T0310839

THE COMPLETE GUIDE TO THE HERSCHEL OBJECTS

Sir William Herschel's contributions to astronomy during the late eighteenth century are unrivalled. His lasting legacy is his dedicated all-sky survey of star clusters and nebulae, and these objects continue to be among the most studied in the night sky. This unique book provides a complete re-examination of Herschel's entire catalogue of nonstellar discoveries, making it the most accurate and up-to-date reference of its kind.

Retrace the footsteps of one of history's greatest astronomers and explore every one of Herschel's landmark discoveries, including those considered to be lost or nonexistent. Read detailed notes about each object's appearance and physical characteristics, and view hundreds of photographs of the most intriguing Herschel objects, along with dozens of sketches of what is visible at the eyepiece. For the reader's convenience, formatted lists of the Herschel objects are available online at www.cambridge.org/9780521768924. This superb book is a must-have for amateur astronomers seeking new and exciting observing challenges, and as the ultimate reference on the Herschel objects.

Mark Bratton has more than two decades of observational experience, and he is one of the few amateur astronomers to have succeeded in observing every one of the Herschel objects. He contributed to *Night Sky: an Explore your World Handbook* (Discovery Books, 1999) and has written articles for several astronomical publications, including *SkyNews* magazine, *Amateur Astronomy* (US), and the *Journal of the Royal Astronomical Society of Canada*. He is a member of the Webb Deep-Sky Society and the William Herschel Society.

The Complete Guide to the

HERSCHEL OBJECTS

Sir William Herschel's
Star Clusters, Nebulae and Galaxies

MARK BRATTON

CAMBRIDGE
UNIVERSITY PRESS

CAMBRIDGE
UNIVERSITY PRESS

Shaftesbury Road, Cambridge CB2 8EA, United Kingdom

One Liberty Plaza, 20th Floor, New York, NY 10006, USA

477 Williamstown Road, Port Melbourne, VIC 3207, Australia

314–321, 3rd Floor, Plot 3, Splendor Forum, Jasola District Centre, New Delhi – 110025, India

103 Penang Road, #05–06/07, Visioncrest Commercial, Singapore 238467

Cambridge University Press is part of Cambridge University Press & Assessment, a department of the University of Cambridge.

We share the University's mission to contribute to society through the pursuit of education, learning and research at the highest international levels of excellence.

www.cambridge.org
Information on this title: www.cambridge.org/9780521768924

First published 2011

A catalogue record for this publication is available from the British Library

Library of Congress Cataloging-in-Publication data
Bratton, Mark, 1955 –
The complete guide to the Herschel objects : Sir William Herschel's star clusters, nebulae, and galaxies / Mark Bratton.
    p. cm.
Includes bibliographical references and index.
ISBN 978-0-521-76892-4
1. Astronomy – Observers' manuals.   2. Stars – Clusters – Observers' manuals.
3. Nebulae – Observers' manuals.   4. Herschel, William, Sir, 1738–1822.   I. Title.
QB64.B736   2011
522–dc22

                                                                2011011263

ISBN   978-0-521-76892-4   Hardback

Additional resources for this publication at www.cambridge.org/9780521768924

Contents

The Complete Guide to the Herschel Objects   41

Preface

This book makes extensive use of images from the Digitized Sky Survey (DSS) and the author wishes to thank Lynn Kozloski, Space Telescope Science Institute, Scott Kardel, California Institute of Technology, and Sue Tritton, Royal Observatory Edinburgh, for their permission to use images originally obtained by the Oschin Schmidt Telescope of Palomar Observatory and the United Kingdom Schmidt Telescope of the Anglo-Australian Observatory in the catalogue portion of this book.

We extend our appreciation and thanks to Andreas Maurer and the Antique Telescope Society for permission to use material from their publication *A Compendium of All Known William Herschel Telescopes* in the chapter entitled 'The telescope maker'. Mr Maurer also graciously provided the photograph of the 10-foot telescope built by William Herschel and presented to Göttingen University, as well as the photograph of the replica of the 25-foot Spanish telescope.

Alison Rowe, Herschel Museum of Astronomy, provided the images of the engraving of Herschel's 20-foot telescope, the Herschel workshop display at the Herschel Museum and the Herschel polishing machine, and for this we express our thanks.

Fred Schlesinger, Treasurer and Membership Secretary of the William Herschel Society, provided assistance in the form of many back issues of the Society's publication, *The Speculum*, and for this the author is grateful.

The catalogue portion of this book makes extensive use of data for the Herschel objects which are now freely and extensively available on the Internet.

It would have been very difficult, if not impossible, to gather these data in any other way.

Our sincere appreciation is extended to:

The Jet Propulsion Laboratory, the California Institute of Technology and the National Aeronautics and Space Administration for providing the NASA/IPAC Extragalactic Database;

Le Centre de Données astronomique de Strasbourg (CDS), Strasbourg, France for providing the SIMBAD database;

L'Université de Lyon, Lyon, France for providing the HyperLEDA database;

The Space Telescope Science Institute, AURA, Inc., the California Institute of Technology, the National Geographic Society, Palomar Observatory, the Royal Observatory Edinburgh and the Anglo-Australian Observatory for making the Digitized Sky Survey possible.

The original Herschel catalogue published in the *Philosophical Transactions of the Royal Society* was not a perfect document. There were occasional errors in position, duplicate (and sometimes multiple) observations of single objects, as well as suspected objects which later proved to be nonexistent. It is unfortunate that Herschel's contemporaries did not try to verify his discoveries independently while he was still alive, as this would have gone a long way towards perfecting what was already an extraordinary document. The first attempt to verify William Herschel's discoveries was made by his son John, but only in the years following his father's death. And John Herschel indicated that he did not observe his father's entire catalogue: more than 600 objects were not observed, and therefore, not confirmed. In the many decades that have followed, astronomers have tried to clear up mysteries and solve identity problems in Herschel's catalogue with considerable success, but some ambiguities remain.

The present volume is my own attempt to provide a guide to the complete catalogue of Herschel's non-stellar discoveries. Any errors in the present volume are my own.

Introduction

IT IS THE EVENING OF 5 SEPTEMBER 1784. As darkness descends on the village of Datchet, the bustling activity of the day slowly draws to a close. A few gentlemen, mostly in pairs, walk the streets at a leisurely pace, taking their evening constitutionals, quietly discussing the day's events. Their way is lit by a few widely spaced lamps, but these will be extinguished shortly; no need to waste precious fuel lighting streets long after everyone has gone to bed. The doors of the village pub open and close with diminishing regularity as the last of the patrons leave and make their way home. By 9:30 p.m. the village is largely quiet, save for the occasional bark of a dog.

It is a beautiful, star-lit night with a slight breeze and the air is crisp, suggesting that the cool days of autumn are not far off. After several days of cloud the moonless skies are clear and the brilliant summer Milky Way arches overhead, seen in all its glory. A short way from the village centre stands a solid two-storey house with a low stone wall extending away down the road. The house is dark save for the light of a single lamp near an open window on the second floor. In the garden the silhouette of a curious structure can be seen rising into the night sky, a peculiar construction of slanted spars and crossbeams, integrated with a pair of ladders on both sides. It appears to be more than 20 feet high and in the centre is a considerable tube, like a cannon but far larger, pointed almost vertically into the sky. There is a small platform off to the side of the tube near the top and one can see that there is a man standing there, leaning over the tube in the darkness. He is silent for a while but then calls out quietly to the ground below and all of a sudden the structure moves ever so slightly, turning on the casters supporting it. There is another man, this one standing at the base of the structure, who next turns a hand crank slowly just above his head. He stops immediately upon a second command and the night is again silent for several minutes.

Suddenly, there is an excited shout from the man on the platform. The assistant, who had been dozing, suddenly springs into action, turning the hand crank slowly one way, then stopping and turning it again in the opposite direction, until he is told to stop again. There is silence, then a request to turn the mighty structure ever so slowly to the west. The man on the platform then calls out in German-accented English towards the lit window in the house. One cannot quite make out what he says, but when he stops a female voice with the same thick accent, repeats what he said, word for word: 'Branching nebulosity, extending in right ascension near one and a half degrees and in polar distance, 52 minutes. The following part divides into several streams uniting again towards the south.'

'That's correct,' says the voice from the platform, 'a large nebula ... fifth class.' A request is made to the assistant below to continue turning the mighty telescope to the west; the object in the field of view is truly an extraordinary one, unlike anything the astronomer on the platform, William Herschel, has ever seen. He studies it for a considerable time before finally telling the assistant to stop turning the structure. The nebula departs the field of view and the crowded star fields again drift slowly through the eyepiece field, studied intently by the astronomer. Herschel has made yet another discovery to go along with the hundreds of nebulae both bright and faint, large and small that he has recorded over the last ten months. This one will one day be known as the Veil nebula.

It is interesting to note that in the final decades of the eighteenth century, the frontiers of astronomical discovery were being breached not by the telescopes of the national observatories of Europe, staffed with astronomers educated at the finest universities but by a single amateur astronomer, recently turned professional, who gave up a successful career as a composer and musician to devote his life to the stars he loved. He was ably assisted in this endeavour by his talented sister, who had even less formal training in astronomy than he had. The advantage that William and Caroline Herschel had was that they were constructing some of the finest, most powerful telescopes the world had ever seen and they possessed the will and determination to exploit this equipment to the fullest. In the process, their discoveries increased the apparent size of the universe many fold and it would be more than a century before astronomers began to probe the universe more deeply than they did.

William Herschel was born on 15 November 1738, the fourth of ten children born to Isaac and Anna Herschel in Hanover, Germany. Isaac was a gardener by profession but his love of music led to him teaching himself to play the violin and later, the oboe, and he eventually embarked on a career in music. He passed on his love of music to most of his surviving children, his son William included, and by the 1750s Isaac and his sons William and Jacob were engaged as musicians in the Hanoverian Guards. This allowed them the opportunity to travel extensively with the army and by 1756 they found themselves in England, where William learned to speak, read and write English. War broke out the following year and the Hanoverian Guards returned to Germany where they saw action. At this point, Isaac encouraged William to leave the military and

in late 1757 William made his way back to England, where, at the age of 19, he embarked on a career as a musician.

Not surprisingly, it took William many years to establish himself in his chosen field and he was itinerant through much of his first decade in England. He spent time in London, then moved about through much of the north of England, finding work where he could, sometimes with military bands, sometimes teaching music, and sometimes as an organist. In 1764, he briefly returned to Hanover to visit his family and did not return there again until 1772.

By 1766, Herschel had arrived at Bath, in the west of England, and it was here that he established himself as a musician, becoming the organist and musical director of the Octagon Chapel. Herschel had been composing music since his arrival in England and he continued this in Bath, many of his musical presentations here being of his own works. He also had a considerable sideline as a musical instructor and much of his time was devoted to teaching music. By 1770, he was well established and making a very comfortable living as a musician. At various times his brothers Dietrich, Jacob and Alexander all visited from Hanover, spending considerable time with William. Alexander established himself as a musician in Bath and spent most of the rest of his life in England.

Herschel's youngest sister, Caroline, was born in 1750 and, though there was a considerable difference in age, he had a soft spot in his heart for her. Unlike her brothers, however, she received little in the way of formal education, a situation not uncommon for women of the era, and the family, particularly her mother and oldest brother Jacob, foresaw and encouraged a life of domestic drudgery for her.

William was aware of the situation back in Hanover, and did not approve of the plans for his sister's future. In 1772, he visited Hanover with the intention of returning to England with his sister in tow. Although Anna and Jacob resisted the idea, William's influence and leverage in the situation were considerable, and the family relented.

Herschel's intention was to give his sister a musical education, training her as a singer. He thought she could also perform domestic and household duties, as well as help to manage Herschel's musical activities, including the scheduled appointments of his many musical students.

Although her training as a singer did not proceed as quickly nor as smoothly as she would like, Caroline's other duties kept her considerably occupied and her days in Bath were happy ones. For the next ten years she was indispensable to her brother in managing and enabling his career as a musician.

Although William Herschel had considerable skills and talent as a musician, his father had ensured that he had a broad education and he was interested in many things, philosophy, politics and science included. An early interest in the stars had blossomed by 1773, when Herschel began acquiring books on mathematics and astronomy. About this time he began to rent small telescopes to observe the heavens, and also started his first rudimentary attempts at constructing telescopes. His interest in astronomy quickly deepened and what started out as an interest quickly became an all-consuming passion.

Early on and indeed for the rest of his life, many aspects of observational astronomy interested him. He took a great interest in observing the moon and all the known planets, especially Saturn, and by the mid-1770s he began regular reviews of the brightest stars in the sky. These stellar surveys required great dedication and discipline of the observer, taking years to complete. Herschel's free time during daylight hours was spent grinding and figuring speculum metal discs, turning them into precision mirrors for his telescopes. He had become a telescope maker when he realized that commercial opticians in England did not build large-aperture telescopes of sufficient optical quality to satisfy his needs. Acquiring the tools, metal discs and other paraphernalia from a gentleman who was giving up the pastime of grinding optics, constant experimenting and practice soon found Herschel making telescopes of unexcelled optical quality with apertures ranging from 4 inches to 12 inches. The 12-inch telescope had a focal length of 20 feet and was one of the largest telescopes in the world. Despite the fact that its mounting was somewhat clumsy and inefficient, it proved to be an effective instrument. Within a few years, astronomy was taking up a considerable amount of Herschel's time, impacting on the time he devoted to his musical career.

During this early period, Herschel was pursuing astronomy and his experiments in optics in virtual obscurity, but this situation soon changed. Although the Herschel household was well established in Bath, William and Caroline frequently changed residence. At various times they made houses at New King Street, River Street and Walcot turnpike their home. Most of the residences had back gardens where Herschel could set up his telescopes and observe the heavens, but the house that they moved into in River Street in December of 1779 had no garden and Herschel had to lease a nearby plot of land in order to set up his largest telescope, a 12-inch reflector.

At the time, Herschel was engaged in a study of the heights of lunar mountains and one evening he set up a small telescope in the street in front of his home in order to get a clear view of the moon. While Herschel was gazing intently at the eyepiece, a gentleman happened by who would alter both Herschel's life and the course of astronomy. His name was William Watson; he was a doctor who moved in the circles of England's most learned men, and when he asked Herschel if he could look through his telescope, Herschel obliged.

Watson returned the next day to thank Herschel and to invite him to join the newly formed Philosophical Society of Bath, to which Herschel readily agreed. From this point forward Herschel's days as an obscure amateur astronomer were at an end. He regularly attended meetings of the Society, writing and reading dozens of papers on all aspects of philosophy and science, including astronomy. He and Watson became good friends and Watson's association with the Royal Society in London introduced Herschel to influential men such as Sir Joseph Banks and the Astronomer Royal, Nevil Maskelyne, who travelled to Bath to meet the amateur astronomer and inspect his telescopes and workshop.

At the beginning of March 1781, the Herschels moved again, this time to No. 19, New King Street, where Herschel would shortly make a discovery that would shake the scientific world. On the evening of 13 March, Herschel was engaged in his latest review of the heavens, involving the inspection of all stars in the heavens brighter than the eighth magnitude, a survey which was turning up hundreds of previously unknown double stars. Moving his telescope through fields near the star Zeta Tauri, Herschel came upon a bright object that was not on his charts. It was not a star, as high magnification showed the object to be a tiny disc and Herschel thought that he might have happened upon a comet, a suspicion which seemed confirmed four nights later when he noted that the object moved amongst the stars. The discovery was communicated to the observatories at Greenwich and Oxford and from there to observatories throughout Europe and it soon became apparent that Herschel had not discovered a comet, but a new planet which eventually became known as Uranus.

The discovery caused a sensation in the worldwide scientific community: who would have suspected that the six planets known throughout history had an unseen associate? Herschel's life changed almost overnight. He became world renowned, certainly in scientific circles, and before the year was out he was elected to membership of the Royal Society, and was introduced to the royal family and King George III. His friends Sir Joseph Banks and William Watson petitioned the King to provide Herschel with a pension sufficient to allow him to give up his career in music and pursue his astronomical studies full time. This was readily agreed to and by the summer of 1782, the Herschels wound up their affairs in Bath and moved to Datchet, near London, where Herschel began his career as a professional astronomer.

In 1783, Herschel began construction of his largest telescope to date, an instrument which would prove to be the most effective of the many telescopes he built. This telescope, of 18.7-inch aperture, became known as the 'large 20-foot telescope' and he used it to conduct his 19-year survey of the universe beyond the solar system.

Free to devote himself full time to astronomy, Herschel's observational and theoretical work soon expanded. So too did his career as a telescope maker. Encouraged by King George III, Herschel was soon producing instruments for other astronomers, institutions in England and Europe and wealthy patrons of science. This sideline greatly augmented the modest pension he received from the King and Herschel's brother Alexander, as well as Caroline, were pressed into service producing instruments. Caroline became an astronomer in her own right during this period, both as an observer and as a valued assistant to William in his astronomical pursuits.

William Herschel's survey of undiscovered nebulae began after the completion of the large 20-foot telescope late in 1783 and continued almost unabated for the next seven years, continuing fitfully through the years after 1790 until its completion in 1802. During this time Herschel continued his studies of the planets, the moon and the sun and among his many discoveries were Titania and Oberon (satellites of Uranus) in 1787 and Mimas and Enceladus (satellites of Saturn) in 1789.

Shortly after completion of the large 20-foot telescope, Herschel began planning an even larger instrument, a 48-inch telescope of 40-foot focal length. Again Banks and Watson petitioned the King for funding for the new project, which was ultimately successful: the King granted £4000 for construction as well as an additional yearly sum for operational expenses. The telescope took the better part of four years to complete but despite the successful completion of the project, the instrument was something of a practical failure and little used by Herschel. The telescope was cumbersome to move and took a long time to prepare for observation. The large speculum metal mirror tarnished quickly and was enormously heavy; it was subject to dewing in the damp English climate and cold weather saw heavy frosts settle on its surface. Nevertheless, the telescope was a wonder of the age and a testament to the skills of the astronomer who built it. It might have been a practical failure, but as a symbol of the state of English astronomy, it was a valuable artefact.

Although Herschel enjoyed great success as an observer, his theoretical work is now largely forgotten. The cameras, spectroscopes and other instruments of a later epoch which would prove invaluable to providing a rational understanding of the nature of the universe were unknown in Herschel's time and his theories were little more than poorly founded speculations based on observations with instruments which were advanced for the times, but not up to the task for which they were built. Nevertheless, Herschel did do some solid work during his career, which included understanding the direction of motion of the solar system through the galaxy and the discovery of infrared radiation.

Besides Herschel's discovery of more than 700 double and multiple stars, the planet Uranus and four planetary satellites, the cataloguing of almost 2500 galactic and extragalactic objects by William and Caroline Herschel is probably their most significant and lasting work. William Herschel was the greatest astronomer of his age; he laid the foundation for the observational work which would follow in the nineteenth century and the significant expansion of our understanding of the universe which began in the early twentieth century and continues to this day.

The telescope maker

BY THE TIME THAT WILLIAM HERSCHEL embarked on his career in astronomy, the telescope had been in use for over 160 years. Progress in improvement and refinement of the instrument had been painfully slow, however, with a combination of factors responsible for the circumstance.

The primary problem with the refractor telescopes of the seventeenth and eighteenth century was spherical and chromatic aberration, a result of the fact that these were single lens instruments that could not overcome their inherent faults. The solution that was found was to increase the focal lengths of the instruments. Where the simple Galilean telescopes of the mid-seventeenth century had apertures of under 2 inches and focal lengths of 2 or 3 feet, instrument makers found that increasing the focal lengths by a factor of 5 or 10 times helped reduce the effects of the aberrations, though they were not nearly eliminated. Over the course of the succeeding decades, astronomers began using telescopes with focal lengths of well over 100 feet and in some cases over 200 feet. Not surprisingly, these were clumsy and cumbersome instruments to use; it was difficult enough aiming the telescope at a selected object, let alone following the object across the sky for any length of time. It was only in the mid-eighteenth century when the first achromatic telescopes were developed that the long refractor telescopes passed into history.

The achromatic telescope used two lenses of crown and flint glass to form an image. The glasses had different refractive indexes and when combined, largely cancelled out chromatic and spherical aberration. The problem was to get glass of sufficient purity, which was difficult for the glass makers of the era to produce and it was rare for a good lens larger than about 3 or 4 inches to be made. Although telescopes returned to much more manageable lengths, their light-gathering abilities were modest at best.

Isaac Newton's invention of the reflecting telescope was a promising avenue in telescope development as reflected light produced no chromatic aberration. Newton's early telescopes were little more than design exercises, however. The apertures were so small that the telescopes were ineffective astronomical instruments.

The credit for building the first effective Newtonian reflector goes to John Hadley, who, along with his brothers George and Henry, constructed a 6-inch, f/10.2 reflector in 1720. The telescope was very well thought out, especially in the way it was mounted, which made it an easy instrument to point and manipulate. Thereafter, London opticians began producing Newtonian reflectors, though the instruments were usually a little smaller than Hadley's reflector.

Except for two brief written references in 1766, one to an observation of Venus on 19 February, the other to an eclipse of the moon on the morning of 24 February, William Herschel's interest in astronomy did not bloom until the spring of 1773. At that time his heavy music and teaching schedule for the season was winding down and Herschel found time to nurture his interest in astronomy. He acquired books on optics, trigonometry and astronomy, then purchased object glasses of 4- and 10-foot focal length, along with tin tubes and eyepieces so that he could construct the first of his telescopes. By June he had rented a small 2-foot reflector and purchased 15-foot and 30-foot focal length objectives for even larger telescopes. The difficulty of managing the long tubes of refractors led him to looking again at reflectors.

In September he rented a 2-foot Gregorian telescope and the ease of use of this instrument whetted his appetite for even larger reflectors. When he tried to acquire a mirror for a 5- or 6-foot reflector, however, he was informed that no mirror for so large a telescope was available commercially. At this point his thoughts turned to making his own mirrors and by a happy coincidence Herschel learned of a Quaker gentleman in Bath who had been making mirrors but had lost interest in the occupation. Herschel acquired his tools and patterns, some unfinished mirrors and instructions for casting metal discs. By October, Herschel's career as a telescope maker had begun.

Initially, Herschel had the metal discs cast for him. The speculum metal mirrors were composed of a mixture of 32 parts copper, 13 parts tin and 1 part of regulus of antimony which produced a bright metal which could be ground and polished relatively easily. Producing the proper alloy was something of a black art: too much tin resulted in a brittle metal, while too much copper produced metal which tarnished quickly.

Herschel's first complete telescope was a Gregorian reflector of 5.5-foot focus but he had great difficulty aligning the optics and the telescope was never used in this configuration. He turned his efforts to producing a Newtonian telescope and he achieved success on 1 March 1774 using his newly made telescope to view Saturn and the Orion Nebula.

For the next two years Herschel contented himself with observing; he found no time to grind mirrors or build telescopes. But by May of 1776 he had evidently made himself an excellent 7-foot reflector, and was completing work on a 10-foot reflector, which used a 9-inch mirror, and an even larger telescope of 20-foot focal length with a 12-inch mirror. The 10-foot reflector is of interest because it was the first of his instruments that he tried to use in the Herschelian rather than Newtonian form. For this, Herschel removed the secondary mirror and tilted the primary mirror slightly so that the focus would come to the edge of the open tube where he placed the eyepiece. This made for a brighter image, as no light was lost in the secondary reflection. On 28 May 1776 he wrote: 'I tried a 10 feet mirror without the small one, looking in at the front of the tube and holding the eye glass in my hand. I liked the method very well.'

Both the 10-foot and 20-foot telescopes were mounted in the same manner. The bottom of the telescope rested on a small platform on the ground where the instrument could pivot. A tall pole was erected near the front of the telescope and a rope and pulley system ran from the top of the pole down to the front of the tube where a wheel could be turned to raise or lower the instrument. The lateral motions of the telescopes were severely limited, however, and the instruments were probably lined up with the meridian to be used to full advantage. The 20-foot telescope, in particular, required a large ladder and the observer had to stand on the ladder and lean in sideways towards the telescope to look through the eyepiece. When observing towards the zenith, the observer was almost 20 feet in the air, leaning over precariously in the dark. When the wind was up, this somewhat flimsy mounting must have resulted in a swaying tube, which would have produced unsteady images. In later years Herschel tried to persuade his sister Caroline to use the 20-foot telescope for her observations, but she had little interest in this.

Each of Herschel's telescopes had several mirrors and because of this, he developed great skill as a mirror grinder and polisher. In the English climate, the speculum metal mirrors tarnished quickly and Herschel's days (and often, the days of his brother, Alexander) were spent switching mirrors in and out of telescopes and repolishing tarnished specula.

This is an example of a polishing machine devised by William Herschel for small speculum metal mirrors. Image provided by the Herschel Museum of Astronomy.

Herschel's experience taught him that there were no hard and fast rules to the art of mirror grinding: each mirror seemed to require a different combination of strokes and grinding techniques to achieve the much desired perfect parabola. He tested his mirrors frequently, trying to achieve the best figure possible and then polished them to an extremely smooth and bright surface. On occasion, he produced exceptional mirrors.

One example is a 7-foot Newtonian constructed in November of 1778 which had a mirror 6.2 inches in diameter. Herschel described it as 'a most capital speculum' and in August of the following year, he used the telescope to begin a survey of the heavens which resulted in his first catalogue of multiple stars.

It was during this survey, on the night of 13 March 1781, that Herschel came upon an object in the heavens which was not on his charts and which he instantly recognized as a possible comet due to its nonstellar appearance. He wrote: 'In the quartile near Zeta Tauri the lowest of two is a curious either nebulous star or perhaps a comet. A small star follows the comet at two thirds of the field's distance.' Four nights later he further remarked: 'I looked for the Comet or Nebulous Star and found that it is a Comet, for it has changed its place.' The news of the discovery was forwarded to the Astronomer Royal, Nevil Maskelyne, and to another astronomer, Thomas Hornsby, but both astronomers had difficulty locating the object due to the inferiority of their instruments. Maskelyne observed it on 3 April, but only after comparing the position of the object with that from earlier observations of the field. He was convinced that it was a planet, which subsequent orbital calculations proved it to be. Hornsby, on the other hand, had even more difficulty and did not realize until 14 April that he had unknowingly located the planet on 30 March.

The news of the discovery of Uranus caused a sensation in both England and the rest of Europe and the amateur astronomer from Bath became something of an overnight sensation. Herschel himself was unfazed by the discovery, feeling that it would have been made sooner or later: 'It has generally been supposed,' he wrote, 'that it was a lucky accident that brought this star to my view; this is an evident mistake. In the regular manner I examined every star of the heavens, not only of that magnitude but many far inferior, it was that night its turn to be discovered. I had gradually perused the great Volume of the Author of Nature and was now come to the page which contained a seventh Planet. Had business prevented me that evening, I must have found it the next, and the goodness of my telescope was such that I perceived its visible planetary disc as soon as I looked at it; and by the application of my micrometer, determined its motion in a few hours.'

In May of 1782, Herschel took his 7-foot reflector to Greenwich where its performance was extensively tested by Maskelyne and others against an achromatic refractor of 46 inches focal length and a 6-foot reflector made by Short. The performance of Herschel's telescope was superior in all respects. Observing with Maskelyne and the wealthy amateur astronomer Alexander Aubert, Herschel was able to resolve double stars with his instrument that could not be split with any of the telescopes of the Royal Observatory.

As good as his small telescopes were, however, Herschel longed for instruments of great aperture. The problem was that foundries in the Bath region, and elsewhere, were not equipped to produce the castings for mirrors of large aperture. Herschel decided to overcome this by making the questionable decision to construct a furnace and melting oven in the basement of the house on New King Street. He intended to cast a mirror for a proposed telescope of 30-foot focal length and 3-foot aperture. Herschel began by experimenting with different metal alloys to produce a mirror that would be highly reflective and at the same time not too brittle. He settled on a ratio of 5 pounds tin to 12 pounds of copper. Because

the mirror would be heavy and difficult to lift onto a polisher, an iron ring 24 inches in diameter would be attached to the back of the mirror. Holes drilled in the ring would allow ropes to be passed through and the whole apparatus could be lifted and moved by a crane. The total weight of the mirror would be almost 470 pounds. A mould made of loam would receive the molten metal for casting and on 11 August 1781, the casting took place. Unfortunately as the metal was poured into the mould, a crack appeared and some of the molten alloy leaked out. As the metal cooled, cracks began appearing and the casting was abandoned.

Thinking that the metal was too brittle, Herschel reduced the percentage of tin in his second melt but as the metal heated it began to leak out of the furnace and into the fire below. The crack in the furnace widened and the metal began to pour out quickly, running over the flagstones of the floor. These subsequently cracked and the stones began to explode, flying through the room in all directions. Herschel and his workmen scattered, narrowly escaping serious injury. For the time being, Herschel abandoned the pursuit of large reflecting telescopes.

After the discovery of the planet Uranus, Herschel's friends William Watson and Joseph Banks of the Royal Society petitioned King George III to provide the now famous astronomer with a posting or perhaps a life-long pension of some kind so that he could abandon the day-to-day necessity of earning a living through music and devote his entire time to the exploration of the heavens. George III readily agreed to this and thereafter furnished him with an annual pension of £200, his only duties being to sometimes entertain the royal family with views of the heavens through his telescopes. Unfortunately, generous though the pension was, it was something of a reduction in income for Herschel and he now turned to making telescopes for sale in order to supplement his income.

George III was amongst his first customers, ordering five 10-foot reflectors with mirrors 8.8 inches in diameter, and with this seeding, Herschel the commercial telescope maker was in business.

Some of William Herschel's telescope-making equipment on display at No. 19 New King Street in Bath. Image provided by the Herschel Museum of Astronomy.

The excellence of Herschel's telescopes was by this time well known and he did not lack an eager clientele. For Herschel, this became a most lucrative enterprise, though a decidedly time-consuming one. Herschel was able to call upon George III's own carpenters to provide the labour to build the tubes and stands for the King's own order of telescopes. For other customers, Herschel tried to utilize local carpenters and smiths but the workmanship left something to be desired and Herschel eventually entrusted them only with simple work. For the rest he relied on his own labour as well as that of his brother, Alexander, who built stands and ground and polished mirrors when not occupied with his own career as a musician. Even Caroline was pressed into service to polish mirrors when she was not otherwise involved in housekeeping or her work as William's astronomical assistant.

Most of the telescopes constructed for sale by Herschel were 7-foot focal length models, generally with mirrors of 6.2 inches aperture, though occasionally smaller or larger mirrors were fitted. Herschel set up a small furnace to cast the mirrors himself. The cost of a complete 7-foot telescope was usually £105, an enormous amount of money at the time, and some 7-foot telescopes with additional eyepieces or other equipment sold for more than £200. (A 7-foot telescope for the Empress of Russia was sold for £525.) By way of comparison, we have already noted that Herschel's yearly salary from the King was £200. The Astronomer Royal's annual salary was £300, while Caroline Herschel, as William Herschel's assistant, was granted £50 per annum. William Herschel's best year as a musician and music teacher netted almost £400.

A Herschel 10-foot reflecting telescope with an aperture of 8.8 inches. It is one of five telescopes commissioned by King George III of the United Kingdom and built between the years 1784 and 1786. This is the telescope personally delivered to Göttingen University by Alexander and William Herschel at the request of King George in the summer of 1786. Photograph courtesy of Andreas Maurer.

Not surprisingly, therefore, Herschel's clientele consisted almost exclusively of wealthy individuals, the nobility and occasionally other astronomers. A list that Herschel compiled from memory which was published in *The Scientific Papers of Sir William Herschel* (Herschel and Dreyer, 1912) lists Prince Usemoff, Baron Hahn, Count Brühl of Denmark, the Duke of Tuscany and Chevallier Forteguerri amongst his noble clientele, while John Pond, Giuseppe Piazzi, Johan Bode and J. H. Schröter were some of the astronomers who owned Herschel telescopes.

Herschel's marriage to the wealthy widow Mary Pitt in 1788 reduced his financial constraints, but he continued to build telescopes commercially, one of the last being a 25-foot reflector with a 24-inch mirror for King Carlos IV of Spain at the turn of the nineteenth century. This telescope was sold for £3150 and Herschel's total income from telescope-making activities over almost two decades was at least £14 743.

Shortly after the move to Datchet near Windsor in 1782 Herschel began construction of his most important telescope, which became known as the 'large' 20-foot reflector. This telescope used a mirror 18.7 inches in diameter and Herschel initially made two mirrors for the telescope so that at least one would always be in a state of high polish.

The telescope featured the most elaborate mounting that Herschel had attempted to date, a massive triangular framework more than 20 feet high set upon a circular platform which could be rotated on a series of metal casters. An elaborate rope and pulley system raised or lowered the telescope and at least one workman, and sometimes two, were required to move the telescope. There are only three known illustrations of this telescope. One is a watercolour painting showing the telescope in its Newtonian configuration in the back garden of the house at Datchet shortly after it was built. The second is a copperplate engraving that was discovered in the 1970s, showing the telescope in its Herschelian form with a large gallery mounted on the front of the telescope framework so that the observer could use it as a front-view instrument. The latter illustration shows an altogether more robustly mounted telescope, though both mountings were constructed along the same general lines. The third illustration is a *camera lucida* drawing by John Herschel of the telescope as it was set up at the Feldhausen estate in South Africa shortly after 1834.

The telescope tube was a massive, eight-sided affair, made up of long wooden planks, braced along each seam in the interior by a series of triangular wedges. The seams along the exterior were covered in fabric attached to the planks by rivets. Two evenly spaced metal straps were wrapped around the tube to further hold it together and the tube was supported at the opening by a roughly worked interior framework of lumber.

The telescope was a cumbersome affair but it was solidly built and far superior to the long telescopes of an earlier era. The tube was heavy enough that the optics did not easily go out of alignment. It took up to 15 minutes to aim the telescope at any given object but considering that its main work was sweeping the skies for nebulae and 'star-gaging' (a project that Herschel conceived of to count stars in field after field over the sky to determine how the stars of the sidereal system were organized), the telescope was admirably suited to both tasks.

The large 20-foot was essentially complete by the autumn of 1783 and became Herschel's favourite instrument; it was used for almost all his observing projects, including planetary studies. The telescope was used to discover the two brightest satellites of Uranus, Titania and Oberon, in 1787 and Saturn's moons, Mimas and Enceladus, in 1789.

As well as the 20-foot reflector worked, Herschel longed for an even larger instrument. The failure of the 30-foot telescope in Bath was only a temporary setback and Herschel made plans for a greater telescope. The main obstacle was financial. He had invested a considerable amount of his own money building the large 20-foot. He could only build a larger telescope with outside financial assistance and, with this in mind, he asked the President of the Royal Society, Sir Joseph Banks, to petition King George III on his behalf for the funds to build a large telescope. The request was successful and Herschel offered the King the choice of either a 30-foot or a 40-foot telescope. Not surprisingly, the King elected to support a 40-foot instrument, an unfortunate choice as things turned out.

In September of 1785 the King granted £2000 for the project and preparations began almost immediately to hire carpenters, smiths and bricklayers to build the foundation, framework and tube for the mighty telescope, which would have a mirror 4 feet in diameter. William Herschel would be the project manager, overseeing every detail of the construction of the telescope, including all of the optical components.

By this time the Herschels had moved to a large property in Slough, in sight of Windsor Castle, where the large 20-foot was already set up. For the next four years the property was overrun by labourers as the telescope and its mounting slowly rose in the garden behind the house.

Like all of his telescopes, the 40-foot had two mirrors and Herschel arranged for a foundry in London to cast the giant metal discs. The work began in late October of 1785; the first casting was of a thin mirror with a total weight of 1023 pounds but it was not entirely successful as a depression in the centre reduced the thickness to less than an inch. Herschel decided to go ahead with grinding the disc anyway, feeling that it would be good practice and anything learned could be applied to a second, thicker mirror.

The mirror fitted in an iron ring which could be suspended over the grinding tool and handles were attached to the ring so that ten men could work the mirror back and forth over the tool. Each man had a number on his back and Herschel would call the numbers out in sequence, commanding the men to either push or pull as they were called out. The work proceeded at a very slow pace but, as the rest of the telescope was not nearly ready, there was no way to test the mirror and no real need to rush the work of grinding the mirror.

By February of 1787 work on the telescope was sufficiently complete that the mirror could be mounted and a test done. Crawling into the tube, Herschel found the focus and the test object, the Orion Nebula, appeared bright though the figure of the mirror was not nearly good enough. The work of figuring and polishing the mirror continued through the spring and summer.

By this time, Herschel was running out of money for the great telescope. Although work on the rest of

A copperplate engraving of Herschel's 'large 20-foot' telescope of 18.7-inch aperture. It is seen here in its front view, Herschelian mode, circa 1787. Image provided by the Herschel Museum of Astronomy.

the telescope was nearing completion, the first mirror was nowhere near finished and a second mirror had yet to be cast. Herschel approached the King again to request additional financing, but the monarch was not amused. Nevertheless, in August of 1787 he granted a further £2000 pounds, as well as a yearly stipend of £200 for operational expenses for the 40-foot telescope. In addition, the King granted Caroline Herschel a lifetime pension of £50 per annum as Herschel's assistant.

While work continued on the first mirror, Herschel proceeded to have the second mirror cast in late January of 1788. The first attempt was not successful as the disc cracked in several places but after reducing the proportion of tin, another melt was performed and on 16 February, a mirror 3.5 inches thick, weighing 2118 pounds was made. Grinding began in July but because of the great weight of the disc, 20 men were now employed

and progress was so unsatisfactory that Herschel decided to build a machine to polish the great mirror. It was constructed in the spring of 1789 and was used to polish the thin mirror. With the work successfully completed, the thicker mirror was polished in August 1789 and Herschel considered the telescope project complete on 28 August.

The mighty instrument was a sight to behold. Like the large 20-foot, the mounting was a triangular framework, this time 50 feet high, with a large gallery for the observer and a small hut at the base from which the telescope tube emerged. Caroline could work as the assistant in the hut, reading the clocks and other instrumentation, and labourers were stationed here as well, controlling the gears and pulleys to turn, raise and lower the telescope. The tube was over 40-feet long and 5-feet in diameter. It was made of sheets of iron which were wrapped with iron straps as well as iron bars, riveted longitudinally along the tube. The words of Oliver Wendell Holmes serve to describe its impact: 'It was a mighty bewilderment of slanted masts, spars and ladders and ropes from the midst of which a vast tube, looking as if it might be a piece of ordnance such as the revolted angels battered the walls of heaven with, according to Milton, lifted its mighty muzzle defiantly to the sky.'

Despite the fact that the mirror was not in a perfect state (some scratches remained on the surface) the instrument was seen as an immediate success as Herschel claimed to have discovered Enceladus, a satellite of Saturn, with it on 27 August. In actual fact, this observation was only a confirmation, as he had observed the satellite with the 20-foot reflector on 19 August. Likewise the seventh satellite of Saturn, Mimas, was first observed with the 20-foot and only followed up later with the 40-foot telescope.

The 40-foot was an instrument at the absolute limits of late eighteenth-century technology and never performed as well as Herschel had hoped. The telescope was far too massive, difficult to prepare for a night's observing and difficult to aim. The mirrors reacted poorly to temperature changes and their great weight meant that flexure was a problem; the definition of the image was rarely better than that of the 20-foot. In cold weather, thick frosts settled over the surface of the mirror, making observation impossible. The mirror surfaces tarnished quickly and the great weight of both mirrors made it a difficult chore to remove the optics and repolish them to their former gloss.

The observing records for the 40-foot are extremely scant. Saturn was a favourite target, but many years went by between observations. Uranus and its satellites were observed only three times. There are records of observations of globular clusters, but these had been discovered previously by Herschel with the large 20-foot. Surprisingly, despite the great light-gathering ability of its mirrors, the telescope was not employed to sweep the sky for nebulae and not a single one was discovered with the great telescope.

The greatest problem with the telescope, however, was that visitors came from far and wide to view the mighty wonder of the age. The telescope was far and away the largest that had ever existed. In about 1780, the Reverend J. Michell of Yorkshire had constructed a 29.5-inch Gregorian reflector, but its performance was abysmal. In addition to the King and members of the royal family, English and European noblemen as well as other astronomers were frequent visitors. Herschel spent more time explaining the workings of the instrument, answering questions and otherwise entertaining his guests than he ever did using it. Herschel never completely gave up on the telescope, however, and he persisted in polishing the second mirror (the first, thin mirror was eventually abandoned as unusable) periodically up to 1809. By then, Herschel was 71 years old and his most active days as an observer were far behind him.

Mention should also be made at this point of the telescopes that Herschel made exclusively for the use of his sister Caroline. In 1782 he felt that Caroline's time could be put to good use by sweeping the heavens for double stars, nebulae and comets and to this end he made her a small, simple refractor which used low-magnification eyepieces to provide a wide field of view. The instrument was quite crude but Caroline enjoyed some success nevertheless, tracking down some of Messier's discoveries and even adding a cluster or two of her own. The following year he built a Newtonian sweeper for her of 4.2-inch aperture and 2-foot focal length. The improved instrument allowed Caroline's pace of discovery to increase, and by the end of 1783, she had located eleven new nebulae, including M110 (NGC205), observed by Messier but not included in his catalogue, and M48 (NGC2548), a 'missing' Messier object that was considered lost for almost two centuries. Despite her success, Caroline's career as an independent observer abruptly ended here; her services as an assistant were required by her brother and Caroline subsequently only observed while William was away travelling or when he otherwise did not require her participation in his work.

Despite the reduced opportunities, Caroline was able to discover her first comet on 1 August 1786 and a second on 21 December 1788. In 1791, Herschel built her an even larger telescope, with an aperture of 9 inches and 5-foot focal length. Caroline discovered eight comets during her observing career, becoming a celebrated and respected astronomer in her own right.

An intriguing telescope that Herschel built later in life was the 'X-feet', a 10-foot focal length telescope which used a mirror 24 inches in diameter and was set up in the Newtonian form. The instrument was ideal for an older man as it did not require climbing up the framework of the telescope, as the large 20-foot and the 40-foot did. The mirror defined celestial objects well and the telescope was used extensively, though not for the discovery of nebulae. It was finally sold to Lucien Bonaparte of France in 1814.

Despite his extensive experience as a telescope maker, William Herschel made no major breakthroughs in optical design during his career; his only real innovation was the adoption of the Herschelian front-view method for reflector telescopes, employed to gain a brighter image due to the limitations of the speculum metal mirrors of the age. Nevertheless, his telescopes were far and away the most effective of the era, due principally to the great care he employed in grinding and polishing his mirrors.

As an instrument of discovery, the large 20-foot of 18.7-inch aperture is arguably one of the most significant telescopes ever built, ranking with Galileo's best telescopes and very likely in the same league as the Hooker 100-inch reflector at Mount Wilson, the Hale 200-inch Palomar reflector and the Hubble Space Telescope, all twentieth-century instruments. The large 20-foot expanded the extent of the known universe many fold. Herschel's 'stargaging' efforts were instrumental in understanding the form of the Milky Way and the almost 2500 star clusters, nebulae and galaxies which he discovered laid the groundwork for understanding the universe beyond the Milky Way. Though the telescope's woodwork began to deteriorate shortly before his death, the framework of the telescope was eventually restored and Herschel instructed his son John in the art of mirror grinding, giving new life to the instrument. It allowed John to resurvey the heavens between 1823 and 1828, confirming the discoveries of his father and adding a further 466 nebulae that had escaped William Herschel's earlier sweeps. The telescope was subsequently dismantled and was taken to South Africa in 1834, where it was erected at Feldhausen and employed by John Herschel to sweep the southern skies which were inaccessible from England. Over the course of four years the telescope was used to catalogue more than 1700 star clusters, nebulae and galaxies in the southern skies. The telescope and Sir John returned to England and though still in an excellent state, the telescope was never mounted again and its effective life was at an end after 55 years of almost continuous service. Today, only the 20-foot tube and three of the mirrors survive, the property of the National Maritime Museum, London.

Happily, many other Herschel telescopes survive, either as complete instruments or partially complete with some components missing. Many of them are smaller instruments which Herschel sold overseas. Often, the original owners donated the telescopes to museums in the UK, Europe, Africa and America, where they are now preserved.

As for Herschel's own telescopes, their fate is a mixed lot. The 40-foot stood in the back garden of Observatory House, Slough, slowly deteriorating until December, 1839, when it was dismantled by John Herschel. One of the earliest known photographs, taken by John Herschel, is a ghostly image of the framework shortly before it was taken down. The iron tube of the telescope rested on the ground for many decades and was partially destroyed by a tree which fell during a violent storm, while the second 4-foot speculum metal mirror hung on a wall in Observatory House until the property was sold in the 1950s. Both of the mirrors, the surviving 10-foot section of the tube and some paraphernalia are now in the possession of the National Maritime Museum and the Science Museum, London.

The replica of the 24-inch aperture Spanish telescope on display at Madrid. It is the only complete example of what a large-aperture William Herschel telescope looked like. Photograph courtesy of Andreas Maurer.

Nothing remains of the small 20-foot telescope of 12-inch aperture; the same is true of the 'X-foot', sold to Lucien Bonaparte. But two 7-foot reflectors of 6.2-inch aperture are complete and survive, one in the possession of Mr John Herschel-Shorland, the other at the Science Museum, London. Of Caroline Herschel's telescopes, only some optical parts of the 2-foot comet sweeper survive at the Historisches Museum am Hohen Ufer Hannover.

Of some interest is the fate of the Spanish 25-foot telescope, aperture 24 inches, delivered to King Carlos IV of Spain in 1802. Herschel ground and polished two mirrors for the telescope in September of 1796 and the telescope was built and erected at Slough for extensive testing shortly thereafter. Herschel must have regretted parting with this telescope, as years later when writing about the 40-foot reflector, he said: 'The difficulty of repolishing its mirror, which is tarnished, and preserving or restoring its figure when lost, is so great that if a larger telescope than a 20 ft. should ever be wanting, I am of opinion that one of 25 ft. with a mirror of 2 feet in diameter, such as I have made and which acted uncommonly well, should be a step between the 20 and 40 feet Instruments.'

The telescope was accompanied by extensive documentation, which included a series of drawings of the entire telescopic apparatus. Upon its arrival in Spain, the telescope was erected near the Madrid Observatory, with observations beginning in August of 1804. This was near the beginning of some dark times in Europe, however, with war spreading rapidly across the continent. Spain was not spared and before the arrival of French troops in Madrid in 1808, many parts of the Spanish telescope, including the mirrors, winches, ropes and drawings were hidden away. Only the wooden framework of the telescope remained and when the troops occupied the observatory, it was dismantled and used as firewood.

In 2002, a project to rebuild the 25-foot telescope was set in motion, and using the preserved documentation, a replica of the telescope was built at a shipbuilding yard in Bilbao, Spain. The project is now complete and the telescope is on display next to the Madrid Observatory, which also houses the two original speculum metal mirrors. The telescope stands in mute testament to a man who, almost singlehandedly, breached the gates of heaven two centuries before.

The deep sky before Herschel

THE NIGHT SKY OF THE PRETELESCOPIC ERA was a decidedly uncomplicated place. The heavens contained the fixed stars, the luminous band of the Milky Way, the moon and the five wanderers or planets, all known since before recorded history. Occasionally a brilliant comet swept across the sky; there were also shooting stars which appeared and disappeared in the blink of an eye. And very rarely, a brilliant 'guest star' would appear where no star had shone before and then slowly fade over the following months, disturbing the otherwise immutable starry vault.

Beyond this were a handful of cloudy spots in the sky, nature unknown, but like the stars they did not move and so could not be atmospheric phenomena. There was the hazy cloud of stars in the tail of Leo, named Coma Berenices, and the compressed grouping of tiny stars known as the Pleiades. There was a hazy patch of light in the Cassiopeia Milky Way which would one day be known as the Double Cluster. And finally there was the luminous patch in Cancer called the Praesepe.

Slowly, additional cloudy spots were noted in the heavens. Ptolemy, in his *Almagest*, mentioned seven, one of which would later be known as the star cluster M7 in the constellation Scorpius. In the tenth century, the Persian astronomer Al-Sûfi made the first recorded mention of the Andromeda Galaxy, and he is probably the first person to note the star cluster surrounding the star Omicron Velorum, now catalogued as IC2391.

Then for the next 500 years no new nebulae were recorded until Portuguese sailors made their way south of the equator and discovered the Clouds of Magellan, no doubt known since time immemorial to the indigenous people south of the equator but unknown to the civilizations of Europe, the Middle East and Asia.

When Galileo and other early astronomers first applied the telescope to astronomy, they understandably concentrated their attention on the moon and planets, but slowly they turned their telescopes to other mysteries of the heavens. Galileo himself observed that the Praesepe was actually a cluster of stars and counted almost 40 telescopic stars involved with the Pleiades. His friend Nicholas Piersec is credited with the discovery of the Orion Nebula in 1610 and the German astronomer Simon Marius was the first to turn a telescope to the Andromeda Galaxy. Otherwise, the exploration of the deep sky beyond the naked-eye stars was a slow and hesitant process.

The first true deep-sky observer was an obscure Italian priest born in Sicily in 1597, Giovanni Battista Hodierna. A man of humble origins, Hodierna developed an interest in astronomy at a young age and by the year 1618 owned a simple Galilean telescope which had a magnification of 20×. With this telescope he observed three comets over the next two years. Ordained as a priest in 1622, he eventually found his way to Palma di Montechiaro, where he served as parish priest, and later became the court astronomer.

Like most astronomers of the era, he observed the planets, including Saturn, and he was in correspondence with the Dutch astronomer Christiaan

Huygens. He produced a few astronomically related pamphlets but his most compelling work, '*De systemate orbis cometici; deque admirandis coeli characteribus*' (Of the systematics of the world of comets, and on the admirable objects of the sky) published in 1654, received little circulation outside his native Sicily and was unknown to the astronomers of the era. Apart from a brief reference to this work by the astronomer Lalande in 1803, this book was unknown until the mid-1980s when it was discovered by G. F. Serio, L. Indorato and P. Nastasi.

The book is remarkable for its catalogue of 40 deep-sky objects or asterisms, including 16 that Hodierna discovered himself with his Galilean telescope. He was the first person to try to classify clusters and nebulae according to their appearance in a telescope and his system divided deep-sky objects into three categories: *luminosae*, stars visible to the naked eye; *nebulosae*, those appearing nebulous to the naked eye but resolved in a telescope; and *occultae*, nebulae appearing as unresolved nebulosity even in a telescope.

Not all the objects recorded by Hodierna were true deep-sky objects; like others of the era he included some naked-eye asterisms that otherwise appeared hazy. Nevertheless, his achievements are remarkable for the times. In the luminosae class, he discovered the stellar association surrounding Alpha Persei, now known as Melotte 20, and the spectacular open cluster NGC6231 in Scorpius (recorded by the Abbé de Lacaille at the Cape of Good Hope in 1751–1752).

His discoveries in the nebulosae class included the Messier objects M6 (later credited to De Chéseaux), M8 (recorded by John Flamsteed in 1680), M36 and M38 (Le Gentil recorded these in 1749) and M37 (located by Charles Messier in 1764).

Other likely discoveries made by Hodierna include M41, M47, NGC752 (a star cluster in Andromeda), M34, NGC2169, NGC2175, NGC2362, NGC2451 and M33, the first galaxy to be discovered in northern skies since Al-Sûfi's description of the Andromeda galaxy almost 700 years before. All these objects remained unknown to astronomers

for decades and in many cases centuries after Hodierna observed them with his small telescope.

In the 90 years following the publication of Hodierna's catalogue, the discovery of deep-sky objects was a very slow process, individual astronomers seldom recording more than one or two new objects. In fact, beyond M8 and M41 (already recorded by Hodierna) only nine nebulae were discovered.

Modest progress was made by the Swiss astronomer Philippe De Chéseaux, who listed a total of 20 nebulae in the years 1745–1746, including eight discoveries of his own (the Messier objects M4, M6 (already recorded by Hodierna), M16, M17, M25, M35 and M71, as well as NGC6633 and IC4665). Unfortunately, as with Hodierna, the news of his discoveries was not widely circulated, and his priority in discovering the aforementioned objects was not generally recognized until the nineteenth century.

For significant advancement in deep-sky astronomy, the work of the Abbé Nicholas-Louis de Lacaille must be recognized. In the years 1751–1752, Lacaille travelled to the Cape of Good Hope to perform a series of scientific programmes which included the first comprehensive catalogue of 9776 southern stars, as well as a catalogue of 42 nebulae and star clusters.

Lacaille's instruments included a mural quadrant equipped with a small refractor telescope, which magnified eight times, as well as simple refractors of 15-foot and 18-foot focal length.

Lacaille's discoveries were organized into three classes: Class I nebulae without stars; Class II nebulous stars in clusters; and Class III stars accompanied by nebulosity. Although his catalogue had some dubious entries, including objects which could not later be positively identified, he discovered at least 22 genuine deep-sky objects, as well as recording Omega Centauri (first identified as a cluster by Edmund Halley in 1677) and NGC6231 (discovered by Giovanni Hodierna in 1654). Lacaille's discoveries include some of the southern skies' most spectacular objects, including the globular cluster NGC104 (47 Tucanae), NGC2070 (the Tarantula

Nebula), M83 (the spectacular spiral galaxy in Hydra) and NGC3372 (the Eta Carinae Nebula).

While Lacaille was at the Cape of Good Hope, a young astronomer in France was beginning to emerge as one of the most celebrated astronomers of the eighteenth century, a young man whose name is well known even to this day.

Charles Messier's career as an astronomer began in 1751 when he began his employment in Paris under the tutelage of Joseph Nicholas Delisle. He was 21 years old and was engaged as a draughtsman. He was also trained in the use of astronomical instruments and by 1754 had become established as an observer at the Marine Observatory at Paris, working with Delisle in a search for comets. At this time, anticipation was high for the return of Halley's comet, which the English astronomer had predicted would return sometime in the year 1758. Messier searched diligently for the comet but was misled by the charts Delisle gave him and credit for recovery of the famous comet went to a German amateur astronomer named Palitzsch, who located it on Christmas night 1758.

This was one of the few times in his career when Messier was not the first to observe a comet; he discovered 16 over a 39-year period ending in 1798, and observed many more. Although renowned in his day as a great comet hunter, it was his catalogue of nebulous objects which brought him lasting fame.

On 28 August 1758, while following the comet of that year as it made its way through the starfields of Taurus, Messier noted what he called a whitish light that was extended and tapered like a flame. This was the Crab Nebula, discovered in 1731 by the English astronomer John Bevis. Two years later, Messier came across another nebulous object, a globular cluster in Aquarius discovered in 1746 by J. D. Maraldi, which he thought resembled the nucleus of a comet.

It is unknown precisely when Messier decided to compile his catalogue of 'false comets', but he made his first discovery on 3 May 1764, when he recorded the globular cluster in Canes Venatici which became known as M3. Five nights later he recorded De Chéseaux's globular cluster near Antares and on 23 May he observed four widely separated objects that had been previously discovered. These were the globular cluster in Serpens discovered by Gottfried Kirch in 1702, the De Chéseaux star cluster near the tail of Scorpius, Ptolemy's star cluster located nearby and John Flamsteed's cluster in Sagittarius discovered in 1680. These objects subsequently became M5, M6, M7 and M8. For much of the rest of the summer and autumn of 1764, Messier reobserved as many previously discovered nebulae as he could and at the same time made many discoveries of his own. By the end of October 1764, his list of nebulous objects had reached 40 objects, including 16 which he could claim for himself.

Somewhat surprisingly, this burst of exploration and discovery was not repeated by Messier in the following year or indeed for several years thereafter. Messier continued his comet hunting campaign, as well as solar observations and observations of transits, occultations and eclipses, so he was certainly not idle. In 1765 he recorded a single star cluster, the previously discovered M41. Nothing was recorded for the next three years but by 1769 he had decided to publish his list and added M42, M43 (the Orion Nebula complex), M44 (the well-known Praesepe) and M45 (the even better known Pleiades).

Messier's catalogue was published in the *Mémoires de l'Académie Royale des Sciences 1771* as well as Lalande's *Ephémérides 1775–1784*. In addition to his list of 45 nebulae and star clusters, Messier also reported on objects that had previously been thought of as nebulous but which he considered, correctly as it turned out, nonexistent.

Messier began recording star clusters and nebulae again in 1771, coming upon M46, M47, M48 and M49 on 19 February of that year. Rather than performing a systematic search for nebulae, Messier recorded objects which he happened to notice while following or searching for comets. One object was discovered in 1772, none in 1773 and two in 1774. The globular cluster M53 was found on 26 February 1777 and the following year two globular clusters in Sagittarius were recorded on the same night, 24 July.

The year 1779 saw quite a bit of activity. Eight objects were recorded though not all were original discoveries. And object number 63, the last discovered that year, was recorded by Pierre Méchain, who would become a close associate of Messier for many years thereafter.

Méchain was fourteen years younger than Messier and was a calculator at the Marine Observatory at Versailles. A capable mathematician and excellent observer, he discovered 8 comets during his career as well as 28 nebulae. Out of deference to Messier, he reported his discoveries to the older astronomer rather than start a catalogue of his own. Messier was always careful to credit Méchain for his discoveries, as he was for the discoveries of others.

In early 1780, Méchain recorded the objects which became known as M65, M66 and M68, along with M64 and M67, which had been discovered by the astronomers Bode and Koehler, respectively. Messier submitted his list of supplemental objects to the French almanac, the *Connaissance des Temps*. This catalogue included not only Messier's list, which now totalled 68, but also the 42 southern objects discovered by Lacaille in 1751–1752.

Through the rest of 1780, the Messier catalogue continued to grow, with 11 more objects recorded, including 7 discoveries by Pierre Méchain. The year 1781 was even more fruitful: 24 objects were added to the list making a total of 103. The complete list was subsequently published in the *Connaissances des Temps* for 1784. Unfortunately, the 102nd object in the catalogue was a duplicate observation of M101, which Méchain admitted in 1783. Objects which later became known as M104, M105, M106 and possibly M108 and M109 were also discovered by Méchain that year and in addition he noted the companion galaxy to M51, subsequently recorded by William Herschel and finally catalogued as NGC5195. The last of the Messier objects discovered by Méchain, M107, a globular cluster, was recorded in April 1782. The object which subsequently became known as M110, a satellite galaxy of M31, was never described by Messier, though he did include it in a drawing of the Andromeda Galaxy and its other satellite, M32, made in 1773.

The Messier catalogue was a remarkable document for the time. It virtually eliminated the 'nebulous asterisms' which had been a part of so many lists that had been published previously. It contained only two newly introduced asterisms, the double star M40 and the coarse grouping of four stars in Aquarius, M73. Three objects were considered 'missing' for many years, M47, M48 and M91, but M47 and M48 were subsequently correctly identified with bright star clusters. Only M91 remains as something of a mystery, though it is now identified as the barred spiral galaxy NGC4548 in Virgo. And the catalogue is unique in being the only catalogue created by a single observer whose nomenclature (in this case, the Messier object number) survives down to the present day.

The catalogue is noteworthy for the accuracy of its positions, its inclusion of some of the sky's most compelling star clusters, nebulae and galaxies, and the detail that Messier strove for in his descriptions of the objects. Unfortunately, Messier was saddled with decidedly mediocre equipment at the Marine Observatory, which compromised his descriptions of many objects, especially bright globular star clusters which he frequently described as 'nebula without star'. Of the dozen or so telescopes that Messier used, some were small, simple refractors of short focal length, generally 3 feet or less. The observatory also had an 8-inch Newtonian with a speculum metal mirror of poor light-gathering ability. The largest telescope that he used was a 7.5-inch aperture Gregorian reflector with a magnification of 104× and he also had access to at least one 3.5-inch aperture achromatic refractor which magnified 120×. Messier always claimed to be satisfied with the equipment available to him, saying he only needed telescopes sufficient to show the comets he sought. But when trying to take measurements of some of the galaxies he discovered, he often commented: 'The least illumination of the micrometer wire causes the nebula to disappear.' It must have been frustrating to work with such equipment.

The Messier catalogue made all other catalogues published up to that point superfluous. The German astronomer Johan Bode compiled his *Complete Catalogue of Nebulous Stars and Star Clusters* in 1777. This was a list of 75 objects, including three of his own discoveries, which Messier subsequently catalogued as M53, M81 and M82.

Unfortunately, Bode observed only two dozen of the objects himself. He perpetuated the inclusion of many 'nebulous asterisms' which Messier's work revealed as nonexistent. His catalogue also contained positional errors and generally does not match the rigour of Messier's catalogue. It is largely forgotten today.

At this point, mention must be made of one last astronomer of the 'pre-Herschel' era who made significant contributions to deep-sky astronomy: Caroline Herschel.

Herschel's sister was employed as his assistant for a number of years. Although her duties were frequently mundane, such as fetching items and equipment for her brother, copying published papers he wished to read, or preparing William's work for publication, by 1782 he began to consider an expanded role for her. He suggested that she might make use of her free time by sweeping the heavens for comets, as well as double and multiple stars and undiscovered nebulae. Originally he suggested she use his small 20-foot reflector of 12-inch aperture to reobserve the double stars he had catalogued to determine if there were any changes in position angle or distance since his measures. But Caroline considered the telescope difficult, if not down right dangerous, to use and was resistant to the suggestion. Her brother then built her a telescope for her exclusive use. This was known as the 'tube with two glasses', a simple refractor not much bigger than a finderscope.

Caroline had little enthusiasm for the task at first and little success to match. Initially, the objects of interest she located were either double stars that her brother had already recorded, or chance recoveries of objects catalogued previously by Messier, such as M13, M27, M36 and M37. After a few months,

however, she started to actually enjoy her time under the stars and success soon followed. On the night of 26 February 1783, she made her first discovery, the star cluster NGC2360. She also evidently came across M47 that same night, one of Messier's 'missing' objects and observed M41 and M93. On 4 March 1783 she recorded the asterism NGC2349 and four nights later, on 8 March, recovered another 'missing' Messier object, M48.

William Herschel took note of the success of his sister and thereupon built her a better telescope, a Newtonian reflector of 2-foot focal length, aperture 4.2 inches with a magnifying power of 24×. Later that summer she swept up the undiscovered cluster NGC6866 in Cygnus, NGC6633 in Ophiuchus (recorded by De Chéseaux in 1746) and independently recovered M110, a satellite of the Andromeda Galaxy. Her success continued into the autumn when she located the Sculptor spiral galaxy NGC253, the Cassiopeia star clusters NGC189, NGC225, NGC659 and NGC7789, and probably rediscovered the star cluster NGC752 in Andromeda, originally catalogued by Giovanni Hodierna.

In an eight-month period, Caroline had discovered 13 deep-sky objects, a pace that clearly was the equal of any deep-sky observer before her and notably better than her illustrious brother, who up to this time had discovered only one deep-sky object, the planetary nebula NGC7009 in Aquarius. There can be little doubt that Caroline's success encouraged her brother to speed up completion of his large 20-foot reflector and begin his systematic sweeps of the skies for nebulae.

The autumn of 1783 saw the end of Caroline's career as an independent observer, at least temporarily. In a few short months she embarked on her career as her brother's indispensable assistant at the large 20-foot reflector, recording by candlelight his discoveries at the eyepiece and copying the previous night's work into the register during the day. Thereafter, Caroline's time at the telescope was restricted to evenings when her brother was away or did not otherwise require her services. As a result, her deep-sky discoveries were infrequent at best. In

May of 1784 she discovered the open cluster NGC6819 in Cygnus and her next, and last, deep-sky object was the cluster and nebula NGC7380 in Cepheus, recorded on 7 August 1787.

It is not surprising that Caroline's record of deep-sky discovery was so meagre after 1783. By this time, the 20-foot reflector was a formidable instrument of discovery. William Herschel would sometimes record dozens of nebulae nightly. The light grasp of his telescope was far superior to that of his sister's. Nevertheless, Caroline now made her mark as an observer by discovering comets.

In the summer of 1786, William and Alexander Herschel were in Europe, the main purpose of the trip being to deliver a telescope that King George III had ordered to Göttingen University. Caroline remained in Slough, working on her brother's catalogue of clusters and nebulae, but on the evening of 1 August while sweeping with the 2-foot Newtonian, she made her first comet discovery, which she confirmed on the following evening. Reporting the discovery to the secretary of the Royal Society and her brother's friend, the amateur astronomer Alexander Aubert, she became only the second female comet discoverer in history (the first, Maria Margarethe Kirch, discovered a comet in 1702). Her next discovery was on 21 December 1788 and two more followed in 1790. Her independent achievements as an observer earned her the respect of astronomers both in England and abroad. She finished her comet-hunting career with eight discoveries and no other woman matched her achievements until the twentieth century. As well, in a male-dominated profession, she is the only female discoverer of objects listed in the *New General Catalogue* (Dreyer, 1888).

Surveyor of the skies

O N THE OCCASION OF WILLIAM HER-
SCHEL'S ELECTION to the Royal Society
on 7 December 1781, his friend William
Watson presented him with a copy of Charles
Messier's 'Catalogue of nebulae and star clusters',
the now famous list of 103 objects which appeared
in the *Connaissance des Temps* for 1784.

At the time, Herschel was preparing to embark on
his third survey of the heavens. The first, carried out
in the years before 1779, was a survey of all stars of
magnitude 4 or brighter, examined with a 7-foot
reflector of 4.5-inch aperture. The second
survey had begun in August 1779, and was an
examination of all stars brighter than magnitude 8
with a 7-foot reflector of 6.2-inch aperture. The
primary result of that survey was the compilation
of a catalogue of double and multiple stars, numbering 269 in all, which was due to be published in the
Royal Society's *Philosophical Transactions* early in
1782. An unexpected byproduct of the survey had
been Herschel's discovery of the planet Uranus, the
event which transformed Herschel into a professional astronomer of world renown.

The third survey was again concerned primarily
with cataloguing double stars and when published
by the Royal Society in 1785 resulted in another
434 double and multiple stars being recorded, bringing Herschel's total to 703.

By the beginning of 1782, William Herschel had
been observing the heavens telescopically for
almost nine years. Yet curiously, in all that time
he had not discovered a single star cluster or nebulae, and had evidently only observed bright

objects like M31, M42 and M13, albeit on a number of occasions. Certainly, he did not lack for
opportunity for there are a number of deep-sky
objects which are quite close to prominent stars in
the sky and many of them should have been bright
enough to be seen in Herschel's fine instruments
during his search for multiple stars. A few examples
are: NGC404, visible to the NW of the bright star
Beta Andromedae; NGC2024, bright emission nebula, immediately E of Zeta Orionis; M109, the galaxy which follows Gamma Ursa Majoris.

Yet it would appear that Herschel's single-minded
pursuit of multiple stars caused him to miss the
opportunity to add to the slowly growing number
of known deep-sky objects in the sky.

Charles Messier was well known to William
Herschel and Watson's gift piqued Herschel's interest in the French comet hunter's catalogue of nebulous objects. In August of 1782, Herschel began
to examine the Messier catalogue with his 7-foot,
10-foot and small 20-foot reflectors.

Over the course of the next year, he had occasion
to observe almost half the entries in the list and was
pleased to note that many of the objects that Messier
had described as 'nebula without star' were either
partially or totally resolved in Herschel's superior
telescopes. The prevailing opinion among astronomers of the age was that the nebulae were most
likely extremely remote clusters of stars, unresolvable with the telescopes available. Herschel shared
this opinion and the results that he was getting with
his own telescopes fuelled his desire for even larger
instruments of greater power.

By happenstance, most of the Messier objects that Herschel observed in 1782 and 1783 were either open or globular clusters. He missed many of the Coma–Virgo galaxies as well as many of the specimens in Leo, Canes Venatici and Ursa Major, so his belief in the resolvability of all nebula might have been tempered somewhat had he had occasion to observe these as well. He had observed the Crab, the Dumbbell and the Ring Nebula, as well as the galaxies M33 and M101 and mistakenly thought they were all on the verge of resolvability. He even thought that the Ring Nebula was in reality a ring of stars. While continuing his survey for multiple stars, Herschel began to realize that the nature of the nebulae might be an interesting field to investigate and with this in mind, he began to build his large 20-foot telescope of 18.7-inch aperture.

Added to this was the success that his sister Caroline was having in her discovery of nebulae. Messier's catalogue and Lacaille's much earlier southern hemisphere catalogue encompassed almost all of the true star clusters and nebulae known up to that time. The total number of discoveries then amounted to 138 and yet Caroline, a completely untrained observer working with very small telescopes, had added about a dozen new clusters and nebulae in less than a year of observing. It was obvious to Herschel that the Messier and Lacaille catalogues were not the final word on the nebulae. As with the multiple stars, it appeared that much remained to be discovered.

Herschel had his own, albeit modest, proof of this on the evening of 7 September 1782. Engaged in his pursuit of double and multiple stars on this occasion, he happened upon a round, bright and small disc while examining the field around the star Nu Aquarii. The location is only about 5° from the ecliptic and one can imagine that Herschel might have briefly thought he had stumbled upon yet another new planet. It was, however, a planetary nebula, now known as NGC7009, the first deep-sky object that Herschel discovered. He returned to it another ten times in the next two years, trying to understand what it was.

Through much of the summer and early autumn of 1783 Herschel constructed his large 20-foot telescope in the back garden at Datchet and by October of 1783 it was in a sufficient state of completion for him to embark on his sweeps of the skies. His first sweep took place on the evening of 23 October 1783 and five nights later he achieved his first success, discovering the galaxy NGC7184 in the constellation Aquarius. Two nights later he came upon two more objects in Sculptor, the galaxies NGC7507 and NGC253, though the latter object had been discovered by his sister Caroline five weeks previously.

Initially, Herschel's technique involved literally 'sweeping' the sky. Perched on a temporary platform, he would point the telescope at a given area of the sky and slowly draw the telescope eastward, covering about 12–14° of sky. He used a magnification of 157×, which provided a field of view a little more than 15.0′ in diameter. If nothing was noted, the telescope was returned to the starting point and raised or lowered by about 10.0′ and another strip of sky was observed. Anywhere from 10 to 30 of these oscillations would constitute a 'sweep'. Any object that he chanced upon would be noted, at which point he would descend from the telescope and head indoors, where he would write down his observations by candlelight and then return to the telescope for another sweep.

There were problems with this observing method, not the least of which was the fact that he was losing valuable observing time climbing up and down the telescope framework, writing his observations down and taking many minutes to recover his dark adaptation after exposing his sight to direct candlelight. Herschel also found the massive telescope heavy to move back and forth and he was concerned about the accuracy of the positional measurements of the objects he was recording. Over the course of almost two months observing in this fashion, Herschel managed to discover only six new star clusters and nebulae and by the forty-first sweep, the impracticality of this method of observing became obvious to him.

He then settled upon a vertical sweeping method, whereby the telescope was raised or lowered slowly by about 2°, allowing the natural, westward progression of the sky through the telescope field to slowly bring new stars and other objects into the field of view. Hiring a workman to operate the vertical motion of the telescope freed Herschel to concentrate on observing and after a few sweeps in this manner, Herschel then pressed his sister Caroline into action as his assistant.

Herschel then spent all of his time at the telescope. If anything of note appeared in the field of view, he would call out his observations to his sister. Caroline was stationed at a nearby window and working by candlelight, wrote down her brother's observations and repeated them back to him. If there were any errors, they could be corrected immediately.

The results of this new method of observing were dramatic. By the end of February 1784, Herschel and his sister recorded a total of 103 new nebulae and star clusters, matching the total number of objects in Messier's celebrated catalogue, which had taken the French observer 23 years to compile. A month later, as Herschel swept through the rich galaxy fields of Leo, Coma Berenices and Virgo, he more than doubled his discoveries to 239.

Through April and May the Herschels continued their furious observing pace, slowing briefly only in June and the early part of July as the brief, and bright, northern summer nights took their toll. By the end of June, 1784, only eight months after beginning their sweeps, the Herschel discoveries stood at 445. As the summer progressed and the nights lengthened, the pace quickened again. There were few nights on which Herschel was unsuccessful in his search, indeed most nights he discovered a dozen or more new nebulae. Special note should be made of the night of 11 April 1785. The sweep that night began in the constellation Leo Minor and progressed eastward into Coma Berenices. Unknown to Herschel, his sweep took him over the distant galaxy cluster Abell 1656 and as the night progressed, field after field of tiny, faint nebulae appeared in his eyepiece. By the end of the night, the Herschels had recorded 74 new nebulae, an absolutely astonishing feat. By 26 April 1785 the

Herschels' total number of discoveries stood at exactly 1000.

While observations continued, the work of the previous 16 months was organized for publication in the *Philosophical Transactions of the Royal Society*. This took the better part of a year to complete. Titled 'Catalogue of one thousand new nebulae and clusters of stars' the paper was read before the society on 27 April 1786. After a brief general introduction describing his telescope and other equipment and the method by which he worked, Herschel presented the catalogue proper. Herschel organized his discoveries into eight classes, the first three classes described nebulae by their apparent brightness, from bright, through faint, to very faint. The fourth class grouped together planetary nebulae as well as nebulae with unusual shapes or appearance, while the fifth class was made up of very large nebulae. The sixth, seventh and eighth classes described resolved star clusters from rich and compressed to large and scattered.

In each class Herschel numbered and ordered his objects by date of discovery. Next the location in the sky of each object was described in reference to a known star, its position in what is now understood as right ascension being described as either preceding or following the reference star in minutes and seconds of time, while its declination was measured as either north or south of the star's position in degrees and minutes. The next column listed the number of times the object had been observed and this was followed by a description of the visual appearance of the object at the eyepiece. In the interest of brevity, Herschel developed a system of abbreviation whereby a single letter, either capitalized or in lower case, would stand for a word or short series of words. This system of abbreviation was adopted and expanded upon by John Herschel and all later observers and finally codified in Dreyer's *A New General Catalogue of Nebulae and Clusters of Stars* (1888). As an example, Herschel described the galaxy NGC4168 as: 'pB. pL. R. vgmbM. r.' (Pretty bright. Pretty large. Round. Very gradually much brighter to the middle. Resolvable.)

The catalogue was an incredible achievement. In a period of only 16 months, William and Caroline Herschel had increased the number of known nebulae and star clusters seven-fold. When Herschel published his two catalogues of multiple stars, it was impossible for contemporary astronomers to confirm his discoveries; the telescopes available could not approach the resolving power of Herschel's instrument. On one occasion, the Astronomer Royal, Nevil Maskelyne, took pains to inform Herschel that the amateur astronomer Alexander Aubert had succeeded in confirming that Polaris was a double star, one of Herschel's early discoveries. It was only when Herschel brought one of his 7-foot reflectors to Greenwich in June 1782 and demonstrated its abilities on double stars to the astronomers present that his discoveries could be well and truly accepted.

The circumstances surrounding his catalogue of nebulae and star clusters were even more extreme. His large 20-foot reflector was far and away the most powerful telescope in the world and his abilities as an observer made it doubly so. Only the brightest of its entries could be observed by other astronomers and the entries in the catalogue were not systematically observed until John Herschel reviewed his father's catalogue 40 years later.

All of the observations for the first catalogue of 1000 nebulae had been conducted from the back garden of the house at Datchet. It was a low-lying area near the river Thames and prone to dampness. Nightly, and especially during the winter months, Herschel was exposed to the cold and damp and he suffered from acute fevers as a result. The house itself was in a poor state of repair and subject to leaks. In June of 1785, therefore, the Herschels moved to a house at Clay Hall in Old Windsor but their stay there was brief as the Herschels found themselves at odds with the landlady. When the Herschels made improvements to the property, the landlady took it as a cue to increase their rent and finding the situation untenable, the Herschels moved again in April of 1786 to the village of Slough and a home that would eventually become

known as Observatory House. It was ʃo
William Herschel spent the rest of his days.

The Herschels continued their observati
nebulae through May of 1785 but suspended ι
observations during the summer months. It w
around this time that Herschel was petitioning the King for funds to build a 40-foot telescope, and he was ultimately successful in his efforts. In the middle of August, the Herschels resumed their sweeps at the same unrelenting pace.

In retrospect one can only be astonished at the relentless drive, determination and energy of Caroline and William Herschel. Not only were they awake every clear night observing, but Herschel's days were occupied supervising workmen in the construction of the 40-foot telescope. He busied himself writing papers for publication, based on his nightly observations and at the same time was engaged in the construction of telescopes for sale to clients in England and abroad. Caroline worked constantly organizing and copying the results of the observations of the nebulae, as well as her brother's 'star-gages' and observations of the planets. She was involved as well in copying her brother's papers for publication and doing the day to day work of running a household. When she found the time, she did a little observing of her own. One may rightly ask when either of them found time to sleep.

As good as the large 20-foot reflector was, a disadvantage of the telescope was the relatively poor efficiency of polished speculum metal in terms of light-gathering ability. Used in the Newtonian form, a considerable amount of light was lost owing to the two reflections necessary to bring the light to the eyepiece. In Herschel's early days as an observer, he had tried using one of his 10-foot reflectors without the secondary mirror, bringing the image to focus at the top end of the tube. He was pleasantly surprised by the resulting brighter image and during the twenty-ninth sweep with the large 20-foot in November of 1783, briefly removed the secondary again. Despite the bright image, the telescope was not convenient to use in this configuration and Herschel went back to the Newtonian cus, with the eyepiece set at an angle of 45° from ıe top of the tube.

In September of 1786, however, during his six hundredth sweep of the heavens, Herschel abandoned the Newtonian configuration for good; thereafter all observations with the large 20-foot and 40-foot reflectors were made with front-mounted eyepieces, which became known as the Herschelian form. This required that the objective mirror be tilted slightly to bring the focus to the edge of the tube, where the focuser was placed. Ordinarily, this would result in a distorted image but the telescope's relatively long, f/12.8 focal ratio meant that the imperfection of the image was minimal and far outweighed by the brighter image that the telescope now provided.

Although their progress was not as rapid as it had been for the first catalogue of nebulae, the Herschels continued to record new nebulae at an impressive rate. Herschel's principal daytime preoccupation from 1786 to 1789 was the construction of the 40-foot telescope, whose framework rose slowly in the back garden of Observatory House. He continued to write papers and build telescopes and in May of 1788 at the age of 49, he married the widow Mary Pitt. It is therefore not surprising that the Herschels took 43 months to discover their next 1000 nebulae, reaching the milestone on 3 December 1788.

It took the Herschels only four months to prepare their data for publication and the paper devoted to the second 1000 new nebulae was read before the Royal Society on 2 April 1789.

The large body of observational evidence that Herschel had acquired over the preceding five years furnished him with the opportunity to speculate both on the nature of the objects that he had seen and the organization of the universe, what he called the 'construction of the heavens'. Despite the superiority of the equipment he used, however, Herschel was at a great disadvantage here. He could see galaxies, star clusters and nebulae but he lacked the sophisticated equipment which might allow him to analyse what he saw. Circumstances forced him to base his theories on assumptions for

which he lacked solid evidence and many of his assumptions were completely wrong.

During the first year of his sweeps, Herschel engaged in an activity which he called 'star-gaging', the purpose of which was to determine the structure of the Milky Way and the sun's place in it. Herschel would simply count the number of stars he could see in successive fields of his telescope, examining representative samples of the sky. He felt that this form of 'sky census' would lead him to determine how the stars of the Milky Way were organized in space. Unfortunately, he based his analysis on three assumptions which would all prove to be false: first, that all stars were roughly of the same brightness; next that they were distributed more or less evenly throughout space; and third that his telescope was powerful enough to allow him to see to the edge of the galaxy in any direction.

His adherence to the first premise is especially puzzling. Even in Herschel's time there was fairly widespread recognition that stars very likely came in a wide variety of sizes and brightnesses. Some were even seen to vary in brightness over time. And Herschel must have realized that he had direct observational evidence of the wide brightness range of stars. He had already discovered dozens of star clusters, presumably with their member stars all at roughly the same distance, whose individual members exhibited wide brightness differences. Using star-gages would therefore tell the observer little about the actual distribution of stars in space. Herschel was nothing if not 'results-oriented', however. He was sometimes quite willing to overlook errors in methodology in order to formulate his theories.

It was during a 'star-gaging' session that Herschel missed the opportunity to discover the dark nebulae, which are some of the largest structures in our galaxy. While counting rich fields of stars near the border between Scorpius and Ophiuchus, he saw that the stars suddenly disappeared and very few stars could be seen for several fields before they suddenly appeared again, richer than ever. Here, he thought, was a 'hole in the heavens', especially as it was next to the bright globular cluster M80, which Herschel had earlier described as one of the richest and most condensed clusters of stars that he had seen. He immediately speculated that all the stars that had previously occupied the 'hole' had been gathered together to form the globular cluster. What he was actually seeing, of course, was a dense cloud of interstellar matter that appeared dark because illuminating stars were buried within and behind the cloud.

Herschel's star-gages led him to assume that the Milky Way was a highly flattened system of stars which he thought had a very irregular outline. He speculated that at least some of the nebulae he was seeing, in particular many of the flat, glowing rays of light, were very probably immense systems of stars like the Milky Way. He even speculated that inhabitants of these distant systems looking back toward the Milky Way would see a flat ray of light.

Up to this time Herschel had seen little evidence that the nebulae were anything but clusters of stars observed at immense distances. He well understood that they were far from evenly distributed in the sky and he thought they were organized into systems which he called 'stratum'. He identified at least two, one of which he called the stratum of Coma Berenices, where he thought that the nebulae which he observed there, presumably all distant clusters of stars, were organized in a great wedge pointed towards the sun, the nearest of which was the Coma Berenices star cluster. The other stratum which he recognized was in Cancer, involving the galaxies there in a system whose nearest member was the Praesepe cluster, M44.

Herschel resisted the idea that true nebulae, clouds that somehow shone, existed. Even the large and bright Orion Nebula, which he observed countless times, was to him simply a huge and extremely distant system of stars, unresolvable in the best of his telescopes.

This assumption on Herschel's part was about to change, however. Through 1789 and 1790, the Herschels continued to catalogue new nebulae and on the evening of 13 November 1790, Herschel

discovered 'a star of about the 8th magnitude, surrounded with a faintly luminous atmosphere, of a considerable extent'. This object was the planetary nebula NGC1514 in Taurus and its visual appearance led to a conclusion that was difficult to ignore. The star and the glowing cloud were undoubtedly connected, yet if the cloud was simply a very distant cluster of stars, the central star would probably be millions of times brighter than its faint neighbours. How could this be so?

Up to this time Herschel had discovered several true planetary nebulae but they were almost invariably the brightest specimens of this class, such as NGC7009 (the Saturn Nebula), NGC6543 (the Cat's Eye) and NGC3242 (the Ghost of Jupiter), all with dense, bright atmospheres surrounding central stars which were often difficult to distinguish. He had seen stars which he described as surrounded by 'milky chevelures' such as NGC2023, but rather than speculate on their possible controversial nature, he simply assigned them to his catch-all Class IV and continued to observe.

Now he rushed a paper into print which discussed his newly discovered true nebula. 'On nebulous stars, properly so called' was published in the *Philosophical Transactions* of 1791; in it, after briefly discussing the appearance of NGC1514, Herschel reexamines many of his Class IV objects, describing the visual appearance of each and speculating whether they are, in fact, true nebulae or not. He did an admirable job of ferreting them out, though some galaxies with bright cores such as NGC3982 and galaxies like NGC6301 with foreground stars involved, were mistakenly classified as nebulous stars. Still, it was a remarkable paper, which laid the groundwork for identifying the characteristics of true nebulae, whose identities and physical properties would not be confirmed until William Huggins' application of the spectroscope to astronomy in the latter half of the nineteenth century.

Curiously at this point, the Herschels' relentless pursuit of new nebulae began to slow. Up until the end of 1790, the Herschels had discovered more than 2300 new nebulae, an average of almost 330 objects per year. In 1791, however, they recorded only 38 and subsequent years up to 1802 were often even less productive.

It is certain that Herschel's marriage to Mary Pitt took up much of his time. By all accounts it was a happy marriage, and William and Mary travelled extensively during this period, an activity which took many weeks due to the slow pace of travel. Herschel's wife was a wealthy woman and the earlier financial pressures that Herschel felt to produce telescopes for sale were alleviated somewhat, though this activity continued. In 1792 the couple had their first and only child together, John, who was doted on not only by his parents but by his aunt Caroline as well.

Herschel's observing interests were not restricted to the nebulae; he spent considerable time observing the planets, Saturn and Uranus in particular, as well as observing the sun. All his observations resulted in papers in the *Philosophical Transactions*, which he contributed to frequently through the 1790s.

As a celebrated astronomer, many demands were made on Herschel's time, especially by royalty and other personages from all over Europe, who would make the pilgrimage to Slough to see the scientific wonder of the age, the 40-foot telescope.

Herschel published a paper in the *Philosophical Transactions* of 1795 titled 'Description of a forty-feet reflecting telescope' in which he went into considerable detail describing the mechanical construction of the great telescope, though tellingly, was completely silent on his method of producing large specula. The paper appeared six years after completion of the project; perhaps it was published in an effort to stem the tide of visitors. Certainly, he never felt the need to describe the workings of the large 20-foot or any of his other telescopes for that matter. At any rate, if the paper was intended to curtail interruptions it did not work; Slough continued to be a popular destination for scientists and the upper classes.

Many of the nebula discoveries in the late 1790s and early 1800s were of north circumpolar objects, an area of the sky relatively infrequently observed

during the early sweeps. The last sweep, number 1112, occurred on 30 September 1802, the final 14 years of observation producing a third catalogue, this time of 500 nebulae, which was published in the *Philosophical Transactions*.

Herschel was then 64 years of age and his many years of almost ceaseless activity, to say nothing of his frequent nights exposed to the cold and damp, had taken their toll. He was an old man and not in the best of health, though he lived for another 20 years. He continued to observe, though fitfully, and produced papers for publication based on his many years of observational activity. Some of his later theoretical work started to veer away from the true nature of the nebulae. He now considered many of the faint nebulae to be solar systems in the making. The majority of them were, in fact, distant galaxies; he was actually closer to the mark when he thought of them as distant, unresolvable star clusters.

The 19-year survey conducted by William and Caroline Herschel was at an end, a stupendous effort which resulted in the discovery of almost 2500 discrete objects, many of them among the most significant in the universe. Though it was William's eye at the telescope, he could not have accomplished his considerable observing feats without the able assistance of his sister, who recorded his thousands of observations virtually without error. She has a share in each one of his discoveries.

As John Herschel grew to manhood he decided to expand upon the work of his father and together they refurbished the large 20-foot telescope. At the same time, William showed his son how to grind and polish the specula for the telescope. After William's death, John Herschel retraced his father's footsteps through the heavens, reviewing his discoveries and adding many more of his own before taking the large 20-foot and a smaller refractor telescope to South Africa, where he extended his father's work on multiple stars and nebulae to the largely unexplored southern skies. The majority of the 7840 entries in J. L. E. Dreyer's *New General Catalogue of Nebulae and Clusters of Stars* (1888) were discovered by Herschels; William, Caroline or John.

In 1813, William Herschel remarked: 'I have looked further into space than ever human being did before me' but even Herschel would have been stunned to know how far into the universe he had penetrated. The most distant galaxies in Herschel's catalogue have radial velocities in excess of 20 000 km/s, implying a distance of almost 900 million light years. Dozens more are at distances in excess of 400 million light years. Astronomers did not routinely investigate objects at these distances until the mid-twentieth century, and then only with state-of-the-art spectroscopes and sensitive cameras attached to massive telescopes with mirrors almost as wide as Herschel's telescope was long.

There can be no doubt that Herschel's observations and speculations on the nature of the universe were indispensable to the cosmology of the twentieth century and laid the groundwork for our current understanding of the universe. He was one of history's greatest astronomers.

The Herschel catalogue

Visual observations

The visual observations for the following guide were carried out by the author during two principal time periods: 1992–2001 and 2006–2010. In 1992, a computer printout of the Herschel catalogue was obtained from David and Brenda Branchett of the Astronomical League. The list had been generated by Father Lucian Kemble to facilitate an ongoing project to observe the entire Herschel catalogue with modern amateur telescopes.

All objects in this guide were located by the starhopping method, whereby the observer uses star charts to proceed from a known location to a new target. The present author's procedure was to record observations on a preprinted form. In addition to identifying the target, the date and time observed, viewing conditions, location, telescope and magnification used, a written description of the object was included and further supported by a field sketch with surrounding stars placed as accurately as possible. Later, each observation was compared with a Digitized Sky Survey (DSS) image of the object in question. If the star patterns and general appearance of the object matched the DSS image, the object was recorded as seen. In the overwhelming majority of cases, the first attempt was successful. However, there were occasional faint objects or objects located in crowded fields that were not successfully recorded the first time. If the DSS image indicated a negative observation, the procedure was repeated until success was attained.

Observing locations and instruments used

While it is always best to strive for uniformity and consistency in any series of observations, many factors including observing site, weather conditions and instruments used have a tendency to vary, especially over the course of a long observing project spanning many years.

The following is a listing of observing locations and instruments used at these locations during the course of compiling the observations for this catalogue.

Mount Sutton, Québec Observations at this location were conducted during 1992. The observing site was the deck of a commercially operated ski lodge situated at the edge of the town of Sutton, Québec. The lodge itself blocked any light coming from the town and especially from the road leading up to the lodge. While there was a certain amount of sky glow in the northwest (where the town was located) observations were never conducted in that part of the sky. Although sky access was otherwise excellent, the upper ridge of the mountain (elevation: approximately 2000 feet) was in the south, effectively blocking access to the sky below 30° S latitude. The air was fairly dry at the site and dew was only occasionally a problem. Limiting magnitude was about magnitude 6.0 on the best nights. Observations of about 40 Herschel objects were made from this location. The telescope used was an 8-inch Schmidt–Cassegrain telescope on a German equatorial mount.

Robertville, New Brunswick This location is a rural village, population about 1000, situated about 6 miles southwest of the city of Bathurst, New Brunswick (population: 12 000). Observations were conducted at this location during the summer of 1992 and the summer of 1999. Although the streets of the village were well lit, the isolated location of the village meant that the sky overhead was quite dark. Care was taken in choosing an observing site so that the worst of the offending streetlights were blocked by buildings and vegetation. Consequently, sky conditions were very good and approached those of a rural location free from light pollution. On the best nights, stars to about magnitude 6.0 were visible, but the site's location near sea level (the village is near the shore of a bay which opens out eventually to the Atlantic Ocean) meant that occasional ground fog and high humidity were a problem. About 50 Herschel objects were recorded at this site. The instruments used were an 8-inch Schmidt–Cassegrain telescope on a German equatorial mount (1992) and a 15-inch Dobsonian reflector (1999).

Ways Mills, Québec   Observations were conducted from this location during the winter and spring of 1993. The site itself was an old, rural farmhouse situated between the villages of Ayer's Cliff and Ways Mills in Québec's Eastern Township region. Sky conditions were excellent at this location and only very small light domes were visible on the horizon, although a larger one from the city of Sherbrooke could be seen 20 miles to the north. The site was quite isolated and no outdoor lighting was visible in any direction. Sky access was excellent and conditions were generally very dry. On the best nights stars to magnitude 6.5 could be detected visually. About 80 Herschel objects were observed here. The instruments used were an 8-inch Schmidt–Cassegrain telescope on a German equatorial mount and a 15-inch Dobsonian reflector.

Sutton, Québec   This site is an isolated rural location situated midway between the town of Sutton and the village of Abercorn in Québec's Eastern Townships. Observations were conducted here from 1993 to 1999 from the deck and the grounds of a small country house. The site was free from local lights except for a yard light situated about 300 yards down the road on the side of a barn. The light was mounted in such a way that it was not directly visible from the observing site and only an indirect glow was noted. Sky access was good, although a high treeline to the south compromised observing somewhat from the deck. A location behind the house furnished limited access to the southern skies to a latitude of about 30° S through a gap in the treeline. The house was situated at the edge of a valley about 3 miles west of Mount Sutton and conditions were such that the site was plagued by high humidity and frequent ground fogs rising from the valley floor, especially in summer. The worst of these fogs would sometimes envelope the observing site, but usually the fogs remained in the valley. High humidity meant that generally the limiting magnitude was about magnitude 5.5 but on the best nights stars fainter than magnitude 6.0 could be seen visually. Approximately 500 Herschel objects were recorded at this location and all observations were made with a 15-inch Dobsonian reflector.

St Chrysostôme, Québec   This site is an isolated farm located about 5 miles from the village and is the location of the observatory of the Montreal Centre of the Royal Astronomical Society of Canada. The site was free of local light; however, it was only 25 miles south of the city of Montreal and the entire northern part of the sky was affected by a light dome which sometimes reached more than 45° above the northern horizon. By necessity, most Herschel observations at this site were of objects in the southern half of the sky. The site was sometimes subject to high humidity and localized ground fogs, though conditions were usually good. The limiting magnitude on most nights was about magnitude 5.2. About 30 Herschel objects were recorded here in 2000 and 2001, using a 12-inch equatorially mounted reflector.

Limerick, Saskatchewan   Observations were carried out at this location between 2006 and 2010. The village has a population of about 150 residents and is surrounded by ranches and farmland. The nearest major population area is the town of Assiniboia (2700 residents), which is located 12 miles to the east. Although the streets of the village are well lit, most of the offending light can be blocked off by buildings, vegetation and artificial barriers. Nevertheless, conditions are variable and the average night sees a limiting magnitude of about 5.5–5.7. The best nights occur from mid-July until the end of October when heavy tree cover blocks most of the local light. On these occasions, stars to about magnitude 6.3 can be detected visually. Sky access is good, though the treeline to the south blocks any observation below 25° S latitude. The site was usually very dry and humidity was only occasionally a problem. Proper dark adaptation was difficult at this site, however, particularly in the late autumn, winter and spring, due to reflection of artificial light sources from buildings, pavement or snow. When critical observations had to be made, they were done from a location about 1 mile southwest of the village in a protected site bordered on the east, north and west by a low treeline which effectively blocked all artificial light. Sky access was excellent here with the southern sky accessible right to the horizon. On the best nights, stars to about magnitude 6.5 were visible. The zodiacal light was frequently seen at both locations, most often during the months of January–March, and August–October. Almost 1600 Herschel objects were recorded at these locations, with all observations made with a 15-inch Dobsonian reflector.

Mayhill, New Mexico   In the autumn of 2008, observations were carried out during a stay at the astronomy enclave New Mexico Skies, which is located in the Lincoln National Forest just west of the village of Mayhill. Over the course of a six-night visit about 90 southern Herschel objects were recorded as part of a more extensive series of observations. Sky conditions here were generally excellent as the site is at a high altitude (about 7300 feet) in a dry climate. The limiting magnitude was usually about magnitude 6.5 visually with the gegenschein visible on about half of the nights around midnight and the zodiacal light was a bright, well defined cone of light in the predawn sky. Except for red light, the location was free of artificial lighting. Observations were made with a 15-inch Dobsonian reflector, although a few selected observations were made with a 25-inch Dobsonian reflector.

San Pedro de Atacama, Chile   A two-week expedition to observe the southern skies was conducted in March 2009 at the Atacama Lodge, which is a rural astronomical enclave located about 5 miles from the town of San Pedro de Atacama. The site is a high-altitude location (elevation about 8000 feet) in the Atacama desert and affords some of the best observing conditions available anywhere in the world. All thirteen nights spent at the Lodge were clear from dusk to dawn with very low humidity and cool though comfortable temperatures. Both the gegenschein and zodiacal light were seen nightly with stars to at least magnitude 6.5 seen visually. Although the primary purpose of the trip was to observe the Magellanic Clouds in detail as well as many of the showpieces of the southern sky, about 70 southern Herschel objects were recorded over a seven-night period. The observations were conducted with a 17.5-inch Dobsonian reflector, although observations of NGC4594 and NGC4038/39 were made with a 24-inch Dobsonian telescope.

Sir William Herschel's original catalogue

Herschel's original catalogue appeared in three instalments in the *Philosophical Transactions of the Royal Society* under the following titles:

1. Catalogue of One Thousand new Nebulae and Clusters of Stars, *Phil. Trans.*, **lxxvi**, 1786

2. Catalogue of a second Thousand of new Nebulae and Clusters of Stars; with a few introductory Remarks on the Construction of the Heavens, *Phil. Trans.*, **lxxix**, 1789

3. Catalogue of 500 new Nebulae, nebulous stars, planetary Nebulae, and Clusters of Stars; with Remarks on the Construction of the Heavens, *Phil. Trans.*, **lxxxxii**, 1802.

The catalogue was subsequently reprinted in *The Scientific Papers of Sir William Herschel*, volume 1 and volume 2, compiled by J. L. E. Dreyer and published in 1912 by The Royal Society and The Royal Astronomical Society. Although the catalogue itself was largely unaltered, the tabular data were accompanied by an additional column which attempted to identify each Herschel object by its corresponding number in the *New General Catalogue*. There was also an extensive and valuable notes section at the end of the catalogue, compiled by J. L. E. Dreyer, in which he tried to clarify positional and identification problems in Herschel's original catalogues.

Herschel approached the task of cataloguing his objects in a way similar to that of a natural historian, believing that classifying them according to appearance would help to bring about an understanding of the nature of these objects. His classification scheme organized the objects into the following eight classes:

Class I	Bright nebulae;
Class II	Faint nebulae;
Class III	Very faint nebulae;
Class IV	Planetary nebulae; stars with burs, with milky chevelure, with short rays, remarkable shapes, etc.;
Class V	Very large nebulae;
Class VI	Very compressed and rich clusters of stars;
Class VII	Pretty much compressed clusters of large or small stars;
Class VIII	Coarsely scattered clusters of stars.

In Herschel's time, the true nature of deep-sky objects was poorly understood and it is not surprising that Herschel's classification scheme was subsequently abandoned, not only by his son John, but by later astronomers as well. Nevertheless, for the modern amateur astronomer working visually the classification scheme does provide some valuable general guidance to the appearance and visibility of the Herschel objects. As more than 75% of the objects in the Herschel catalogue are galaxies, the classes can provide a quick and handy reference to most of the objects.

Class I: Bright nebulae

Generally speaking these are bright, nearby galaxies of high surface brightness and no particular type, though most of them are elliptical, lenticular or oblique spiral galaxies. A handful of globular clusters can be found in this class as well, usually remote, compact systems that are bright but difficult to resolve, for example, NGC5694 in Hydra or NGC7006 in Delphinus.

Class II: Faint nebulae

This class is made up almost exclusively of galaxies of all types whether elliptical, lenticular, spiral and sometimes irregular form. Typically, these galaxies are within 100 million light years of the Milky Way, though some are high-surface-brightness galaxies in the 100–200 million light year range. Most of the members are normal-sized galaxies, though some relatively nearby, dwarfish galaxies are included. In spring skies, many Virgo supercluster galaxies are members of this class and in autumn skies constellations such as Pegasus, Pisces, Eridanus and Cetus are rich in this class of object. Observers using moderate-sized instruments (generally apertures of 12–18 inches) will find most of these objects easy to see.

Class III: Very faint nebulae

This class features a wide variety of galaxies which can generally be characterized in the following manner: small, lower-surface-brightness objects within 100 million light years of the Milky Way, which are frequently the fainter members of small groups of galaxies or minor galaxies in major clusters such as the Virgo cluster. From about 100–300 million light years they are usually normal-sized galaxies, principally elliptical, lenticular or spiral, and are often the principal members of small galaxy groups or major clusters. From 300 million out to about 900 million light years (which is probably the distance of the most remote of the Herschel objects) they are usually large, massive, high-surface-brightness galaxies, the dominant members of groups and large clusters. This class contains many challenging and difficult objects, particularly for observers using small and even moderate-sized instruments.

Class IV: Planetary nebulae; stars with burs, with milky chevelure, with short rays, remarkable shapes, etc

As this class is often referred to simply as 'planetary nebulae', it is a class that is sometimes misunderstood by modern amateur astronomers, who are surprised that galaxies could be mistaken for 'planetary nebulae'. This class was something of a 'catch-all' grouping of objects that must have puzzled Herschel. Although there are many true planetary nebulae in this class, one also finds small emission and reflection nebulae here, especially if they are involved with one or more bright stars. Good examples are NGC2261 (Hubble's Variable Nebula) and NGC2170 in Monoceros or NGC2023 in Orion. The class also contains galaxies with intensely bright cores, such as the Seyfert galaxy NGC4051 in Ursa Major or galaxies which appear to be involved with foreground stars in our own galaxy, examples being NGC3507 and NGC3662 in Leo.

Class V: Very large nebulae

This class contains many true emission or reflection nebulae, but it is also the repository of large nearby galaxies, sometimes bright (NGC253 in Sculptor and NGC4565 in Coma Berenices) but often quite low in surface brightness (such as NGC4236 in Draco and NGC4395 in Canes Venatici).

Class VI: Very compressed and rich clusters of stars

Most of the objects here are made up of bright, rich open clusters or bright globular clusters that Herschel considered resolvable.

Class VII: Pretty much compressed clusters of large or small stars

This is perhaps Herschel's most consistent and coherent class of object. Every member is a moderately rich grouping of stars, usually with a fairly large range in brightness of the members. The clusters are usually easy to recognize with all sizes of amateur telescopes.

Class VIII: Coarsely scattered clusters of stars

The members of this class are typically large, scattered open clusters; however, there are a number of asterisms or condensed Milky Way patches. Many of the so-called 'nonexistent' star clusters are found in this class because the groupings were hard to recognize when

examined on the Palomar Sky Survey prints. Though some of the groupings were later found to exist, some of these Herschel objects are truly nonexistent.

Herschel's 'revised' catalogue

Although there were nominally 2500 objects in the original catalogue, Sir John Herschel listed an additional eight objects observed by his father in an appendix which appeared in his *Results of Astronomical Observations Made During the years 1834–1838 at the Cape of Good Hope* (1847) (hereinafter the *Cape Observations*). These objects were identified as: $H908^2$ = NGC2650, $H909^2$ = NGC3066, $H979^3$ = NGC3210, $H980^3$ = NGC3212, $H981^3$ = NGC3215, $H982^3$ = NGC2629, $H983^3$ = NGC2641 and $H79^4$ = NGC3034 = M82.

The likely reason that they were not included in the third instalment is that the number of objects included in that listing was already 500 and Sir William probably intended to prepare a fourth instalment, but never did. He was already 64 years of age when the third instalment was published and never systematically swept the heavens again, although he did occasionally observe until the year 1815.

All the 'appendix' items were included in the revised Herschel catalogue compiled by J. L. E. Dreyer and published in *The Scientific Papers of Sir William Herschel* (1912) (hereinafter *The Scientific Papers*) except for M82, which was already well known as a discovery by Johan Bode and included in Charles Messier's catalogue of nebulae. While its exclusion is understandable under the circumstances, there are a number of Messier objects which were included in Sir William Herschel's catalogues, though he usually made a point of avoiding objects already catalogued by the French comet hunter. As M82 is already a part of the Astronomical League's Herschel 400 observing project, it is included here as a late addition to Herschel's catalogue.

Dreyer himself included a number of objects observed by Sir William Herschel but not originally included in his catalogue. They are: $H1b^8$ = NGC2319, recorded on 18 December 1783 and included in Caroline Herschel's Zone Catalogue; $H910^2$ = NGC4646, recorded on 24 March 1791; H984[3] = NGC7810, recorded on 17 November 1784; and H985[3] = NGC4695, recorded on 24 March 1791. It should be noted, however, that H985[3] is probably a duplicate observation of a Herschel object found in 1789, H796[2]. The total number of objects catalogued by Sir William Herschel should therefore be 2511.

Herschel's catalogue (circa 1990)

The computer generated list supplied by Father Lucian Kemble contains a total of 2478 Herschel objects, not 2511 as one might expect.

The reason for the discrepancy is not hard to find. Sir William Herschel, observing as he did at the very limits of his equipment as well as his capabilities as an observer, occasionally made mistakes, as every cataloguer of the heavens before him did.

One of the problems for Herschel was that some areas of the sky were swept multiple times, while other areas were observed only once. Since his telescope was not equatorially mounted his method of identifying the location of a new nebula was to indicate its offset position from a known star. This was done in minutes and seconds of time for right ascension (a nebula being either preceding or following the reference star) and degrees and minutes (either north or south) for declination. Problems sometimes occurred because Herschel would use a given star in his initial sweep and in a subsequent sweep (sometimes conducted years later) would offset from an entirely different star. It was left to Sir John Herschel and other astronomers in the decades and centuries that followed to resolve the errors. For this reason, some Herschel objects will have two (and sometimes more) catalogue numbers.

Occasionally Herschel, working at the limits of his equipment, would see a nebula where none existed or would perceive a faint star as nebulous. On other occasions, human error in reading the offsets correctly placed an existing nebula where none could subsequently be found.

Another source of errors is the fact that many of Herschel's star clusters are not discrete objects but rather asterisms made up of unrelated stars or condensed patches of stars in crowded Milky Way fields. One of Herschel's theories of cluster formation was that, in a region of space where stars were otherwise evenly distributed, the effects of gravitation would cause stars to be drawn together over time, meaning that loose collections of stars were relatively young while compressed clusters were much older. Especially during the early days of his sweeps, Herschel was on the lookout for these forming clusters and he would frequently identify perceived groupings as actual clusters.

Amongst the 2478 objects appearing in Father Lucian Kemble's observing list, 107 objects are identified as either 'nonexistent' or 'not documented'. Since the list was compiled, the identities of many of these objects have been resolved. Others which were identified as single or multiple stars have been identified as nonstellar objects, usually galaxies, while some galaxies have turned out to be double or multiple stars. A significant number are multiple observations of the same object, while some are single stars which were presumed to be nebulous. The best clearing house for identification problems in the Herschel catalogue as well as the *New General Catalogue* is the NGC/IC Project, available on the Internet at: www.ngcicproject.org.

For the purposes of this book, every object whether documented or not, has been searched for visually and the results compared with research already conducted. In the following guide, every Herschel object has been listed, whether it exists and can be identified, or not.

Based on observational evidence conducted by the author and the best research on errors and corrections in the *New General Catalogue*, the actual number of Herschel objects appears to be 2435.

Organization of this guide

The following is a complete guide to the entire catalogue of star clusters, nebulae and galaxies discovered by Sir William Herschel. Entries are organized by constellation, which appear in alphabetical order in this guide. Within each constellation, entries are ordered by their New General Catalogue number, from lowest to highest, irrespective of their position in right ascension.

Each catalogue entry is presented as a data heading, followed by a descriptive paragraph. The information presented is ordered in the following manner:

1. **NGC number** The number of the object as it is listed in the New General Catalogue (NGC). Three Herschel objects were not listed in the original NGC and appear here under their Index Catalogue (IC) number.

2. **Herschel number** The number and class originally assigned to the object in Herschel's catalogue. H171^3 would be the 171st object in Herschel's Class III Very faint nebulae.

3. **Location** The object's position in the sky, listed by right ascension and declination in epoch 2000.0 coordinates.

4. **Type of object** The object type is presented in abbreviated form. The types are as follows:

SS	single star
DS	double star
TS	triple star
OC	open cluster
AST	asterism
MWP	Milky Way patch
ST ASS	stellar association
GC	globular cluster
PN	planetary nebula
EN	emission nebula
RN	reflection nebula
SNR	supernova remnant
GLX	galaxy

5. **Classification** The classification of the object within its type. Classifications vary depending on the type of object. The classification schemes used in this guide are well-known standard systems which are explained below.

6. **Physical dimensions** The apparent size of the object, measured in minutes of arc, as it appears in the sky. The first dimension is the major axis, the second (if applicable) is the minor axis. These data are drawn primarily from the NASA/IPAC Extragalactic Database (NED) or from the SIMBAD database, and are finally compared with the apparent size as it appears on the DSS image.

7. **Magnitude** The apparent magnitude of the object, as it appears in standard catalogues. Values which are photographic (blue) magnitudes are followed by the letter 'B'. Values which are visual magnitudes (where available) are followed by a 'V'. For galaxies, the primary source was the HyperLEDA database, with the SIMBAD database used as a secondary source. For all other classes of deep-sky objects, the SIMBAD database was used as a primary source. The interested observer is also directed to *The Deep Sky Field Guide to Uranometria 2000.0* (Cragin, Rappaport and Lucyk, 1993), which provides visual magnitudes for almost all the Herschel objects.

8. **Description** Two descriptions are usually presented for each object. The first is a description of its appearance on the DSS image. This description is followed by a description of the object as it appears visually through a moderate-aperture telescope (12-inch to 18-inch aperture). There are some cases where the photographic and visual appearances of an object are quite similar, for instance an elliptical galaxy. In this case, only a visual description is supplied. Where it is found appropriate, information of a historical nature may be presented with the description, for instance the date of discovery and William Herschel's original description of the object. Also presented are identifications of other objects which may be seen in the immediate vicinity of the Herschel object, with their positions and apparent distances from the Herschel object indicated in minutes of arc. The primary source for these data is the NASA/IPAC Extragalactic Database (NED). Other information that may be presented, if available, includes spectral type, age or distance. For all galaxies listed in this

catalogue, data on the galactocentric radial velocity in kilometers per second are presented, as well as estimates of the galactocentric distance in millions of light years and an estimate of the actual size of the galaxy, based on its apparent size and distance. Unless otherwise noted, this information is derived from the NED, which uses the *Third Reference Catalogue of Bright Galaxies* (de Vaucouleurs, *et al.*, 1991) as its primary source. Although an object's radial velocity can be accepted as a rough indicator of distance, gravitational effects can sometimes skew radial velocity measurements in groups or clusters of galaxies. There are many instances where several galaxies appear together in a small area of the sky, forming an obvious grouping and yet the measured radial velocities of each member seem to indicate that each object is separated by a considerable distance, often tens of millions of light years. In this guide, no attempt has been made to correct these discrepancies. In certain instances, the interested observer may choose to take the average of all measured distances of individual galaxies in a group to represent the distance of the whole.

Open cluster classification

In this catalogue, open cluster classification is based on the standard three-part system developed by R. J. Trumpler, which is organized as follows:

Concentration

I. Detached: strong concentration towards the centre.

II. Detached: weak concentration toward the centre.

III. Detached: no concentration towards the centre.

IV. Not well detached from surrounding star field.

Range in brightness

1. Small range in brightness.

2. Moderate range in brightness.

3. Large range in brightness.

Richness

p Poor (less than 50 stars).

m Moderately rich (50 to 100 stars).

r Rich (more than 100 stars).

As an example, the open cluster NGC2129 in Gemini is classified I3m – a detached cluster showing a strong concentration to the centre, a large range in brightness of its component stars and moderately rich membership.

Globular cluster classification

The classification of globular clusters in this catalogue is based on the concentration class system developed by Harlow Shapley and Helen Sawyer-Hogg. It is a twelve-step system, whereby the smaller the number, the more highly concentrated towards the centre the cluster is. In this catalogue the number will appear with the prefix 'CC' (concentration class). As an example, the globular cluster NGC6717 in Sagittarius is classified CC8.

Planetary nebula classification

The classification of planetary nebula in this catalogue is based on the Vorontsov–Velyaminov system, which is divided into the six following types:

1. Stellar image.

2. Smooth disc (a – brighter towards the centre; b – uniform brightness; c – traces of ring structure).

3. Irregular disc (a – very irregular brightness distribution; b – traces of ring structure).

4. Ring structure.

5. Irregular form similar to a diffuse nebula.

6. Anomalous form.

As an example, the planetary nebula NGC6772 in Aquila is classified 3b+2 – an irregular disc with traces of ring structure plus a smooth disc.

Galaxies

Galaxy classification in this catalogue is based on the extensive system developed by Gérard de Vaucouleurs and his collaborators, the most recent iteration being the one appearing in the *Third Reference Catalogue of Bright Galaxies* (1991) and the interested reader is directed to that resource for a complete exposition of the system. The classification system also appears in *The Deep Sky Field Guide to Uranometria 2000.0* (Cragin, Rappaport and Lucyk, 1993) while an earlier iteration of the system can be found in *The Observing Handbook and Catalogue of Deep-Sky Objects* (Luginbuhl and Skiff, 1989).

Briefly, the system identifies five major classes of galaxies: ellipticals, lenticulars, spirals, irregulars and peculiars, with various subgroupings refining the classification further. Ellipticals are type 'E', lenticulars type 'S0', spirals type 'S', irregulars type 'I' and peculiars are identified as 'Pec'. If there is no bar involved, a lenticular is type 'SA0' while a spiral is type 'SA'. If a bar is involved, the lenticular becomes 'SB0', while the spiral becomes 'SB'. A mixture of barred and nonbarred features produces an 'SAB0' type for lenticulars and an 'SAB' for spirals. A lower case 'r' indicates an inner ring structure, a capital 'R', an outer ring structure. If there is doubt about a type, this is indicated by a '?', uncertainty by a ':'.

In this scheme, the galaxy NGC7592 in Aquarius is classified S0$^+$ pec: – a late stage lenticular galaxy with peculiarities, though the classification is uncertain.

Abbreviations in the text

Abbreviations have been kept to a minimum in the descriptive paragraphs but in an effort to minimize redundancy, the following have been employed:

DSS = Digitized Sky Survey;
mly = million light years;
ly = light years.

Illustrations

All drawings which appear in this guide are field sketches created at the eyepiece by the author. Details regarding telescope and magnification used to produce the sketch are included with the drawing.

The original sketches were pencil and ink drawings (black on white background) and all sketches are oriented with north up and west to the right.

All photographs appearing in this guide are DSS images reproduced to the same scale: $1' = 7.5$ mm and are oriented with north up and west to the right.

Acknowledgments

Use of northern starfield images reproduced from the the DSS Survey © AURA are courtesy of the Palomar Observatory and the DSS Sky Survey created by the Space Telescope Science Institute, operated by AURA, Inc. for NASA and are reproduced here with permission from AURA/STScI.

Use of southern starfield images is courtesy of the UK Schmidt Telescope (copyright of which is owned by the Science and Technology Facilities Council of the UK and the Anglo-Australian Telescope Board) and the DSS created by the Space Telescope Science Institute, operated by AURA, Inc, for NASA, and are reproduced here with permission from the Royal Observatory Edinburgh.

This research has made use of the NASA/IPAC Extragalactic Database (NED) which is operated by the Jet Propulsion Laboratory, California Institute of Technology, under contract with the National Aeronautics and Space Administration (http://nedwww.ipac.caltech.edu).

I acknowledge the use of the HyperLEDA database (http://leda.univ-lyon1.fr): Paturel G., Petit, C., Prugniel, Ph., *et al.*, HYPERLEDA. I. Identification and designation of galaxies. *Astron. Astrophys.*, **412**, 45–55 (2003).

This research has made use of the SIMBAD database, operated at CDS, Strasbourg, France (http://simbad.u-strasbg.fr/simbad/).

This research has made use of data available at the NGC/IC Project to clear up identifications of a number of New General Catalogue entries (http://www.ngcicproject.org/).

The Complete Guide to the

HERSCHEL OBJECTS

Andromeda

Herschel recorded 40 objects in this constellation between the years 1784 and 1790. Two were brought to his attention by Caroline, the elliptical galaxy NGC205 (seen by Charles Messier but not originally included in his catalogue) and the star cluster NGC752 (probably first recorded by Giovanni Hodierna before the year 1654). The fine galaxies NGC891 and NGC7640 can be found here as well as the bright planetary nebula NGC7662. However, the galaxies associated with the NGC68 group, the Hickson 10 compact galaxy group and the Abell 262 galaxy cluster are all interesting targets.

NGC13

H866[3]	0:08.8	+33 26	GLX
(R) Sab:	2.3′ × 0.5′	14.08B	

The DSS image shows a highly inclined spiral galaxy with an odd, triangular-shaped central region. The spiral arms are very smooth with low surface brightness. Visually, only the core of this elongated galaxy is visible. The core is a small, round, well condensed spot with well defined edges immediately N of a faint field star. Radial velocity: 4990 km/s. Distance: 223 mly. Diameter: 149 000 ly.

NGC29

H853[2]	0:10.8	+33 21	GLX
SAB(s)bc:	1.7′ × 0.8′	13.37B	

This is a highly inclined barred spiral galaxy with two principal spiral arms and a faint field star embedded in the arm N of the core. Telescopically, the galaxy is moderately bright and well seen at high magnification as an opaque, extended streak of light. The edges are fairly well defined and the galaxy is a little brighter along its major axis; it is elongated NW–SE. Radial velocity: 4950 km/s. Distance: 221 mly. Diameter: 109 000 ly.

NGC39

H861[3]	0:12.3	+31 03	GLX
SA(rs)c	1.1′ × 1.0′	14.34B	

Photographs show a probably face-on spiral galaxy with a poorly defined spiral pattern, a small, brighter core and a faint field star at the edge of the disc S of the core. Visually, the galaxy is dim but nevertheless well seen as a round, nebulous patch of light with a brighter, star-like core. The envelope is faint but well defined and even in brightness. Radial velocity: 5033 km/s. Distance: 225 mly. Diameter : 72 000 ly.

NGC68

H16[5]	0:18.3	+30 04	GLX
SA0⁻	1.3′ × 1.2′	14.41B	

The brightest member of the NGC68 group of galaxies, this object has long been accepted as the galaxy that Herschel saw. The DSS image shows this compact galaxy group well; more than

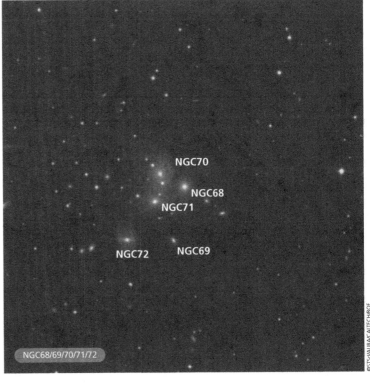

NGC68/69/70/71/72

two dozen galaxies can be counted in a 15.0′ field, with a mix of elliptical/lenticulars and spiral galaxies. Surprisingly, this galaxy group is easily visible as a nebulous patch at low magnification in a moderate-aperture telescope. The largest patch is NGC70, a spiral galaxy whose envelope contains two faint field stars, one NNE and one S of the core. NGC68 precedes to the WSW as a round patch of light with a brighter core and fairly well defined edges. Also visible to the S of NGC70 is NGC71. Herschel described entry H16[5] as 'Extremely faint. 5 or 6′ diameter.' Since NGC68 is itself only 1.3′ × 1.2′ in size, and since Herschel entered this object in his 'Class V: very large nebulae', it is probable that Herschel originally saw the unresolved images of three galaxies: NGC 68, NGC70 and NGC71. The discovery of NGC70 and NGC71 are credited to Lord Rosse at Parsontown in 1855. Radial velocity: 5907 km/s. Distance: 263 mly. Diameter: 100 000 ly.

NGC108

H148[3]	0:25.9	+29 13	GLX
(R)SB(r)0[+]	2.0′ × 1.6′	13.34B	

The DSS image reveals a face-on barred ring galaxy with faint, thin bars which broaden and brighten where they attach to a round inner ring. A second, very faint outer ring is visible, unattached to the galaxy. The major axis of this ring is 2.5′, somewhat larger than the published value and increasing the actual diameter of the galaxy considerably. Telescopically, the galaxy is moderately bright, condensed and brighter to the centre, where a round core is visible. The outer envelope is oval and extended roughly E–W with well defined edges. Radial velocity: 4904 km/s. Distance: 219 mly. Diameter: 128 000 ly.

NGC160

H476[3]	0:36.1	+23 57	GLX
(R)SA[+] pec	2.3′ × 1.2′	13.61B	

Photographs show an inclined lenticular galaxy with a brighter

elongated core embedded in a faint inner disc. An extensive outer ring has some characteristics of spiral structure and is particularly thin and bright along the eastern flank. Visually, it is a moderately bright galaxy but hindered somewhat by the presence of a magnitude 7 field star 4.2′ to the NNE. It is seen as a roundish patch of light that is fairly well defined and slowly brighter to the middle. A faint triangle of stars to the W points towards the galaxy. In a medium-magnification field the galaxy pair NGC169/IC1559 can be seen 10.8′ to the ENE. The very dim galaxy NGC162 is ESE of NGC160 and is far beyond the capabilities of a medium-aperture telescope. Radial velocity: 5405 km/s. Distance: 241 mly. Diameter: 211 000 ly.

NGC205

H18[5]	0:40.4	+41 41	GLX
E5 pec	21.9′ × 11.0′	8.89B	8.1V

Records indicate that this galaxy was seen by Charles Messier in 1773 but not included in his original catalogue, though today it is best known as M110. It appears, clearly labeled ('small, faint nebula, Messier 1773'), on a drawing Messier did of the Andromeda Galaxy which was published in 1807. It was independently discovered by Caroline Herschel on 27 August 1783 and recorded in William Herschel's own catalogue on 5 October 1784, where he credits Caroline with the discovery. This Andromeda Galaxy companion is a very elongated dwarf elliptical galaxy with a bright, slightly elongated core and fainter extensions. It was first resolved into stars by Walter Baade in 1944, his red-sensitive plates resolving the brightest members at visual magnitude 22. Visually and photographically the appearances are quite similar, though photographs bring out two dust patches bordering the core to the E. Telescopically, this galaxy is large, though fairly bright and moderately diffuse. The centre is brighter, slightly elongated N–S and broadly brighter to a nonstellar middle. There is a moderately sharp drop off in brightness to the much

fainter and diffuse outer disc, which fades slowly to ill defined edges. Radial velocity: 62 km/s in approach. Distance: 2.4 mly. Diameter: 18 500 ly.

NGC206

©STScI/AURA/CALTECH/ROE

NGC206

H36[5]	0:40.6	+40 44	ST ASS

The DSS image shows a well-resolved cluster of faint stars elongated N–S in a spiral arm of the Andromeda Galaxy, situated SW of the core. There may be nebulosity involved though it is faint and the resolved stars are generally brighter than the galaxy stars visible nearby. Visually, the cluster appears as a relatively bright, triangular-shaped nebulous patch in the SW spiral arm of M31. It is best seen in a very dark sky, which brings out the faint outer spiral arm where NGC206 is located. In brighter skies, the stellar association is difficult to locate as it appears to be isolated from M31. Distance: 2.2 mly.

NGC214

H209[2]	0:41.5	+25 30	GLX
SAB(r)c	1.9′ × 1.4′	12.94B	

Photographs show a slightly inclined multi-armed spiral galaxy with a large bright core and knotty spiral arms. The two main spiral arms emerge from the bright inner disc; the arm on the NW flank is considerably longer and brighter. Visually, this is quite a bright galaxy, fairly large but with slightly diffuse edges. It appears almost round with a small but nonstellar, condensed and brighter core, well seen at high magnification. Radial velocity: 4686 km/s. Distance: 209 mly. Diameter: 116 000 ly.

NGC233

H149[3]	0:43.4	+30 35	GLX
E?	1.1' × 1.0'	13.91B	

In photographs this galaxy appears as an elliptical of somewhat peculiar structure. A spur borders the bright core to the S and a faint condensation, or perhaps a small companion galaxy borders the galaxy's faint outer envelope to the NW. A possible compact companion is about 5.0' to the ESE. Telescopically, this galaxy is faintly visible as a small, moderately condensed patch of light with a gradually brighter core. The object appears round with little indication of fainter extensions. Radial velocity: 5583 km/s. Distance: 249 mly. Diameter: 80 000 ly.

NGC252

H609[2]	0:48.0	+27 38	GLX	
(R)SA(r)0+:	1.6' × 1.3'	13.34B	12.5V	

Images reveal a slightly inclined possible spiral galaxy with a large, bright, elongated core and a prominent ring-like feature in its disc. The ring is complete for about three quarters of a revolution, fading out along the southern flank. A faint, compact galaxy is 3.1' NE while the brighter NGC260 is 8.25' NNE. Visually, the galaxy is bright and well seen at high magnification; it is a fat oval galaxy brighter to a broad middle. The outer envelope is fairly well defined and elongated E–W. It forms a flat triangle with two field stars, one ENE, the other WNW. Also faintly visible is NGC260, a hazy, ill defined patch of light SW of a field star. Radial velocity: 5143 km/s. Distance: 230 mly. Diameter: 107 000 ly.

NGC280

H477[3]	0:52.5	+24 20	GLX
SB?	1.7' × 1.1'	14.65B	

Photographically, this is an inclined spiral galaxy, very probably barred, with two principal spiral arms, somewhat asymmetrical in structure emerging from the central region. A peculiar oval brightening WSW of the core and between two spiral arms is the probable satellite galaxy UGC534. Telescopically, the galaxy is quite faint but fairly well seen at medium magnification as a diffuse patch of light, 0.8' W of a faint field star. The galaxy is elongated E–W and is very slightly brighter towards the middle with ill defined edges. Radial velocity: 10 249 km/s. Distance: 458 mly. Diameter: 226 500 ly.

NGC393

H54[1]	1:08.6	+39 40	GLX
S0-:	1.7' × 1.4'	13.61B	

Photographs show this lenticular type galaxy has a large round core and an extensive, oval-shaped outer envelope, oriented NNE–SSW. Also in the field is NGC389 located 3.25' to the NNW. Telescopically, the galaxy is small and moderately bright and is situated in a fairly rich field of stars. Fairly well defined at the edges, the envelope is round and bright and slightly brighter to the middle. Radial velocity: 6266 km/s. Distance: 280 mly. Diameter: 138 000 ly.

NGC404

H224[2]	1:09.4	+35 43	GLX
SA(s)0-:	4.7' × 4.7'	11.19B	

This dwarfish lenticular galaxy lies just beyond the presumed boundaries of the Local Group and is not considered a member. The DSS image shows a round galaxy, broadly brighter to the middle with a mottled texture suggesting threshold resolution into stars. Telescopically, this moderately bright galaxy is greatly hindered by the presence of Beta Andromedae 6.0' to the SSE and high magnification is necessary to keep the bright star out of the field of view. It is round and fairly large with a brighter core; the edges are ill defined and the envelope appears mottled.

NGC404

Radial velocity: 109 km/s. Distance:
4.9 mly. Diameter: 6500 ly.

NGC477

H577[3]	1:21.3	+40 29	GLX
SAB(s)c	2.2′ × 1.2′	14.26B	

This is an inclined, barred spiral galaxy
with a slightly brighter core and a very
weak bar. Photographs show the two
principal spiral arms form a tight
pseudo-ring around the bar and branch
to form another two spiral arms after a
half revolution. Two fainter galaxies are
in the field, MCG+7-3-31 is 2.4′ WSW
and a much fainter edge-on galaxy is
4.3′ WSW. Telescopically, NGC477 is a
very faint and difficult object, visible
only at medium magnification as a
small, diffuse, ethereal blur, small
and elongated roughly SSE–NNW
with a faint field star bordering the
galaxy to the S. The edges are hazy
and ill defined. Radial velocity:
6035 km/s. Distance: 270 mly.
Diameter: 173 000 ly.

NGC513

H169[3]	1:23.8	+33 49	GLX
Sb/c	0.7′ × 0.3′	13.38B	

Photographs show a galaxy with a
bright, lenticular-shaped envelope with
a very slightly brighter core. Several
faint companion galaxies are in the field
including NGC512 8.2′ to the NW. The
galaxy is a possible outlying member of
the NGC507 galaxy group. Visually, it is
quite small and moderately faint,
though it is well seen at high
magnification as an oval patch of light,
elongated ENE–WSW. The envelope is
opaque and even in surface brightness,
though a little brighter to the middle.
The edges are well defined and a
magnitude 10 field star is 3.25′ SSW.
Radial velocity: 6005 km/s. Distance:
268 mly. Diameter: 55 000 ly.

NGC523

H170[3]	1:25.3	+34 02	GLX
Pec	3.2′ × 0.9′	13.50B	

This galaxy has also been catalogued as
Arp 158 and the DSS image reveals a

NGC523

galaxy with a very peculiar morphology,
very possibly the result of a recent
merger of two galaxies. The principal
galaxy would seem originally to have
been a barred spiral: an elongated core
seems to have a bar involved with two
disturbed and asymmetrical spiral
extensions attached to the bar. A bright
oval condensation bordering the bar to
the E appears to be the core of a smaller
galaxy. A round condensation bordering
the bar on the W side is probably a faint

foreground star. The arm emerging from
the bar from the W and extended ESE has
a very faint and broad plume of material
attached to it. The morphology is very
similar to the peculiar galaxy NGC520
located in Pisces. Telescopically, the
galaxy is faint but readily visible at
medium magnification and is fairly
distinct at high power. A diffuse,
rectangular smudge of light, it is
elongated E–W and is moderately well
defined at the edges. The main envelope
is fairly opaque and even in surface
brightness. The galaxy is listed as
NGC537 in *The Scientific Papers* and is
probably an outlying member of the
NGC507 galaxy group. Radial velocity:
4904 km/s. Distance: 219 mly.
Diameter: 204 000 ly.

NGC536

H171[3]	1:26.4	+34 43	GLX
SB(r)b	3.0′ × 1.1′	13.22B	

The DSS photograph shows a highly
inclined, theta-shaped barred spiral

NGC531/536/542

galaxy with an elongated core perpendicular to a weak bar. The bar is attached to a ring and two very faint and large, smooth-textured spiral arms emerge from the ring. In a medium-aperture telescope, only the core of this elongated galaxy is visible, which is brighter to a stellar core. At high magnification a faint star is visible immediately N of the galaxy core. This galaxy is a member of the Hickson 10 compact galaxy group, the other members are NGC529, NGC531 and NGC542. All stand out well in a medium magnification field. Radial velocity: 5336 km/s. Distance: 238 mly. Diameter: 208 000 ly.

NGC551

H560[3]	1:27.6	+37 11	GLX
SBbc	1.8′ × 0.8′	13.76B	

Images show an inclined four-branch spiral galaxy with a brighter core. Two arms emerge from the central region and quickly divide into separate branches. In a moderate-aperture telescope the galaxy is best seen at medium magnification as an extended, diffuse patch of light, elongated SE–NW with a fairly smooth envelope. Two faint field stars are to the NW and are aligned with the galaxy's major axis. The brighter of the two stars is 1.0′ from the core just beyond the outer envelope. Radial velocity: 5347 km/s. Distance: 239 mly. Diameter: 125 000 ly.

NGC679

H175[3]	1:49.7	+35 47	GLX
S0⁻:	1.2′ × 1.1′	13.55B	

This lenticular-type galaxy presents a similar appearance both photographically and visually. In moderate-aperture telescopes, this is a moderately bright galaxy, well defined and fairly well condensed to a brighter core. Averted vision at high magnification occasionally brings out a faint stellar nucleus. The galaxy is round. It is SW of the main concentration of the Abell 262 galaxy cluster of which it is a member. Herschel discovered the galaxy on 13 September

1784, calling it: 'Stellar.' Radial velocity: 5186 km/s. Distance: 232 mly. Diameter: 81 000 ly.

NGC687

H561[3]	1:50.6	+36 21	GLX
S0	1.4′ × 1.4′	13.58B	13.3V

This outlying member of the Abell 262 galaxy cluster appears much the same photographically and visually. At high magnification, this lenticular galaxy is a small, round, well-condensed spot growing brighter to the middle. A small, faint secondary envelope is visible with averted vision. A somewhat fainter galaxy, UGC1308, is 6.5′ SE. Radial velocity: 5228 km/s. Distance: 233 mly. Diameter: 95 000 ly.

NGC703

H562[3]	1:52.8	+36 09	GLX
S0⁻:	0.9′ × 0.5′	14.35B	

This galaxy, along with NGC704, NGC705 and NGC708 are at the core of the Abell 262 galaxy cluster, a rich and relatively nearby cluster of galaxies which probably forms an association with two other rich galaxies clusters in the constellation Pisces: the NGC507 group and the Pisces Cloud, whose principal member is NGC383. Photographs show NGC703 is a lenticular type galaxy, an elongated oval, oriented NE–SW with a faint secondary envelope. Visually, it is the second brightest of the four galaxies, only NGC708 is brighter. It appears small, round and condensed. Radial velocity: 5716 km/s. Distance: 255 mly. Diameter: 67 000 ly.

NGC704

H563[3]	1:52.7	+36 06	GLX
S0	0.5′ × 0.3′	14.34B	

This galaxy is elongated E–W and is at the core of the Abell 262 galaxy cluster; the DSS image shows that this lenticular type galaxy is part of a galaxy pair with

NGC703/704/705/708/709/710

a slightly fainter companion galaxy immediately S. The bright cores of the two galaxies appear to be almost in contact and they seem to share a common envelope but there is no sign of tidal interaction between them. The radial velocities are virtually identical so it is very likely that they form a physical pair. Visually, NGC704 is difficult, very small, round and condensed and the companion is not seen. However, 1.2' to the WSW the slightly fainter galaxy CGCG522-033 can be detected. Radial velocity: 4864 km/s. Distance: 217 mly. Diameter: 32 000 ly.

NGC705

H564[3]	1:52.8	+36 08	GLX
S0/a	1.2' × 0.3'	14.53B	

This Abell262 member is seen in photographs as an edge-on, spindle-shaped galaxy with a large bright core and a small companion galaxy immediately WSW of the core. It is the most striking of the four principal Herschel galaxies at the centre of the cluster. At the eyepiece only the core of this edge-on galaxy is visible; it is fairly difficult, small and round. Radial velocity: 4650 km/s. Distance: 208 mly. Diameter: 73 000 ly.

NGC708

H565[3]	1:52.8	+36 10	GLX
cD;E	3.0' × 2.5'	13.69B	

This giant elliptical-type galaxy is the largest and brightest of the four Herschel objects at the core of the Abell 262 galaxy cluster. In photographs it has a round, bright core and an extensive fainter envelope, elongated NE–SW. The four galaxies are contained within an area of 2.75' × 2.0' and the minimum separation between NGC703 and NGC704 would be 178 000 ly at the presumed distance. Herschel recorded all four galaxies on 21 September 1786 and described them together as: 'Four, stellar, unequal. Three in a row, and the fourth making a rectangle with them. That at the angle is much larger.' This is a reference to NGC708 itself, which visually is the

brightest and largest of the group; round, slightly brighter to the middle with hazy edges and a distinct secondary envelope. In a medium-magnification field eight galaxies can be seen including the four Herschel objects. The others are CGCG522-033, NGC710, NGC714 and NGC717. Radial velocity: 4991 km/s. Distance: 223 mly. Diameter: 194 000 ly.

NGC752

H32[7]	1:57.8	+37 41	OC
II2r	75.0' × 75.0'	6.5B	5.7B

Credit for discovery of this cluster is usually given to Caroline Herschel, who recorded it on 29 September 1783, but the object was evidently first recorded by the obscure Italian astronomer Giovanni Battista Hodierna who probably observed it before the year 1654 with a 20× Galilean telescope and recorded it in his catalogue as object Ha IV.4 (nebulosae – an object appearing as a nebula to the naked eye but resolved into stars in a telescope). William Herschel entered it in his catalogue on 21 September 1786 and described it as: 'A very large coarse scattered cluster of very large stars, irregularly round, very rich. Takes up a half degree, like a nebulous star to the naked eye.' This is a large, scattered cluster; more than 70 stars are accepted as members and are visible in a low-magnification field. It is readily apparent to the naked eye under dark skies and is a compelling sight in binoculars. It loses much of its impact in a telescope. The

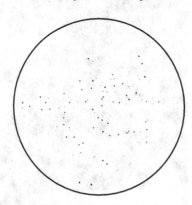

NGC752: 15-inch Newtonian 48×

cluster members are uniform in brightness for the most part, predominantly magnitude 10–12. A few cluster members are magnitude 8–9 and bordering the cluster to the S are a wide pair of magnitude 6 stars. Spectral type is A4. Age: 1.1 billion years. Radial velocity: 4 km/s in recession. Distance: 1180 ly.

NGC797

H566[3]	2:03.4	+38 07	GLX
SAB(s)a	1.9' × 1.4'	13.60B	13.1V

This is an almost face-on spiral galaxy, probably barred, with a large, bright core and very faint, smooth-textured spiral arms. In the DSS image, a fainter companion galaxy is located at the tip of the southern spiral arm, 0.75' WSW of the core. Though the radial velocities of the two galaxies are almost identical, there does not appear to be any interaction between them. At magnitude 16.9 and 0.15' × 0.15' in size, this compact elliptical may be detectable in large amateur telescopes. Visually, NGC797 is moderately bright, though only the core is visible. It is round and steadily brighter to the middle; the edges are moderately well defined and a faint field star is visible 0.75' WNW of the core. Radial velocity: 5788 km/s. Distance: 259 mly. Diameter: 143 000 ly.

NGC818

H604[2]	2:08.7	+38 47	GLX
SABc:	3.0' × 1.3'	13.19B	

Photographs show an inclined spiral galaxy with a slightly brighter core and two strong spiral arms. Telescopically, the galaxy is moderately bright and fairly large; it is an oval, smooth-textured patch of light, elongated ESE–WNW. The edges are well defined and the brightness even across the disc. Radial velocity: 4376 km/s. Distance: 196 mly. Diameter: 171 000 ly.

NGC828

H605[2]	2:10.2	+39 12	GLX
Sa: pec	2.9' × 2.2'	13.14B	

The DSS image shows a peculiar galaxy, possibly a spiral, with a large,

NGC828

©STScI/AURA/CALTECH/ROE

bright core. A thick dust cloud borders the core to the SW and seems to broaden as it crosses in front of a faint plume of spiral structure WNW of the core. A broad envelope emerges S of the core, while an irregular extension projects from the core towards the ESE. Telescopically, the galaxy is a fairly bright but somewhat diffuse oval patch elongated SE–NW. The main envelope is broadly brighter to the middle and a bright wide pair of field stars is 2.8′ to the ESE. Herschel recorded the galaxy on 18 October 1786, noting its irregular form. Radial velocity: 5507 km/s. Distance: 246 mly. Diameter: 208 000 ly.

NGC834

H567[3]	2:11.0	+37 40	GLX
S?	1.1′ × 0.5′	13.82B	

The DSS image shows a peculiar galaxy of uncertain morphology; it is irregular and fairly bright with four large knots or condensations in the envelope surrounding the presumed core plus one very faint, detached condensation to the SE. Visually, the galaxy is moderately bright though very small; it is a roundish, condensed galaxy, brighter to the middle and is quite well defined. In a medium-magnification field the similarly bright galaxy NGC841 is also visible to the SSE and the two galaxies together with NGC845 possibly form a physically related group. Radial velocity: 4722 km/s. Distance: 211 mly. Diameter: 68 000 ly.

NGC845

H604[3]	2:12.3	+37 29	GLX
Sb	1.7′ × 0.3′	14.35B	

This edge-on spiral galaxy has a large, oval core and thin tapered spiral arms; the DSS image shows a probable dust lane extending along the major axis. Morphologically, the galaxy looks like a distant twin of the well-known edge-on spiral NGC4565. Visually, however, the galaxy is quite faint and only the central region is visible as a very small oval patch, slightly elongated SSE–NNW. Radial velocity: 4557 km/s. Distance: 203 mly. Diameter: 101 000 ly.

NGC891

H19[5]	2:22.6	+42 21	GLX	
SA(s)b? sp	13.5′ × 2.5′	10.84B	9.9V	

Credit for discovery of this galaxy was erroneously assigned to Caroline Herschel by Sir William in his original *Philosophical Transactions* publication, but corrected in the reprint which appeared in *The Scientific Papers* (Herschel had meant NGC205). The DSS image shows a breathtaking edge-on spiral galaxy with a slightly brighter central bulge and a heavy dust lane that is particularly thick and prominent against the bulge. The dust lane obscures a significant percentage of the galaxy's light and visually it can be something of a challenge for small-aperture telescopes. In a medium-aperture instrument this large galaxy is well seen at medium magnification as a smooth-textured,

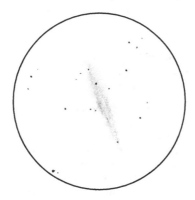

NGC891: 15-inch Newtonian 146×

very flat patch of light which bulges gradually towards the centre. The envelope is well defined and tapers to fairly sharp points. Three field stars are involved: one near the tip in the SSW, one NNE from the core and the last near the NNE tip. The dust lane in front of the core is well seen and the galaxy is oriented NNE–SSW. The galaxy is situated in a rich field of stars and is one of the prominent members of the NGC1023 group of galaxies. Herschel discovered the galaxy on 6 October 1784 and described it as: 'Considerably bright, much extended above 15′ long, 3′ broad. A black division 3 or 4′ long in the middle.' Radial velocity: 660 km/s. Distance: 30 mly. Diameter: 116 000 ly.

NGC898

H570[3]	2:23.3	+41 57	GLX
Sab	1.9′ × 0.3′	13.64B	

The visual and photographic appearances of this westernmost bright member of the Abell 347 galaxy cluster are quite similar. At high magnification, this is an almost edge-on spiral galaxy with a high surface brightness; it is oriented almost due N–S with a prominent diamond-shaped asterism visible to the SE. The bright, condensed core is situated in a well defined, extended envelope which tapers to points. A probable dust lane along the eastern flank is visible on the DSS image. Radial velocity: 5626 km/s. Distance: 251 mly. Diameter: 139 000 ly.

NGC910

H571[3]	2:25.4	+41 50	GLX
E+	1.9′ × 1.7′	13.25B	

This is the most prominent of the galaxies in the Abell 347 cluster. Visually, it is round and well condensed with sharply defined extremities and even surface brightness. On the DSS image a faint star or possible compact companion is involved immediately to the SE and an even fainter star or companion borders the galaxy on the W. Several fainter galaxies circle the galaxy, the brightest of which, NGC912, is visible 4.0′ SE. Radial velocity: 5337 km/s. Distance: 238 mly. Diameter: 132 000 ly.

NGC980

H572[3]	2:35.3	+40 56	GLX
S0	1.0′ × 0.5′	14.15B	

The DSS image shows an inclined lenticular type galaxy with a bulging core and a bright inner region embedded in a fainter outer envelope. The Herschel object NGC982 is located 3.75′ SSE and is probably a physical companion; the minimum core-to-core separation between the two galaxies would be about 285 000 ly at the presumed distance. Visually, both galaxies can be seen together in a high-magnification field. NGC980 is a quite faint oval, elongated ESE–WNW, and is moderately well defined with even brightness across its disc. Radial velocity: 5856 km/s. Distance: 262 mly. Diameter: 76 000 ly.

NGC982

H573[3]	2:35.4	+40 52	GLX
Sa	1.5′ × 0.6′	13.05B	

This is an inclined spiral galaxy with an elongated bright core and moderately bright spiral arms. The galaxy appears much larger than NGC980 on the DSS image, though the published values for its apparent dimensions suggest it is smaller. Telescopically, NGC982 is larger and brighter, a well defined sliver of light, fairly bright along its major axis, elongated SE–NW with a star-like, brighter core. Radial velocity: 5860 km/s. Distance: 262 mly. Diameter: 114 000 ly.

NGC7640

H600[2]	23:22.1	+40 51	GLX
SB(s)c	10.5′ × 2.0′	11.61B	

Photographs show a highly inclined spiral galaxy with a small, brighter core and two curving spiral arms. The arms are dusty with a thin dust lane crossing the core. High-resolution images show the galaxy well resolved into HII regions and stars, with the brightest being about blue magnitude 21.5. Visually, the galaxy is well seen at medium magnification and is moderately bright, bracketed by a prominent triangle of field stars. The galaxy is much elongated N–S but the extensions are very poorly

NGC7640

© STScI/AURA/CALTECH/ROE

defined and fade uncertainly into the sky background. The envelope is smooth textured along the major axis. Herschel recorded the galaxy on 17 October 1786, though his description indicates that he had seen the galaxy two years before: 'Pretty bright, much extended north preceding, south following but nearly meridial. A little brighter to the middle, resolvable, 5′ long, 1.5′ broad. Also observed 1784.' Radial velocity: 576 km/s. Distance: 26 mly. Diameter: 78 000 ly.

NGC7662

H18[4]	23:25.9	+42 33	PN
4+3	0.6′ × 0.5′	9.20B	8.3V

The DSS image is overexposed and shows an opaque, very slightly elliptical nebula with a blunt spur emerging in the NE and a slightly sharper spur in the SSW. The high surface brightness of this planetary nebula allows very high magnification with moderate-aperture telescopes. There is a strong bluish-white colour and the envelope is quite round. The main body is a thick, bright, quite mottled ring with a slight drop off in brightness to a fainter core. The ring is surrounded by a fainter, fuzzy secondary envelope. The central star, HD220733, is magnitude 12.5 but is not seen even at high magnification. Herschel recorded the object on 6 October 1784, recognizing it as a planetary nebula. His description seems to indicate that he used one of his smaller telescopes to make the observation: 'Bright, round, a planetary pretty well defined disc. 15″ diameter with a 7 feet reflector.' Expansion velocity: 26 km/s. Radial velocity: 13 km/s in approach. Distance: 3900 ly.

NGC7686

H69[8]	23:30.2	+49 08	OC or AST
III2p	15.0′ × 15.0′	6.8B	5.6V

This group of stars is poorly seen on the DSS image. In a small telescope, this coarse, scattered cluster of stars surrounds a magnitude 7 yellow star, designated HD 221246. In all, about two dozen stars are visible, including about five magnitude 9 stars; the rest are fainter. This may not be a cluster. About 80 stars are suspected of being members. The spectral type is A0. Distance estimated at about 3300 ly.

NGC7707

H579[3]	23:34.8	+44 18	GLX
S0⁻:	1.3′ × 1.1′	14.48B	

This lenticular galaxy's appearance is quite similar both visually and photographically. The galaxy is located 0.75′ SE of a magnitude 10 field star which hinders observation

somewhat. It is a round, hazy patch of light, broadly brighter to the middle with ill defined edges. Radial velocity: 5750 km/s. Distance: 257 mly. Diameter: 97 000 ly.

Antlia

Although this southern constellation appeared quite low on the horizon from Herschel's location, he was nevertheless able to catalogue one object in the constellation, the bright classic spiral galaxy NGC2997.

NGC2997

H50[5]	9:45.6	−31 11	GLX
SA(s)c	9.6′ × 6.8′	10.04B	9.4V

This is a classic grand-design spiral, well seen in the DSS image. The two main spiral arms emerge from the small, bright circular core. There are several knotty condensations along the spiral arms as well as narrow dust lanes delineating the spiral arms. The arms fragment into smaller segments after about half a revolution; at least eight of these can be seen in the outer disc. High-resolution photographs reveal several very thin dust lanes spiralling outward from the core. Unfortunately, for observers at mid-northern latitudes this beautiful galaxy is something of a disappointment. It is a very faint and diffuse, E–W elongated, large, hazy patch of light, broadly brighter to the middle with a small, though nonstellar patch a little brighter than the main envelope, embedded in the middle. A faint field star is in the outer envelope to the WSW. Recorded on 4 March 1793, Herschel called the galaxy: 'Very faint, very small, a little extended 15 degrees south preceding, north following. A little brighter to the middle. 8′ l, 5 or 6′ broad.' Radial velocity: 869 km/s. Distance: 39 mly. Diameter: 109 000 ly.

NGC2997

©STScI/AURA/CALTECH/ROE

Aquarius

This large, but decidedly faint constellation is where Herschel began his sweeps of the heavens in 1783 and the first object discovered was the spiral galaxy NGC7184. Herschel noted 37 objects in the constellation, all galaxies apart from the exceptional planetary nebula NGC7009 (recorded before he began his formal sweeps in September 1782), the faint and remote globular cluster NGC7492 and an asterism NGC7526. There are few spectacular objects here, though NGC7606 is an impressive spiral and some of the Herschel objects mark interesting fields of faint galaxies. Also, there are a high number of galaxies with peculiar morphologies, including the well-known NGC7252.

NGC6962

H426[2]	20:47.3	+0 19	GLX
SAB(r)ab	2.9′ × 2.3′	13.0B	12.4V

In photographs this is a slightly inclined, two-armed spiral galaxy. The large, bright, oval core has a very faint star involved to the SE. The spiral arms are faint and smooth with a few subtle condensations in the eastern arm. Visually, this galaxy is the largest and brightest of a small, compact group of galaxies including NGC6959, NGC6961, NGC6964, NGC6965 and NGC6967. All are visible in a medium-magnification field; NGC6964 is only 1.75′ to the SE and at the apparent distance, the core-to-core separation between the two galaxies would be about 100 000 ly. NGC6962 is a well defined, round galaxy with a brighter, star-like core. Radial velocity: 4368 km/s. Distance: 195 mly. Diameter: 164 000 ly.

NGC6964

H427[2]	20:47.4	+0 18	GLX
E+ pec:	1.7′ × 1.3′	13.8	

Photographs show this elliptical galaxy as slightly elongated with a faint outer envelope. There is a faint condensation in the N towards NGC6962 and the outer envelope appears to extend towards the larger galaxy though there is no suggestion of interaction. Visually, this galaxy follows NGC6962 to the SE. It appears a little fainter and much smaller, but is quite concentrated and well defined. A magnitude 13 star follows the galaxy to the ESE. Radial velocity: 3961 km/s. Distance: 177 mly. Diameter: 87 500 ly.

NGC7009

H1[4]	21:04.2	−11 22	PN
4+6	1.0′ × 0.6′	8.30B	8.0V

This object has the distinction of being the earliest entry in Herschel's catalogue, recorded on 7 September 1782, nearly 14 months before the next entry (for the galaxy NGC7184) was made. The DSS image is burned out and shows no detail, owing to the brightness of this planetary nebula, but high-resolution photographs reveal a nebula of complex structure,

NGC6962/6964

©STScI/AURA/CALTECH/ROE

featuring a dense, elliptical ring surrounding the central star and embedded in a fainter, secondary shell. The defining features are the ansae, faint extensions which give rise to the popular name for this object, the Saturn Nebula, which was bestowed by Lord Rosse, based on observations he made with his 72-inch reflector. Herschel did not remark upon the ansae in his discovery description, saying only: 'Very bright, nearly round planetary, not well defined disc.' In a moderate-aperture telescope, the object is very distinctly elongated ENE–WSW and sharply defined to the N and S. A subtle but definite pale blue colour is detectable. The nebulosity appears coarse and grainy, very dense and opaque and the edge is a little less well defined towards the E. The ansae cannot be detected with certainty, but are suspected as blunt and subtle extensions to the ENE and WSW. They are visible with large-aperture telescopes, however, the present author having seen them distinctly with a 25-inch reflector. The central star, HD200516, is magnitude 11.5 but is difficult to detect owing to the brightness of the gaseous shell. Its spectrum is continuous. Radial velocity in approach: 46 km/s. Expansion velocity of the nebulous shell: 21 km/s. Distance: 2900 ly.

NGC7081

H859[3]	21:31.4	+2 30	GLX
Sb pec?	1.3′ × 1.3′	14.10B	

Images show a galaxy with a peculiar morphology. It is viewed face-on and has a bright core and half a dozen brighter condensations arrayed around the core. An asymmetrical, faint ring surrounds the core with a faint plume emerging and connecting to a ring in the NNE. The galaxy UGC11760 is 4.25′ ESE. Telescopically, this is a very small galaxy but the core is fairly well condensed and it takes magnification fairly well. It is somewhat elongated N–S in a hazy, ill defined envelope. Two very faint field stars are immediately S. Radial velocity: 3435 km/s. Distance: 153 mly. Diameter: 58 000 ly.

NGC7165

H930[3]	21:59.5	−16 31	GLX
SB(r)ab	1.0′ × 1.0′	14.03B	13.5V

Photographically, this face-on barred spiral galaxy has a small, bright core and a faint bar embedded in an oval inner envelope. Two S-shaped spiral arms emerge from the bar; they fade and narrow as they spiral outward. Telescopically, only the bright inner region is seen, so the galaxy appears as a faint, slightly oval glow, elongated E–W with well defined edges. The galaxy is situated just N of a ESE–WNW line joining two faint field stars. Radial velocity: 5363 km/s. Distance: 240 mly. Diameter: 70 000 ly.

NGC7171

H692[3]	22:01.0	−13 16	GLX
SB(rs)b	2.6′ × 1.5′	13.05B	

The DSS image shows an inclined spiral galaxy with two principal arms of the grand-design type and a brighter core. The bright inner arms emerge from a short bar and form a partial ring. Visually, this is a very faint and diffuse galaxy, appearing almost round and only slightly brighter to the middle. The edges are poorly defined and fade very gradually into the sky background. A magnitude 10 chain of three equally-spaced field stars is visible immediately NE. Radial velocity: 2825 km/s. Distance: 126 mly. Diameter: 96 000 ly.

NGC7180

H693[3]	22:02.3	−20 33	GLX
S0°?	1.5′ × 0.7′	13.54B	

The visual and photographic appearances of this lenticular galaxy are quite similar. Telescopically, it is a distinct oval of light with a bright envelope, brightening to a condensed, nonstellar core and is fairly well defined and elongated ENE–WSW. A slightly fainter galaxy, NGC7185, is located 9.5′ ENE and is visible with NGC7180 in a medium-magnification field, though it was missed by Herschel. Discovered on

©STScI/AURA/CALTECH/ROE

NGC7183

11 September 1787, he described NGC7180 as 'Very faint, very small 360 (magnification) confirmed it.' Radial velocity: 1318 km/s. Distance: 59 mly. Diameter: 26 000 ly.

NGC7183

H595[2]	22:02.4	−18 56	GLX
S0+ pec sp	3.8′ × 1.1′	13.00B	

The DSS image shows a lenticular galaxy with a brighter core and three prominent dust lanes: one crossing the core, a short, narrow one to the NE, and a third in the SW taking a notch out of the outer envelope. Visually, the galaxy forms a triangle with two field stars to the S. Despite the dominating presence of dust along the disc, the galaxy is a moderately bright, though diffuse, oval patch elongated almost due E–W with fairly even surface brightness and moderately well defined edges. Radial velocity: 2720 km/s. Distance: 122 mly. Diameter: 135 000 ly.

NGC7184

H1[2]	22:02.7	−20 49	GLX
SB(r)c	6.0′ × 1.3′	11.69B	11.2V

The entry for this bright spiral galaxy marks the true beginning of Herschel's systematic sweeps of the heavens in search of the nebulae. Recorded on 28 October 1783, Herschel described this galaxy as: 'Faint, considerably large, much extended, brighter in the middle. Easily resolvable.' Photographs show this highly inclined spiral galaxy has a moderately large core tilted somewhat

NGC 7184

©STScI/AURA/CALTECH/IROE

compared to the plane of the galaxy. This core is surrounded by a bright ring of overlapping inner spiral arms and the outer spiral arms are broader with knotty condensations. In a medium-magnification field, this bright spiral galaxy is much extended ENE–WSW with a prominent core. The envelope in the ENE extends almost to a magnitude 11 field star. Although the extensions are fairly well defined, they are not bright and appear patchy and irregular. Radial velocity: 2698 km/s. Distance: 121 mly. Diameter: 210 000 ly.

NGC7218

H897[2]	22:10.2	−16 40	GLX
SB(r)c	2.5′ × 1.1′	12.40B	

The DSS image shows a slightly inclined spiral galaxy with a brighter core and a possible bar. Two spiral arms emerge, and several bright knots are prominent, particularly in the western spiral structure. In moderate apertures this

fairly bright galaxy is elongated NNE–SSW and is bordered on the E and NNE by a pair of magnitude 13 field stars. The core is moderately bright and the major axis is bright and well defined. Averted vision brings out a fainter outer envelope perpendicular to the major axis so that the galaxy appears more oval in form. Discovered on 6 September 1793, Herschel described the galaxy as: 'Pretty bright, little extended, resolvable. 1.5′ long, 1.25′ broad.' Radial velocity: 1751 km/s. Distance: 78 mly. Diameter: 57 000 ly.

NGC7230

H931[3]	22:14.3	−17 04	GLX
SA(s)bc	0.9′ × 0.8′	14.44B	

This face-on galaxy appears to be a two-arm spiral. Photographs show one short and stubby arm emerging from the N; the other arm which emerges from the S is longer with a bright knot, after which the arm continues much fainter,

encircling the galaxy and ending at a faint condensation. Visually, the galaxy is faint but takes magnification well. It is a roundish, diffuse, nebulous patch of light with fairly well defined edges and is slightly brighter to the middle. Radial velocity: 4481 km/s. Distance: 200 mly. Diameter: 52 000 ly.

NGC7246

H932[3]	22:17.5	−15 32	GLX
(R)SA(r)a:	1.7′ × 1.0′	13.67B	

Photographs show an inclined spiral galaxy with a brighter core, two arms and a faint knotty outer structure. A very faint star is located immediately NE of the core. Visually, a prominent triangular asterism marks the location of this diffuse galaxy, which is situated immediately S of the westernmost star in the asterism. The galaxy is a hazy, oval patch of light, elongated N–S, and is very slightly brighter to the middle. The edges are poorly defined and the proximity of the star interferes with the view. Recorded on 6 September 1793, Herschel commented: 'Extremely faint, small, little extended, south of a star to which it seems to be attached, but is free from it. The star is the first of three, making a small triangle.' Radial velocity: 4223 km/s. Distance: 188 mly. Diameter: 93 000 ly.

NGC7251

H933[3]	22:20.4	−15 46	GLX
(R')SA(rs)a?	1.9′ × 1.7′	13.50B	

Photographically, this face-on, multi-arm spiral has a large, bright core and thin, tightly wound and faint spiral arms. At medium magnification, the galaxy is readily visible but diffuse, quite round and broadly brighter to the middle with somewhat well defined edges. Radial velocity: 4959 km/s. Distance: 221 mly. Diameter: 122 000 ly.

NGC7252

H458[3]	22:20.7	−24 41	GLX
(R)SA(r)°:	1.9′ × 1.6′	12.79B	12.1V

This very peculiar galaxy has been catalogued as Arp 226 and also goes by the prosaic name 'Atoms for Peace'. High-resolution photographs give some idea why: part of the outer structure consists of looping spiral arms encircling the central region of the galaxy, looking very much like the orbits of electrons around the core of an atom. The unusual structure of this galaxy is presumed to be the result of a merger of two galaxies. The DSS image shows a bright, lenticular core with a fainter envelope. Two condensations border the core on the W. The faint and peculiar outer structure features loops of matter and two very faint extensions: one is to the NW while the other lies to the east. Telescopically, much of this intriguing outer structure is lost. Although faint, the galaxy takes magnification well; it is a roundish, diffuse, nebulous patch of light with fairly well defined edges and is slightly brighter to the middle. The published apparent size does not take the extensions into consideration; the

extension to the NW continues for 4.5′ from the core before fading while the one to the E extends for about 2.7′. At the apparent distance, the NW plume would extend for at least 283 000 ly, while the eastern one would be at least 170 000 ly in length. Radial velocity: 4852 km/s. Distance: 217 mly. Diameter: 120 000 ly.

NGC7284

H469[2]	22:28.6	−24 51	GLX
SB(s)0° pec	1.2′ × 0.7′	13.35B	

This galaxy forms an interacting pair with NGC7285 and together they are also catalogued as Arp 93. In the DSS image both galaxies appear as barred spirals and the outer reaches of NGC7284 seem to blend into the spiral arms of NGC7285. The western arm of the latter galaxy extends counterclockwise for almost a full revolution and seems to enclose both

NGC7284

galaxies. An extensive, faint plume emerges from this arm on the E side and extends for about 3.7′ towards the S. Telescopically, the pair are located in a star-poor field; they are faint but high magnification resolves the two bright cores and reveals separate galaxies involved in a diffuse, common envelope. The core of NGC7285 is located immediately ENE of NGC7284, which appears somewhat brighter. Discovered on 26 October 1785, it is unclear whether Herschel saw both galaxies. His description reads: 'Faint. Pretty small, little extended, easily resolvable, some of the stars visible.' Radial velocity: 4739 km/s. Distance: 212 mly. Diameter: 74 000 ly.

NGC7302

H31[4]	22:32.4	−14 07	GLX
SA(s)0⁻:	1.8′ × 1.1′	13.15B	

Photographs show that this lenticular galaxy has a bright, slightly oval core in a fainter, but distinct, outer envelope. In moderate apertures, although small and moderately bright, this galaxy stands out well and is easy to locate 2.8′ N of a magnitude 9 field star. Only the core of this galaxy is visible, appearing round, well defined and brighter to the middle to a faint, stellar core. Herschel placed this galaxy in his Class IV, recording the observation on 3 October 1785 and calling the galaxy: 'Faint, small, stellar, with a pretty large chevelure.' Radial velocity: 2799 km/s. Distance: 125 mly. Diameter: 66 000 ly.

NGC7252

NGC7309

H476[2]	22:34.3	−10 21	GLX
SAB(rs)c	2.0′ × 1.8′	13.04B	

The DSS image shows a small, face-on, three-branch spiral with a bright core. The three branches are bright but short and farther out the arms fade quickly, becoming broader and more diffuse. Visually, the galaxy is fairly dim, visible as a faint, unconcentrated, almost round patch, with poorly defined extremities. A magnitude 9 field star is located 6.25′ N. Radial velocity: 4114 km/s. Distance: 184 mly. Diameter: 107 000 ly.

NGC7364

H442[2]	22:44.5	−0 07	GLX
S0/a pec:	1.7′ × 1.1′	13.43B	

The visual and photographic appearances of this galaxy are quite similar. Bracketed by two bright, widely separated field stars, this is a moderately bright galaxy with an opaque oval envelope, elongated ENE–WSW. There is a brighter, star-like core and the edges are moderately well defined. Radial velocity: 5003 km/s. Distance: 223 mly. Diameter: 110 000 ly.

NGC7371

H477[2]	22:46.1	−11 00	GLX
(R)SA(r)0/a:	2.0′ × 2.0′	12.69B	

The DSS image shows a face-on multi-arm spiral with a large core and narrow, tightly wound spiral arms. Higher-resolution photographs resolve a faint bar in the core and show the spiral arms to be studded with HII regions. Visually, the galaxy is small but fairly bright, with a bright, oval envelope, possibly evidence of the bar, and fairly even surface brightness. Radial velocity: 2786 km/s. Distance: 125 mly. Diameter: 72 000 ly.

NGC7377

H598[2]	22:47.8	−22 19	GLX
SA(s)0+	3.2′ × 2.6′	12.10B	11.6V

The DSS image reveals a lenticular galaxy of somewhat peculiar

morphology. The extensive secondary envelope shows several discrete dust patches; higher-resolution images show a definite spiral aspect to the envelope, delineated by subtle dust lanes, some of which emerge from the bright inner disc. Telescopically, only the bright central region is seen; it is small but fairly bright, well condensed and brighter to a nonstellar core with fairly well defined edges. Radial velocity: 3401 km/s. Distance: 152 mly. Diameter: 141 000 ly.

NGC7391

H443[2]	22:50.7	−1 31	GLX
E:	1.7′ × 1.3′	13.12B	

The classification of this galaxy is uncertain and it may be a lenticular type as it has an extensive, faint outer envelope. Visually, the galaxy is fairly bright, featuring a bright, star-like core in a slightly fainter, condensed inner envelope. This is surrounded by a faint, ill defined outer envelope. The galaxy appears round, with a prominent field star 1.1′ N. Radial velocity: 3180 km/s. Distance: 142 mly. Diameter: 70 000 ly.

NGC7392

H702[2]	22:51.8	−20 36	GLX
(R′:)SB(rs)ab	2.1′ × 1.3′	12.58B	11.8V

Photographs show an inclined spiral galaxy with a brighter, elongated central region and two principal spiral arms. High-resolution images show no evidence of a bar, despite the classification, and also reveal a knotty, multi-arm structure. Visually, this bright and fairly well defined, galaxy appears oval and well condensed and is elongated ESE–WNW. The core area is large and is elongated along the major axis. Radial velocity: 3259 km/s. Distance: 145 mly. Diameter: 89 000 ly.

NGC7393

H453[2]	22:51.7	−5 33	GLX
SB(rs)c pec	2.0′ × 07′	13.53B	

Photographs show a peculiar galaxy with a bright core surrounded by a

bright ring that is broken in the E. There is a faint nebulous extension off the break and a hazy outer envelope to the W. At high magnification the galaxy is fairly faint and quite diffuse. Broadly brighter near the middle, it is elongated E–W with the eastern portion a little brighter with well defined edges. The western portion is fainter and fades gradually into the sky background. Radial velocity: 3883 km/s. Distance: 173 mly. Diameter: 101 000 ly.

NGC7443

H450[2]	23:00.1	−12 48	GLX
SB(s)0+: sp	1.5′ × 0.6′	13.88B	

This galaxy forms a visual pair with NGC7444, which is situated 1.6′ to the S. Although there is a difference of about 600 km/s in the radial velocities of the two galaxies, they very probably form a physical pair and the minimum separation between the two galaxies would be about 75 000 ly at the suspected distance of NGC7443. The two galaxies appear similar both visually and photographically. They can be seen together in a high-magnification field. NGC7443 is moderately bright and is well seen at high magnification. It is an elongated streak of light oriented NE–SW with well defined edges and has a small, brighter core embedded in the bright envelope. Radial velocity: 3600 km/s. Distance: 161 mly. Diameter: 70 000 ly.

NGC7444

H451[2]	23:00.1	−12 50	GLX
SB(r)0°? sp	1.5′ × 0.6′	13.86B	

Photographs show an elongated lenticular galaxy with a bright central region and tapered extensions. This galaxy appears a little fainter visually than its neighbour, NGC7443, but their forms are quite similar. It is an elongated streak oriented N–S; the edges are well defined and a small, brighter core is visible in the bright envelope. Radial velocity: 3012 km/s. Distance: 135 mly. Diameter: 59 000 ly.

NGC7443

NGC7492

H558[3]	23:08.4	−15 37	GC
CC12	4.2′ × 4.2′	11.2V	

This is a very loosely structured globular cluster, poorly concentrated to the middle. The DSS image resolves the cluster well and its total population of stars is quite low. Telescopically, this globular cluster is extremely faint and is best seen at medium magnification as

NGC7492

an irregularly round, ill defined patch of light, very slightly brighter to the middle. No resolved stars are visible and the cluster is only slightly brighter than the sky background. The brightest members of the cluster are magnitude 15.5. Radial velocity in approach: 189 km/s. Distance: 82 000 ly.

NGC7526

H470[3]	23:13.9	−9 12	AST
0.9′ × 0.5′			

This is an asterism involving four faint stars, three of them in a N–S line, the fourth a little separated towards the WNW. It appears as a very faint, nebulous streak at medium magnification, while high magnification brings out the four stars which appear fainter than magnitude 14. Recorded on 28 November 1785, Herschel commented: 'Extremely faint, very small, 240 (magnification) left doubtful.'

NGC7576

H454[2]	23:17.4	−4 44	GLX
SA(r)0[+]	1.2′ × 1.0′	13.81B	

In photographs this lenticular galaxy displays a bright core surrounded by a fainter outer envelope; the division between the two is very sharp and sudden. In a medium-magnification field, NGC7576 is visible with NGC7585 (see next entry). Herschel only came across this galaxy a year after discovering NGC7585. Smaller and fainter than its neighbour, it has a brighter and more prominent stellar nucleus. Though the core is round and well defined, the outer envelope is diffuse. Radial velocity: 3687 km/s. Distance: 165 mly. Diameter: 58 000 ly.

NGC7585

H236[2]	23:18.0	−4 39	GLX
(R')SA(s)0[+] pec	3.3′ × 2.7′	12.44B	

This lenticular galaxy has a very faint, asymmetrical outer envelope, more extensive towards the N and has also been catalogued as Arp 223. Photographs show the structure of the outer envelope is somewhat peculiar with a small, round dust patch visible and the distortions are presumed to have been caused by an interaction or merger with another galaxy in the distant past. Visually, the galaxy is quite bright and well condensed at the core with a brighter nucleus. It appears round and is surrounded by a hazy secondary envelope. A meandering chain of magnitude 11 or fainter field stars points toward the galaxy from the NNW. The galaxy is probably physically related to NGC7576, which is in the field to the SW. Radial velocity: 3652 km/s. Distance: 163 mly. Diameter: 157 000 ly.

NGC7592

H186[3]	23:18.4	−4 25	GLX
S0[+] pec:	1.2′ × 0.9′	14.50B	

Discovered on the same night as NGC7585, 20 September 1784. This is actually a triple interacting galaxy

©STScI/AURA/CALTECH/ROE

NGC7526

NGC7592

system with a peculiar morphology. Two galaxies with bright cores are in contact; the western galaxy is ejecting a plume to the N while the eastern galaxy is ejecting a plume through a bright condensation to the SW. The condensation is actually the third, and smallest, galaxy, located immediately S between the two cores. Herschel described it as 'extremely faint, very small' and visually it is a very faint galaxy, not resolved into separate components and a little brighter to the middle. Oval in shape, the system is oriented E–W. A magnitude 10 field star follows to the NE. Radial velocity: 7441 km/s. Distance: 332 mly. Diameter: 116 000 ly.

NGC7600

H431[2]	23:18.9	−7 35	GLX
S0⁻	2.5′ × 1.1′	12.91B	

Photographically and visually, this lenticular galaxy presents a similar appearance. It is a moderately bright galaxy, well condensed towards the centre with a brighter core. The outer envelope is grainy and the galaxy is extended in an ENE–WSW direction. Radial velocity: 3539 km/s. Distance: 158 mly. Diameter: 116 000 ly.

NGC7606

H104[1]	23:19.1	−8 29	GLX
SA(s)b	5.3′ × 2.0′	11.58B	

Images reveal a moderately inclined spiral galaxy of the multi-arm, filamentary type. The small, bright core is embedded in a fairly tightly wound, symmetrical spiral pattern. Visually, the galaxy is large and moderately bright; it is an elongated oval oriented SE–NW with a bright, elongated core surrounded by a disc that fades gradually to somewhat ill defined edges. Radial velocity: 2331 km/s. Distance: 104 mly. Diameter: 161 000 ly.

NGC7665

H438[3]	23:27.2	−9 24	GLX
0.7′ × 0.6′	14.23B		

This galaxy is unclassified, though on the DSS image it appears that it may be a barred spiral of peculiar structure. The bright core has two tight, bright extensions in a faint, asymmetrical envelope. Visually, this small galaxy is a bright, well defined patch of light, irregularly round and slightly brighter to the middle. A bright field star is 2.75′ ESE. Radial velocity: 4903 km/s. Distance: 219 mly. Diameter: 45 000 ly.

NGC7721

H432[2]	23:38.8	−6 31	GLX
SA(s)c	3.5′ × 1.4′	12.25B	

This is a highly inclined, three-branch spiral with an elongated core and knotty spiral arms. Visually, the galaxy appears somewhat diffuse but large and moderately bright: it is an elongated patch of light with rather ill defined edges and is slightly brighter along its major axis. The envelope has a grainy texture and is elongated NNE–SSW. Radial velocity: 2113 km/s. Distance: 94 mly. Diameter: 96 000 ly.

NGC7723

H110[1]	23:38.9	−12 58	GLX
SB(r)b	3.3′ × 2.2′	11.93B	

This is a four-branch barred spiral viewed almost face-on. Images reveal a bright

NGC7606

NGC7606: 15-inch Newtonian 293×

core with a bright bar; the inner spiral arms are tightly wound and bright, while the outer arms are faint and symmetrical. In a moderate-aperture telescope the galaxy is quite bright and best seen at medium magnification, but it takes higher magnification well. It is an oval patch of light with ill defined edges and a grainy texture to the envelope. A small, bright

core is visible and the galaxy is elongated ENE–WSW. Radial velocity: 1951 km/s. Distance: 87 mly. Diameter: 84 000 ly.

NGC7725

H189[3]	23:39.3	−4 32	GLX
0.4′ × 0.4′	14.74B		

This galaxy is unclassified but is most probably elliptical. Photographs show a round, bright disc with no outer

envelope. Visually, it is very small, almost round and quite faint and is difficult to hold with direct vision. The surface brightness is quite even and the edges are well defined. Radial velocity: 6057 km/s. Distance: 270 mly. Diameter: 31 500 ly.

NGC7727

H111[1]	23:39.9	−12 18	GLX
SAB(s)a pec	6.0′ × 3.5′	11.57B	

This peculiar galaxy exhibits outer structural similarities with NGC7252, the 'Atoms for Peace' galaxy and has also been catalogued as Arp 222. The core is bright and elongated with what appears to be a stubby bar oriented perpendicular to the core. There is very faint outer structure, including a curving band S of the core and a large, broad band sweeping from the core in the E. This may be a case of two galaxies merging: short exposure images seem to reveal a double nucleus near the centre. Visually, the galaxy is small but bright, with a bright, grainy envelope elongated E–W and a bright, star-like core. The edges are well defined. Radial velocity: 1946 km/s. Distance: 87 mly. Diameter: 151 000 ly.

Aquila

This bright constellation dominates the equatorial region of the summer sky, with the Milky Way as a backdrop. Casual sweeping with a telescope reveals field after field of faint stardust as well as numerous dark dust clouds blotting out the myriad stars beyond. Despite the large size of the constellation, this is a meagre hunting ground for Herschel objects: ten are recorded in all. Three are planetary nebulae, five are open clusters and two are galaxies.

NGC6755

H19[7]	19:07.8	+4 14	OC
II2r	15.0′ × 15.0′	8.6B	7.5V

Small telescopes show this cluster as a gathering of magnitude 11 or fainter stars, with a very slightly nebulous central region, indicating the presence of unresolved stars. Photographs show this nebulous patch is actually a clutch of about a dozen stars, though not all of them are necessarily cluster members. Otherwise, the cluster is somewhat scattered but the surroundings are relatively starless, so the group stands out well. The cluster comprises about 160 stars. The spectral type is B2 and the age 35 million years. Distance: 5500 ly.

NGC6756

H62[7]	19:08.7	+4 41	OC
I1m	4.0′ × 4.0′	10.6V	

In a small-aperture telescope this is a very small and nebulous cluster, which is readily picked up at low magnification and is well seen at medium magnification. The central region is an unresolved haze, oriented E–W with two or three magnitude 12 stars resolved. To the S is a smaller unresolved haze with a magnitude 12 star shining through. About six other magnitude 12–13 stars are visible surrounding the central region. In photographs there is a conspicuous ring of stars starting at the compressed cluster to the N, curving E, S, then W. There is also a second chain of stars in a broad arc from E to W over the top of the compressed knot. Forty stars are accepted as members, the spectral type is B3 and the age about 47 million years. Distance: 4900 ly.

NGC6837

H18[8]	19:53.5	+11 41	OC
IV1p	3.0′ × 3.0′	12.0V:	

Neither this cluster nor NGC6840 are plotted in the first edition of *Uranometria 2000.0* (Tirion, Rappaport and Lovi, 1987). Photographs show a small, scattered grouping of stars slightly brighter than the surrounding Milky Way field and difficult to separate from the background. Telescopically, the general field is rich in Milky Way stars and examination with a low magnification reveals two coarse knots of stars which may be the objects.

NGC6756

NGC6837 is visible as a coarse knot of stars at low magnification. About 18 magnitude 11 or fainter stars are seen in an 8.0′ × 8.0′ area. The area to the SSW is marked by a diamond-shaped asterism, the area to the NNE by a row of magnitude 11 stars oriented ESE–WNW.

NGC6772

H14[4]	19:14.6	−2 42	PN
3b+2	1.4′ × 1.2′	14.2B	12.7V

The DSS image shows a bright planetary nebula. It is an elongated ring which broadens to the WSW and narrows and brightens considerably in the E. The central star (magnitude 18.2) is faint but visible in a narrow, elongated clearing at the centre. In a moderate-aperture telescope the planetary nebula is visible as a fuzzy out-of-focus star at low magnification. At high magnification, it is a moderately large, dim, mottled patch of light with hazy edges. An OIII filter does not substantially improve the view. Herschel's 21 July 1784 discovery entry correctly surmises that this is a planetary nebula, calling it: 'Very faint of equal light, resolvable 1′ in diameter in the midst of numberless stars of the Milky Way.' Distance: about 4200 ly.

NGC6781

H743[3]	19:18.4	+6 33	PN
3b+3	2.5′ × 2.1′	11.8B	11.8V

The DSS image shows a very bright, ring-shaped planetary nebula. The ring is most pronounced in the W, curving S, then E, where it narrows and brightens considerably. It continues through the N, fainter but unbroken. There is a faint extension of material outside the ring here. The centre of the nebula is fainter and several stars are embedded in it. The central star is faint but visible. The visual appearance of this planetary nebula is fairly pale, yet bright and obvious even at low magnification. It is quite large, about 2.0′ across and holds magnification well. It is fairly featureless and appears quite round, though it is decidedly brighter along its southern edge and a little hazy and ill defined in

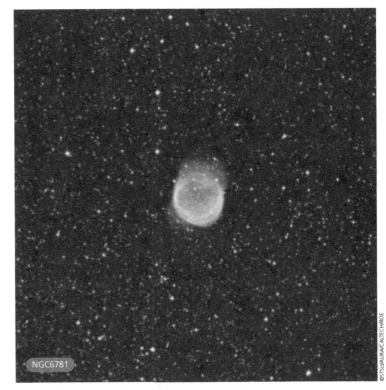

NGC6781

©STScI/AURA/CALTECH/ROE

the N; it is smooth textured. At magnitude 16.2, the central star is not seen visually. Herschel recorded this object as a very faint nebula on 20 July 1788, calling it: 'Considerably faint, irregularly round, resolvable, 3.0′ or 4.0′ in diameter.' Radial velocity in recession: 4 km/s. Expansion velocity: 12 km/s. Distance: 2600 ly.

NGC6804

H38[6]	19:31.6	+9 13	PN
4+2	1.0′ × 0.9′	13.4B	12.0V

The DSS image reveals a very bright planetary nebula, with a bright core surrounded by an oval doughnut-shaped ring. This ring is surrounded by a paler outer shell that is brighter to the NNW and the SSE. In a moderate-aperture telescope this is a moderately bright, although tenuous, almost round nebula with poorly defined edges. A magnitude 13 field star borders the nebula on the NE side. Here, the shell appears slightly indented. Averted vision reveals the

magnitude 14.4 central star and at high magnification the interior is less bright than the surrounding shell. The spectral type of the central star is O9. Herschel discovered this object on 25 August 1791 and curiously placed it in his Class VI – very compressed and rich clusters of stars. His description reads: 'Considerably bright, small, irregular form. Easily resolvable. Some of the stars are visible.' Radial velocity in approach: 12 km/s. Distance: 4200 ly (estimated).

NGC6814

H744[3]	19:42.7	−10 19	GLX
SAB(rs)bc	3.0′ × 3.0′	12.10B	

Images reveal a bright, face-on galaxy in a fairly rich star field with a large core and two principal spiral arms emerging. After about half a revolution, both arms branch into a broader, multi-arm spiral pattern. Visually the galaxy is moderately large but faint, diffuse and almost round with an even-surface-brightness disc. The edges are diffuse

NGC6804

©STScI/AURA/CALTECH/ROE

fainter than magnitude 12 is situated immediately N and E of the bright star, well separated from the sky background. The bright star is about 1° S of Altair.

NGC6840

H19[8]		19:55.3	+12 07	OC
4.0′ × 4.0′				

In a rich Milky Way field, photographs show a small, scattered grouping of about 15 brighter stars. Telescopically, the group appears as a compressed, diamond-shaped asterism in a Milky Way field. It looks slightly nebulous at low magnification, while high magnification reveals only six stars in a 4.0′ × 4.0′ area, with a brightness range about magnitude 11–12. This grouping may just be an asterism.

NGC6926

H142[3]		20:33.1	−2 01	GLX
SB(s)bc pec		1.9′ × 1.3′	13.31B	

The DSS image shows a two-armed spiral galaxy with the faint core offset towards the N and a dust patch immediately S. The spiral structure is weak and somewhat fragmentary with a brighter, knotty spiral arm emerging on the E side and curving to the N. Visually, this galaxy is a moderately large, though hazy patch of light, oval in form and oriented N–S. It is very ill defined with fairly even surface brightness. Also visible in the field is NGC6929, located 4.0′ almost due E. Radial velocity: 6029 km/s. Distance: 269 mly. Diameter: 149 000 ly.

and poorly defined and a tiny star-like core is visible at high magnification. Several field stars are in the immediate vicinity. Radial velocity: 1678 km/s. Distance: 75 mly. Diameter: 65 000 ly.

NGC6828

H73[8]		19:50.4	+7 55	OC?

Recorded on 30 July 1788, Herschel described this object as: 'A cluster of coarse scattered stars with one pretty bright star in the middle.' The DSS photograph shows a scattered group of bright stars with a wide range of brightnesses in a fairly rich field. Visually, there appears to be a cluster or asterism in the field; with the seventh magnitude star centred in the field, a scattered group of seven slightly fainter stars is located to the S, while a more compressed group of about a dozen stars

Aries

Herschel recorded the 27 objects discovered in this constellation between the years 1784 and 1786. All are galaxies except NGC1240, which is a faint pair of stars. For the most part, the objects here are nondescript, but the spiral galaxy NGC772 is a bright and interesting specimen.

NGC673

H589[2]	1:48.4	+11 32	GLX
SAB(s)c	2.5′ × 1.9′	13.22B	

Photographs show a spiral galaxy seen face-on with a small, brighter core. The two principal spiral arms split into fragmentary arms after half a revolution and there are several condensations in the arm N of the core. The galaxy is fairly bright visually, large and situated 3.1′ WSW of a bright field star. It is a moderately well defined oval of light, elongated N–S and broadly brighter to the middle. Radial velocity: 5261 km/s. Distance: 235 mly. Diameter: 171 000 ly.

NGC678

H228[2]	1:49.4	+22 00	GLX
SB(s)b: sp	4.5′ × 0.8′	13.35B	

This edge-on spiral galaxy forms a contrasting pair with NGC680 5.25′ to the ESE. They almost certainly form a physical pair and the core-to-core separation may be as little as 200 000 ly. Both galaxies are part of the NGC691 group of galaxies. The DSS photograph shows a spiral galaxy with a large, bright core embedded in low-surface-brightness spiral arms. A prominent but peculiar dust lane extends along the major axis, bending slightly as it crosses in front of the core. Telescopically, only the core of this galaxy can be seen under average conditions. The core brightens to the middle and the edges are fairly well defined. Radial velocity: 2941 km/s. Distance: 131 mly. Diameter: 171 000 ly.

NGC678/680

©STScI/AURA/CALTECH/ROE

NGC680

H229[2]	1:49.8	+21 58	GLX
E⁺ pec:	1.9′ × 1.6′	12.9B	

This elliptical galaxy has a large, bright core and a peculiar, asymmetrical outer envelope. In a moderate-aperture telescope it appears very slightly brighter than its neighbour, NGC678; it is a grainy, roundish patch of light with moderately well defined edges and a brighter core. Radial velocity: 2905 km/s. Distance: 130 mly. Diameter: 72 000 ly.

NGC691

H617[2]	1:50.7	+21 46	GLX
SA(rs)bc	3.5′ × 2.6′	12.28B	

Photographs show a slightly tilted, low-surface-brightness, multiple-arm spiral galaxy with a small, slightly brighter core. Visually, the galaxy is moderately faint and best seen at medium magnification. It is quite diffuse and gradually brighter to the middle with very poorly defined edges, oval in shape and slightly elongated E–W. A bright, tight double star with equal components borders the galaxy 1.6′ NE. Radial velocity: 2769 km/s. Distance: 124 mly. Diameter: 126 000 ly.

NGC695

H618[2]	1:51.2	+22 35	GLX
S0? pec	0.6′ × 0.5′	13.78B	

The DSS photograph shows a very small, bright peculiar galaxy with possible small dust patches in the outer envelope to the W. Visually, the galaxy is very faint, a small nebulous patch of light, almost round and fairly even in surface brightness with well defined edges. A faint field star precedes the galaxy 0.6′ W. Radial velocity: 9841 km/s. Distance: 440 mly. Diameter: 77 000 ly.

NGC697

H179[3]	1:51.3	+22 21	GLX
SAB(r)c:	4.7′ × 1.6′	12.79B	

This is a highly inclined spiral galaxy. Photographs show an elongated, patchy central region with symmetrical spiral arms emerging. The galaxy is moderately bright visually; it is a well defined oval of light, much extended and lens-shaped, oriented ESE–WNW and very broadly brighter to the middle, though it lacks a brighter core. Radial velocity: 3222 km/s. Distance: 144 mly. Diameter: 197 000 ly.

NGC772

H112[1]	1:59.3	+19 01	GLX	
SA(s)b	7.2′ × 4.3′	10.71B	10.3V	

This is a very large and bright galaxy, well seen on the DSS photograph as a tilted, multi-arm spiral galaxy with a small core and dusty spiral arms. The structure is somewhat asymmetrical: the principal spiral arm is decidedly brighter and broad, narrowing to a long, thin and fairly straight arm after half a revolution around the core. This arm is studded with star-forming

regions while the rest of the multi-arm pattern appears fainter and fairly smooth-textured, defined by discrete dust patches and lanes. A very faint and apparently detached arc extends around the galaxy from the SSE to the NW. A small elliptical companion galaxy is situated 3.4′ SSW; this is NGC770 and together with the larger galaxy the system is also catalogued as Arp 78. They form a physical pair: the minimum core-to-core separation would be about 113 000 ly at the presumed distance. Telescopically, NGC772 is a very bright galaxy, well seen at high magnification as a large, elongated patch of light. The bright envelope, which is oriented ESE–WNW, brightens to an ill defined and nonstellar core, while the edges of the galaxy are diffuse and fade uncertainly into the sky background. NGC770 is a small, concentrated patch of light a little extended NNE–SSW and it is curious that Herschel overlooked this object. Radial velocity: 2564 km/s. Distance: 114 mly. Diameter: 240 000 ly.

NGC774

H214[3]	1:59.4	+14 01	GLX
S0	1.2′ × 1.0′	14.25B	

Photographically this lenticular galaxy is elongated N–S with a bright core, a faint disc and a very faint outer envelope. A dim spiral galaxy, UGC1468, is 3.9′ SSW. The galaxy is moderately bright visually, appearing round, fairly well defined and a little brighter to the middle. Radial velocity: 4673 km/s. Distance: 209 mly. Diameter: 73 000 ly.

NGC781

H215[3]	2:00.1	+12 38	GLX
Sb(f)	1.5′ × 0.4′	13.91B	

The DSS shows an edge-on lenticular galaxy with a large, bright, bulging core and bright extensions oriented NNE–SSW. Only the core of this edge-on lenticular galaxy is detected visually, but it is fairly bright and well seen at high magnification as a condensed, well

NGC772

defined disc with a small, bright core embedded. Radial velocity: 3557 km/s. Distance: 159 mly. Diameter: 69 000 ly.

NGC794

H207[3]	2:02.5	+18 22	GLX
S0⁻:	1.2′ × 1.0′	13.82B	

This lenticular galaxy presents similar appearances visually and photographically. At high magnification it is moderately faint but is well seen as a round patch of light with even surface brightness and well defined edges. Radial velocity: 8313 km/s. Distance: 371 mly. Diameter: 130 000 ly.

NGC803

H208[3]	2:03.8	+16 02	GLX
SA(s)c: sp	3.0′ × 1.3′	13.03B	

In photographs this highly tilted, multi-arm spiral has a very small, bright core, while the symmetrical spiral arms are grainy though fairly continuous. A bright field star situated 0.8′ to the WSW hinders observation visually. The galaxy appears diffuse at medium magnification; it is a hazy, slightly concentrated patch of light, very slightly elongated N–S. Radial velocity: 2182 km/s. Distance: 98 mly. Diameter: 85 000 ly.

NGC821

H152[1]	2:08.4	+11 00	GLX
E6?	2.6′ × 2.0′	11.76B	11.3V

This is an example of the relatively rare E6 elliptical galaxy, which features a bright, elongated central region and tapered extensions which almost suggest the beginnings of edge-on spiral structure. A considerable faint envelope surrounds the bright core and disc-like extensions, and is well seen on the DSS image. Visually, the galaxy can be seen 1.0′ SE of a bright field star. It is a bright galaxy that is well seen at high magnification but the star hinders the view somewhat. The very small bright core is embedded in a well defined, round disc which has a grainy texture. Radial velocity: 1799 km/s. Distance: 80 mly. Diameter: 61 000 ly.

NGC871

H201[3]	2:17.2	+14 33	GLX
SB(s)c:	1.2′ × 0.5′	14.15B	13.6V

Photographs show an irregular, wedge-shaped galaxy which is bright along its major axis with a star or compact companion galaxy immediately E. Three prominent, but faint galaxies are in the field and may be companions, including NGC870, only 1.5′ S. Visually, NGC871 is a small, bright galaxy; it is a well condensed and well defined oval, oriented N–S. Its surface brightness is fairly even and it is visible in a medium-magnification field with NGC877. Radial velocity: 3808 km/s. Distance: 170 mly. Diameter: 59 000 ly.

NGC877

H246[2]	2:18.0	+14 33	GLX
SAB(rs)bc	2.4′ × 1.8′	12.53B	

This is the brightest member of the NGC877 group of galaxies, which comprise eight members including NGC870, NGC871 and NGC876. The DSS image shows what appears to be a four-branch spiral galaxy seen almost face-on, possibly with a bar. The core is a little brighter and the spiral arms are knotty and tightly wound with a prominent dust lane along the western flank. Visually, this is a bright galaxy; it is an elongated oval of light oriented NW–SE and brighter to the middle. The main envelope is quite grainy and the edges fade gradually into the sky background. A magnitude 12 field star lies just outside the SE edge and the companion galaxy NGC876, a dim edge-on galaxy with a prominent dust lane, is only 2.0′ SW. At the apparent distance, the core-to-core separation of the two galaxies, would be about 103 000 ly. High-contrast images indicate a faint plume of material connects the two galaxies. Radial velocity: 3981 km/s. Distance: 178 mly. Diameter: 124 000 ly.

NGC871

NGC877

NGC924

H474[3]	2:26.8	+20 30	GLX
S0	2.3′ × 1.3′	13.72B	

While this galaxy is classified as lenticular, the DSS image reveals an object with a bright, elongated core embedded in a very faint disc which is seen obliquely. This disc appears to involve dust and there are suspicions of a spiral pattern. A small, fainter barred spiral galaxy, MCG+3-7-13, is 3.0′ to the NE. Telescopically, NGC924 is a small, somewhat concentrated patch of light with well defined edges; only the bright core is visible. Radial velocity: 4540 km/s. Distance: 203 mly. Diameter: 135 000 ly.

NGC932

H489[2]	2:27.9	+20 20	GLX
SAa	1.6′ × 1.6′	13.77B	

This galaxy is erroneously plotted in the first edition of *Uranometria 2000.0* as

NGC930. Photographs show a face-on spiral galaxy with a bright, round core and faint, narrow spiral arms. A faint field star is immediately SE of the core while there is a bright condensation in the outer spiral arm to the NE, probably a satellite galaxy. Visually, the galaxy is quite faint and is visible at medium magnification as a round patch of light, evenly bright across its disc with well defined edges. Herschel's 29 November 1785 discovery description reads: 'Faint, small, a little extended, contains 3 stars unconnected.' In addition to the star SE of the core, there is a slightly brighter star at the edge of the galaxy further to the SE but there is no third star. There is a slight possibility that Herschel may have detected the satellite galaxy to the NE. Another explanation, though an unlikely one, for the third star is that it may have been a supernova. Radial velocity: 4156 km/s. Distance: 186 mly. Diameter: 86 000 ly.

NGC972

H211[2]	2:34.2	+29 19	GLX
Sab	3.3′ × 1.7′	12.16B	

This is a highly tilted spiral galaxy with a bright core and a complex system of dust patches and lanes bordering the core. The dust lanes may not be coincident with the plane of the galaxy. Photographs show that beyond the bright inner region the outer disc is smooth-textured and does not reveal an easily recognizable spiral pattern. At high magnification the galaxy is bright and is bordered on the S by a curving chain of five prominent field stars. The galaxy is roughly oval and elongated SSE–NNW with a grainy and mottled envelope. The edges are moderately well defined, with the ENE edge being a little more sharply defined. Radial velocity: 1640 km/s. Distance: 73 mly. Diameter: 70 000 ly.

NGC990

H557[3]	2:36.4	+11 39	GLX
E	1.6′ × 1.3′	13.49B	

In photographs this elliptical galaxy has a bright core and an extensive outer envelope, elongated NE–SW. Visually, it is a small, condensed galaxy, well seen at high magnification as a round, well defined object with a brighter, stellar core. Radial velocity: 3555 km/s. Distance: 159 mly. Diameter: 74 000 ly.

NGC1012

H152[3]	2:39.3	+30 09	GLX
S0/a?	2.4′ × 0.9′	13.26B	

A highly inclined spiral galaxy with an extended, bright inner region and spiral

NGC972

arms. The DSS image shows bright condensations in the spiral arms to the NNE and SSW as well as evidence of dust patches; it is very unlikely that this is an S0 system. Telescopically, this is a moderately large and bright galaxy, a little brighter along its major axis and fairly well defined. A prominent, though faint, field star borders the galaxy immediately to the ESE of the core. Radial velocity: 1083 km/s. Distance: 48 mly. Diameter: 34 000 ly.

NGC1024

H592[2]	2:39.2	+10 51	GLX
(R')SA(r)ab	4.7′ × 1.8′	13.22B	

This is a highly tilted spiral galaxy of unusual structure that has also been catalogued as Arp 333. Photographically, the bright core is surrounded by a prominent inner ring with much fainter and larger spiral arms attached. The arms emerge at an angle to the ring; they are smooth textured except for some small, faint knots. A prominent field star is 0.6′ NE of the core, and there are several fainter galaxies in the immediate field, including NGC1129, 6.75′ ESE. In a moderate-aperture telescope NGC1024 is the brightest of the three galaxies visible in the field. Located immediately SW of the field star, it is a round, mottled glow, ill defined at the edges with a detached, sizable brighter core. NGC1029 is smaller and only slightly fainter; it is a well defined streak oriented ENE–WSW. NGC1028 is 3.0′ to the N, and is a threshold object only intermittently visible as a tiny, nebulous patch east of a diamond-shaped asterism. Radial velocity: 3574 km/s. Distance: 160 mly. Diameter: 218 000 ly.

NGC1030

H581[3]	2:39.9	+18 01	GLX
S0	1.6′ × 0.7′	14.36B	

This highly inclined spiral galaxy has a morphology which appears quite similar to that of NGC3628 and a more accurate classification might be SAb pec. Photographs show an edge-on galaxy with a small core which is partially obscured by a thick dust lane, slightly inclined to the plane of the galaxy. The overall structure is decidedly rectangular with the south extension appearing slightly disturbed. There are fainter anonymous galaxies 4.3′ ENE and 3.5′ WNW. The galaxy is quite faint visually and is a little more distinct at medium magnification. It is elongated N–S with moderately well defined edges and fairly even surface brightness along its major axis. Radial velocity: 8615 km/s. Distance: 385 mly. Diameter: 179 000 ly.

NGC1036

H475[3]	2:40.5	+19 18	GLX
Pec?	1.2′ × 0.8′	13.74B	

This galaxy, probably a dwarf, appears to be an inclined irregular galaxy, with an elongated bright core and a faint irregular outer envelope. The DSS image shows several faint knots which may trace an incipient spiral pattern, particularly N of the core. Telescopically, the galaxy is quite faint and best seen at medium magnification as a fairly well defined, small, oval patch of light, elongated N–S and a little brighter to the middle. Radial velocity: 857 km/s. Distance: 38 mly. Diameter: 13 000 ly.

NGC1056

H584[3]	2:42.9	+28 33	GLX
Sa	2.3′ × 1.1′	13.33B	

In photographs, this is a highly inclined spiral galaxy with a large bright core and a dense dust lane, which is narrow to the S and broadens quickly to the N. Telescopically, the galaxy is moderately faint, but visible as a small, opaque, oval patch of light, slightly elongated SSE–NNW, brightening to a small, prominent core. The edges are fairly well defined and the galaxy takes higher magnification well. Radial velocity: 1635 km/s. Distance: 73 mly. Diameter: 49 000 ly.

NGC1088

H582[3]	2:47.0	+16 11	GLX
S0⁻a	1.1′ × 0.6′	14.95B	

The DSS image shows a close, unequal pair of galaxies; the principal galaxy is lenticular, elongated E–W, while the satellite is a very small, compact object, possibly elliptical, situated 0.25′ ENE of the core. There does not seem to be tidal interaction between the galaxies and the minimum separation between the two would be about 24 000 ly at the presumed distance. NGC1088 is an extremely small and difficult galaxy visually, situated 1.1′ ESE of a faint field star. It is a round, somewhat condensed patch of light, fairly well defined and even in surface brightness across the disc. Radial velocity: 7264 km/s. Distance: 324 mly. Diameter: 104 000 ly.

NGC1134

H254[2]	2:53.6	+13 00	GLX
S?	2.5′ × 0.9′	13.27B	

Photographs show a peculiar, inclined spiral galaxy which has also been catalogued as Arp 200. It has a small, bright core and the bright inner disc shows spiral structure with dust involved and many knotty condensations, indicating a region of high star formation. The outer envelope is much fainter and appears less active. A faint plume of material extends westward towards the faint, irregular galaxy UGC2362, 7.0′ to the W. Visually, only the bright interior of the galaxy is visible and the core of the galaxy is 0.75′ WSW of a field star. The galaxy is quite bright and is well seen at high magnification as a mottled, round and moderately well defined object with a small, brighter core. Radial velocity: 3680 km/s. Distance: 164 mly. Diameter: 119 000 ly.

NGC1156

H619[2]	2:59.7	+25 14	GLX	
IB(s)m	3.3′ × 2.5′	12.06B	11.7V	

The DSS image shows a dwarf Magellanic galaxy with a bar-like

NGC1156

©STSCI/AURA/CALTECH/ROE

diffuse at the edges with a field star bordering the outer envelope immediately N. The galaxy is a little brighter along a thin band running the length of the major axis and is oriented NNE–SSW. Herschel discovered the galaxy on 13 November 1786 and called it: 'Pretty bright, considerably large, pretty much extended in the meridian, resolvable, 1′ south of a star.' Radial velocity: 446 km/s. Distance: 20 mly. Diameter: 20 000 ly.

NGC1240

H164[3]	3:13.5	+30 30	DS

Herschel recorded this object on 12 September 1784, his description reading: 'Extremely faint, very small. 240 (magnification) left a doubt.' In his sweep notes (sweep 264) he was more specific and closer to the truth: 'Suspected, 240 (magnification) left a doubt; extremely faint and very small, most probably two close stars; between two stars.' This is, in fact, a close pair of stars appearing visually as a faint, fairly equal pair of field stars, about magnitude 13–14 and only resolved at high magnification. It does not appear nebulous and is clearly a pair of stars which can be seen on a line joining two stars oriented ESE–WNW and another slightly fainter pair of stars aligned NNE–SSW.

central region and a very small, star-like nucleus embedded. The envelope is irregular in form with many bright, knotty condensations. High-resolution photographs begin to resolve individual stars in the galaxy at blue magnitude 21.

Visually, the galaxy is faintly visible at low magnification, is quite bright and large at high magnification and is situated in a rich starfield. The galaxy appears as a broad, elongated patch of light that is quite bright but somewhat

Auriga

Except for the star cluster NGC 1896 (recorded in January of 1784) all of the objects discovered in this constellation were recorded between the years 1786 and 1793. There are 13 Herschel objects here in all, star clusters and nebulae for the most part, though the very remote galaxy NGC 2387 provides a difficult challenge for observers.

NGC1664

H59[8]	4:51.1	+43 42	OC
III1p	18.0′ × 18.0′	8.0B	7.6V

In photographs, this cluster appears as an oval-shaped grouping of bright stars in a rich starfield. Visually, this is a very delicate cluster of predominantly magnitude 11 stars; one magnitude 8 star to the SE is probably not a member. About 20 stars are resolved, beginning with a N–S chain W of the bright field star and the rest are scattered towards the NW. Up to 100 stars may be members of this cluster. The spectral type is B8 and the age about 300 million years. Radial velocity in recession: 35 km/s. Distance: 3670 ly.

NGC1778

H61[8]	5:08.1	+37 03	OC
III2p	8.0′ × 8.0′	8.2B	7.7V

Photographs show a scattered cluster of bright stars elongated NW–SE with a 'dog-leg' in the SE where the cluster turns towards the SW. Though not particularly rich visually, this is an attractive cluster of moderately bright stars, well resolved and well separated from the sky background. The brightness range is moderate and the cluster appears elongated NW–SE with an attractive double star near the SE edge. About 25 stars are involved, though faint unseen cluster members increase the membership to more than 100. The brightest member is magnitude 10.06 and the spectral type of the cluster is B6, leading to an age of 160 million years. Distance: 4820 ly.

NGC1857

H33[7]	5:20.2	+39 21	OC
I3m	10.0′ × 10.0′	8.1B	7.0V

On the DSS photograph this is a very pretty, compressed cluster of faint stars surrounding three bright field stars oriented NW–SE. Most of the cluster members are faint and arrayed below the middle of the three field stars. Visually, it is a fairly compressed cluster of bright and faint stars, well resolved at high magnification and dominated by one magnitude 8 star. Thirty-two stars are resolved in the field and in all about 40 stars are considered members. The spectral type is A. Distance: 6200 ly.

NGC1883

H34[7]	5:25.9	+46 33	OC
II1m	5.0′ × 5.0′	12.0V	

The DSS image reveals a faint, triangular grouping of stars, well separated from the surrounding field. In a medium-magnification eyepiece, this is a faint, subtle cluster located E of Capella. Most resolved cluster members are magnitude 14 or fainter. It is visible as a misty patch NNE of a magnitude 10 field star at low magnification; resolution begins at higher powers. A round concentration of stars is visible just W of the centre and high magnification brings out a few more stars. Recorded on 11 December 1786, Herschel called this object 'A cluster of very faint and very small stars, pretty compressed but not rich. Irregular form. 3.0′ in diameter.' About 30 stars are probable cluster members.

NGC1896

H4[8]	5:25.7	+29 20	OC?
20.0′ × 20.0′			

The DSS photograph shows an irregular grouping of brighter stars in an otherwise moderately rich field. Telescopically, the group is well seen at medium magnification as a somewhat scattered and well resolved group of about two dozen stars with a fairly wide brightness range. It was not identified in the first edition of *Uranometria 2000.0* but is coincident with a small triangle of stars plotted between Beta Tauri and the star cluster Be19. Recorded on 16 January 1784, Herschel described the group as: 'A cluster of coarse and irregularly scattered pretty large stars.'

NGC1907

H39[7]	5:28.0	+35 19	OC
I1m n	5.0′ × 5.0′	8.9B	8.2V

Photographs show a compressed grouping of moderately bright stars, well separated from the background field of stars. At medium magnification, this is a faint, though very attractive cluster, lozenge-shaped and oriented NNW–SSE. Resolution is complete at high magnification, though much of the cluster's impact is lost. Most of the cluster members are magnitude 12–14. Two magnitude 10 field stars border the cluster to the S. Total cluster membership is in excess of 100 stars with the brightest cluster member about magnitude 11. Spectral type is

NGC1857

©STScI/AURA/AC.ALTECH/ROE

B3 and the age about 440 million years. Distance: 4335 ly.

NGC1931

H261[1]	5:31.4	+34 15	EN/RN+OC
6.0′ × 6.0′	10.1V		

This complex object is well seen in the DSS image, which reveals a bright nebula, quite intense at its core, with fainter nebulosity arrayed to the N, W and S. The whole nebula complex is roughly crescent-shaped. Visually, the nebula is quite subtle and is brightest around a triangular asterism of fairly bright stars which have been catalogued as ADS4112. A star NE of the triangle and one to the SW indicate the visual extent of the nebulosity, which is broadly brighter to the middle and poorly defined. Total membership of the cluster is believed to be 20 stars though the DSS image shows literally dozens of faint stars in the less intense portions of the nebula. Herschel recorded this object on 4 February 1793, describing it as: 'Very bright, irregularly round, very gradually brighter to the middle. 5′ in diameter. Seems to have one or two stars in the middle, or an irregular nucleus; the chevelure diminishes very gradually.'

NGC1931

member of the cluster. One stream leads to the WSW, while the other goes to the SW. There are other stars scattered to the S of the bright star. Telescopically, the cluster is a faint scattering of stars with several tight pairs involved. The cluster members show a very narrow brightness range with about two dozen resolved, all fainter than magnitude 13. About 40 stars are probable members, with the brightest magnitude 13. Distance: 4600 ly.

NGC2192

H57[7]	6:15.2	+39 51	OC
II2m	5.0′ × 3.0′	10.9V	

Photographs show a compressed group of faint stars, irregularly round with a narrow range of brightness. It is located 5.0′ S of a magnitude 8 field star. Visually, it is a small, compressed cluster of very faint stars, well separated from the sky background. There is resolution into individual stars at high magnification but these stars are faint and quite similar in brightness. The cluster is oval and elongated NE–SW. About 20 stars are resolved, but there are probably 45 members of the cluster. The brightest stars are magnitude 14.

NGC2240

H49[8]	6:33.2	+35 15	OC

Situated E of a wide pair of magnitude 7 field stars oriented N–S, this is a slightly compressed grouping of about two dozen stars with a moderate brightness range. Fairly distinct and well separated from the sky background, the cluster is almost round.

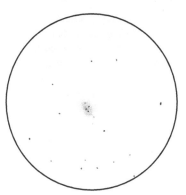

NGC1931 15-inch Newtonian 272x

positively seen at high magnification as a very small, nebulous spot with a tiny star shining through, offset a bit towards the SSE.

NGC2126

H68[8]	6:03.0	+49 54	OC
III2m	6.0′ × 6.0′		

Photographs show two principal streams of faint stars leading away from a magnitude 6 field star which is not a

NGC1985

H865[3]	5:37.7	+32 00	EN/RN?

Photographs show a faint, round nebula with the brightest portion to the SE around a magnitude 10 or 11 star. A dark patch is in the nebula SSW of the star, with two further dark patches in the fainter portion of the nebula N of the star. Visually, this nebula is only

NGC1985

NGC2281

H71[8]	6:49.3	+41 04	OC
I3m	25.0′ × 25.0′	6.0B	5.4V

Photographs show a very bright grouping of about 25–30 stars, magnitude 7–11, although many other much fainter stars are in the field. The brighter members in the group are oriented NW–SE and are easily separated from the stellar background. Visually, the cluster is well seen and is

NGC2281

At least 40 stars are resolved. Almost 120 stars are thought to be members, the brightest of which is magnitude 7.3. The spectral type is A0 and the age about 300 million years. Radial velocity in recession: 21 km/s. Distance: 1500 ly.

NGC2387

H820[2]	7:29.0	+36 52	GLX
Spiral	0.5′ × 0.4′	15.71B	16.5V:

This galaxy may be the most remote object in the Herschel catalogue and it is certainly a very challenging object visually. On the DSS image, it is a very small and very dim two-branch spiral galaxy with a slightly brighter core viewed face-on. An even smaller field galaxy is 0.5′ SE. Telescopically, this is an extremely faint and very small galaxy, visible only intermittently as a roundish, hazy patch of light. It is diffuse and occasionally displays a brighter core. Herschel discovered the galaxy on 10 March 1790, describing it as: 'Pretty bright, small, stellar.' The description is puzzling, as the galaxy is far and away one of the faintest objects in Herschel's catalogue, but his offsets from the star 65 Aurigae lead directly to the galaxy's position and there are no alternative nebulous candidates nearby. Radial velocity: 20 028 km/s. Distance: 895 mly. Diameter: 130 000 ly.

very attractive at low magnification. It is bright and well resolved, with the majority of the members around magnitude 9–10; overall the cluster exhibits a moderate brightness range. Eighteen of the stars are quite bright and dominate the cluster and there are some attractive pairs scattered throughout.

Boötes

This area of the sky was a rewarding hunting ground for Herschel, who returned to it almost every spring between the years 1784 and 1802. Herschel recorded 113 objects in this constellation, all galaxies except for the loosely structured globular cluster NGC5466. Two entries can definitely be classed as nonexistent: the pair of stars NGC5621 (it is extremely unlikely that this is the object Herschel thought he saw) and the single star NGC5856 (erroneously thought to involve nebulosity). There is also the case of the pair of stars NGC5922, which is thought by some to be a Herschel object, though the present author feels that the nearby galaxy NGC5923 is more likely to be the object that Herschel recorded. While most of the objects in the constellation are visually unremarkable, dedicated observers will find much here to challenge their observing skills and there are some interesting objects, including NGC5248, NGC5529 and NGC5673, to name a few.

NGC5239

H101[3]	13:36.5	+7 22	GLX
SB(rs)bc	1.8′ × 1.8′	14.54B	

Photographs show a face-on spiral galaxy with a small, bright core and an oval bar surrounded by a broken inner pseudo ring. Two faint and narrow spiral arms are attached and increase the apparent size of the galaxy to about 2.7′ × 2.7′. The galaxy is fairly large but quite diffuse visually and is best seen at medium magnification as a round, fairly faint glow, very broadly brighter to the middle with moderately well defined edges. It is situated within a triangle of field stars. Radial velocity: 7012 km/s. Distance: 313 mly. Diameter: 246 000 ly.

NGC5248

H34[1]	13:37.5	+8 53	GLX
(R)SB(rs)bc	6.2′ × 4.5′	10.88B	

The DSS image shows a slightly tilted two-arm spiral galaxy. The large, bright core generates two broad, bright spiral arms featuring many knotty condensations and thin dust lanes. The inner spiral structure is peculiarly blunted and abruptly transforms to extremely faint traces of broader outer spiral arms. There are three of these fragments. Visually, this is a large, bright, but somewhat diffuse galaxy with an extensive envelope which is elongated ESE–WNW. The core is bright and concentrated; the outer envelope is broad and poorly defined. Hints of spiral structure are suspected with slight curves to the SE and NW. Radial velocity: 1128 km/s. Distance: 51 mly. Diameter: 91 000 ly.

NGC5248

©STScI/AURA/C.ALTECH/ROE

NGC5249

H72[3]	13:37.6	+15 58	GLX
S0?	1.3′ × 0.8′	14.32B	

Though the classification is uncertain, photographs show a probable lenticular galaxy with a bright core and extensive, elongated outer disc. A dust lane runs E–W just N of centre, perpendicular to the major axis and partially obscures the core. Visually, this galaxy is small and quite faint; it is seen as a slightly oval patch of light, well defined and brighter in the middle, oriented N–S. Radial velocity: 7668 km/s. Distance: 342 mly. Diameter: 130 000 ly.

NGC5251

H369[3]	13:37.5	+27 25	GLX
S?	1.0′ × 0.7′	14.68B	

The DSS image shows an almost face-on spiral galaxy with a large, bright core and two fainter spiral arms. A very faint field star or possible compact companion is seen against the spiral arm NNE of the core. Although faint visually, this galaxy is well seen at medium magnification as a small, round, fairly well defined patch of light that is gradually brighter towards the middle to a compressed core. Radial velocity: 11 022 km/s. Distance: 492 mly. Diameter: 143 000 ly.

NGC5293

H6[5]	13:46.9	+16 16	GLX
SA(r)c	1.9′ × 1.5′	13.73B	

In photographs this galaxy appears as an almost face-on spiral with a star-like core. Two principal spiral arms emerge from the central disc, branching to a dusty multi-arm pattern after less than half a revolution. The galaxy is fairly dim visually; it is almost round but large and very slightly brighter to the middle. The edges are poorly defined. Radial velocity: 5789 km/s. Distance: 259 mly. Diameter: 143 000 ly.

NGC5416

H56[3]	14:02.1	+9 27	GLX
Scd:	1.4′ × 0.8′	13.98B	

This galaxy is part of a fairly prominent group of more than a dozen NGC galaxies and many fainter galaxies located in a 1 square degree field. Herschel recorded three of them (NGC5416, NGC5438 and NGC5463) but several others were certainly within the range of his telescope and were overlooked. In photographs, NGC5416 is a slightly inclined multiple-arm spiral with a small, bright core and at least four bright arms in a fainter halo. Companion galaxy NGC5409 is 6.2′ to the WNW and a small, compact galaxy is 4.7′ to the NE, situated between two field stars. Telescopically, NGC5416 is a moderately bright galaxy and appears as a round, somewhat ill defined disc which is gradually brighter to the middle. A medium-magnification field easily shows NGC5423 and NGC5424 following to the E. Radial velocity: 6233 km/s. Distance: 278 mly. Diameter: 113 000 ly.

NGC5417

H11[3]	14:02.1	+8 03	GLX
Sa	1.6′ × 0.7′	13.66B	

Photographs show an inclined spiral galaxy with a bright slightly elongated core in an unevenly bright outer envelope. Though the spiral structure is not well seen on the DSS image, dust patches are suspected on the E and W sides and a brighter nebulous knot is near the edge to the WNW. The galaxy is moderately bright visually, very slightly oval and oriented ESE–WNW with a hazy-edged disc and a small, star-like brighter core. Radial velocity: 4868 km/s. Distance: 217 mly. Diameter: 101 000 ly.

NGC5438

H57[3]	14:03.8	+9 37	GLX
0.8′ × 0.8′	15.92B		

This galaxy was originally designated NGC5446 by Dreyer but nothing was seen at the position by the French observer Guillaume Bigourdan. Herschel's object is very likely NGC5438, which is at the same declination as Herschel gave but about 1.5′ in time further W. The galaxy is unclassified but is probably a face-on lenticular: photographs show a bright round core in an extensive, fainter outer disc. The galaxy is part of a fairly compact group of at least ten NGC galaxies and the edge-on spiral NGC5436 is 2.8′ SW, while NGC5437 is 5.8′ S. Visually, NGC5438 is fairly faint and seen as a round patch of light, which is broadly brighter to the middle, with hazy but fairly well defined edges. NGC5436 can be seen to the SW but only the core is visible, appearing round, smaller and fainter than NGC5438. Radial velocity: 6939 km/s. Distance: 310 mly. Diameter: 72 000 ly.

NGC5463

H58[3]	14:06.1	+9 22	GLX
S?	1.2′ × 0.5′	13.99B	

The classification of this object is uncertain but photographs show an edge-on lenticular galaxy with a brighter core and tapered extensions. It forms a likely physical system with NGC5463B situated 0.7′ to the NE, though there appears to be no interaction between the two. The minimum core-to-core separation between the two would be about 65 000 ly at the presumed distance. Visually, only the larger galaxy is visible, appearing as a small but moderately bright and well defined object, seen as an elongated, opaque disc oriented NE–SW with tapered ends and a bright star-like core. Radial velocity: 7173 km/s. Distance: 320 mly. Diameter: 112 000 ly.

NGC5466

H9[6]	14:05.5	+28 32	GC
CC12	9.0′ × 9.0′	10.5B	

Herschel recorded this loosely structured globular cluster on 17 May 1784, describing it as: 'A cluster of extremely small and compressed stars 6′ or 7′ in diameter, many of the stars visible, the rest so small as to appear

NGC5466

round; it is an opaque, condensed object with a brighter core and well defined edges. Radial velocity: 2244 km/s. Distance: 100 mly. Diameter: 52 000 ly.

NGC5482

H59[3]	14:08.4	+8 56	GLX
S0	1.2′ × 0.9′	13.98B	

Photographs show this lenticular galaxy has a brighter core with tapered extensions in an overall very faint halo. Telescopically, this is a somewhat faint object; it is a little diffuse and is seen as an elongated patch of light, of even brightness, oriented E–W. Radial velocity: 7095 km/s. Distance: 317 mly. Diameter: 111 000 ly.

NGC5490

H32[3]	14:10.0	+17 33	GLX
E	1.7′ × 1.3′	13.07B	

This elliptical galaxy is moderately bright and well condensed with a faint, star-like core embedded in an opaque envelope which has well defined edges. Photographs show an extensive but very faint outer envelope. Faint companions NGC5490B 1.5′ ENE and NGC5490C 4.75′ NNE are not visible in a moderate-aperture instrument. Radial velocity: 4878 km/s. Distance: 218 mly. Diameter: 108 000 ly.

NGC5492

H876[2]	14:10.5	+19 37	GLX
Sb pec?	1.7′ × 0.3′	13.75B	

Photographs show an edge-on galaxy of uncertain classification. Visually, the galaxy is moderately bright and is visible as a hazy spot at low magnification. Medium magnification shows the galaxy's elongated form well; it is oriented NNW–SSE. The major axis is well defined and the extremities taper to soft points. High magnification brings out a slightly brighter, elongated core. Radial

nebulous.' Photographs resolve the cluster completely to the core, though it is a little more compressed to the middle. In a moderate-aperture telescope, it is a large, faint globular cluster, poorly condensed and only very gradually brighter to the middle. Resolution begins at low magnification, as the brightest stars are about magnitude 13.8. High magnification brings out dozens of faint stars down to a magnitude of about 15 over a nebulous haze. Radial velocity: 120 km/s in recession. Distance: 51 500 ly.

NGC5481

H693[2]	14:06.7	+50 43	GLX
E+	1.8′ × 1.4′	13.28B	

The DSS photograph shows an elliptical galaxy, with a faint secondary envelope and a very faint outer envelope. Companion galaxy NGC5480 (in Ursa Major) is 3.2′ W;

the constellation boundary between Ursa Major and Boötes separates the two galaxies. They are probably physically associated and the minimum core-to-core separation would be about 93 000 ly at the presumed distance. The two galaxies are bright and well seen together in a high-power field. Although smaller, NGC5481 is the brighter and is almost

NGC5480/5481

velocity: 2300 km/s. Distance: 103 mly. Diameter: 51 000 ly.

NGC 5500

H674[3]	14:10.2	+48 33	GLX
E	1.10′ × 0.9′	14.49B	

Presenting quite similar appearances visually and photographically this is a fairly small, round but moderately bright galaxy which is well seen at high magnification. The opaque envelope, which is well defined at the edges, has a brighter, intermittently visible core. Radial velocity: 2097 km/s. Distance: 94 mly. Diameter: 30 000 ly.

NGC 5513

H877[2]	14:13.2	+20 26	GLX
S0	1.9′ × 1.1′	13.92B	

This slightly inclined lenticular galaxy has a bright oval core in a fainter envelope which is irregular in outline and uneven in brightness. Photographs show a very faint plume extending towards the WNW and a small, edge-on lenticular galaxy is located 1.25′ SW of the core. Although moderately bright, only the core of this lenticular galaxy is detected visually. It is best seen at medium magnification as a sharp stellar core embedded in a round, well defined envelope. Radial velocity: 5026 km/s. Distance: 225 mly. Diameter: 124 000 ly.

NGC 5515

H685[3]	14:12.5	+39 18	GLX
Sab	1.3′ × 0.7′	14.05B	

Photographs show an inclined spiral galaxy with a large, bright elongated core and an extensive, bright inner disc surrounded by a faint outer envelope. The galaxy is moderately bright visually; it is a well defined oval of light, oriented almost due E–W with a bright envelope of even surface brightness. A small, brighter core is visible and is well seen at high magnification. Radial velocity:

7806 km/s. Distance: 349 mly. Diameter: 132 000 ly.

NGC 5519

H12[3]	14:14.3	+7 31	GLX
Sa	1.6′ × 1.0′	14.44B	

This galaxy was originally identified as NGC 5570 by J. L. E. Dreyer but there is nothing at the designated position. However, NGC 5519 further to the W matches Herschel's 23 January 1784 description perfectly: 'Very faint, forming an arch with 3 stars.' The DSS image shows an inclined spiral galaxy which may be barred, with a bright core and two faint spiral arms. One arm is short and passes through a field star immediately E of the core. The other spiral arm is very long and sweeping, emerging from the core in the E and swinging counterclockwise to the S encircling the galaxy for more than a full revolution. Visually, the galaxy is fairly faint and is seen as a small, oval and ill defined patch of light, oriented E–W with a faint field star involved E of the core. The other two stars of Herschel's arc are 3.3′ WSW and 5.0′ WSW of the core. Radial velocity: 7449 km/s. Distance: 333 mly. Diameter: 155 000 ly.

NGC 5520

H676[3]	14:12.3	+50 22	GLX
Sb	2.0′ × 1.1′	13.31B	

This highly inclined spiral galaxy has a slightly brighter core. Photographs show two bright spiral arms in the NE. Less distinct spiral structure is seen to the SW. Telescopically, the galaxy is quite bright with a well defined, extended envelope, oriented ENE–WSW. A small, bright core is embedded in an opaque but grainy envelope. Radial velocity: 1991 km/s. Distance: 89 mly. Diameter: 52 000 ly.

NGC 5522

H644[3]	14:14.8	+15 09	GLX
Sb	1.9′ × 0.4′	14.07B	

Images show a highly inclined spiral galaxy with a brighter, much elongated

central region. There are knots in the spiral arms, particularly in the NE. Visually, this is a moderately bright galaxy, very well defined at the edges with a bright, opaque disc, oriented NE–SW, a little brighter to the middle with tapered extensions, oriented NE–SW. Radial velocity: 4591 km/s. Distance: 205 mly. Diameter: 113 000 ly.

NGC 5523

H134[3]	14:14.8	+25 19	GLX
SA(s)cd:	4.6′ × 1.3′	13.03B	

Photographs show an almost edge-on spiral galaxy with a bright, elongated core in a bright inner disc with a knotty, spiral filament W of the core. The outer spiral structure is much fainter and ill defined. Visually, this spiral galaxy is moderately faint at medium magnification. The envelope is rather smooth and elongated E–W. Though the galaxy is moderately bright along its major axis, the edges are poorly defined. Averted vision betrays hints of a condensation W of the core. Radial velocity: 1088 km/s. Distance: 49 mly. Diameter: 65 000 ly.

NGC 5529

H414[3]	14:15.6	+36 13	GLX
Sc	6.3′ × 0.6′	12.76B	

The DSS image shows a beautiful edge-on spiral galaxy, very flat with a very small, brighter core. A thin dust lane bisects the galaxy crossing just N of the core. The spiral arm in the WNW is a little longer and slightly disturbed, curving slightly to the W. Several very small, dim galaxies are nearby, possibly dwarf satellites. Telescopically, this is one of the more attractive galaxies in the constellation; it is a flat, elongated object which has fairly high surface brightness and is well seen at high magnification. It is fairly even in brightness along its major axis, is quite well defined and is oriented ESE/WNW. The faint extensions taper

NGC5529

©STScI/AURA/CALTECH/ROE

and smooth-textured; one arm is clearly seen emerging from the N of the bar and curving into E. The arm emerging from S of the bar and curving W is difficult to detect. Visually, the galaxy is small and faint and is best seen with averted vision as a condensed, almost round patch of light, with well defined edges and no sign of the bar. The galaxy NGC5541 is 5.4′ NNE but its higher radial velocity implies that there is no physical connection between the two galaxies. Radial velocity: 5937 km/s. Distance: 265 mly. Diameter: 116 000 ly.

NGC5541

H732³	14:16.4	+39 35	GLX
Sc	0.9′ × 0.9′	13.89B	

Photographs show a very bright, slightly inclined, two-armed spiral, with extremely faint outer spiral arms. A smaller galaxy with a bright core and two faint plumes of material oriented ENE and ESE seems to be impacting the N spiral arm of the brighter galaxy which appears disturbed. A faint field galaxy is 3.8′ NNW. NGC5541 and NGC5536 can be seen together in a medium-magnification field. NGC5541 can easily be seen with direct vision but is still somewhat faint. It is bright and condensed along its major axis, with well defined edges and is elongated N–S. Radial velocity: 7787 km/s. Distance: 348 mly. Diameter: 91 000 ly.

NGC5544

H419²	14:17.0	+36 34	GLX
(R)SB(rs)0/a	1.0′ × 1.0′	14.25B	

Photographs show a face-on spiral, perhaps barred, with a brighter core, an inner ring of material and a fainter outer ring. It is involved with a highly inclined spiral to the E, NGC5545, and the pair are also catalogued as Arp 199. Visually, NGC5545 is the more prominent of the two galaxies, appearing as a quite faint, elongated streak, oriented

to blunt points. Radial velocity: 2955 km/s. Distance: 132 mly. Diameter: 242 000 ly.

NGC5532

H47³	14:16.9	+10 48	GLX
S0	1.3′ × 1.0′	12.96B	

Photographs show a lenticular galaxy with a brighter core and faint outer envelope. A small, edge-on galaxy is 0.5′ S, while NGC5531 is 5.0′ NNW. Visually, NGC5532 is a poorly defined, elongated patch of light oriented NNW–SSE with a brighter, nonstellar core. The envelope is mottled and the lenticular galaxy NGC5531 is readily visible in a high-magnification field. Herschel recorded NGC5532 on 15 March 1784, his description reading: 'Very faint, resolvable, 2 or 3 stars in it.' Radial velocity: 7413 km/s. Distance: 331 mly. Diameter: 125 000 ly.

NGC5533

H418²	14:16.1	+35 21	GLX
SA(rs)ab	3.1′ × 1.9′	12.68B	

The DSS image shows a slightly inclined spiral galaxy with a brighter elongated core. The spiral pattern is faint, asymmetrical and more extensive to the SW. Visually, the galaxy appears elongated NNE–SSW with a faint, star-like core intermittently visible. The envelope is smooth and even in surface brightness. Radial velocity: 3944 km/s. Distance: 176 mly. Diameter: 159 000 ly.

NGC5536

H731³	14:16.3	+39 30	GLX
SBa	1.5′ × 1.3′	14.62B	

The DSS image shows a two-armed barred spiral galaxy seen face-on, with a bright round core in a bright bar. The spiral arms are very faint, narrow

ENE–WSW. There is a slight brightening in the envelope to the WSW and this may be the core of NGC5544. Herschel is presumed to have discovered NGC5544 on 1 May 1785, but his description only reads: 'Faint, pretty large.' Individually, both galaxies are actually quite small and it seems unlikely that Herschel would have picked up NGC5544 without seeing the brighter galaxy. Radial velocity: 3121 km/s. Distance: 140 mly. Diameter: 41 000 ly.

NGC5546

H551[3]	14:18.1	+7 34	GLX
E	1.5′ × 1.2′	13.32B	

This elliptical galaxy has a large oval core in an extensive inner envelope with a faint outer envelope. The slightly fainter galaxies NGC5538, NGC5542 and NGC5543 are all in the immediate field to the W. Visually, NGC5546 is moderately bright and is seen as a round disc, broadly brighter to the middle with a grainy texture and fairly well defined edges. It follows NGC5542, which is a N–S oriented, well defined oval of light. Radial velocity: 7321 km/s. Distance: 327 mly. Diameter: 226 000 ly.

NGC5548

H194[2]	14:18.0	+25 08	GLX
(R′)SA(s)0/a	1.4′ × 1.3′	13.20B	

This face-on spiral is a Seyfert galaxy and features a bright core in a very faint outer envelope. Visually, it is very small, quite condensed and suddenly brighter to a nonstellar core. The edges are hazy and ill defined. Radial velocity: 5199 km/s. Distance: 232 mly. Diameter: 94 000 ly.

NGC5557

H99[1]	14:18.4	+36 30	GLX
E1	2.3′ × 1.9′	11.93B	

Visually and photographically similar, this elliptical galaxy is bright at medium magnification, almost round with a small, bright core and fairly bright

condensed inner envelope. The envelope is surrounded by a faint, ill defined haze. A faint field star is immediately SSE of the core in the outer envelope. Radial velocity: 3295 km/s. Distance: 147 mly. Diameter: 99 000 ly.

NGC5559

H347[3]	14:19.1	+24 48	GLX
SBb	1.5′ × 0.4′	14.61B	

This highly inclined barred spiral galaxy has a bright core and two spiral arms visible in photographs. Visually, it is a difficult galaxy in a moderate aperture and less than perfect skies. Situated in a star-poor field, the galaxy is a small, oval, diffuse object, elongated ENE–WSW. It is very gradually brighter to the middle but the core, which is prominent in photographs, is not visible. Radial velocity: 5216 km/s. Distance: 233 mly. Diameter: 102 000 ly.

NGC5579

H415[3]	14:20.5	+35 11	GLX
SABcd	2.0′ × 1.4′	14.54B	

This galaxy is the faintest of three that can be seen in a medium-magnification field: the others are NGC5589 and NGC5590. Photographs show a face-on spiral galaxy, with a faint, very small, elongated core, possibly a bar. Three principal spiral arms emerge: they are asymmetrical and faint with many nebulous knots. The galaxy has also been catalogued as Arp 69. Visually, the galaxy is quite difficult; it is an ill defined blur bracketed by three faint field stars. It is only intermittently seen and exhibits even surface brightness across the envelope and ill defined edges. Radial velocity: 3682 km/s. Distance: 164 mly. Diameter: 96 000 ly.

NGC5582

H754[2]	14:20.7	+39 42	GLX
E	5.3′ × 2.7′	12.50B	

Despite the classification, the DSS image shows a possible lenticular or elliptical

galaxy with a bright core embedded in a fainter inner disc. The disc is surrounded at a considerable distance by a thin outer ring which does not appear to be in contact with the inner disc. Visually, the galaxy appears small but fairly bright at medium magnification with a small, bright core embedded in a bright envelope with fairly well defined edges; it is elongated NNE–SSW. Radial velocity: 1454 km/s. Distance: 65 mly. Diameter: 101 000 ly.

NGC5587

H110[3]	14:22.2	+13 55	GLX
S0/a	3.0′ × 1.0′	13.79B	

Images show a highly inclined spiral galaxy with a small, slightly elongated core. The grainy spiral structure is bright near the core and fades outwards with dust lanes involved throughout. Telescopically, this is a fairly bright and quite well defined object: it is a sliver of light oriented SSE–NNW, even in surface brightness with a brighter core. A bright field star is 3.1′ to the SSE. Radial velocity: 2322 km/s. Distance: 104 mly. Diameter: 90 000 ly.

NGC5589

H416[3]	14:21.5	+35 15	GLX
SBa	1.1′ × 0.9′	14.25B	

The DSS image shows a face-on, S-shaped spiral galaxy, possibly barred. The two spiral arms are very symmetrical with quite high surface brightness. The lenticular galaxy NGC5590 is 4.5′ SSE. Both galaxies have similar radial velocities, so they probably form a physical pair. The core-to-core separation of the two galaxies would be about 203 000 ly at the projected distance. Visually, only the core of NGC5589 is visible as a pale, diffuse spot, fairly even in surface brightness. Both this galaxy and NGC5590 can be observed in the same high-magnification field. Radial velocity: 3477 km/s. Distance: 155 mly. Diameter: 50 000 ly.

NGC5590

H417[3]	14:21.7	+35 11	GLX
S0	1.1′ × 1.0′	13.56B	

The photographic and visual appearances of this lenticular galaxy are quite similar. Visually, the galaxy appears round and the extremities are moderately well defined, the light growing gradually brighter to the middle. Radial velocity: 3288 km/s. Distance: 147 mly. Diameter: 47 000 ly.

NGC5594

H135[3]	14:23.2	+26 15	GLX
E/S0?	1.0′ × 0.5′	14.80B	

This remote galaxy is possibly an elliptical or S0, with a bright core and very faint elongated outer envelope oriented SSE–NNW. Visually, it is located NNW from a magnitude 9 field star; it appears as a faint, almost stellar spot at medium magnification. It is round with well defined edges and an even surface brightness across the envelope. Discovered on 19 May 1784, Herschel used a magnification of 240× to verify its nonstellar nature. Radial velocity: 11 291 km/s. Distance: 504 mly. Diameter: 147 000 ly.

NGC5596

H418[3]	14:22.5	+37 07	GLX
S0	1.1′ × 0.7′	14.44B	

The visual and photographic, appearance of this lenticular galaxy are identical. At medium magnification it is small and moderately faint with a condensed central region, moderately well defined and slightly elongated E–W. Radial velocity: 3208 km/s. Distance: 143 mly. Diameter: 46 000 ly.

NGC5598

H733[3]	14:22.5	+40 19	GLX
S0	1.2′ × 0.9′	14.07B	

The DSS image shows four galaxies together in the field: NGC5598, NGC5601, NGC5603 and NGC5603B.

NGC5598 is a face-on, multi-arm spiral galaxy and is the faintest in the field. It is lenticular and slightly elongated. Visually, it is a small, but moderately bright, oval, condensed and opaque galaxy with a bright, star-like core. The other three NGC galaxies can all be seen in a high-magnification field, though NGC5601 is very faint and difficult. The four galaxies probably form a physical system. Radial velocity: 5561 km/s. Distance: 248 mly. Diameter: 87 000 ly.

NGC5600

H177[2]	14:23.8	+14 38	GLX
Sc pec	1.5′ × 1.5′	12.92B	

The DSS image reveals a peculiar galaxy, possibly spiral, with a disturbed morphology. The core region is bright and somewhat crescent-shaped, with a spur to the S. Two bright condensations are in the W and a narrow spiral arm emerges from the NW, curving counterclockwise for half a revolution. South of the core is a broad envelope. Visually, the galaxy is bright and irregular in form with a brighter oval central region embedded in a hazy envelope. Herschel thought this galaxy was resolvable. Radial velocity: 2342 km/s. Distance: 105 mly. Diameter: 46 000 ly.

NGC5602

H694[2]	14:22.3	+50 31	GLX
Sa	1.4′ × 0.8′	13.31B	

The DSS image shows an inclined ring galaxy with a bright elongated core and a smooth, fainter disc, which forms a broad ring around the core; a slightly darker zone borders the core. A faint field star is immediately ENE of the core. Telescopically, the galaxy is quite bright with an oval, slightly extended envelope elongated almost due N–S. The envelope is well defined and a small,

NGC5600

bright core is embedded. Radial velocity: 2348 km/s. Distance: 105 mly. Diameter: 43 000 ly.

NGC5603

H734[3]	14:23.1	+40 22	GLX
S0	1.0′ × 0.9′	13.98B	

This lenticular galaxy is part of a group of four that includes NGC5603B, which is 2.5′ to the NNW. Both in photographs and visually, NGC5603 appears larger and brighter than the similar galaxy NGC5598 6.9′ to the WSW. It is a well defined galaxy and very broadly brighter to the middle. Radial velocity: 5759 km/s. Distance: 257 mly. Diameter: 75 000 ly.

NGC5608

H673[2]	14:23.3	+41 46	GLX
Im:	2.6′ × 1.3′	14.25B	

Classified as an irregular type, this dim galaxy nevertheless seems to show something of a spiral character in photographs. The small core is embedded in a weak and ill defined three-branch spiral pattern with many knots in the arms. The structure is asymmetrical with an extensive, very faint envelope to the east. Telescopically, the galaxy is quite faint and diffuse, and is best seen at medium magnification as a slightly oval, dim patch of light, elongated E–W and only very gradually brighter to the middle. Radial velocity: 761 km/s. Distance: 34 mly. Diameter: 26 000 ly.

NGC5610

H136[3]	14:24.3	+24 36	GLX
SB(s)ab	2.0′ × 0.7′	14.22B	

Photographically, this is a highly inclined barred spiral with a bright central region and a bright bar. Two spiral arms are bright where they emerge from the bar, thereafter they are faint and smooth-textured. Situated within a 15′ triangle of magnitude 9 field stars, this is a small, though

moderately bright galaxy visually, extended due E–W and a little brighter to the middle. The envelope appears grainy along the major axis, and the extremities are fairly well defined. Radial velocity: 5115 km/s. Distance: 229 mly. Diameter: 133 000 ly.

NGC5614

H420[2]	14:24.1	+34 52	GLX
SA(r)ab pec	2.5′ × 2.0′	12.57B	

This galaxy is part of a system of three galaxies including NGC5613 and NGC5615 that is also catalogued as Arp 178. The DSS image shows that NGC5614 has a bright core, offset to the S in a bright envelope with a suspected curved dust patch to the N. A very faint secondary envelope is offset to the S and E. NGC5615 is a round compact galaxy in the outer halo to the NW. NGC5613 is 1.9′ almost due N. High-resolution images show the offset spiral pattern of NGC5614 well, with several HII regions embedded. A faint plume of material emerges where NGC5615 contacts the larger galaxy, while NGC5613 features a bright inner ring and fainter outer ring. Visually, NGC5614 is a fairly bright galaxy; it is round with a bright core and is quite condensed to the middle. The inner envelope is quite bright and mottled; a very faint, sizable secondary envelope, is visible with averted vision. NGC5615 is not visible, but NGC5613 can be seen as an ethereal, broadly condensed patch of light. Radial velocity: 3973 km/s. Distance: 177 mly. Diameter: 129 000 ly.

NGC5616

H419[3]	14:24.2	+36 26	GLX
Sbc	2.3′ × 0.4′	14.46B	

Photographs show an edge-on spiral galaxy with a small, bright core. The spiral pattern is asymmetrical and a dust lane is E of the core, broadening as it extends to the S. Visually, the galaxy is a moderately faint, well defined, small, flat streak of light, with even surface brightness along its

major axis, which is oriented NNW–SSE. Radial velocity: 8522 km/s. Distance: 380 mly. Diameter: 255 000 ly.

NGC5621

H14[3]	14:27.8	+8 14	DS (nonexistent)

Although identified today with a pair of magnitude 16 stars, joined by a magnitude 19 one to the N forming a triangle, this has nothing to do with Herschel's suspected object, which he described as: 'Extremely faint, very large. Not verified.' An object matching this description has not been found. As a Herschel object, this can be safely classified as nonexistent.

NGC5622

H677[3]	14:26.2	+48 33	GLX	
Sb	1.6′ × 0.9′	14.09B	13.4V

Photographs reveal an inclined spiral galaxy of the grand-design type with a bright round core. Two very long, high-surface-brightness spiral arms emerge from the core with two prominent condensations. Visually, however, this is an extremely faint galaxy, just a little brighter than the sky background. Fairly diffuse and just a little brighter to the middle, it is slightly elongated E–W. Radial velocity: 3977 km/s. Distance: 178 mly. Diameter: 83 000 ly.

NGC5623

H329[2]	14:27.2	+33 14	GLX	
E	1.6′ × 1.1′	13.64B	12.5V

The classification of this galaxy may more correctly be S0 due to its extensive outer envelope and the three-step brightness gradient. It is elongated NNE–SSW but visually appears as a very small, but moderately bright galaxy with a hazy envelope and a bright, star-like core. Well seen at high magnification, it appears almost round and is moderately well defined. A bright field star is 4.25′ S. Radial velocity: 3434 km/s. Distance: 153 mly. Diameter: 71 000 ly.

NGC5630

©STScI/AURA/CALTECH/ROE

NGC5630

H674[2]	14:27.6	+41 16	GLX
Sdm:	2.1′ × 0.7′	13.73B	

The DSS photograph reveals a highly inclined spiral, possibly barred. There is no core visible and the bright bar has two bright spurs emerging towards the E on each end. There is a bright condensation at the end of the southern spur and fragmentary structure to the spiral arms. Visually, the galaxy is somewhat faint but is visible at high magnification as an elongated streak of light which is fairly opaque along its major axis but a little ill defined. The galaxy is oriented E–W. Radial velocity: 2754 km/s. Distance: 123 mly. Diameter: 75 000 ly.

NGC5633

H185[1]	14:27.5	+46 09	GLX	
(R)SA(rs)b	1.6′ × 1.0′	13.11B	12.9V	

The DSS photograph shows a slightly inclined spiral galaxy, with a small, brighter core in a bright inner disc surrounded by two faint spiral arms. Visually, the galaxy is fairly small but has high surface brightness. Well defined at the edges, it is oval and slightly elongated NNE–SSW with an opaque but somewhat grainy envelope, fairly even in surface brightness. Radial velocity: 2445 km/s. Distance: 109 mly. Diameter: 51 000 ly.

NGC5635

H132[3]	14:28.5	+27 25	GLX
S pec	2.7′ × 1.7′	13.53B	

Photographically, this is a highly inclined spiral galaxy with a bright, elongated core. Two major spiral arms emerge, with the ENE arm disrupted and curving to the N. A dust lane borders the core to the S. Visually, the galaxy is moderately bright with a brighter core and occasional hints of a faint stellar nucleus. The major axis is oriented ENE–WSW and fairly even in surface brightness, while the edges fade suddenly into the sky background. Radial velocity : 4379 km/s. Distance 196 mly. Diameter: 154 000 ly.

NGC5637

H357[2]	14:28.9	+23 11	GLX
Sc	0.9′ × 0.5′	14.79B	

The DSS photograph shows a small, bright two-armed spiral elongated N–S and seen obliquely with a peculiar, bright spur emerging from the spiral arm on the eastern flank. The galaxy is quite faint visually but is somewhat concentrated and well defined at the edges. It appears very broadly brighter to the middle and almost round. Radial velocity: 5298 km/s. Distance: 237 mly. Diameter: 62 000 ly.

NGC5642

H126[3]	14:29.2	+30 01	GLX
E	3.2′ × 1.3′	13.93B	

The DSS image shows an elliptical galaxy with a bright core in a fainter disc with a tenuous and asymmetrical outer envelope, oriented SE–NW and much more extensive on the SE side. At high magnification, it is fairly bright visually and is located 1.4′ almost due W of a bright field star. Initially, it appears as a galaxy with a double nucleus owing to the presence of a faint field star 0.25′ ESE of the true core. The core is a little fainter than the star and the envelope is grainy and very slightly extended ESE–WNW. Large-aperture telescopes may detect a field galaxy 3.0′ to the N. Herschel's discovery description on 16 May 1784 reads: 'Two small stars with suspected nebulosity almost verified 240 (magnification).' Radial velocity: 4377 km/s. Distance: 196 mly. Diameter: 182 000 ly.

NGC5653

H330[2]	14:30.2	+31 13	GLX
(R')SA(rs)b	1.7′ × 1.3′	12.90B	

Photographs show a peculiar spiral with a bright core containing bright

NGC5654

©STScI/AURA/CALTECH/ROE

small, a little brighter to the middle, between 2 very faint stars, 300 (magnification).' This is clearly what the DSS image shows for NGC5655, as the galaxy is situated between two field stars oriented SSE–NNW and separated by 3.0'. The face-on spiral galaxy NGC5648/NGC5649 is 6.2' to the WNW and the present author would argue that this is the object that John Herschel discovered, not NGC5655. The first edition of *Uranometria 2000.0.* plots the N preceding galaxy as NGC5648, the S following one as NGC5649 and assigns NGC5655 to a very faint and small galaxy 6.8' E of NGC5648. Visually, NGC5655 is small, but moderately well defined and moderately bright. It is round and even in surface brightness and is situated between two field stars. Radial velocity: 5234 km/s. Distance: 234 mly. Diameter: 61 000 ly.

NGC5656

H421[2]	14:30.4	+35 19	GLX
SAab	1.9' × 1.5'	13.13B	12.4V

This inclined spiral galaxy has a broad, bright core and faint spiral arms. High-resolution images reveal that the spiral pattern is of the multi-arm-type. In a moderate-aperture telescope this is a fairly bright galaxy, oval in shape and elongated NE–SW. A faint, star-like core is consistently visible, embedded in a moderately bright, grainy envelope with fairly well defined extremities. Radial velocity: 3249 km/s. Distance: 145 mly. Diameter: 80 000 ly.

NGC5660

H695[2]	14:29.8	+49 37	GLX
SAB(rs)c	2.8' × 2.6'	12.38B	

Photographs show a face-on spiral galaxy with a small core and two main high-surface-brightness spiral arms emerging from the core, fading and broadening after about half a revolution. A Magellanic dwarf companion is 2.5' WNW and may be

---

condensations and with a spur emerging from the core in the N. The faint, smooth-textured spiral arms are asymmetrical and more extensive to the SSE. Visually, only the bright inner disc of the galaxy is seen. It is a quite bright, slightly oval, well defined and grainy disc, oriented E–W, surrounding a brighter, elongated core which also appears extended E–W. Radial velocity: 3636 km/s. Distance: 162 mly. Diameter: 80 000 ly.

NGC5654

H420[3]	14:29.9	+36 21	GLX
S?	1.4' × 0.9'	14.14B	

The classification of this peculiar object is uncertain but it is unlikely to be a spiral. The DSS image reveals what appear to be two high-surface-brightness galaxy cores in contact in a faint, asymmetrical envelope aligned NNW–SSE. In a moderate-aperture telescope this is a faint

object, though well seen at high magnification. The envelope is fairly well defined and elongated with a faint, star-like core. Radial velocity: 8677 km/s. Distance: 387 mly. Diameter: 158 000 ly.

NGC5655

H645[3]	14:30.9	+13 58	GLX
S?	0.9' × 0.9'	14.50B	

The DSS photograph reveals two face-on spiral galaxies in the field; the N preceding one carries two NGC numbers: NGC5648 and NGC5649. The S following galaxy is now identified as NGC5655. J. L. E. Dreyer originally identified H645[3] as NGC5649 and assigned the number NGC5655 to the S following object, presuming this was the object discovered in a subsequent sweep by John Herschel. However, William Herschel's original description of H645[3] reads: 'Extremely faint, very

visible in large-aperture telescopes. Visually, the galaxy is well seen at medium magnification; it is fairly bright but rather diffuse, gradually brighter to a faint stellar core, which is only intermittently visible. The outline is oval but only slightly so and is oriented NNE–SSW with hazy edges. Radial velocity: 2448 km/s. Distance: 109 mly. Diameter: 89 000 ly.

NGC5665

H27[2]	14:32.4	+8 05	GLX
SAB(rs)c pec?	2.0′ × 1.3′	12.72B	

This peculiar, one-armed spiral galaxy has also been catalogued as Arp 49. The galaxy is bright but poorly resolved on the DSS image, though the one principal spiral arm, which emerges in the S and curves counterclockwise, is well seen. High-resolution photographs reveal the complex inner region of the galaxy, which is studded with bright HII knots, though no core or bar is visible. The outer spiral disc is fairly smooth-textured, indicating little if any star formation. In a moderate-aperture telescope this galaxy is quite bright and appears as an elongated brighter spine in an otherwise roundish and diffuse disc. The bright spine is elongated N–S and the edges of the disc are poorly defined. Radial velocity: 2236 km/s. Distance: 100 mly. Diameter: 58 000 ly.

NGC5669

H79[2]	14:32.7	+9 53	GLX
SAB(rs)cd	4.1′ × 2.7′	12.73B	

This galaxy is well seen on the DSS photograph and is an almost face-on barred spiral galaxy with a short, bright bar, no core and two spiral arms. The northern arm branches quickly into two arms defined by bright knots. The arms quickly expand into a fainter, broader pattern. A small, compact possible satellite galaxy is 6.2′ to the WNW. Telescopically, NGC5669 is quite large but fairly

diffuse, moderately bright and round. The disc is fairly even in surface brightness and the edges are a little hazy. Radial velocity: 1385 km/s. Distance: 62 mly. Diameter: 74 000 ly.

NGC5672

H310[3]	14:32.6	+31 40	GLX
Sb?	0.5′ × 0.3′	14.13B	

Photographs show a small, faint, nearly edge-on spiral galaxy with a brighter core. A swarm of very faint galaxies is in the immediate field, mostly to the S but they are probably unrelated to the larger object. Visually, the galaxy is very small; it is a well defined circular patch of light, even in surface brightness, with a bright, tight double star visible 3.8′ to the WSW. Radial velocity: 3610 km/s. Distance: 161 mly. Diameter: 24 000 ly.

NGC5673

H696[2]	14:31.5	+49 58	GLX
SBc? sp	2.4′ × 0.5′	12.98B	

Images reveal a highly inclined spiral galaxy with a small, elongated core and grainy spiral arms, seen almost edge-on. Visually, this galaxy is the faintest and smallest of the three field galaxies, the other two being NGC5660 and NGC5676. It has a fairly prominent core which is best seen at medium magnification. It is fairly bright along its major axis, well condensed and elongated NW–SE. The envelope appears spindle-shaped and tapers slightly at the extremities. A magnitude 13 field star is visible off its NW tip. Radial velocity: 2203 km/s. Distance: 98 mly. Diameter: 69 000 ly.

NGC5675

H422[2]	14:32.6	+36 17	GLX
S?	2.7′ × 0.7′	13.82B	

The DSS photograph shows a highly inclined spiral galaxy with a broad, bright core. A thin dust lane is tilted

with respect to the major axis and the spiral arm appears disturbed in the SE. Visually, it is a moderately faint galaxy; at high magnification it is an elongated blur with even surface brightness and is fairly well defined to the SW and NE but with SE–NW oriented extensions which fade slowly into the sky background. Radial velocity: 4062 km/s. Distance: 181 mly. Diameter: 143 000 ly.

NGC5676

H189[1]	14:32.8	+49 28	GLX
SA(rs)bc	4.0′ × 1.7′	11.87B	

This is the principal galaxy of the NGC5676 galaxy group, which comprises 11 galaxies, including probable members NGC5660 and NGC5673. In photographs, this is a bright, inclined spiral galaxy with a bright core. The inner spiral arms have high surface brightness and broaden to a multi-branch pattern with many condensations and a decidedly asymmetrical structure. Visually, the galaxy is bright and greatly elongated NE–SW. The outer envelope is broad and well defined and brightens to a prominent core. A sharp stellar nucleus is visible. Although fairly smooth-textured, the outer envelope appears brightest to the NE. Radial velocity: 2235 km/s. Distance: 100 mly. Diameter: 116 000 ly.

NGC5676
©STScI/AURA/CALTECH/ROE

NGC5677

H283[3]	14:34.2	+25 27	GLX
S?	0.9′ × 0.6′	14.69B	

Although the classification is uncertain, photographs show a small, faint, three-armed spiral galaxy seen face-on with a bright core. Telescopically, the galaxy is quite dim and small; it is situated ESE of a flat triangle of brightish field stars which hinder visibility. A diffuse blur, the object is somewhat well defined at the edges. Radial velocity: 4900 km/s. Distance: 219 mly. Diameter: 57 000 ly.

NGC5684

H421[3]	14:35.8	+36 32	GLX
S0	1.5′ × 1.0′	13.60B	

Photographs show a lenticular galaxy with a faint outer envelope. A possible dust lane is visible to the W and a condensation or companion galaxy is seen in silhouette SW from the core. NGC5686 is 3.3′ to the SE. In a moderate-aperture telescope both galaxies are visible. NGC5684 is a faint, fairly well defined galaxy with a brighter, round, nonstellar core. Faint extensions are visible oriented E–W. Radial velocity: 4174 km/s. Distance: 186 mly. Diameter: 81 000 ly.

NGC5687

H808[2]	14:34.9	+54 29	GLX
S0⁻?	2.5′ × 1.9′	12.59B	

This lenticular galaxy presents a similar appearance both visually and photographically. Located 2.0′ N of a bright field star which hinders observation somewhat, this is a bright galaxy, small but well seen at high magnification as an oval, condensed object with a very bright core, elongated E–W with three stars involved. The star immediately W is easy; the star immediately E is only glimpsed, as is the faint star immediately S of the W star. Herschel's 24 April 1789 description reads: 'Pretty bright, small, irregular form, easily resolved, mixed with some pretty large stars, which may

perhaps belong to it.' Radial velocity: 2248 km/s. Distance: 100 mly. Diameter: 73 000 ly.

NGC5689

H188[1]	14:35.5	+48 45	GLX
SB(s)0/a:	4.3′ × 0.9′	12.74B	

This is the brightest member of a small group of five galaxies about 1° SE of the NGC5676 group. Their similar radial velocities suggest that the two groups may be physically related; the core-to-core separation at the presumed distance would be about 1.75 mly. In photographs this is a bright, almost edge-on lenticular galaxy with an elongated central region and a thin dust lane bisecting the core. High-resolution photographs show that the galaxy actually has three dust lanes running parallel to each other: the one visible in the DSS image is the most northerly one. In moderate-aperture telescopes, the galaxy is well condensed and much elongated E–W. Well defined at the edges, the core is bright and well condensed and suddenly brighter to the middle. To the SE the companion galaxy NGC5693 is a dull, hazy patch of light, round and just a little brighter to the middle. NGC5682 and NGC5683 are about 8.0′ to the WSW and may be detected in large-aperture telescopes. Radial velocity: 2280 km/s. Distance: 102 mly. Diameter: 127 000 ly.

NGC5695

H423[2]	14:37.3	+36 34	GLX
SBb	1.2′ × 0.8′	13.68B	

The DSS image shows what may be two spiral galaxies in contact. The galaxy image to the SE is larger and brighter. The image of the NW galaxy has a brighter core and possibly a detached spiral arm. Visually, the galaxy is a round, well condensed object with a brighter, though nonstellar core. The envelope is quite mottled with moderately well defined edges. Herschel discovered the galaxy on 1 May 1785 and remarked on its irregular form.

Radial velocity: 4318 km/s. Distance: 193 mly. Diameter: 67 000 ly.

NGC5696

H648[2]	14:37.0	+41 49	GLX
Sbc	2.0′ × 1.5′	13.99B	

This galaxy probably forms a physical system with NGC5697, which is located about 9.0′ to the SSW. In photographs it is an inclined barred spiral, with a bright, elongated core, prominent bar and two faint spiral arms attached. Visually, the galaxy is located in a star-poor field and is moderately faint. It appears as a broad, oval patch of light, is moderately well defined with an envelope broadly brighter to the middle and is oriented NNE–SSW. Radial velocity: 5576 km/s. Distance: 249 mly. Diameter: 145 000 ly.

NGC5697

H675[2]	14:36.6	+41 41	GLX
S?	1.0′ × 0.4′	14.61B	

The classification of this galaxy is uncertain though it is probably a spiral. Photographs show that a bright core is embedded in a slightly fainter envelope with two bright streaks, probably spiral arms, in the envelope S of the core. In a moderate-aperture telescope the galaxy is fairly faint and is located E of a bright, four-star asterism. The galaxy is oriented NNE–SSW with a fairly well defined envelope and is opaque with a small, round core embedded. Radial velocity: 5548 km/s. Distance: 248 mly. Diameter: 72 000 ly.

NGC5698

H700[2]	14:37.2	+38 28	GLX
SBb	2.3′ × 0.8′	13.88B	

This highly inclined barred spiral galaxy features two strong spiral arms and is situated in an elongated triangle of field stars. Visually, the galaxy is a faint oval of light, best seen at medium magnification. Very smooth textured and only very gradually brighter to the middle, it is elongated ENE–WSW with the extremities fading very gradually into the sky background. Radial

NGC5698

©STScI/AURA/CALTECH/ROE

NGC5708

H649[2]	14:38.3	+40 27	GLX
Sdm:	1.8′ × 0.60′	13.83B	

Originally identified as NGC5704, it is not certain that this galaxy is H649[2] but Herschel's discovery description on 18 March 1786 is a good match for this galaxy. He described it as: 'Faint, small, extended nearly in the meridian. Resolvable.' Photographs show an inclined spiral galaxy with a small, brighter core in a fairly bright spiral disc. Visually, this galaxy is moderately bright, fairly large and is well seen at high magnification. It is an extended, N–S oriented oval with fairly well defined edges and even surface brightness across the disc. A faint field star is located 0.6′ S of the core, just off the edge of the galaxy. Radial velocity: 2854 km/s. Distance: 127 mly. Diameter: 67 000 ly.

NGC5709

H128[3]	14:38.8	+30 27	GLX
SBa	1.7′ × 0.4′	14.39B	

This highly inclined, barred spiral galaxy is the likely candidate for the nonexistent NGC5703, described under the entry for NGC5706 above. Photographs reveal an inclined galaxy with a bright core and faint bar embedded in a spiral disc. Telescopically, NGC5709 is far brighter and larger than its neighbour NGC5706. It is a small, elongated oval patch of light that is difficult at high magnification; it is elongated ESE–WNW and very broadly brighter to the middle. Radial velocity: 3785 km/s. Distance: 169 mly. Diameter: 83 000 ly.

NGC5710

H895[3]	14:39.3	+20 03	GLX
E	1.3′ × 1.1′	14.01B	

Photographs show that this elliptical galaxy is condensed with a faint outer envelope. NGC5711, a spiral galaxy and a likely physical companion, is 3.5′ to the SSE. At the presumed distance, the minimum core-to-core separation

velocity: 3776 km/s. Distance: 169 mly. Diameter: 113 000 ly.

NGC5702

H894[3]	14:38.8	+20 31	GLX
S0	1.2′ × 0 9′	14.28B	

This galaxy is classified as an S0; the DSS image shows a bright-cored object which may have a bar and very possibly spiral structure in its faint outer envelope. Visually, the galaxy is faint and quite blurry; it is a slightly oval patch of light, extended E–W and slightly brighter to the middle. Radial velocity: 5462 km/s. Distance: 244 mly. Diameter: 85 000 ly.

NGC5706

H127[3]	14:38.6	+30 28	GLX
0.3′ × 0.3′	15.74B		

This unclassified galaxy is one of the most remote in the Herschel catalogue. Originally identified as NGC5699, this

galaxy and NGC5703 are listed as nonexistent, though probably only because of a positional error made by the Herschels. The galaxies NGC5706 and NGC5709, subsequently discovered by the French astronomer Édouard Stephan, match Herschel's 16 May 1784 discovery description, which reads: 'Two. 3′ distant on the parallel. The following very faint, very small, irregularly round. The preceding extremely faint, very small, verified 240 (magnification).' On the DSS image the galaxy appears to be possibly a lenticular with a nebulous patch immediately SSW of the core. A faint field star is immediately N and NGC5709 is located 2.1′ ESE. Visually, NGC5706 is extremely difficult, only intermittently suspected as a very small, round patch of light NW of NGC5709 and about mid-way towards a wide pair of faint field stars. Radial velocity: 21 363 km/s. Distance: 954 mly. Diameter: 83 000 ly.

between the two galaxies would be 414 000 ly. Telescopically, although fairly faint, NGC5710 is quite well seen at medium and high magnification as a condensed, round patch of light, steadily brighter to the middle, though there is no star-like core. The edges are well defined and NGC5711 is also visible in the field as a small, slightly elongated blur SE of an attractive double star which separates the two galaxies. Radial velocity: 9108 km/s. Distance: 407 mly. Diameter: 154 000 ly.

NGC5714

H675[3]	14:38.2	+46 38	GLX
Sc	3.0′ × 0.3′	14.04B	

The DSS image shows an edge-on galaxy with a small, bright core embedded in an extended inner disc. The extended envelope is a little fainter and very probably spiral, though no structure is visible. There are several fainter, smaller galaxies in the field including NGC5717 4.6′ to the ENE. Telescopically, the galaxy is located about 1.0′ SSW of a wide pair of faint field stars and is quite faint; it is a very flat, elongated streak of light, broadly brighter along its major axis. Oriented almost due E–W, the companion galaxy NGC5717 is also visible in the field as a faint patch of light to the ENE. Radial velocity: 2354 km/s. Distance: 105 mly. Diameter: 92 000 ly.

NGC5730

H657[3]	14:40.3	+42 45	GLX
Im:	1.8′ × 0.4′	14.37B	

This galaxy forms a physical pair with NGC5731 located 3.75′ to the NE. Classified as a Magellanic irregular, the DSS image suggests a possible barred spiral seen almost edge-on. The brighter, elongated core is surrounded by two dusty, knotted spiral arms. A bright condensation is at the tip of the spiral arm in the E. This may be a field star, but seems to be disturbing the spiral arm: a faint tuft of material emerges from the arm to the N. Visually, both galaxies can be seen in a high-magnification field with NGC5730

being quite faint; it is an E–W elongated patch of light, somewhat ill defined but with an even surface brightness envelope. It appears larger than NGC5731, though fainter. The minimum core-to-core separation between the two galaxies would be about 129 000 light years at the presumed distance. Radial velocity: 2642 km/s. Distance: 118 mly. Diameter: 62 000 ly.

NGC5731

H658[3]	14:40.0	+42 46	GLX
S?	1.7′ × 0.4′	14.15B	

This is possibly a nearly edge-on spiral galaxy. Photographs show a bright, irregularly-shaped central region embedded in an asymmetrical envelope that is a little more extensive to the ESE. Visually, the galaxy is moderately bright and only the bright central region is visible. It is a small, thin sliver of light elongated ESE–WNW, fairly well defined and even in surface brightness. Herschel recorded NGC5730 and NGC5731 on 9 April 1787, describing them as: 'Two. Both very faint, very small, extended in different directions. 2′ or 3′ distance on the parallel. Each south of a small star.' Radial velocity: 2646 km/s. Distance: 118 mly. Diameter: 59 000 ly.

NGC5732

H686[3]	14:40.6	+38 39	GLX
Sbc	1.5′ × 0.7′	14.32B	

Photographs show an inclined spiral galaxy with two principal, knotty arms emerging from a small core. Only the bright inner region is detected visually. It appears as a roundish glow, moderately concentrated towards a faint stellar nucleus and is best seen at medium magnification. Radial velocity: 3848 km/s. Distance: 172 mly. Diameter: 75 000 ly.

NGC5735

H133[3]	14:42.4	+28 44	GLX	
SB(rs)bc	2.4′ × 1.9′	13.78B	13.4V	

The DSS image shows a face-on, four-armed, barred spiral galaxy. The

brighter core and bar are surrounded by tightly wound, knotty spiral arms forming a pseudo-ring before expanding to a more relaxed and broad spiral structure. Telescopically, the galaxy is moderately bright but fairly diffuse and is best seen at medium magnification as a round object with an even-surface-brightness envelope and somewhat hazy edges. Radial velocity: 3817 km/s. Distance: 171 mly. Diameter: 119 000 ly.

NGC5737

H896[3]	14:43.2	+18 53	GLX
SBb	1.2′ × 0.8′	14.63B	

This is a face-on, barred spiral galaxy with a brighter, elongated core and a narrow bar. A pseudo-ring surrounds the bar, made up of the tightly wound inner spiral arms which relax after about half a revolution into a broader, fainter outer spiral pattern. A condensation or possible companion galaxy is seen in silhouette against the spiral arm in the SSE. At high magnification the galaxy appears as an irregularly round, uncondensed patch of light, very slightly brighter to the middle with edges that are fairly hazy. Radial velocity: 9565 km/s. Distance: 427 mly. Diameter: 149 000 ly.

NGC5739

H171[1]	14:42.5	+41 50	GLX
SAB(r)0[+]:	2.3′ × 2.1′	13.21B	

Photographs show a slightly inclined, probable spiral galaxy, with a bright, elongated core and fainter outer envelope. Dust patches in the outer envelope N and E of the core hint at a possible spiral structure. An extremely faint outer structure is seen to the SW. Visually, the galaxy is quite bright and is located immediately W of a small asterism of five faint stars. The galaxy appears almost round; a small, round brighter core is embedded in a fainter secondary envelope. Radial velocity: 5485 km/s.

Distance: 245 mly. Diameter: 164 000 ly.

NGC5747

H48[3]	14:44.4	+12 09	GLX
Sb	1.0′ × 0.9′	14.43B	

This face-on galaxy, possibly barred, is part of a probably interacting system, with photographs showing a second galaxy in the outer halo immediately S of the core. The inner spiral pattern is bright but indefinite and only one arm is seen with certainty. The outer spiral pattern is faint and the beginnings of two 'antennae' are seen to the S. The companion is bright but its morphology is difficult to determine. Telescopically, the galaxy is quite faint and is seen as a hazy, very slightly oval patch of light oriented N–S. The envelope is fairly well defined, with even surface brightness and is only very slightly brighter to the middle. Radial velocity: 9103 km/s. Distance: 407 mly. Diameter: 118 000 ly.

NGC5751

H809[2]	14:43.9	+53 25	GLX
Scd:	1.5′ × 0.8′	13.88B	

Photographs show an inclined spiral galaxy with a slightly brighter core and possibly a short bar. Telescopically, the galaxy is moderately faint but is fairly well seen at high magnification. It is an oval patch of light, fairly even in brightness, moderately well defined and elongated NE–SW. Radial velocity: 3420 km/s. Distance: 153 mly. Diameter: 67 000 ly.

NGC5754

H687[3]	14:45.4	+38 47	GLX
SB(rs)b	2.3′ × 1.8′	13.93B	

This is a face-on barred spiral galaxy with two narrow spiral arms and a small companion galaxy (NGC5752) situated 1.0′ W. Visually, NGC5754 is a round, ill defined patch of light, broadly concentrated to the centre. High magnification brings out a faint

nucleus, though the outer envelope is rendered very tenuous. Herschel discovered this galaxy on 16 May 1787, describing it as: 'Considerably faint, pretty small. Another suspected 2′ north at 300×.' This suspected object is NGC5755, located 2.75′ NNE from NGC5754. Though uncatalogued, it is clear that Herschel has priority here in discovering this galaxy. A fourth galaxy, NGC5753, is 4.4′ N of NGC5754. Radial velocity: 4663 km/s. Distance: 208 mly. Diameter: 139 000 ly.

NGC5760

H885[3]	14:47.7	+18 31	GLX
Sa	1.5′ × 0.6′	14.10B	

This is a highly inclined spiral galaxy with an elongated core. The DSS photograph suggests a likely multi-arm spiral pattern. The fainter companion galaxy IC4507 is 2.75′ to the S. Telescopically this is a

moderately faint galaxy, but fairly well seen at high magnification as an oval, elongated patch of light, oriented E–W and broadly brighter to the middle, with a slightly brighter, stellar core. The edges are pretty well defined. Radial velocity: 5955 km/s. Distance: 266 mly. Diameter: 116 000 ly.

NGC5771

H129[3]	14:52.2	+29 51	GLX
E?	0.8′ × 0.7′	14.83B	

This elliptical galaxy with a faint outer envelope probably forms a physical pair with NGC5773, located 4.0′ to the ESE. Visually, both galaxies can be seen in a medium-magnification field and NGC5771 is very slightly the fainter. It appears round with an opaque, even-surface-brightness envelope. The minimum core-to-core separation between the two galaxies would be 361 000 ly at

NGC5754

the presumed distance. Radial velocity: 6948 km/s. Distance: 310 mly. Diameter: 72 000 ly.

NGC5773

H130[3]	14:52.5	+29 48	GLX
Sab	0.9′ × 0.9′	14.82B	

Photographs show a possible spiral galaxy with a dust lane bordering the core to the SE. Telescopically, this galaxy is very slightly brighter and larger than its neighbour NGC5771, appearing round, faint and opaque with a well defined disc. Radial velocity: 7053 km/s. Distance: 315 mly. Diameter: 82 000 ly.

NGC5784

H676[2]	14:54.2	+42 33	GLX
S0	1.9′ × 1.8′	13.70B	

This galaxy is classified as lenticular but may possibly be spiral as the DSS images show traces of dust lanes in the faint outer envelope. The core is very bright and is elongated E–W. Visually, the galaxy is round with a mottled envelope and poorly defined edges. It brightens suddenly to a bright core and the companion galaxy NGC5787 can be seen in the same medium-magnification field. A fainter, anonymous spiral galaxy is 4.4′ to the W and may be visible in large-aperture telescopes. Radial velocity: 5486 km/s. Distance: 245 mly. Diameter: 135 000 ly.

NGC5787

H677[2]	14:55.3	+42 30	GLX
S?	1.0′ × .0.8′	· 14.08B	

Photographs reveal a disc galaxy with an elongated core and a fainter, well defined envelope. There is possible spiral structure in the envelope but it is fairly smooth-textured and the classification may more likely be S0. In medium-aperture instruments only the bright core is visible, appearing almost round and well

defined. Radial velocity: 5601 km/s. Distance: 250 mly. Diameter: 73 000 ly.

NGC5789

H976[3]	14:56.6	+30 13	GLX
Sdm	1.1′ × 1.0′	14.07B	

Classified as a spiral, photographs show an irregular galaxy with chaotic structure and knotty condensations in a well defined envelope. The spiral structure, if it exists, is quite fragmentary in nature. Visually, the galaxy is very faint, diffuse and very slightly brighter to the middle with hazy, ill defined edges. It appears almost round and is best seen at medium magnification. Radial velocity: 1892 km/s. Distance: 84 mly. Diameter: 27 000 ly.

NGC5797

H678[3]	14:56.3	+49 42	GLX
S0/a	2.4′ × 1.8′	13.66B	

Located 4.1′ N of a magnitude 6 field star, this galaxy is the brightest of a

group of six galaxies visible on the DSS photograph. Despite its classification the galaxy shows some characteristics of a spiral: the bright inner disc is surrounded by faint, irregular spiral arms or perhaps a pseudo-ring, which brings the apparent size of the galaxy to about 2.4′ × 1.8′. NGC5794 is 5.3′ WNW, while NGC5804 is 7.1′ ESE and they would seem to form a physical system. The first edition of *Uranometria 2000.0* plots four galaxies in the field and misplots and misidentifies each of them except for NGC5794. Visually, the three brightest galaxies, NGC5794, NGC5797 and NGC5804, can be seen together in a medium-magnification field forming an ESE–WNW oriented line of galaxies, but the bright field star interferes somewhat. NGC5797 is bright and well condensed, small and almost round with well defined edges. Radial velocity: 4110 km/s. Distance: 183 mly. Diameter: 128 000 ly.

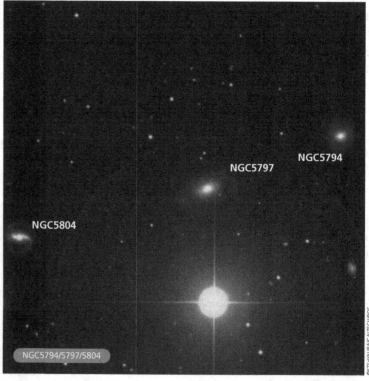

NGC5794

NGC5797

NGC5804

NGC5794/5797/5804

NGC5798

H131[3]	14:57.7	+29 58	GLX
Im:	1.4′ × 0.9′	13.39B	

Photographs show an irregular galaxy with knotty structure in its envelope and a brighter bar oriented NE–SW. Two faint field stars border the disc to the W. Telescopically, the galaxy is a moderately bright but 'blurry' object, more distinct at medium magnification. It is an oval patch of light, somewhat well defined at the edges and elongated NE–SW and is broadly brighter to the middle with a faint field star 0.9′ NE of the core. Radial velocity: 1875 km/s. Distance: 84 mly. Diameter: 34 000 ly.

NGC5804

H679[3]	14:55.7	+49 39	GLX	
SB(s)b	1.3′ × 1.2′	13.91B	13.3V	

This is an attractive, S-shaped barred spiral galaxy viewed face-on. Photographs show a bright, round core in a bright bar which curves into narrow, faint and smooth spiral arms. The bar is longer to the W than to the E and there is a bright condensation involved in the eastern part of the bar. Visually, the galaxy is bright, though only the core and part of the bar are visible; it appears slightly elongated E–W with a star-like core. The sixth magnitude field star 7.5′ to the WSW hinders visibility. Radial velocity: 4268 km/s. Distance: 191 mly. Diameter: 72 000 ly.

NGC5820

H756[2]	14:58.7	+53 53	GLX
S0	1.8′ × 1.4′	13.21B	

This lenticular galaxy is also known as Arp 136 due to the presence of a faint, though extensive plume of material emerging from the disc in the SE. There is a compact, high-surface-brightness galaxy in the field 1.6′ S; the plume extends roughly towards this galaxy. There are several faint galaxies in the field, either dwarfish companions or faint background galaxies. NGC5821 is

3.6′ ENE. In moderate-aperture telescopes the galaxy is a bright, E–W elongated oval, very opaque and very well defined with a bright, condensed core. The wide, bright pair of field stars 8.0′ ESE greatly hinders the view. Radial velocity: 3475 km/s. Distance: 155 mly. Diameter: 81 000 ly.

NGC5821

H811[3]	14:58.9	+53 55	GLX
S?	1.7′ × 0.9′	14.71B	

Photographs show an inclined spiral with a brighter core. Telescopically, the galaxy is quite faint and is seen as a roundish patch of light, broadly brighter to the middle with hazy edges. The galaxy is probably physically associated with NGC5820; the minimum core-to-core separation between the two galaxies at the presumed distance would be 165 000 ly. The bright pair of field stars 6.0′ to the ESE interferes with the view. Radial velocity: 3516 km/s. Distance: 157 mly. Diameter: 78 000 ly.

NGC5851

H886[3]	15:06.9	+12 52	GLX
S?	1.0′ × 0.2′	14.97B	

Situated 3.0′ SE of a bright field star, this spiral galaxy is part of a triple system which includes NGC5852 1.0′ to the SE and CGCG077-007 located 1.75′ to the WSW. The DSS photograph shows an almost edge-on spiral galaxy with a bright, round core and a disturbed spiral pattern; the arm to the NE narrows and curves towards the N. Visually, only the NGC galaxies are visible and they are quite faint; NGC5851 is smaller and fainter and seen as a very small, hazy patch of light with a bright core intermittently visible with averted vision. The minimum separation core-to-core between NGC5851 and NGC5852 would be 85 000 ly at the presumed distance, while that between NGC5851 and CGCG077-007 would be about 148 000 ly. Radial velocity: 6516 km/s. Distance: 291 mly. Diameter: 85 000 ly.

NGC5852

H887[3]	15:07.0	+12 51	GLX
S?	1.1′ × 0.6′	14.66B	

The morphology of this galaxy is a little difficult to determine: it may be a spiral with faint arms or perhaps a lenticular with a bright oval core and a faint outer disc. It forms a tight grouping with NGC5851 and CGCG077-007, with the minimum core-to-core separation between NGC5852 and CGCG077-007 about 212 000 ly. At high magnification NGC5852 is a very slightly oval glow with hazy edges, somewhat mottled and with a star-like core occasionally visible. Radial velocity: 6687 km/s. Distance: 299 mly. Diameter: 96 000 ly.

NGC5856

H71[4]	15:07.3	+18 27	SS
nonexistent			

Herschel's 21 May 1791 discovery descriptions reads: 'A star 7.6 magnitude enveloped in extensive milky nebulosity. Another star 7 magnitude is perfectly free from such appearance.' The star in question is unremarkable and this object can be safely classed nonexistent.

NGC5857

H751[2]	15:07.5	+19 36	GLX
SB(s)b	1.2′ × 0.6′	13.85B	

This highly inclined spiral galaxy forms a physical pair with NGC5859

NGC5857/5859

situated 1.8′ to the ESE. Photographs show the galaxy has a large, bright, elongated core and faint spiral arms. The two galaxies form a fairly bright pair in a high-magnification field. NGC5857 is smaller but features a bright core embedded in an opaque, well defined envelope, very slightly elongated NE–SE. Radial velocity: 4832 km/s. Distance: 216 mly. Diameter: 75 000 ly.

NGC5859

H752[2]	15:07.6	+19 35	GLX
SB(s)bc	2.7′ × 0.7′	13.25B	

The DSS photograph shows a highly inclined spiral galaxy, probably barred, with narrow spiral arms featuring knotty condensations. The galaxy has an almost identical radial velocity to its companion NGC5857, and faint plumes of material extend from both galaxies. The minimum core-to-core separation between the two galaxies would be 113 000 ly at the estimated distance. Telescopically, it is seen as an elongated patch of light, bright along its major axis and quite well defined at the edges, though there are hints of a fainter secondary envelope. Radial velocity: 4829 km/s. Distance: 216 mly. Diameter: 170 000 ly.

NGC5875

H755[2]	15:09.2	+52 32	GLX
SAb:	2.3′ × 1.1′	13.35B	

Photographs show an inclined spiral galaxy with a small, round core and two principal spiral arms. Telescopically, the galaxy is quite bright with four bright field stars to the SE. It is a well defined oval galaxy, oriented SE–NW, fairly large and even in surface brightness. Radial velocity: 3669 km/s. Distance: 164 mly. Diameter: 110 000 ly.

NGC5888

H659[3]	15:13.1	+41 16	GLX
SB(s)bc	1.4′ × 0.8′	14.18B	

This galaxy is the brightest of a triplet which includes NGC5886 4.5′ to the WSW and NGC5889 4.2′ to the NNE.

Photographs show an inclined barred spiral galaxy with a round core, a fainter bar, and two spiral arms attached. Telescopically, this galaxy is somewhat faint and is best seen at medium magnification as a roundish glow, very slightly brighter towards the middle with the edges fading gradually into the sky background. NGC5886 is also visible as a smaller but higher-surface-brightness object; it is a condensed, roundish glow that is fairly well defined at the edges. A bright field star is 3.5′ SSW of NGC5888. Radial velocity: 8853 km/s. Distance: 395 mly. Diameter: 161 000 ly.

NGC5893

H678[2]	15:13.6	+41 58	GLX
SB(r)b	1.8′ × 1.3′	14.15B	

This face-on, possibly barred, spiral galaxy has a round core and two main spiral arms that form a pseudo-ring before they branch out and broaden. Photographs show NGC5895, a faint edge-on companion 4.0′ to the NE, and NGC5896 immediately N of the edge-on galaxy, but both of these objects may be visible only in large-aperture telescopes. Visually, NGC5893 is a faint, moderately well defined object visible with direct vision. The envelope is smooth, round and does not brighten to the middle. Radial velocity: 5505 km/s. Distance: 246 mly. Diameter: 129 000 ly.

NGC5899

H650[2]	15:15.0	+42 03	GLX
SAB(rs)c	3.2′ × 1.2′	12.56B	

Images reveal a highly inclined, two-armed grand-design spiral galaxy with an elongated core and narrow, bright, knotty spiral arms expanding to a broad outer envelope. In a moderate-aperture instrument it is a bright, well defined object, much elongated NNE–SSW with a sharp stellar core. The envelope is smooth in texture and in a medium-magnification field the companion galaxy NGC5900 can be seen as well. Radial velocity: 2687 km/s. Distance: 120 mly. Diameter: 111 000 ly.

NGC5900

H660[3]	15:15.1	+42 13	GLX
Sb: sp	1.5′ × 0.5′	14.75B	

Images show an almost edge-on spiral galaxy, with a dust lane along the galaxy's NE flank, slightly tilted to the plane. The galaxy is elongated SE–NW. This object is quite faint in a moderate aperture, appearing as a poorly defined patch of light, a little brighter to the middle, small and round with no evidence of faint extensions. Radial velocity: 2637 km/s. Distance: 118 mly. Diameter: 51 000 ly.

NGC5902

H737[3]	15:14.4	+50 21	GLX
1.1′ × 1.0′	14.68B		

This galaxy is unclassified but may be a spiral. Photographs show faint arcs E and W of the core and two bright condensations are visible, one adjacent to the bright core to the NNW, the other in the outer envelope to the NE. Telescopically, the galaxy is small but quite distinct at high magnification; it is almost round with a star-like brighter core embedded in a bright, opaque and well defined envelope. A wide, equally bright pair of stars is 1.6′ to the E while a bright field star is 5.5′ to the SW. Radial velocity: 11 241 km/s. Distance: 502 mly. Diameter: 161 000 ly.

NGC5922

H661[3]	15:21.2	+41 40	DS

Observed only once, Herschel's 9 April 1787 discovery description for this object called it: 'Extremely faint, small' and NGC5922 is now associated with a very faint pair of stars located 3.4′ SSW of the face-on spiral galaxy NGC5923, which was discovered by John Herschel. It is difficult to imagine William Herschel detecting what is a very difficult pair of stars while overlooking the faint galaxy in the field to the N. Visually, NGC5923 is quite faint, a roundish glow, even in surface brightness and poorly defined at the edges. The pair of stars is very faint and difficult to resolve in a moderate-aperture telescope and does not appear nebulous. The present author has examined the field on a number of occasions and the galaxy is always seen while several attempts were necessary to correctly identify the faint pair of stars. It was well within William Herschel's capabilities as an observer to detect the galaxy and, guardedly, the present author suggests that NGC5923 = H661[3]. The galaxy is type SAB(s)bc, its dimensions are $1.8' \times 1.8'$ and it shines at blue magnitude 13.77. Radial velocity: 5696 km/s. Distance: 254 mly. Diameter: 133 000 ly.

NGC5930

H651[2]	15:26.1	+41 41	GLX
SAB(rs)b pec	$1.9' \times 0.8'$	13.23B	

Photographs show this galaxy forming an interacting pair with NGC5929 0.4′ SW and the pair is also catalogued as Arp 90. NGC5930 is an oblique barred spiral galaxy with a bright core and a bright, compressed S-shaped spiral pattern emerging from the bar. The spiral pattern quickly fades to a broader, smooth spiral pattern and the arm on the western flank appears in direct contact with NGC5929. A faint, disturbed, possible companion galaxy is 5.0′ NE. Visually, NGC5929 and NGC5930 form an interesting pair, and at medium magnification they definitely appear to be in contact. NGC5930 is the larger and brighter galaxy; it is a roundish, mottled patch of light with a much brighter core. Its extremities are moderately well defined. NGC5929 is located immediately SW. It is curious that Herschel did not note the galaxy as a separate object; however, in the 18 March 1787 entry for NGC5930, Herschel called it: 'Pretty bright, pretty large, irregularly extended. Easily resolvable.' This may indicate that he saw the two components but thought they were one object. Radial velocity: 2746 km/s. Distance: 123 mly. Diameter: 68 000 ly.

NGC5966

H634[3]	15:35.8	+39 47	GLX
E	$1.3' \times 0.7'$	13.14B	

Photographs show that this elliptical galaxy should probably be classified as

NGC5992/5993

E4. Larger telescopes may pick up IC4563 4.25′ to the NE and a fainter field galaxy is 2.7′ almost due N. Visually, this is a small though moderately bright galaxy, a well defined oval elongated E–W with even brightness across the envelope and a brighter star-like core. A faint field star is immediately E and the presence of two bright field stars to the NNE greatly hinders visibility. Radial velocity: 4605 km/s. Distance: 206 mly. Diameter: 78 000 ly.

NGC5992

H635[3]	15:44.4	+41 05	GLX
SBb	$0.9' \times 0.7'$	14.54B	

Photographs show a peculiar, face-on spiral galaxy with a very bright core and two spiral arms. A broad, bright southward curving spiral arm has two condensations. At the condensations, the spiral arm becomes extremely faint. Companion galaxy NGC5993 is 2.25′ NNE and they probably form a physical pair; the core-to-core

separation at the presumed distance would be 284 000 ly. In a moderate-aperture telescope both galaxies can be seen in the same high-magnification field. NGC5992 appears almost round and well defined but faint. Radial velocity: 9656 km/s. Distance: 431 mly. Diameter: 113 000 ly.

NGC5993

H636[3]	15:44.5	+41 07	GLX
SB(r)b:	$2.0' \times 1.0'$	14.18B	

Images show a face-on spiral galaxy with a somewhat peculiar morphology. The core is bright and the spiral structure is asymmetrical, being brighter to the S. The N curving spiral arm is 'sheared off', then resumes, more faintly, N of the galaxy. Visually, this object is a roundish, fairly well defined patch of light, slightly brighter to the middle, and is larger and brighter than its neighbour NGC5992. Radial velocity: 9703 km/s. Distance: 433 mly. Diameter: 252 000 ly.

Camelopardalis

Although this constellation only contains 12 Herschel objects (NGC2253 is very likely nonexistent), many of them have interesting structural features including the galaxies NGC1569, NGC1961, NGC2403 and NGC2655. The planetary nebula NGC1501 and the open cluster NGC1502 are also very interesting to observe. All the objects were recorded in the years between 1787 and 1802.

NGC1501

H53[4]	4:07.0	+60 55	PN
3	0.9′ × 0.9′	15.2B	11.4V

Herschel discovered this object on 3 November 1787 and correctly surmised that it was a planetary nebula. His description reads: 'A pretty bright planetary nebula, near 1′ diameter. Round, of uniform light and pretty well defined, with 360 magnified in proportion; but still the borders pretty abruptly defined, and a little elliptical.' Photographs show a very slightly oval planetary nebula, elongated E–W with a complex inner structure of bright and dark zones. The outer ring is a little brighter than the interior and particularly bright to the N and S. The central star is spectral type WC7 and magnitude 14.45. In a small telescope, the nebula is a somewhat uniform glow but irregularities are suspected in the envelope and the central star is not visible. It appears slightly brighter along the N and E rim. Expansion velocity: 18 km/s. Radial velocity in approach: 16 km/s. Distance: about 3900 ly.

NGC1502

H47[7]	4:07.7	+62 20	OC
I3m	20.0′ × 20.0′	7.5B	6.9V

This attractive cluster of stars presents a similar appearance visually and photographically. It is well isolated and well resolved and exhibits a wide range of brightness. Very pretty at low magnification, the brightest stars are for the most part arranged in wide pairs in a formation that is elongated E–W and to some extent resembles a tuning fork. High magnification picks out a number of threshold stars. Twenty-nine stars are resolved in all. The central bright pair of stars is HD25638/39 and the cluster is a possible member of the Camelopardalis OB1 association. More than 60 stars are associated with the cluster, the brightest being magnitude 6.93. The spectral type is B0, the age about 20 million years. The cluster shows a radial velocity of 16 km/s in approach. Distance: 2650 ly.

NGC1569

H768[2]	4:30.8	+64 51	GLX
IBm	3.6′ × 1.8′	11.79B	11.0V

This is an unusual dwarf galaxy of the Magellanic-type which may be a

NGC1501

neighbour of our own Local Group of galaxies; indeed, the heliocentric radial velocity is 104 km/s in approach. A starburst galaxy, it has also been catalogued as Arp 210. It is well seen on the DSS image as an inclined irregular galaxy with a bright, ragged central region and a chaotic, knotty outer envelope. A faint extension emerges from the galaxy and extends to the SW and a bright field star is 1.0' N of the galaxy. High-resolution images of the galaxy reveal resolved stars at blue magnitude 20 and at least two objects that are probably super star clusters of blue magnitude 15.7 and 16.5. Spectra of the galaxy reveal emission lines indicative of active star formation. Visually, the galaxy is small but very bright and is seen immediately S of a bright, unequal pair of stars. The high-surface-brightness disc is quite well defined and elongated ESE–WNW, with a mottled texture. The disc is brightest and widest to the WNW. Herschel

recorded the galaxy on 4 November 1788, calling it: 'Pretty bright, small, a little extended, a bright nucleus just south of a pretty bright star.' Radial velocity: 25 km/s. Distance: 5.22 mly. Diameter: 6200 ly.

NGC1961

H747[3]	5:42.1	+69 23	GLX
SAB(rs)c	4.6' × 3.0'	11.79B	11.0V

This inclined spiral galaxy is also catalogued as Arp 184 and displays somewhat peculiar structure in the outer spiral arms. The small, brighter core may be involved with a weak bar. Two moderately well defined spiral arms displaying large amounts of dust emerge from the central region and a faint and very long curving extension to the northern spiral arm emerges perpendicularly in the E. This extension is very dusty and displays several knotty HII regions as the arm sweeps S of the core. A large and disturbed dust

NGC1961

patch is located SW from the core. Telescopically, this galaxy is visible even at low magnification; it is a quite bright and large galaxy situated in a rich star field. Elongated E–W, the inner region is oval and is broadly brighter to the middle, while the outer envelope appears as a very faint, indefinite and ill defined haze surrounding the brighter core. A faint field star is visible 0.4' SSE of the core. Radial velocity: 4059 km/s. Distance: 181 mly. Diameter: 243 000 ly.

NGC2253

H54[7]	6:42.4	+66 20	nonexistent

NGC2347

H746[3]	7:16.1	+64 43	GLX
(R')SA(r)b:	2.1' × 1.7'	13.31B	

This is an inclined spiral galaxy with a large, bright, elongated core. Photographs show the outer envelope is faint and involves tightly wound spiral arms. Telescopically, the galaxy is located 4.2' S of a magnitude 7 field star. It is a moderately bright, roundish, well defined patch of light that is a little brighter to the middle; the faint outer envelope is not visible. A magnitude 10 field star precedes the galaxy to the W and the similarly bright galaxy IC2179, located 13.0' N is also visible. Radial velocity: 4516 km/s. Distance: 202 mly. Diameter: 123 000 ly.

NGC1569

NGC2366

H748[3]	7:28.9	+69 13	GLX
IB(s)m	8.1′ × 3.3′	11.57B	10.8V

This dwarfish, irregular galaxy is a part of the M81/NGC2403 group of galaxies. The DSS image shows a large, very faint, irregular galaxy, elongated with small, faint knots along its main envelope. Two bright HII regions in the SW are the brightest features of the galaxy. The one to the W is the larger and measures about 18″ in diameter. About 1.0′ to the W of this HII region is a dwarf irregular companion galaxy measuring little more than 2600 ly across. The brightest stars in the galaxy resolve at blue magnitude 19. Visually, the galaxy is very dim and difficult, located immediately S of a magnitude 12 triangle of field stars. The main body of the galaxy is an extremely hazy streak of light, poorly defined and oriented NNE–SSW. The two HII regions are brighter than the rest of the galaxy and look like a hazy double star at the SW edge of the galaxy; they are

separately designated NGC2363. Herschel came upon this object on 3 December 1788 and thought the HII regions were the principal object, describing the galaxy as: 'Very faint, very small, has a very faint branch north following.' Radial velocity: 190 km/s. Distance: 8.5 mly. Diameter: 21 000 ly.

NGC2403

H44[5]	7:36.9	+65 36	GLX
SAB(s)cd	21.9′ × 12.3′	8.82B	8.5V

This is one of the principal galaxies in the NGC2403/M81 group of galaxies, a scattered group which is one of the nearest to our own Local Group of galaxies. Morphologically, the galaxy shares many characteristics with M33, including a small core and diffuse, loosely wound spiral arms. In photographs it appears as a very large and bright, inclined, two-armed spiral galaxy with a slightly brighter central region. There is some dust in the spiral arms and very many condensations

which are star clusters and/or regions of nebulosity which appear as sizable patches of diffuse light. In deep photographs, blue supergiant stars begin to resolve at blue magnitude 18, while red supergiants begin to appear at visual magnitude 19.5. Telescopically, the galaxy is large but fairly faint, with several foreground stars involved. An E–W chain of three stars is prominent, the easternmost star is involved with the core which brightens immediately to the N; a fourth star farther to the E marks the eastern edge of the nebulosity. The galaxy is irregular with poorly defined edges and decidedly elongated WNW–ESE. Dust patches are suspected W of the core. On the best nights, something of its spiral structure may be detected in mid–large-sized instruments. Recorded on 1 November 1788, Herschel described the galaxy as: 'Considerably bright, round, very gradually brighter to the middle. Bright nucleus, 6′ or 7′ in diameter with a faint branch extending a great way to the north preceding side; not less than half a degree and to the north or north following the nebulosity diffused over a space, I am pretty sure, not less than a whole degree.' Radial velocity: 227 km/s. Distance: 10.14 mly. Diameter: 64 000 ly.

NGC2655

H288[1]	8:55.6	+78 13	GLX
SAB(s)0/a	6.0′ × 5.3′	10.98B	

The DSS image shows a galaxy with a large, bright and elongated central region surrounded by a very faint, smooth-textured outer spiral structure. High-resolution images show a fine pattern of dust bordering the bright central region. The galaxy is quite bright visually and is well seen at high magnification as a slightly oval, prominent disc with a grainy texture and hazy edges. The central region is bright and elongated E–W with a tiny, star-like core embedded in an extended brighter bar. This object was the last bright nebula observed by Herschel, recorded on 26 September 1802 and described as: 'Very bright, considerably large, a little extended, suddenly much

NGC2366

NGC2403

©STScI/AURA/CALTECH/ROE

brighter to the middle.' Radial velocity: 1537 km/s. Distance: 69 mly. Diameter: 120 000 ly.

NGC3901

H970[3]	11:43.0	+77 24	GLX
Scd:	1.8′ × 0.8′	14.55B	

Photographs show a slightly inclined spiral galaxy with a small, brighter core and patchy spiral structure. The galaxy is quite faint visually and seen as an oval, diffuse glow of even surface brightness, oriented N–S, with fairly well defined edges. Radial velocity: 1829 km/s. Distance: 81 mly. Diameter: 43 000 ly.

NGC4127

H279[1]	12:08.4	+76 48	GLX
SAc?	2.2′ × 1.0′	13.58B	

This is a slightly inclined spiral galaxy. Photographs show a brighter, oval core and a star-like nucleus in a grainy outer envelope. Only the brighter, central region of this galaxy

is visible telescopically; it appears as a small, oval patch of light, slightly brighter to the middle and fairly well defined at the edges. The surface brightness of this galaxy is fairly low and it is surprising that Herschel classified it as a bright nebula. Recorded 12 December 1797, he described it as: 'Considerably bright, considerably large, a little extended, brighter to the middle.' Radial velocity: 1961 km/s. Distance: 88 mly. Diameter: 56 000 ly.

NGC5295

H946[3]	13:38.6	+79 27	GLX
0.3′ × 0.3′	14.78B		

This faint galaxy presents a similar appearance visually and photographically and, though unclassified, would appear to be an elliptical or lenticular type. In a moderate-aperture telescope, this is a very faint galaxy that is a little easier to see at medium magnification. It is a round nebulous spot, a little brighter

to the middle with hazy, ill defined edges, not much brighter than the sky background. Radial velocity: 7095 km/s. Distance: 317 mly. Diameter: 28 000 ly.

NGC5640

H949[3]	14:20.8	+80 06	GLX
0.7′ × 0.5′	15.50B		

This galaxy is unclassified but the DSS image reveals an elegant, barred spiral galaxy viewed face-on with a bright, slightly elongated core, and a fainter, slightly curved bar which makes a smooth transition into two narrow, smooth-textured spiral arms. Visually, the galaxy is very faint, an ill defined patch of light requiring averted vision to detect. It is a roundish, nebulous patch that is a little brighter to the middle in a star-poor field. A similarly bright field galaxy is 7.0′ almost due W, but is not detected in a moderate-aperture telescope. Radial velocity: 14 371 km/s. Distance: 642 mly. Diameter: 131 000 ly.

Cancer

The objects recorded by Herschel in Cancer were observed between the years 1784 and 1789. A total of 34 Herschel objects are found in this constellation, all of them galaxies except for one asterism (NGC2678). The present author is suspicious of the likelihood that the extremely faint and remote galaxy NGC2843 was observed by Herschel and is an actual Herschel object; however, it is included here and offered as a challenge to visual observers with moderate–large-aperture telescopes. Except for the bright galaxy NGC2775, all of the galaxies here were classed by Herschel as faint or very faint.

NGC2507

H554[2]	8:01.6	+15 43	GLX
S0/a pec	2.5′ × 1.8′	13.84B	

Images show a distorted spiral galaxy, almost face-on with a bright core and a slightly fainter inner envelope. A thin dust lane borders the core to the NNE and extends to a large dust patch S of the core. The faint, smooth spiral arms are irregular with a nebulous condensation in the arm 0.5′ ENE of the core and a faint outer arm emerging from the S and curving E–N. Telescopically, the galaxy is moderately large and fairly diffuse; it is well seen at high magnification as a roundish patch of light, broadly brighter to the middle with ill defined edges. Radial velocity: 4463 km/s. Distance: 200 mly. Diameter: 145 000 ly.

NGC2512

H605[3]	8:03.1	+23 24	GLX
SBb	1.4′ × 0.9′	13.90B	

Photographs show a two-armed barred spiral galaxy, viewed almost face-on. The bright, elongated core extends into a bright bar. The beginnings of the spiral extensions off the bar are bright but fade quickly into faint, smooth and open spiral arms. Telescopically, the galaxy is moderately faint and somewhat brighter to the middle. It is small and slightly oval in shape, oriented ENE–WSW with the edges fading gradually into the sky background. Radial velocity: 4631 km/s. Distance: 207 mly. Diameter: 84 000 ly.

NGC2513

H512[3]	8:02.5	+9 25	GLX
E	2.1′ × 1.7′	12.62B	

Photographs show this elliptical galaxy is the brightest of a small group of galaxies which include NGC2510 situated 5.6′ to the NW and NGC2511 located 2.6′ to the WSW. Telescopically, NGC2513 is a moderately bright object, steadily brighter to the middle and fairly well defined with a somewhat grainy texture. The galaxy has a star-like, brighter core and the disc is elongated slightly N–S. Radial velocity: 4543 km/s. Distance: 203 mly. Diameter: 124 000 ly.

NGC2530

H752[3]	8:07.9	+17 49	GLX
SB(s)d	1.4′ × 1.0′	14.7B	

This galaxy is erroneously plotted as NGC2529 in the first edition of *Uranometria 2000.0*. Photographs show a face-on, probably barred, spiral galaxy with ill defined arms in a faint disc. Telescopically, the galaxy is quite faint and diffuse, situated 0.75′ S of a faint field star. The disc is mottled and irregularly round, there is no brightening to the middle and the edges are poorly defined. Radial velocity: 4936 km/s. Distance: 220 mly. Diameter: 90 000 ly.

NGC2545

H627[2]	8:14.2	+21 21	GLX
(R)SB(r)ab	2.6′ × 1.3′	13.21B	

The DSS image shows an almost face-on spiral galaxy, with a small, brighter, round core. Deeper photographs reveal that the bright inner envelope is a high-surface-brightness ring attached to a very faint bar, with two spiral arms emerging from the ring. Visually, this is a fairly faint galaxy, small, though much elongated N–S. The envelope is fairly diffuse and fades gradually into the sky background. The core area is fairly bright and features a star-like nucleus. A faint field star is visible to the NNW and is comparable in brightness to the nucleus. A brightening near the core may be a brighter portion of the ring. Interestingly, during the sweep in which this galaxy was discovered, on 11 January 1787, Herschel observed the planet Uranus and discovered the satellites Titania and Oberon. Radial velocity: 3305 km/s. Distance: 148 mly. Diameter: 111 000 ly.

NGC2554

H303[2]	8:17.9	+23 28	GLX
SAB(rs)ab	3.2′ × 2.3′	12.94B	

Photographs reveal a barred spiral galaxy with a bright core and a fainter bar attached to a very faint and relaxed two-armed spiral structure. Visually, only the core and a portion of the bar are visible; the galaxy is quite bright and small but with a well defined, grainy envelope that brightens to a small core. It is very slightly extended SSE–NNW with one star 1.25′ SSE and another 1.3′ NE. Radial velocity: 4914 km/s. Distance: 219 mly. Diameter: 170 000 ly.

NGC2558

H606[3]	8:19.4	+20 30	GLX
SAB(rs)ab	1.6′ × 1.1′	13.87B	

Photographs show a slightly inclined spiral galaxy, possibly barred, with a brighter, elongated core and two smooth-textured spiral arms. In a medium-magnification eyepiece this galaxy is moderately faint and small, though readily seen. The envelope appears smooth and quite round and fades gradually into the sky background and a bright, star-like nucleus is embedded in it. A very faint field star is WSW from the galaxy. Radial velocity: 4914 km/s. Distance: 219 mly. Diameter: 102 000 ly.

NGC2562

H607[3]	8:20.4	+21 08	GLX
S0/a:	0.9′ × 0.5′	14.00B	

Photographs reveal that this lenticular galaxy with a faint outer envelope is involved in a rich field of galaxies, including several anonymous ones as well as others in the *New General Catalogue*. With a large-aperture telescope, some of these galaxies are revealed. The Herschel object NGC2563 is 4.75′ SE and is visible in the same high-magnification field in moderate-aperture telescopes. The surface brightness of NGC2562 is fairly high and it displays a well-developed core in a round envelope. The faint extensions, oriented N–S in photographs, are not seen visually. Radial velocity: 4917 km/s. Distance: 220 mly. Diameter: 58 000 ly.

NGC2563

H634[2]	8:20.6	+21 04	GLX	
S0°:	1.5′ × 1.2′	13.34B	12.4V	

Photographs reveal this lenticular galaxy has an extensive faint outer envelope and is oriented ENE–WSW. An anonymous galaxy is 5.75′ W and may be visible in large amateur telescopes. In moderate apertures, NGC2563 is moderately bright with a well-developed core and a faint, diffuse outer envelope.

Radial velocity: 4398 km/s. Distance: 196 mly. Diameter: 86 000 ly.

NGC2577

H259[2]	8:22.7	+22 33	GLX
S0⁻:	1.4′ × 0.7′	13.43B	

The visual and photographic appearances of this lenticular galaxy are very similar. Visually, it is small but fairly bright, elongated ESE–WNW with a bright, condensed core. The outer envelope is bright and fairly well defined and the envelope has a grainy texture. The spiral galaxy UGC4375 is 9.2′ to the NE and may be visible in large amateur instruments. Radial velocity: 1980 km/s. Distance: 88 mly. Diameter: 36 000 ly.

NGC2582

H753[3]	8:25.3	+20 20	GLX
(R′)SAB(s)ab	1.2′ × 1.0′	13.97B	

The DSS image shows a highly symmetrical, two-armed, barred spiral galaxy, viewed face-on. The round, bright core is embedded in a fainter, but well defined bar. Two strong, smooth spiral arms are involved in a fainter envelope and a very faint field star is located E of the core at the tip of the northern spiral arm. Visually, the galaxy is moderately bright, visible with direct vision at medium magnification. An irregularly round patch of light, it is just a little brighter to the middle and moderately well defined. Radial velocity: 4354 km/s. Distance: 194 mly. Diameter: 68 000 ly.

NGC2592

H315[2]	8:27.1	+25 58	GLX
E	1.7′ × 1.4′	13.61B	

Fairly similar visually and photographically, this elliptical galaxy is small, bright and round, well condensed with a sharply defined, opaque envelope, brightening to a small, bright core. Photographs show it extended NE–SW with a small, bright, compact neighbour 5.75′ SSE. Radial velocity: 1982 km/s. Distance: 89 mly. Diameter: 44 000 ly.

NGC2595

H599[3]	8:27.7	+21 29	GLX
SAB(rs)c	1.8′ × 1.4′	13.16B	

Photographs show a face-on barred spiral galaxy with a small, round, bright core. The weak bar is connected to a spiral arm or faint ring, broken to the W. A small detached portion is SE and a large, faint, curving detached arm is in the W. One spiral arm emerges from the N, curving E with a faint field star immediately NE. On the DSS image, the galaxy appears about 2.5′ × 2.5′. A bright field star is 2.25′ WSW. Visually, the galaxy appears moderately bright at medium magnification. It is faint and fairly large, with even surface brightness except for a slightly brighter, star-like core that is visible intermittently. The field star to the NE is visible. The outer envelope is not well defined. Radial velocity: 4249 km/s. Distance: 190 mly. Diameter: 100 000 ly.

NGC2599

H234[3]	8:32.2	+22 34	GLX
SAa	1.9′ × 1.7′	13.19B	

Photographs show this is a face-on spiral galaxy with a bright core embedded in a fainter inner envelope, surrounded by very faint and thin spiral arms which are difficult to trace. Only the core is seen visually; it is fairly bright but small with a condensed, round nucleus, slightly brighter to the middle with fairly well defined edges. Radial velocity: 4664 km/s. Distance: 208 mly. Diameter: 115 000 ly.

NGC2604

H292[3]	8:33.3	+29 33	GLX	14.9V
SB(rs)cd	2.1′ × 2.1′	13.82B		

This is a faint barred spiral galaxy viewed face-on. Photographs show a small, round core in a brighter bar. Two spiral arms emerge from the bar; they are faint and broaden into a diffuse envelope. The very dim companion galaxy NGC2604B is 3.5′ SE. Telescopically, NGC2604 is visible as a moderately large, but quite diffuse glow that is broadly brighter to the middle.

NGC2595

©STScI/AURA/CALTECH/ROE

It is almost round and the edges are hazy and fade gradually into the sky background. The galaxy is more distinct at medium magnification. Radial velocity: 2027 km/s. Distance: 91 mly. Diameter: 55 000 ly.

NGC2608

H318²	8:35.3	+28 28	GLX
SB(s)b:	2.0′ × 1.3′	12.96B	

In photographs, this is an inclined spiral galaxy, probably barred, with a smaller, brighter, round core; two grainy spiral arms emerge, the one on the W side branches further into two fainter subarms. A very faint field star is immediately NNW of the core.
At high magnification the galaxy is moderately bright with a brighter core and a mottled disc which appears elongated ENE–WSW and moderately well defined. Radial velocity: 2081 km/s. Distance: 93 mly. Diameter: 54 000 ly.

NGC2619

H319²	8:37.6	+28 43	GLX
Sbc	2.3′ × 1.4′	13.40B	

Images show an inclined, multi-arm spiral galaxy, possibly barred, with a bright oval core. A faint bar is attached to an inner, partial ring with spiral structure emerging from the ring. Visually, the galaxy is moderately bright, oval in form and gradually brighter to the middle. It is slightly elongated NE–SW with diffuse, ill defined edges and is a little more distinct at medium magnification. Radial velocity: 3420 km/s. Distance: 153 mly. Diameter: 102 000 ly.

NGC2628

H235³	8:40.5	+23 33	GLX
SAB(r)c?	1.2′ × 1.0′	14.10B	

This is an almost face-on spiral galaxy with a possible bar and a round, bright core. The DSS photograph shows one very long, knotty spiral arm emerging

from the core, forming a pseudo-ring and wrapping 1.5 times around the core. A second spiral arm emerges from the first in the NNE. This arm also appears attached to the main spiral arm in the SE. Telescopically, the galaxy is fairly faint and is best seen at medium magnification. A hazy, oblong patch of light, it is broadly brighter to the middle with a faint field star beyond the envelope 1.0′ to the NNE. Radial velocity: 3548 km/s. Distance: 158 mly. Diameter: 55 000 ly.

NGC2648

H49³	8:42.7	+14 17	GLX
S(s)b	3.2′ × 1.1′	12.77B	11.8V

This galaxy forms an interacting pair with the much fainter edge-on galaxy CGCG060-036, which is situated 2.5′ to the ESE. Together the pair has also been catalogued as Arp 83. Photographs show a highly inclined spiral galaxy, with a very bright, large and elongated core and two faint, smooth-textured spiral arms. A plume of material extends from the SE edge of the galaxy towards the smaller companion, which itself has a disturbed outer arm to the E. At the presumed distance, the minimum core-to-core separation would be about 64 000 ly. Visually, only the brighter core of NGC2648 is visible, but it holds high magnification well and is seen as a bright and well defined oval, oriented SSE–NNW with a tiny, star-like core. The companion may be visible in large-aperture telescopes. Radial velocity: 1951 km/s. Distance: 87 mly. Diameter: 81 000 ly.

NGC2661

H50³	8:46.0	+12 38	GLX
Scd:	1.7′ × 1.6′	13.84B	

Photographs show a face-on spiral galaxy with a very small core in a fragmentary spiral structure. There is a brighter, 'dog-leg' spiral arm in the S. This galaxy is very dim visually and is seen at medium magnification only as a diffuse, roundish patch of light 1.0′ to the ESE of a magnitude 10 field star which hinders observation. The surface

NGC2648

©STScI/AURA/CALTECH/ROE

photographs show a possible barred spiral galaxy. The bright core has faint extensions running N–S in a very faint outer envelope. Two small, bright stars or compact companion galaxies are in the envelope immediately E of the core. Visually, this is a small but fairly bright galaxy with what appear to be twin cores clearly visible at high magnification. They are embedded in a common, hazy and circular envelope. The edges of the envelope are moderately well defined. Recorded on 13 March 1785, Herschel called the object: 'Very faint, very small, round, brighter to the middle, large stellar.' Radial velocity: 2004 km/s. Distance: 90 mly. Diameter: 42 000 ly.

NGC2743

H608[3]	9:05.0	+25 00	GLX
Sdm:	1.5′ × 1.0′	14.17B	

Photographs show a slightly inclined spiral or irregular galaxy with a small, brighter core. There is knotty, chaotic structure in the outer envelope which appears made up of spiral arm fragments. Visually, the galaxy is quite faint, difficult at medium magnification, but is visible as a small, oblong patch of light, broadly brighter to the middle and extended E–W with hazy edges. Radial velocity: 2925 km/s. Distance: 131 mly. Diameter: 57 000 ly.

NGC2744

H60[3]	9:04.7	+18 28	GLX
SBab	1.6′ × 1.0′	13.85B	

This galaxy is the brighter of a physical pair, appearing on photographs as a

brightness is even across the disc and the edges are somewhat ill defined. Radial velocity: 3994 km/s. Distance: 178 mly. Diameter: 88 000 ly.

NGC2672

H48[2]/H80[2]	8:49.3	+19 04	GLX
E1-2	2.5′ × 2.5′	12.64B	

This galaxy was recorded twice by Herschel and forms an interacting pair with NGC2673 located immediately E. Their centres are separated by 33″ and the pair is also catalogued as Arp 167. NGC2672 is a moderately bright galaxy; it is fairly condensed and brighter to a nonstellar core, very slightly extended SE–NW. The projected separation of NGC2672 and NGC2673 is a little over 30 000 ly. J. L. E. Dreyer erroneously considered H48[2] a separate object, identified as NGC2677, but Herschel's description of the object, 'pretty bright, pretty large, a little brighter to the middle, contains one star' does not match the characteristics of NGC2677,

which is small and faint. Radial velocity: 4252 km/s. Distance: 190 mly. Diameter: 138 000 ly.

NGC2678

H10[8]	8:50.2	+11 20	AST

This is a fairly bright and loose asterism, plotted on *Uranometria 2000.0* as three magnitude 8 stars situated SW of M67. Eight stars are readily visible, having a moderate brightness range and appearing well isolated. There is also an elongated diamond-shaped asterism to the SSW made up of five stars that may be considered part of Herschel's group. His 15 March 1784 entry for this object reads: 'A cluster of very coarse scattered stars not rich.'

NGC2679

H294[3]	8:51.5	+30 52	GLX
SB0:	1.6′ × 1.3′	14.03B	

Plotted as NGC2679/80 in the first edition of *Uranometria 2000.0*,

NGC2679

©STScI/AURA/CALTECH/ROE

peculiar, three-branch spiral with the S branch attached to a larger condensation, evidently the fainter companion. Visually, the galaxy is quite faint and can be seen in the same medium power field as NGC2749, a brighter galaxy missed by Herschel, which is located to the ESE. NGC2744 is a diffuse patch, slightly extended E–W with hazy edges. The surface brightness appears even across the disc and a very faint field star is visible 0.8′ E. Radial velocity: 3331 km/s. Distance: 149 mly. Diameter: 69 000 ly.

NGC2750

H291[3]	9:05.7	+25 26	GLX
Sc	1.9′ × 1.5′	12.77B	

The DSS image shows an almost face-on spiral galaxy with a small, elongated, brighter core in a fainter, inner disc. Four knotty spiral arm fragments emerge from the disc, two from the ENE, two from the WSW, unwinding for a half revolution, though the largest extends well past three-quarters of a revolution. Visually, the galaxy is a moderately faint, roundish, ill defined patch of light with a small, bright, condensed core embedded in a mottled envelope. Radial velocity: 2607 km/s. Distance: 116 mly. Diameter: 64 000 ly.

NGC2764

H236[3]	9:08.3	+21 27	GLX
S0:	1.5′ × 0.5′	13.65B	

This galaxy is highly inclined, possibly spiral, with a bright, elongated core. Deep photographs show heavy dust lanes along the eastern flank and clumpy spiral structure, indicating that the type is more probably Sb (pec). Visually, it is moderately faint and bracketed by two magnitude 10 field stars. It is elongated NNE–SSW and the outer regions are moderately well defined; averted vision helps in viewing the core, which is very gradually brighter to the middle. Radial velocity: 2638 km/s. Distance: 116 mly. Diameter: 51 000 ly.

NGC2774

H61[3]	9:10.6	+18 42	GLX
0.8′ × 0.8′	14.73B		

This galaxy is unclassified, but images indicate that it is probably elliptical or lenticular, featuring a brighter core in an extensive, fainter envelope. Visually, the galaxy is a very faint and small, hazy patch, almost round with a slightly brighter core. Radial velocity: 8673 km/s. Distance: 387 mly. Diameter: 90 000 ly.

NGC2775

H2[1]	9:10.3	+7 02	GLX
SA(r)ab	4.5′ × 3.5′	11.14B	10.5V

The DSS image shows a bright, slightly inclined spiral galaxy with a large, brighter core. The spiral pattern is tightly wound with a multi-arm flocculent texture that is very symmetrical. Dust patches are seen throughout the inner spiral pattern and a prominent, narrow dust lane extends along the western flank separating the

brighter inner spiral pattern from the much fainter outer spiral structure. The galaxy is fairly bright visually; it is a round concentrated disc that is much brighter to the middle with a bright nucleus. The outer disc is hazy and ill defined and fades gradually into the sky background. This galaxy was Herschel's second Class I discovery, recorded on 19 December 1783. Radial velocity: 1220 km/s. Distance: 54 mly. Diameter: 71 000 ly.

NGC2783

H295[3]	9:13.6	+29 59	GLX
S0+:	2.1′ × 1.0′	13.55B	

This is the brightest member of the Hickson 37 compact galaxy group. Photographs show an elongated, lenticular galaxy with a fainter outer envelope and four very close companions, the brightest of which is the edge-on galaxy NGC2783B. The companions are all to the NW, with the nearest being a small, condensed

NGC2775

NGC2783

DSS image suggests that it may be a two-armed spiral, though this feature is overwhelmed by the bright core. At the eyepiece the galaxy is extremely faint, a bright field star to the SSW hindering the view. It is a very hazy smudge, elongated SE–NW and not resolved as two separate galaxies; it appears very tenuous with hazy, ill defined edges. The minimum separation between the two galaxies at the assumed distance is about 85 000 ly. Radial velocity: 8645 km/s. Distance: 386 mly. Diameter: 79 000 ly.

NGC2803

H63[3]	9:16.7	+18 58	GLX
S?	1.2′ × 1.2′	14.3B	

This physical companion of NGC2802 is presumed to be a spiral but photographically it appears elliptical with a faint outer envelope. Visually, the two galaxies may be difficult to resolve at medium magnification, though Herschel had success using high magnification: his 21 March 1784 discovery description states: 'Two, nearly in the meridian. Both very faint, pretty small, round, a little brighter to the middle, resolvable with 240 (magnification), considerably large.' Radial velocity: 8814 km/s. Distance: 393 mly. Diameter: 137 000 ly.

NGC2843

H64[3]	9:20.5	+18 56	GLX
0.5′ × 0.3′	16.50B		

Herschel evidently saw this object only once and recorded it on 21 March 1784,

satellite in the outer halo. The most distant member of the quintet is only 3.0′ to the NNW. NGC2783 is fairly bright visually and is seen as an oval, elongated glow, oriented N–S and broadly brighter to the middle. The texture of the envelope is a little grainy and the edges are moderately well defined. Two bright field stars are seen to the SW. With care, the edge-on galaxy may be visible in moderate-aperture telescopes. All five galaxies may be seen in large-aperture telescopes. Radial velocity: 6696 km/s. Distance: 298 mly. Diameter: 183 000 ly.

NGC2796

H296[3]	9:16.7	+30 55	GLX
SA?	1.6′ × 1.0′	14.62B	

This is a highly tilted spiral galaxy, possibly a 'Sombrero' type. The DSS image shows an obliquely tilted galaxy with a prominent, dark dust lane adjacent to the core and a dust lane in the outer

disc. A peculiar jet of matter or perhaps an edge-on galaxy is visible perpendicular to the plane of the galaxy NW of the core. The galaxy appears to be at least a triple system with possible companions 1.0′ to the ESE and 1.1′ to the WSW. At the presumed distance, the minimum separation of both these objects from the principal galaxy would be about 90 000 ly. NGC2796 is a moderately faint object visually, though it is well seen as a round patch of light, very slightly brighter to the middle and well defined at the edges. Neither satellite galaxy is visible in a moderate aperture, though they should be visible in large-aperture telescopes. Radial velocity: 6931 km/s. Distance: 309 mly. Diameter: 144 000 ly.

NGC2802

H62[3]	9:16.7	+18 58	GLX
S?	0.7′ × 0.5′	15.20B	

This galaxy forms a physical pair with NGC2803 located 0.75′ to the SE and the

NGC2796

probably after recording NGC2802 and NGC2803 above. The position was taken from Delta Cancri and described as 36 minutes of time following and 52.0′ N of the star. His description reads: 'Extremely faint. 240 (magnification) shewed some small star(s) with suspected nebulosity.' The notes for sweep 181 state: 'A suspected nebula, but 240 (magnification) shewed some small star(s) with suspected nebulosity, probably a deception from want of light and power.' Herschel used the abbreviation 'st.' both times in the descriptions, which could mean either a single star or more than one star. Presumably, he would have used the word 'a' instead of 'some' if a single star was indicated. The object was searched

for but not found by the French observer Guillaume Bigourdan (using the 12-inch refractor of the Paris Observatory). And John Herschel did not observe it either, though it was included in his *General Catalogue* (1864) and described as: 'Small star and nebulae.' Dreyer has identified the object as NGC2843 and on the DSS image this galaxy is a very small and faint object, possibly a barred spiral galaxy with a small, brighter core, immediately S of a faint field star. As Herschel was using his reflector in its Newtonian configuration, an object as faint as NGC2843 would have been extremely difficult to detect, especially when so close to a field star. The present author believes that it is doubtful that this is the object that Herschel allegedly

observed, particularly as his description mentions '*some* small star(s) with suspected nebulosity'. There is nothing of that description in the vicinity of NGC2843 and Herschel's alleged object was recorded as 2.0′ N of the position of NGC2802/NGC2803 and not 1.0′ S, which is the location of NGC2843. Visually, several unsuccessful attempts were made to observe NGC2843 with a 15-inch reflector under good to very good conditions. The galaxy was never convincingly detected. Guardedly, we would suggest that although NGC2843 certainly exists, it is not likely to be H64[3] which should probably be considered nonexistent. Radial velocity: 16 931 km/s. Distance: 756 mly. Diameter: 110 000 ly.

Canes Venatici

Although it is not a particularly large constellation, Canes Venatici is a rich repository of Herschel objects. We list 144 Herschel objects here, all galaxies, except for NGC4401, an HII region in the large spiral galaxy NGC4395. Herschel began exploring this area of the sky in the spring of 1785 and returned to it often in the years that followed. The constellation is notable for the large number of relatively nearby galaxies within its borders, many of which show

rich detail in medium- and large-aperture telescopes. There are a good number of remote systems as well, small and faint, which will present a challenge to observers; among these we may number NGC5003, NGC5009 and NGC5096. Exploring the galaxies in this corner of the sky is much like taking a core sample into the universe, from nearby objects only a few million light years distant to star systems greater than 500 million light years away.

NGC4111

H195[1]	12:07.1	+43 04	GLX
SA(r)0⁺: sp	4.6′ × 1.0′	11.67B	10.7V

The DSS reveals a very bright lenticular galaxy with a slightly bulging, elongated core and bright extensions. A small spiral galaxy is 4.7′ SW and a bright, unequal pair of field stars is 3.6′ ENE. Visually, the galaxy is well seen in small-aperture telescopes and is a striking object in medium apertures. The bright core bulges slightly with a

sharp stellar nucleus embedded and the high-surface-brightness extensions, oriented SSE–NNW, are well defined and taper to points. Radial velocity: 848 km/s. Distance: 38 mly. Diameter: 50 000 ly.

NGC4117

H708[3]	12:07.8	+43 08	GLX
S0°:	1.7′ × 0.6′	14.10B	

Photographs show a highly inclined galaxy, probably a lenticular, with a

large oval core and faint extensions. A magnitude 8 field star is 5.0′ WSW. The lenticular galaxy NGC4118 is 1.4′ ESE, while NGC4111 is 8.25′ WSW. Visually, NGC4117 is a small and faint object, though it is readily visible as a slightly elongated patch of light, brighter to the middle with fairly well defined edges. The galaxy is oriented NNE–SSW. Radial velocity: 975 km/s. Distance: 44 mly. Diameter: 22 000 ly.

NGC4138

H196[1]	12:09.5	+43 41	GLX
SA(r)0+	2.6′ × 1.7′	12.29B	11.3V

This galaxy appears as a lenticular galaxy on the DSS image, but higher-resolution photographs suggest a possible spiral galaxy with a thin dust lane along the ESE flank. A faint star is superimposed on the core. Well seen at high magnification, the galaxy is a very well defined, bright oval with mottled texture and a brighter core embedded, which may be the foreground star. The major axis is aligned with a bright field star 2.0′ to the NNW. Radial velocity: 932 km/s. Distance: 42 mly. Diameter: 32 000 ly.

NGC4143

H54[4]	12:09.6	+42 32	GLX
SAB(s)0°	2.3′ × 1.4′	11.82B	

This bright lenticular galaxy presents a similar appearance visually and photographically. Telescopically, the galaxy is a well defined, moderately condensed object with a slightly brighter core. High magnification reveals an oval disc, oriented SE–NW, with a grainy texture. A bright field star is 4.8′ SW. Radial velocity: 998 km/s. Distance: 45 mly. Diameter: 30 000 ly.

NGC4145

H169[1]	12:10.0	+39 53	GLX
SAB(rs)d	6.6′ × 4.3′	11.78B	11.3V

This is a probable member of the Ursa Major Cluster. Photographs shows a bright, inclined spiral galaxy, with a brighter, squarish core involving a faint bar. Two spiral arms emerge from the core, branching into two additional arms after about half a revolution. There are many knots and evidence of dust in the arms and stars begin to resolve at about blue magnitude 22. Visually, the galaxy is best seen at medium magnification as a large but faint and very diffuse oval patch of light elongated E–W with a broadly brighter core. The edges are very poorly defined and a very faint field star borders the galaxy to the SSW. Radial velocity: 1047 km/s. Distance: 47 mly. Diameter: 90 000 ly.

NGC4151

H165[1]	12:10.5	+39 24	GLX
(R′)SAB(rs)ab:	9.3′ × 8.6′	11.36B	11.9V

This bright spiral galaxy was one of the first Seyfert galaxies to be recognized; the class features very bright cores which exhibit great activity, sometimes including variability in light output. Photographs show the galaxy has a very bright, round core. The inner spiral arms are dusty and knotty and two extremely faint, thin outer spiral arms emerge well away from the main body of the galaxy, each extending for a little more than a half revolution around the core. The outer arms begin in the NW and SE regions of the inner spiral arms, at the point where suspected HII regions are at their brightest. A small bright companion to the NW, NGC4156, is probably not physically related as its radial velocity is far higher. Visually, NGC4151 has a very bright, star-like core embedded in a slightly fainter central region. The envelope surrounding the core is extremely faint, elongated SE–NW and is best seen at high magnification. The brightness of the core interferes with the envelope's visibility. Radial velocity: 1025 km/s. Distance: 46 mly. Diameter: 124 000 ly.

NGC4156

H642[2]	12:10.8	+39 28	GLX
SB(rs)b	1.3′ × 1.2′	13.98B	

Photographs show a bright face-on barred spiral galaxy with a bright, elongated core, a short bar and two principal spiral arms attached. A third subarm emerges from the N. Despite its evident great distance, the galaxy is bright and well resolved. Visually, the galaxy is best seen at medium

NGC4145

©STScI/AURA/CALTECH/ROE

NGC4151

magnification as a faint, round patch of light that is fairly well defined. Radial velocity: 6785 km/s. Distance: 303 mly. Diameter: 115 000 ly.

NGC4163

H399[3]	12:12.2	+36 10	GLX
IAm	2.4′ × 1.3′	14.46B	

Photographs reveal an irregular galaxy with a round central region in a grainy, N–S elongated envelope, gradually brighter to the middle with a few brighter knots N of centre. The radial velocity is very low and it is probably a dwarf galaxy which may be physically related to NGC4212, situated 43.2′ to the ENE. Visually, the galaxy is quite faint, and is seen as a diffuse, moderately large, roundish patch of light with very poorly defined edges. It is situated mid-way between a bright pair of stars to the SW and a single bright field star to the NE. Radial velocity:

184 km/s. Distance: 8.2 mly. Diameter: 5600 ly.

NGC4183

H697[3]	12:13.3	+43 42	GLX
SA(s)cd? sp	7.3′ × 0.7′	12.83B	

Photographs show an edge-on spiral galaxy with a slightly brighter, elongated central region and grainy extensions. A faint field star is in the envelope, SSE from the core. The galaxy is a member of the Ursa Major Cluster with its brightest stars resolving at about blue magnitude 20. Best seen at medium magnification, the galaxy is faint but readily seen as an elongated, flat streak of light oriented NNW–SSE. It is well defined with an even surface brightness along its major axis, though very slightly brighter at the core. A faint field star is beyond the tip immediately S. Radial velocity: 976 km/s. Distance: 44 mly. Diameter: 93 000 ly.

NGC4187

H813[2]	12:13.4	+50 44	GLX
E	1.3′ × 1.0′	14.21B	

Photographs show this elliptical galaxy has bright spurs to the SE and NW in an extensive, fainter outer envelope. The classification may more correctly be S0. Many smaller, fainter galaxies are in the immediate vicinity, principally to the S, W and N. In a moderate-aperture telescope this is a very faint galaxy, small, almost round with poorly defined edges and a fairly prominent core. Radial velocity: 9207 km/s. Distance: 411 mly. Diameter: 155 000 ly.

NGC4190

H409[2]	12:13.7	+36 38	GLX
Im pec	1.6′ × 1.5′	13.86B	13.2V

Like NGC4163, this dwarfish irregular galaxy is probably a physical companion of NGC4214, which is 30.0′ to the SE. Photographs show a brighter, 'wasp-waisted' inner region with many knotty condensations that is oriented NNE–SSW and embedded in a fainter, irregular halo. The galaxy is moderately faint and diffuse visually; it is squarish and very slightly brighter along its major axis, which is extended NNE–SSW. The edges are hazy and poorly defined. A field star is 6.25′ N. Radial velocity: 250 km/s. Distance: 10.4 mly. Diameter: 5300 ly.

NGC4214

H95[1]	12:15.6	+36 20	GLX
IAB(s)m	8.5′ × 6.6′	10.21B	

This member of the Canes Venatici I Cloud of galaxies is an irregular,

NGC4190

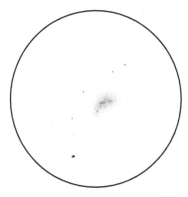

NGC4214 15-inch Newtonian 272x

Magellanic galaxy with stubby spiral structure. Photographically, the major axis features a bright bar embedded in a grainy envelope with many detached brighter condensations, particularly to N, SE and SW. The brightest stars resolve at blue magnitude 18.8. Visually, this is one of the most rewarding galaxies in Herschel's catalogue to view. The galaxy is bright with a complex, unusual structure. The outer envelope is tenuous and ill defined with a vaguely crescent-shaped structure. The central region is bright but very fragmentary in nature. Two condensations appear fused into a bar shape. In the SE a third condensation appears detached with a dark channel separating it from the core. Herschel's 28 April 1785 discovery description reads: 'Considerably bright, considerably large, extended north preceding, south following. Brighter to the middle. 4.0′ long, 3.0′ broad.' Radial velocity: 312 km/s. Distance: 14 mly. Diameter: 35 000 ly.

NGC4217

H748[2]	12:15.8	+47 06	GLX
SAb sp	6.0′ × 1.5′	12.05B	

The DSS photograph shows a classic, edge-on spiral galaxy, which is very flat with a brighter central region which is nevertheless mostly hidden by a thick, prominent dust lane. A faint field star lies

NGC4217

immediately N of the galaxy, brighter field stars lie a little farther N and W. In a moderate-aperture telescope the galaxy is faint and ethereal, visibility being hindered somewhat by the bright, nearby field stars. The galaxy is an elongated sliver of light, broadly brighter along its major axis; it is oriented NE–SW and is well defined along its edges. NGC4226 is 7.2′ to the ESE, appearing as a faint, roundish patch of light, a little brighter to the middle. Radial velocity: 1085 km/s. Distance: 49 mly. Diameter: 85 000 ly.

NGC4218

H718[3]	12:15.8	+48 08	GLX
Sa?	0.9′ × 0.6′	13.71B	

The DSS image shows a squarish irregular galaxy, possibly a dwarf, with a brighter central bar oriented along the major axis and a fainter outer envelope. The galaxy is small and only moderately bright visually; it is much more distinct at medium magnification. It appears as an elongated patch of light that is a little brighter to the middle, quite well defined and oriented SE–NW. Radial velocity: 791 km/s. Distance: 35 mly. Diameter: 9000 ly.

NGC4220

H209[1]	12:16.2	+47 53	GLX
SA(r)0+	3.8′ × 1.2′	12.23B	

The DSS photograph reveals a highly inclined spiral galaxy with a very bright, elongated central region and faint spiral structure in the outer envelope. The

NGC4214

brighter central region is dusty in its outer reaches, particularly to the SE and NNW. This is a large and bright galaxy visually, greatly elongated SE–NW with a very well defined envelope which is bright along its major axis. A small, bright core is embedded and the galaxy is very well seen at high magnification. Radial velocity: 975 km/s. Distance: 44 mly. Diameter: 48 000 ly.

NGC4227

H518[2]	12:16.5	+33 31	GLX
SAB0°:	1.5′ × 0.9′	13.96B	

This galaxy forms a probable physical pair with NGC4229 situated 2.5′ to the NNE. Photographs show an inclined lenticular galaxy, possibly barred, with a bright core in a disc surrounded by a very faint outer envelope. The pair is somewhat faint visually, with NGC4227 being a little larger, brighter and more distinct. It is a slightly oval patch of light, broadly brighter to the middle with somewhat hazy edges. Radial velocity: 6462 km/s. Distance: 289 mly. Diameter: 126 000 ly.

NGC4229

H519[2]	12:16.7	+33 34	GLX
S0/a	1.3′ × 0.8′	14.46B	

Photographs reveal a very slightly inclined lenticular galaxy with a small, bright core and an extensive, fainter outer envelope. It is quite possibly a physical companion of NGC4227: the minimum core-to-core separation between the two galaxies would be approximately 210 000 ly at the presumed distance. Visually, the galaxy appears as a slightly elongated blur, oriented N–S with moderately well defined edges and even surface brightness across its disc. Radial velocity: 6716 km/s. Distance: 300 mly. Diameter: 113 000 ly.

NGC4231

H719[3]	12:16.8	+47 27	GLX
SA0+ pec?	1.0′ × 0.8′	14.83B	

Photographs show a small galaxy, probably a spiral, seen face-on. It has a bright core with very faint suspected spiral arms. It is located N of companion galaxy NGC4232 and there is a broad, faint patch extending towards the southern galaxy. They are almost certainly a physical pair and the apparent separation of 1.1′ indicates a core-to-core separation of approximately 107 000 ly at the suspected distance. Visually, the galaxy is difficult to see and appears as an indistinct smudge of light; it is best seen with averted vision. Radial velocity: 7460 km/s. Distance: 333 mly. Diameter: 97 000 ly.

NGC4232

H720[3]	12:16.8	+47 26	GLX
SBb pec:	1.3′ × 0.6′	14.57B	

This is an inclined barred spiral galaxy with a bright core, a very short bar running N–S and two broad spiral arms. Images reveal a very faint field star located adjacent to the spiral arm in the S. Visually, the galaxy is a little more distinct than its companion NGC4231, appearing as a slightly oval patch of light broadly brighter to the middle. Radial velocity: 7315 km/s. Distance: 327 mly. Diameter: 124 000 ly.

NGC4242

H725[3]	12:17.5	+45 37	GLX
SAB(s)dm	5.0′ × 3.8′	11.59B	10.8V

Along with the nearby galaxies NGC3949, NGC3675 and others, this may be a member of the Canes Venatici II Cloud of galaxies. The DSS image shows a large, loose-structured spiral galaxy with a star-like brighter core. The somewhat tentative spiral pattern is embedded in a fainter envelope, with three spiral fragments visible. The envelope and spiral arms contain many small, nebulous concentrations. This is a large and very diffuse galaxy visually, with a very slightly oval envelope that is poorly defined at the edges. The galaxy is very broadly brighter to the middle and set within an equilateral triangle of field stars. Radial velocity: 560 km/s. Distance: 25 mly. Diameter: 36 000 ly.

NGC4244

H41[5]	12:17.5	+37 49	GLX
SA(s)cd: sp	19.4′ × 2.1′	10.46B	

Photographs show this Canes Venatici I Cloud member as a very thin, edge-on galaxy with a very small, star-like core. A small, thin dust lane passes just above the core; many small, brighter condensations are seen along the spiral extensions plus small dust patches. In a moderate-aperture telescope this galaxy is a splendid sight. A very bright, edge-on galaxy with a well defined central plane with faint, intermittently visible extensions. The core of the galaxy is elongated along the major axis; the galaxy is oriented NE–SW. A very faint stellar nucleus is intermittently visible at high magnification. Radial velocity: 271 km/s. Distance: 12.1 mly. Diameter: 68 000 ly.

NGC4242

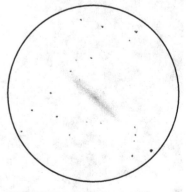

NGC4244 15-inch Newtonian 146x

NGC4244

The envelope runs N–S and in the northern portion it is slightly brighter along its eastern flank. To the S, it is bright along the western flank. Otherwise, the envelope is quite diffuse and poorly defined. Radial velocity: 508 km/s. Distance: 23 mly. Diameter: 124 000 ly.

NGC4288

H726[3]	12:20.6	+46 17	GLX
SB(s)dm	2.2′ × 1.6′	13.30B	

Images show an irregular galaxy with a curved, central bar and a slightly brighter, small core. The grainy, irregularly shaped outer envelope has two brighter condensations immediately SSE of the central bar. A small, compact galaxy to the SSE, designated NGC4288A, is evidently a background object of high radial velocity and may be visible in large amateur telescopes. Visually, NGC4288 is quite faint and is a little more distinct at medium magnification. A diffuse, oval patch of light, it is a little brighter along its N–S axis, where a thin, bright streak is suspected. The edges are poorly defined and the galaxy is situated within a diamond-shaped asterism of four field stars. Radial velocity: 577 km/s. Distance: 26 mly. Diameter: 16 000 ly.

NGC4346

H210[1]	12:23.5	+47 00	GLX
SB0	3.3′ × 1.3′	12.14B	

Photographs show a lenticular galaxy with a very bright, oval core and bright extensions. Although small visually, this galaxy is bright and well condensed; it is a very well defined, edge-on galaxy with a high-surface-brightness disc and very slightly tapered extensions. A round, slightly bulging core, which is brighter to the middle, is embedded and the galaxy is elongated E–W. Radial velocity: 830 km/s. Distance: 37 mly. Diameter: 36 000 ly.

NGC4357

H743[2]	12:24.0	+48 47	GLX
SAbc	3.9′ × 1.3′	13.46B	

This is a highly inclined multi-arm spiral galaxy with a small, bright and

NGC4248

H742[2]	12:17.8	+47 25	GLX	12.5V
I0 sp	3.0′ × 1.1′	13.14B		

Photographically, this is a highly inclined irregular, or possibly barred, spiral galaxy. The very flat, bright, elongated core has faint extensions displaying a possible spiral pattern, particularly off the E extension. A faint field star borders the core in the outer envelope to the W. This galaxy is the western companion of NGC4258, which is located 12.5′ SE; they are very likely physically related. Visually, this is an elongated patch of light, moderately bright and extended ESE–WNW. It is brighter to a small, star-like core with fairly well defined edges. NGC4231 and NGC4232 can be seen in the same medium magnification field to the WNW. Radial velocity: 544 km/s. Distance: 24 mly. Diameter: 21 000 ly.

NGC4258

H43[5]	12:19.0	+47 18	GLX	8.4V
SAB(s)bc	18.6′ × 7.2′	9.12B		

The DSS image reveals a very large, inclined spiral galaxy with a bright, complex inner region and a brighter core. The core is elongated – it is possibly a stubby bar – with two strong, narrow spiral arms emerging. There are many dust patches in the inner region and the inner spiral arms display many bright condensations. The galaxy is surrounded by complex, fainter outer spiral arms which almost triple the extent of the galaxy. These spiral arms are dotted with brighter condensations and star-forming regions. The discovery of this object is credited to Pierre Méchain and it is also catalogued as M106. Visually, the full extent of the galaxy is very difficult to see and the bright inner region dominates. It is a large and fairly bright galaxy with a short, brighter, bar-like core, elongated roughly E–W with a star-like nucleus.

NGC4258

round core. Photographs show that strong, grainy spiral arms emerge from the core with decidedly symmetrical structure. A slight positional error led Dreyer to originally designate this object NGC4381 but there is no object at that location. It is plotted as NGC4357 in the correct position in the first edition of *Uranometria 2000.0*. At the eyepiece the galaxy is fairly faint; it is an oval patch of light elongated almost due E–W and is

gradually brighter to the middle but no core is visible. Radial velocity: 4188 km/s. Distance: 187 mly. Diameter: 196 000 ly.

NGC4369

H166[1]	12:24.6	+39 23	GLX
(R)SA(rs)a	2.1′ × 2.0′	12.40B	

Images show a round disc galaxy, or possibly a spiral, with a very bright

round core and a round outer envelope. There is evidence of a dust patch in the core and a small, bright condensation is attached to the core in the E. A possible dust patch, seen as a dark arc in the outer envelope, is visible to the W. Visually, the galaxy appears small but it is quite bright and is well seen at high magnification. It appears round and well condensed with a small, bright core and a very faint outer halo suspected. Radial velocity: 1080 km/s. Distance: 48 mly. Diameter: 30 000 ly.

NGC4389

H749[2]	12:25.6	+45 41	GLX
SB(rs)bc pec:	2.5′ × 1.6′	12.55B	

The DSS image shows an inclined barred spiral galaxy; high-resolution photographs reveal that the bar is made up of a series of HII regions. There is no core and the spiral structure is difficult to trace, appearing as grainy fragments in an oval outer envelope. Telescopically, this galaxy can be seen in the same medium-magnification field as NGC4392 though they are not physically related, as NGC4392 has a far higher radial velocity. NGC4389 is a bright elongated galaxy featuring a bright elongated bar embedded in a bright but diffuse envelope. It appears large and a little hazy at the edges and is oriented ESE–WNW. Radial velocity: 775 km/s. Distance: 35 mly. Diameter: 25 000 ly.

NGC4392

H729[3]	12:25.2	+45 52	GLX
1.0′ × 0.90′	14.35B		

This galaxy is unclassified and has a peculiar form as seen on the DSS image. The small, moderately bright, irregularly oval disc has a bright core and possible bar-like extensions elongated E–W. The galaxy is surrounded by a very faint and ill defined halo, elongated SSE–NNW and appearing to be about 4.0′ × 2.3′ on the DSS image. At the apparent distance, this halo would be about 380 000 ly in extent. Possible companion galaxies are 5.75′ ENE and 5.5′ N. At medium magnification, NGC4392 is located NNW of NGC4389; it is smaller and fainter but well defined and well condensed. It is a slightly elongated patch of light, evenly bright across its disc. Radial velocity: 7290 km/s. Distance: 326 mly. Diameter: 95 000 ly.

NGC4395

H29[5]	12:25.8	+33 33	GLX	
SA(s)m:	13.2′ × 11.0′	10.84B	10.1V	

This is a member of the nearby Canes Venatici I Cloud of galaxies, which includes NGC4190, NGC4214 and NGC4449 amongst others. Photographs show a large, complex but very faint spiral galaxy with a possible central bar. A very small, star-like nucleus is embedded in the central bar and six distinct, identifiable though weak, spiral arms emerge from the bar. Small condensations are seen throughout the spiral arms with bright star-forming regions SSW and SE of the centre of the galaxy. Stars begin to resolve at about blue magnitude 18. Recorded on 2 January 1786, Herschel observed a bright HII region ESE of the core as well, which was included in the *New General Catalogue* as NGC4401. Herschel's description of NGC4395 reads: 'Extremely faint, very large, very little brighter to the middle. Resolvable. 10′ long, 8 or 9′ broad.' In moderate-aperture instruments this is a difficult object; it is a large, diffuse and very faint galaxy, best seen at low magnification. There is a faint concentration of light toward the centre; with averted vision this appears as a bar oriented ESE–WNW. The envelope is extremely tenuous and fades very gradually into the sky background. NGC4399, NGC4400 and NGC4401 are HII regions in NGC4395. NGC4401 is the brightest portion of the galaxy; at medium magnification it appears as an oval glow extended ESE–WNW and is smaller than the galaxy's central region which is 2.0′ to the WNW. Radial velocity: 335 km/s. Distance: 15 mly. Diameter: 57 000 ly.

NGC4401

H29[5]	12:25.8	+33 31	EN

This is an HII region in the spiral galaxy NGC4395.

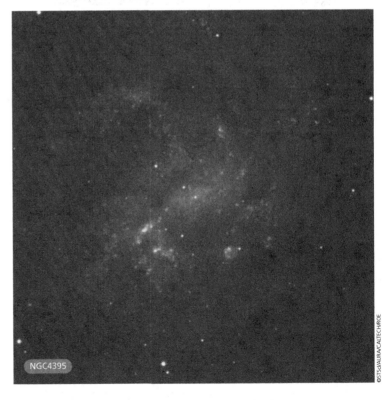

NGC4395

NGC4449

H213[1]	12:28.2	+44 06	GLX
IBm	6.2′ × 4.4′	9.50B	

This bright, starburst galaxy is a member of the Canes Venatici I Cloud and is well seen on the DSS photograph as a large, bright, irregular galaxy with a complex morphology. It is possibly a Magellanic galaxy: the large brighter core is triangular in shape and embedded in a squarish outer envelope. There are several bright and complex star-forming regions to the N and another bright region S of the core. Hydrogen-alpha streamers are common and the brightest stars resolve at blue magnitude 19. Telescopically, it is a large and very bright galaxy, but quite irregular in morphology. It is very bright and mottled along its major axis with some brighter condensations visible, as well as a very faint field star in the outer envelope E of the core. A fainter and ill defined envelope surrounds the bright inner envelope and the galaxy is oriented NE–SW. Herschel recorded the galaxy on 27 April 1788 and described it as: 'Very brilliant, considerably large, extended south preceding north following, with difficulty resolvable. Has 3 or 4 bright nuclei.' Radial velocity: 259 km/s. Distance: 11.6 mly. Diameter: 20 600 ly.

NGC4449

NGC4460

H212[1]/H750[2]	12:28.8	+44 52	GLX
SB(s)0$^+$? sp	4.1′ × 1.2′	12.38B	

Recorded twice by Herschel, on 10 April and 17 April 1788, this is an edge-on

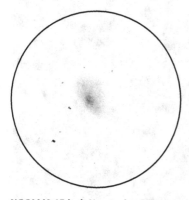

NGC4449 15-inch Newtonian 272x

lenticular galaxy with a bright, elongated core and a fainter extended envelope. High-resolution images reveal a grainy texture to the disc, particularly in the central region. A probable member of the Canes Venatici II Cloud, this object is quite bright visually and is located 7.6′ ENE of a well-resolved, bright double star. The galaxy is a well defined streak of light, quite flat with a brighter core, oriented NE–SW. Radial velocity: 545 km/s. Distance: 24 mly. Diameter: 29 000 ly.

NGC4485

H197[1]	12:30.5	+41 42	GLX
IB(s)m pec	2.3′ × 1.6′	12.34B	11.9V

This galaxy forms a bright, interacting pair with NGC4490 and has a peculiar morphology. Photographically, the galaxy is decidedly irregular in form, though there are traces of incipient spiral structure, particularly to the S, where

the thin arm is studded with bright HII regions which seem to reach back towards the larger galaxy. Visually, the galaxy is smaller and fainter that its companion and appears as a narrow streak of light, elongated N–S with somewhat ill defined edges but is brighter and mottled along its major axis. Radial velocity: 538 km/s. Distance: 24 mly. Diameter: 16 000 ly.

NGC4490

H198[1]	12:30.6	+41 38	GLX
SB(s)d pec	6.3′ × 3.1′	9.76B	9.8V

This galaxy and NGC4485 form a bright pair of interacting galaxies and are among the most spectacular objects in Herschel's catalogue. Probable members of the Canes Venatici II Cloud, the galaxies have also been catalogued as Arp 269. Herschel discovered the pair on 14 January 1788, describing them as: 'Two. The south very bright, very large, a little extended. The north bright, pretty small, irregularly formed, distance 1.5′.' The likely close encounter between the two galaxies has resulted in a tremendous amount of star formation in both galaxies and the brightest stars can be resolved at visual magnitude 18. The DSS image shows a very bright galaxy with a peculiar morphology. The core is bright and elongated and two irregular, short, but bright, stubby arms can be seen in a fainter extended envelope. There are many HII knots, particularly along the bright major axis. The western spiral arm extends towards the companion galaxy NGC4485 and a very faint bridge of material seems to connect the galaxies. There is a great deal of dust involved in the spiral pattern of NGC4490, though it is not well seen photographically due to the high surface brightness of the galaxy. Telescopically, NGC4490 is very large and bright, elongated WNW–ESE with a bright, elongated core embedded in a slightly fainter envelope. A faint, curving patch

in the WNW is suspected, spreading towards NGC4485. The core-to-core separation between the two galaxies is 3.5′ which suggests a minimum separation of about 28 000 ly. Radial velocity: 611 km/s. Distance: 27 mly. Diameter: 51 000 ly.

NGC4534

H410[2]	12:34.1	+35 31	GLX
SA(s)dm:	4.1′ × 3.0′	12.98B	

This is a four-branch spiral galaxy viewed almost face on. Photographs show a broadly brighter central region with low-surface-brightness spiral arms emerging. The arms are heavily mottled with many knotty condensations. Visually, the galaxy is a quite large but diffuse patch of light and is fairly well seen at high magnification as a somewhat dim patch of light, round and very broadly brighter to the middle. The edges are hazy and ill defined. The galaxy is a possible member of the Canes

Venatici II Cloud of galaxies. Radial velocity: 828 km/s. Distance: 37 mly. Diameter: 44 000 ly.

NGC4583

H495[3]	12:38.0	+33 28	GLX
1.1′ × 1.1′	14.56B		

This galaxy is unclassified and on the DSS image displays a peculiar morphology. A small, brighter, elongated core is embedded in a slightly elongated outer envelope. A narrow, slightly tapered dark dust lane intrudes from the N with the apex of the dust lane immediately adjacent to the core. A complementary, though blunter, dust patch is visible S of the core. A fainter anonymous galaxy is 4.6′ to the WSW. Visually, NGC4583 is a small and somewhat ill defined patch of light immediately E of a faint triangle of field stars. It is slightly extended E–W with a slightly brighter middle. Radial velocity: 6942 km/s. Distance: 310 mly. Diameter: 99 000 ly.

NGC4617

H744[2]	12:41.1	+50 26	GLX
Sb	3.0′ × 0.4′	14.03B	

Images show a very flat, almost edge-on spiral galaxy with a brighter, elongated core and faint spiral extensions. Visually, only the core of this galaxy is visible as an oval patch elongated N–S; excellent conditions may help to bring out the fainter extensions. Radial velocity: 4732 km/s. Distance: 211 mly. Diameter: 184 000 ly.

NGC4618

H178[1]/H179[1]	12:41.5	+41 09	GLX
SB(rs)m	4.2′ × 3.4′	11.31B	

Along with NGC4625 8.25′ to the NNE, this galaxy has also been catalogued as Arp 23. It is part of the widely dispersed Canes Venatici II Cloud and the DSS image shows it as a bright irregular galaxy with a peculiar morphology. It is dominated by a bright, well-resolved bar in a fainter envelope. Several HII patches can be made out here and in the one spiral arm which emerges from the

NGC4490

©STScI/AURA/CALTECH/ROE

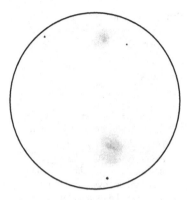

NGC4618 NGC4625
15-inch Newtonian 272x

about 63 000 ly at the presumed distance. The galaxy is rewarding to observe visually and can be seen with NGC4625 in the same high-magnification field. The bright E–W bar is well seen, offset towards the N where the fainter outer envelope is fairly well defined. The envelope is broader and diffuse towards the S with a grainy texture. A prominent field star is 3.75′ to the S. Herschel discovered the pair on 9 April 1787, but recorded NGC4618 as two separate objects, calling it: 'Two. The north very bright, very much brighter to the middle. The south pretty bright. Their nebulosity run together.' Radial velocity: 593 km/s. Distance: 26 mly. Diameter: 32 000 ly.

NGC4619

H411[2]	12:41.7	+35 04	GLX
SB(r)b pec?	1.3′ × 1.3′	13.62B	

This is a multi-arm spiral galaxy viewed face-on with a small, round, bright core.

NGC4625

Images show a strong, bright spiral structure with an asymmetrical form; most of the spiral structure is W of the core. Visually, the galaxy is fairly bright and is well seen at high magnification as a slightly oval, grainy patch of light, well defined and fairly even in surface brightness with a very slightly brighter core. A bright field star 1.9′ to the ESE does not significantly affect the view. Radial velocity: 6955 km/s. Distance: 311 mly. Diameter: 117 000 ly.

NGC4625

H660[2]	12:41.9	+41 16	GLX
SAB(rs)m pec	1.8′ × 1.6′	13.03B	

This galaxy is well seen on the DSS image and has a peculiar form that is very similar to that of its companion galaxy NGC4618. It has a bright, elongated core with a possible short bar emerging at an angle from the core. One spiral arm emerges from the NE and almost completely encircles the core of the galaxy. Telescopically, it is smaller but only very slightly fainter than NGC4618, almost round with a brighter core offset to the N in a diffuse envelope. Radial velocity: 658 km/s. Distance: 29 mly. Diameter: 16 000 ly.

NGC4627

H659[2]	12:42.0	+32 34	GLX
E4 pec	2.6′ × 1.8′	13.01B	

This is a small, dwarf elliptical companion galaxy to NGC4631; both galaxies are part of the Canes Venatici II

bar on the W and encircles the galaxy. It seems to form a partial ring with a second, much fainter spiral arm which emerges N of the bar on the W side. There was probably interaction between the two galaxies in the past; the apparent separation on the sky indicates a minimum core-to-core separation of

NGC4618

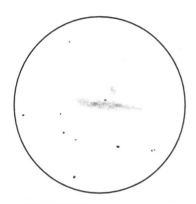

NGC4627 NGC4631
15-inch Newtonian 146x

Cloud. The two galaxies have also been catalogued as Arp 281; high-contrast images show a faint connection between the two galaxies. Telescopically, the galaxy is faint, small and oriented NNE–SSW with a smooth envelope and moderately well defined edges. Medium magnification brings out a faint stellar nucleus. It is situated only 3.0′ from the core of NGC4631 and the core-to-core separation at the presumed distance would be about 22 000 ly. Radial velocity: 562 km/s. Distance: 25 mly. Diameter: 19 000 ly.

NGC4631

H42[5]	12:42.1	+32 32	GLX
SB(s)d	15.5′ × 2.7′	9.52B	9.2V

This galaxy is the principal member of the Canes Venatici II Cloud of galaxies, an association of 14 galaxies whose membership also includes NGC4656. Photographically, this is a very bright, large, spiral galaxy viewed almost edge-on. A small, brighter core is partially obscured by dust patches. An extremely complex structure of dust lanes runs along the major axis with several knotty star-forming regions. The structure is asymmetrical: the eastern spiral arm is short and thick, while the western spiral arm is thinner and longer with the brightest stars resolving at about blue magnitude 19. In moderate-aperture telescopes, this galaxy is a rewarding sight. Extremely bright and elongated

E–W, the western extension tapers to a point, while the eastern one is blunt. There is much mottling along the major axis. A magnitude 10 field star is located immediately N of the galaxy centre. About 1.5′ ESE of the star, a bright condensation is visible in the envelope. A similar, but fainter one is visible 2.0′ WSW. Herschel recorded the galaxy on 20 March 1787, describing it as: 'Very bright, much extended south preceding, north following but near the parallel. Much brighter middle. 16′ long.' Radial velocity: 626 km/s. Distance: 28 mly. Diameter: 127 000 ly.

NGC4655

H661[2]	12:43.6	+41 02	GLX
0.8′ × 0.7′	14.99B		

Although unclassified, photographs show this galaxy is probably an elliptical: it is very small and round with a fainter outer envelope. At high magnification it is small and quite faint

and is visible as a round, well defined patch of light of even surface brightness. A faint field star is 0.75′ ESE. Radial velocity: 7274 km/s. Distance: 325 mly. Diameter: 76 000 ly.

NGC4656

H176[1]	12:44.0	+32 10	GLX
SB(s)m pec	10.8′ × 2.0′	9.83B	10.5V

In photographs this probable Magellanic-type galaxy is viewed almost edge-on and displays a peculiar morphology, with a bright, complex core featuring several star-forming regions. A very faint, curving extension sweeps towards the SW with a brighter extension to the NE involving knotty condensations. It fades gradually until connecting to a bright, hook-shaped extension with many bright, star-forming knots (NGC4657). It is a member of the Canes Venatici II Cloud of galaxies and there is evidence of a past encounter between this galaxy and

NGC4631

©STScI/AURA/CALTECH/ROE

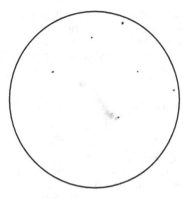

NGC4656 NGC4657
15-inch Newtonian 272x

NGC4631, located about 31.0′ to the NW. At the presumed distance, the core-to-core separation would be about 267 000 ly. Herschel originally saw NGC4656 as two separate objects and assigned it the numbers H176[1] and H177[1]. This was copied into the *New*

General Catalogue as NGC4656 and NGC4657. Recorded on 20 March 1787, he described it as: 'Two. The south considerably bright, elongated, much brighter to the middle. The north pretty bright, elongated south preceding, north following. Both join and form the letter S.' Visually, the galaxy is fairly bright and takes magnification well. It is oriented NE–SW and the brightest concentration is in the SW with a stubby arm curving to the W. At the tip of this branch is a very faint star. A faint bridge of light curves up to the N where a second, fainter concentration is located. This is NGC4657. Radial velocity: 666 km/s. Distance: 30 mly. Diameter: 93 000 ly.

NGC4657

H177[1]	12:44.2	+32 12	GLX
SB(s)m pec	2.0′ × 1.0′	9.83B	

This bright knot of material in the NE portion of the galaxy NGC4656 was

catalogued as a separate object by Herschel. Visually, it is a mottled and patchy object, elongated E–W and clearly connected to the galaxy by a faint bridge of material.

NGC4662

H643[2]	12:44.5	+37 07	GLX
SB(rs)bc	2.3′ × 1.8′	13.53B	

This is a face-on barred spiral galaxy with a brighter, slightly elongated core and short, fainter bar. There are three principal spiral arms, two tangent to the bar. They are grainy textured and narrow but well formed. Visually the galaxy is moderately bright and is best seen at medium magnification as a round patch of even-surface-brightness light with a small, slightly brighter core. The edges are fairly well defined and the disc is somewhat mottled. Radial velocity: 7021 km/s. Distance: 314 mly. Diameter: 210 000 ly.

NGC4704

H662[2]	12:48.8	+41 55	GLX
Sbc pec	1.0′ × 0.9′	14.52B	

Photographs show a three-branch, barred spiral galaxy. A slightly elongated, brighter core with a fainter bar has two principal spiral arms emerging; the third arm detaches from the principal arm in the NE. Visually, the galaxy is small and quite faint, requiring high magnification to confirm it. It appears as a slightly oval patch of light, oriented NE–SW, and is very slightly brighter to the middle with well defined edges. Radial velocity: 8188 km/s. Distance: 366 mly. Diameter: 106 000 ly.

NGC4707

H815[3]	12:48.4	+51 10	GLX
Sm:	2.2′ × 1.1′	13.43B	

The DSS image shows a probable dwarf irregular galaxy, or possibly a spiral; it is faint, oval and face-on with a faint star at the centre mimicking a brighter core. The outer envelope is brighter to the NW and several small, faint knots are distributed throughout. Visually, the

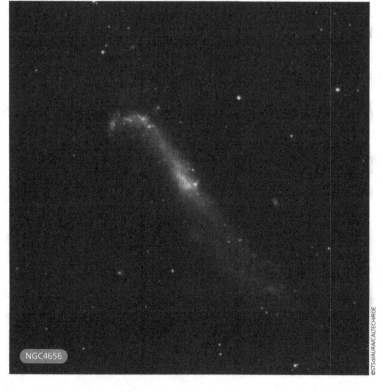

NGC4656

galaxy is well seen at both medium and high magnification as a diffuse and slightly oval patch of light, elongated N–S, with hazy edges and with the faint star in the middle. Radial velocity: 551 km/s. Distance: 25 mly. Diameter: 16 000 ly.

NGC4711

H412[2]	12:48.9	+35 21	GLX
Sb:	1.5′ × 0.9′	14.24B	

This is an inclined spiral galaxy with a slightly brighter core. Images show bright, grainy spiral arms and while the structure is fairly symmetrical, the arms are brighter to the NE. This is a fairly bright galaxy in a moderate-aperture telescope and is best at medium magnification. An oval patch of light, it is quite well defined, oriented NE–SW, fairly opaque and very gradually brighter to the middle. A bright field star follows 6.25′ to the ESE. Radial velocity: 4095 km/s. Distance: 183 mly. Diameter: 80 000 ly.

NGC4719

H424[3]	12:50.1	+33 09	GLX
SB(s)b	1.6′ × 1.4′	13.99B	

The DSS image shows a small, faint, S-shaped barred spiral galaxy with a small, bright, round core and a curved bar in a faint, elongated envelope. Two faint, smooth-textured spiral arms emerge from the bar. The galaxy is a quite small and somewhat difficult object visually, appearing as a round, hazy disc, with a star-like brighter core. Radial velocity: 7117 km/s. Distance: 318 mly. Diameter: 148 000 ly.

NGC4737

H496[3]	12:50.9	+34 09	GLX
0.5′ × 0.4′	14.92B		

This galaxy is unclassified but is probably lenticular. Photographs show a bright, round core and a fainter, elongated outer envelope. It is very faint and small visually, almost round and gradually brighter to the middle with fairly well defined edges. Radial velocity: 6929 km/s. Distance: 309 mly. Diameter: 45 000 ly.

NGC4741

H721[3]	12:50.9	+47 40	GLX
Scd:	1.1′ × 0.7′	14.38B	

Photographs show a slightly inclined, four-branch spiral galaxy with a bright, small, round core and strong, bright spiral structure. The galaxy is quite faint visually and best seen at medium magnification as an oval fairly well defined patch of light, very broadly brighter to the middle and oriented N–S. Radial velocity: 8958 km/s. Distance: 400 mly. Diameter: 128 000 ly.

NGC4774

H618[3]	12:53.2	+36 49	GLX
S pec (Ring:)	0.6′ × 0.4′	14.82B	

The DSS image shows a galaxy pair with a peculiar morphology. The principal galaxy is an oval ring with a brighter, round core on the northern extremity of the galaxy. A short, slightly fainter bar emerges, oriented E–W connected to the ring located S of the core. A faint plume of matter is directed N towards a smaller, fainter companion galaxy. Only the larger, brighter galaxy is seen visually and it is quite a dim object, best seen at medium magnification as a roundish patch of light, even in surface brightness with hazy, poorly defined edges. Radial velocity: 8412 km/s. Distance: 376 mly. Diameter: 66 000 ly.

NGC4800

H211[1]	12:54.6	+46 32	GLX
SA(rs)b	1.6′ × 1.2′	12.33B	

The DSS photograph shows a slightly inclined spiral galaxy with a large, bright, slightly oval core in a bright, compressed spiral pattern. Dust lanes are well seen in the outer spiral pattern and nebulous knots are prominent in the inner spiral structure. Well seen at high magnification, the galaxy is large and quite bright visually; it is an oval, opaque disc, quite well defined and oriented NNE–SSW with a condensed, bright core which is best seen at medium magnification. The disc is a little grainy-textured and a field star is

immediately W of the core. Distance: 43 mly. Diameter: 20 000 ly.

NGC4834

H817[3]	12:56.4	+52 18	GLX
0.9′ × 0.3′	15.20B		

Though unclassified, this remote galaxy is probably a highly tilted SB spiral; the DSS image shows an elongated, elliptical disc galaxy, with a bright, slightly elongated core in a fainter, extended envelope. Visually, this galaxy is very faint, little more than a roundish, diffuse patch of light with ill defined edges. Radial velocity: 10 430 km/s. Distance: 466 mly. Diameter: 122 000 ly.

NGC4861

H30[4]	12:59.0	+34 52	GLX
SB(s)m:	4.0′ × 1.5′	12.86B	

This galaxy has also been catalogued as Arp 266 due to its peculiar morphology. In the DSS image it is a faint irregular galaxy, very elongated and brightening along its major axis. The envelope is grainy and leads to a very bright core of nebulous material at the extreme SSW portion of the galaxy. This HII region has been catalogued as UGC8098. The galaxy is very much a threshold object visually; it is very dim and a little more distinct at medium magnification. It appears as a gradually brightening ray extending from the bright HII region in the SSW almost to a field star in the NNE, 1.4′ from the centre of the galaxy. The HII region resembles a fuzzy, out-of-focus star and is irregularly round. Herschel recorded the galaxy on 1 May 1785 and described it as: 'Two stars, distance 3′, connected with a very faint, narrow nebulosity.' Radial velocity: 869 km/s. Distance: 39 mly. Diameter: 45 000 ly.

NGC4868

H644[2]	12:59.1	+37 19	GLX
SAab?	1.6′ × 1.5′	12.96B	

Photographs show a face-on, two-armed spiral galaxy with a bright, round core. The spiral arms feature

NGC4861 15-inch Newtonian 272x

bright, curved, stubby roots leading to fainter, but prominent outer spiral arms. The galaxy is a probable physical companion of NGC4914, situated 20.0' to the E. Visually, the galaxy is moderately bright but somewhat diffuse; the envelope is mottled but moderately well defined with a slightly

brighter core. A field star is 1.0' N, while a slightly fainter one is 0.4' WSW of the core. Radial velocity: 4709 km/s. Distance: 210 mly. Diameter: 98 000 ly.

NGC4914

H645[2]	13:00.7	+37 19	GLX
E[+]/S0	3.5' × 2.1'	12.51B	

The appearance of this elliptical galaxy is very similar photographically and visually. At high magnification, it is a moderately bright galaxy, an extended oval of light with a grainy-textured envelope and a bright, small core. Well defined at the edges, the galaxy is oriented SSE–NNW. Radial velocity: 4708 km/s. Distance: 210 mly. Diameter: 214 000 ly.

NGC4932

H818[3]	13:02.6	+50 27	GLX
SA(r)c	1.5' × 1.3'	14.63B	

Photographs show a face-on, barred spiral galaxy with a bright, elongated

core and a fainter bar attached. The inner spiral structure forms a pseudo-ring around the core and is bright and narrow, extending to broader, fainter outer spiral arms. A field star is 1.6' S while a small possible companion galaxy is 2.4' W. Telescopically, the galaxy is best seen at medium magnification though it is very faint, appearing as a roundish blur of light, very broadly brighter to the middle, with edges fading into the sky background. Radial velocity: 7174 km/s. Distance: 320 mly. Diameter: 140 000 ly.

NGC4956

H413[2]	13:05.1	+35 11	GLX
S0	1.5' × 1.5'	13.26B	

This lenticular galaxy presents a similar appearance visually and photographically. It is a moderately bright galaxy, well seen at high magnification as a round, fairly opaque object, well defined at the edges and broadly brighter to the middle, where there is a bright, star-like core. Radial velocity: 4790 km/s. Distance: 214 mly. 93 000 ly.

NGC4963

H663[2]	13:05.8	+41 43	GLX
S?	0.9' × 0.8'	14.26B	

Images show that this is possibly a face-on spiral galaxy with a bright, round core embedded asymmetrically in a fainter envelope, which is a little more extensive to the N. The outer envelope is extremely faint to the W and N with dust patches suspected. Visually, the galaxy is moderately faint but is well seen at high magnification as a round, condensed, well defined patch of light, with a tiny, brighter core. A faint field star is 0.8' SSE. Radial velocity: 7211 km/s. Distance: 322 mly. Diameter: 84 000 ly.

NGC4985

H654[3]	13:08.1	+41 41	GLX
S0	1.0' × 0.7'	14.74B	

This is classified as a lenticular galaxy. Photographs show that this remote

NGC4861

©STScI/AURA/CALTECH/ROE

system has a very faint and somewhat asymmetrical outer envelope. Telescopically, it is a very small, round, well defined patch of light with even surface brightness across the disc. A faint field star follows 2.0′ to the ESE. Radial velocity: 8483 km/s. Distance: 379 mly. Diameter: 110 000 ly.

NGC4986

H401[3]	13:08.4	+35 12	GLX
SB(r)b	1.6′ × 0.9′	14.06B	

Images show a barred spiral galaxy with a bright, round core and a fainter bar. A ring or pseudo-ring surrounds the bar and two extremely faint and thin spiral arms emerge from the ring at the bar. A third thin arm branches from the northern spiral arm on the inside. A faint field star is in the ring immediately S of the core. Visually, the galaxy is quite faint, slightly oval and fairly even in surface brightness with moderately well defined edges. Radial velocity: 4864 km/s. Distance: 217 mly. Diameter: 101 000 ly.

NGC4987

H815[2]	13:08.0	+51 56	GLX
E	1.2′ × 0.7′	14.40B	

Photographs show an elliptical galaxy with an elongated core in a fainter, elongated envelope. The diffuse, face-on spiral galaxy UGC8222 is 5.0′ to the NNE, while the bright lenticular galaxy MCG+9-22-20 is 8.0′ to the NE. It is similar in size and brightness to NGC4987 and visually they form a visual pair in a medium-magnification field. NGC4987 appears as a round and fairly condensed patch of light, even in brightness and well defined at the edges. Radial velocity: 4817 km/s. Distance: 215 mly. Diameter: 75 000 ly.

NGC4998

H819[3]	13:08.2	+50 40	GLX
0.8′ × 0.7′	14.92B		

Though NGC4998 is unclassified, the DSS photograph shows a face-on spiral galaxy with a small, bright, round core

and two strong spiral arms emerging from the core. Telescopically, the galaxy is fairly faint, but is seen at medium magnification as a roundish and diffuse patch of light, fairly well defined and even in brightness. Radial velocity: 9003 km/s. Distance: 402 mly. Diameter: 94 000 ly.

NGC5003

H655[3]	13:08.6	+43 44	GLX
Sa	1.1′ × 0.8′	15.00B	

Photographs show a possible face-on spiral galaxy with somewhat asymmetrical structure. The brighter, round core is offset slightly to the SW in the disc and dust patches impact on the bright core in the W and E. Telescopically, the galaxy is small and very faint, requiring high magnification to confirm. Situated within a faint triangle of field stars, the galaxy is a round well defined patch of light. Radial velocity: 10 733 km/s. Distance: 479 mly. Diameter: 153 000 ly.

NGC5005

H96[1]	13:10.9	+37 03	GLX
SAB(rs)bc	5.8′ × 2.6′	10.54B	

This is a large, bright, highly inclined spiral galaxy, with a large, bright, elongated core. Photographs show tightly wound, bright, spiral structure with prominent, narrow dust lanes primarily along the NNW flank, which appears to be the near side. This galaxy and NGC5033 are both principal members of the cloud of galaxies known as the Canes Venatici Spur. Telescopically, this galaxy is large and very bright, featuring a large and bright core with two inner arms, one on the NNW side curving to the W, the other SSE curving towards the E. This is surrounded by a fainter, but grainy-textured outer envelope which is diffuse and somewhat ill defined. The disc is elongated ENE–WSW. Radial velocity 995 km/s. Distance: 44 mly. Diameter: 75 000 ly.

NGC5005

NGC5033

©STScI/AURA/CALTECH/ROE

medium magnification as a large, flat streak of light, quite well defined and elongated NNE–SSW. Radial velocity: 478 km/s. Distance: 21 mly. Diameter: 46 000 ly.

NGC5025

H649[3]	13:12.8	+31 48	GLX
Sb	2.3′ × 0.6′	14.38B	

This almost edge-on spiral galaxy has an elongated, brighter central region in photographs which may indicate a bar. The spiral pattern appears to be of the multi-arm type and seems to be quite symmetrical. Telescopically, this is a fairly faint galaxy; it is a slightly elongated patch of light oriented ENE–WSW, fairly well defined with a very faint field star 0.5′ NNE of the core. Radial velocity: 6376 km/s. Distance: 285 mly. Diameter: 190 000 ly.

NGC5033

H97[1]	13:13.4	+36 36	GLX
SA(s)c	10.5′ × 5.1′	10.70B	

This Canes Venatici Spur galaxy is a large, bright, highly inclined spiral galaxy with a brighter, elongated core. The DSS image shows at least five principal spiral arms, which are narrow, well defined by brighter condensations and loosely wound around the central disc. Along the western flank of the bright inner disc are distinct dust lanes. Telescopically, however, only the brighter inner region of this galaxy is detected; it is seen as a N–S elongated, fairly bright envelope, broadly brighter to an extended core. Radial velocity: 924 km/s. Distance: 41 mly. Diameter: 125 000 ly.

NGC5009

H820[3]	13:10.8	+50 05	GLX
SBb	1.1′ × 0.6′	15.21B	

Photographs show a barred spiral galaxy with a bright core and a fainter, short bar attached to two strong, smooth-textured spiral arms. Although small, this galaxy is seen at both medium and high magnification as an irregularly round, nebulous patch of light, fairly well defined and evenly bright. A prominent diamond-shaped asterism is in the field to the NE. Radial velocity: 9490 km/s. Distance: 424 mly. Diameter: 136 000 ly.

NGC5014

H414[2]	13:11.5	+36 17	GLX
Sa? sp	1.6′ × 0.6′	13.48B	

The DSS image shows an edge-on spiral or possibly an irregular galaxy with a bright major axis and faint extensions. A very narrow dust patch, perpendicular to the major axis, is near the core. Faint plumes of matter emerge to the NNW and N, presenting a morphological type similar to a distant M82. This is a bright galaxy visually, and is well seen at high magnification as a well defined, elongated streak of light oriented roughly E–W with slightly tapered extensions and a small, star-like core. Radial velocity: 1173 km/s. Distance: 53 mly. Diameter: 24 000 ly.

NGC5023

H664[2]	13:12.2	+44 02	GLX
Scd	7.3′ × 0.9′	12.81B	

The DSS image shows a faint, large edge-on galaxy, very likely a spiral. Very even in surface brightness along the major axis, it fades slightly to the extremities. There is no brighter core and little evidence of dust or star-forming regions. Visually, the galaxy is somewhat faint but is well seen at

NGC5040

H816[2]	13:13.6	+51 16	GLX
	1.2′ × 0.6′	15.01B	

Though NGC5040 is unclassified, photographs show a probably lenticular galaxy with an elongated core and fainter extended outer envelope. Visually, the galaxy is a moderately faint, round patch of light, even in surface brightness and well defined at the edges.

Radial velocity: 7606 km/s. Distance: 340 mly. Diameter: 119 000 ly.

NGC5074

H309[3]	13:18.4	+31 28	GLX
SAb pec?	0.7′ × 0.6′	14.43B	

Photographs show a galaxy with a somewhat chaotic structure. A brighter, irregular bar, oriented SSE–NNW has slight flares at both ends with a partial ring or possibly a single spiral arm attached. A suspected curving dust patch crosses in front of the bright bar to the E; the structure may be the result of the merger of two galaxies. The galaxy is small and very faint visually, a roundish ill defined glow, fairly even in surface brightness. The galaxy may be associated with NGC5056, NGC5057 and NGC5065 about 30.0′ to the SW. Radial velocity: 5639 km/s. Distance: 252 mly. Diameter: 51 000 ly.

NGC5093

H633[3]	13:19.6	+40 23	GLX
Sa	1.1′ × 0.6′	14.44B	

Images reveal an inclined spiral galaxy with a bright, elongated core and a fainter outer envelope. Visually, only the bright core is seen; at high magnification it is a quite bright, condensed oval patch elongated ESE–WNW in a very faint, roundish secondary envelope and is quite diffuse. Radial velocity: 7215 km/s. Distance: 322 mly. Diameter: 103 000 ly.

NGC5096

H650[3]	13:20.2	+33 05	GLX
E1	0.7′ × 0.5′	15.17B	

This galaxy is the principal member of a compact triple system of elliptical galaxies which appear to be interacting, as faint plumes connect both smaller galaxies to the largest one. In addition, the DSS image shows a pair of edge-on galaxies 0.9′ to the ENE as well as NGC5098, a pair of elliptical galaxies 3.5′ to the NNE. Photographs show that NGC5096 has an elongated core in a fainter outer envelope. One companion is NW of the core, while the other is NNE

of it. This latter galaxy has a short, faint spike of material emerging from its core to the N. This group cannot be resolved in medium-aperture telescopes, and appears as an oval blur broadly concentrated to the middle, elongated E–W and ill defined at the edges. NGC5098 to the NNE is even more difficult, a haze barely brighter than the sky background. The quoted physical dimensions for NGC5096 are generous: it does not appear to be much more than 0.4′ × 0.3′ on the DSS image. Radial velocity: 11 810 km/s. Distance: 528 mly. Diameter: 61 000 ly.

NGC5103

H665[2]	13:20.5	+43 04	GLX
Sab	1.1′ × 0.5′	13.47B	

Photographs show an edge-on lenticular galaxy with a bright, bulging core and bright extensions. Telescopically, the galaxy is located 1.8′ SSE of a bright field star and a fainter field star is 0.6′ to the NE. It takes magnification well but the bright field star hinders observation. A high-surface-brightness object visually, it is elongated NW–SE with very well defined edges. It is bright along its major axis and a little brighter to the middle. Herschel's 9 April 1787 description reads: 'Pretty bright, considerably small, extended. 300 (magnification) showed it like a star with burrs.' Radial velocity: 1348 km/s. Distance: 60 mly. Diameter: 19 000 ly.

NGC5107

H619[3]	13:21.4	+38 32	GLX
SB(s)d? sp	2.3′ × 0.5′	13.94B	

This is an almost edge-on galaxy, possibly a spiral or an irregular. Images show it is quite bright along its major axis with a fainter extension to the NW. A perpendicular, very narrow dust patch is in the main envelope, offset from the centre to the NW, while an extensive but extremely faint plume extends to the ESE. The morphology is similar to M82. At medium magnification the galaxy is faint but well seen as an elongated streak of light

with a brighter central region. It is fairly well defined and oriented SE–NW, and the extremities taper slightly to soft points. Radial velocity: 1005 km/s. Distance: 45 mly. Diameter: 30 000 ly.

NGC5112

H646[2]	13:21.9	+38 44	GLX
SB(rs)cd	4.0′ × 2.8′	12.63B	

The DSS photograph reveals a slightly inclined barred spiral galaxy with no core and a central, narrow bar surrounded by a bright, multi-arm spiral pattern. The arms feature several knotty condensations; they start out narrow and bright, then broaden and fade slightly as they spiral outward. Visually, the galaxy is large and moderately bright but quite diffuse. It is very slightly elongated SE/NW and broadly brighter to the middle and the edges are poorly defined with a field star 1.9′ SE of the core. Radial velocity: 1024 km/s. Distance: 46 mly. Diameter: 53 000 ly.

NGC5123

H666[2]	13:23.2	+43 04	GLX
Scd:	1.2′ × 1.1′	13.63B	

This is a three-branch, face-on spiral galaxy, with a small, round, bright core. Photographs show the spiral arms are bright and symmetrical. Telescopically, the galaxy is moderately faint but well seen at high magnification as a round patch of light with moderately well defined edges and an envelope which is

NGC5112

broadly brighter to the middle. Radial velocity: 8331 km/s. Distance: 372 mly. Diameter: 130 000 ly.

NGC5127

H328[2]	13:23.8	+31 34	GLX
E pec	2.4′ × 1.7′	13.61B	

This elliptical galaxy has a bright, elongated core in a fainter outer envelope. Photographs show a brighter condensation, or a possible companion galaxy, in the outer envelope to the ENE. The galaxy CGCG161-041 is 4.75′ to the N and a compact edge-on galaxy is 3.0′ S. Visually, NGC5127 is a moderately bright, small, well-condensed oval of light oriented ENE–WSW, somewhat grainy in texture with a star-like core and well defined edges. Radial velocity: 4900 km/s. Distance: 219 mly. Diameter: 153 000 ly.

NGC5141

H402[3]	13:24.9	+36 23	GLX
S0	1.3′ × 1.0′	13.79B	

This lenticular galaxy forms a triple system with NGC5142 2.25′ to the ENE and NGC5143 4.0′ to the NNE. Photographs show that it has a bright, elongated core in a fainter, outer envelope. Visually, only NGC5141 and NGC5142 are seen in a moderate-aperture telescope. NGC5141 is very slightly the brighter of the two, an irregularly round patch of light, well defined and brighter to the middle. Radial velocity: 5264 km/s. Distance: 235 mly. Diameter: 89 000 ly.

NGC5142

H403[3]	13:25.0	+36 24	GLX
S0	1.0′ × 0.7′	14.19B	

This lenticular galaxy has a bright core and fainter extensions elongated N–S and may be physically associated with NGC5141 although its radial velocity is somewhat higher. At the presumed distance the minimum core-to-core separation between the two galaxies would be 166 000 ly. Visually, it appears a little smaller than NGC5141; it is a roundish patch of light with a

bright, star-like core. Radial velocity: 5671 km/s. Distance: 253 mly. Diameter: 74 000 ly.

NGC5145

H667[2]	13:25.2	+43 15	GLX
S?	2.0′ × 1.8′	13.61B	

The morphology of this starburst galaxy is somewhat peculiar but it seems to be a face-on spiral. It appears slightly inclined with an intensely bright, elongated core. A bright wedge of material emerges from the core in the NE and two short spiral fragments lie outside the central core, one to the E and one to the W. Several peculiar spokes of matter emerge from the bright core in the W and S. Extremely faint spiral structure surrounds the core; three branches are visible but they are ill defined and fragmentary. Visually, the galaxy is small and moderately faint and only the central region is visible. It appears as a well defined disc, slightly

extended E–W and even in surface brightness. A tiny, star-like core is brighter than the disc. Radial velocity: 1299 km/s. Distance: 58 mly. Diameter: 34 000 ly.

NGC5149

H404[3]	13:26.1	+35 56	GLX
SBbc	1.7′ × 0.9′	13.75B	

This slightly inclined, barred spiral galaxy forms a likely physical pair with NGC5154, located 5.75′ to the NE, and both galaxies may be associated with the NGC5141 triplet 30.0′ to the NNW. Photographs show NGC5149 has an elongated core running into a short bar with two bright, well defined S-shaped spiral arms attached. The pair of galaxies is quite faint and diffuse telescopically, although NGC5149 is a little brighter and easier to see; it is an elongated oval, even in surface brightness and well defined at the edges.

NGC5149/5154

©STScI/AURA/CALTECH/ROE

Radial velocity: 5705 km/s. Distance: 254 mly. Diameter: 126 000 ly.

NGC5154

H405³	13:26.4	+36 00	GLX
Scd:	1.3′ × 1.3′	14.57B	

Photographs show a face-on, two-branch spiral galaxy with a small, round core and knotty spiral arms, symmetrical in structure. It is probably physically associated with NGC5149 to the SW and the minimum core-to-core separation between the two galaxies would be 420 000 ly at the presumed distance. Visually, it is a roundish, diffuse patch of light that is a little brighter to the middle with hazy edges. Radial velocity: 5632 km/s. Distance: 252 mly. Diameter: 96 000 ly.

NGC5157

H651³	13:27.4	+32 01	GLX
SAB(r)a	1.4′ × 1.1′	14.17B	

The DSS image shows an inclined, theta-shaped barred galaxy. The elongated core leads into a fainter bar, which is connected to a ring or pseudo-ring. Two very faint spiral arms are suspected to emerge from the ring in a tight pattern. The galaxy is quite faint visually, a somewhat roundish patch of light that is a little brighter to the middle; the edges are quite hazy. Radial velocity: 7364 km/s. Distance: 329 mly. Diameter: 134 000 ly.

NGC5173

H672³	13:28.4	+46 36	GLX
E0:	1.0′ × 0.9′	13.38B	

It is likely that the classification of this elliptical galaxy should be E1 or E2, as it appears slightly oval and oriented E–W on DSS images. Telescopically, the galaxy is moderately bright; it is a round, condensed object with a tiny, star-like core. The edges are well defined and the dim, inclined spiral galaxy NGC5169 is visible in the same field 5.4′ to the NNW. Radial velocity: 2504 km/s. Distance: 112 mly. Diameter: 32 000 ly.

NGC5187

H652³	13:29.9	+31 07	GLX
Sb	0.9′ × 0.7′	14.77B	

This is a slightly inclined spiral galaxy. Photographs show a large, very bright, elongated core. The spiral pattern is very faint, involving two or perhaps three branches with a suspected dust patch and a narrow dust lane crossing the core, oriented SW–NE. The galaxy is quite faint visually, but is fairly well seen at high magnification as a round patch, broadly brighter to the middle with well defined edges and a field star 1.5′ to the NNW. Radial velocity: 7212 km/s. Distance: 322 mly. Diameter: 84 000 ly.

NGC5195

H186¹	13:30.0	+47 16	GLX	
SB01 pec	5.8′ × 4.6′	10.50B	9.6V	

The DSS image shows a large, bright galaxy with a suspected theta-shaped barred ring morphology. Dust from the outer spiral arm of M51 obscures the view. The core is elongated E–W, while the broad bar is oriented almost due N–S. Small dust patches are in the envelope W of the core and are faintly seen in silhouette as they pass over the western portion of the ring. Beyond the main envelope of the galaxy, there is a chaotic halo of matter, especially prominent to the E, N and W, probably the result of gravitational interaction with M51, although the larger galaxy appears relatively unperturbed. The two galaxies are also catalogued as Arp 85. Visually, NGC5195 is quite large and bright in a medium-magnification field. The core is bright, embedded in the suspected bar and is elongated N–S in the somewhat diffuse outer envelope, which is irregularly round and somewhat ill defined. Herschel officially recorded the galaxy in his catalogue on 12 May 1786 though he described it earlier on 17 September 1783, when using a 7-foot telescope, magnification of 57 (and a likely aperture of 6.2 inches) as: 'Two nebulae joined together; both suspected of being stars. Of the most north [H I.186, NGC5195] I

have hardly any doubt. 7 feet, about 150 (magnification). A strong suspicion next to a certainty of being stars. I make no doubt the 20 ft. will resolve them clearly, as they want light and prevent my using a higher power with this instrument.' Three nights later, using the 20-foot telescope, he said: 'Most difficult to resolve, yet I do no longer doubt it. In the southern nebula [NGC 5194] I saw several stars by various glimpses, in the northern [NGC5195] also three or four in the thickest part of it, but never very distinctly. Evening very bad.' Credit for the discovery of NGC5195 goes to Pierre Méchain who observed it in 1781. It was later incorporated into Charles Messier's description of M51, published in the *Connaissance des Temps*, where he stated: 'Elle est double, ayant chacune un centre brillant, éloigné l'un de l'autre de 4′ 35″. Les deux atmosphères se touchent. L'une est plus faible que l'autre. Revue plusieurs fois.' (It is double, each having a brilliant centre and separated by 4 minutes 35 seconds. The two nebulae touch. One is fainter than the other. Seen several times.) Radial velocity: 553 km/s. Distance: 25 mly. Diameter: 42 000 ly.

NGC5198

H689²	13:30.2	+46 40	GLX
E1-2:	1.5′ × 1.3′	12.70B	

Visually and photographically, this elliptical galaxy appears quite similar, although photographs show that the galaxy is slightly elongated NNE–SSW. At medium magnification, the galaxy is small and roundish, fairly concentrated with a brighter, compressed core. A magnitude 12 star is visible to the NNE and a magnitude 14 star is visible to the W. Radial velocity: 2619 km/s. Distance: 117 mly. Diameter: 51 000 ly.

NGC5199

H406³	13:30.6	+34 19	GLX
Compact	0.9′ × 0.9′	14.71B	

Though the classification of this galaxy is uncertain, the DSS photograph indicates that it is probably an elliptical or lenticular galaxy seen face-on. Visually, the galaxy is very faint, a diffuse, roundish patch of light, broadly brighter to the middle with ill defined edges. Radial velocity: 7218 km/s. Distance: 322 mly. Diameter: 84 000 ly.

NGC5214

H656[3]	13:32.8	+41 52	GLX
Scd:	1.1′ × 0.9′	14.35B	

The DSS image reveals a four-branch spiral galaxy, probably barred, viewed almost face-on, with a slightly brighter, elongated core and a probable bar. A smaller, edge-on galaxy involved in the outer envelope to the S is designated NGC5214A and is probably a physical companion. At magnitude 16, it may be detected in very large amateur telescopes. Telescopically, NGC5214 is a roundish, quite dim galaxy, best seen with medium magnification. It is a hazy patch of light, broadly brighter to the middle and its edges are moderately well defined. Radial velocity: 8151 km/s. Distance: 364 mly. Diameter: 116 000 ly.

NGC5223

H407[3]	13:34.4	+34 42	GLX
E	1.5′ × 1.2′	13.96B	

This elliptical galaxy is part of a triplet which includes NGC5228 and NGC5233. Although its radial velocity is somewhat lower than that of the other two, the galaxies may be physically related. Visually, the three galaxies form an attractive grouping in a medium-magnification eyepiece with NGC5223 being fairly bright, round and brighter to the middle with a faint field star 0.6′ to the WSW. Radial velocity: 7258 km/s. Distance: 324 mly. Diameter: 141 000 ly.

NGC5225

H822[3]	13:33.3	+51 30	GLX
S?	0.7′ × 0.7′	14.38B	

The DSS image shows a disc, or possibly spiral, galaxy viewed face-on with a small, bright, round core in a high-surface-brightness, round envelope. Two suspected S-shaped spiral arms,

which are slightly brighter than the disc, emerge from the core. Telescopically, the galaxy is moderately bright and fairly well defined; it is visible as a round, opaque galaxy that is brighter to a very small core. Radial velocity: 4719 km/s. Distance: 211 mly. Diameter: 43 000 ly.

NGC5228

H408[3]	13:34.5	+34 46	GLX
S0⁻:	1.0′ × 0.9′	14.28B	

Situated 5.3′ NNE of NGC5223 this face-on lenticular galaxy has a bright core and a fainter outer envelope. The DSS image shows several fainter galaxies nearby, including an edge-on spindle 5.1′ to the WSW, so NGC5223, NGC5228 and NGC5233 are probably the brightest members of a small galaxy group. NGC5228 is the second brightest galaxy in a medium-magnification field; it is fairly bright, almost round and brighter to the middle with faint field stars flanking the galaxy to the NNE and SSW. Radial velocity: 7760 km/s. Distance: 347 mly. Diameter: 101 000 ly.

NGC5233

H425[3]	13:35.1	+34 40	GLX
Sab	1.1′ × 0.5′	14.45B	

Photographs show an almost edge-on spiral galaxy with a bright, elongated core and fainter extensions. A prominent dust lane crosses N of the core, extending for almost the entire length of the disc. The galaxy was recorded by Herschel on 3 May 1785, two nights after he discovered NGC5223 and NGC5228. It is considerably fainter than these other two galaxies visually, an ill defined patch of light that is a little brighter to the middle. Radial velocity: 7990 km/s. Distance: 357 mly. Diameter: 114 000 ly.

NGC5238

H823[3]	13:34.7	+51 37	GLX
SAB(s)dm	1.7′ × 1.0′	13.85B	

The DSS image reveals a dwarf irregular or possible spiral galaxy viewed face-on with a small, round, brighter core in a faint, grainy outer envelope. A smaller, barred spiral

NGC5238

galaxy with a small, round core is involved to the SSE. It is unclear whether the smaller galaxy is seen in projection or is interacting, but they seem to share a common envelope. These galaxies are probably distant satellites of M101. Visually, this is a very faint object, only intermittently visible at medium magnification as a small, round, very diffuse patch of light, only a little brighter to the middle and not well defined. Radial velocity: 336 km/s. Distance: 15 mly. Diameter: 7300 ly.

NGC5240

H409[3]	13:35.9	+35 34	GLX
SB(s)cd?	1.9′ × 1.4′	13.84B	

Photographs show an almost face-on barred spiral galaxy with a small, brighter core and slightly curved bar. The two narrow spiral arms open to a broader, fragmentary pattern away from the core. Telescopically, the galaxy is quite faint and difficult; it is a hazy, ill defined and roundish patch of light that is very broadly brighter to the middle. Radial velocity: 2291 km/s. Distance: 102 mly. Diameter: 57 000 ly.

NGC5243

H620[3]	13:36.2	+38 21	GLX
S	1.5′ × 0.4′	14.47B	

This is a highly inclined spiral galaxy with a brighter, elongated core. Photographs show the core is offset towards the NW and the spiral structure is somewhat asymmetrical. Visually, the galaxy is fairly faint and is seen as an elongated patch of light oriented ESE–WNW with fairly well defined edges. It is a little brighter along its major axis though there is no core. Radial velocity: 4270 km/s. Distance: 191 mly. Diameter: 83 000 ly.

NGC5263

H370[3]	13:40.0	+28 24	GLX
S[+]	1.6' × 0.4'	14.41B	

This is probably a highly inclined spiral galaxy. Photographs show that it is very bright throughout the entire envelope and several dust patches are involved; a bright, narrow probable spiral arm is to the SE. A bright field star is 3.1' S. In a moderate-aperture telescope the galaxy is faint though readily visible as a narrow, flat streak of light, even in surface brightness along its major axis, with very well defined edges and elongated NNE–SSW. Radial velocity: 4867 km/s. Distance: 217 mly. Diameter: 101 000 ly.

NGC5265

H410[3]	13:40.1	+36 51	GLX
Irr	0.6' × 0.5'	14.64B	

Despite its classification, the DSS photograph shows a face-on spiral galaxy with two broad principal spiral arms, though the structure is somewhat asymmetrical with a broader, fainter pattern to the E. A faint field star is 0.4' to the NNE and a slightly brighter one is 1.6' to the NNW. Visually, the galaxy is a faint and difficult object, a roundish, diffuse glow very slightly brighter to the middle with ill defined edges. Radial velocity: 5818 km/s. Distance: 260 mly. Diameter: 45 000 ly.

NGC5273

H98[1]	13:42.1	+35 39	GLX
SA(s)0°	2.2' × 2.1'	12.53V	

The DSS photograph shows a face-on disc galaxy with a bright core and an extensive, fainter outer envelope. Although it is faint, a spiral pattern can be seen in the outer envelope and a thin dust lane borders the core on the E side.

Telescopically, NGC5273 is quite bright and is well seen at high magnification; it is a round and fairly large galaxy, with an opaque envelope and a brighter middle. The edges are ill defined and fade gradually into the sky background. The likely companion galaxy NGC5276 can be seen 3.25' to the ESE as a very small but moderately bright object that is very slightly extended roughly N–S and quite concentrated. The minimum core-to-core separation between the two galaxies would be 48 000 ly at the presumed distance. Radial velocity: 1124 km/s. Distance: 50 mly. Diameter: 32 000 ly.

NGC5289

H668[2]	13:45.1	+41 30	GLX
(R)SABab: sp	2.2' × 0.6'	13.01B	

This is an almost edge-on galaxy, possibly a spiral or ring. Photographs show a bright, elongated core, perhaps a bar and a ring-like fainter outer

envelope. Visually, this galaxy can be seen in the same medium-magnification field as NGC5290 and they probably form a physical pair. The minimum core-to-core separation is about 423 000 ly. NGC5289 is well defined and elongated almost due E–W with a brighter core; a field star follows it to the ESE. Radial velocity: 2605 km/s. Distance: 116 mly. Diameter: 74 000 ly.

NGC5290

H170[1]	13:45.3	+41 43	GLX	
Sbc: sp	3.6' × 0.6'	13.29B	13.1V

Images show a bright, large, almost edge-on spiral galaxy with a brighter, elongated core and very dusty spiral arms. Telescopically, it is bright and much extended E–W with sharply defined edges and a brighter core. The extensions taper slightly. Radial velocity: 2652 km/s. Distance: 118 mly. Diameter: 124 000 ly.

NGC5289

NGC5290

©STScI/AURA/CALTECH/ROE

The galaxy is elongated SSE–NNW and a pair of bright field stars are located to the S. Radial velocity: 1595 km/s. Distance: 71 mly. Diameter: 77 000 ly.

NGC5303

H681[3]	13:47.8	+38 18	GLX
Pec	1.0′ × 0.5′	12.83B	

Photographs show an irregular galaxy with an oval, bright, dense core and a very faint extension emerging from the E and curving N. The companion galaxy NGC5303B is 2.7′ S and is difficult to see in moderate-aperture telescopes. Visually, NGC5303 is a fairly bright oval patch of light elongated E–W and brighter to a small core. It is sharply defined at the edges and the surface brightness of the envelope is even with a grainy texture. Radial velocity: 1490 km/s. Distance: 67 mly. Diameter: 19 000 ly.

NGC5305

H621[3]	13:47.9	+37 50	GLX
SB(r)b	1.5′ × 1.0′	14.42B	

Situated 6.0′ SE of a bright field star, photographs show a slightly inclined, theta-shaped barred spiral galaxy. A bright core is embedded in a fainter bar with a suspected ring attached. The spiral structure is faint with three principal arm fragments visible. A likely companion galaxy, UGC8724, is 4.75′ SSW. Telescopically, NGC5305 is very faint and is best seen at medium magnification as an ill defined, diffuse blur of light, slightly elongated N–S. Radial velocity: 5728 km/s. Distance: 256 mly. Diameter: 112 000 ly.

NGC5311

H710[2]	13:49.0	+40 00	GLX
S0/a	2.6′ × 2.2′	13.44B	

Images reveal an inclined spiral galaxy with a bright, elongated core in a fainter inner envelope. The outer spiral structure is extremely faint and is defined by a prominent dust lane bordering the inner envelope to the S in an otherwise featureless outer disc. In a medium-magnification eyepiece, both this galaxy

NGC5297

H180[1]	13:46.4	+43 52	GLX
SAB(s)c: sp	5.6′ × 1.3′	12.40B	

This galaxy forms a close physical pair with NGC5296, an S0 galaxy 1.5′ to the SW. The DSS photograph shows an almost edge-on spiral galaxy, possibly barred, with a bright, elongated core and dusty spiral arms. The arms are thin with many brighter condensations throughout and there is some evidence of interaction between the two galaxies. A thin spiral fragment from the larger galaxy curves back towards NGC5296, while faint plumes of material extend from the fainter galaxy towards the larger galaxy's core. The core-to-core separation between the two galaxies would be a minimum of 49 000 ly at the presumed distance. Visually, this is a very attractive pair at high magnification. NGC5297 is a large, bright, much extended galaxy with a large, bright core and bright extensions which taper to points. A bright field star precedes it to the ENE and the northern extension points directly at a slightly fainter field star. NGC5296 is fairly difficult, appearing as a small, round, nebulous patch SW of the larger galaxy's core. Radial velocity: 2495 km/s. Distance: 112 mly. Diameter: 182 000 ly.

NGC5301

H688[2]	13:46.4	+46 06	GLX
SA(s)bc: sp	3.7′ × 0.7′	13.34B	

The DSS image shows this galaxy as an almost edge-on spiral with a bright, elongated core and dusty spiral arms. A prominent dust lane runs almost the length of the galaxy NE of the core and the arms are studded with many brighter condensations. At high magnification the galaxy is quite bright; it is a fairly well defined object with a slightly brighter, elongated core and a star-like nucleus which is best seen at medium magnification. The extensions are fairly well defined and taper slightly to points.

and NGC5313 can be observed; the separation is 9.4′ due E–W. Although moderately bright, only the central portion of NGC5311 can be observed; it is well condensed with well defined extremities, slightly elongated E–W. This is a possible member of the NGC5353 group of galaxies. Radial velocity: 2774 km/s. Distance: 124 mly. Diameter: 94 000 ly.

NGC5312

H422[3]	13:49.8	+33 38	GLX
0.8′ × 0.5′	14.73B		

This galaxy is W of a small group of galaxies including NGC5318, NGC5319 and NGC5321 that are probably physically related. The galaxy is unclassified but on the DSS image appears to be lenticular with a bright elongated core in a fainter outer envelope. Visually, the galaxy can be seen in a medium-magnification field along with NGC5318 and NGC5321 as a small, somewhat concentrated spot slightly elongated NNE–SSW with well defined edges. Radial velocity: 4366 km/s. Distance: 195 mly. Diameter: 45 000 ly.

NGC5313

H711[2]	13:49.7	+39 59	GLX
Sb?	2.0′ × 1.0′	12.77B	

This is a high-surface-brightness galaxy. Photographs show an inclined spiral of the grand-design type with a brighter, elongated core and two bright, narrow, knotty spiral arms involved in a dusty outer envelope. Visually, this companion to NGC5311 is a bright galaxy, elongated NE–SW with a bright, well defined envelope, broadly brighter along its major axis. Like its companion, this is a possible member of the NGC5353 group of galaxies. Radial velocity: 2615 km/s. Distance: 117 mly. Diameter: 68 000 ly.

NGC5318

H423[3]	13:50.5	+33 43	GLX
S0?	1.2′ × 0.8′	13.64B	

Photographs show this lenticular galaxy has a bright core and short extensions and is the brightest of a group of at least seven galaxies in the field, including NGC5319 (3.4′ NNE) and NGC5321 (4.6′ SSW). Two very faint galaxies are immediately N of NGC5318; the first is a very small, face-on two-arm spiral (0.8′ N), the second is an inclined two-arm spiral 1.8′ N. Visually, NGC5318 is the largest and brightest of three galaxies in a medium-magnification field: the others are NGC5312 and NGC5321. It is slightly oval and elongated N–S with a brighter core and fairly well defined edges. The barred spiral NGC5321 to the S is not a Herschel object and appears as a very small, concentrated spot forming a triangle with two field stars. Radial velocity: 4388 km/s. Distance: 196 mly. Diameter: 68 000 ly.

NGC5320

H669[2]	13:50.3	+41 22	GLX
SAB(rs)c:	3.4′ × 1.7′	13.35B	

In photographs, this is an inclined, dusty spiral galaxy with a small, slightly brighter core and knotty spiral arms. At high magnification the galaxy is moderately faint and somewhat diffuse, but is well seen as a broad, oval patch of light, broadly brighter to the middle with somewhat ill defined edges and is elongated NNE–SSW. Radial velocity: 2700 km/s. Distance: 121 mly. Diameter: 119 000 ly.

NGC5326

H712[2]	13:50.8	+39 34	GLX
SAa:	2.2′ × 1.1′	12.89B	

Images show a highly inclined spiral galaxy with a bright, large, lens-shaped core in a fainter envelope. An extensive, prominent dust lane to the NE borders the inner lens. In moderate apertures at high magnification, this is a fairly bright, moderately condensed galaxy; it is very small but displays a very bright core. The edges are moderately well defined and the galaxy is clearly elongated SE–NW. It is a possible member of the NGC5353 group of galaxies. Radial velocity: 2596 km/s. Distance: 116 mly. Diameter: 74 000 ly.

NGC5336

H670[2]	13:52.1	+43 15	GLX
Scd:	1.3′ × 1.0′	13.82B	

Photographs show an inclined spiral galaxy, with a large, brighter, oval core and four thin, knotty spiral arms. Telescopically, the galaxy is moderately bright, appearing as an oval patch of light, elongated E–W, and slightly brighter to the middle. The edges are fairly well defined and two field stars follow the galaxy to the E. Radial velocity: 2423 km/s. Distance: 108 mly. Diameter: 41 000 ly.

NGC5337

H698[3]	13:52.5	+39 42	GLX
S?	1.7′ × 0.8′	13.44B	

The DSS image shows a highly inclined galaxy, probably a spiral, with a bright, elongated core in a fainter outer envelope. In a medium-magnification field, this galaxy can be seen as part of a wide pair which includes NGC5346, 9.8′ to the SW. Although faint, NGC5337 is a small, moderately well defined streak, oriented NNE–SSW and a little brighter to the middle. It may be associated with the NGC5353 group. Radial velocity: 2242 km/s. Distance: 100 mly. Diameter: 50 000 ly.

NGC5347

H424[2]	13:53.3	+33 29	GLX
(R′)SB(rs)ab	1.8′ × 1.3′	13.44B	

Photographs reveal a slightly inclined, theta-shaped barred galaxy. The elongated, brighter core is embedded in a fainter bar. Two tightly wound spiral arms emerge, forming a pseudo-ring which extends out into a faint, short spiral arm with a faint spur extending to the south. This object is fairly bright at medium magnification and less distinct at high magnification. It is a moderately large, diffuse, fat oval of light of fairly even surface brightness except for a slightly brighter, nonstellar core. The edges are moderately well defined and the galaxy is elongated ESE–WNW. A prominent field star is 8.25′ to the SW.

Radial velocity: 2395 km/s. Distance: 107 mly. Diameter: 56 000 ly.

NGC5350

H713[2]	13:53.4	+40 22	GLX
SB(r)b	3.2′ × 2.3′	12.28B	

This is the largest member of the NGC5353 group of galaxies. It is a barred spiral galaxy with a spiral pattern intermediate between the grand-design and multi-arm spiral class. Photographically, the core is round and bright with a prominent bar. The two main arms emerge from the bar but are tightly wound before branching out into a fragmentary, four-branch pattern. The extensive outer structure involves faint extensions of the spiral arms. Visually, it is the largest member of the group, though not the brightest; the core is large and a little brighter than the diffuse envelope, which is slightly elongated NNE–SSW. The edges are poorly defined. A bright field star is 2.8′ SW. Radial velocity: 2401 km/s. Distance: 107 mly. Diameter: 100 000 ly.

NGC5351

H697[2]	13:53.5	+37 55	GLX
SA(r)b:	2.7′ × 1.5′	13.08B	

In photographs this is an inclined spiral galaxy with a brighter, round core in a possible inner ring. The inner spiral arm structure is bright with several knots and expands to an extensive and fainter multi-arm pattern. Telescopically, the galaxy is large and fairly bright, but somewhat diffuse and elongated E–W. The disc is even in surface brightness and only very slightly brighter to the middle. The edges are diffuse, blending into an ill defined, very faint outer envelope. A likely companion galaxy NGC5349 is visible 3.5′ to the WSW. The galaxy is a barred spiral but is faint visually and only the core and bar, oriented E–W, are visible. At the presumed distance, the minimum core-to-core separation between the two galaxies would be 168 000 ly. Radial velocity: 3683 km/s. Distance: 165 mly. Diameter: 130 000 ly.

NGC5352

H415[2]	13:53.6	+36 09	GLX
S0⁻:	1.1′ × 0.9′	14.05B	

Photographs show a slightly inclined lenticular galaxy with a bright, elongated core and a fainter outer envelope. The galaxy is moderately faint visually, though well seen at high magnification as a roundish patch of light, a little brighter to the middle with fairly well defined edges. Radial velocity: 8036 km/s. Distance: 359 mly. Diameter: 115 000 ly.

NGC5353

H714[2]	13:53.5	+40 17	GLX
S0	2.2′ × 1.1′	12.00B	

This is the principal galaxy of the NGC5353 group, also catalogued as Hickson 68. In addition to a number of smaller, dwarfish galaxies in the immediate vicinity, the wider field contains several prominent galaxies which probably make up a larger group

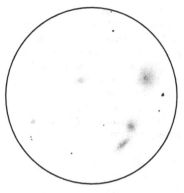

NGC5353 group
15-inch Newtonian 272x

or cloud of galaxies, as the radial velocities are all fairly similar. Photographically, NGC5353 is a very bright lenticular galaxy with a bright core and short, bright extensions. There is an extensive, fainter outer envelope which blends into the outer envelope of NGC5354, which is 1.4′ N, but there is

NGC5350/5353/5354/5355/5358

no evidence of interaction between the two galaxies. In medium-aperture telescopes, the five principal members of the group are all easily visible and can be seen together in the field of a high-magnification eyepiece. NGC5353 is the brightest; it is a large, elongated object, very bright along its major axis, well defined and oriented NW–SE. Herschel observed the four brightest members of the group, cataloguing them on 14 January 1788. Radial velocity: 2404 km/s. Distance: 107 mly. Diameter: 69 000 ly.

NGC5354

H715[2]	13:53.5	+40 18	GLX
SA0 sp	2.1′ × 1.6′	12.34B	

Images show this lenticular galaxy has a bright core and an extensive outer envelope and is decidedly elongated E–W. It is the second brightest member of the NGC5353 group of galaxies. Visually, the galaxy is a little smaller than its immediate neighbour, NGC5353, almost round and well defined; it is very much brighter to the middle, though no stellar core is visible. Radial velocity: 2659 km/s. Distance: 119 mly. Diameter: 73 000 ly.

NGC5355

H699[3]	13:53.8	+40 21	GLX
E3	1.1′ × 0.7′	14.04B	

This is an elliptical galaxy with a brighter, oval core and a faint outer envelope. The visual and photographic appearances of this NGC5353 group member are quite similar. Though faint, the galaxy is readily seen as a round, well defined spot at high magnification. Radial velocity: 2424 km/s. Distance: 108 mly. Diameter: 35 000 ly.

NGC5361

H682[3]	13:54.7	+38 27	GLX
S?	0.8′ × 0.4′	14.91B	

The classification of this galaxy is uncertain but photographs show a small, inclined galaxy, probably a spiral, with a disc that is fainter in the E, possibly due to a dust patch. Visually,

the galaxy is extremely faint and difficult, and is only intermittently visible as a very small oval patch of light, 1.0′ WSW of a field star. It is elongated ENE–WSW and exhibits even surface brightness. Radial velocity: 5724 km/s. Distance: 256 mly. Diameter: 59 000 ly.

NGC5362

H671[2]	13:54.9	+41 19	GLX
Sb? pec	2.3′ × 1.0′	13.31B	

In the DSS photograph this is a highly inclined spiral galaxy with a small, brighter, elongated core in a fainter envelope. It is a three-branch spiral with knotty condensations, including a large and fainter spiral arm to the E. A peculiar jet of material made up of two knotty condensations emerges from the galaxy in the S. This could also be a fainter companion galaxy. Visually, the galaxy is moderately faint and best at medium magnification, appearing as an elongated oval patch of light that is a little brighter to the middle and precedes a faint field star. The edges are well defined and the disc is oriented E–W. Radial velocity: 2258 km/s. Distance: 101 mly. Diameter: 67 000 ly.

NGC5371

H716[2]	13:55.7	+40 28	GLX
SAB(rs)bc	4.9′ × 3.5′	11.28B	

The radial velocity of this galaxy suggests that it may be a member of the NGC5353 group of galaxies. Photographs show a large, bright, slightly inclined spiral galaxy with a

NGC5371

©STScI/AURA/CALTECH/ROE

possible bar and a very slightly elongated brighter core. The four-branch spiral structure has narrow arms featuring many brighter knots. At the eyepiece, this is a large, bright galaxy and is very rewarding to view. The outer envelope is almost round and is diffuse with poorly defined edges. The envelope's texture is quite smooth with a brighter core surrounding a sharp, bright stellar nucleus. Several field stars immediately surround the galaxy, including a bright one 2.5′ NE and another 5.0′ NNE. Radial velocity: 2639 km/s. Distance: 118 mly. Diameter: 168 000 ly.

NGC5375

H125[3]	13:56.9	+29 10	GLX
SB(r)ab	3.7′ × 2.5′	12.82B	

This galaxy was originally designated NGC5396 by Dreyer, but there is nothing at Herschel's position and Dreyer notes the likelihood that the John Herschel discovery h1711 is, in fact, William Herschel's object. The DSS image shows an almost face-on multiple-arm barred spiral galaxy with an oval central region and a short bar attached to narrow, knotty spiral arms. There are peculiar dark gaps in the spiral structure both E and W of the core. Visually, this galaxy is moderately bright and is seen at high magnification as a hazy oval of light, oriented N–S, broadly brighter to the middle with a dim, secondary envelope which is ill defined. Radial velocity: 2435 km/s. Distance: 109 mly. Diameter: 117 000 ly.

NGC5377

H187[1]	13:56.3	+47 14	GLX
(R)SB(s)a	5.0′ × 2.7′	12.17B	

The DSS image shows this galaxy well, revealing its somewhat peculiar morphology. It is a highly inclined spiral galaxy with a large, bright core. Two very broad spiral arms emerge from the core, embedded in a faint oval envelope. Extremely faint and broad spiral arms emerge from the tips of the galaxy's major axis, forming something of a theta-shaped, ring structure seen

NGC5377

©STScI/AURA/CALTECH/ROE

galaxy is almost round, with hazy, poorly defined edges and a faint double star is 1.1′ ENE of the core. Radial velocity: 2356 km/s. Distance: 105 mly. Diameter: 73 000 ly.

NGC5394

H191[1]	13:58.6	+37 27	GLX
SB(s)b pec	1.7′ × 1.0′	13.62B	13.1V

This galaxy forms a physically interacting pair with the larger NGC5395 1.8′ to the SSE and the pair have also been catalogued as Arp 84. High-resolution photographs show a bright, two-armed spiral galaxy with a small, round core surrounded by three bright, inner spiral arcs. The outer, S-shaped spiral arms seem to emerge from the arcs; the arms are thin and taper as they spiral outward and the S arm appears to be in contact with the larger galaxy to the S. The minimum core-to-core separation between the two galaxies would be 83 000 ly at the presumed distance. Visually, only the core of this galaxy is visible; it is seen as a small, blurry patch slightly elongated NE–SW and is pretty well defined. Radial velocity: 3546 km/s. Distance: 158 mly. Diameter: 78 000 ly.

NGC5395

H190[1]	13:58.6	+37 25	GLX
SA(s)b pec	2.7′ × 1.3′	12.27B	11.8V

This inclined spiral galaxy is the dominant member of the interacting pair which includes NGC5394. Photographs show that the galaxy

obliquely. Telescopically, the galaxy is moderately large; it is bright and well condensed along its major axis, which is oriented NE–SW. The edges are fairly well defined and the central region is moderately large and bright, slightly elongated N–S. A bright star-like core is embedded in the bright central region. Radial velocity: 1892 km/s. Distance: 84 mly. Diameter: 123 000 ly.

NGC5380

H698[2]	13:56.9	+37 37	GLX
SA0⁻	1.7′ × 1.7′	13.28B	

The DSS image shows a face-on, prototypical lenticular galaxy with a bright core surrounded by a slightly fainter inner disc and a much fainter, broad outer envelope. Visually, the galaxy is moderately bright, a roundish patch of light, broadly brighter to the middle with ill defined edges. NGC5378, 11.0′ to the N can be seen in the field and is similarly bright and a

little larger with a brighter, nonstellar core. Curiously, Herschel never recorded NGC5378. Radial velocity: 3247 km/s. Distance: 145 mly. Diameter: 72 000 ly.

NGC5383

H181[1]	13:57.1	+41 51	GLX
(R′)SB(rs)b: pec	2.4′ × 2.4′	12.22B	

Well seen on the DSS image, this is a slightly inclined, bright and large, barred spiral galaxy with a bright, oval core. A prominent, straight dust lane is in the NW of the bar, while patchy dust is in the SE of the bar with knotty, star-forming regions. Two short spiral arms emerge from the bar, a narrow one from the NW of the bar, a much broader arm from the SE of the bar. A very faint, face-on barred spiral galaxy lies 3.25′ to the S. Visually, only the central part of this galaxy is visible; it is moderately bright though slightly diffuse and broadly brighter to the middle. The

NGC5383

©STScI/AURA/CALTECH/ROE

NGC5394/5395

and fainter extensions. Photographs show that a faint field star lies just outside the core to the WNW. Visually, the galaxy is quite faint, but it is well seen at medium magnification as an elongated, well defined oval oriented almost due E–W with a tiny, star-like core. Radial velocity: 3828 km/s. Distance: 171 mly. Diameter: 80 000 ly.

NGC5403

H683³	13:59.9	+38 11	GLX
SB(s)b: sp	3.2′ × 0.7′	14.38B	

The DSS photograph reveals an edge-on spiral galaxy with a curious structure. The large, brighter core is embedded in a small, bulging central region, which is diminished somewhat by a dust lane which bisects the galaxy. Curiously, the SE spiral extension shows no material below the dust lane, while the NW extension shows no material above the dust lane. A small, bright galaxy, NGC5403A, is 1.5′ to the NE. They may form a physical pair: the core-to-core separation between the two galaxies would be a minimum of 55 000 ly at the presumed distance. Visually, NGC5403 is quite faint, and is best seen at medium magnification as an oval, diffuse patch of light broadly brighter to the middle, elongated NNW–SSE and hazy at the edges. The surface brightness of the companion galaxy NGC5403A is higher and it is seen intermittently as a condensed, very small patch of light. Radial velocity: 2823 km/s. Distance: 126 mly. Diameter: 117 000 ly.

appears to be a one-armed spiral with the arm emerging from the W of the small, bright core. It is thin and brightens as it curves N and then down the eastern flank of the galaxy. It fades slightly in the S then broadens as it heads N and loops around the core again. The arm can be traced for almost two turns around the core. There are two considerable dust lanes long the western flank which appear to rise above the plane of the galaxy in the N. Telescopically, NGC5395 is by far the larger and brighter of the pair. It is a quite-high-surface brightness object with some structure detected. Elongated N–S and fairly well defined, it is brightest along its eastern flank with a slightly darker gap along the major axis in the middle. The spiral arm on the western flank appears as a thin, slight brightening in the envelope and is sharply defined. Radial velocity: 3565 km/s. Distance: 159 mly. Diameter: 125 000 ly.

NGC5399

H411³	13:59.5	+34 47	GLX
S?	1.5′ × 0.3′	14.60B	

The DSS image shows an almost edge-on galaxy, probably a spiral, with a broad, brighter core in an elongated outer envelope. A large, elongated dust patch is visible in the envelope extension to the W. The galaxy is small and fairly faint visually, and is best seen at medium magnification as an elongated patch of light oriented E–W and situated WSW from a pair of field stars. The surface brightness is even and the edges are fairly well defined. Radial velocity: 3732 km/s. Distance: 167 mly. Diameter: 73 000 ly.

NGC5401

H412³	13:59.7	+36 15	GLX
Sa	1.6′ × 0.4′	14.39B	

This is an elongated, lens-shaped galaxy with a bright, extended central region

NGC5403

©STScI/AURA/CALTECH/ROE

NGC5406

H699[2]	14:00.3	+38 55	GLX
SAB(rs)bc	2.0′ × 1.1′	13.12B	

This is a face-on, four-branch barred spiral galaxy with a bright, elongated core and a fainter bar. Photographs show the inner spiral arms form a pseudo-ring with arm fragments emerging after less than a quarter turn. Visually, the galaxy can be seen in the same medium-magnification field as NGC 5407 about 15.0′ to the NNE and is probably a neighbour galaxy. NGC5406 appears round, moderately well defined and broadly brighter to the middle. A magnitude 7 field star is 6.6′ N. Radial velocity: 5281 km/s. Distance: 236 mly. Diameter: 137 000 ly.

NGC5407

H684[3]	14:00.9	+39 09	GLX
S0	1.2′ × 0.7′	14.2B	

The appearance of this lenticular galaxy is similar visually and photographically. Located in a field of bright stars, this galaxy is quite small but fairly high in surface brightness with a star-like core embedded in a fairly well defined envelope. Very slightly elongated E–W, the galaxy is 1.8′ NNE of the centre of a bright triangle of field stars and 8.75′ NNE of a magnitude 7 field star. Radial velocity: 5483 km/s. Distance: 245 mly. Diameter: 85 000 ly.

NGC5410

H672[2]	14:00.7	+41 00	GLX
SB?	1.7′ × 0.9′	14.37B	

This highly inclined irregular galaxy, possibly a barred spiral, has a brighter, elongated core. Photographs show an outer ring of material with many knots defining a fainter inner disc. A smaller companion, elongated and with disturbed structure, is visible 1.0′ to the NNE. Telescopically, this is quite a faint galaxy which is a little more distinct at medium magnification. It is a small, slightly oval patch of light, very slightly brighter to the middle and moderately well defined. It is diffuse and slightly elongated ENE–WSW. Radial velocity: 3823 km/s. Distance: 171 mly. Diameter: 84 000 ly.

NGC5433

H653[3]	14:02.5	+32 31	GLX
Sdm:	1.6′ × 0.4′	14.05B	

Photographs show a highly inclined, almost edge-on spiral galaxy with a bright, elongated core and a suspected disturbed spiral arm to the N appearing as a slightly brighter, nebulous patch. This may be a satellite galaxy seen in silhouette against the outer spiral arm. Visually, the galaxy is quite faint and is seen as an elongated, fairly well defined patch of light oriented N–S. The surface brightness is fairly even along the major axis. A small edge-on galaxy of fairly high surface brightness, CGCG191-037, is 4.75′ SW. CGCG191-037's radial velocity is quite low, however, and it is very possibly a dwarf galaxy as its estimated diameter is only 5000 ly. Radial velocity: 4416 km/s. Distance: 197 mly. Diameter: 92 000 ly.

NGC5440

H416[2]	14:03.0	+34 46	GLX
Sa	3.1′ × 1.2′	13.18B	

This highly inclined spiral galaxy has a bright, slightly elongated core and a faint outer envelope. The DSS image hints at a possible multi-arm spiral structure. The very dim NGC5441 is 4.75′ SSE, while UGC8955 is 7.5′ NW. Telescopically, only the bright central region of NGC5440 is visible; the galaxy appears as a small oval of light, has quite high surface brightness and is well seen at high magnification. The edges are well defined and a tiny, star-like core is visible. A field star is 1.25′ WSW. Radial velocity: 3758 km/s. Distance: 168 mly. Diameter: 152 000 ly.

NGC5444

H417[2]	14:03.4	+35 08	GLX
E+:	2.4′ × 2.1′	12.78B	

This elliptical galaxy has a large, bright core and an extensive outer envelope. It forms a physical pair with NGC5445 6.5′ to the SSE. At the presumed distance the minimum core-to-core separation between the two galaxies would be 336 000 ly. Both galaxies are well seen at high magnification in moderate-aperture telescopes. NGC5444 is larger and brighter, a roundish, quite bright object which is broadly brighter to the middle with a well defined central region embedded in a blurry, very faint outer envelope. Radial velocity: 4018 km/s. Distance: 179 mly. Diameter: 125 000 ly.

NGC5445

H413[3]	14:03.5	+35 02	GLX
S0?	1.7′ × 0.5′	13.96B	

Photographs show this lenticular galaxy has a bright core and fainter extensions which suggest possible ring-like structure seen obliquely. Visually, only the bright core is seen at high magnification; the galaxy appears very small but moderately bright with a slightly elongated envelope, it is well defined and is oriented NNE–SSW. The tiny, star-like core is well seen and a faint field star is 1.0′ SSW. Radial velocity: 3971 km/s. Distance: 177 mly. Diameter: 88 000 ly.

Canis Major

The 12 objects observed here by Herschel were all recorded in the years 1784 and 1785. There is one galaxy, the rest of the objects are star clusters, asterisms or nebulae. Though some of the clusters are quite attractive, the most spectacular object is NGC2359, a Wolf–Rayet star involved in extensive nebulosity.

NGC2204

H13[7]	6:15.7	−18 39	OC
II2r	10.0′ × 10.0′	9.3B	8.6V

Photographs show a rich cluster of stars with a relatively wide range in brightness. Two chains of prominent stars, one running from ENE to WSW, with a magnitude 10 field star marking the WSW end, the other almost due N–S from a magnitude 9 field star dominate the central region, with many fainter stars sprinkled throughout. Visually, the two star chains produce a distorted 'X' pattern with most of the stars magnitude 13 or fainter. At least 25 stars are resolved; the brighter stars at the periphery are evidently not associated with the cluster. The total membership of the cluster is probably more than 350 stars, with the age estimated at 3 billion years. The group is quite remote at a distance of 14 000 ly.

NGC2283

H271[3]	6:45.9	−18 14	GLX
SB(s)cd	3.5′ × 2.6′	12.95B	

In the DSS image this is a face-on barred spiral galaxy with a weak core and weak bar. One narrow spiral arm emerges from the S before curving N. The spiral arm from the N bar splits into two heading S and is much broader. Even in moderate apertures this is a very difficult object to view. The galaxy appears to be a faint, hazy patch of light involved with the westernmost star of a flat, faint triangle of stars. It is best seen at medium magnification and appears poorly defined with irregular edges. It is

8.25′ ENE of a magnitude 8 field star. Herschel observed the galaxy on 7 February 1785, and saw it as: '3 or 4 small stars with nebulosity, very faint, verified at 240 (magnification).' Radial velocity: 662 km/s. Distance: 30 mly. Diameter: 30 000 ly.

NGC2318

H14[7]	6:59.5	−13 42	AST

Recorded on 8 February 1785 and described by Herschel as: 'A cluster of coarse scattered stars 20′ in diameter', this is a loose grouping of stars of moderate brightness range arrayed around a magnitude 8.3 star catalogued as HD52133. About 20 stars are visible, moderately well separated from the sky background, but this is probably not an actual star cluster.

NGC2327

H25[4]	7:04.3	−11 18	RN?
2.0′ × 1.75′	6.2B	6.2V	

Recorded on 31 January 1785, Herschel described this object as: 'A pretty considerable star with very faint and very small milky chevelure, irregular form.' Photographs show a small, bright, irregularly round nebula with ragged extensions to the E, N and SE. Faint stars are involved, in particular a chain of four very faint ones on the western flank. The central double star (ADS5761) is burned out on the DSS image and the nebula is silhouetted against a larger dark nebula running roughly N–S. It is probably a part of the much larger emission nebula complex

IC2177, which is immediately E. Visually, NGC2327 is a bright, irregular glow around a double star in a region of extremely faint, formless nebulosity in a N–S band to the E of the bright nebula.

NGC2352

H15[7]	7:13.6	−24 06	OC/AST?
7.0′ × 4.0′			

The DSS image shows an elongated grouping of stars, a little brighter than the surrounding star field, though it does not stand out well. Telescopically, at the indicated position is a somewhat compressed group of stars, fairly well separated from the sky background and featuring a N–S elongated group with stars of moderate brightness range. A chain of four stars oriented NNW–SSE marks the northern extension with a faint triangle near the centre. Altogether about 20 stars seem to be involved. Herschel recorded the object on 6 March 1785 and called it: 'A small cluster of pretty compressed stars, not very rich.'

NGC2327

©STScI/AURA/CALTECH/ROE

NGC2354

H16[7]	7:14.3	−25 44	OC
III2r	18.0′ × 18.0′	7.3B	6.5V

This cluster is difficult to see on the DSS image, owing to the richness of the Milky Way field and the faintness of the cluster members. Visually, however, the cluster is fairly easy to see, though the cluster members are faint and narrow in brightness range. There are two distinct groups of stars: the first is central and appears as a broad, elongated stream oriented N–S, while the second group of stars appears to the W. The cluster is well resolved with at least 30 stars visible. The total membership is close to 300 stars, while the spectral type is A0 and the age about 180 million years. Distance: 5850 ly.

NGC2358

H45[8]	7:16.8	−17 03	MWP

This object is not apparent on the DSS image and, visually, this grouping is not initially obvious, but sweeping shows a detached field of stars of moderate brightness range at the position indicated. The stars are quite scattered, however, and there is little to indicate that it is a coherent group. At least 30 stars are in the field, which is probably just a more condensed Milky Way patch. The field matches Herschel's 31 December 1785 description of: 'A coarse scattered cluster of stars, not rich.'

NGC2359

H21[5]	7:18.6	−13 12	EN?
25.0′ × 17.0′			

In recent years this nebula has become known as Thor's Helmet. The DSS image shows it as a bright nebula with complex structure and extensive, fainter tendrils. The central region is dominated by a bubble of nebulosity which is sharply defined to the W and more diffuse in the E. To the S is the brightest portion, a short spur running WSW which becomes faint when it straightens out towards the W. The fainter portions of the nebula are three discrete spurs, one running NW, the second to the E

and the last emerging below the eastern spur before curving S. Much fainter, detached nebulosity is visible to the NE of the main nebula. In medium- and large-aperture telescopes, the brightest portions of the nebula are well seen so that the visual size is at least 10.0′ × 10.0′. It is a large, bright and extremely complex nebula, irregularly bright throughout. To the S there is a prominent bar oriented ENE–WSW with a field star embedded to the ENE. A broad, comma-shaped nebulosity emerges from the bar to the N, then curves to the ENE. From the N a much fainter, hazy band curves towards the NW; this patch is very ill defined. A very faint, hazy, detached patch is ESE of the main nebula. Herschel recorded the object on 31 January 1785 and described it as: 'A broad extended nebulosity. Forms a parallelogram with a ray southwards; the parallelogram 8′ long, 6′ broad, very faint.' The illuminating star is the Wolf–Rayet star HD56925 of spectral type WN4. The

nebulosity is very probably composed of debris shells blown off by the star.

NGC2360

H12[7]	7:17.8	−15 37	OC
I3r	14.0′ × 14.0′	7.6B	7.2V

This object is well seen in photographs as a compressed grouping of the stars in a rich Milky Way field. In a small

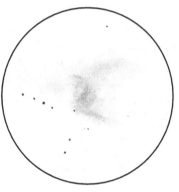

NGC2359 15-inch Newtonian 94x

NGC2359

NGC2360

©STScI/AURA/CALTECH/ROE

of 25 million years. Radial velocity in recession: 33 km/s. Distance: 5100 ly.

NGC2367

H27[8]	7:20.1	−21 56	OC
II3m	5.0′ × 3.0′	8.0B	7.9V

Photographs show a small, coarse cluster of bright stars emerging from a bright star in the S and forming two parallel rows of stars oriented N–S. This impression is reinforced visually, the cluster being pretty distinct and moderately bright with a moderate range in brightness of the member stars. The cluster is fairly compressed and made up of two principal chains of stars running roughly parallel and oriented N–S that come together to a brightish star in the S. About 15 stars are resolved, although the total membership of the cluster is about 30 stars. The spectral type is B3 with the brightest star magnitude 9.39. Distance: 9320 ly.

NGC2374

H35[8]	7:24.0	−13 16	OC
IV2p	12.0′ × 12.0′	8.5B	8.0V

Photographs show a grouping of stars that are brighter than the surrounding stars in a rich field. Herschel's description of this cluster, discovered on 31 January 1785, reads: 'A cluster of pretty large scattered stars, pretty rich, about 20′ long, crooked figure.' In small telescopes there is a distinctive N–S chain of stars with a fainter spur of stars beginning in the S and angling to the NW. The cluster is faint, however, and only about two dozen members are resolved. More than 70 stars are actual members and the age of the cluster is 320 million years. Distance: 3950 ly.

telescope, this is an absolutely spectacular cluster of faint star dust. It is a well-resolved but very compressed grouping, elongated E–W with a conspicuous spur angling off to the SE. The stars are tightly packed and a strong background haze of unresolved members is evident. About 50 stars are resolved. This is one of Caroline Herschel's discoveries, located on 26 February 1783. Herschel himself included it in his catalogue on 4 February 1785, saying: 'A beautiful cluster of pretty compressed stars near one half degree diameter. C. H. [Caroline Herschel].' Membership of the cluster is at least 90 stars, the spectral type is B8 and the age of the cluster is 1.3 billion years. Distance: 3725 ly.

NGC2362

H17[7]	7:18.8	−24 57	OC
I3r	6.0′ × 6.0′	4.1V	

Photographs show a rich and compressed cluster of stars, which are very similar in brightness, arrayed around the bright star Tau Canis Majoris (magnitude 4.39), which is considered a member of the cluster. This is a bright cluster for all telescopes, with about 30 stars of about magnitude 8–10 and about 20 fainter ones visible. Most of the brighter members are E, S and W of Tau Canis Majoris and they are tightly grouped around the bright star. About 60 stars are considered members of the cluster, the spectral type is O8 and the cluster is extremely young with an age

Canis Minor

This small constellation is a barren ground for Herschel objects. The two galaxies here are quite faint and the two asterism are nondescript. Herschel recorded all four in 1784 and 1785.

NGC2394

H44[8]	7:28.5	+7 05	AST?
9.0?V			

This is something of a controversial object and is listed as nonexistent in the *Revised New General Catalogue*. Herschel discovered this object on 28 December 1785, describing it as: 'A cluster of very coarse, scattered, large stars form a cross. Not rich'. It is listed in Archinal and Hynes' catalogue *Star Clusters* (2003), with a position and the notation 'cl??'. Luginbuhl and Skiff provide a detailed description in their *Observing Handbook and Catalogue of Deep-Sky Objects* (1989). The present author observed this object on 12 March 1994, describing it as:

'Labeled nonexistent in the Herschel list, the eyepiece reveals a loose asterism north and east from Eta Canis Minoris. About fifteen magnitude 9 and fainter stars are visible in a 10.0′ area, well separated from other field stars. No hint of an unresolved, nebulous background.' There is fair agreement between the author's field sketch and the DSS image.

NGC2402

H19[3]	7:30.8	+9 39	GLX
E	0.8′ × 0.8′	15.15B	

The DSS photograph reveals a small, faint elliptical galaxy with a brighter core and faint outer envelope. A faint spiral companion galaxy is 0.5′ NE with two faint field stars between the two galaxy images. The radial velocities are similar and the two form a probable physical pair with a minimum core-to-core separation of about 34 000 ly at the presumed distance. Visually, the galaxy is a hazy, ill defined nebulous patch of light, poorly concentrated and hazy. Better seen at medium magnification, it is oval and roughly E–W in orientation. Radial velocity: 5186 km/s. Distance: 231 mly. Diameter: 54 000 ly.

NGC2459

H479[3]	7:52.0	+9 33	AST
0.8′ × 0.8′			

This asterism is well seen at high magnification and appears as a small, hazy patch of light with three stars resolved. Herschel recorded it on 26 December 1785 with the description: 'Suspected. Extremely faint, very small, a little extended.' J. L. E. Dreyer in *The Scientific Papers* further noted it as: 'Only a small cluster of very faint stars.'

NGC2508

H7[3]	8:02.0	+8 34	GLX
E?	1.8′ × 1.4′	14.13B	

Similar in appearance visually and photographically, this galaxy is moderately bright and is seen as an irregularly round and fairly well defined object which is even in surface brightness except for a small, brighter core. It precedes two field stars which point directly at the galaxy. Radial velocity: 4253 km/s. Distance: 190 mly. Diameter: 99 000 ly.

NGC2394

©STScI/AURA/CALTECH/ROE

Capricornus

This relatively large constellation, which is most prominent late in the summer and into the early part of autumn, follows the rich starfields of Sagittarius and Scutum. Despite its size, however, this is a very barren area of the sky and Herschel managed to locate only one galaxy,

NGC6q07. There are others in the region but the constellation's low position on the southern horizon from Herschel's English location, combined with the dimming effects of a thickening atmosphere must have contributed to his lack of success in the area.

NGC6907

H141[3]	20:25.1	−24 49	GLX
SB(s)bc	3.6′ × 2.6′	11.92B	11.3V

The DSS image shows an inclined barred spiral galaxy with a brighter, elongated core and an extended, oval-shaped bar. Two S-shaped spiral arms start out very strong and bright before fading, with knotty condensations defining the arms. Telescopically, the galaxy is moderately bright; it is a mottled oval elongated E–W with somewhat ill defined edges and a prominent stellar core. The spiral arms are too faint to detect visually. Radial velocity: 3258 km/s. Distance: 145 mly. Diameter: 152 000 ly.

NGC6907

Cassiopeia

Embedded in the starfields of the autumn Milky Way, Herschel listed 19 deep-sky objects within the confines of Cassiopeia. Most are open star clusters; the exceptions are the gaseous nebulae NGC896 and NGC7635 and the galaxies NGC185, NGC278 and NGC1343. Interestingly, three of Caroline Herschel's early discoveries are here: NGC225, NGC659 and the beautiful star cluster NGC7789.

NGC129

H79[8]	0:29.9	+60 14	OC
III2m	12.0′ × 12.0′	7.3B	6.5V

Photographs show a fairly bright, though scattered and ill defined cluster. At its core is a bright, 4.0′ diameter triangle of magnitude 9 stars with most cluster members magnitude 10 or fainter. Visually, the cluster is set in an attractive field of stars making it difficult to separate from the background. The brightness range is moderate and the group is well resolved with the central triangle of brighter stars dominating. At least 60 stars are seen, many forming triangles and short chains. As many as 200 stars may be members of the group, the spectral type is B5 and the age 150 million years. Radial velocity in approach: 14 km/s. Distance: 5200 ly.

NGC129 15-inch Newtonian 85x

NGC136

H35[6]	0:31.5	+61 32	OC
II1p	1.5′ × 1.5′		

Photographs show a very small and compressed grouping of stars with a narrow range of brightness, well detached from the sky background. At medium magnification this small, compressed cluster is a round, nebulous patch in a star-rich field. High magnification helps to resolve about a dozen cluster members, particularly on the western flank, but the hazy background persists. Around 20 stars are accepted as members of the cluster, the brightest being magnitude 13.0. Herschel recorded the object on 26 November 1788 and described it as: 'A small cluster of very faint and extremely compressed stars about 1′ in diameter. The next step to an easily resolved nebula.' Distance: 13 300 ly.

NGC185

H707[2]	0:39.0	+48 20	GLX
dSph/dE3	11.7′ × 10.0′	10.14B	9.1V

This dwarf elliptical galaxy holds an important place in history as one of the galaxies that Walter Baade resolved into stars, helping him to establish his 'Population' concept of the organization of stars in galaxies. The galaxy has been well studied with RR Lyrae variable stars, planetary nebulae and globular clusters all identified in the system. Photographically, the galaxy features a small, slightly brighter core in a faint secondary envelope and a very faint outer envelope; it has a very grainy texture with hints of resolution into individual stars. A small dust patch is visible immediately W of the core. Visually, it is a fairly large, diffuse glow in a rich star field. It appears slightly extended NNE–SSW and is very gradually brighter to the middle with a very smooth brightness rise. Herschel recorded the galaxy on 30 November 1787, but was unable to locate the very similar NGC147 to the W. His description reads: 'Pretty bright, very large, irregularly round. Very gradually much brighter to the middle. Resolvable. 5′ or 6′ in diameter.' Radial velocity in approach: 15 km/s. Distance: 2.15 mly. Diameter: 9800 ly.

NGC225

H78[8]	0:43.4	+61 47	OC
III1p n	15.0′ × 15.0′	7.4B	7.0V

This is one of Caroline Herschel's discoveries: she first observed it on

NGC185

27 September 1783 and recorded it a second time on 23 February 1784. The brightest stars in the cluster form a very coarse grouping about magnitude 9–10 with a small range in brightness. It stands out well at low magnification but is less impressive as magnification increases. To the E, a bright jagged N–S line of five bright and three faint stars is isolated from the main group, which is rather scattered. At least 30 stars are visible in all; and at least 75 stars are thought to be members of the group. William Herschel did not himself catalogue this cluster until 26 November 1788, more than five years after his sister first observed it. His description reads: 'A cluster of very coarsely scattered large stars take up 15 or 20′. Caroline Herschel discovered it 1784.' Spectral type B8. Age: 140 million years. Distance: 2000 ly.

NGC278

H159[1]	0:52.1	+47 33	GLX
SAB(rs)b	2.4′ × 2.4′	11.50B	

This face-on spiral galaxy is very poorly seen on the DSS image as the bright core and bright internal spiral structure are overexposed. High-resolution photographs show a compact, knotty inner spiral structure with two large HII regions which measure about 5.0″ in diameter (about 880 ly at the presumed distance). Visually, this is a bright, fairly well defined galaxy with a grainy, high-surface-brightness disc. There is a sizable, round, brighter and nonstellar core, which is well defined and clearly brighter than the disc, while the surrounding field is rich in stars. Radial velocity: 808 km/s. Distance: 36 mly. Diameter: 25 000 ly.

NGC381

H64[8]	1:08.3	+61 35	OC
III1m	7.0′ × 7.0′	9.3V	

This is a small and very compressed cluster, visible as a circular haze at low magnification. At medium magnification it is well resolved; about

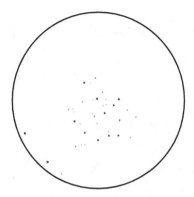

NGC381 15-inch Newtonian 177x

35 stars of magnitude 10 and fainter are visible. Total group membership is about 50 stars and the spectral type is A2.

NGC436

H45[7]	1:15.6	+58 49	OC
I2m	5.0′ × 5.0′	8.0B	8.8V

This is a small, rather tight cluster of stars, visible at low magnification as a partially resolved haze. At high magnification it is a well-resolved group of predominantly magnitude 13 stars, though a few are as bright as magnitude 10. The cluster is elongated ESE–WNW, with about 50 stars recognized as members. Spectral type: B5. Age: 79 million years. Distance: 6900 ly.

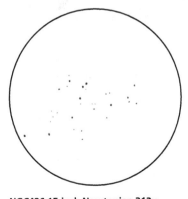

NGC436 15-inch Newtonian 313x

NGC457

H42[7]	1:19.1	+58 20	OC
II3r	20.0′ × 20.0′	7.0B	6.4V

This is a large, rich cluster of stars with a large range of brightness: the main body is elongated SE–NW and the stars Phi-1 and Phi-2 Cassiopeiae are in the SE. Two spurs of stars off the main body, one oriented to the NE, the other to the SW, give the cluster its characteristic shape and name, the Nightowl. The cluster is bright and the brightest members are well resolved even in small-aperture telescopes but there are many fainter members, well seen in photographs, which escape detection at the eyepiece. Actual membership of the cluster probably exceeds 200 stars. Spectral type: B2. Age: approximately 25 million years. Distance: 10 000 ly.

NGC559

H48[7]	1:29.5	+63 18	OC
I1m	7.0′ × 7.0′	9.9B	9.5V

Visible as a faint haze at low power with a handful of members resolved, the cluster is well seen at medium magnification. The three brightest members are near the centre, the rest are magnitude 13–14 or fainter. About 20 stars are resolved and the cluster is elongated NE–SW. About 120 stars are recognized as members. Spectral type: B7. Age: about 1.3 billion years. Distance: 3700 ly.

NGC637

H49[7]	1:42.9	+64 00	OC
I2m	3.0′ × 3.0′	8.6B	8.2V

Very similar both visually and photographically, this is a compressed, though bright cluster with about 20 stars visible at high magnification. The range of brightness is moderate and the brighter stars are aligned NE–SW with the cluster well separated from the sky background. The bright double star at the centre is ADS1342. More than 50 stars are probably members of the cluster. Spectral type: B0. Radial velocity in approach: 14 km/s. Distance: 7800 ly.

NGC457

NGC654

H46[7]	1:44.1	+61 53	OC
II2r	6.0′ × 6.0′	7.3B	6.5V

Photographs show a rich, compressed cluster of stars of moderate brightness range, roughly triangular in shape and with ill defined and very faint detached nebulosity to the SSW. In small telescopes, this is an interesting cluster, consisting of an irregularly round, nebulous haze with about a dozen faint stars shining through. A conspicuous chain of magnitude 7–11 stars running roughly E–W borders the cluster to the S and E. Moderate-aperture telescopes fully resolve the cluster at medium–high magnification. The brightness range of the member stars is moderate. The cluster is somewhat triangular in form and is well separated from the sky background. Total membership is probably more than 80 stars. It is a possible member of the Cassiopeia OB8 association. Spectral type: B0. Age: 15 million years. Radial velocity in approach: 31 km/s. Distance: 6500 ly.

NGC659

H65[8]	1:44.2	+60 42	OC
I2m	6.0′ × 6.0′	8.4B	7.9V

Photographs show a cluster of bright and faint stars which are not well separated from the sky background with

NGC637

most of the bright stars arrayed to the NE. The discovery of this cluster is credited to Caroline Herschel, who observed it on 27 September 1783, though William Herschel did not record it in his catalogue until 3 November 1787. Most of the member stars of this cluster are quite faint and in a small aperture they form a very tenuous nebulous haze with about half a dozen

faint stars arrayed in a rough circle shining through. Medium-aperture telescopes at high magnification pick up some of the fainter members but the nebulous background haze remains. The total membership of the cluster may be close to 200 stars, the spectral type is B5 and the age about 20 million years. Distance: 8200 ly.

NGC663

H31[6]	1:46.0	+61 15	OC
II3r	15.0′ × 15.0′	7.8B	7.1V

Small-aperture telescopes show a rich and fairly compressed cluster of stars, with a wide brightness range, although the majority are fainter than about magnitude 10. Many of these fainter members are arranged in a narrow column oriented N–S. The membership of the cluster totals at least 100 stars. Spectral type: B1. Age: about 22 million years. Radial velocity in approach: 31 km/s. Distance: about 7200 ly.

NGC896

H695[3]	2:24.8	+61 54	EN
eF-1-R	20.0′ × 20.0′		

Herschel classified this object as a very faint nebula; it is the brightest portion of a nebula complex which also involves IC1795 to the ENE. On the DSS image, NGC896 is a large, wedge-shaped nebula of intermingled light and dark patches; the brightest portion is a 3.0′ × 3.0′ irregular patch in the W central portion of the nebula. Although it is not bright, telescopically this nebula is readily seen at low and medium magnification as an indistinct haze which permeates a moderately rich field of faint stars. It is brightest in a small patch surrounding a faint field star in the W. From here it flows back faintly towards the E, then curves N with at least one dark channel suspected. The extensive nebula complex IC1805 is immediately ESE.

NGC1027

H66[8]	2:42.7	+61 33	OC
II3m n	15.0′ × 15.0′	7.3B	6.7V

In photographs, this is a fairly rich cluster of stars of moderate brightness range surrounding a bright central star (HD16626, magnitude 7.0), which is not a cluster member. Visually, this is a rich cluster, the brightness range is moderate and the members are arrayed in subtle groups of three or four stars each. The cluster is moderately well separated from the background with 60 well-resolved members and no background glow of fainter stars. The total membership of the cluster may be as high as 200 stars with the brightest at magnitude 9.0. Spectral type: B3. Age: 350 million years. Distance: 3950 ly.

NGC1343

H694[3]	3:37.8	+72 34	GLX
SAB(s)b: pec	2.6′ × 1.6′	13.66B	

The DSS photograph shows a spiral galaxy of very unusual structure. A small, bright core is surrounded by an equally bright, round inner ring with a

NGC654

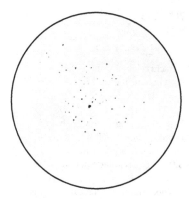

NGC1027 15-inch Newtonian 85x

NGC7635

prominent dark patch immediately E of the core. The outer spiral structure is faint; two dim spiral arms are seen in a very faint envelope with a very faint, dwarfish galaxy seemingly involved with one of the spiral arms NE of the core. Visually, only the bright core of this peculiar galaxy is visible. The core is round, moderately well defined and brightens gradually to the middle. A field star is 0.9′ NNW of the core. Radial velocity: 2371 km/s. Distance: 106 mly. Diameter: 80 000 ly.

NGC7635

H52[4]		23:20.7	+61 12	EN
15.0′ × 8.0′				

The DSS image shows a spectacular nebula with bright and faint components. It is centred on a bubble of gas slightly elongated E–W; the bubble is brightest to the N with a bright star involved (HD220057, magnitude 6.93, spectral type B2IV) immediately E of the nebula's brightest portion. There is a bright patch of nebulosity immediately N of the bubble, extending towards the NW. Fainter clouds of gas are visible S of the bubble, where there is a brighter arc of gas, which eventually crosses a bright field star. Beyond this the field is bathed in very faint nebulosity and the size quoted above refers only to the nebula's brightest part. In a moderate-aperture telescope the brightest portions of the nebula are clearly visible, making it appear elongated N–S with a 'wasp-waisted'

form, where the bright northern component contacts the brightest portion of the 'bubble' portion of the nebula. Herschel recorded this nebula on 3 November 1787 and called it: 'A star of 9th magnitude with very faint nebulosity of small extent about it.'

NGC7789

H30[6]	23:57.0	+56 44	OC
II2r	25.0′ × 25.0′	7.7B	6.7V

This spectacular cluster is one of Caroline Herschel's discoveries. She came across the cluster on 30 October 1783; William Herschel did not record it in his catalogue until 18 October 1787, describing it as: 'A beautiful cluster of very compressed small stars. Very rich.' Even in a small aperture this is a compelling object. Low magnification reveals a faint haze which is tolerably round and well separated from the sky background. The brightness range of

the cluster members is very narrow and there is a very slight compression towards the centre. Literally hundreds of resolved members are visible and very definite dark zones or channels wind their way through the cluster. One in particular crosses the S portion of the cluster before curving to the E. These channels are well seen in photographs and appear to be actual structure of the cluster members rather than obscuring dust patches or lanes. More than 580 stars are members of this cluster. Spectral type: B9. Age: about 1.6 billion years. Distance: 6000 ly.

NGC7790

H56[7]	23:58.4	+61 13	OC
II2m	5.0′ × 5.0′	9.2B	8.5V

In photographs, this cluster is a grouping of slightly brighter stars in a star-rich field. In small telescopes, it is a small grouping of brighter stars with a

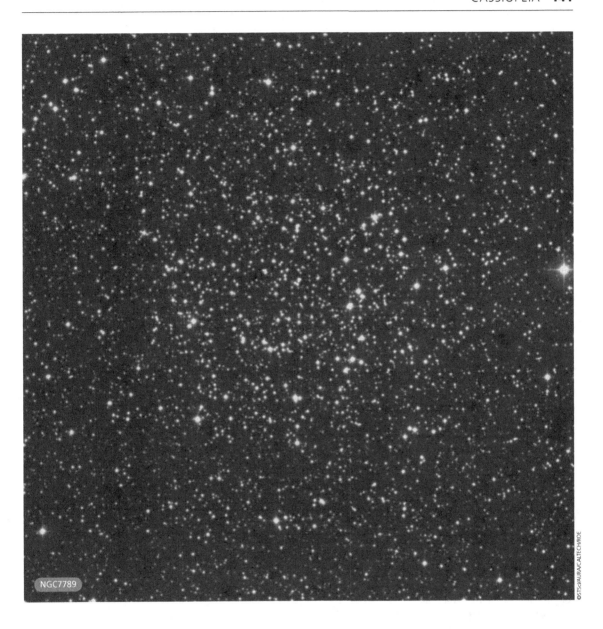

NGC7789

narrow range of brightness and is oriented roughly E–W. It is moderately well separated from the background field of stars; resolution is somewhat difficult and a nebulous background haze is apparent at low magnification. Moderate apertures at high magnification resolve the cluster well. More than 130 stars are suspected of being members, including the variables CE and CF Cassiopeaie. Age: about 78 million years. Distance: 9700 ly.

Centaurus

This sprawling constellation is an area rich in deep-sky objects, but it was too far South to observe from Herschel's location, though he did manage to pull the bright galaxy NGC5253 out of the horizon murk in the spring of 1787.

NGC5253

H638[2]	13:39.9	−31 39	GLX
Im pec	5.0′ × 1.9′	10.80B	10.4V

This galaxy is part of the nearby Centaurus group of galaxies, the principal member of the group being the spectacular galaxy NGC5128, one of the brightest in the sky, though it was too far S to be seen by Herschel. NGC5253 is one of the most southerly entries in his catalogue and the only one in the constellation Centaurus. It appears quite similar to a lenticular galaxy both visually and in photographs and two bright supernovae have been recorded here. The first in 1895 was seen before supernovae were understood as a separate class of object and was classified as a variable star (Z Centauri). The second object was discovered in 1972 and probably reached a maximum blue magnitude of 8.5. On the DSS image, several small, nebulous patches can be seen in the outer halo, particularly SW of the core; they are probably globular clusters. Visually, this is a bright and beautiful galaxy featuring a large, bright, elongated core in an opaque and well defined envelope. The envelope is somewhat grainy and fades slightly to the edges where there is a sharp drop in brightness. The galaxy is elongated NE–SW with several prominent field stars in the immediate vicinity. Radial velocity: 274 km/s. Distance: 12 mly. Diameter: 18 000 ly.

NGC5253

©STScI/AURA/CALTECH/ROE

Cepheus

All 19 of the objects recorded here by Herschel were observed between 1787 and 1798. Most of the objects recorded are either nebulae or star clusters, though the spectacular spiral galaxy NGC6946 is well worth examining on nights of exceptional transparency. Many of the clusters and nebulae are rewarding objects to observe and one of them, NGC7380, was discovered by Caroline Herschel.

NGC40

H58[4]	0:13.0	+72 32	PN
3b+3	1.2′ × 0.8′	11.27B	10.6V

This nebula is overexposed on the DSS image, though its irregular form is well seen. It appears as a very bright and opaque shell of gas, very slightly oval and elongated NNE–SSW. The shell is slightly indented to the N and the S with a very faint elongated patch of nebulosity N of the main nebula. This planetary nebula is a superb and fairly bright object visually, surrounding the bright central star HD826, which is magnitude 11.6 and spectral type WC8. The nebula is round, a little diffuse and certainly not opaque, giving the impression there is a delicate, almost filamentary structure around the periphery of the nebula, making the edges rather well defined. Herschel observed this object on 25 November 1788 and described it as: 'A star of the 9th magnitude surrounded with very faint milky nebulosity. The star is either double or not round. Less than 1′ in diameter.' Distance: 2900 ly.

NGC1184

H704[2]	3:16.6	+80 48	GLX
S0/a	3.2′ × 0.6′	13.06B	

Photographs show an edge-on lenticular galaxy with a large, elongated core and bright extensions which fade gradually into a fainter outer envelope. The appearance is quite similar visually: it is a fairly bright and elongated galaxy which appears as a streak of light oriented NNW–SSE. It is well defined along its edges with a small, brighter core and is well seen at high magnification. Radial velocity: 2510 km/s. Distance: 112 mly. Diameter: 105 000 ly.

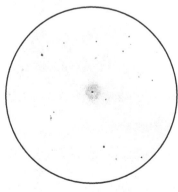

NGC40 15-inch Newtonian 293x

NGC1184

NGC6939

H42[6]	20:31.4	+60 38	OC
II1r	10.0′ × 10.0′	8.8B	7.8V

This was the last entry in Herschel's Class VI: very compressed and rich clusters of stars. Recorded on 9 September 1798, Herschel described the cluster as: 'A beautiful compressed cluster of small stars. Extremely rich, of an irregular figure. The preceding part of it is round and branching out on the following side, both towards the north and towards the south. 8′ or 9′ in diameter.' Small telescopes show this cluster well as a very rich object with an unresolved background. The brightness range of the member stars is narrow, almost all are magnitude 12 or fainter. Averted vision emphasizes the unresolved mass and suppresses the resolved members. The cluster is slightly elongated SW–NE. Resolution is almost complete in moderate-aperture instruments at high magnification. The cluster is well separated from the background field of stars. The total membership of the cluster is about 300 stars. Spectral type: B8. Age: 1.8 billion years. Distance: 3900 ly.

NGC6946

H76[4]	20:34.8	+60 09	GLX
SAB(rs)cd	11.5′ × 10.0′	9.75B	

The DSS image shows a large and bright four-branch spiral galaxy with a small, brighter core obscured somewhat by a dust patch. Four spiral arms are distinctly seen; they are marked by many condensations indicating star clusters or star-forming regions. The arms are very dusty as well. Photographically, the brightest stars are resolved at blue magnitude 21 but owing to the galaxy's low galactic latitude, considerable light absorption by dust in the plane of our own Milky Way may be dimming the galaxy. The galaxy has also been catalogued as Arp 29. In a medium-aperture telescope, this galaxy takes on a ghostly appearance, quite large and very gradually brighter to the middle. At least two of the spiral arms can be detected at medium

magnification under a dark sky. The arm emerging from the core and curving NE is most distinctly seen, while the combined light of two spiral arms curving towards the W forms a stubby extension which is intermittently visible. Large-aperture telescopes show the spiral pattern well and it is a breathtaking sight located in a rich star field. Herschel recorded the galaxy on 9 September 1798, placing it in his Class IV. His description reads: 'Considerably faint, very large, irregular form, a sort of bright nucleus in the middle. The nebulosity 6′ or 7′. The nucleus seems to consist of stars, the nebulosity is of the milky kind. It is a pretty object.' There have been a large number of supernovae detected in the galaxy over the years, including events in 1917, 1939, 1948, 1968, 1969, 1980, 2002, 2004 and 2008. For supernova hunters, the galaxy bears watching. This is one of the nearest galaxies to the Local Group. Radial velocity:

275 km/s. Distance: 12.25 mly. Diameter: 40 500 ly.

NGC7023

H74[4]		21:00.5	+68 10	RN
18.0′ × 18.0′	7.2B			

The DSS image shows a distinct nebula with a bright central region and a mix of bright dust and darker cavities. The bright centre is irregularly round with a slightly fainter extension to the S. The surrounding region is slightly fainter and squarish, with darker pockets especially towards the WSW. Beyond this, the field is bathed in fainter, ill defined nebulosity. The nebula is opaque and few stars shine through except in the darker regions. The star cluster Collinder 427 is W of the nebula but is not seen visually. Telescopically, the nebula is best seen at low magnification as a very large, though very pale nebula associated

NGC6946

©STScI/AURA/CALTECH/ROE

NGC7023

with the magnitude 6.8 star HD200775. The nebula is brightest immediately surrounding the star and covers a considerable area. There is little detail visible and the edges are ill defined and not regular. Dark gaps or channels are suspected to the N. Herschel recorded this object on 18 October 1794, calling it: 'A star, seventh magnitude very much affected with nebulosity, which more than fills the field. It seems to extend to at least a degree all around; smaller stars, such as ninth or tenth magnitude of which there are many, are perfectly free from this appearance. A star 7.8 magnitude is perfectly free from this appearance.' The spectral type of the illuminating star is B5e.

NGC7076

H936[3]	21:26.3	+62 53	PN
3b	0.95′ × 0.95′	14.50B	

Curiously, this planetary nebula is listed in *Sky Catalogue 2000.0*

(Hirshfeld and Sinnott, 1985) and *Planetary Nebulae* (Hynes, 1991) only as PK101+8.1 (Hynes also has Abell 75). It is also listed in *Sky Catalogue 2000.0* under its NGC designation, but in the 'Bright Nebulae' section with the comment: 'Planetary?' In the DSS image, this faint planetary nebula is annular in form and almost round. A narrow, brighter inner ring begins in the NW and continues counterclockwise to the S. A fainter shell of nebulosity is outside this inner ring and completes a circle. In addition to the central star, two very faint field stars are projected against the nebula, one to the NNW, the other to the ENE. Visually, this is a faint nebula, best seen at medium magnification as a poorly defined, small, round glow. Three magnitude 14–15 stars lie immediately N. At high magnification, the nebula occasionally appears stellar. Herschel's 15 October 1794 description reads: 'Very faint. Easily resolvable.' The

central star is magnitude 17.4 and not seen visually. Distance: about 6200 ly.

NGC7129

H75[4]	21:41.3	+66 06	OC+EN/RN
IV2p n	8.0′ × 8.0′	11.50B	

The DSS image shows a bright, irregular nebula with a much brighter central region, almost round. Three bright stars are involved to the N and a fainter outer shell, annular in shape, surrounds the core, offset and elongated towards the SW. Additional very faint nebulous patches surround stars to the N and NE. Visually, medium magnification reveals a coarse grouping of bright stars, dominated by four stars of about magnitude 8. The associated nebulosity is quite bright and surrounds the three SW stars. The nebula is considerably brighter to the middle and roughly triangular in shape. Herschel recorded this object on 18 October 1794; his description reads: 'Three stars about ninth magnitude involved in nebulosity. The whole takes up a space of about 1.5′ in diameter. Other stars of the same size are free from nebulosity.' Ten stars are considered members of the star cluster.

NGC7139

H696[3]	21:45.9	+63 39	PN
3b	1.3′ × 1.3′	13.0B	

Photographs show a bright, round, fairly well defined shell of gas with

NGC7076

slightly darker cavities both within the shell surrounding the central star and also towards the NE and SW. About 15 extremely faint field stars shine through the shell or border it closely. In a moderate-aperture telescope this object appears almost round, displaying low contrast with the sky background. The edges appear fuzzy and poorly defined and the surface brightness is relatively flat. The central star is magnitude 18.1 and not visible. Radial velocity in approach: 54 km/s. Distance: about 3900 ly.

NGC7142

H66[7]	21:45.9	+65 48	OC
I2r	12.0′ × 12.0′	10.4B	9.3V

In photographs this is quite a rich cluster of faint stars. The surrounding field is fairly rich but narrow, dark starless zones on all sides seem to border the cluster, separating it from the background. In moderate-aperture telescopes this is quite a rich cluster of faint stars with a very narrow range of brightness except for three slightly brighter stars on the E side, which may not be cluster members. At least 100 stars are resolved in a medium-magnification field and the cluster is fairly well separated from the sky background. This was one of Herschel's last Class VII discoveries, recorded on 18 October 1794. The total population of the cluster is about 185 stars. Spectral type: F3. Age: about 4 billion years. Distance: 9600 ly.

NGC7160

H67[8]	21:53.7	+62 36	OC
I3p	5.0′ × 5.0′	6.3B	6.1V

Photographs show a bright, coarse grouping of stars in a rich stellar field. There is a wide range in the brightness of the components and the cluster is elongated E–W. Visually, it is an interesting cluster even in a small-aperture telescope. About 20 stars are visible with the two brightest stars near the E edge. A broad crescent-shaped

asterism of magnitude 12–13 stars is on the W side and the overall form of the cluster is strongly elongated. Cluster membership is about 60 stars. Spectral type: B1. Radial velocity in approach: 25 km/s. Age: about 10 million years. Distance: 2600 ly.

NGC7235

H63[8]	22:14.5	+57 16	OC
II3m	6.0′ × 6.0′	8.6B	7.7V

Herschel recorded this object on 16 October 1787 and described it as: 'A small cluster of pretty large stars.' It was later entered into the *New General Catalogue* as NGC7234. But the position given by Herschel leads to a very small and nondescript grouping of seven faint stars of small brightness range in a rich field. And none of these stars can be called 'large' (bright). The credit for discovery of NGC7235 goes to John Herschel, who examined his father's position for NGC7234 and found nothing. A likely positional error seems to indicate that William Herschel saw and described NGC7235 but recorded it at an erroneous position, so it is very likely that NGC7235 = H63[8]. The DSS image shows a compressed grouping of fairly bright stars of moderate brightness range and visually this is a very attractive cluster, well separated from the rich field of background stars. It is a bright and coarse cluster, oval in shape and elongated E–W with the brighter stars towards the E; the rest of the resolved members are magnitude 10–12. About 100 stars are accepted as members, the brightest being magnitude 8.8. Spectral type: B0. Age: 2 million years. Distance: 10 300 ly.

NGC7354

H705[2]	22:40.4	+61 17	PN
4+3b	0.4′ × 0.4′	12.90B	

Photographically, this is a bright, opaque planetary nebula with a bright, inner, diamond-shaped shell of gas surrounded by a slightly fainter round

shell with darker regions near the extremities. Small, faint spurs emerge from the outer shell in the N and the S. In small apertures the planetary nebula is a dim, roundish smudge, with fairly even surface brightness and well defined edges. Moderate apertures show a bright, mottled disc which is broadly brighter to the middle. Herschel discovered the nebula on 3 November 1787 but placed it in his Class II: faint nebulae, saying: 'Pretty bright, small, irregularly round, easily resolvable almost equally bright.' The central star's spectrum is continuous but at visual magnitude 16.1 the central star is far beyond detection in most amateur telescopes. The expansion velocity of the gaseous shell is 26 km/s. Radial velocity in approach: 42 km/s. Distance: about 5200 ly.

NGC7380

H77[8]	22:47.0	+58 06	OC
III2m n	20.0′ × 20.0′	7.6B	7.2V

This is one of Caroline Herschel's discoveries, made on 7 August 1787, which Herschel catalogued on 1 November 1788. Photographs show a rich star field and the cluster does not stand out well. There is also considerable nebulosity involved, which Herschel did not see but which has also been designated as NGC7380. In a small telescope, only the star cluster is seen; this comprises about 30 fairly faint stars of moderate brightness range, organized into a roughly triangular form which is fairly compressed. Altogether, about 125 stars are involved and the group is a possible member of the Cepheus OB1 association. It is a young cluster, only 3.8 million years of age. Spectral type: O6. Radial velocity in approach: 38 km/s. Distance: 9700 ly.

NGC7419

H43[7]	22:54.3	+60 50	OC
I2m	6.0′ × 6.0′	13.0V	

The DSS photograph shows a rich cluster of faint stars with a moderate range in brightness and quite

NGC7419

pretty large, irregular figure, easily resolvable.'

NGC7510

H44[7]	23:11.5	+60 34	OC
II3r n	7.0′ × 7.0′	8.77B	7.9V

Photographs show a rich and compressed cluster of fairly bright stars with a moderate range of brightness. It is impressive even in small apertures: the brightest members form an elongated group of at least three star chains, oriented ENE–WSW. The cluster is embedded in a rich background of stars, but is not difficult to separate. Herschel's description of this cluster reads: 'A cluster of pretty compressed, pretty large stars, considerably rich. The stars arranged chiefly in lines from south preceding to north following.' About 75 stars are considered members of the cluster. The cluster is quite young, about 10 million years of age. Spectral type: O9. Radial velocity in approach: 66 km/s. Distance: 10 100 ly.

NGC7423

compressed. The brighter stars are located towards the SW and are organized in an elongated grouping oriented SE–NW with a brighter field star about 1.5′ from the edge of the cluster. The background field of stars is not rich, so the cluster easily stands out. Telescopically, this cluster is small and well separated from the sky background. It is nebulous and poorly resolved at low and medium magnification, and about a dozen faint and perhaps three brighter stars are seen at high magnification, mostly clumped together in groups. The faint stars appear superimposed over an unresolved, very faint background haze and a bright field star is just beyond the cluster boundary to the NW. In total, 40 stars are accepted as members but the DSS image shows many faint stars involved and the actual total is probably higher than this. Distance: 6300 ly.

NGC7423

H745[3]	22:55.3	+57 08	OC
II2m	5.0′ × 5.0′		

This is a fairly rich cluster of stars on the DSS image; it is slightly compressed towards the middle, with a moderate range of brightness. The stars seem organized into seven chains which all converge towards the centre like spokes of a wheel. A curious small nebulous patch is located 4.3′ to the NE. In a moderate-aperture telescope this is a very faint cluster, best seen at medium magnification as a hazy patch of light immediately E of a triangle of field stars. The haze is elongated SE–NW and is gradually brighter to the middle where a couple of very faint stars are resolved. The brightest members of this cluster are magnitude 15 and 40 stars are accepted as members. This is one of the few star clusters recorded in Herschel's Class III. Discovered on 1 November 1788, Herschel described it as: 'Very faint,

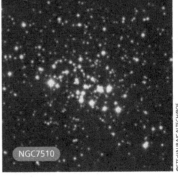

NGC7510

NGC7538

H706[2]	23:13.5	+61 31	EN
10.7′ × 7.3′			

This is a fairly bright nebula in a moderately rich star field. Images reveal a brighter central portion with several darker channels and two bright stars embedded near the centre. Fainter nebulosity extends broadly towards the NE and the boundaries of the bright portion of the nebula are fairly well defined. Telescopically, the nebula is readily visible at medium magnification as a glow engulfing two moderately bright field stars. The nebula takes magnification well and appears oval in shape with ill defined edges; it is elongated NE–SW.

NGC7708

H62[8]	23:36.5	+72 51	AST?
30.0′ × 30.0′	10.0V		

The DSS image shows a large and scattered cluster of stars of wide brightness range, including HD221774, which at magnitude 7.3 is the brightest star to the NW. Telescopically, it is a fairly coarse and scattered grouping of stars. It is quite well separated from the sky background and involves three magnitude 8–9 stars which define the edges to the N and S. In all about two dozen stars are involved with a moderate brightness range. The group is probably not a cluster. Recorded on 19 September 1787, Herschel called the group: 'A cluster of coarse scattered large stars, not rich but the stars are brilliant, one 7th magnitude.'

NGC7762

H55[7]	23:49.8	+68 02	OC
II2m	15.0′ × 15.0′	10.0V	

This cluster is quite similar in appearance visually and photographically. It is a fairly large cluster, well separated from the sky background, but the member stars are quite faint. The cluster has about 40 members, all of similar brightness, with the brightest being about magnitude 11. The cluster is somewhat scattered and is elongated SE–NW. Distance: 3000 ly.

Cetus

The majority of the 105 Herschel objects in this sprawling constellation were recorded between the years 1783 and 1787. Five were recorded in 1790 (NGC132, NGC173, NGC192, NGC196 and NGC201) and two (NGC163 and NGC270) in 1798. All the objects in the constellation are galaxies except NGC246, a planetary nebula, and the asterism NGC7826. Although many of the galaxies are faint and/or remote, there is much of interest here, including several outstanding objects such as NGC246, NGC247 and NGC908. Amongst the faint objects, some are members of small galaxy groups and are therefore interesting to examine in moderate–large-aperture telescopes.

NGC132

H855[2]	0:30.2	+2 06	GLX
SAB(s)bc	2.1′ × 1.9′	13.74B	

Photographs show this two-armed barred spiral galaxy has a faint star-like core. Clumpy structure is seen in the inner spiral arms, which are bright for less than half a rotation from the bar before broadening and fading. Visually, this moderately faint galaxy is well seen at high magnification as an oval glow slightly elongated N–S and located WSW from a faint field star. The surface brightness is fairly even and the edges of the galaxy are poorly defined. Herschel thought the galaxy was resolvable.

Radial velocity: 5464 km/s. Distance: 244 mly. Diameter: 149 000 ly.

NGC151

H478[2]	0:34.0	−9 42	GLX	
SB(r)bc	3.7′ × 1.7′	12.27B	11.6V	

This is a two-armed barred spiral galaxy, slightly tilted to our line of sight. Photographs show a very bright core and bar with weaker, loosely wound spiral arms. The pattern is somewhat asymmetrical with each arm having a second, fragmentary arm emerging. There is a bright condensation at the end of the eastern spiral arm leading to a bright field star. Telescopically, this is a fairly bright and moderately large galaxy; it is a fat oval of light elongated E–W with a very diffuse and faint secondary envelope. The inner envelope is quite grainy in texture, surrounding a small, bright core. A prominent field star is beyond the outer envelope to the ENE. Radial velocity: 3807 km/s. Distance: 170 mly. Diameter: 183 000 ly.

NGC154

H467[3]	0:34.1	−12 39	GLX
1.0′ × 0.9′	15.02B		

Unclassified, but very likely an E1 or E2 elliptical galaxy, visually this galaxy is fairly faint and small, very slightly oval and extended E–W. It is fairly sharply defined at the edges and a little brighter to the middle. Radial velocity: 8080 km/s. Distance: 361 mly. Diameter: 105 000 ly.

NGC157

H3[2]	0:34.8	−8 24	GLX
SAB(rs)bc	4.2′ × 2.5′	11.07B	

The DSS image shows a bright two-armed spiral galaxy of the grand-design type with massive, dusty spiral arms of high surface brightness. The spiral arms fragment as they unwind outward. Visually, the galaxy is a large and bright, broad, oval patch of light which is moderately well defined at the edges and slightly brighter to the centre. Elongated NE–SW, the envelope appears grainy and a faint field star borders the galaxy to the NE. Radial velocity: 1716 km/s. Distance: 77 mly. Diameter: 94 000 ly.

NGC157

NGC163

H954[3]	0:36.0	−10 07	GLX
E0	1.5′ × 1.2′	13.6B	13.68V

The visual and photographic appearances of this elliptical galaxy are similar. Visually, it is a small, condensed object which is fairly well defined and very slightly brighter to the middle. Good conditions bring out NGC165, a slightly fainter barred spiral about 7.0′ to the ENE. Radial velocity: 6040 km/s. Distance: 270 mly. Diameter: 118 000 ly.

NGC173

H871[3]	0:37.2	+1 56	GLX
SA(rs)c	3.2′ × 2.6′	13.41B	

Photographs show this spiral galaxy has evidently low mass and faint spiral arms as well as a small bright core. In moderate-aperture telescopes, this is a moderately faint, very ill defined galaxy, located between two field stars of about magnitude 12, oriented NE–SW in the field. It is best seen at medium magnification as a very slightly oval patch of light of even surface brightness. The fainter galaxy NGC170 is visible in the field, 7.25′ to the WSW. Radial velocity: 4460 km/s. Distance: 199 mly. Diameter: 185 000 ly.

NGC175

H223[3]	0:37.4	−19 56	GLX
SB(r)ab	2.1′ × 1.9′	12.98B	

This bright barred spiral galaxy has two spiral arms emerging from the bar. Photographs show the arms are tightly wound and form a pseudo-ring after half a revolution around the core. The arms then fade as they broaden. Visually, the galaxy is a fairly bright, oval object elongated ESE–WNW. It is broadly brighter along its major axis with diffuse edges and the core cannot be differentiated from the bar and is seen as a broadening in the middle. This galaxy and NGC171 are the same object. Radial velocity: 3935 km/s. Distance: 176 mly. Diameter: 107 000 ly.

NGC191

H479[2]	0:39.0	−9 00	GLX
SAB(rs)c: pec	1.4′ × 1.2′	14.19B	

This barred spiral has a bright, compact lenticular companion galaxy, IC1563, situated 1.0′ to the SSE and together they have also been catalogued as Arp 127. Photographs show that the principal galaxy is irregularly round and has asymmetrical spiral arms, suggesting interaction with the smaller galaxy, which also appears slightly disturbed. The elongated core is perpendicular to the short, faint bar. Telescopically, NGC191 is a faint and quite diffuse galaxy, best seen at medium magnification as a roundish, unconcentrated spot with a low surface brightness. The edges are ill defined and uncertain; a bright triangle of stars lies to the NNW. Herschel's 28 November 1785 discovery description suggests that he actually saw the unresolved image of both galaxies, as his description states: 'Pretty bright, much extended in the meridian, 2′ long.' Radial velocity: 6135 km/s. Distance: 274 mly. Diameter: 112 000 ly.

©STScI/AURA/CALTECH/ROE

NGC191

NGC192

H872[3]	0:39.3	+0 51	GLX
(R')SB(r)a:	1.7' × 0.8'	13.43B	

Together with NGC196 and NGC201, which are Herschel objects, and NGC197, which is not, this is a member of a compact group of four galaxies catalogued as Hickson 7; it is the brightest of the four. On the DSS image the galaxy is a highly inclined spiral galaxy with a bright core and what appears to be a pseudo-ring made up of two, tightly wound spiral arms. Visually, high magnification shows NGC192 as bright and well defined, much extended NNW–SSE and a little brighter to the middle. The entire group follows a magnitude 8 field star located about 8.0' to the NW. Core-to-core distances to the other member galaxies are: NGC197, 2.1' NE; NGC196, 3.0' NNE; NGC201, 5.3' E. At the apparent distance, the core-to-core separations would be: NGC197, 115 000 ly; NGC196, 164 000 ly; NGC201, 290 000 ly. Radial velocity: 4223 km/s. Distance: 189 mly. Diameter: 93 000 ly.

NGC196

H860[2]	0:39.3	+0 54	GLX
SB0 pec:	0.5' × 0.3'	14.44B	

The visual and photographic appearances of this lenticular galaxy are quite similar. At high magnification, this galaxy is just a little fainter than NGC192; it is well condensed and slightly elongated N–S. The companion galaxy NGC197 is just 1.25' SSE; at the presumed distance the core-to-core separation would be about 68 000 ly. Radial velocity: 4345 km/s. Distance: 194 mly. Diameter: 28 000 ly.

NGC201

H873[3]	0:39.7	+0 51	SAB(r)c
1.8' × 1.6'	13.66B		

The DSS image shows this face-on, three-armed barred spiral well. The bar is very faint and is attached to a small bright core. Visually, the galaxy appears moderately bright but very diffuse; it is round with even surface brightness across the disc. Part of the Hickson 7 galaxy group, it is the largest of the four galaxies in the field. The fourth member of the group, NGC197, is extremely faint but can be seen in moderate-aperture telescopes under good conditions. The group appears brightest at medium magnification. Radial velocity: 4505 km/s. Distance: 201 mly. Diameter: 105 000 ly.

NGC210

H452[2]	0:40.6	−13 52	GLX
SAB(s)b	5.0' × 3.3'	11.72B	

On the DSS image this is a two-armed spiral galaxy, possibly barred, with thin, narrow spiral arms studded with brighter patches. After about half a revolution, the arms broaden but still form a tightly wound, well defined pattern. The interior appears to be a very bright, large, massive and elongated dusty core, but high-resolution photographs reveal this core to have a separate, tightly wound spiral structure which is overexposed on the DSS image. Two compact galaxies are situated within 7.0', one to the NE, the other to the SE. In a moderate-aperture telescope, NGC210 is quite a bright galaxy located ENE from two field stars. The envelope is elongated N–S and is fairly well defined, grainy in texture with a small, bright core. The outer spiral structure is not

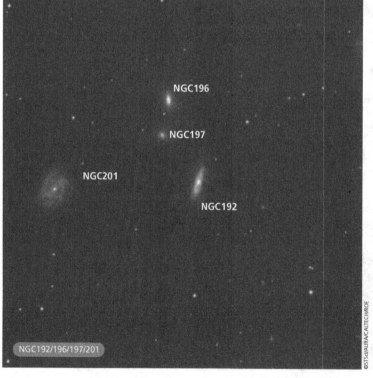

NGC196

NGC197

NGC201

NGC192

NGC192/196/197/201

NGC210

©STScI/AURA/CALTECH/ROE

magnification as a moderately bright galaxy with a roundish, opaque envelope. It is fairly well defined with a prominent, star-like core. Radial velocity: 5378 km/s. Distance: 240 mly. Diameter: 112 000 ly.

NGC244

H485[3]	0:45.8	−15 36	GLX
S0 pec?	1.3′ × 1.0′	13.79B	

Photographs reveal a peculiar galaxy with asymmetrical outer structure, including one large bright condensation in a single, curving, possibly spiral arm in the E. The galaxy has a large, bright core. This may be a case of two separate galaxies interacting. Telescopically, this galaxy is quite faint and is visible as a round, nebulous patch which is gradually brighter to the middle and located NNW from a moderately bright field star. The edges are slightly hazy but fairly well defined. Radial velocity: 976 km/s. Distance: 44 mly. Diameter: 17 000 ly.

NGC245

H445[2]	0:46.1	−1 44	GLX
SA(rs)b pec?	1.4′ × 1.2′	12.91B	

The DSS image shows a spiral galaxy with clumpy, asymmetrical arms and a small, bright and round core. Three spiral branches emerge from the core; they are initially faint and brighten as they wind away from the central region. In a moderate-aperture telescope the galaxy is moderately bright but diffuse with a very mottled, round envelope. It is moderately well defined with a brighter core offset towards the S. A wide pair of faint field stars is located 1.75′ to the SSW. Radial velocity: 4157 km/s. Distance: 185 mly. Diameter: 76 000 ly.

NGC246

H25[5]	0:47.0	−11 53	PN
3b	2.75′ × 2.75′	8.00B	11.8V

Discovered on 27 November 1785, Herschel placed this planetary nebula in his Class V: very large nebulae. He described it as: 'Four or five pretty large

detected with any certainty and a large aperture is required to see it. Radial velocity: 1679 km/s. Distance: 75 mly. Diameter: 109 000 ly.

NGC216

H244[3]	0:41.4	−21 03	GLX
S0°? sp	1.8′ × 0.7′	13.71B	13.1V

Photographs show a lenticular galaxy with a small, bright core and asymmetrical extensions, the one to the SW being brighter than that to the NE. Visually, this is a moderately faint galaxy, but is fairly opaque along its major axis and evenly bright. Well defined at the edges, it is oriented NNE–SSW. Radial velocity: 1563 km/s. Distance: 70 mly. Diameter: 37 000 ly.

NGC217

H480[2]	0:41.5	−10 02	GLX
S0/a: sp	2.8′ × 0.6′	13.35B	12.1V

This is a highly inclined galaxy. Photographs show that it is probably a

spiral with a large, bright central region which may be a bar seen obliquely. A probable spiral pattern is most prominent to the ESE, defined by thin dust lanes. The surface brightness is high and a small and very faint galaxy, perpendicular to the larger galaxy's spiral arm, is situated on the WNW tip. Telescopically, the galaxy is small but quite bright and attractive. It is an elongated, well defined object, with a brighter core which bulges slightly and an outer envelope which tapers to points. The galaxy is oriented ESE–WNW. Radial velocity: 4031 km/s. Distance: 180 mly. Diameter: 147 000 ly.

NGC227

H444[2]	0:42.6	−1 32	GLX
S0⁻ pec:	1.6′ × 1.3′	13.12B	

This somewhat remote galaxy has also been classified as a possible E5 system and is well seen at high

NGC246

stars forming a trapezium of about 5′ diameter. The inclosed space is filled up with faintly terminated milky nebulosity. The stars seem to have no connexion with the nebulosity.' Seven magnitude 10 or fainter field stars mark the location of this large and somewhat faint planetary nebula. Four stars are seen to be involved, including the magnitude 11.9 central star, which is easily visible. Averted vision brings out much detail, including a dark zone immediately NE of the central star and a large dark zone to the E. The edges are brighter than the central region, but overall the nebula is quite diffuse. Spectral type of the central star: OVI+K. Distance: about 1300 ly.

NGC247

H20[5]	0:47.1	−20 46	GLX
SAB(s)d	21.4′ × 6.9′	9.67B	9.1V

Photographically, this highly tilted spiral galaxy with low-mass spiral arms

NGC247

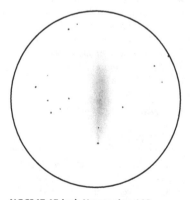

NGC247 15-inch Newtonian 146x

appears to be an almost edge-on twin of M33, the Triangulum galaxy. It has many small bright condensations in the spiral arms as well as a very large dust cloud in the N. Deep photographs show many resolved stars, star clusters and HII regions. The core is very small and faint. Telescopically, it is a very large, fairly bright, though diffuse, galaxy. It is much elongated N–S; a magnitude 10 field star indicates its southern extremity and a magnitude 12 field star is involved 3.0′ to the N. The galaxy is brighter to the middle but the core is much offset to the N. The southern extension tapers to a sharp point at the magnitude 10 field star, but the northern edge is much blunter and more diffuse. The dust cloud and northern loop visible in the photograph are not visible at the eyepiece. Discovered on 30 October 1784, Herschel described the galaxy as: 'A streak of light, nearly in the direction of the meridian, 26' long, 3' or 4' broad, pretty bright.' This galaxy is probably a member of the South Polar Group, which includes the nearby galaxies NGC55, NGC300, NGC7793 and the Herschel object NGC253. Radial velocity: 172 km/s. Distance: 7.7 mly. Diameter: 46 000 ly.

NGC255

H472²	0:47.8	−11 28	GLX
SAB(rs)bc	3.0' × 2.5'	12.40B	

Photographically, this two-armed barred spiral has a bright bar and core with bright inner spiral structure, studded with HII regions and extremely faint outer spiral structure. Visually, this is a moderately large and fairly bright galaxy, though quite diffuse. It is a roundish glow which is just gradually brighter to the middle with no core visible. Though bright, it is best at moderate magnification, becoming diffuse at high magnification. Radial velocity: 1632 km/s. Distance: 73 mly. Diameter: 63 000 ly.

NGC259

H621²/H703²	0:48.1	−2 47	GLX
Sbc: sp	2.9' × 0.6'	13.70B	

Herschel inadvertently recorded this galaxy twice, observing it on 13 December 1786 and 11 September 1787. Photographs show a highly inclined spiral galaxy with a bright inner region and faint outer structure. Telescopically, the galaxy is situated within a triangle of moderately bright field stars. It is fairly bright and elongated SE–NW; it is quite opaque along its major axis with even surface brightness and no evidence of a brighter core. The edges are sharply defined

with tapering tips. Radial velocity: 4119 km/s. Distance: 184 mly. Diameter: 155 000 ly.

NGC268

H463³	0:50.2	−5 12	GLX
SB(s)bc:	1.5' × 1.0'	13.54B	

High-resolution photographs show a slightly tilted spiral galaxy with two principal spiral arms which fragment into a multi-arm pattern. The surface brightness of the arms is quite high and there are several bright condensations in the pattern. Telescopically, the galaxy is faint and diffuse and is best seen at medium magnification as a slightly oval object, elongated E–W. The diffuse envelope is weakly concentrated to a brighter middle and the edges are poorly defined. The galaxy precedes two bright field stars: the one to the N is a little fainter than the one to the S. Radial velocity: 5559 km/s. Distance: 248 mly. Diameter: 108 000 ly.

NGC270

H955³	0:50.7	−8 39	GLX
S0⁺	2.6' × 2.5'	13.81B	

Despite the classification, the DSS image shows a probable spiral galaxy with a bright, large, elongated central region and very faint, low-mass, asymmetrical spiral arms. Only one arm is clearly visible and is brightest to the SW. Visually, only the bright centre is visible; the galaxy is moderately bright and is well seen at high magnification as a slightly oval, condensed and well defined patch of light, fairly even in surface brightness and elongated NNE–SSW. Radial velocity: 3827 km/s. Distance: 171 mly. Diameter: 129 000 ly.

NGC271

H446²	0:50.8	−1 53	GLX
(R')SB(rs)ab	2.6' × 2.1'	13.18B	

Photographs reveal a barred spiral galaxy with a small, brighter core and a bar which characteristically flares in brightness where the spiral arms begin. The inner arms form a pseudo-ring,

which expands and broadens to two, faint outer spiral arms. Visually, the galaxy is located 1.6' WNW of a bright field star which hinders observation. Only the core and bar are visible; the galaxy is moderately faint, elongated N–S and brighter to the middle and its edges are fairly well defined. Radial velocity: 4204 km/s. Distance: 188 mly. Diameter: 142 000 ly.

NGC273

H430³	0:50.8	−6 53	GLX
S0	2.0' × 0.6'	13.86B	

This lenticular galaxy is situated about 10.0' NNW of the interesting pair NGC274–NGC275 and in telescopes can be seen in the same medium-magnification field. Photographs show it as a high-surface-brightness lens-shaped galaxy with faint extensions. Visually, it is well seen at high magnification as a small, elongated patch of light with a slightly brighter core. The opaque envelope has well defined edges and is elongated ESE–WNW. Radial velocity: 4800 km/s. Distance: 214 mly. Diameter: 125 000 ly.

NGC274

H429³	0:51.0	−7 03	GLX
SAB(r)0⁻ pec	1.5' × 1.5'	13.40B	

This looks like a bright S0 galaxy seen face on. The DSS image shows it forms a

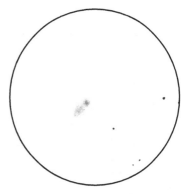

NGC274 NGC275
15-inch Newtonian 272x

NGC274

©STScI/AURA/CALTECH/ROE

structure. An extremely faint outer ring increases the apparent size of the galaxy to $3.0' \times 2.4'$. Telescopically, the galaxy is moderately bright with a hazy, though fairly well defined envelope and a brighter, round core. The galaxy appears very slightly elongated SSE–NNW. Radial velocity: 3951 km/s. Distance: 176 mly. Diameter: 154 000 ly.

NGC337

H433[2]	0:59.8	−7 35	GLX
SB(s)d	$2.9' \times 1.8'$	12.01B	

This odd galaxy displays three short but massive spiral arms on the DSS image but no evidence of a brighter core or classical central region. High-resolution photographs reveal three or four clumpy condensations near the centre, delineated by dust patches, while knotty condensations in the outer envelope are probably star-forming regions. Telescopically, this is a large and bright galaxy, well seen at high magnification. It is an elongated, though not quite oval galaxy, and there seems to be a slight indent in the envelope to the SW. The envelope is quite opaque, bright and very gradually brighter towards the middle. The edges are well defined and the galaxy is elongated ESE–WNW. Herschel discovered the galaxy on 10 September 1785, and he described it as: 'Pretty bright, pretty large, brighter to the middle. An irregular parallelogram in the meridian.' Radial velocity: 1700 km/s. Distance: 76 mly. Diameter: 64 000 ly.

NGC352

H191[3]	1:02.1	−4 15	GLX
(R')SB(rs)b?	$2.3' \times 0.8'$	13.54B	

In photographs this is a highly inclined spiral galaxy with a bright central region, which may include a bar, and faint, tightly wound spiral arms. Although somewhat faint, this galaxy is fairly well seen at high magnification as a well defined, even-surface-brightness streak of light, elongated SSE–NNW. A tiny, slightly brighter core is embedded

close pair with NGC275 and the two galaxies have also been catalogued as Arp 140. It is unclear if the two galaxies are interacting: NGC274 shows absolutely no disturbance in its structure, but NGC275 appears to be a bright, possibly barred spiral galaxy and high-resolution images show many bright knots along both inner spiral arms and in the fainter envelope as well, suggesting intense star formation. The radial velocities of the two galaxies are very similar but we may be seeing one galaxy projected in front of the other. The core-to-core separation is only 0.75', suggesting a minimum separation of about 18 000 ly at the presumed distance. At high magnification both galaxies are clearly seen as a pair in contact. NGC274 is the brighter; it is a round, concentrated object that quickly brightens to the middle with well defined edges. NGC275 is oval and more diffuse, larger and elongated SE–NW; intermittently a

slightly brighter, star-like core is visible embedded in the bright, well defined envelope which is somewhat grainy in texture. Herschel is credited with the discovery of NGC274, but it seems likely that he saw both galaxies as an unresolved object. Recorded on 10 September 1785, his description reads: 'Very faint, pretty small, extended.' Credit for the discovery of NGC275 goes to John Herschel, who recorded the galaxy in 1828. Radial velocity: 1809 km/s. Distance: 81 mly. Diameter: 35 000 ly.

NGC279

H439[3]	0:52.3	−2 12	GLX
(R')SAB(r)0+ pec:	$1.5' \times 1.2'$	13.95B	

It is difficult to classify this galaxy from the DSS photograph but elements of a lenticular ring galaxy are faintly seen. The small, round core is offset to the W in the bright inner region, which is surrounded by a fainter, partial ring in the envelope. Radial velocity: 5345 km/s. Distance: 239 mly. Diameter: 160 000 ly.

NGC357

H434[2]	1:03.4	−6 20	GLX
SB(r)0/a:	2.2′ × 1.7′	13.16B	

Photographs show a barred spiral galaxy with a large, round, bright core and a fainter very short bar which flares in brightness where the smooth-textured spiral arms emerge. On the DSS image, several very faint galaxies are in the field, mostly W and SW of the bright galaxy. Visually, the galaxy appears small but it is fairly bright and well seen at high magnification as a bright, round core surrounded by a small, hazy and irregularly illuminated envelope. The envelope is poorly defined and a faint field star borders the galaxy to the ENE. Radial velocity: 2460 km/s. Distance: 110 mly. Diameter: 70 000 ly.

NGC426

H592[3]	1:12.9	−0 17	GLX
E[+]	1.1′ × 0.8′	13.84B	

The DSS image shows this E3 elliptical galaxy forming a triangle with two other galaxies: NGC429 3.8′ to the SE and NGC430 3.4′ to the NE. All three have similar radial velocities and probably form a physical system. The minimum separation between NGC426 and NGC 429 would be about 260 000 ly at the presumed distance while that between NGC426 and NGC430 would be not less than 233 000 ly. Visually, the trio is quite faint but it is well seen in a high-magnification field. NGC426 is a small, slightly oval patch of light that is a little brighter to the middle. Radial velocity: 5266 km/s. Distance: 235 mly. Diameter: 75 000 ly.

NGC428

H622[2]	1:12.9	+0 59	GLX
SAB(s)m	4.1′ × 3.1′	11.95B	

Photographs show a very slightly inclined, three-branch barred spiral galaxy with a slightly brighter, knotty core. The bar is very weak and there is a heavy concentration of bright knots at the root of the spiral arm, which begins in the NW. The spiral structure is somewhat chaotic and the arms to the S display several condensations. Visually, the galaxy is moderately bright and large, oval in shape with hazy, indefinite boundaries. Broadly brighter to the middle, the envelope is somewhat grainy and the galaxy is elongated ESE–WNW. It is situated within a faint triangle of field stars: the star to the S is actually a close pair. Radial velocity: 1223 km/s. Distance: 54 mly. Diameter: 65 000 ly.

NGC429

H593[3]	1:13.0	−0 20	GLX
S0°: sp	1.4′ × 0.3′	14.58B	

This highly inclined lenticular galaxy has a bright core and faint extensions photographically and forms a group with NGC426 and NGC430. Visually, it is the faintest of the three galaxies and is seen as a thin sliver of light oriented NNE–SSW. Radial velocity: 5692 km/s. Distance: 254 mly. Diameter: 104 000 ly.

NGC430

H447[2]	1:13.1	−0 15	GLX
E:	1.3′ × 1.0′	13.52B	

The classification of this galaxy is uncertain but if it is indeed an elliptical it is probably an E4. Forming a physical system with NGC426 and NGC429, it is the brightest of the three, a small, slightly oval patch of light that is a little brighter to the middle and slightly elongated SSE–NNW. Radial velocity: 5366 km/s. Distance: 240 mly. Diameter: 91 000 ly.

NGC450

H440[3]	1:15.5	−0 52	GLX
SAB(s)cd:	3.0′ × 2.2′	12.74B	

The DSS image shows a multi-arm barred spiral galaxy with a small, weak bar and no evident core. The spiral arms are fairly faint but there are several bright HII regions in the spiral structure on the E side. A small, inclined spiral galaxy, UGC807, is to the ENE and appears embedded in the outer spiral structure, though it is presumed to be a background object. Its spiral structure is slightly warped on the NE side, however, and it is curious that all the bright HII regions of NGC450 are on the side nearest the small galaxy. Visually, NGC450 is a large, though fairly faint galaxy, best seen at medium magnification. It is a fat oval of light which is broadly brighter towards the middle with hazy, indefinite edges. A triangle of field stars to the S points towards the galaxy, which is elongated E–W. UGC807 may be visible in large-aperture telescopes. Radial velocity: 1824 km/s. Distance: 82 mly. Diameter: 71 000 ly.

NGC493

H594[3]	1:22.2	+0 57	GLX
SAB(s)cd:	3.4′ × 1.0′	12.96B	

This is a highly inclined multi-arm spiral galaxy with a slightly brighter central region. Photographs show clumpy spiral arms involved with prominent dust patches and lanes. In a moderate-aperture telescope, the galaxy is somewhat faint, moderately large and extended, and is oriented NE–SW. It is a little brighter towards the centre along the major axis with grainy texture across the envelope and with the edges fading slowly into the sky background. Radial velocity: 2403 km/s. Distance: 107 mly. Diameter: 106 000 ly.

NGC521

H461[2]	1:24.6	+1 44	GLX
SB(r)bc	3.2′ × 2.9′	12.62B	

Photographs show a barred spiral galaxy viewed face-on, with a round core and a thin, short bar. The spiral pattern is extremely complex: the arms are thin and somewhat fragmentary, with short fragments connecting major spiral arcs. A bright, small, compact elliptical galaxy is 2.9′ WNW. Visually, both this galaxy and NGC533 can be seen in the same medium-magnification field. NGC521 appears small, faint, fairly well condensed and almost round with a faint stellar core. It is embedded in a small envelope with ill defined edges. Radial velocity: 5083 km/s. Distance: 227 mly. Diameter: 212 000 ly.

NGC533

H462[2]	1:25.5	+1 46	GLX
cD;E3:	3.8′ × 2.3′	12.49B	14.1V

Images show an elliptical galaxy with an extensive, faint outer envelope that is elongated NE–SW. In a moderate-aperture telescope, this is a fairly bright elliptical galaxy which appears almost round and is diffuse with a brighter, stellar core. Some graininess is visible in the fairly bright envelope. The DSS image shows many very small, elliptical and lenticular galaxies surrounding NGC533. Radial velocity: 5614 km/s. Distance: 251 mly. Diameter: 277 000 ly.

NGC545

H448[2]	1:26.0	−1 20	GLX
SA0⁻	2.4′ × 1.6′	13.34B	13.3V

The DSS image shows a lenticular galaxy with an extensive outer envelope, which is oriented ENE–WSW. At least 18 galaxies are visible in a 15.0′ field. NGC545 and NGC547, which is 0.5′ to the SE, are two of the brightest galaxies in the Abell 194 galaxy cluster, although the principal galaxy, NGC541 (not a Herschel object), lies 4.4′ to the SW. The cluster is classified as type 'L' in the Rood–Sastry classification system, owing to the dominance of a chain of the principal members in the distribution of galaxies in the cluster. NGC545 and NGC547 are also catalogued as Arp 308, as their outer envelopes overlap, although there is no indication of disruption in either galaxy's structure. The minimum core-to-core separation would be about 35 000 ly at the presumed distance. Visually, it is difficult to separate the two galaxies into separate objects, but averted vision helps. The core is bright with a hazy envelope extending to NGC547. Herschel discovered the galaxies on 1 October 1785, and listed them with a common

NGC535/541/543/545/547

description: 'Two. Both stellar, within 1′ distance. Nebulosities run together.' Radial velocity: 5394 km/s. Distance: 241 mly. Diameter: 168 000 ly.

NGC547

H449[2]	1:26.0	−1 21	GLX
E1	1.3′ × 1.3′	13.13B	13.3V

The DSS image shows an elliptical galaxy with an outer envelope which blends into the outskirts of NGC545. Visually, averted vision brings out a bright round core in a diffuse outer envelope, located immediately SE of NGC545 and appearing within that galaxy's outer halo. In addition to NGC541 to the SW, very large amateur instruments can be used to try to locate Minkowski's Object, a disrupted galaxy fragment 0.8′ NE of NGC541. Radial velocity: 5523 km/s. Distance: 247 mly. Diameter: 93 000 ly.

NGC550

H463[2]	1:26.7	+2 01	GLX
SB(s)a?	1.5′ × 0.6′	13.64B	

Photographs show a highly inclined spiral galaxy with faint outer arms and a thin dust lane bordering the core along the NNE flank. Visually, this galaxy is a hazy, extended oval of light; it is fairly well condensed, is extended WNW–ESE and is fairly even in surface brightness across the envelope. Radial velocity: 5894 km/s. Distance: 263 mly. Diameter: 115 000 ly.

NGC560

H441[3]	1:27.4	−1 55	GLX
S0°	2.0′ × 0.3′	13.89B	

This lenticular galaxy has a bright, slightly elongated core, embedded in a slightly fainter inner envelope, and faint extensions. The DSS image shows some slight brightenings at the edge of the inner halo, which may indicate that we are seeing a barred lenticular galaxy edge on. NGC558 is 4.25′ to the SW, while NGC564 is 6.1′ ENE, and all are outlying members of the Abell 194 galaxy cluster. Visually, all three galaxies can be seen in a high-power field of a moderate-aperture telescope. Only the bright core and central region of NGC560 are detected; the galaxy is seen as a well defined, oval, slightly elongated patch of light, oriented N–S. NGC558 is the smallest and faintest of the three galaxies; it follows a faint field star and is elongated ESE–WNW. Radial velocity: 5563 km/s. Distance: 248 mly. Diameter: 144 000 ly.

NGC564

H442³	1:27.8	−1 53	GLX
E	1.4′ × 1.2′	13.52B	

This elliptical galaxy should probably be classified E3. Photographs show a bright core and an extensive outer envelope. It is part of a curving E–W chain of four galaxies, the easternmost galaxy being IC120, which is situated 6.8′ to the ESE, while to the W are NGC560 (6.1′ WSW) and NGC558. Visually, NGC564 is the brightest of the four galaxies. It appears as a condensed, opaque patch of light, oval in shape with a tiny, bright core. Radial velocity: 5889 km/s. Distance: 263 mly. Diameter: 107 000 ly.

NGC584

H100¹	1:31.3	−6 52	GLX
E4	4.2′ × 2.3′	11.33B	10.5V

This is the first of a 2° long, loose chain of Herschel galaxies in Cetus, including NGC586, NGC596, NGC600, NGC615 and NGC636. The galaxies have similar radial velocities and appear to be about 85 mly distant. Presumably, they form a gravitationally bound group. NGC584 is a very bright galaxy, and is well condensed with a much brighter core. A grainy envelope surrounds the core and a much fainter and diffuse secondary envelope is also visible. Oval in shape, the galaxy is oriented ENE–WSW and deep photographs show characteristics of S0 galaxies. It is paired with NGC586, which follows 4.25′ ESE. There is also a

fainter, compact elliptical galaxy preceding NGC584, 7.0′ WNW. Radial velocity: 1837 km/s. Distance: 82 mly. Diameter: 68 000 ly.

NGC586

H431³	1:31.6	−6 54	GLX
SA(s)a:?	1.6′ × 0.8′	14.08B	13.2V

Although it is much fainter than NGC584, Herschel discovered this galaxy on the same night: 10 September 1785. Photographs show a highly inclined spiral galaxy with an extensive, bright inner region and a fainter outer disc; thin dust lanes are suspected along the galaxy's western flank. Visually, this highly inclined spiral galaxy is much elongated with a brighter core and is oriented N–S. The extensions are fairly diffuse. Radial velocity: 2025 km/s. Distance: 90 mly. Diameter: 42 000 ly.

NGC596

H4²	1:32.9	−7 02	GLX
E⁺ pec:	3.2′ × 2.1′	11.5B	

Interestingly, Herschel discovered this galaxy two years before the previous two entries, on 13 December 1783. Photographs show a large, round core embedded in an extensive, fainter envelope oriented NNE–SSW. Telescopically, this is a very bright, well-condensed galaxy; the core is quite round with well defined edges and traces of a brighter nucleus. At high magnification, the outer envelope is faint and grainy and the galaxy is visible in a medium-magnification field with NGC600. Radial velocity: 1910 km/s. Distance: 85 mly. Diameter: 79 000 ly.

NGC599

H473²	1:32.8	−12 11	GLX
SAB0⁻ pec:	1.4′ × 1.3′	14.15B	

On the DSS image this otherwise normal looking face-on lenticular galaxy has two faint plumes emerging from the outer envelope, one extended ESE, the other WNW. A small, elongated possible companion galaxy is situated at the edge of the outer envelope, 0.5′ NNW of

the core. It may be interacting with the larger galaxy and be the source of the faint plumes. Visually, the galaxy is round and moderately bright with fairly well defined edges and an evenly bright disc. Radial velocity: 5514 km/s. Distance: 246 mly. Diameter: 100 000 ly.

NGC600

H432³	1:33.1	−7 19	GLX
(R′)SB(rs)d	3.6′ × 3.5′	13.01B	

Herschel discovered this three-branch, face-on barred spiral galaxy on the same evening as NGC584 and NGC586. Photographically, the galaxy has a bright bar and no core is visible. There are several condensations in weak and somewhat fragmentary spiral arms. In a medium-magnification field, this galaxy is located SSE from NGC596. It is a round, faint and diffuse glow poorly concentrated to the centre. It appears to be about the same size as NGC596 but is much fainter. Radial velocity: 1875 km/s. Distance: 84 mly. Diameter: 87 000 ly.

NGC615

H282²	1:35.1	−7 20	GLX
SA(rs)b	3.6′ × 1.4′	12.47B	

In photographs, this is a highly inclined spiral galaxy with a dense, opaque, elliptical inner region and fainter but strongly structured outer spiral arms; deep photographs show HII knots. Visually, it is a rewarding object to view: it is a very bright and well defined galaxy, appearing highly tilted with a brighter core. The galaxy is oriented NNW–SSE with very well defined extremities. The outer envelope shows arms that taper to points. A small stellar core is visible at high magnification. Averted vision brings out traces of an outer halo. The galaxy follows a magnitude 9 field star located 5.25′ WSW. Herschel discovered this galaxy and the nearby NGC636 on the same evening: 10 January 1785. Radial velocity: 1879 km/s. Distance: 84 mly. Diameter: 88 000 ly.

NGC622

H454[3]	1:36.0	+0 40	GLX
SB(rs)b	1.6′ × 1.1′	14.20B	

Photographs reveal a barred-spiral galaxy with a round core and two tightly wound spiral arms attached to the bar. The arms form a pseudo-ring and after half a revolution broaden and relax slightly into an open spiral pattern. The galaxy is a faint and quite difficult object visually. It is a very small, roundish patch of light of even surface brightness and is moderately well defined, suggesting that only the core is visible. Radial velocity: 5216 km/s. Distance: 233 mly. Diameter: 108 000 ly.

NGC624

H471[3]	1:35.7	−10 00	GLX
(R′)SB(r)b pec	1.32′ × 0.66′	14.0B	

Photographs show a barred spiral galaxy seen obliquely with a bright bar surrounded by a pseudo-ring and faint, somewhat chaotic and clumpy outer spiral structure. The fainter galaxy MCG−2-5-9 is 2.0′ WNW. Telescopically, this galaxy is moderately faint and is situated between two field stars oriented NNE–SSW, with the galaxy closer to the S star. It is an oval patch of well defined light, elongated E–W and slightly brighter to the middle. Radial velocity: 5903 km/s. Distance: 264 mly. Diameter: 101 000 ly.

NGC636

H283[2]	1:39.1	−7 31	GLX
E3	2.7′ × 2.2′	12.35B	

This galaxy is visually and photographically similar in appearance. Telescopes show a fairly bright galaxy, small but well-condensed with a bright core at high magnification. The galaxy appears slightly elongated NE–SW, and the envelope appears mottled with well defined edges. It appears just slightly W of a line between two magnitude 9 field stars oriented NE–SW. Radial velocity: 1888 km/s. Distance: 84 mly. Diameter: 66 000 ly.

NGC681

H481[2]	1:49.2	−10 26	GLX
SAB(s)ab	2.7′ × 1.6′	12.77B	

Discovered on 28 November 1785, this is an almost edge-on spiral galaxy of the 'Sombrero' (M104) type, with a large, bright core region and a thin, well defined dust lane. Visually, it is a bright galaxy with a well defined central region. A small secondary envelope is visible as a diffuse, ill defined glow. The core is fairly bright but is not star-like and the overall form is rather boxy. Five magnitude 11 field stars are visible to the W, including one star within the outer envelope immediately to the NW. Radial velocity: 1773 km/s. Distance: 79 mly. Diameter: 62 000 ly.

NGC682

H501[2]	1:49.0	−14 59	GLX
SA0−	1.4′ × 1.1′	13.0B	

The appearance of this lenticular galaxy is quite similar visually and photographically. It is a fairly bright and condensed object, broadly brighter to the middle, slightly oval and elongated E–W with fairly well defined edges. Radial velocity: 5614 km/s. Distance: 251 mly. Diameter: 102 000 ly.

NGC701

H62[1]	1:51.0	−9 42	GLX
SB(rs)c	2.5′ × 1.2′	12.81B	

This bright, highly inclined, multi-armed spiral galaxy was recorded by Herschel on 10 January 1785. Visually, it is a fairly bright and well defined object, much extended in a SW–NE direction. The major axis is quite bright and the galaxy's edges are well defined. The overall texture is quite smooth. The compact spiral IC1738 is 5.25′ S. Radial velocity: 1845 km/s. Distance: 82 mly. Diameter: 60 000 ly.

NGC681

NGC702

H192[3]	1:51.3	−4 03	GLX
SB(s)bc pec + comp	1.5′ × 1.1′	13.97B	

This interacting pair of galaxies has also been catalogued as Arp 75. The DSS image is difficult to interpret as the merged image of the two galaxies is poorly resolved; high-resolution images suggest three galaxies are involved as there are three bright condensations. The first is the core of the principal galaxy, the second is an elongated condensation emerging perpendicularly to the core in the W and the third is a bright condensation seen in silhouette against the spiral arm in the N. This arm extends into a fainter 'ring-tail' spur on the western side, a structure that is often seen as the aftermath of a merger. Visually, the galaxy is small and moderately faint, but is quite well seen at high magnification as a roundish, condensed patch of light, even in surface brightness and very slightly elongated SSE–NNW. It was first observed on 20 September 1784, when Herschel described the object as: 'Extremely faint, small, verified 240 (magnification) with difficulty.' Radial velocity: 10 615 km/s. Distance: 474 mly. Diameter: 207 000 ly.

NGC720

H105[1]	1:53.0	−13 44	GLX
E5	4.6′ × 2.5′	11.15B	

This elliptical galaxy presents similar appearances visually and photographically. In a moderate-aperture telescope it is quite bright and fairly large; it is seen as an extended, almost oval patch of light, oriented NW–SE and is quite well defined with an elongated brighter central region and a small, bright core. Radial velocity: 1745 km/s. Distance: 78 mly. Diameter: 104 000 ly.

NGC723

H460[3]	1:53.9	−23 46	GLX
SA(r)bc	1.5′ × 1.3′	13.57B	

Photographs show a multi-arm spiral galaxy viewed face-on, with a very bright, large core and extremely faint, weak spiral arms. Located 2.25′ N of a magnitude 10 field star, the galaxy is bright visually and is well seen at high magnification. Only the bright core is seen and this is almost round and brightens to a large, nonstellar core. The edges are fairly well defined. Radial velocity: 1454 km/s. Distance: 65 mly. Diameter: 29 000 ly.

NGC731

H266[3]	1:54.9	−9 01	GLX
E⁺:	1.7′ × 1.7′	13.07B	

The visual and photographic appearances of this galaxy are identical. It is a round and moderately bright galaxy and is fairly well defined at the edges. The envelope is even in brightness and quite opaque. An edge-on field galaxy is about 7.0′ to the ESE and may be visible in larger-aperture telescopes. Radial velocity: 3845 km/s. Distance: 172 mly. Diameter: 85 000 ly.

NGC748

H193[3]	1:56.4	−4 28	GLX
(R′)SA(r)b?	2.3′ × 1.1′	13.42B	

This is an inclined two-arm spiral galaxy. Photographs show a bright central region with a bright inner pseudo-ring made up of two principal spiral arms which expand to fainter, thin outer spiral arms. A field star is 2.0′ to the NW. NGC748 is a fairly bright and largish galaxy visually. It exhibits even surface brightness and is very slightly elongated SE–NW. The edges are well defined and no core is visible. Radial velocity: 5342 km/s. Distance: 239 mly. Diameter: 16 000 ly.

NGC755

H265[3]	1:56.4	−9 04	GLX
SB(rs)b	3.4′ × 1.1′	13.10B	

Photographs show a highly inclined spiral galaxy with a bright, though irregular inner disc and a much fainter outer envelope. The galaxy is fairly bright visually and is elongated NE–SW. The main envelope is fairly well defined and even in surface brightness; no core is visible and the extremities taper slightly. Radial velocity: 1651 km/s. Distance: 74 mly. Diameter: 73 000 ly.

NGC762

H464[3]	1:57.1	−5 26	GLX
(R′)SB(rs)a	1.3′ × 1.1′	13.93B	

Images show an S-shaped two-arm spiral galaxy with a bright core and a possible short bar which curves at the end into the fainter, outer spiral pattern. The spiral arms extend for half a revolution before turning back in towards the bar, forming a pseudo-ring. The galaxy is small and fairly faint visually; it is elongated N–S and is quite well defined with an even-surface-brightness disc. Radial velocity: 4922 km/s. Distance: 220 mly. Diameter: 83 000 ly.

NGC773

H468[3]	1:59.0	−11 31	GLX
SAB(r)a pec	1.3′ × 0.7′	13.91B	

The morphology of this galaxy is difficult to determine on the DSS image but a brighter, round core is embedded in an oval ring or pseudo-ring and a slightly fainter outer envelope. The galaxy is of moderate brightness visually, oval in outline and oriented N–S. The edges are well defined and the disc is even in brightness. Radial velocity: 5435 km/s. Distance: 243 mly. Diameter: 92 000 ly.

NGC779

H101[1]	1:59.7	−5 58	GLX
SAB(r)b	4.0′ × 1.2′	11.97B	

This is a highly inclined spiral galaxy of the multi-arm type. Photographs show a bright, elongated core with a tightly wound inner spiral pattern resembling a ring, expanding to a grainy and fragmentary outer spiral pattern. The galaxy is bright and attractive visually and is well seen at high magnification as a much extended, lenticular-shaped galaxy, broader at the middle with extensions tapering to points. It is well defined at the edges with a large, brighter central region. Radial velocity:

1411 km/s. Distance: 63 mly. Diameter: 74 000 ly.

NGC788

H435[2]	2:01.1	−6 49	GLX
SA(s)0/a:	2.0' × 1.6'	13.06B	

The DSS image shows a probable lenticular galaxy with a bright, slightly oval core surrounded by a narrow, fainter inner disc and with a very faint, smooth-textured, possible spiral arm, most prominently seen to the ESE. Visually, the galaxy is fairly bright, quite round and well defined, with a small, bright core. Radial velocity: 4094 km/s. Distance: 183 mly. Diameter: 106 000 ly.

NGC790

H433[3]	2:01.5	−5 23	GLX
SA(r)0°?	1.3' × 1.3'	13.98B	

The form of this lenticular galaxy appears fairly similar visually and photographically. Telescopically, it is somewhat faint and quite small; the envelope is round and fairly condensed with a tiny brighter core embedded in the well defined disc. Radial velocity: 5360 km/s. Distance: 239 mly. Diameter: 91 000 ly.

NGC827

H227[3]	2:08.8	+7 58	GLX
S?	2.2' × 0.8'	13.81B	

The DSS image confirms that this is quite obviously a highly inclined spiral galaxy. The core is bright with extensive, but very faint, outer spiral structure. There is a probable dust lane beyond the inner core to the W. At medium magnification, this is a fairly distinct galaxy; it is somewhat bright and is well defined, elongated almost due E–W and a little brighter to the middle with ends that taper. The main envelope is well condensed and quite opaque. Herschel observed this object on 7 November 1784, but his description is somewhat perplexing: 'Two or three small stars with nebulosity rather confirmed 240 (magnification).' The notes in *The Scientific Papers* indicate that his position

is correct and coincides with NGC827; there is not a nebulous group of stars anywhere nearby. Radial velocity: 3513 km/s. Distance: 157 mly. Diameter: 100 000 ly.

NGC833

H482[2]	2:09.3	−10 08	GLX	
(R')Sa:pec	1.7' × 0.7'	13.71B	11.4V	

Together with NGC835, NGC838 and NGC839, this forms a compact group of galaxies that has been catalogued as Hickson 16. The quartet is also known as Arp 318 and probably forms an actual physical group, as their radial velocities are all quite similar. Photographically, NGC833 is a bright, inclined galaxy, probably a spiral, which displays a faint, warped outer disc. There is evidence of interaction with NGC835, which is located 0.8' to the E. All four galaxies display disturbed and/or peculiar structure and the core-to-core separation between NGC833

and NGC839 is only about 331 000 ly at the presumed distance. In a moderate-aperture telescope each member of the group is well seen and the galaxies bracket a magnitude 9 field star located to the S. Visually, NGC833 is bright, sharply defined and very small, appearing very slightly oval in an E–W direction. Herschel discovered all four galaxies on 28 November 1785, his description reading: 'Four. The preceding two, both faint, extended, small, within 1.0' distant on the parallel. The following two, both pretty faint, pretty small, extended, about 2.0' distant and nearly in the direction of the meridian.' Radial velocity: 3865 km/s. Distance: 173 mly. Diameter: 88 000 ly.

NGC835

H483[2]	2:09.4	−10 08	GLX
SAB(r)ab: pec	1.3' × 1.0'	13.16B	

On the DSS image this appears as an almost face-on ring-shaped spiral galaxy

NGC833/835/838/839

NGC833
NGC835
NGC838
NGC839

with a large, bright, circular core. High-resolution photographs show a prominent dust patch or lane in the ring on the NE side and a faint plume of material extending back to NGC833, as well as a plume curving towards NGC838, 3.5′ to the ESE. The core-to-core separation between NGC833 and NGC835 would only be about 40 000 ly at the presumed distance. Visually, NGC835 is the brightest of the group; it is well defined, almost round and displays a star-like core. Radial velocity: 4073 km/s. Distance: 182 mly. Diameter: 69 000 ly.

NGC838

H484[2]	2:09.6	−10 09	GLX
SA(rs)0° pec:	1.1′ × 0.9′	13.53B	

Photographs reveal a possible spiral galaxy with an extensive outer envelope, asymmetrically elongated towards the E, as well as a narrow dust lane in the outer envelope S of the core. Deep images show the E extension is somewhat disturbed. Telescopically, this galaxy is a little fainter than NGC833 and NGC835 to the W; the galaxy is well defined and slightly elongated E–W. Radial velocity: 3851 km/s. Distance: 172 mly. Diameter: 55 000 ly.

NGC839

H485[2]	2:09.7	−10 11	GLX
S pec sp	1.4′ × 0.6′	13.97B	11.4V

Photographs show that this may be a lenticular galaxy, despite the classification. Deep, high-contrast images show the faint outer envelope is somewhat disturbed and curves slightly towards NGC838, 2.5′ to the NNW. Visually, this is the faintest of the four members of Hickson 16, but the galaxy is well condensed to the core and quite elongated E–W. A magnitude 13 field star is visible to the NW. Radial velocity: 3874 km/s. Distance: 173 mly. Diameter: 70 000 ly.

NGC850

H259[3]	2:11.2	−1 30	GLX
SAB(s)0⁺	1.3′ × 1.0′	14.07B	

Situated 5.6′ NNW of a bright field star, this lenticular galaxy is seen face-on and

photographs show a bright, round core in an extensive outer envelope slightly elongated E–W. At high magnification the galaxy is a small, somewhat faint object, but is well defined with even brightness across its disc. Radial velocity: 8190 km/s. Distance: 366 mly. Diameter: 138 000 ly.

NGC853

H486[2]	2:11.8	−9 18	GLX
Sm pec?	1.6′ × 1.2′	15.86B	

The DSS photograph reveals an inclined S-shaped galaxy, possibly a spiral, with three bright condensations involved in a fainter outer envelope. Visually, the galaxy appears oval and fairly bright and is oriented ENE–WSW with fairly even surface brightness and well defined edges. Radial velocity: 1533 km/s. Distance: 69 mly. Diameter: 32 000 ly.

NGC863

H260[3]	2:14.6	−0 46	GLX
SA(s)a:	1.1′ × 1.0′	13.93B	13.8V

Images show a possible face-on spiral galaxy with a very faint outer structure and suspected dust patches involved. Telescopically, the galaxy is small, round and is quite well seen at high magnification as a condensed, well defined patch of light with a small, brighter core. Radial velocity: 7936 km/s. Distance: 354 mly. Diameter: 113 000 ly.

NGC864

H457[3]	2:15.5	+6 00	GLX
SAB(rs)c	4.7′ × 3.5′	11.62B	

The DSS image shows a three-branch spiral. Two of the arms are narrow and bright where they emerge from the central region which may be a faint bar. The third arm is broad at its base, emerging from the central region in the S. The core is small, round and condensed. A very faint outer arm to the N increases the extent of the galaxy to almost 5.5′. At high magnification, this is an interesting galaxy with a magnitude 10 field star embedded in the

NGC864

outer envelope ESE of the core that almost swamps the galaxy's light. The envelope is quite tenuous and diffuse; it is roughly round but appears slightly elongated E–W. The central region is a little brighter and there is evidence of an E–W bar. With averted vision, the core appears round and moderately bright. Radial velocity: 1607 km/s. Distance: 72 mly. Diameter: 98 000 ly.

NGC873

H474[2]	2:16.6	−11 20	GLX
Sc pec:	1.6′ × 1.3′	13.17B	12.7V

This slightly inclined spiral galaxy is fairly bright on the DSS image and has a peculiar structure. The core is bright but irregular with fragmentary spiral arms emerging; one arm is bright and located ENE of the core. The outer envelope is asymmetrical, being more extensive towards the S. Telescopically, the galaxy is fairly bright, irregularly round with a mottled envelope, though fairly even in surface brightness. Radial velocity: 4005 km/s. Distance: 179 mly. Diameter: 83 000 ly.

NGC875

H2[3]	2:17.1	+1 15	GLX
S0⁺:	1.3′ × 1.3′	14.14B	

This galaxy was originally documented by J. L. E. Dreyer in the *New General Catalogue* as NGC867, but at the position provided by Herschel (about 13 minutes of time following 60 Ceti with no offset provided for declination), nothing could be found by subsequent observers. It is now identified with NGC875, which was discovered by Heinrich d'Arrest in 1865. On the DSS image, the galaxy appears as a face-on lenticular galaxy with the edge-on galaxy IC218 situated 2.4′ to the NNE. Telescopically, NGC875 is a very faint galaxy; it is round and somewhat well defined with an evenly bright disc. Radial velocity: 6412 km/s. Distance: 286 mly. Diameter: 108 000 ly.

NGC881

H436[2]	2:18.7	−6 35	GLX
SAB(r)c	2.2′ × 1.5′	13.23B	12.4V

This is a tightly wound, slightly inclined two-armed spiral galaxy with a bright core and thin, symmetrical spiral arms. The DSS image shows several compact galaxies in a chain running towards a bright field star 5.1′ WNW, which may be an interesting target for amateurs with very large telescopes. Visually, NGC881 and NGC883 can be seen together in a medium-magnification field. NGC881 is the larger and brighter, though it is moderately faint and fairly diffuse with a faint pair of field stars located immediately N. An arc of three bright field stars to the NW aids in locating the galaxy. Radial velocity: 5268 km/s. Distance: 235 mly. Diameter: 151 000 ly.

NGC883

H437[2]	2:19.0	−6 45	GLX
SA(s)0⁻:	1.7′ × 1.4′	13.60B	

This lenticular galaxy may be physically related to NGC881 as they have similar radial velocities. Visually, the galaxy is located 4.0′ N of a bright, unequal pair of stars and appears as a faint, condensed patch of light which is well defined and round. Radial velocity: 5444 km/s. Distance: 243 mly. Diameter: 120 000 ly.

NGC887

H486[3]	2:19.6	−16 04	GLX
SAB(rs)c	2.3′ × 1.8′	13.38B	

Photographs show a symmetrical face-on spiral galaxy, probably barred, with two bright, well-developed spiral arms which fade suddenly to very faint outer spiral structure. A faint field star is 0.3′ E of the core. Telescopically, the galaxy is small and situated just NNE of a line joining two field stars oriented ESE–WNW. Roundish and ill defined, the galaxy is slightly brighter to the middle. Radial velocity: 4292 km/s. Distance: 192 mly. Diameter: 128 000 ly.

NGC895

H438[2]	2:21.6	−5 31	GLX
SA(s)cd	3.6′ × 2.6′	12.30B	11.7V

This is an inclined spiral galaxy. Photographs show two thin, bright spiral arms that originate in the small, bright core. The arms remain bright for about three-quarters of a revolution before fading to fragmentary outer spiral structure which defines a faint outer envelope. Visually, the galaxy appears large and quite bright, but is very diffuse and ill defined. The envelope is grainy and broadly brighter to a nonstellar core which is offset towards the E. The edges are very hazy and ill defined and a threshold star is noted 1.8′ ENE of the core. Radial velocity: 2294 km/s. Distance: 102 mly. Diameter: 107 000 ly.

NGC907

H224[3]	2:23.0	−20 43	GLX
SBdm? sp	1.9′ × 0.6′	13.18B	

The DSS image shows a highly inclined spiral galaxy, very likely barred which would probably present an S-shaped spiral pattern if it could be viewed face-on. The bar is curved and on the W side is involved with two small and one very large dust patches. The slightly fainter spiral arms appear tightly wound and are unresolved but grainy in texture. Visually, the galaxy is somewhat faint but well defined and well seen as an elongated, even-surface-brightness

NGC895

©STScI/AURA/CALTECH/ROE

object, oriented E–W with tapered extensions. Radial velocity: 1614 km/s. Distance: 72 mly. Diameter: 40 000 ly.

NGC908

H153[1]	2:23.1	−21 14	GLX
SA(s)c	6.3′ × 3.1′	10.82B	10.2V

This magnificent, inclined spiral galaxy is very well seen on the DSS image. The slightly brighter core is encircled by an inner disc from which four major spiral arms emerge. The arms are bright and massive, featuring many large HII regions or stellar associations, and are well defined by intricate dust lanes. This galaxy is the principal member of a small group of eight galaxies, including NGC907 located 31′ N. The minimum separation between the two galaxies would be about 588 000 ly at the presumed distance. Telescopically, this is a large and quite bright galaxy; it is a grainy oval of light, broadly brighter along its major axis and moderately well

defined at the edges. It is elongated ENE–WSW and a field star borders the galaxy to the E. It is a curious fact that, despite their proximity on the sky, Herschel discovered the fainter galaxy NGC907 exactly 23 months before he recorded this brighter one. Radial velocity: 1467 km/s. Distance: 66 mly. Diameter: 120 000 ly.

NGC936

H23[4]	2:27.6	−1 09	GLX
SB(rs)0+	4.7′ × 4.1′	11.19B	

This is the principal member of a seven-galaxy group that includes the Herschel objects NGC941 and NGC955. The DSS image shows a prototypical barred lenticular galaxy with a large, bright core and a prominent bar terminating well within the disc of the galaxy. A faint inner ring structure is attached to the bar, brightening where it is in contact, and the smooth outer envelope is fairly bright and betrays hints of spiral

structure. Telescopically, the galaxy is extremely bright and quite large; the bar is well seen at medium magnification as an E–W extension and the core is very small but bright. The secondary halo is a fairly bright but diffuse oval and extends E–W before fading. NGC941 can be seen in the same field 12.6′ to the E. Radial velocity: 1446 km/s. Distance: 65 mly. Diameter: 88 000 ly.

NGC941

H261[3]	2:28.5	−1 09	GLX
SAB(rs)c	2.5′ × 1.8′	13.04B	

The DSS image shows a patchy, irregular galaxy with a small, round core involved in a brighter ellipse. The outer envelope is knotted and dusty and the spiral structure is fragmentary and difficult to trace. Individual stars and HII regions begin to resolve at visual magnitude 22. It is a likely member of the NGC936 group of galaxies and its minimum separation from the parent galaxy would be about 237 000 ly at the presumed distance. Visible in the same medium-magnification field as NGC936, it is smaller and fainter, a large diffuse patch of light extended N–S, oval and broadly brighter to the middle with hazy, ill defined edges. Radial velocity: 1623 km/s. Distance: 72 mly. Diameter: 53 000 ly.

NGC945

H487[2]	2:28.6	−10 32	GLX
SB(rs)c	2.4′ × 2.0′	12.80B	

This face-on barred spiral galaxy is the principal member of a loose grouping of seven galaxies which includes the Herschel object NGC977 as well as the smaller, neighbouring barred spiral NGC948, which is situated 2.5′ to the ENE and a good target for large-aperture telescopes. Photographs show NGC945 as a three-armed, barred spiral with a bright core, curved bar and thin, grainy spiral arms. Visually, the galaxy is quite diffuse, oval in shape and elongated N–S; it is faint and broadly brighter to the middle. A faint field star is 1.4′ SE and a brighter field star is 5.25′ S. Radial

NGC908

velocity: 4467 km/s. Distance: 200 mly. Diameter: 140 000 ly.

NGC955

H278[2]	2:30.6	−1 07	GLX
Sab: sp	2.8′ × 0.7′	12.93B	12.0V

On the DSS image this galaxy appears as a very thin lenticular with faint extensions and a bulging, elongated core, but high-resolution photographs show that it is an almost edge-on spiral with a thin dust lane that is seen in silhouette against the core and a morphology very similar to that of NGC2683 in Lynx. It is a member of the NGC936 group of galaxies. Telescopically, the galaxy is bright and quite large and is well seen at high magnification as an elongated, sharply defined streak of light. Very bright along its major axis, it is oriented NNE–SSW. There is a very slight bulge at the centre but no core is seen; the outer extensions taper slightly. Radial velocity: 1501 km/s. Distance: 67 mly. Diameter: 55 000 ly.

NGC958

H237[2]	2:30.7	−2 57	GLX
SB(rs)c:	2.8′ × 0.8′	12.91B	

This is a highly inclined, two-armed spiral with bright, symmetrical, knotty arms of the grand-design type. It appears barred on the DSS photograph but high-resolution images seem to show the arms winding directly into the core. Visually, this is a large and moderately bright galaxy; it exhibits a slightly oval form with even surface brightness. It is quite well defined at the edges, is elongated N–S and is well seen at high magnification. Radial velocity: 5746 km/s. Distance: 257 mly. Diameter: 209 000 ly.

NGC977

H472[3]	2:33.0	−10 44	GLX
(R')SAB(r)a:	1.9′ × 1.6′	14.22B	

This spiral galaxy has a large, bright core and a possible short bar; photographs show it surrounded by a large but very faint, S-shaped spiral

pattern which forms almost a complete ring around the galaxy. Visually, only the core and central region are visible and they are faint, quite diffuse, evenly bright across the disc and slightly elongated E–W. There is a scattered grouping of field stars centred about 5.0′ to the NNE, which Herschel noted in his discovery description of 28 November 1785. Radial velocity: 4562 km/s. Distance: 204 mly. Diameter: 113 000 ly.

NGC991

H434[3]	2:35.5	−7 09	GLX
SAB(rs)c	2.7′ × 2.4′	13.01B	

The DSS image shows a low-surface-brightness galaxy with an oval central region and a fragmentary, asymmetrical spiral pattern emerging. High-resolution photographs show a minute, faint nucleus and many HII regions scattered along the spiral arms. Visually, the galaxy is a faint and diffuse, roundish patch of light and is ill defined at the edges with a faint field star situated 1.6′ S. Radial velocity: 1524 km/s. Distance: 68 mly. Diameter: 53 000 ly.

NGC1022

H102[1]	2:38.5	−6 40	GLX
(R')SB(s)a	2.4′ × 2.0′	12.11B	11.3V

This barred spiral galaxy is bright and displays some unusual structural features, some of which are seen in the DSS image. High-resolution photographs show a bright bar with two bright, tightly wound spiral arms emerging which form a pseudo-ring. The

NGC1022
©STScI/AURA/CALTECH/ROE

interior of this pattern is laced with filamentary dust patches which also cross the bar on the W side. Two faint plumes emerge from the NE, curving in opposite directions to encircle the galaxy. Telescopically, this is a bright galaxy, slightly oval in shape and well defined, oriented E–W with the bright bar seen as a sizable knot elongated E–W. Radial velocity: 1445 km/s. Distance: 65 mly. Diameter: 45 000 ly.

NGC1032

H5[2]	2:39.4	+1 06	GLX
S0/a	3.5′ × 1.0′	12.65B	

Photographs show a spectacular, edge-on lenticular galaxy with a bulging core and tapered extensions. A thin dust lane bisects the core and can be traced to the edges of the inner disc. Visually, only the bright inner lens of this galaxy is visible, though it is quite bright and opaque, well defined at the edges, oriented ENE–WSW and broadly brighter to the middle. Radial velocity: 2708 km/s. Distance: 121 mly. Diameter: 123 000 ly.

NGC1035

H284[2]	2:39.5	−8 08	GLX
SA(s)c?	2.2′ × 0.7′	12.94B	

This is one of a loose grouping of galaxies centred around NGC1052. Images show that this highly inclined spiral galaxy has bright arms and many dust patches along the NE flank, which is the near side of the galaxy. Telescopically, the galaxy is moderately bright but somewhat ghostly, an elongated object oriented SSE–NNW with extensions that taper slightly. It is smoothly and very slightly brighter to the middle; a faint field star can be seen in the outer envelope to the SSE. Radial velocity: 1228 km/s. Distance: 55 mly. Diameter: 35 000 ly.

NGC1044

H228[3]	2:41.1	+8 44	GLX
E/S0	0.5′ × 0.5′	14.42B	

The DSS image reveals two elliptical or lenticular galaxies which seem to share

NGC1032

©STScI/AURA/CALTECHROE

a common envelope and may be tidally interacting; each galaxy appears to have a very faint spiral 'wisp' attached. The minimum core-to-core separation between the two galaxies would be less than 25 000 ly at the presumed distance. A smaller anonymous companion is 0.9′ to the WNW, MCG +1-7-20 is 9.25′ to the NW, UGC2167 is 9.0′ SSW and NGC1046 2.0′ ESE. Except for NGC1046, these other galaxies are targets for large amateur telescopes. Visually, NGC1044 and NGC1046 are seen together in a medium-magnification field. NGC1044 is larger and slightly brighter; the two components are unresolved and appear as a hazy, elongated glow oriented SE–NW. Herschel recorded both NGC1044 and NGC1046 on 7 November 1784 and described them together as: 'Two about 1′ distant. The preceding extremely faint, very small verified 240 (magnification). The following extremely faint, extremely small. 240 (magnification) doubtful.'

Radial velocity: 6272 km/s. Distance: 280 mly. Diameter: 41 000 ly.

NGC1045

H488[2]	2:40.4	−11 18	GLX
SA0⁻ pec?	2.0′ × 1.3′	13.42B	

Photographs show this lenticular galaxy has a bright, elongated central region and an extensive outer envelope. The galaxy is moderately bright visually and appears well defined with a brighter, condensed core and a disc which is gradually elongated along an ENE–WSW axis. Radial velocity: 4622 km/s. Distance: 206 mly. Diameter: 120 000 ly.

NGC1046

H229[3]	2:41.2	+8 43	GLX
S0⁻:	0.3′ × 0.3′	14.70B	

This lenticular galaxy is 2.0′ ESE of NGC1044 and is probably physically related, as it displays a similar radial

velocity. The minimum core-to-core separation between the two galaxies would be about 158 000 ly at the presumed distance. Visually, it is seen as a round glow with a slightly brighter stellar core intermittently visible and is best seen at medium magnification. Radial velocity: 6080 km/s. Distance: 272 mly. Diameter: 24 000 ly.

NGC1052

H63[1]	2:41.1	−8 15	GLX
E4/S0	3.0′ × 2.1′	11.45B	10.5B

This is the brightest member of a group of galaxies which includes NGC991, NGC1022, NGC1033, NGC1035, NGC1042, NGC1047, NGC1048, NGC1051 and possibly NGC1084 in Eridanus. Similar in appearance both visually and photographically, in a moderate-aperture telescope the galaxy is quite bright with a small, prominent, bright core embedded in a grainy inner disc. This disc is surrounded by a roundish, fainter outer envelope with somewhat hazy edges. NGC1047 can be seen to the NW as a small, hazy patch of light of even surface brightness in a medium-magnification field. Radial velocity: 1495 km/s. Distance: 67 mly. Diameter: 58 000 ly.

NGC1055

H1[1]/H6[2]	2:41.8	+0 26	GLX
SBb: sp	7.6′ × 2.7′	11.46B	10.6V

This galaxy was evidently recorded twice by Herschel on consecutive nights in December of 1783, though it is uncertain if Herschel actually saw the galaxy on the first night. Recorded as H6[2] on 18 December, his description reads: 'Small, cometic, between 2 large and 1 small star.' But in his note section, he further states: 'This has probably been a telescopic comet, as I have not been able to find it again, notwithstanding the assistance of a drawing which represents the telescopic stars in its neighbourhood.' To confuse matters further, no measured offsets from the star Delta Ceti are given. A more definite sighting was made on 19 December, when he recorded the

NGC1055

©STScI/AURA/CALTECH/ROE

galaxy as the first entry in his bright nebulae class: 'Considerably bright, considerably large, irregular form, brighter to the middle.' In photographs, the galaxy is an almost edge-on spiral, with a very large central region and a heavy dust lane extending along the entire length of the plane of the disc. It is reminiscent of a fainter version of the Sombrero galaxy, NGC4594. Visually, the galaxy is very dim; it is a large, though uncertain patch of light, ill defined at the edges, quite diffuse and elongated ESE–WNW. The surface brightness brightens very slowly towards the middle. A triangle of field stars is immediately NNW. Radial velocity: 1004 km/s. Distance: 45 mly. Diameter: 100 000 ly.

NGC1070

H273[2]	2:43.3	+4 58	GLX
Sb	2.4′ × 2.0′	12.80B	

Photographs show a tightly wound, multi-arm spiral galaxy viewed almost face-on. The large, bright core is oval and the fainter spiral structure is defined by dust and many small, faint knots. At high magnification, this is a fairly bright galaxy in a star-poor field. Broadly concentrated to the middle, the main envelope is surrounded by a faint, tenuous glow. Two field stars are nearby to the S. Radial velocity: 4111 km/s. Distance: 184 mly. Diameter: 128 000 ly.

NGC1073

H455[3]	2:43.7	+1 23	GLX
SB(rs)c	4.9′ × 4.5′	11.68B	

The DSS image reveals a classic barred spiral galaxy of the SBc type. The elongated core is located in a long, narrow, asymmetrical bar, which is brighter to the SW. Two wide, open spiral arms emerge from the bar; the arms are quite faint with many knots. They broaden and fade after about a half revolution around the bar. In deep, high-resolution photographs, the brightest stars start to resolve at about blue

magnitude 21. The galaxy is a member of the NGC1068 group. Visually, the galaxy is fairly large but exceedingly dim, located NE of a large magnitude 11 triangle of stars. It is very broadly brighter to the middle, where the bar is suspected as a brightening oriented ENE–WSW. The envelope is slightly oval in shape and is very ill defined at the edges. Recorded on 9 October 1785, Herschel's description reads: 'Very faint, very large, a little brighter to the middle. Easily resolvable, 6 or 7′ in diameter.' Radial velocity: 1220 km/s. Distance: 54 mly. Diameter: 93 000 ly.

NGC1087

H466[2]	2:46.4	−0 30	GLX
SAB(rs)c	3.7′ × 2.2′	11.52B	

In photographs this is an exceptionally interesting galaxy, with features combining irregular and spiral structure. There appears to be a very tiny, brighter nucleus embedded in a multi-arm spiral pattern with many bright knots and small dust patches. The spiral structure is fragmentary and much brighter to the N, the southern envelope is extensive but much fainter. The DSS field contains many faint, background galaxies. Visually, this is a fairly bright galaxy, elongated N–S with a broadly concentrated core. The envelope is smooth and well defined. Two magnitude 10 field stars flank the galaxy to the E. To the N in the same medium-magnification field is NGC1090. Herschel discovered both galaxies on the night of 9 October 1785. Radial velocity: 1522 km/s. Distance: 68 mly. Diameter: 73 000 ly.

NGC1090

H465[2]	2:46.6	−0 15	GLX
SB(rs)bc	4.0′ × 1.7′	12.60B	

This is a highly inclined, three-branch spiral with a faint bar; deep photographs show thin dust lanes crossing the bright, elliptical core, while the spiral arms are knotty and dusty. Telescopically, it is a bright galaxy, oriented E–W, oval and elongated. The moderately well defined envelope is smooth in texture and there is a gradual brightening to the middle. Although they are similarly bright and show comparable resolution in photographs, NGC1090 has a much higher radial velocity than NGC1087, implying that they are not a physical pair. Radial velocity: 2765 km/s. Distance: 124 mly. Diameter: 144 000 ly.

NGC1094

H462[3]	2:47.5	−0 17	GLX
SAB(s)ab	1.6′ × 1.0′	13.87B	

This galaxy was discovered on 7 November 1785. Photographs show a highly inclined, symmetrical spiral galaxy with a bright core embedded in a bright envelope. Two faint, tightly wound spiral arms are attached. A smaller, edge-on companion galaxy is 1.0′ N. Telescopically, this is a faint galaxy located due S of two bright field

stars, one magnitude 9, the other magnitude 10. It is an oval, diffuse haze, oriented E–W and slightly brighter to the centre. Radial velocity: 6469 km/s. Distance: 289 mly. Diameter: 135 000 ly.

NGC1153

H274[2]	2:58.1	+3 22	GLX
S0[+]?	1.3′ × 1.1′	13.68B	

Photographs show a condensed lenticular or elliptical galaxy with a broad but very faint outer envelope. A very faint star, or perhaps a compact companion, is in the halo to the SW. In moderate apertures, this is a small galaxy which nevertheless stands out well from the sky background. It features a bright, star-like core surrounded by a slightly fainter, diffuse envelope, which is only moderately well defined. Round.

Radial velocity: 3134 km/s. Distance: 140 mly. Diameter: 53 000 ly.

NGC7826

H29[8]	0:05.2	−20 44	AST
14.0′ × 9.0′			

This object is well seen on the DSS photograph as a scattered group of bright stars. Telescopically, it appears as a coarse scattering of moderately bright stars and is well seen at low magnification. The main body of the asterism consists of about 15 moderately bright stars and is elongated N–S. Towards the SW there are another eight fairly bright stars and it is unclear whether Herschel considered these stars to be part of the group. Herschel recorded the group on 9 December 1784, calling it: 'A cluster of a few coarse scattered large stars.'

Coma Berenices

Herschel recorded 153 objects in this constellation, all of them galaxies except for the globular clusters NGC4147 and NGC5053. There are a number of bright, nearby galaxies in this constellation including NGC4725, NGC4559 and NGC4565 and many others which are associated with the Virgo galaxy cluster. Many are more remote, however, and the constellation is notable for the

presence of the Abell 1656 galaxy cluster, by far the richest cluster accessible in moderate-sized amateur telescopes. Almost all of the objects discovered here were recorded by Herschel between 1783 and 1785, while a handful were discovered in the winter of 1787 and 1788. Also note that NGC4501 = H118[2], an object that Herschel thought followed the galaxy M88, is nonexistent.

NGC4014

H3[3]	11:58.6	+16 11	GLX
S0/a	1.7′ × 1.0′	14.31B	

In photographs, this inclined spiral galaxy has a bright, elongated core with fainter spiral arms and was originally identified as NGC4028 by Dreyer. The galaxy is quite bright visually, appearing as an elongated patch, oriented ESE–WNW with fairly well

defined edges. The disc is a little grainy with a brighter core. Radial velocity: 3706 km/s. Distance: 166 mly. Diameter: 82 000 ly.

NGC4017

H369[2]	11:58.7	+27 26	GLX
SABbc	2.0′ × 1.7′	13.04B	

This galaxy and the peculiarly structured barred spiral galaxy

NGC4016 have also been catalogued together as Arp 305. The DSS image shows a barred spiral galaxy with a brighter, round core and a very short, fainter bar. Two strong, narrow spiral arms with condensations and grainy texture are attached. NGC4016 is situated 5.8′ to the NNW and the DSS image shows an elongated nebulous condensation almost half way between the two galaxies and oriented with

NGC4017's eastern spiral arms, indicating possible interaction between the two galaxies. At the presumed distance, the minimum core-to-core separation between the two galaxies would be about 259 000 ly and both NGC4016 and NGC4017 are probably members of the NGC4008 group of galaxies. Visually, NGC4017 is moderately faint and somewhat diffuse, appearing as an oval patch of light elongated E–W with a broadly brighter envelope and fairly well defined edges. Radial velocity: 3437 km/s. Distance: 154 mly. Diameter: 89 000 ly.

NGC4032

H404[2]	12:00.6	+20 04	GLX
Im:	1.9′ × 1.3′	12.86B	

Photographs reveal an irregular galaxy, with possible incipient spiral structure, dominated by a brighter irregular core in a faint envelope. A detached chain of matter to the E is oriented NNW–SSE and has two condensations involved. A similar chain is visible in the outer envelope to the W, and is a little shorter with three distinct condensations. There is an overall grainy texture to both the core and the outer envelope of the galaxy. The galaxy is moderately bright visually and is well seen at high magnification as a roundish, very slightly diffuse disc which is a little grainy in texture and very slowly brighter to the middle. The edges are moderately well defined. Radial velocity: 1226 km/s. Distance: 55 mly. Diameter: 30 000 ly.

NGC4037

H77[3]	12:01.4	+13 24	GLX
SB(rs)b:	2.6′ × 2.2′	13.31B	

Photographs show a very faint, probable spiral galaxy with a brighter, round, star-like core. A faint bar is oriented N–S with a partial ring of matter attached to the bar in the N and S. This ring has several faint condensations, particularly in its northern region. Visually, this is a faint

and diffuse patch of light, slightly extended N–S, with a faint but distinct stellar core. The edges are ill defined and a bright field star is located 5.25′ E. Radial velocity: 867 km/s. Distance: 39 mly. Diameter: 30 000 ly.

NGC4049

H390[3]	12:03.0	+18 46	GLX
I?	0.9′ × 0.6′	14.28B	

This is very probably a dwarf irregular galaxy. Photographs show a faint object with a mottled envelope, oriented NE–SW. The galaxy is moderately faint visually, appearing as an irregularly round patch of light, a little brighter to the middle with hazy edges. Recorded on 27 April 1785, Herschel described this object with one word: 'Suspected.' Radial velocity: 783 km/s. Distance: 35 mly. Diameter: 9200 ly.

NGC4061

H394[3]	12:04.1	+20 13	GLX
E:	1.0′ × 0.9′	13.81B	

This elliptical galaxy is very probably a member of the NGC4065 group of galaxies, which is situated immediately W of a similar tight grouping of galaxies arrayed around NGC4092. Although there is a considerable spread in the radial velocities of the galaxies in these two groups, it is probable that both groups are physically related. The first edition of *Uranometria 2000.0* plots 19 NGC galaxies here in a 30′ × 30′ area. The DSS image reveals many more anonymous members. Herschel discovered nine objects associated with the two groupings but sorting out the identities can be something of a chore. Photographs show NGC4061 is an elliptical galaxy with a bright, round core, and a fainter envelope. Visually, it is round, well defined and condensed and forms a bright E–W pair with NGC4065, which is 1.2′ to the ENE. Radial velocity: 7163 km/s. Distance: 320 mly. Diameter: 93 000 ly.

NGC4065

H395[3]	12:04.2	+20 13	GLX
E	1.0′ × 0.9′	13.58B	

This is the principal galaxy of the NGC4065 group of galaxies, though it has one of the lowest radial velocities of the galaxies in the field. Photographs show this elliptical galaxy has a bright, elongated core and a fainter envelope oriented SE–NW. Visually, it forms a pair with NGC4061 and appears round, condensed and well defined at the edges. Radial velocity: 6286 km/s. Distance: 281 mly. Diameter: 82 000 ly.

NGC4066

H392[3]	12:04.2	+20 21	GLX
E	1.1′ × 1.0′	14.13B	

J. L. E. Dreyer associated this Herschel object with NGC4069, but modern sources identify NGC4069 with a small, faint lenticular object 1.75′ SSW of NGC4066. Herschel discovered the members of the NGC4065 group of galaxies on 27 April 1785, describing them as: 'Six nebulae. The places belong to the three first, which are very faint, very small. The other three are 10 or 12′ more south, but there was not time to take their places, more suspected.' Unfortunately, the three measured places cannot be made to correspond with the actual physical relationship of any three galaxies in the northern part of the group, but the three brightest and likeliest candidates are NGC4070, NGC4066 and NGC4074. The three southern objects are NGC4061, NGC4065 and NGC4076. Photographs show NGC4066 as an elliptical galaxy with a bright, round core and a fainter envelope. Like the other bright members of this group, the galaxy appears round, condensed and well defined visually and NGC4069 can be seen as a small, faint patch of light to the SSW. Radial velocity: 7325 km/s. Distance: 327 mly. Diameter: 105 000 ly.

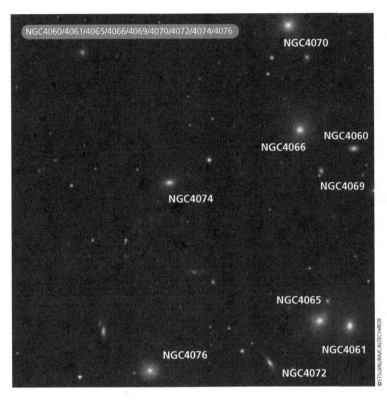

NGC4060/4061/4065/4066/4069/4070/4072/4074/4076

NGC4070

NGC4060
NGC4066

NGC4069

NGC4074

NGC4065

NGC4076

NGC4061

NGC4072

©STScI/AURA/CALTECH/ROE

outer envelope. Radial velocity:
6174 km/s. Distance: 276 mly.
Diameter: 80 000 ly.

NGC4080

H355[3]	12:05.0	+27 00	GLX
Im?	1.2′ × 0.5′	14.04B	

Images reveal an inclined galaxy,
possibly a dwarf irregular or a spiral,
with a slightly brighter core and a
bright, very knotty envelope.
Telescopically, the galaxy is moderately
faint and is seen as an elongated oval of
even surface brightness with well
defined edges and is oriented
ESE–WNW. Radial velocity: 551 km/s.
Distance: 25 mly. Diameter: 9 000 ly.

NGC4092

H382[3]	12:05.9	+20 29	GLX
Sa	1.0′ × 1.0′	14.24B	

This galaxy is the principal member of
the NGC4092 group, which comprises
six galaxies including NGC4089,
NGC4091, NGC4093, NGC4095 and
NGC4098. Photographs show an
almost face-on spiral with a bright,
round core in an oval, slightly fainter
disc with very faint spiral arms
emerging from the disc. Herschel
observed three members of the group,
which he discovered on 26 April 1785
and described as: 'Three. The place is of
the last which is very faint, small. The
other two are south preceding,
extremely faint, very small.' J. L. E.
Dreyer has suggested NGC4093,
NGC4095 and NGC4098 for the
galaxies seen by Herschel, but the four
brightest in a medium-magnification
field are NGC4089, NGC4092,
NGC4095 and NGC4098, while
NGC4093 is the faintest. Herschel
probably observed NGC4092, which is
the largest and one of the brightest
galaxies in the field. Visually, NGC4092
is a round, somewhat diffuse patch of
light, broadly brighter to the middle and
a little hazy at the edges. A field star
0.75′ to the NW interferes somewhat
with visibility. Radial velocity:
6681 km/s. Distance: 298 mly.
Diameter: 87 000 ly.

NGC4070

H391[3]	12:04.2	+20 26	GLX
E	1.0′ × 1.0′	14.03B	

This is the northernmost bright member
galaxy associated with the NGC4065
group. Photographs show a bright
elliptical galaxy with a bright core
embedded in a fainter disc and an
extensive outer envelope. Visually, the
galaxy forms a wide pair with
NGC4066, which is 3.75′ to the S and
appears as a moderately bright, round
and well-condensed object. Radial
velocity: 7174 km/s. Distance: 320 mly.
Diameter: 93 000 ly.

NGC4074

H393[3]	12:04.6	+20 20	GLX
S0: pec	0.4′ × 0.2′	15.11B	

This member of the NGC4065 galaxy
group appears in photographs as an
inclined lenticular galaxy with a bright,

elongated core in a fainter envelope.
Possible dust patches are suspected to
border the core. Visually, the galaxy
appears somewhat isolated from the
other members of the group,
appearing round, condensed and 1.5′
ESE of a faint field star. Radial velocity:
6689 km/s. Distance: 299 mly.
Diameter: 35 000 ly.

NGC4076

H396[3]	12:04.6	+20 12	GLX
Sa	1.0′ × 0.9′	14.25B	

The DSS image shows this member of
the NGC4065 galaxy group is a spiral
galaxy viewed face-on with a brighter,
round core and a fainter outer envelope.
The core is offset slightly towards the W
in the spiral structure. The galaxy seems
to be the largest of the seven group
members visible in a medium-
magnification field, appearing as a small
condensed core surrounded by a diffuse

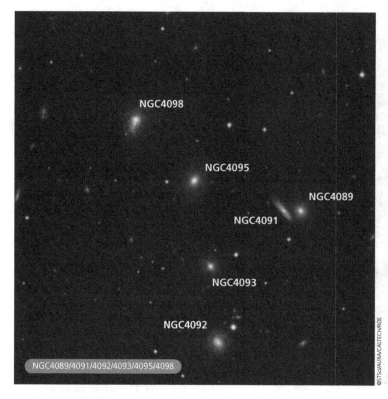

NGC4098

NGC4095

NGC4089

NGC4091

NGC4093

NGC4092

NGC4089/4091/4092/4093/4095/4098

©STScI/AURA/CALTECH/ROE

galaxy is quite faint and is seen as a hazy, irregularly round patch, brighter to a very small core, with the whole embedded in a hazy, ill defined secondary envelope. Radial velocity: 6377 km/s. Distance: 285 mly. Diameter: 83 000 ly.

NGC4104

H370[2]	12:06.6	+28 10	GLX
S0	2.6′ × 1.5′	13.13B	13.6V

Photographs show this lenticular galaxy has a bright core, an elongated inner disc and a faint envelope. There are several smaller galaxies in the immediate vicinity, including MCG+5-29-15 2.75′ SW, and NGC4104 may be a very large galaxy in a small cluster. It is moderately bright visually. The main envelope is fairly well defined and even in surface brightness; it is squarish in form and elongated NNE–SSW. Radial velocity: 8442 km/s. Distance: 377 mly. Diameter: 285 000 ly.

NGC4126

H68[3]	12:08.7	+16 09	GLX
S0	0.8′ × 0.6′	14.65B	

The visual and photographic appearances of this very small, S0 galaxy are similar. Visually, it is a small, faint, oval patch of light which is fairly well defined at the edges, and elongated N–S. The envelope is broadly brighter to the middle. Radial velocity: 6742 km/s. Distance: 301 mly. Diameter: 70 000 ly.

NGC4131

H356[3]	12:08.7	+29 18	GLX
Sb	1.6′ × 0.7′	14.18B	13.3V

Images reveal an inclined, possibly spiral galaxy with a bright extended core and a fainter outer envelope, oriented ENE–WSW. Companion galaxies NGC4132 and NGC4134 are 4.4′ SE and 9.0′ SSE, respectively. The three galaxies can be seen together in a medium- or high-magnification field; NGC4131 appears well condensed, round and virtually stellar at medium magnification and its galactic nature is confirmed at higher power. Herschel

NGC4095

H383[3]	12:06.0	+20 35	GLX
E	0.8′ × 0.7′	14.38B	

Photographs show that this member of the NGC4092 group is an elliptical galaxy with a bright, elongated core, and a fainter envelope oriented SE–NW. Visually, the galaxy appears as a small, condensed, round patch 1.0′ W of a faint star. Radial velocity: 7110 km/s. Distance: 318 mly. Diameter: 74 000 ly.

NGC4098

H384[3]	12:06.1	+20 37	GLX
E	1.2′ × 0.6′	14.47B	

This is the most northerly member of the NGC4092 group and may be an interacting pair of galaxies. On the DSS image the principal galaxy is an elliptical with a round core and a fainter envelope. A possible spiral companion is

0.25′ S and may be situated behind the larger elliptical, as its northern spiral extension is not visible. The minimum separation between the two galaxies would be about 24 000 ly at the presumed distance. Visually, only the brighter elliptical galaxy is seen as an irregular patch of light which is brighter to the middle in a hazy, ill defined envelope. Radial velocity: 7282 km/s. Distance: 325 mly. Diameter: 114 000 ly.

NGC4101

H326[3]	12:06.2	+25 34	GLX
S0/a	1.0′ × 0.8′	14.49B	

Photographs show a lenticular galaxy with a bright, squarish central region and a fainter outer envelope, oriented NE–SW. A fainter extension, or more probably an elongated background galaxy, is projected against the outer envelope to the NW. Visually, the

found this trio on 11 April 1785, and described the fainter members as: 'Two of three. The place is that of II. 371 (NGC4134). Both very faint, much elongated. A fourth suspected.' Herschel does not mention the location of this fourth object but there is a faint S-shaped spiral in the field, 2.3′ NE of NGC4132 that might be Herschel's object. Radial velocity: 3830 km/s. Distance: 171 mly. Diameter: 80 000 ly.

NGC4132

H357[3]	12:09.0	+29 14	GLX
Sab	1.1′ × 0.3′	14.72B	14.0V

This inclined spiral has a bright, very elongated core and disturbed outer spiral structure, particularly the northern extension. A non-NGC open spiral galaxy to the NE may have been seen by Herschel but was not catalogued. At medium magnification NGC4132 is small and very faint but well defined and oriented NNE–SSW. Radial velocity: 4008 km/s. Distance: 179 mly. Diameter: 57 000 ly.

NGC4134

H371[2]	12:09.2	+29 11	GLX
Sb	2.2′ × 0.9′	13.82B	13.1V

Photographs show an inclined spiral galaxy, possibly with a small bar, since the core is bright and elongated. The spiral pattern is complex and multi-armed, with a possible inner ring. The galaxy very probably forms a physical system with NGC4131 and NGC4132 as they have similar radial velocities. Visually, this galaxy is the largest and brightest of three visible in a high-magnification field. Oval and moderately well defined, it is oriented SE–NW and is a little brighter to the middle. Radial velocity: 3819 km/s. Distance: 169 mly. Diameter: 109 000 ly.

NGC4136

H321[2]	12:09.3	+29 56	GLX
SAB(r)c	4.0′ × 3.7′	12.05B	

The DSS image shows a face-on spiral galaxy with a small core and a possible bar surrounded by a complete inner ring. The spiral pattern emerges from this ring with three likely principal arms and at least two additional spiral fragments. High-resolution images begin to resolve stars at blue magnitude 22.0 and there are several bright HII regions involved. Visually, the galaxy is large and moderately bright but quite diffuse and broadly brighter to the middle. The disc appears round with hazy, ill defined edges. The galaxy is a likely member of the Coma I group, whose principal member is NGC4062. Radial velocity: 605 km/s. Distance: 27 mly. Diameter: 31 000 ly.

NGC4146

H327[3]	12:10.3	+26 26	GLX
(R)SAB(s)ab:	1.6′ × 1.5′	13.99B	

This is a face-on barred spiral galaxy with a bright core and a bar attached to a pseudo-inner-ring made up of the two principal spiral arms. Photographs show a peculiar spur that seems to extend northward from the core to the inner ring, perpendicular to the bar. The galaxy is a hazy, roundish patch of light visually and is moderately well defined with a star-like core. Radial velocity: 6504 km/s. Distance: 291 mly. Diameter: 135 000 ly.

NGC4147

H11[1]/H19[1]	12:10.1	+18 33	GC
CC6	4.8′ × 4.8′	11.4B	10.9V

Photographs show this globular cluster has a compressed, bright core and an extensive outer halo of stars, very slightly elongated NNE–SSW. It is moderately bright visually and is well seen at high magnification as an irregularly round disc with ragged edges and very uneven surface brightness across the disc. It is brighter to the middle and grainy textured with several faint stars resolved, particularly near the edge of the disc. Averted vision brings out a faint outer halo. The cluster was not resolved by Herschel. The brightest stars shine at magnitude 14.5 with the horizontal branch visual magnitude at 16.9. Herschel

inadvertently recorded this cluster as two separate objects, observing it on 15 February and 14 March 1784. Spectral type: F2. Radial velocity in recession: 182 km/s. Distance: 61 000 ly.

NGC4150

H73[1]	12:10.6	+30 24	GLX
SA(r)0°?	2.2′ × 1.6′	12.49B	11.6V

This lenticular galaxy has a bright central region and a slightly fainter outer envelope. High-resolution photographs show the characteristic three brightness zones of the lenticular class as well as faint dust lanes along the western flank. Visually, the galaxy is a small but fairly bright oval of light, elongated SSE–NNW. Well defined along the edges, the disc is a little grainy in texture with a bright, star-like core. The radial velocity suggests this galaxy is relatively nearby, which indicates that it may be a dwarf. It may possibly be a member of the Coma I galaxy group. Radial velocity: 224 km/s. Distance: 10 mly. Diameter: 6500 ly.

NGC4152

H83[2]	12:10.6	+16 02	GLX
SAB(rs)c	2.2′ × 1.7′	12.89B	

Photographs show an almost face-on, three-branch spiral galaxy with a bright core and a probable bar attached to a ring. The spiral arms emerge from the ring, and broaden into a fainter, multi-arm pattern. Visually, the galaxy is moderately bright and broadly brighter to the middle. The main envelope is well defined, but there are hints of a faint secondary envelope, elongated SSE–NNW. The galaxy may be an outlying member of the Virgo Cluster of galaxies, possibly on the far side. Radial velocity: 2116 km/s. Distance: 95 mly. Diameter: 61 000 ly.

NGC4158

H405[2]	12:11.2	+20 11	GLX
SA(r)b:	1.9′ × 1.7′	13.08B	

Images show a slightly inclined multi-arm spiral galaxy with a large, elongated core and very faint spiral arms. The entire disc appears mottled. Visually, this is a moderately bright galaxy, appearing as a slightly oval glow, broadly brighter to the middle and oriented ENE–WSW with slightly hazy edges. Radial velocity: 241 km/s. Distance: 108 mly. Diameter: 60 000 ly.

NGC4162

H353[2]	12:11.9	+24 07	GLX
(R)SA(rs)bc	2.3′ × 1.4′	12.46B	

This is an inclined spiral with a brighter core and complex spiral structure. Photographs show the high-surface-brightness spiral arms are knotty, with many condensations. Telescopically, the galaxy is situated mid-way between two field stars, oriented ENE–WSW. The galaxy is bright and very slightly elongated N–S; the envelope is bright and quite mottled and a little brighter along its western flank. The edges are moderately well defined and the galaxy is broadly brighter to the middle. Radial velocity: 2546 km/s. Distance: 114 mly. Diameter: 76 000 ly.

NGC4169

H358[3]	12:12.2	+29 10	GLX	
S0/a	1.8′ × 0.9′	13.16B	12.3V	

This lenticular galaxy is the brightest of a compact group of four galaxies which has been catalogued as Hickson 61. It is also part of a larger complex of 20 galaxies collectively known as the NGC4169 group. Photographs show a bright elongated core and a fainter outer envelope, oriented NNW–SSE. NGC4173 is 1.75′ N, NGC4174 is 2.4′ SE and NGC4175 is 2.6′ ESE. At the presumed distance, this would result in core-to-core separations of about 89 000 ly to NGC4173, 122 000 ly to NGC4174 and 132 000 ly to NGC4175. However, NGC4173 shows a much lower radial velocity than the other members of the group, implying that it may be a foreground object. There are a few scattered galaxies in the area with similar, though lower, radial velocities to NGC4173, so it is possible that we are seeing two separate galaxy groups at widely different distances in the same

NGC4173 ⬝ NGC4175 ⬝ NGC4169 ⬝ NGC4174

NGC4169/4173/4174/4175

general field. Visually, NGC4169 is a well-condensed, elongated object with a bright core. Herschel discovered this tight grouping of galaxies on 11 April 1785 and described NGC4169, NGC4174 and NGC4175 together as: 'Three of a quartile. The place is that of II. 372 (NGC4173). All very faint, very small and all within 3′.' Radial velocity: 3779 km/s. Distance: 169 mly. Diameter: 88 000 ly.

NGC4173

H372[2]	12:12.3	+29 11	GLX	
Sdm	5.0′ × 0.7′	13.59B	13.3V	

Images show this edge-on irregular or spiral galaxy is oriented NW–SE. The core is hidden by a spiral arm which crosses the central region at an angle to the major axis and appears slightly disturbed at its SE extremity with many knotty condensations throughout. High-contrast images reveal a faint bridge apparently connecting the

galaxy to NGC4175 but the galaxies show highly dissimilar radial velocities, implying that they are not at the same distance. At medium magnification this is the largest galaxy in the compact galaxy group Hickson 61; it appears as a faint flat streak of light with even surface brightness and moderately well defined edges. Herschel's description reads: 'One of four. The most northerly of the preceding side of a quartile. Faint. Small.' Radial velocity: 1122 km/s. Distance: 50 mly. Diameter: 73 000 ly.

NGC4174

H359[3]	12:12.4	+29 08	GLX	
S0	0.8′ × 0.2′	14.35B	13.6V	

Images show an edge-on lenticular galaxy with a bright, slightly elongated core and bright extensions surrounded by a much fainter halo. Oriented NE–SW, the galaxy is a member of the compact galaxy group Hickson 61. In moderate-aperture telescopes it is the smallest of the group and appears virtually stellar at all magnifications; it is a compact, well-condensed object. Radial velocity: 4040 km/s. Distance: 180 mly. Diameter: 42 000 ly.

NGC4175

H360[3]	12:12.5	+29 09	GLX
Sbc	1.8′ × 0.4′	14.19B	13.4V

In photographs this is an edge-on spiral galaxy with a bright, elongated core. A prominent dust lane is visible along much of the length of the major axis. A large dust patch is in the spiral extension immediately adjacent to the core in the NW and there is possible interaction with NGC4173. Oriented SE–NW, it is a member of the Hickson 61 compact galaxy group. Visually, only the core of this galaxy is visible; it is small and well condensed. Radial velocity: 4007 km/s. Distance: 179 mly. Diameter: 94 000 ly.

NGC4185

H373[2]/H375[2]	12:13.4	+28 31	GLX
Sbc	2.6′ × 1.9′	13.48B	

Photographically, this is a slightly inclined spiral galaxy with a very small, round and bright core. There is a possible inner ring, with narrow spiral arms expanding to a broad, faint, outer spiral pattern. A faint field star is in the inner ring immediately E of the core. Visually, the galaxy is a large, ill defined hazy patch of the light which is a little brighter to the middle. The edges fade unevenly into the sky background. This galaxy may have been inadvertently recorded twice on the same night, 11 April 1785, by Herschel, but this is uncertain. There is no object at the position recorded for H375[2]. Radial velocity: 3897 km/s. Distance: 174 mly. Diameter: 132 000 ly.

NGC4189

H106[2]	12:13.8	+13 26	GLX
SAB(rs)cd?	2.4′ × 1.7′	12.58B	11.7V

This is an almost face-on spiral galaxy, with a very short and bright bar surrounded by four spiral arms. The spiral structure is asymmetrical with the core offset towards the S. The galaxy is a probable member of the NGC4168 group. Visually, the galaxy is somewhat faint and very slightly elongated E–W with a smooth-textured envelope that is fairly well defined. Radial velocity: 2057 km/s. Distance: 92 mly. Diameter: 64 000 ly.

NGC4196

H374[2]	12:14.4	+28 25	GLX
S0?	1.8′ × 1.2′	13.92B	

Photographs show a probable S0 galaxy with a bright, slightly elongated core. There is a peculiar, fainter outer envelope with very short, stubby extensions to the W and SW. A suspected narrow but extremely faint plume emerges from the core and extends to the NE. Visually, this galaxy is small but moderately bright and is quite condensed to the middle. It is round with moderately well defined edges. Radial velocity: 3968 km/s. Distance: 178 mly. Diameter: 93 000 ly.

NGC4203

H175[1]	12:15.1	+33 12	GLX
SAB0⁻:	4.0′ × 3.0′	11.73B	10.9V

This lenticular galaxy has a very slightly elongated bright core, and photographs show an extensive, fainter outer envelope. The galaxy is fairly bright visually and is well seen at high magnification as a round, opaque ball of light, gradually, then suddenly brighter to the middle to a bright core. The envelope is grainy in texture and the edges are moderately well defined. A bright field star is 3.75′ NNW. Radial velocity: 1096 km/s. Distance: 49 mly. Diameter: 57 000 ly.

NGC4204

H397[3]	12:15.2	+20 39	GLX
SB(s)dm	3.6′ × 2.9′	14.03B	

The DSS image shows a dim, face-on barred spiral galaxy with a slightly brighter bar and two weak spiral arms emerging. The arms are brightest near the bar and are dotted with nebulous knots but fade after less than quarter of a revolution to a broader, ill defined outer structure. The galaxy is quite dim visually and only the brighter bar is visible as an elongated patch of light oriented ESE–WNW which is very broadly brighter along the major axis and somewhat ill defined. Radial velocity: 823 km/s. Distance: 37 mly. Diameter: 39 000 ly.

NGC4212

H107[2]/H108[2]	12:15.7	+13 54	GLX
SAc:	3.2′ × 1.9′	11.78B	

This galaxy was inadvertently recorded as two different objects by Herschel on the night of discovery, 8 April 1784. Photographically, it is an inclined spiral galaxy with a small, round, brighter core and diffuse, three-branch spiral pattern. There are many knotty condensations in the spiral arms and envelope. It is bright even in small-aperture instruments and appears as an oval elongated ENE–WSW which fades very gradually into the sky background. The central region is well defined but only gradually brighter to the middle and a magnitude 12 field star is 2.3′ due S. Located about 4° from the centre of subcluster A in the Virgo Cluster, the galaxy actually shows an approach radial velocity of 137 km/s. Radial velocity: −137 km/s. Distance: 55 mly (estimate). Diameter: 51 000 ly (estimate).

NGC4213

H354[2]	12:15.7	+23 59	GLX
E	1.4′ × 1.1′	13.51B	

Photographically and visually, this small, bright elliptical presents a similar appearance. Located 9.0′ WNW from the bright field star 7 Comae, this is a high-surface-brightness galaxy, small, round and well condensed with well defined edges. It is surprisingly easy to see at medium magnification, even with 7 Comae in the field. Several fainter companions to the W may be visible in very-large-aperture telescopes. Radial velocity: 6718 km/s. Distance: 300 mly. Diameter: 122 000 ly.

NGC4222

H109[2]	12:16.4	+13 19	GLX
Sc	3.2' × 0.4'	13.90B	

Images show an edge-on spiral with no core visible; the minor axis is very flat and small dust patches are seen along the major axis. Located on the Coma–Virgo border, the galaxy forms an attractive triplet with NGC4206 and NGC4216 and is presumably at about the same distance. A bright field star is 4.4' almost due E. Seen in a medium-magnification field, this galaxy is considerably fainter than either NGC4206 or NGC4216; it is little more than a hazy flat streak of light oriented ENE–WSW, though it is fairly well defined. Radial velocity: 172 km/s. Distance: 55 mly (estimate). Diameter: 51 000 ly (estimate).

NGC4237

H11[2]	12:17.2	+15 19	GLX
SAB(rs)bc	2.2' × 1.4'	12.38B	

Photographically, this inclined spiral galaxy has a very small, brighter core and diffuse spiral structure with many knotty condensations. Telescopically, the galaxy is a moderately bright though hazy oval disc, which is slightly brighter to the centre with a grainy texture and well defined edges. Radial velocity: 817 km/s. Distance: 37 mly. Diameter: 23 000 ly.

NGC4245

H74[1]	12:17.6	+29 36	GLX
SB(r)0/a:	3.3' × 2.6'	12.34B	

Photographs show a slightly inclined, theta-shaped barred galaxy. The large, bright core is slightly elongated with a fainter bar attached. The diffuse spiral structure is tightly wound and appears to form a ring surrounded by a faint and smooth-textured disc. Visually, the galaxy is bright but small and well defined. It is a grainy oval patch of light elongated SE–NW with a tiny, star-like brighter core. This may be an outlying member of the Coma I galaxy group. Radial velocity: 814 km/s. Distance: 36 mly. Diameter: 35 000 ly.

NGC4251

©STScI/AURA/CALTECH/ROE

NGC4251

H89[1]	12:18.1	+28 10	GLX
SB0? sp	3.9' × 2.6'	11.63B	

Images reveal a very bright lenticular galaxy viewed almost edge-on with a bright, elongated central bulge and short, bright extensions along the major axis. The galaxy is surrounded by a faint but extensive, diffuse outer envelope, oriented ESE–WNW. In a small telescope, this is a bright object with a concentrated, almost stellar core surrounded by a fainter, elongated inner envelope which holds magnification very well. Radial velocity: 1065 km/s. Distance: 48 mly. Diameter: 54 000 ly.

NGC4253

H702[3]	12:18.4	+29 49	GLX
(R')SB(s)a:	0.9' × 0.8'	14.0B	13.73V

This is a slightly inclined barred galaxy, which photographs show probably involves a theta-shaped ring. The bar is very bright and is attached to a fainter outer envelope which is opaque on its eastern side with two dark zones in the envelope, one to the N of the bar, the other to the S, visible on the western side. The galaxy is small and moderately faint visually; it is a fat oval slightly extended E–W, well defined and broadly brighter to the middle. Radial velocity: 3876 km/s. Distance: 173 mly. Diameter: 45 000 ly.

NGC4262

H110[2]	12:19.5	+14 53	GLX
SB(s)0−?	1.9' × 1.7'	12.40B	11.5B

Photographs show a lenticular galaxy with a bright, elongated core and small, bright extensions, probably a bar, oriented NNE–SSW. It is surrounded by a fainter outer envelope, is slightly extended and is oriented perpendicular to the core. Telescopically, only the bright core is detected with certainty, appearing as an oval disc with a brighter core. It is well defined but a faint, ill defined haze is suspected. Radial velocity: 1308 km/s. Distance: 58 mly. Diameter: 32 000 ly.

NGC4272

H299[3]	12:19.8	+30 20	GLX
E/S0[−]	1.0′ × 0.9′	14.24B	

In a field sprinkled with very faint and small galaxies, photographs show that this elliptical galaxy has a brighter, elongated core and faint outer envelope. Visually, the galaxy is quite faint and is best seen at medium magnification as a small, round, well defined patch of light with even surface brightness. Radial velocity: 8454 km/s. Distance: 378 mly. Diameter: 110 000 ly.

NGC4274

H75[1]	12:19.8	+29 37	GLX
(R)SB(r)ab	7.3′ × 2.6′	11.33B	10.4V

The DSS image shows a highly inclined, theta-shaped spiral galaxy that may be barred. The bright, elongated central region may contain a bar oriented NNE–SSW. A straight dust lane is seen in the bulge on the W side. Two dusty spiral arms are tightly wound and overlap to form a pseudo-ring. Beyond is the fainter outer spiral pattern, which contains extensive dust patches. The galaxy may be a member of the Coma I galaxy group. Visually, this galaxy is bright and detailed in moderate-aperture telescopes. At high magnification the brightest inner portion of the galaxy is visible, featuring a well defined ring elongated almost due E–W. The core is bright and small, though not star-like and is embedded in a bright bar oriented NNE–SSW which attaches to the ring. Radial velocity: 930 km/s. Distance: 41 mly. Diameter: 89 000 ly.

NGC4275

H376[2]	12:19.9	+27 37	GLX
S?	1.1′ × 1.0′	13.61B	

Photographs show a small, poorly resolved galaxy which has a high surface brightness. It appears to be a

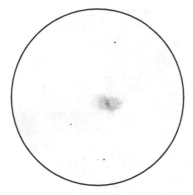

NGC4274 15-inch Newtonian 272x

spiral, possibly barred, with a brighter core and four tightly wound spiral arms. A bright condensation or possible small companion galaxy is located at the end of a spiral arm to the N. Telescopically, the galaxy is quite faint, small and round with even surface brightness and fairly well defined edges.

A faint field star follows 1.2′ to the ENE. Radial velocity: 2305 km/s. Distance: 103 mly. Diameter: 33 000 ly.

NGC4278

H322[2]/H90[1]	12:20.1	+29 17	GLX
E1-2	4.9′ × 4.5′	11.04B	10.2V

This elliptical galaxy is the brightest of a linear triplet that includes NGC4283 and NGC4286 which may form a physical group. The DSS image shows the galaxy has a round, bright core with an extensive, fainter outer envelope. The many bright knots in and around the outer halo are the system's globular clusters. The three galaxies are aligned ENE–WSW with NGC4283 3.4′ ENE and NGC4286 8.6′ ENE. All three galaxies can be seen in the same high-magnification field with NGC4278 easily the largest and brightest. It is a round, grainy, opaque galaxy brightening steadily to the middle to a

NGC4274

NGC4278/4283/4286

large core. The envelope is well defined but embedded in a very faint secondary halo. The three galaxies were first recorded by Herschel on 13 March 1785, using the star 13 Canes Venaticorum as the reference star (H322[2], H323[2], H300[3]). NGC4278 and NGC4283 were subsequently recorded a month later on 11 April 1785, this time using the star 14 Comae as the reference. The faintest galaxy was not seen on this occasion and Herschel recorded NGC4278 as H90[1] and NGC4283 as H377[2] believing they were undiscovered objects. Radial velocity: 648 km/s. Distance: 29 mly. Diameter: 41 000 ly.

NGC4283

H323[2]/H377[2]	12:20.3	+29 19	GLX	
E0	1.5′ × 1.5′	12.99B	12.1V	

This is the second galaxy in a chain that includes NGC4278 and NGC4286. Photographs show that this elliptical galaxy has a round, bright core in a faint envelope. Despite its location, this galaxy may not be physically related to the other two galaxies as its radial velocity is about 300 km/s higher. Telescopically, it appears as a small, round, bright and well defined object with even surface brightness. Radial velocity: 983 km/s. Distance: 44 mly. Diameter: 19 000 ly.

NGC4286

H300[3]	12:20.7	+29 21	GLX
SA(r)0/a:	1.7′ × 1.1′	14.72B	

This galaxy has a similar radial velocity to NGC4278 and is probably physically related. The core-to-core separation between the two galaxies at the presumed distance would be about 72 000 ly. Photographs show a slightly inclined, low-surface-brightness galaxy, probably a dwarf lenticular or elliptical with a very small, round, brighter core embedded in an inner disc surrounded

by a faint outer envelope. Visually, the galaxy is by far the dimmest of the three galaxies; it is a hazy, dim patch of light and is difficult to see but it can be held steadily with averted vision. The surface brightness appears even and the galaxy is a little extended SSE–NNW. Radial velocity: 643 km/s. Distance: 29 mly. Diameter: 14 000 ly.

NGC4293

H5[5]	12:21.2	+18 23	GLX
(R)SB(s)0/a	5.6′ × 3.0′	11.05B	

This bright galaxy displays a peculiar morphology. Images show a highly inclined galaxy, probably barred, with a brighter core involved with a very elongated possible ring structure, which is defined by several large, massive dust patches in the core and in the disc. This inner disc is surrounded by a much fainter outer envelope, which is highly inclined to the major axis of the inner ring and displays some characteristics of spiral arms. The outer envelope is fairly smooth-textured with evidence of dust to the W and S. Telescopically, the galaxy is fairly dim but is well seen at medium and high magnification as a large, moderately well defined disc, elongated ENE–WSW. Bright along its major axis, it is somewhat mottled in texture with bright field stars bordering the galaxy along its northern flank. Radial velocity: 855 km/s. Distance: 38 mly. Diameter: 62 000 ly.

NGC4293

NGC4298/4302

©STScI/AURA/CALTECH/ROE

NGC4298

H111[2]	12:21.5	+14 36	GLX	
SA(rs)c	3.3′ × 1.3′	12.06B	11.3V	

The DSS image shows an inclined, multi-armed spiral galaxy with a very small, round, brighter core. The diffuse spiral pattern has knotty condensations and dusty patches and is somewhat asymmetrical, being broader to the SE. A field star is ENE from the core. NGC4298 may form a physical pair with NGC4302 2.4′ to the E but there is no evidence of tidal distortion in either galaxy, despite the very similar radial velocities. If they are at the same distance, the minimum core-to-core separation between the two galaxies would be about 35 000 ly. This pair of galaxies is a beautiful sight in a moderate-aperture telescope. NGC4298 is a large oval galaxy; it is moderately well defined and gradually brighter to the middle. The galaxy is elongated SE–NW and the faint field star is plainly visible to the ENE. Herschel discovered this pair on 8 April 1784 and his combined description reads: 'Two, about 2′ distant. The first round, resolvable. The second, extended, resolvable.' Radial velocity: 1084 km/s. Distance: 49 mly. Diameter: 47 000 ly.

NGC4302

H112[2]	12:21.7	+14 36	GLX	
Sc:	5.4′ × 0.6′	12.50B	11.6V	

The DSS image reveals a large, very thin spiral galaxy, edge-on with a brighter core, though there is no bulge. The core is largely obscured by a narrow, prominent dust lane which runs almost the entire length of the galaxy's disc. A faint field star is projected against the tip of the dust lane in the S. Visually, the galaxy is a bright, flat streak oriented N–S and is very ethereal with a field star at its NW tip and another a little farther to the NE. The dust lane is not visible in small telescopes, though large amateur instruments may show it. Radial velocity: 1099 km/s. Distance: 49 mly. Diameter: 77 000 ly.

NGC4310

H378[2]	12:22.4	+29 12	GLX	
(R')SAB(r)0+?	2.3′ × 1.1′	13.57B		

This highly inclined spiral galaxy has a large, bright, elongated core in a fainter disc. Photographs show evidence of a dust lane bordering the core along the western flank. Visually, the galaxy is moderately faint but is well seen at medium magnification as a well defined oval of even surface brightness, elongated SSE–NNW. Radial velocity: 912 km/s. Distance: 41 mly. Diameter: 27 000 ly.

NGC4312

H628[2]	12:22.5	+15 32	GLX	
SA(rs)ab: sp	4.2′ × 1.1′	12.52B	11.7V	

In photographs this is a highly inclined spiral galaxy, oriented very nearly N–S with a thin, very elongated core and a star-like nucleus. The fainter outer envelope shows some evidence of spiral structure, in particular a dust lane which borders the central core to the W. The visual appearance of this galaxy is quite similar to the photographic one; it is a faint sliver of light with even surface brightness along its major axis and well defined edges. A wide, nearly equal pair of stars is 2.5′ ESE of the core. The galaxy may be a satellite of M100, which is nearby, and this may help to explain its extraordinarily low radial velocity, which may be a result of high-speed orbital motion around the principal galaxy. It is difficult to ascertain an accurate size and distance for the galaxy. Radial velocity: 106 km/s. Distance (estimate): 68 mly. Diameter (estimate): 83 000 ly.

NGC4314

H76[1]	12:22.6	+29 53	GLX	
SB(rs)a	4.4′ × 4.1′	11.42B	10.6V	

The DSS image reveals a bright, face-on barred spiral galaxy with a possible theta-shaped ring or smooth disc. The core and bar are both bright; the core is

NGC4314

elongated, while the bar fades only slightly towards the inner ring or disc. The ring is very faint and smooth-textured, as are the two S-shaped spiral arms, which emerge where the bar meets the ring. The galaxy is bright and quite large visually and is greatly extended SSE–NNW with a field star to the NNW and a very faint star 0.6′ SE of the core. The structure is complex; it is a smooth, elongated, oval envelope with a large, bright, elongated core, which features a tiny, star-like nucleus. Bright extensions on either side of the core are the brighter portions of the bar. Radial velocity: 965 km/s. Distance: 43 mly. Diameter: 55 000 ly.

NGC4317

H324[2]	12:22.6	+31 03	nonexistent

NGC4328

H84[2]	12:23.3	+15 48	GLX
SA0⁻:	1.3′ × 1.2′	14.20B	

This is a small, faint, lenticular galaxy which is very similar in appearance visually and photographically. At medium magnification it is a very small, hazy, unconcentrated patch of light located 6.0′ E of M100. It is round and is a little brighter to a small core with a hazy envelope. This may be a dwarf galaxy; its association with M100 is uncertain. Radial velocity: 433 km/s. Distance: 19.4 mly. Diameter: 7400 ly.

NGC4336

H406[2]	12:23.6	+19 27	GLX
SB0/a	2.0′ × 0.9′	13.42B	

Photographs show an inclined galaxy, possibly a dwarfish barred spiral, with a small, brighter core and brightenings in the disc E and W of the core. These are probably the attachment points of the bar to the spiral structure. At the eyepiece, the galaxy is moderately faint; medium magnification reveals a diffuse, elongated patch of light, oriented N–S and broadly brighter to the middle. The edges are somewhat ill defined. Radial velocity: 998 km/s. Distance: 45 mly. Diameter: 26 000 ly.

NGC4340

H85[2]	12:23.6	+16 43	GLX
SB(r)0⁺	3.5′ × 2.8′	12.10B	

Photographs show an inclined barred galaxy, with a theta-shaped ring structure. The bright elongated core is embedded in a fainter inner envelope.

The faint bar brightens where it joins the otherwise smooth-textured ring. The bright Herschel galaxy NGC4350 is 5.6′ to the ESE but there is some difference in the radial velocities of the two galaxies and they may not form a physical pair. At the presumed distance of NGC4340, the core-to-core separation would be about 66 000 ly. Visually, both galaxies can be seen in a medium- or high-magnification field and form an attractive contrasting pair. NGC4340 appears slightly fainter, smaller and almost round with a small, brighter core. The edges are fairly well defined but there is little evidence of the bar or the outer ring. Radial velocity: 908 km/s. Distance: 40 mly. Diameter: 41 000 ly.

NGC4344

H31[3]	12:23.6	+17 32	GLX
SpN/BCD	1.7′ × 1.6′	13.20B	

The DSS image shows a face-on galaxy, but the resolution is insufficient to

NGC4340/4350

determine its morphology, although it does have a peculiar, squarish core. There are condensations in the core, particularly to the SE and SW, which are brighter than the central region. The core is surrounded by a much fainter, round envelope. Visually, this is a small galaxy, well condensed to the core and irregularly round with ragged, poorly defined edges. Radial velocity: 1102 km/s. Distance: 49 mly. Diameter: 24 000 ly.

NGC4350

H86[2]	12:24.0	+16 42	GLX
SA0	2.7′ × 1.0′	11.86B	11.1V

This is an almost edge-on lenticular galaxy; the visual and photographic appearances are very similar. At medium magnification, the galaxy is a bright, sharply defined object which is bright along its major axis with a prominent, star-like core. The galaxy is elongated NNE–SSW. Radial velocity: 1158 km/s. Distance: 52 mly. Diameter: 41 000 ly.

NGC4359

H648[3]	12:24.2	+31 31	GLX
SB(rs)c? sp	3.6′ × 1.0′	13.92B	

The DSS image shows a very thin, almost edge-on spiral galaxy, with an elongated, brighter inner region but no bulge. The faint outer envelope has some evidence of spiral structure including a dust lane S of the central region and a prominent dust patch in the western spiral arm. The galaxy is large and somewhat faint visually and is best seen at medium magnification as an elongated streak of light that is fairly distinct along the major axis and oriented ESE–WNW. The NNE flank is a little more sharply defined and the ends taper slightly. Radial velocity: 1262 km/s. Distance: 56 mly. Diameter: 59 000 ly.

NGC4375

H379[2]	12:25.0	+28 33	GLX
SB(r)ab pec:	1.4′ × 1.0′	14.0B	

This is a face-on spiral galaxy. Photographs show a bright core, which may be embedded in a bar, and two bright, symmetrical spiral arms. Visually, the galaxy is a faint patch of light which is fairly even in surface brightness, roundish and somewhat ill defined at the edges. Radial velocity: 9058 km/s. Distance: 405 mly. Diameter: 165 000 ly.

NGC4377

H12[1]	12:25.2	+14 46	GLX
SA0⁻	1.7′ × 1.4′	12.70B	

This almost face-on lenticular galaxy has a very slightly elongated, bright core and a fainter outer envelope. The DSS image reveals that there are three brighter knots involved in the outer envelope, which are very probably background galaxies. At medium magnification NGC4377 is a small, though moderately bright galaxy with a bright, well defined envelope. A small, bright, round core is visible; this is well defined with the brightness at its edges dropping off sharply as it merges with the main envelope. Radial velocity: 1327 km/s. Distance: 59 mly. Diameter: 29 000 ly.

NGC4379

H87[2]	12:25.2	+15 36	GLX
S0⁻ pec:	1.9′ × 1.6′	12.64B	11.7V

This lenticular galaxy presents very similar appearances visually and photographically. At medium magnification it is a bright and well-condensed galaxy with a bright inner envelope, brightening to a small, bright core. The edges are well defined but a hazy outer envelope is suspected and the galaxy is elongated E–W. Radial velocity: 1024 km/s. Distance: 46 mly. Diameter: 25 000 ly.

NGC4393

H361[3]	12:25.8	+27 33	GLX
SABd	4.0′ × 3.3′	13.84B	

The DSS image reveals a faint, almost face-on, barred spiral galaxy. It is only very slightly concentrated to the middle with a fainter, tapered bar running N–S. There is grainy texture to the central region from which weak spiral arms emerge. Many knotty condensations are seen in the spiral arms, including the brightest, an E–W brightening in the spiral arm to the S. Visually, this is a very faint, diffuse patch of light NNE of a magnitude 10 field star. It appears almost round, is poorly defined at the edges and is not much brighter to the middle. Radial velocity: 750 km/s. Distance: 34 mly. Diameter: 39 000 ly.

NGC4394

H55[2]	12:25.9	+18 13	GLX
(R)SB(r)b	3.6′ × 3.2′	11.7B	

This face-on barred spiral galaxy has a bright, elongated core and a slightly fainter bar. Faint, but broad, spiral arms form a pseudo-ring around the bar before broadening and fading outward. This galaxy is a companion of NGC4382 (M85) 7.75′ WSW and a probable member of subcluster A of the Virgo Cluster. At high magnification, NGC4394 is visible in the same high power field as M85. Only the core is visible and the galaxy is much brighter towards the middle. A magnitude 12 field star is 2.8′ S. Herschel thought the galaxy was resolvable. Radial velocity: 886 km/s. Distance: 40 mly. Diameter: 42 000 ly.

NGC4405

H88[2]	12:26.1	+16 11	GLX
SA(rs)0/a:	1.8′ × 1.1′	12.96B	

Photographs show a slightly inclined galaxy, possibly a barred spiral, with a bright, star-like nucleus embedded in a slightly fainter central core. The core appears a little brighter along a N–S axis, possibly the bar, which is surrounded by a prominent inner envelope and a fainter, extensive outer envelope. In a moderate-aperture telescope, this is a moderately bright, oval galaxy, brightening broadly to the middle. Elongated almost due N–S, the faint outer envelope is fairly well defined at the edges. Radial velocity: 1704 km/s. Distance: 76 mly. Diameter: 40 000 ly.

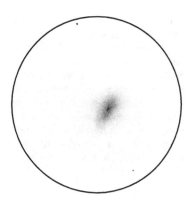

NGC4414 15-inch Newtonian 272x

NGC4414

H77[1]	12:26.4	+31 13	GLX	
SA(rs)c?	4.9′ × 3.5′	10.99B	10.1V	

This galaxy is probably a member of the Canes Venatici II Cloud of galaxies. Photographs show a bright, inclined spiral galaxy with an elongated, brighter central region. The symmetrical multi-arm spiral structure is dusty with many knots and only slightly fainter than the central region. It expands to a much fainter and extensive outer spiral structure. Visually, this is a large and very bright galaxy; high magnification reveals an oval, high-surface-brightness object, very grainy in texture and steadily brighter to the middle. Elongated SSE–NNW, the envelope is fairly well defined and a brighter core is visible. Radial velocity: 724 km/s. Distance: 32 mly. Diameter: 46 000 ly.

NGC4419

H113[2]	12:26.9	+15 03	GLX	
SB(s)a	3.3′ × 1.1′	11.99B	11.2V	

The DSS image shows a highly inclined spiral galaxy with a brighter, elongated disc and a slightly fainter outer envelope. The spiral structure is not well seen. A very prominent dust lane borders the core to the NE, broadening E of the core before continuing to the SE. Visually, this is a large galaxy, high magnification revealing an oval, high-surface-brightness object, very grainy in

texture, and increasing in intensity to the middle. The envelope is fairly well defined and a brighter core is visible. The galaxy is oriented SE–NW with a faint field star 1.9′ SSW of the core. The galaxy is suspected to be member of subcluster A in the Virgo Cluster and has a radial velocity of 307 km/s in approach, one of a handful of galaxies in the Virgo Cluster to do so. Distance: 55 mly (estimate). Diameter: 53 000 ly (estimate).

NGC4421

H89[2]	12:27.0	+15 28	GLX	
SB(s)0/a	2.5′ × 1.4′	12.34B	11.6V	

Photographs show a probable barred spiral galaxy with a large, bright, elongated core and an unevenly bright bar extending N–S. The core and bar are embedded in a faint, ill defined haze. The visual appearance is quite similar to the photographic one. At medium magnification the galaxy is bright and the bar is well seen with a

bright and very small core embedded. The hazy, ill defined secondary envelope is quite evident. A magnitude 10 field star is 2.25′ WNW. Radial velocity: 1558 km/s. Distance: 69 mly. Diameter: 50 000 ly.

NGC4448

H91[1]	12:28.2	+28 37	GLX	
SB(r)ab	4.1′ × 1.5′	12.04B		

The DSS image shows a highly inclined galaxy, possibly barred with a bright, broad core elongated N–S, connected to an inner ring. The fainter outer envelope appears dusty. The galaxy is well seen in medium apertures at high magnification as a very bright and well-concentrated galaxy. The central region is bright and condensed, rising to the middle but no stellar core is visible. A faint secondary envelope, best seen at medium magnification, is elongated due E–W. The edges of the envelope are moderately well defined and very mottled. Herschel's discovery

NGC4419

description on 11 April 1785 states: 'Very bright, elongated on the parallel. Pretty bright, large nucleus and 2 branches'. Radial velocity: 661 km/s. Distance: 30 mly. Diameter: 35 000 ly.

NGC4450

H56²/H90²	12:28.5	+17 05	GLX
SA(s)ab	5.2′ × 3.9′	10.84B	10.1V

This is a magnificent inclined spiral galaxy with a large, bright, slightly elongated core. The DSS photograph shows that the surrounding spiral arms are massive and smooth textured, highlighted by thin, extensive dust lanes, particularly near the core on the eastern flank. A very broad dark patch, possibly dust, is in the outer spiral structure on the E as well. Telescopically, the galaxy is large and bright and easily seen as a hazy patch at low magnification. At medium and high magnification the core is large, bright and slightly elongated and is embedded in a grainy and extensive disc elongated N–S. A thin dust lane is suspected along the eastern flank; the faint outer halo fades gradually into the sky background. A bright field star is 3.75′ SW. The galaxy appears to have been recorded twice by Herschel, on 14 March and 21 March 1784, but J. L. E. Dreyer raises the possibility that Herschel may have recorded a faint comet in addition to the galaxy on 21 March. Radial velocity: 1915 km/s. Distance: 85 mly. Diameter: 129 000 ly.

NGC4455

H355²	12:28.7	+22 49	GLX
SB(s)d? sp	2.8′ × 0.8′	13.29B	

Photographs show a highly inclined galaxy, possibly a spiral with fragmentary structure. There is a gradual brightening to an irregularly shaped central region. There are many brighter, knotty condensations in both the core and the fainter extensions. Visually, the galaxy is best seen at medium magnification. It is a moderately bright, elongated streak that is brighter along its major axis with the extensions tapering to soft points. Fairly well defined along its flanks, the galaxy is oriented NNE–SSW and no core is detected. Radial velocity: 618 km/s. Distance: 28 mly. Diameter: 22 000 ly.

NGC4459

H161¹	12:29.0	+13 59	GLX
SA(r)0⁺	3.5′ × 2.7′	11.46B	

Photographs show a lenticular galaxy with a bright core and an extensive outer disc. In a medium-magnification field, this galaxy appears WSW of NGC4468 and NGC4474 in Markarian's Chain. This lenticular galaxy is large and bright, slightly elongated E–W and brighter to the middle with hazy extremities. A magnitude 8 field star is 2′ SE. Deep photographs show a very thin dust lane encircling the core and separating it

from the fainter, outer envelope. Radial velocity: 1161 km/s. Distance: 52 mly. Diameter: 53 000 ly.

NGC4468

H630²	12:29.5	+14 03	GLX
SA0⁻?	1.3′ × 1.0′	13.83B	

Photographs show a lenticular galaxy with a small core and an extensive outer envelope. Visible in the field with NGC4459 and NGC4474, this is a very small galaxy, possibly a dwarf, and appears as a faint oval patch of light; it is poorly defined with even surface brightness across the envelope. Radial velocity: 861 km/s. Distance: 38 mly. Diameter: 15 000 ly.

NGC4473

H114²	12:29.8	+13 26	GLX
E5	3.7′ × 2.4′	11.07B	10.2V

Photographs show a bright elliptical galaxy with a large, elongated core in an extensive and very ill defined outer envelope. It is located 12.0′ S of the pair NGC4477 and NGC4479 in Markarian's Chain. Visually, a small, bright core is embedded in an extended, fainter outer envelope, elongated E–W. Radial velocity: 2194 km/s. Distance: 98 mly. Diameter: 106 000 ly.

NGC4474

H117²/H629²	12:29.9	+14 04	GLX
S0 pec:	2.0′ × 1.0′	12.49B	

This galaxy was recorded twice by Herschel: first on 8 April 1784 and again on 14 January 1787. In a medium-magnification field, this Markarian Chain galaxy can be seen with NGC4459 and NGC4468. Its visual and photographic appearances are similar. Visually, the galaxy is a small, elongated streak elongated almost due E–W. The core is bright and the envelope is even in surface brightness and well defined. Photographs show a very large central bulge. A faint pair of stars lies 2.1′ NNE. Radial velocity: 1540 km/s. Distance: 69 mly. Diameter: 40 000 ly.

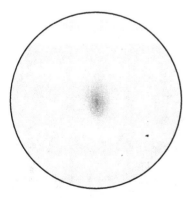

NGC4450
15-inch Newtonian 272x

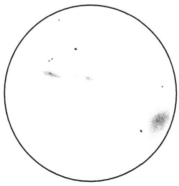

NGC4459 NGC4464 NGC4468
15-inch Newtonian 146x

NGC4475

H362[3]	12:29.8	+27 15	GLX
SAbc	1.7′ × 1.0′	14.56B	

The DSS image shows a very slightly inclined spiral galaxy, which may be barred as an E–W extension is visible emerging from the very bright core. The bar connects to an inner ring with four spiral arms emerging, two from the ring. The spiral structure is relaxed and quite symmetrical. The galaxy is very faint visually, however, and is visible as a hazy, irregularly round patch of light of even surface brightness with ill defined edges. Radial velocity: 7389 km/s. Distance: 330 mly. Diameter: 163 000 ly.

NGC4477

H115[2]	12:30.0	+13 38	GLX	
SB(s)0:?	3.8′ × 3.5′	11.32B	10.4V	

Photographically, this is a bright, face-on galaxy with a round, bright core and a short, fainter, tapered bar oriented N–S. They are embedded in a slightly oval and very faint outer envelope which displays an ill defined spiral pattern in deep photographs. Visually, the inner region appears bright and well defined and brightens to a well condensed core. The lenticular galaxy NGC4479 is 5.0′ SE. Both galaxies are part of Markarian's Chain in subcluster A of the Virgo Cluster of galaxies. Radial velocity: 1306 km/s. Distance: 58 mly. Diameter: 65 000 ly.

NGC4479

H116[2]	12:30.3	+13 35	GLX	
SB(s)0°:?	1.5′ × 1.3′	13.51B	12.4V	

Photographs show a barred lenticular galaxy with a bright core and a moderately bright bar, oriented NNE–SSW, embedded in a very faint and ill defined envelope. Only the core of this lenticular galaxy is visible in moderate-aperture telescopes. It appears round, small and faint and is moderately brighter to the middle. The edges are poorly defined and the outer envelope fades first quickly and then more gradually into the sky background. This neighbour of NGC4477 has a much

smaller radial velocity and may not be physically related. Radial velocity: 827 km/s. Distance: 37 mly. Diameter: 16 000 ly.

NGC4489

H91[2]	12:30.8	+16 45	GLX
E	1.7′ × 1.6′	12.84B	

The visual and photographic appearances of this dwarfish elliptical galaxy are quite similar, though photographs show an extensive outer envelope. Visually, the galaxy is small, moderately bright and almost round with even surface brightness and well defined edges. Medium magnification shows it and NGC4498 in the same field. Radial velocity: 929 km/s. Distance: 41 mly. Diameter: 21 000 ly.

NGC4494

H83[1]	12:31.4	+25 47	GLX	
E1-2	4.8′ × 3.5′	10.68B	9.7V	

Photographs show a large, bright, elliptical galaxy with a bright core, slightly elongated N–S, and a fainter outer envelope. The galaxy is large and bright visually, appearing almost round with a somewhat ill defined disc which is grainy in texture and broadly brighter to the middle to a suddenly brighter, nonstellar core. Radial velocity: 1336 km/s. Distance: 60 mly. Diameter: 84 000 ly.

NGC4495

H301[3]	12:31.3	+29 09	GLX
Sab	1.5′ × 0.8′	14.26B	

Photographs reveal a highly inclined galaxy, probably a spiral, with a bright elongated core in a brighter inner envelope, with faint, slightly warped outer arms. Telescopically, it is visible only as a very faint, roundish patch of light, best seen at medium magnification. Radial velocity: 4554 km/s. Distance: 203 mly. Diameter: 89 000 ly.

NGC4498

H69[3]	12:31.7	+16 51	GLX
SAB(s)d	3.0′ × 1.6′	12.77B	

This inclined barred spiral galaxy has an extensive bar but no core is visible.

Photographs show fragmentary spiral structure with many dust patches and brighter condensations in an asymmetrical outer envelope. Visually, the galaxy is a hazy oval patch, elongated ESE–WNW, and broadly concentrated towards the middle with fairly well defined edges. Radial velocity: 1469 km/s. Distance: 66 mly. Diameter: 58 000 ly.

NGC4502

H92[2]	12:32.1	+16 42	GLX
Scd:	1.1′ × 0.6′	14.59B	

Images show an inclined, dwarfish galaxy, possibly spiral or irregular, with a brighter central region. The outer envelope is somewhat wedge-shaped with short flares extending away from the central region towards the N. Visually, the galaxy is small and extremely faint and is visible only intermittently with averted vision. It appears as a hazy, oval patch of light elongated NE–SW. Radial velocity: 1588 km/s. Distance: 71 mly. Diameter: 23 000 ly.

NGC4506

H631[2]	12:32.2	+13 25	GLX
Sa pec?	1.8′ × 1.1′	13.70B	

This is a small, slightly inclined galaxy, possibly a dwarf spiral. Photographs show a bright, small and round core and a squarish inner envelope surrounded by a fainter, oval envelope oriented ESE–WNW. A prominent dust patch, oriented ENE–WNW, passes in front of

NGC4498

the core. The galaxy is moderately faint visually, quite diffuse and a little more distinct at medium magnification. It follows a fairly bright field star and appears as a hazy, roundish patch of light, a little brighter to the middle with ill defined edges. Radial velocity: 688 km/s. Distance: 31 mly. Diameter: 16 000 ly.

NGC4514

H302[3]	12:32.5	+29 44	GLX
Sbc	1.2′ × 1.0′	14.47B	14.0V

Photographs show a small, face-on, multi-arm spiral galaxy with a round, bright core in a fainter envelope with two weak spiral arms emerging from the bar. Secondary arms emerge after about a half revolution around the core. The galaxy is quite faint visually, visible as a low-surface-brightness, small, round patch of light, and is moderately well defined with an evenly bright disc. Radial velocity: 8095 km/s. Distance: 362 mly. Diameter: 126 000 ly.

NGC4515

H93[2]	12:33.0	+16 15	GLX
S0⁻:	1.3′ × 1.1′	13.28B	

This galaxy is similar visually and photographically. Moderate apertures reveal a very small, faint, almost round galaxy, with even surface brightness and well defined edges. The galaxy is likely a dwarf lenticular. Radial velocity: 912 km/s. Distance: 41 mly. Diameter: 16 000 ly.

NGC4516

H78[3]	12:33.1	+14 34	GLX
SB(rs)ab?	1.9′ × 0.8′	13.70B	

Images show a highly inclined barred galaxy with a bright, much elongated core and bar, with two fainter, slightly curved spiral arms emerging into a still fainter outer envelope. Visually, the galaxy is a faint, elongated patch of light, quite even in surface brightness along its major axis and well defined along the edges. No core is visible and the envelope is elongated N–S. On its discovery on 8 April 1784, Herschel

commented: 'Very faint, resolvable by moonlight.' Radial velocity: 872 km/s. Distance: 39 mly. Diameter: 22 000 ly.

NGC4525

H325[2]	12:33.8	+30 17	GLX
Scd:	2.7′ × 1.5′	13.30B	

This low-surface-brightness galaxy is classified as a nonbarred spiral but the DSS image reveals a short, E–W oriented band that may be a bar. The galaxy is somewhat inclined and though the galaxy is symmetrical, the fragmentary spiral structure is brighter N of the bar. This galaxy is fairly faint visually but is distinctly seen as a fairly well defined oval of light which is smooth-textured, a little brighter to the middle and elongated NE–SW. Radial velocity: 1181 km/s. Distance: 53 mly. Diameter: 41 000 ly.

NGC4529

H26[3]	12:32.6	+20 11	GLX
Scd:	2.1′ × 0.4′	15.52B	

This object was observed by Herschel, on 12 March and 17 March 1784, and again on 16 March 1790, though none of the positions that he supplied falls directly on an actual galaxy. On 12 March he commented: 'Suspected a large, extremely faint nebula, but tho' I looked at it a good while I could not verify the suspicion, nor could I convince myself that it was a deception.' Although the identity is somewhat uncertain, the present author feels that NGC4529 = UGC7697: photographs show a very flat, almost edge-on spiral galaxy, with a small, elongated, brighter core and faint, grainy extensions, oriented almost due E–W. The plotted position in the first edition of *Uranometria 2000.0* would then be incorrect; the Herschel object is about 25.0′ SE of the indicated position. Although very faint visually, this galaxy is seen as a blunt, narrow, elongated patch of light, oriented E–W, fairly well defined and even in brightness. The field agrees with both the DSS image and the NASA/IPAC Extragalactic Database

listing. Radial velocity: 2510 km/s. Distance: 112 mly. Diameter: 69 000 ly.

NGC4540

H94[2]/H119[2]	12:34.8	+15 33	GLX
SAB(rs)cd	1.9′ × 1.5′	12.47B	

The DSS image shows a slightly inclined spiral galaxy with a small core with two ill defined spiral arms emerging and forming a pseudo bar, before expanding to a clumpy, diffuse outer structure. A very faint field star is involved in the outer envelope 0.3′ NW of the core and IC3528, a small, high-surface-brightness face-on spiral, is 1.5′ NE. Visually, NGC4540 is a hazy, though moderately bright galaxy, very broadly concentrated to the middle. It appears almost round with poorly defined edges. Herschel recorded the galaxy as separate objects on two occasions, 21 March and 8 April 1784. Radial velocity: 1246 km/s. Distance: 56 mly. Diameter: 31 000 ly.

NGC4548

H120[2]	12:35.4	+14 30	GLX
SBb(rs)	5.4′ × 4.9′	10.94B	13.6V

Photographically, this is a beautiful, almost face-on barred spiral galaxy. It has a bright, elongated core with a slightly fainter bar. The spiral arms emerge from the bar, very faintly at first, then brighten considerably after about half a rotation around the central bar. At this point, narrow dust lanes bordering nebulous knots (stellar associations and HII regions) appear.

NGC4540

NGC4548

Fainter, outer spiral arms extend outward to the S and to the N. The southern spiral arm is broad and features small, knotty condensations. Higher-resolution photographs show complex, sinuous dust lanes in the bar on both sides of the core. This galaxy has been identified as the missing Messier object M91. Visually, little more than the central region of this galaxy is visible. The nucleus is very bright and well condensed and is surrounded by a fainter outer envelope with poorly defined edges and elongated ENE–WSW. It is presumably a Virgo Cluster member: its radial velocity is only 442 km/s. Distance: 55 mly (estimate). Diameter: 87 000 ly (estimate).

NGC4555

H343[2]	12:35.7	+26 31	GLX
E	1.7′ × 1.3′	13.50B	

Photographs show an elliptical galaxy with a bright, elongated core in a fainter outer envelope, oriented ESE–WNW. Visually, the galaxy is moderately bright, fairly condensed, round and steadily brighter to the middle with well defined edges. Radial velocity: 6680 km/s. Distance: 298 mly. Diameter: 148 000 ly.

NGC4556

H380[2]	12:35.8	+26 54	GLX
E[+]:	1.2′ × 1.0′	14.22B	

This elliptical galaxy is the brightest of nine galaxies in a 10′ × 10′ field that includes NGC4558 and NGC4563. The field should be a rewarding one for observers with large-aperture telescopes. On the DSS image the galaxy has a bright, elongated core in a fainter outer envelope, oriented E–W. A compact galaxy lies immediately W. In a moderate-aperture telescope only NGC4556 is visible, appearing somewhat faint and round, with an evenly bright disc and well defined edges. Radial velocity: 7489 km/s. Distance: 335 mly. Diameter: 117 000 ly.

NGC4559

H92[1]	12:36.0	+27 58	GLX
SAB(rs)cd	10.5′ × 4.4′	10.28B	10.0V

This is a highly inclined barred spiral galaxy, oriented NW–SE. The DSS image shows a small, narrow bar embedded in dusty spiral arms with many knotty condensations. The inner arms are quite bright but fade suddenly to a fainter outer structure which involves several resolvable HII regions. The brightest stars of this relatively nearby galaxy begin to appear at blue magnitude 22. This galaxy may form a group with NGC4565 and NGC4725, two other large, bright spirals, though its radial velocity is a little lower. Telescopically, this large and very bright galaxy is seen as an elongated, long oval of light, broadly brighter along its major axis and fairly well defined at the edges. The disc is a little mottled and a triangular group of field stars is involved, one star E of the core, the second to the SE of the core and the last S of the core. Herschel saw a fourth star which is just W of the southernmost star and a little fainter. Discovered on 11 April 1785, Herschel's description reads: 'Very bright, very large, much extended north preceding, south following, 10 or 12′ long. 4 stars in it.' Radial velocity: 818 km/s. Distance: 37 mly. Diameter: 111 000 ly.

NGC4559

NGC4561

H407[2]	12:36.1	+19 20	GLX
SB(rs)dm	1.5′ × 1.3′	13.02B	

The DSS image shows an irregular galaxy with a brighter, curving bar oriented E–W. A bright condensation is visible near the eastern end of the bar and there is a similar, though fainter, condensation in the W. The fainter outer envelope is squarish and is oriented NNE–SSW with a grainy texture. Telescopically, this galaxy is moderately bright and shows some structure at high magnification. A bright, elongated bar oriented ESE–WNW is embedded in a diffuse and ill defined outer envelope which is irregularly round. Radial velocity: 1407 km/s. Distance: 62 mly. Diameter: 27 000 ly.

NGC4565

H24[5]	12:36.3	+25 59	GLX
SA(s)b? sp	14.8′ × 2.1′	10.06B	

This magnificent galaxy is a classic edge-on spiral galaxy with a bright, compact, elongated central bulge. It is well seen on the DSS image as a well defined object with a complex dust lane along the length of the major axis. The dust lane helps to show that the galaxy is very slightly tilted from the edge-on position with the side along the SW flank tilted towards us. Herschel recorded the galaxy on 6 April 1785, calling it: 'A lucid ray 20′ long or more. 3 or 4′ broad, north preceding, south following, very bright to the middle, a beautiful appearance.' This galaxy is well seen even in small-aperture telescopes and never fails to generate an exclamation of wonder when it drifts into the field of view. The dust lane is best seen as it passes over the central bulge. In a moderate aperture, on the best nights the galaxy extends for more than half the field of a medium-magnification eyepiece. Although the surface brightness is a little low, the galaxy is well defined with a bright, oval central bulge and a sharply defined dust lane which crosses the core. Averted vision helps to trace the dust lane across

a third of the galaxy with the faint, slightly mottled extensions tapering slowly to points. The galaxy is oriented NW–SE and a faint field star is 1.3′ NE of the core. Radial velocity: 1230 km/s. Distance: 55 mly. Diameter: 234 000 ly.

NGC4571

H602[3]	12:36.9	+14 13	GLX
SA(r)d	3.6′ × 3.2′	11.97B	

On the DSS image this face-on spiral galaxy has a detailed, complex structure. Morphologically quite similar to bright, multi-arm spirals like M101, the galaxy has a small, bright, round core embedded in a fainter, ring-like inner envelope. The spiral arms are not well defined, being condensations in the overall outer envelope. Many small, dusty patches and knotty condensations are seen throughout. A faint field star is in the outer envelope 1.2′ W of the core and a magnitude 8 field star is 2.5′ to the NE. In a moderate-aperture telescope,

the galaxy is a round, very diffuse patch of light with the outer edges reaching almost to the bright field star. The surface brightness is even, though very slightly brighter towards the middle and the edges are moderately well defined. The radial velocity measurement leads to a suspiciously low distance and size for this galaxy. Radial velocity: 298 km/s. Distance: 55 mly (estimate). Diameter: 57 000 ly (estimate).

NGC4595

H632[2]	12:39.9	+15 18	GLX
SAB(rs)b?	1.8′ × 1.1′	12.92B	

Photographs show a slightly inclined galaxy, possibly a dwarf barred spiral. The bright, slightly elongated core has faint extensions oriented E–W. Three spiral arms emerge into a knotty outer envelope. Visually, the galaxy is a moderately bright oval oriented ESE–WNW with moderately well defined edges. The envelope is bright and broadly brighter to the middle. The

NGC4565

NGC4571

ENE–WSW. The bright oval core has two bright, spiral arms, which in higher-resolution images emerge from an inner ring. Further out, the spiral arms are fainter and thin, with many knotty condensations and dust patches, as well as one strong dust lane in the outer envelope to the SE. This galaxy is also catalogued as Arp 189 because of an extremely faint, straight jet of matter emerging from the galaxy, extending ENE and ending in a curving arc of matter elongated N–S; this jet is very faintly seen on the DSS image. The apparent length of this plume is about 3.5′ and would be about 35 000 ly in actual length at the presumed distance of the galaxy. At high magnification this is a large and bright galaxy; it is a fat oval with a bright, dense inner envelope, brighter to the middle though no core is visible. The edges of the inner envelope are fairly well defined, but a very faint secondary envelope is visible. Herschel thought the galaxy resolvable. Radial velocity: 772 km/s. Distance: 35 mly. Diameter: 40 000 ly.

galaxy is probably a member of subcluster A of the Virgo Cluster. Radial velocity: 592 km/s. Distance: 26 mly. Diameter: 14 000 ly.

NGC4634

H603[3]	12:42.7	+14 18	GLX
SBcd: sp	2.6′ × 0.7′	13.16B	

Photographs show a large, bright, edge-on spiral galaxy with a bright, elongated main envelope. A thin dust lane runs almost the length of the main envelope but is very slightly inclined to it. An extensive fainter envelope broadens at the SSE and NNW extremities. Visually, this is a moderately bright galaxy, much extended SSE–NNW and fairly flat. The surface brightness is even along the major axis and the edges are moderately well defined. NGC4633 is 3.75′ to the NNW and is visible as a round, hazy patch of light. The radial velocity of NGC4634 is quite low, leading to low estimates of both its distance and size and in the sky the two galaxies are near

the outskirts of the main mass of the Virgo Cluster. Both may in fact be dwarf galaxies. Radial velocity: 257 km/s. Distance: 11.5 mly. Diameter: 8700 ly.

NGC4651

H12[2]	12:43.7	+16 24	GLX
SA(rs)c	4.0′ × 2.6′	11.38B	

The DSS image reveals a large, bright, slightly inclined spiral galaxy, elongated

NGC4634

NGC4659

H127[2]	12:44.5	+13 31	GLX
S0/a	1.5′ × 1.2′	13.07B	

Images show a face-on lenticular galaxy, with a bright core elongated ENE–WSW in a fainter elongated envelope, oriented N–S. There are hints of a weak spiral arm in the outer envelope to the W. In a moderate-aperture telescope, the galaxy is bright and takes magnification well. The

NGC4651

envelope is small, dense and grainy and is moderately well defined, with a very small, bright core. The galaxy is almost round and there is a magnitude 10 field star 1.3′ to the SSW. Herschel considered this galaxy resolvable. Radial velocity: 438 km/s. Distance: 20 mly. Diameter: 8500 ly.

NGC4670

H328[3]	12:45.3	+27 08	GLX
SB(s)0/a pec:	1.4′ × 1.1′	13.15B	

This dwarfish lenticular galaxy has been catalogued as Arp 163 because of its peculiar morphology. The galaxy is elongated E–W with a bright, bulging central region and bright, stubby extensions. The western extension curves slightly to the N. The faint outer envelope is extensive and appears disturbed, streaming irregularly away from the galaxy to the S and to the N, mostly just from the western extension of the galaxy. The Herschel object NGC4673 is 5.5′ to the SE but has a much higher radial velocity and is not likely to be a companion. Visually, both galaxies can be seen in a high-magnification field. NGC4670 is the larger and brighter of the two; it is small with a fairly high-surface-brightness, well defined envelope elongated E–W, with a star-like core embedded. A bright field star follows 4.3′ to the E. Radial velocity: 1073 km/s. Distance: 48 mly. Diameter: 19 000 ly.

NGC4673

H329[3]	12:45.6	+27 04	GLX
E1-2	1.0′ × 0.8′	14.89B	

Photographs show an elliptical galaxy, slightly elongated N–S with a fainter outer envelope. Telescopically, it is smaller and fainter than NGC4670, which can be seen in the same field to the NW. It is a very small, round, concentrated patch of light, even in surface brightness with a well defined envelope. Radial velocity: 6856 km/s. Distance: 306 mly. Diameter: 89 000 ly.

NGC4676AB

H326[2]	12:46.1	+30 44	GLX
S0 pec? ‾	2.0′ × 0.3′–	15.09B	
	1.7′ × 0.7′		

This pair of galaxies is amongst the most famous of the interacting galaxies in the sky; it is commonly known as 'The Mice' and has also been catalogued as Arp 242. Component A is the more northerly of the pair and the one displaying the peculiar 'tadpole' tail structure. The core of the galaxy is elongated N–S with a bright, collimated stream of matter at least four times longer than the host galaxy, directed to the N. Component B is situated 0.6′ to the SSE with the core elongated NE–SW. The southern galaxy has what might be a dust patch located to the NE as well as an elongated patch of matter oriented NW–SE. Two broad streams of faint matter emerge from the core of the southern galaxy and are directed S. A faint bridge connects the cores of the galaxies and they are undoubtedly physically connected. The

minimum core-to-core separation between the two galaxies would be 52 000 ly at the presumed distance. Telescopically, only the cores of this pair of galaxies are visible; they are both very faint and blurry, irregularly round, even in surface brightness and moderately well defined. The 'tadpole' tail may be visible in larger amateur telescopes. Herschel observed these galaxies on 13 March 1785 and was unable to resolve them, though he may have seen the northern extension; he described the pair as: 'Faint, much extended in the meridian.' Radial velocity: 6626 km/s. Distance: 296 mly. Diameter: north component, 172 000 ly; south component, 146 000 ly.

NGC4685

H398[3]	12:47.1	+19 28	GLX
S0‾:	2.3′ × 1.3′	13.60B	

The DSS image shows a lenticular galaxy with a peculiar morphology. The

NGC4676

©STScI/AURA/CALTECH/ROE

NGC4685

bright, elongated core has a faint outer envelope which consists of at least five broad extensions of matter, three directed towards the N, two directed towards the S. Telescopically, this is a small but moderately bright galaxy, seen as a thin sliver of light elongated SSE–NNW; it is fairly well defined with a tiny star-like core. Radial velocity: 6729 km/s. Distance: 301 mly. Diameter: 201 000 ly.

NGC4689

H128[2]	12:47.8	+13 46	GLX
SA(rs)bc	6.0′ × 5.3′	11.58B	

Photographs show a dusty, multi-arm spiral galaxy viewed almost face-on. The small, brighter, round core has tightly wound spiral arms emerging directly from it. Many knots and condensations mark the spiral arms and dust patches and lanes delineate the arms. There is an extensive and tenuous outer spiral pattern. Visually, the galaxy is large, somewhat faint and very diffuse. It appears to be almost round and is slightly brighter to the middle, but no core is visible. The edges are poorly defined and the galaxy is about 3.0′ S of a pair of magnitude 10 field stars. Herschel considered the galaxy resolvable. Radial velocity: 1577 km/s. Distance: 70 mly. Diameter: 123 000 ly.

NGC4692

H381[2]	12:47.9	+27 13	GLX
E[+]:	1.2′ × 1.1′	13.70B	

Visually and photographically this elliptical galaxy's appearance is quite

similar. It is a moderately faint, slightly oval patch of light extended E–W, and is well defined at the edges, with an even-surface-brightness envelope. A bright field star is 6.8′ SSE. Radial velocity: 8011 km/s. Distance: 358 mly. Diameter: 125 000 ly.

NGC4710

H95[2]	12:49.6	+15 10	GLX
SA(r)0[+]? sp	4.9′ × 1.2′	11.85B	

The DSS image reveals an edge-on lenticular galaxy with a broad, bright central bulge and bright extensions. A thin, prominent dust lane crosses the core. Higher-resolution images show that the dust lane curves as it wraps itself around the core and a compact bright patch obscures the dust lane SSW of the core. In a moderate-aperture telescope, this bright galaxy is quite flat and elongated NNE–SSW. It is very bright along its major axis and a little brighter to the middle. The edges are

very well defined and the tips taper to points. A magnitude 10 field star is located 1.5′ to the E. Radial velocity: 1091 km/s. Distance: 49 mly. Diameter: 69 000 ly.

NGC4725

H84[1]	12:50.4	+25 30	GLX
SAB(r)ab pec	11.0′ × 7.6′	9.92B	

This is a large, bright, barred spiral galaxy with a brighter, elongated core and fainter bar. Photographs show the bar displays small patches and streaks of dust. The inner spiral arms are brighter than the bar and form an apparently complete ring. The structure is very complex with sinewy dust lanes involved with brighter HII regions. The outer spiral structure is fainter, broader and for the most part smooth textured with occasional brightenings, particularly to the NNE and SSW. The galaxy is large and quite bright visually; it is a very fat oval, quite

NGC4710

NGC4725

©STScI/AURA/CALTECH/ROE

smooth-textured but very ill defined at the edges and oriented NE–SW. Averted vision increases the size of the disc considerably but there is no sign of the bar. The brightness of the disc is fairly even and a much brighter central region has a star-like, bright core. NGC4712 is seen 12.0′ to the WSW; it is a fainter, well defined, elongated N–S oval patch of light that was missed by Herschel. Radial velocity: 1207 km/s. Distance: 54 mly. Diameter: 172 000 ly.

NGC4747

H344[2]	12:51.8	+25 47	GLX
SBcd? pec sp	3.5′ × 1.2′	12.99B	

This galaxy has also been catalogued as Arp 159. The DSS photograph shows an irregular galaxy, possibly of the barred type, with a brighter, knotty bar offset towards the SW. Dust patches are involved with the bar, which is embedded in a fainter outer envelope. A faint plume emerges from the bar and broadens as it extends towards the ENE

for almost 7.0′ (almost 100 000 ly) before looping back towards the galaxy. Although fairly large, this galaxy is a little dim visually and is best seen at medium magnification. It is elongated NNE–SSW and is a little brighter along its major axis, with moderately well defined edges. Radial velocity: 1192 km/s. Distance: 53 mly. Diameter: 54 000 ly.

NGC4752

H82[3]	12:51.5	+13 47	GLX
0.8′ × 0.4′	15.30B	14.5V	

This galaxy has been suggested as the object that Herschel saw on 12 April 1784 and described as: 'Very faint, small, extended, resolvable.' It is NW of the position he indicated and the French observer Guillaume Bigourdan was unable to find anything at Herschel's location. Photographs show a very small and faint spiral galaxy, oriented NNW–SSE with a brighter core and two well defined spiral arms. This object is

not plotted in the first edition of *Uranometria 2000.0*. Visually, the galaxy is extremely dim and is seen only intermittently as a roundish, hazy patch of light with a slightly brighter core in an ill defined envelope. High magnification is needed to confirm it. The present author feels that it is doubtful that Herschel observed this object. Radial velocity: 11 231 km/s. Distance: 501 mly. Diameter: 117 000 ly.

NGC4758

H70[3]	12:52.7	+15 51	GLX
Im:	3.1′ × 0.7′	13.56B	

Images show a faint, edge-on, probably spiral galaxy, oriented NNW–SSE. The brighter, much extended central region is offset towards the NNW and the extensions are asymmetrical with the SSE one much longer. There is a prominent, elongated dust lane in the spiral arm to the S and a dust lane crosses the core at a very slight angle to the major axis. A faint field star is 0.6′ N of the core. Visually, this is a very faint, nebulous patch of light, fairly even in surface brightness and extended SSE–NNW. The edges are moderately well defined. Radial velocity: 1213 km/s. Distance: 54 mly. Diameter: 49 000 ly.

NGC4789

H345[2]	12:54.3	+27 04	GLX
SA0:	1.4′ × 1.1′	13.09B	12.1V

This lenticular galaxy is elongated N–S with a bright core and a slightly fainter outer envelope. The DSS image shows a small, compact galaxy 1.1′ to the NE which may be a physical companion. The edge-on lenticular galaxy NGC4787 is 2.9′ W while the large, and very faint dwarf irregular galaxy NGC4789A is 5.5′ to the NW. Telescopically, NGC4789 is fairly bright and lies 0.6′ S of a bright field star; it is very slightly elongated N–S with a bright, somewhat ill defined envelope and bright stellar core. NGC4787 is also visible but is very faint, appearing as a diffuse patch of light. NGC4789 is probably a member of the Abell 1656 galaxy cluster, though its radial velocity is considerably higher than the cluster average. Radial velocity: 8373 km/s. Distance: 374 mly. Diameter: 152 000 ly.

NGC4793

H93[1]	12:54.6	+28 56	GLX	
SAB(rs)c	2.8′ × 1.5′	12.30B	11.9V	

The DSS image shows a bright, inclined spiral galaxy with a bright, elongated core. Bright, knotty spiral arms emerge from the core and transition to a faint, outer spiral pattern. Visually, this is quite a bright galaxy, an elongated oval with a bright envelope and slightly brighter, elongated core. The edges are somewhat diffuse. The galaxy is oriented ENE–WSW and the envelope is a little mottled. Radial velocity: 2499 km/s. Distance: 112 mly. Diameter: 91 000 ly.

NGC4798

H382[2]	12:55.0	+27 25	GLX
S0⁻:	1.0′ × 0.6′	14.19B	

This galaxy is also identified as NGC4797, though it is plotted in the first edition of *Uranometria 2000.0* as NGC4798 and identified as NGC4798 in *The Scientific Papers*. Photographs show an inclined lenticular galaxy with a narrow dust lane bordering the bright core along the NW flank of the galaxy. Visually, the galaxy is faint but readily visible as an oval patch of light, even in surface brightness and fairly well defined. In a medium-magnification field, NGC4807 is also visible to the NE, smaller and slightly fainter; it is a round, well defined patch of light. Radial velocity: 7873 km/s. Distance: 351 mly. Diameter: 102 000 ly.

NGC4816

H383[2]	12:56.2	+27 45	GLX	
S0⁻:	1.3′ × 1.1′	13.64B	12.8V	

Photographs show this lenticular galaxy has an elongated core, oriented E–W, and a broad, fainter outer envelope. This galaxy is in the western portion of the Abell 1656 galaxy cluster. The surrounding field is peppered with

faint galaxies and the compact galaxy CGCG160-023 is 1.6′ ENE. Visually, NGC4816 is a relatively bright, well defined, roundish patch of light of fairly even surface brightness with a faint field star 0.5′ to the NE. Radial velocity: 6926 km/s. Distance: 309 mly. Diameter: 117 000 ly.

NGC4819

H346[2]	12:56.5	+26 59	GLX
(R′)SAB(r)a:	1.1′ × 0.8′	14.13B	

This is a possible barred spiral galaxy: photographs show a slightly inclined object with a bright, lens-shaped core oriented ESE–WNW, in a featureless outer envelope oriented SSE–NNW. This is a likely member of the Abell 1656 galaxy cluster, though it is S of the main mass. The fainter galaxy NGC4821 is 3.2′ S. NGC4819 is moderately bright visually, appearing round and very broadly brighter to the middle with moderately well defined edges. NGC4821 is visible as a smaller, round and condensed object and is only a little fainter. Radial velocity: 6472 km/s. Distance: 289 mly. Diameter: 93 000 ly.

NGC4827

H384[2]	12:56.7	+27 11	GLX
S0:	1.3′ × 1.1′	13.89B	

This Abell 1656 cluster member is S of the main mass and somewhat isolated from the main cluster but its membership appears to be definite. Photographs show a small lenticular galaxy with a bright core, very slightly elongated NE–SW and a fainter outer envelope. Visually, the galaxy is quite bright, small and fairly round with well defined edges and even surface brightness except for the brighter, star-like core. Radial velocity: 7640 km/s. Distance: 341 mly. Diameter: 129 000 ly.

NGC4839

H386[2]	12:57.4	+27 30	GLX	
cD;SA0	4.0′ × 1.9′	13.07B	12.1V	

This lenticular galaxy appears to be one of the principal members of the

Abell 1656 galaxy cluster, though it is situated along its southern outskirts. Photographs show a bright, slightly elongated core, oriented ENE–WSW, with an extensive fainter outer envelope. A small compact companion galaxy is in the halo immediately SW of the core. NGC4840 is 7.0′ to the NNE, while NGC4842 2.5′ to the ESE and the immediate field is peppered with faint galaxies. Telescopically, NGC4839 is a moderately bright, slightly oval galaxy of even surface brightness with quite well defined edges. NGC4842 is just a little fainter and smaller and is a concentrated star-like spot. Radial velocity: 7373 km/s. Distance: 329 mly. Diameter: 383 000 ly.

NGC4840

H385[2]	12:57.5	+27 37	GLX
E1:/SAB0	0.7′ × 0.7′	14.70B	

Photographs show that this elliptical galaxy is slightly elongated almost due E–W with a faint outer envelope. NGC4839 is in the field, 7.0′ to the SSW. Visually, NGC4840 is a little fainter and smaller than NGC4839 and is a slightly oval well defined object of even surface brightness. Radial velocity: 6099 km/s. Distance: 272 mly. Diameter: 55 000 ly.

NGC4841

H387[2]	12:57.5	+28 29	GLX
E⁺ pec	1.3′ × 1.1′	13.5B	

This galaxy is usually designated NGC4841A and forms a physical pair with NGC4841B located 0.7′ to the NE. The pair are on the northern edge of the Abell 1656 galaxy cluster. The minimum core-to-core separation between the two galaxies would be 62 000 ly at the presumed distance. Photographs show a small, elliptical galaxy, very slightly elongated almost due E–W with a fainter outer envelope, situated 3.25′ NNW of a field star. The companion is smaller and almost round. Telescopically, the pair of galaxies are barely resolved and look more like an elongated single object but high magnification brings out the 'double nucleus' impression; the SW galaxy has a particularly bright core. Herschel did not resolve two separate objects: his 11 April 1785 description of the object: 'Faint, pretty large.' implies that he saw both galaxies as a single, extended object. Radial velocity: 6790 km/s. Distance: 303 mly. Diameter: 115 000 ly.

NGC4869

H388[2]	12:59.4	+27 55	GLX
E3	0.7′ × 0.7′	14.37B	

The DSS image shows a small, elliptical galaxy, very slightly elongated almost due N–S with a fainter outer envelope. A compact companion galaxy lies immediately to the NNW. This galaxy is located 4.0′ SW of NGC4874, one of the two dominant galaxies in the Abell 1656 galaxy cluster. Visually, NGC4869 is very small, compact and round with well defined edges. Radial velocity: 6873 km/s. Distance: 307 mly. Diameter: 62 000 ly.

NGC4872

H389[2]	12:59.6	+27 57	GLX
SB0	0.6′ × 0.4′	14.55B	

Photographs show that this small, elliptical galaxy is round with an extremely faint outer envelope oriented ESE–WNW and is located 0.9′ SSW of NGC4874. It is an enduring mystery why William Herschel did not discover NGC4874 even though he is credited with NGC4872 and NGC4869, both much more feeble objects. Herschel's 11 April 1785 entry describes them as: 'Two. The time taken between them.' The next entry is for the massive galaxy NGC4889 to the E, so NGC4874 would have had to have been at the edge of the telescopic field, or just out of view, to be overlooked. John Herschel missed it as well; Heinrich d'Arrest is credited with the discovery of NGC4874 in 1864. The present author strongly suspects that William Herschel is much more likely to have observed NGC4874 rather than NGC4872. Telescopically, NGC4872 is intermittently visible as a concentrated fuzzy 'star' SSW of NGC4874. Radial

velocity: 7223 km/s. Distance: 322 mly. Diameter: 56 000 ly.

NGC4889

H391[2]	13:00.1	+27 58	GLX
cD;E4;Db	2.9′ × 1.9′	12.49B	12.2V

Photographically and visually this galaxy presents similar appearances. It is a bright, elliptical galaxy, with a bright, elongated core in an extensive, fainter envelope, oriented almost due E–W. In photographs, several smaller, fainter galaxies are embedded in the outer envelope. Patience and superb skies will reveal many faint, compact galaxies in the field. Along with NGC4874, this is the dominant galaxy of the Abell 1656 galaxy cluster, which is one of the richest clusters in the sky for amateur instruments, Abell 1656 is classified as type 'B' in the Rood–Sastry classification system, owing to the dominance of the giant galaxies NGC4889 and NGC4874 in the field. Radial velocity: 6509 km/s. Distance: 291 mly. Diameter: 245 000 ly.

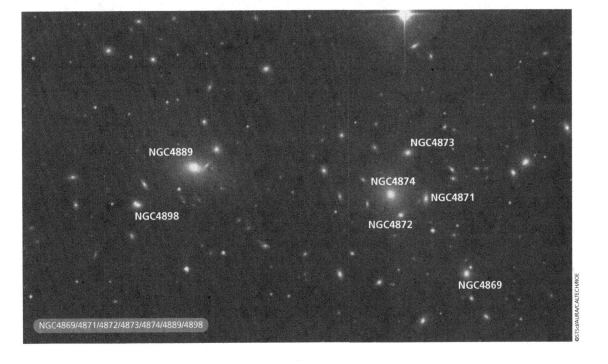

NGC4869/4871/4872/4873/4874/4889/4898

NGC4892

H390[2]	13:00.0	+26 54	GLX
Sb	1.3′ × 0.3′	14.53B	

Photographs show this Abell 1656 member as an edge-on galaxy, probably a spiral, with a small, brighter core and slightly fainter extensions. Two field stars are immediately ENE of the galaxy. It is a moderately faint object visually and best seen at medium magnification as an extended oval of light, broadly brighter along the major axis; it is fairly well defined and oriented NNE–SSW. Radial velocity: 5918 km/s. Distance: 265 mly. Diameter: 100 000 ly.

IC4051

H363[3]	13:00.9	+28 00	GLX
E5	1.0′ × 0.9′	14.22B	

This is the southernmost galaxy in a chain of four, which begins with NGC4907. Mistakenly equated with entry h1510 in John Herschel's *Catalogue of Nebulae and Clusters of Stars* (Herschel, 1864), the position was subsequently found to agree closely with the galaxy IC4051, recorded by the French observer Bigourdan at the Paris Observatory. The DSS image shows an elliptical galaxy with a bright, elongated core and a fainter outer envelope, oriented E–W. The 15′ DSS image shows no bright galaxies to the E of IC4051, but 20 objects are seen to the W. Visually, the galaxy appears as a slightly oval patch of light that is well concentrated and a little brighter to the middle. Radial velocity: 8808 km/s. Distance: 394 mly. Diameter: 114 000 ly.

NGC4911

H392[2]	13:00.9	+27 47	GLX	
SAB(r)bc	1.4′ × 1.3′	13.66B	13.2V	

This Abell 1656 galaxy cluster member is a face-on spiral galaxy; photographs show a bright, oval core, slightly offset to the NW. The outer envelope features chaotic spiral structure with breaks in the N and the S. A small, elongated companion galaxy is 0.5′ to the SW. Visually, NGC4911 is a fairly bright, round, diffuse patch of light of even

surface brightness, and is fairly well defined at the edges. Radial velocity: 7999 km/s. Distance: 357 mly. Diameter: 145 000 ly.

NGC4921

H393[2]	13:01.4	+27 53	GLX
SB(rs)ab	2.4′ × 2.1′	13.49B	

There are relatively few spiral galaxies in the Abell 1656 galaxy cluster but this galaxy is a large and impressive example of one. Photographs show a large barred spiral galaxy viewed face-on. The small, bright, round core is in a fainter bar oriented ESE–WNW, connected to two fainter spiral arms in an extensive outer envelope. Prominent dust patches and lanes are to the ESE, S and WNW of the core. Visually, the galaxy can be seen in the same medium-magnification field as NGC4923, NGC4927 and IC4051, which are all Herschel objects. It is the largest in the field, a round, diffuse patch of light that is fairly well defined with a small,

brighter core. Radial velocity: 5497 km/s. Distance: 246 mly. Diameter: 171 000 ly.

NGC4923

H394[2]	13:01.5	+27 51	GLX
(R′)SA(r)0⁻?	0.8′ × 0.8′	14.69B	

Situated 2.5′ SSE of NGC4921, photographs show that this elliptical galaxy has a slightly elongated core in a fainter outer envelope, oriented almost due E–W. It is very likely physically associated with NGC4921, which is larger. Visually, it appears small, round and concentrated and is well condensed with well defined edges. Radial velocity: 5499 km/s. Distance: 246 mly. Diameter: 57 000 ly.

NGC4927

H364[3]	13:02.0	+28 00	GLX
SA0⁻:	0.8′ × 0.6′	14.79B	

In photographs, this lenticular galaxy, located 9.75′ NE of NGC4921, has an

NGC4921

©STScI/AURA/CALTECH/ROE

elongated core in a fainter outer envelope, oriented NNE–SSW. Visually, it is a small, concentrated galaxy with a moderately bright envelope; it is almost round and is well defined with a star-like, brighter core and a faint field star immediately NNE. It is an Abell 1656 galaxy cluster member but evidently is not associated with the large spiral to the SW. Radial velocity: 7779 km/s. Distance: 348 mly. Diameter: 81 000 ly.

NGC4944

H395[2]	13:03.8	+28 11	GLX
SA0	1.8′ × 0.6′	13.8B	

This is a small lenticular galaxy viewed almost edge-on. Photographs show a bright, elongated core with slightly fainter extensions. A compact possible companion galaxy is 1.1′ to the E while a field star is 1.75′ NNE. This galaxy is located on the eastern outskirts of the Abell 1656 galaxy cluster. Visually, it is moderately bright; it appears as a well defined, extended oval oriented E–W with a bright envelope which brightens to a stellar core. Radial velocity: 7006 km/s. Distance: 313 mly. Diameter: 164 000 ly.

NGC4952

H396[2]/H303[3]	13:05.0	+29 08	GLX
E	1.8′ × 1.1′	13.77B	13.0V

The visual and photographic appearances of this Abell 1656 galaxy cluster member are quite similar. It is a fairly bright galaxy visually, a slightly oval patch of light, fairly well defined and oriented NNE–SSW with a bright, star-like core. It was observed twice by Herschel, first on 13 March 1785 and again on 11 April 1785. Radial velocity: 5988 km/s. Distance: 267 mly. Diameter: 140 000 ly.

NGC4957

H397[2]	13:05.2	+27 34	GLX
E3	1.1′ × 0.9′	14.00B	

This galaxy is situated ESE of the main mass of the Abell 1656 galaxy cluster

but is very likely a member. Photographs show an elliptical galaxy with an elongated core in a fainter outer envelope, oriented E–W. Visually, this galaxy can be seen in the same medium-magnification field as NGC4960 and NGC4961. It is fairly bright and appears as an oval patch of light, which is brighter to a stellar core, with fairly well defined edges. Radial velocity: 6925 km/s. Distance: 309 mly. Diameter: 99 000 ly.

NGC4961

H398[2]	13:05.8	+27 44	GLX
SB(s)cd	1.6′ × 1.1′	13.06B	

Images show a slightly inclined spiral galaxy with a large, bright, elongated core. The spiral pattern appears slightly disturbed: the spiral arms are fragmentary and asymmetrical with knotty condensations and dust patches. At medium magnification, the galaxy is a little larger than, but similar in brightness to NGC4957, which is visible to the SSW. It appears as an oval patch of light, broadly brighter to the middle and elongated E–W and is a little diffuse at the edges. Radial velocity: 2551 km/s. Distance: 114 mly. Diameter: 53 000 ly.

NGC4966

H304[3]	13:06.3	+29 04	GLX
Sa	1.2′ × 0.7′	14.02B	13.5V

Situated 2.4′ NE of a bright field star, photographs reveal a slightly inclined spiral galaxy with an oval, brighter core. One spiral arm emerges from the core in the SE and appears to terminate on a brighter condensation. After a gap another condensation marks the continuation of the arm as it curves N, then E around the core of the galaxy. A much smaller, probable companion galaxy is 0.5′ to the W. NGC4966 is somewhat hard to see visually owing to the presence of the bright field star. It appears diffuse and somewhat ghostly, broadly brighter to the middle with hazy edges. This is probably a member of the Abell 1656 galaxy cluster. Radial

NGC4966

©STScI/AURA/CALTECH/ROE

velocity: 7059 km/s. Distance: 315 mly. Diameter: 110 000 ly.

NGC4979

H346[3]	13:07.6	+24 49	GLX
Sbc	1.7′ × 1.3′	15.69B	

The DSS image clearly shows a face-on barred spiral galaxy. The small, bright, round core is embedded in a fainter bar, oriented E–W. The bar is attached to a ring-like inner spiral pattern with four faint arms emerging, three on the W flank. Visually, the galaxy is fairly faint and diffuse, and is seen as a round, evenly bright patch of light with ill defined edges. It is best seen at medium magnification and is 4.7′ WSW of a magnitude 8 field star which greatly hinders visibility. Radial velocity: 6341 km/s. Distance: 283 mly. Diameter: 140 000 ly.

NGC4983

H365[3]	13:08.4	+28 19	GLX
S0/a	1.1′ × 0.6′	14.90B	

Photographs show a highly inclined galaxy, probably barred, with a bright elongated core and a slightly fainter, curved bar in a faint outer envelope, oriented ESE–WNW. It is very faint visually, a small, ill defined patch of light, irregularly round and best seen with averted vision. The surface brightness is even and the edges are diffuse. It is a possible member of the Abell 1656 galaxy cluster. Radial velocity: 6552 km/s. Distance: 292 mly. Diameter: 94 000 ly.

NGC5000

H366[3]	13:09.8	+28 54		GLX
SB(rs)bc	2.3′ × 1.3′	13.99B	13.4V	

Images show a face-on barred spiral galaxy with a round, brighter core and an E–W oriented bar. The bar is attached to a partial ring structure which brightens around the point of attachment. The ring expands to a relaxed spiral pattern with the southern arm terminating at a brighter condensation. Telescopically, NGC5000 is quite a faint galaxy, difficult to see and requiring averted vision. It appears as an extended oval, oriented E–W and is situated between two faint field stars. It is fairly even in surface brightness and is somewhat well defined; only the core, bar and inner disc are visible. Radial velocity: 5630 km/s. Distance: 251 mly. Diameter: 168 000 ly.

NGC5004

H305[3]	13:11.1	+29 38		GLX
S0	1.5′ × 1.2′	14.16B	13.2V	

This galaxy is the brightest of a triple system which includes IC4210 (aka NGC5004B) 5.2′ to the NNW and NGC5004C 3.5′ to the S. Photographs show that NGC5004 is a lenticular galaxy with a bright, elongated core and a fainter outer envelope. Although there is a considerable spread in the radial velocity of the triplet, the minimum core-to-core separation across the group would be about 770 000 ly at the presumed distance of NGC5004. Visually, the galaxy is moderately faint, an oval patch of light, oriented almost due N–S, well defined and with a star-like, brighter core. The three galaxies may be visible together in large-aperture telescopes. Radial velocity: 7069 km/s. Distance: 316 mly. Diameter: 138 000 ly.

NGC5012

H85[1]	13:11.6	+22 55	GLX
SAB(rs)c	2.7′ × 1.6′	12.79B	

This inclined spiral galaxy has a slightly brighter core and images show a bright well defined internal spiral structure.

The spiral pattern broadens to fainter outer structure involving five distinct spiral branches. The arms are grainy with evidence of dust patches and lanes. A field star is visible at the tip of a spiral arm 0.5′ N of the core. Telescopically, this is a large and bright galaxy; it is well seen at medium and high magnification as an extended oval of light, oriented N–S, a little brighter along the major axis and hazy at the edges. The envelope is somewhat mottled and is also quite opaque. Radial velocity: 2623 km/s. Distance: 117 mly. Diameter: 92 000 ly.

NGC5016

H356[2]	13:12.1	+24 06	GLX
SAB(rs)c	1.7′ × 1.3′	13.13B	

Photographs show a slightly inclined spiral galaxy with a large, bright central region and fainter, multi-arm spiral structure. The knotty arms are defined by dust lanes throughout the outer disc. The galaxy is moderately bright visually, slightly oval in form and extended ENE–WSW. It is very broadly brighter to the middle, there is no core visible and the edges are hazy and ill defined. A faint field star is 1.25′ to the N. Radial velocity: 2620 km/s. Distance: 117 mly. Diameter: 58 000 ly.

NGC5032

H367[3]	13:13.4	+27 48	GLX
SB(r)b	2.0′ × 1.2′	13.70B	

Photographs reveal a slightly inclined barred spiral galaxy with a small, slightly elongated brighter core. The faint bar is attached to an inner pseudo-ring which expands to a relaxed spiral structure, elongated N–S. Companion galaxy NGC5032B is 2.4′ to the S and may be visible in moderately large-aperture telescopes. Situated about 3° E of the centre of the Abell 1656 galaxy cluster, the radial velocity suggests that both galaxies are probably outlying members of that cluster. Although somewhat faint, NGC5032 holds magnification well and is seen as a round, rather diffuse patch of light, brighter to the middle and fading uncertainly at the edges. The galaxy is

bracketed by two faint field stars, the first 1.25′ to the E the other 1.25′ to the SW. Radial velocity: 6426 km/s. Distance: 286 mly. Diameter: 167 000 ly.

NGC5053

H7[6]	13:16.4	+17 42	GC
CC11	10.0′ × 10.0′	9.0V	

This large, faint globular cluster is slightly elongated E–W. It is well resolved in photographs but has a low concentration of stars. It is very loosely structured with little concentration to the centre. Visually, this is a large, though very faint, ephemeral glow which is almost round in form with poorly defined extremities. The surface brightness is uniform across the face of the globular cluster and even at medium magnification many faint cluster members can be resolved. The brightest cluster members are magnitude 13.8. The cluster is ESE of the much brighter globular M53. Herschel discovered the cluster on 14 March 1784, and remarked: 'An extremely faint cluster of extremely small stars with resolvable nebulosity. 8.0′ or 10.0′ in diameter. Verified 240 (magnification) beyond doubt.' Distance: 52 500 ly.

NGC5056

H306[3]	13:16.2	+30 56	GLX
Scd:	1.7′ × 1.0′	13.71B	

This galaxy forms a probable physical pair with NGC5057, situated 5.75′ to the NNE and is probably associated with NGC5065 20.0′ to the ENE as well. Images reveal a slightly inclined spiral galaxy with a large, bright core. Three principal spiral arms emerge from the core, fading to a broader spiral pattern after less than half a revolution. Telescopically, NGC5056 and NGC5057 can be seen in the same field. Located 3.25′ N of a bright field star, the galaxy is a moderately faint, slightly oval patch of light, elongated N–S and broadly brighter to the middle. The edges are hazy and not well defined. Radial velocity: 5624 km/s. Distance: 251 mly. Diameter: 124 000 ly.

NGC5053

©STScl/AURA/CALTECH/ROE

an elongated and very slightly curved patch of light, moderately well defined at the edges. The surface brightness of the envelope is fairly even except for a brighter, star-like spot which is probably a faint field star located 0.25′ WSW of the core. Radial velocity: 2173 km/s. Distance: 97 mly. Diameter: 65 000 ly.

NGC5116

H368[3]	13:22.9	+26 59	GLX
SB(s)c:	2.2′ × 0.7′	13.36B	

This is a highly inclined galaxy, probably a barred spiral. Photographs show a brighter central region connected to a curving bar which expands into two spiral arms in a fainter outer envelope. The galaxy is somewhat faint visually but is well seen at medium and high magnification as an elongated streak of light oriented NE–SW which is a little brighter along the centre line of the major axis. The ends taper slightly and the edges are quite well defined. Radial velocity: 2915 km/s. Distance: 130 mly. Diameter: 83 000 ly.

NGC5057

H307[3]	13:16.5	+31 01	GLX
S0	1.3′ × 1.2′	14.44B	

The DSS image reveals a face-on ring galaxy with a round, bright core and a slightly fainter, N–S elongated envelope. Surrounding this is a broad, faint, smooth ring which does not appear connected to the central region. Visually, the galaxy is round and appears a little larger and brighter than NGC5056. It is fairly well defined at the edges with a star-like, brighter core intermittently visible. Radial velocity: 5915 km/s. Distance: 264 mly. Diameter: 100 000 ly.

NGC5065

H308[3]	13:17.6	+31 04	GLX
Sd	1.4′ × 0.8′	14.26B	

This is an inclined spiral galaxy. Photographs show it is possibly barred, with an elongated, brighter core and fainter, fragmentary spiral arms, elongated E–W. The galaxy is quite faint visually; it is an irregularly round patch of light, even in surface brightness with hazy edges. Radial velocity: 5583 km/s. Distance: 249 mly. Diameter: 102 000 ly.

NGC5089

H327[2]	13:19.6	+30 15	GLX
Sb	2.3′ × 0.9′	11.13B	

Images show a highly inclined galaxy, probably a spiral, with a very bright, irregular and elongated core. The central region has a number of bright condensations around its periphery, particularly to the E and W. Very weak and faint spiral arms emerge from the N and S. The northern arm features a number of condensations and the southern arm extends about 0.7′ to the ESE. The galaxy is moderately faint visually, appearing as

NGC5180

H71[3]	13:29.4	+16 49	GLX
S0?	1.4′ × 1.0′	13.97B	

Situated 5.9′ SW of a magnitude 7 field star, this elliptical or lenticular galaxy has a bright core and a fainter outer envelope, elongated NE–SW. A faint star or companion galaxy is involved in the outer envelope immediately S of the core. A field star is 0.25′ to the SE and is faintly visible in a moderate-aperture telescope. The galaxy is faint and almost round with even surface brightness and a stellar core. The edges are fairly well defined. Herschel came across this object on 21 March 1784 and commented: 'Three small stars with suspected nebulosity. 240 (magnification) left some doubt.' Radial velocity: 6706 km/s. Distance: 300 mly. Diameter: 122 000 ly.

Corona Borealis

This small but distinct constellation is home to only ten Herschel objects, nine of which are Class III: very faint nebulae. These nine are also fairly remote objects and show little in the way of detail. All are very likely situated 400–500 mly away, so they pose something of a challenge to owners of moderate-aperture telescopes.

NGC5958

H399[2]	15:34.7	+28 40	GLX
S?	1.2′ × 1.0′	13.29B	

Photographs show a small face-on galaxy, probably a spiral, with a bright, irregular core. A suspected spiral arm emerges from the core in the E and curves to the N. North of the core are three bright condensations in a line oriented ESE–WNW. A fainter, grainy outer envelope surrounds the brighter inner region. Visually, the galaxy is faint but readily seen with direct vision as a round patch of light, broadly brighter to the middle. The edges are fairly well defined and the galaxy is framed by a triangle of faint field stars. Radial velocity: 2113 km/s. Distance: 94 mly. Diameter: 33 000 ly.

NGC6001

H371[3]	15:47.7	+28 38	GLX
Sc	1.2′ × .1.2′	14.34B	

This is a face-on spiral galaxy, probably barred. Photographs show a small, slightly elongated core with a possible fainter bar. Two spiral arms emerge from each end of the bar branching off quickly to give two principal spiral arms and at least two fainter spiral fragments. A faint, edge-on field galaxy is located 1.0′ to the SW. Telescopically, the galaxy is moderately faint and almost round and a little brighter to the middle. The smooth, opaque envelope is fairly well defined at the edges. Radial velocity: 10 074 km/s. Distance: 450 mly. Diameter: 157 000 ly.

NGC6038

H622[3]	16:02.6	+37 21	GLX
Sc	1.2′ × 1.0′	14.22B	

This small, almost face-on spiral galaxy has a small, round and bright core. Images show two spiral arms emerge from the slightly fainter central region, each of which branches to give an additional spiral arm after about a half revolution. Telescopically, the galaxy is situated 1.3′ NW of a prominent field star. It is moderately faint and visible as a round patch of light, even in surface brightness with well defined edges. Radial velocity: 9532 km/s. Distance: 426 mly. Diameter: 149 000 ly.

NGC6089

H889[3]	16:12.7	+33 02	GLX
Compact	0.6′ × 0.6′	14.88B	

The DSS image reveals a close pair of very faint galaxies. The principal one

NGC6089

©STScI/AURA/CALTECH/ROE

is a small, face-on galaxy, probably a spiral, with a small, bright, round core well offset to the N in a faint, oval envelope, elongated N–S. The companion is a smaller spiral attached to the NE. The companion has a small, bright round core but only one spiral arm, curving away to the NE, is visible, implying that it may be situated behind the larger galaxy. In a moderate-aperture telescope the galaxy is small and very faint, occasionally betraying a faint stellar core at high magnification. The outer envelope is poorly defined and seems oval and slightly elongated NNE–SSW. This may be the combined image of both galaxies. Radial velocity: 9711 km/s. Distance: 434 mly. Diameter: 76 000 ly.

NGC6103

H888[3]	16:15.7	+31 57	GLX
S?	1.0′ × 0.6′	14.45B	

Images reveal a small, slightly inclined, probably spiral galaxy with a small core only a little brighter than the surrounding envelope which has two knotty spiral arms in a fainter, broader disc. It is oval and oriented almost due E–W. Visually, the galaxy is very faint and broadly concentrated to the middle. It appears almost round with moderately well defined edges. It is situated between two faint field stars. Radial velocity: 9557 km/s. Distance: 427 mly. Diameter: 124 000 ly.

NGC6104

Photographs show the main envelope is bright, oval and oriented NNE–SSW. Attached to the WNW is a bright, compact object, possibly a companion galaxy, which bulges slightly from the envelope of the main galaxy. Opposite the companion, a faint plume of matter emerges from NGC6120 in the ESE while a broader, very faint plume is ejected towards the N. Several galaxies are in the field, including NGC6119 2.3′ to the NNW and NGC6122 4.4′ to the ENE. Visually, NGC6120 is the brightest of the NGC trio visible in a high-magnification field. It is small, condensed and opaque with a star-like core which appears offset towards the NE. Radial velocity: 9334 km/s. Distance: 417 mly. Diameter: 73 000 ly.

NGC6129

H891[3]	16:21.8	+37 59	GLX
0.8′ × 0.7′	14.76B		

This small, elliptical galaxy has a bright, elongated core and a fainter outer envelope oriented ESE–WNW. It is unclassified but may be an S0 galaxy. The DSS image reveals half a dozen fainter galaxies in the field, the most prominent being MCG+6-36-36 located 2.0′ to the WNW. Visually, NGC6129 is small and very faint, a round spot showing little concentration to the centre, while the edges are fairly well defined. Radial velocity: 10 113 km/s. Distance: 452 mly. Diameter: 105 000 ly.

NGC6104

H688[3]	16:16.5	+35 42	GLX
S(R)pec/Pec	1.0′ × 0.8′	14.34B	

The DSS image shows a small face-on galaxy with a peculiar morphology. The small, bright, round core is embedded in a bright ring, bordered on the inside by an evident dust lane. There is a bright condensation attached immediately E of the core. The condensation may be the core of a second galaxy; there are hints of a spiral arm projecting obliquely from the plane of the principal galaxy in the N, suggesting a possible merger or near encounter between two separate objects. The smaller galaxy MCG+6-36-12 is 3.8′ ESE. Visually, this faint galaxy is brighter to the middle, though no core is noted. It is moderately well defined and very slightly elongated E–W. The MCG galaxy is visible as a faint, round spot to the ESE. Radial velocity: 8573 km/s.

Distance: 383 mly. Diameter: 111 000 ly.

NGC6120

H623[3]	16:19.8	+37 47	GLX
Pec	0.6′ × 0.5′	14.31B	

This is a small, slightly inclined galaxy with a peculiar morphology.

NGC6120

NGC6137

H624[3]	16:23.1	+37 55	GLX
E	1.9′ × 1.2′	13.42B	

The DSS image shows an elongated lenticular galaxy with a large, bright, oval core and a faint outer envelope. The classification should probably therefore be about E5, or very possibly S0. The neighbouring galaxy NGC6137B is 1.75′ NNW and is probably an Sb spiral. The galaxies are likely physically related and the core-to-core separation at the presumed distance would be about 215 000 ly. There are several other smaller, fainter galaxies in the field. Visually, NGC6137 can be seen in the same medium-magnification field as NGC6129 and appears brighter and larger with a faint star-like core visible with averted vision. The galaxy is elongated N–S and very faint field stars border it to the E and NW. A faint, hazy stellar spot occasionally seen to the NNW may be NGC6137B. Herschel's discovery description of 17 March 1787, reads:

'Very faint, small, brighter to the middle, discovered with 300 [magnification].' Radial velocity: 9454 km/s. Distance: 422 mly. Diameter: 233 000 ly.

NGC6142

H892[3]	16:23.4	+37 16	GLX
Sb	1.7′ × 0.6′	14.71B	

Photographs show a highly inclined spiral galaxy with a bright, small, round core. A dust patch is immediately W of the core and overall the arms are dusty. Visually, this galaxy is a faint and rather diffuse object, more readily seen at medium magnification. A little brighter to the middle, it is elongated NNW–SSE. Radial velocity: 10 382 km/s. Distance: 464 mly. Diameter: 230 000 ly.

Corvus

This springtime constellation, located immediately south of Virgo, lay relatively low on the horizon for Herschel. Nevertheless, he enjoyed some success tracking down nebulae in the region. There are 22 entries in his catalogue for the constellation, all of them galaxies except for the bright planetary nebula NGC4361, which is a very interesting object visually. Also of note is the spectacular interacting pair NGC4038 and NGC4039, which are well seen in moderate-aperture telescopes and are very rewarding in a large aperture on the best of nights.

NGC4024

H295[2]	11:58.5	−18 21	GLX
SB0⁻	1.8′ × 1.5′	12.70B	

Photographs show a lenticular galaxy with a very bright, oval core and bright, stubby extensions. This area is surrounded by a slightly fainter, almost round outer envelope, very slightly elongated ENE–WSW. There are several bright stars in the field and three much smaller and fainter galaxies are located to the NNE. Visually, only the bright central region is noted; the galaxy is small but moderately bright, a well defined elongated oval oriented ENE–WSW with a small, brighter core. Radial velocity: 1537 km/s. Distance: 68 mly. Diameter: 34 000 ly.

NGC4027

H296[2]	11:59.5	−19 16	GLX
SB(s)dm	3.2′ × 2.7′	11.58B	11.7V

Photographs show a face-on spiral galaxy, probably barred, and this interesting galaxy along with the Magellanic-type galaxy NGC4027A located 4.0′ to the S has also been catalogued as Arp 22. The bright core is elongated E–W and a stubby spiral arm with two blunt branches emerges from the E and curves to the S with several bright, knotty condensations. A broader, grainy spiral arm emerges from the W and curves N, giving the galaxy a very asymmetrical appearance. The whole galaxy is embedded in a fainter outer envelope. Two field stars are in the envelope to the E. At the eyepiece this is a rewarding object, appearing as a bright, though peculiarly structured galaxy. The disc is quite diffuse and mottled, brightening to a rather large core. Patchy condensations are

NGC4027

©STScI/AURA/CALTECH/ROE

suspected in the outer envelope and the envelope appears crescent shaped, cradling the magnitude 14 field star located immediately ENE from the galaxy's core. Radial velocity: 1512 km/s. Distance: 67 mly. Diameter: 63 000 ly.

NGC4033

H508[2]	12:00.6	−17 51	GLX
E6/SA:0⁻	2.6′ × 1.1′	12.65B	

The classification of this galaxy is somewhat uncertain but photographs show an inclined galaxy with a bright, oval core in an extensive, elongated envelope, oriented NE–SW, indicating that the galaxy is a lenticular type. Telescopically, this is a bright and opaque galaxy; it is a flattened disc with a brighter core and tapered, well defined extensions. Radial velocity: 1462 km/s Distance: 65 mly. Diameter: 49 000 ly.

NGC4035

H279[3]	12:00.5	−15 57	GLX
(R')SAB(rs)bc pec	1.2′ × 1.1′	14.08B	

Photographs show a face-on spiral galaxy with a slightly elongated, bright core. One bright spiral arm emerges from the E and curves N with several bright condensations along its length. A fainter, broader spiral arm curves to the S. The galaxy is quite dim visually, appearing as a very slightly oval object, a little extended ENE–WSW and fairly even in surface brightness with hazy edges. Radial velocity: 1429 km/s. Distance: 64 mly. Diameter: 22 000 ly.

NGC4038

H28[4]	12:01.9	−18 52	GLX
SB(s)m pec	8.0′ × 2.1′	10.85B	

This is the northern component of the well-known interacting pair which is popularly known as the Ringtail Galaxy. Along with NGC4039 they are also catalogued as Arp 244. Photographically, NGC4038 is large, bright, irregular in structure and elongated E–W. A dense chain of matter with several brighter condensations emerges from the centre of the galaxy,

curving S, W, then N. The galaxy is surrounded by a fainter, smooth outer envelope. A long plume of matter emerges from the E and curves S. A disconnect occurs before the plume curves S, then SSW. Interestingly, Herschel classified this and NGC4039 as Class IV objects on discovering the system on 7 February 1785. He described it as: 'Pretty bright, large, opening with a branch, or two nebulae very faintly joined. The south is smallest.' In a moderate-aperture telescope, NGC4038 is to the N, larger and very slightly brighter than NGC4039. Both components are smooth and featureless at first glance, with NGC4038 much broader across its minor axis. A dark cavity, opening to the W, separates the two galaxies. They appear to be in contact to the E. In a large-aperture telescope, however, both galaxies are transformed into rich and detailed objects with NGC4038 displaying a slightly brighter core and a mottled arm curving W then N and E over the top of the core. A brighter condensation is seen in the arm to the E before the arm thins and connects with NGC4039. Neither plume of ejected matter is visible, however. There are several Herschel objects in a loose grouping here, as well as a number of fainter galaxies that are of similar radial velocity and may therefore be part of a group or cloud of galaxies. They include NGC3956, NGC3957 and NGC3981 in Crater, and NGC4024, NGC4027 and NGC4033 in Corvus. Projected on the sky, these galaxies cover an area of about 2.5° × 2.0°. At the presumed distance, the group would cover about 3 mly. Radial velocity: 1485 km/s. Distance: 66 mly. Diameter: 155 000 ly (including plume).

NGC4039

H28[4]	12:01.9	−18 53	GLX
SA(s)m pec	8.0′ × 2.5′	11.04B	

This is the southern component of Arp 244. Photographs show a large, bright,

NGC4038/4039

irregular structure, elongated NE–SW
with an extensive, fainter outer
envelope which is smooth textured.
Dense condensations of matter and
several dust patches are offset towards
the N in the envelope. A long plume of
matter emerges from the E and curves
N. Visually, the galaxy is visibly smaller,
narrower and a little fainter than its
companion to the N. It appears a little
brighter along a thin strip tracing its
major axis. In a large aperture it
broadens considerably into a long oval
disc oriented ENE–WSW with mottled
texture along the major axis. The edges
are well defined but not sharply so.
Radial velocity: 1484 km/s. Distance:
66 mly. Diameter: 155 000 ly.

NGC4050

H509[2]	12:02.9	−16 22	GLX
SB(r)ab	4.4′ × 2.5′	13.04B	

Photographs show this inclined spiral
galaxy may have a bar or simply an
elongated inner disc with a brighter
core. Three grainy spiral arms emerge,
one on the W, two on the E, expanding
to much fainter outer spiral structure. A
bright field star is about 5.0′ to the SSE.
The galaxy is well seen at medium
magnification, but somewhat faint,
appearing as a large, fat oval patch of
light, fairly even in surface brightness
with moderately well defined edges and
oriented E–W. Radial velocity:
1610 km/s. Distance: 72 mly. Diameter:
92 000 ly.

NGC4114

H533[3]	12:07.2	−14 11	GLX
(R')SAB(rs)a	2.0′ × 1.0′	13.94B	

Images show an inclined spiral galaxy,
probably barred, with a bright,
elongated, oval core. The two spiral
arms appear smooth and well defined
with possible dust patches E and W of
the core. Visually, only the bright
central region is seen; the galaxy is quite
faint and appears as a wedge-shaped,
elongated patch with even surface
brightness, oriented SE–NW. Radial
velocity: 3952 km/s. Distance: 176 mly.
Diameter: 103 000 ly.

NGC4177

H534[3]	12:12.8	−14 00	GLX
SA(rs)c pec:	1.7′ × 1.2′	13.27B	

This inclined barred spiral galaxy has a
brighter core and a stubby bar.
Photographs show a partial ring
structure attached to the bar extending
to two spiral arms. There is a bright
condensation in the spiral arm to the
WSW and very grainy texture
throughout. Telescopically, the galaxy
is somewhat faint, but it is seen as a well
defined oval of light with a slightly
brighter core and mottled envelope,
oriented ENE–WSW. Herschel recorded
the galaxy on 27 March 1786, and
noted the mottling, describing the
galaxy as: 'Very faint, pretty large, of
unequal light.' Radial velocity:
3942 km/s. Distance: 176 mly.
Diameter: 87 000 ly.

NGC4263

H535[3]	12:19.7	−12 15	GLX
SAB(rs)b pec	1.2′ × 0.6′	13.92B	

This small, bright galaxy displays a
somewhat peculiar morphology.
Photographs show a bright, elongated
central region, oriented ENE–WSW. The
spiral structure is as bright as the
central bar and core and forms an
almost complete ring, being open only
in the WNW, where the envelope is
broad and fainter. Despite its high
surface brightness in photographs, the
galaxy appears small and faint visually;
it is very slightly elongated ESE–WNW,
and is moderately well defined with even
surface brightness. Radial velocity:
4038 km/s. Distance: 180 mly.
Diameter: 63 000 ly.

NGC4361

H65[1]	12:24.5	−18 48	PN	
3a+2	2.3′ × 2.1′	10.30B	10.9V	

This planetary nebula is well seen in the
DSS photograph, which reveals the
magnitude 13.2 central star in an oval
inner nebulosity which has curiously
spiral ansae emerging in the NNE and
SSW as well as a fainter outer shell.
Telescopically, it is a detailed object, a

large and bright planetary nebula, well
seen at high magnification as an
irregularly round disc surrounding a
prominent, though not bright, central
star. The envelope is brightest around
the star, slightly oval and extended E–W
with diffuse extensions to the S and N.
A hazy secondary disc is visible at the
edges, which are a little ill defined.

NGC4462

H764[3]	12:29.3	−23 10	GLX	
SB(r)ab	3.2′ × 1.5′	12.79B	11.5V	

This highly inclined barred spiral
galaxy has a bright, very elongated
core and a short, slightly fainter bar.
Images show that the bar is attached
to a probable ring which brightens at
the point of attachment. The spiral
arms are faintly visible in the outer
envelope and there is a large, broad
dust lane along the galaxy's
southwestern flank. A bright optical
double star is 2.0′ to the ESE.

Telescopically, only the brighter inner region of the galaxy is visible, appearing as a small but moderately bright object. An extended oval disc involves a brighter core and is oriented ESE–WNW. Radial velocity: 1639 km/s. Distance: 73 mly. Diameter: 66 000 ly.

NGC4714

H536[3]	12:50.3	−13 18	GLX
SAB(rs)0⁻?	1.9′ × 1.2′	14.12B	

Photographs show a lenticular galaxy with a bright, elongated core in a fainter outer envelope. It is moderately bright visually and is seen as an oval disc oriented NNW–SSE, fairly well defined at the edges and brightening to a nonstellar core. Radial velocity: 4174 km/s. Distance: 186 mly. Diameter: 103 000 ly.

NGC4724

H280[3]	12:50.9	−14 20	GLX
SB0⁻:	0.9′ × 0.5′	13.70B	

This galaxy precedes the larger and brighter NGC4727 by 0.8′ and there are at least six fainter galaxies in the immediate field seen in the DSS image. NGC4724 is a bright, compact lenticular galaxy elongated E–W and is very likely a physical companion of NGC4727 as there is a faint plume of material apparently connecting the two galaxies; the minimum separation at the presumed distance would be 77 000 ly. Visually, it is one of three galaxies which are visible in a medium-magnification field, the others being NGC4727 and NGC4726 to the NNW. NGC4727 appears as a condensed patch with a diffuse outer envelope that seems in contact with NGC4724 to the E, while NGC4726 is seen as a small patch of light. Distance: 329 mly (estimate). Diameter: 86 000 ly (estimate).

NGC4727

H298[2]	12:51.0	−14 20	GLX
SAB(r)bc pec:	1.4′ × 1.1′	12.64B	

This is probably the principal galaxy of a small, distant group which

NGC4727

©STScI/AURA/CALTECH/ROE

includes NGC4724 lying immediately to the W. The DSS image shows a face-on spiral galaxy with a broad and bright central region. There are two short, stubby spiral arms, one in the N which extends to the W, the other in the SE which curves to the W and terminates at a compact object, probably a dwarf companion galaxy. NGC4726 is 4.4′ to the NW. Telescopically, NGC4727 is a bright oval patch of light; it is elongated E–W and a little brighter to the middle with diffuse edges. Radial velocity: 7374 km/s. Distance: 329 mly. Diameter: 134 000 ly.

NGC4748

H537[3]	12:52.2	−13 24	GLX
Sa	1.2′ × 0.8′	14.37B	

Images reveal a pair of galaxies, possibly interacting, though there is no sign of obvious tidal interaction between the two. The principal galaxy is a slightly inclined spiral galaxy with a large, bright core embedded in a fainter disc elongated E–W with faint condensations to the SSE and WNW. The core of the smaller galaxy is 0.25′ to the NNE and it is also embedded in a fainter disc. A bright field star lies 3.0′ to the SSE. Visually, the pair is not resolved but the galaxy appears moderately bright and displays an extended oval envelope oriented N–S, fairly well defined with a bright, star-like core. The apparent N–S disc is probably the merged image of the two galaxies. Radial velocity: 4268 km/s. Distance: 191 mly. Diameter: 66 000 ly.

NGC4756

H281[3]	12:52.9	−15 25	GLX
SAB(s)0°?	2.0′ × 1.3′	13.39B	

The DSS image shows a lenticular galaxy with a bright, elongated core in a fainter outer envelope, oriented NE–SW. It sits in a field swarming with dozens of

NGC4782/4783

©STScI/AURA/CALTECH/OE

a third Herschel object, NGC4794, 8.7′ to the ESE. NGC4782 is the brighter and larger of the close pair, a round object with a grainy disc and a condensed, brighter core. Radial velocity: 4514 km/s. Distance: 202 mly. Diameter: 106 000 ly.

NGC4783

H136[1]	12:54.6	−12 33	GLX
E0 pec	1.8′ × 1.7′	12.82B	12.8V

This elliptical galaxy, paired with NGC4782 immediately to the S, appears round with a bright core and fainter envelope in photographs. Visually, it appears slightly smaller and a little fainter than its apparent companion and the galaxies appear to be in contact, sharing a nebulous haze. The galaxy is well seen at high magnification, which accentuates the graininess of the bright core and surrounding envelope. Radial velocity: 3867 km/s. Distance: 173 mly. Diameter: 91 000 ly.

NGC4794

H538[3]	12:55.2	−12 37	GLX
SB(rs)a	1.9′ × 0.8′	14.64B	13.0V

Photographs show this inclined barred spiral galaxy has a large, bright core and a bright bar connected to a partial pseudo-ring structure. The ring is bright where it connects to the bar and fades slowly after about quarter of a revolution to a pair of broader, symmetrical spiral arms. The galaxy is elongated SSE–NNW and is situated between two field stars, one immediately E, the other immediately WSW of the core. The galaxy has a similar radial velocity to NGC4783 and the two are very probably physically related. Telescopically, NGC4794 can be seen in a high-magnification field with the bright pair NGC4782 and NGC4783, though only the core and bar are well seen, appearing as a small, E–W elongated, well defined glow with even surface brightness. Radial velocity: 3853 km/s. Distance: 172 mly. Diameter: 95 000 ly.

small, faint galaxies which are members of the rich galaxy cluster Abell 1631. Telescopically, NGC4756 is a moderately bright object; it is oval, well defined and gradually brighter to the middle. A faint field star follows the galaxy immediately to the ENE and a small galaxy is visible to the SW, possibly MCG−2-33-36, which is one of a compact group of three galaxies situated about 9.0′ to the SW. Radial velocity: 3954 km/s. Distance: 177 mly. Diameter: 103 000 ly.

NGC4763

H489[3]	12:53.4	−17 00	GLX
SB(r)a:	1.6′ × 1.0′	13.24B	

Photographs show a slightly inclined spiral galaxy with a broad, bright, oval core and three fainter spiral arms. Telescopically, the galaxy is moderately bright though quite small, very slightly oval in shape and oriented ESE–WNW. The edges are hazy and the envelope is

gradually brighter to the middle. Radial velocity: 3918 km/s. Distance: 175 mly. Diameter: 81 000 ly.

NGC4782

H135[1]	12:54.6	−12 34	GLX
E0 pec	1.8′ × 1.7′	12.24B	12.8V

The DSS photograph shows an elliptical galaxy with a round, bright core and a fainter envelope, paired with NGC4783, 0.7′ to the N. There is a considerable spread in the radial velocities of the two galaxies and they may not form a physical pair, though they appear to share a common envelope and there is some apparent blending of the bright cores of the galaxies. If they form a physical pair, the minimum core-to-core separation between the two would be about 33 000 ly at the presumed distance. The lenticular galaxy NGC4792 is 8.0′ to the ENE. Visually, the pair of galaxies form an attractive triplet with

NGC4802

H40[4]	12:55.9	−12 03	GLX
SA(r)0?	2.4′ × 1.6′	12.39B	

This lenticular galaxy was designated NGC4804 by Dreyer but a positional error in declination recorded by Herschel led to the galaxy being considered nonexistent. Observed on 27 March 1786, Herschel's description reads: 'Suspected. A pretty bright star with a seeming brush to it north preceding, may be a very small nebula close to it. No time to verify it.' This is a perfect description of NGC4802, which lies 1° N of Herschel's position and was recorded by Édouard Stephan in 1864. Photographs show a bright lenticular galaxy with an extensive outer envelope. Visually, the galaxy is immediately W of a bright field star, which hinders observation. The star and galaxy appear to be in contact and the galaxy is a round, ghostly glow, very broadly brighter to the middle and fairly well defined. Averted vision helps bring the object out and there is a brighter field star 1.9′ to the WNW. Radial velocity: 901 km/s. Distance: 40 mly. Diameter: 28 000 ly.

Crater

All 25 objects recorded here by Herschel are galaxies and while none are spectacular specimens, some are interesting objects. Most of the objects were recorded in the years 1785–1787, while two were recorded in 1790 and one as late as 1794.

NGC3456

H29[4]	10:54.1	−16 02	GLX
SB(rs)c:	1.9′ × 1.3′	13.29B	

Photographs reveal a slightly inclined spiral galaxy, possibly barred, with a brighter, elongated core and two spiral arms emerging. The pattern is fragmentary with at least two other spurs visible. Three field stars are immediately E of the galaxy and the brightest was noted by Herschel. A faint spiral galaxy is located 5.5′ SSW. Visually, the galaxy appears moderately faint, fairly diffuse and elongated E–W with a magnitude 13 field star bordering the galaxy to the ENE. Herschel's 8 February 1785 description led him to place the galaxy in his Class IV: 'A small star with an extremely faint brush preceding, perceived by gaging, otherwise I should certainly have overlooked it. Verified 240 (magnification).' Radial velocity: 4089 km/s. Distance: 183 mly. Diameter: 101 000 ly.

NGC3508

H507[2]	11:02.9	−16 18	GLX
SA(r)b pec?	1.1′ × 0.8′	13.61B	

In photographs this galaxy appears as a slightly inclined spiral with a large, bright central region and two short spiral arms embedded in a faint outer envelope. The galaxy is moderately bright visually, well condensed and slightly oval, oriented almost due N–S with a very faint field star involved immediately NNE. Radial velocity: 3724 km/s. Distance: 166 mly. Diameter: 53 000 ly.

NGC3511

H39[5]	11:03.4	−23 05	GLX
SAB(s)c	5.0′ × 1.8′	11.54B	

This is a large, highly inclined, spiral galaxy with a small, elongated, brighter core. Photographs show that grainy spiral arms emerge from the core with some small condensations. The arms are very dusty with a short dust lane S of the core and another large dust patch to the WSW. The galaxy forms a likely physical pair with the barred spiral NGC3513, located about 10.0′ to the SSE. The separation between the two galaxies would be a minimum of about 120 000 ly at the presumed distance. Visually, both galaxies are well seen in a medium-magnification field. NGC3511 is a large and quite bright object, though somewhat diffuse; it is a much extended but fairly well defined galaxy, oriented ENE–WSW with a mottled disc and a brighter elongated core. A faint field star can be seen bordering its eastern tip. Radial velocity: 920 km/s. Distance: 41 mly. Diameter: 60 000 ly.

NGC3513

H40[5]	11:03.4	−23 15	GLX
SB(s)c	3.0′ × 2.5′	11.94B	

This is a face-on barred spiral galaxy. Photographs show a small, slightly brighter core in a long, thin bar which brightens where it attaches to the two main spiral arms. There is grainy texture

NGC3511

NGC3513

somewhat mottled. The much fainter galaxy NGC3529 is also visible in the same high-magnification field as a small, round, concentrated patch situated 5.0′ almost due S. Radial velocity: 3491 km/s. Distance: 156 mly. Diameter: 118 000 ly.

NGC3571

H819[2]	11:11.5	−18 17	GLX
(R)SAB:(rs)a	3.0′ × 0.9′	13.03B	

This is a highly inclined spiral galaxy with a bright, slightly elongated core in a grainy outer envelope. The DSS photograph shows that a very long, prominent dust lane almost encircles the galaxy, beginning NNW of the core and circling E, then S, then W below the core before terminating in the W. This is quite a bright galaxy visually and is fairly large, appearing as an oval, quite well defined object, elongated E–W with a fairly bright disc and bright, star-like core. Radial velocity: 3562 km/s. Distance: 159 mly. Diameter: 139 000 ly.

NGC3591

H529[3]	11:14.1	−14 04	GLX
(R')SB(r)0+?	1.3′ × 0.8′	14.37B	

Photographs show an inclined lenticular galaxy with an elongated, bright core, oriented N–S, in a fainter envelope which is oriented NNW–SSE. The galaxy is quite faint visually and only the bright inner region is seen; it is fairly even in surface brightness, moderately well defined at the edges and oval in shape. Radial velocity: 5259 km/s. Distance: 238 mly. Diameter: 90 000 ly.

NGC3636

H550[2]	11:20.4	−10 17	GLX
E0	1.3′ × 1.3′	13.33B	

This elliptical galaxy has a round, bright core and a fainter outer envelope. It is situated 1.75′ WNW of a bright field star and on the DSS image NGC3637, a likely physical companion, is 3.75′ to the ENE. The minimum separation between the two galaxies would be about 77 000 ly at the presumed distance. Visually, both galaxies exhibit high surface brightness and can be seen in the same field, but the bright field star

to the spiral arms and several condensations; high-resolution photographs resolve the brightest stars at about blue magnitude 21.5. Faint field stars bracket the galaxy immediately E, W and SSW. Visually, the galaxy is smaller and a little fainter than NGC3511 and is quite diffuse and a little hazy at the edges. Almost round, it is broadly brighter to the middle and the brightest of the three field stars can be seen to the E. Radial velocity: 1005 km/s. Distance: 45 mly. Diameter: 39 000 ly.

NGC3528

H824[3]	11:07.3	−19 28	GLX
SA(s)°	2.6′ × 1.5′	13.00B	

The DSS photograph reveals an inclined spiral galaxy with a large, bright, elongated central region in a fainter outer envelope. A narrow dust lane in the outer envelope borders the core along the NNW flank. Visually, this is a fairly bright galaxy, a pretty well defined oval of light, quite opaque and a little brighter to the middle. Oriented ENE–WSW, the disc is

NGC3636/3637

hinders observation. NGC3636 is round, small, condensed, even in surface brightness and well defined at the edges. Radial velocity: 1586 km/s. Distance: 71 mly. Diameter: 27 000 ly.

NGC3637

H551[2]	11:20.7	−10 16	GLX
(R)SB(r)0°	1.7′ × 1.6′	13.51B	

Images reveal a ring galaxy with a bright, slightly elongated core

surrounded by a very faint ring. It is located 3.0′ NE of the bright field star which also hinders NGC3636. Visually, only the bright core of this galaxy is visible, appearing small, round and condensed with well defined edges and just a little larger than NGC3636. Radial velocity: 1692 km/s. Distance: 76 mly. Diameter: 37 000 ly.

NGC3660

H635[2]	11:23.6	−8 40	GLX
SB(r)bc	2.7′ × 2.2′	12.74B	

The DSS photographs show a bright, barred spiral galaxy with an elongated core leading to a short bar. The bar is attached to two bright inner spiral arms forming a pseudo-ring, which is complex in structure and includes fainter spurs emerging from the N. After half a revolution the spiral arms are fainter and grainy in texture with several condensations involved. Several compact objects immediately surround the galaxy, especially to the NE. Visually, the galaxy appears very faint with an elongated, even-surface-brightness envelope oriented ESE–WNW with several brightish field stars to the E and one 1.75′ to the NE. Radial velocity: 3534 km/s. Distance: 158 mly. Diameter: 124 000 ly.

NGC3661

H530[3]	11:23.7	−13 48	GLX
SA(r)0/a: sp	1.6′ × 0.6′	14.26B	

Images show a highly inclined galaxy, probably a spiral, with a brighter core and elongated outer envelope, oriented SE–NW. Three bright field stars, oriented NE–SW are located about 2.0′ to the S. Visually, this galaxy is small and faint, appearing slightly oval in shape and oriented SE–NW. The galaxy NGC3667 can be seen in the same medium-magnification field 9.5′ to the ESE, though their radial velocities suggest they are not physical companions. Radial velocity: 6527 km/s. Distance: 291 mly. Diameter: 136 000 ly.

NGC3667

H531[3]	11:24.3	−13 49	GLX
(R')SA(rs)ab:	1.5′ × 1.0′	13.86B	

This inclined spiral galaxy has a bright, slightly elongated core and photographs show a fainter outer envelope with two faint spiral arms, one to the N, the other to the S. A likely physical companion galaxy, NGC3667B, is located 1.0′ E and has a similar radial velocity to the principal galaxy. Though NGC3667 appears normal, the companion galaxy, which is spindle shaped, has a disturbed arm with a faint plume on the NE side of the core. The minimum separation at the apparent distance would be about 68 000 ly. Visually, only the principal galaxy is seen as a small, faint, round and fairly well defined disc. Radial velocity: 5189 km/s. Distance: 232 mly. Diameter: 101 000 ly.

NGC3672

H131[1]	11:25.0	−9 48	GLX
SA(s)c	4.2′ × 2.1′	12.07B	

This is a large, bright, highly inclined spiral galaxy with a bright, elongated core. High-resolution photographs show four principal spiral arms emerging from the central core. The

NGC3660

arms are well defined with several bright, knotty condensations. Telescopically, this is a large, bright and well defined galaxy, elongated N–S and evenly bright along its major axis, except for a broad, gradual brightening to the middle. One faint star immediately precedes the core to the W. Radial velocity: 1711 km/s. Distance: 76 mly. Diameter: 94 000 ly.

NGC3693

H532[3]	11:28.2	−13 10	GLX
(R')SA(r)b:	3.0′ × 0.8′	13.91B	

This highly inclined spiral galaxy, possibly barred, has a bright, elongated core in a slightly fainter inner envelope. Images reveal the fainter, grainy outer envelope has evidence of dust patches and lanes, particularly to the E and W. The galaxy is very faint visually and only the core and inner disc are well seen as an elongated patch of light, oriented E–W, with hazy edges. Radial velocity: 4794 km/s. Distance: 214 mly. Diameter: 187 000 ly.

NGC3715

H562[2]	11:31.5	−14 12	GLX
(R')SB(rs)bc:	1.3′ × 0.9′	13.44B	

This is a slightly inclined spiral galaxy, probably barred, with a broad, bright core and two bright, narrow spiral arms, one emerging from the NE and curving W, the other from the SW and curving E. An anonymous compact elliptical galaxy is 7.0′ to the WSW. NGC3715 is small and very faint visually; it is an oval patch of light, oriented ESE–WNW, even in surface brightness and fairly well defined. Radial velocity: 1964 km/s. Distance: 88 mly. Diameter: 33 000 ly.

NGC3732

H552[2]	11:34.2	−9 51	GLX
SAB(s)0/a:	1.2′ × 1.2′	12.96B	

Its classification is uncertain but images show a small, bright, face-on galaxy, probably a spiral, with a very bright and large central region. The fainter outer envelope has two suspected faint spiral

arms emerging, one from the N, the other from the S. Visually, this is a quite bright, round and opaque object; it is a grainy textured, well defined disc with a small, brighter core. A field star is 1.0′ to the SE. Radial velocity: 1572 km/s. Distance: 70 mly. Diameter: 24 000 ly.

NGC3734

H935[3]	11:34.7	−14 06	GLX
(R')SAB(rs)bc:	1.1′ × 1.0′	14.65B	

Photographs show a faint spiral galaxy, seen face-on, with a brighter, round core and a fainter inner ring. There is a dim and larger outer ring, but this is probably a manifestation of the outer spiral structure. The galaxy is slightly elongated NNE–SSW. Telescopically, this galaxy is moderately faint but well seen as a round, well defined disc, evenly bright and quite small. Radial velocity: 8985 km/s. Distance: 401 mly. Diameter: 128 000 ly.

NGC3791

H609[3]	11:39.8	−9 23	GLX
(R)SAB pec:	0.6′ × 0.6′	14.37B	

Images show a round lenticular galaxy with a fainter outer envelope. A small, bright spur emerging from the core to the WNW suggests a compact companion superimposed on the main galaxy and a bright field star is located 6.0′ almost due S. The galaxy is small but quite bright visually, appearing as an opaque, slightly elongated disc, oriented E–W, even in surface brightness and well defined at the edges. Radial velocity: 5369 km/s. Distance: 240 mly. Diameter: 42 000 ly.

NGC3887

H120[1]	11:47.1	−16 51	GLX
SB(r)bc	3.5′ × 2.8′	11.42B	

Images show a slightly inclined, barred spiral galaxy with a broad, bright, elongated core in a slightly fainter bar. Four distinct and massive spiral arms

NGC3887

emerge with many bright knots throughout and the galaxy is elongated N–S. The faint outer structure visible on the DSS image increases the dimensions of the galaxy to about 4.1′ × 3.2′. Telescopically, this is a large and fairly bright galaxy, appearing round and a little brighter to the middle with a mottled disc. The edges are a little hazy and the galaxy is situated within a flattened triangle of field stars. Radial velocity: 1049 km/s. Distance: 47 mly. Diameter: 56 000 ly.

NGC3892

H553[2]	11:48.0	−10 58	GLX
SB(rs)0+	3.1′ × 2.9′	12.2B	

The DSS image shows a bright barred galaxy with an elongated and bright core. The short, tapered, slightly fainter bar, oriented E–W, is embedded in a grainy disc which is surrounded by a fainter outer halo. The galaxy is quite bright visually and the bar is visible as a brighter enhancement in a fainter, though fairly well defined oval disc. A triangle of field stars is visible immediately W. Radial velocity: 1643 km/s. Distance: 73 mly. Diameter: 66 000 ly.

NGC3955

H623[2]	11:54.0	−23 10	GLX	
S0/a pec	2.9′ × 0.9′	12.59B	11.9V	

The DSS reveals a highly inclined galaxy, probably a spiral, with a peculiar morphology. The bright elongated core is partially obscured by a large dust patch to the S. A small but intense dust patch takes a notch out of the outer envelope N of the core. Telescopically, this galaxy is small and moderately faint, but is well seen at medium magnification as an even-surface-brightness, smooth-textured and well defined streak of light, oriented SSE–NNW. Radial velocity: 1321 km/s. Distance: 59 mly. Diameter: 50 000 ly.

NGC3956

H290[3]	11:54.0	−20 34	GLX
SA(s)c	3.5′ × 1.0′	12.79B	

This highly inclined spiral galaxy has a small, brighter central region, which may be a bar, and grainy-textured spiral arms with small condensations including a small bright knot or faint field star WSW of the core. There appears to be a large dust lane along the outer envelope to the N and a faint, dwarfish, possible companion galaxy, ESO572-11, is 4.5′ to the SSW, while a bright field star is 4.5′ to the NNW. Visually, NGC3956 is a diffuse and moderately faint galaxy which is broadly brighter to the middle. The envelope has a mottled texture and is elongated ENE–WSW with moderately well defined edges. Radial velocity: 1480 km/s. Distance: 66 mly. Diameter: 67 000 ly.

NGC3957

H294[2]	11:54.0	−19 34	GLX
SA0+: sp	3.1′ × 0.7′	12.94B	

This edge-on lenticular galaxy has a bright, elongated core and bright extensions and is very probably a member of the group of galaxies dominated by the Ringtail galaxies NGC4038 and NGC4039 in Corvus. High-resolution photographs show a thin dust lane crossing in front of the core. A faint field star is immediately SSW. Visually, this is a bright, almost edge-on, moderately well defined galaxy oriented almost exactly N–S. The brightness is fairly uniform along its

NGC3955

©STScI/AURA/CALTECH/ROE

NGC3957

©STScI/AURA/CALTECH/ROE

major axis except for a slight brightening of the central region. Herschel thought this object resolvable. Radial velocity: 1475 km/s. Distance: 66 mly. Diameter: 60 000 ly.

NGC3962

H67[1]	11:54.7	−13 58	GLX
E1	3.6′ × 2.8′	11.61B	

Images show an elliptical galaxy with a bright, very slightly elongated core in an extensive outer envelope, oriented NNE–SSW, with two bright field stars about 2.5′ S. This is a fairly bright galaxy visually and the disc is quite opaque and brighter to the middle. It is

fairly well defined, though surrounded by a very faint secondary envelope. The central region is a little mottled and almost round. Radial velocity: 1666 km/s. Distance: 74 mly. Diameter: 78 000 ly.

NGC3981

H274[3]	11:56.1	−19 54	GLX
SAB(s)bc pec	7.0′ × 2.7′	12.14B	

The DSS image reveals a highly inclined spiral galaxy with a bright, extended core. The spiral pattern is of the grand-design type: two narrow, bright, grainy spiral arms with a few brighter knots involved. There is a sudden drop off in

the brightness of the spiral arms and an abrupt change in angle before the arms broaden, fade and become smoother textured. An extremely faint, detached plume of material oriented NNE–SSW can be seen W of the galaxy. Bright field stars are 4.5′ NW and 5.0′ SSW. At moderate magnification this companion galaxy to NGC3957 appears a little brighter and larger but with diffuse, less well defined edges. It is much elongated and oriented NNE–SSW. Though a little brighter to the middle, no core has been detected. Two threshold stars lie to the E of the galaxy's central region. Radial velocity: 1561 km/s. Distance: 70 mly. Diameter: 142 000 ly.

Cygnus

This large constellation, embedded in a rich and detailed section of the Milky Way, is home to 27 Herschel objects and representatives of every class of deep-sky object can be found here except for globular clusters. Herschel observed this region extensively beginning in 1784 and many spectacular objects are among his

discoveries here, including NGC7000 (the North American Nebula), NGC6960, NGC6979 and NGC6992 (all individual portions of the Veil Nebula) and NGC6826 (the Blinking Planetary). The planetary nebulae NGC6894 and NGC7008 are also rewarding objects in moderate-aperture telescopes.

NGC6824

H878[2]	19:43.7	+56 07	GLX
SA(s)b	2.7′ × 1.8′	12.95B	

The DSS image reveals a bright, inclined spiral galaxy with a bright and broad inner region. Knotty spiral arms wind around the core and extensive dust patches and lanes are silhouetted against the very faint, smooth-textured outer envelope, particularly to the N and NW. A bright, unequal pair of stars lies 3.5′ almost due N. Visually, this is a bright galaxy with a well defined outer envelope. The galaxy is oriented NE–SW with a faint stellar core visible at high magnification. A magnitude 14 star lies

immediately S. Radial velocity: 3611 km/s. Distance: 161 mly. Diameter: 127 000 ly.

NGC6826

H73[4]	19:44.8	+50 31	PN	
3a+2	2.25′ × 2.25′	10.02B	10.1V	

The DSS image shows that the inner shell of this planetary nebula is perfectly round, completely opaque and about 40″ in diameter. This shell is overexposed and no detail is visible. A fainter shell surrounds the inner nebula; this is also round and is well defined along its edges by a thin, slightly brighter, continuous ring. This object is

popularly known as the Blinking Planetary, referring to an effect seen in small telescopes whereby the observer concentrates on the bright central star, then looks away to see the nebulosity surrounding the star pop into view. With moderate-aperture telescopes, this effect is not noted as the inner gas shell is visible at all times as a grainy, sharply defined object with the central star prominent and easily visible. The outer shell may be visible with large apertures or with an OIII filter. Herschel came upon this object on 6 September 1793 and his detailed descriptions reads: 'A bright point, a little extended, like two points close to one another; as bright as a star of the 8.9 magnitude, surrounded by a very bright milky nebulosity suddenly terminated, having the appearance of a planetary nebula with a lucid centre; the border however is not very well defined. It is perfectly round, and I suppose about half a minute in diameter. It is of a middle species, between the planetary nebulae and nebulous stars, and is a beautiful phenomenon.' The expansion velocity of the gaseous shell is 13 km/s. The central star, catalogued as HD186924, is magnitude 10.6 with a spectral type of O6fp. Radial velocity in approach: 6 km/s. Distance: 3200 ly.

NGC6834

H16[8]	19:52.2	+29 25	OC
II2m	6.0′ × 6.0′	8.5B	7.8V

Photographs show a compressed grouping of bright stars in a rich field, poorly separated from the sky background. Visually, small apertures show an attractive cluster of magnitude 9 or fainter stars, much extended in an E–W direction. The cluster is defined by a chain of five magnitude 9–11 stars; the fainter members are grouped around these luminaries. At lower magnification, the background appears slightly nebulous. The brightest star is at the centre and is designated HD332843. Herschel came upon this cluster on 17 July 1784, describing it as: 'A cluster of not very compressed stars closest in the middle. It may be called (if the expression be allowed) a forming cluster or one that seems to be gathering.' Total membership in this cluster is about 130 stars. Spectral type: B2. Age: 79 million years. Distance: 7100 ly.

NGC6857

H144[3]	20:01.9	+33 31	EN/RN?
1.0′ × 1.0′			

The DSS image shows a small, faint nebula in a rich field of stars. It is roughly triangular in shape and more opaque to the E, S, and W. A star is visible through the faint cavity in the NW but it is unclear whether this star is involved with the nebulosity. Visually, this nebula is visible in a very rich Milky Way field. Two faint field stars on an ENE–WSW line bracket the nebula and the 'central star' associated with the nebula gives the appearance of a planetary nebula. At low magnification, the three stars and the nebula are seen as a nebulous blur in the field, attracting attention. At high magnification, the nebula is roundish and offset somewhat towards the S of the star, fading gradually into the sky background. Herschel's discovery description of 5 September 1784 states: 'Some extremely small stars with nebulosity, irregularly extended. Verified 240 (magnification).'

NGC6866

H59[7]	20:03.7	+44 00	OC
II2r	15.0′ × 15.0′	8.0B	7.6V

Photographs show a triangular grouping of brighter stars, well separated from the immediate star field. This cluster was first recorded by Caroline Herschel on 23 July 1783; her brother did not list it in his catalogue until 11 September 1790. In small telescopes, this is a rich and very compressed cluster of predominantly magnitude 11–12 stars. A bright, wide pair, elongated NW–SE, is in the S; the central portion features a ragged N–S chain of stars and there is a prominent subgroup branching off towards the E and S. In total 45–50 stars are visible. The actual cluster membership is about 130 stars. Spectral type: A2. Age: 230 million years. Distance: 4200 ly.

NGC6874

H86[8]	20:07.8	+38 14	OC
IV1m n	7.0′ × 7.0′	7.7V	

This cluster is plotted as Basel 6 in the first edition of *Uranometria 2000.0*. On photographs it is a scattering of bright stars in a rich Milky Way field, but visually it appears as a distinct cluster in a low- to medium-magnification field. The most prominent stars are slightly brighter than the surrounding field and give the cluster a distinctive, broadly triangular shape. To the E, a tight diamond of four magnitude 12 stars appears to be separated from the main body. A prominent chain of about six magnitude 12–13 stars lies to the W and stars to about magnitude 15 are arrayed around these. About 30 stars are visible, although only 20 are considered members. Spectral type: B4. Distance: 6850 ly.

NGC6888

H72[4]	20:12.0	+38 21	EN?
16.0′ × 12.0′	7.44B		

Popularly known as the Crescent Nebula, and more recently as Van Gogh's Ear, the origin of this object is somewhat unclear. It is powered by a magnitude 7 Wolf–Rayet star, HD192163, which is clearly visible somewhat off-centre towards the W in the nebulosity. The nebula may be a shell of gas ejected by the star: Wolf–Rayet stars are known to be hot and massive and to produce violent stellar winds, frequently blowing off their outer layers into space. Herschel discovered the nebula on 15 September 1792, describing it as: 'A double star of the 8th magnitude, with a faint south preceding milky ray joining to it, 8′ long, 1.5′ broad.' The double star mentioned by Herschel is ADS13515 and borders the brightest portion of the nebula. Visually, this object is difficult to find without an OIII filter. Arrayed around a large, bright, ·

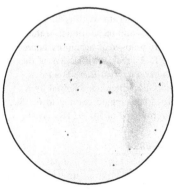

**NGC6888 15-inch Newtonian 90×
(O-III filter)**

NGC6888

©STScI/AURA/CALTECH/ROE

diamond-shaped asterism, the nebula is easy to locate. With a filter it is brightest and most distinct in the NW, where a finger of nebulosity curves to the E. A break is visible SW of the double star, then there is an oval-shaped brightening, followed by a broad faint region curving to the S. Spectral type of HD192163: WR6.

NGC6894

H13[4]	20:16.4	+30 34	PN
4+2	0.9′ × 0.9′	14.40B	12.3V

Images reveal an annular planetary nebula with a faint central star. There is faint nebulosity within the bright ring. The ring brightens as it expands outwards, except for a very bright patch on the inner ring in the NW. The bright ring is surrounded by a thin, fainter outer shell and the object is situated in a rich star field. Visually, the planetary nebula is bright and well seen at high magnification. It is round with fuzzy edges and high magnification brings out the slightly darker centre. The ring is unevenly bright: it appears a little brighter to the N. The magnitude 17.6 central star is not visible. Radial velocity in approach: 58 km/s. Distance: 4600 ly.

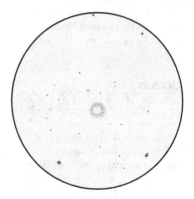

NGC6894 15-inch Newtonian 272×

NGC6894

©STScI/AURA/CALTECH/ROE

NGC6895

H83[8]	20:16.4	+50 14	MWP?
15.0′ × 7.0′			

Recorded on 30 September 1790, Herschel's description of this object reads: 'A cluster of scattered stars, above 15′ diameter, pretty rich, joining to the milky-way or a projecting part of it.' Visually, there is nothing to suggest that this is an actual star cluster. Located SE of a magnitude 7 field star, this is a scattering of stars in a fairly rich field that stands out slightly from the sky background.

NGC6910

H56[8]	20:23.1	+40 47	OC
l3m n	10.0′ × 10.0′	7.1B	7.4V

This cluster appears well resolved in small telescopes and is an elongated group extended NNW–SSE. About 35 stars are resolved, centred on a bright, distorted Y-shaped asterism. There is a wide range in the brightnesses of the cluster members. Located near Gamma Cygni, the cluster is embedded in the extensive nebulosity in the field, though it is too faint to easily detect visually. Photographs show the cluster in the field bathed in faint nebulosity. Herschel did not note any nebulosity when he recorded the cluster on 17 October 1786. His description reads: 'A small cluster of coarse scattered stars of various sizes. Extended like a forming one.' More than 60 stars are suspected to be members. The cluster is part of the Cygnus OB9 association. Spectral type: O8. Age:

about 10 million years. Radial velocity in approach: 33 km/s. Distance: 4900 ly.

NGC6960

H15[5]	20:45.7	+30 43	SNR
70.0′ × 10.0′			

This is the western portion of the Veil Nebula, marked by the bright field star 52 Cygni, which is seen on the western flank of the nebula. The entire nebula complex is a supernova remnant; the progenitor star has never been identified and it is believed that the nebula shines due to the shock front between the expanding shell of gas and a dense interstellar medium. Photographs show an extremely complex filamentary structure, elongated N–S; there are dozens of long, narrow, individual streamers, most of them oriented N–S, involving excited gas as well as dark nebulosity. Herschel observed this segment of the Veil Nebula on 7 September 1784, describing it as: 'Extended; passes through k (52) Cygni. By the Newtonian view above one degree long. By the Front-view (Herschelian) near two degrees long. See note.' This last comment refers to Herschel's explanation of the superiority of the front-view configuration of his reflector, which primarily gave a brighter image of objects. Visually, the nebula is well seen in moderate-aperture telescopes in good skies, but an OIII filter significantly enhances the view. The portion N of 52 Cygni extends for close to a full degree and is moderately bright but smooth textured. It is very well defined and the entire western edge is brighter and more opaque, giving the impression that this is the shock front meeting the interstellar medium. OIII filters impart a distinctly warm yellow tint to the nebula. The northern tip of the nebula ends abruptly in a curve, pointed towards the NNE. The nebula seems to fade in the vicinity of 52 Cygni, no doubt due to the glare from the star, which is a foreground object. The southern portion extends for about 40′ and is a little fainter than the northern section. About 10′ S of 52 Cygni the nebula forks into

two branches that continue to diverge and widen until they fade into the sky background. Distance: about 1300 ly.

NGC6979

H206[2]	20:51.0	+32 09	SNR
25.0′ × 15.0′			

This forms the north central portion of the Veil Nebula complex, NGC6974 being the southern continuation. The DSS image reveals that NGC6979 is fainter than NGC6960 or NGC6992, is not as complex in structure and is oriented NW–SE. Visually, this is a very difficult object to see; it is an extremely faint, ill defined and elongated glow. Only the northernmost portion is visible, with a wide SE–NW oriented pair of field stars, separated by 0.75′, marking the brightest portion. It is broadest to the N, is elongated roughly N–S and tapers slightly towards the S. The star field is very rich. Herschel observed it on the same night as he discovered NGC6960, calling it: 'Faint, small, crookedly elongated. Resolvable.'

NGC6989

H82[8]	20:54.1	+45 17	MWP
10.0′ × 10.0′			

Situated N of an E–W, very broad triangle of three bright field stars in the 'Quebec' portion of the North American Nebula, this is simply a slightly brighter Milky Way patch involved with some nebulosity and a hazy, unresolved field. About 15 faint stars are involved.

NGC6991

H76[8]	20:56.6	+47 25	OC
III2m n	25.0′ × 25.0′		

This object is plotted incorrectly in the first edition of *Uranometria 2000.0*; the symbol should be larger and plotted W of the bright field star involved with the nebula IC5076, as stated in *Star Clusters* (Archinal and Hynes, 2003). Telescopically, this cluster is a scattered grouping of moderately bright stars, with the greatest concentration SW of the bright field star HD 199478. Though the field is relatively rich, the cluster members stand out well. There is a moderate range in brightness and about 35 stars are visible in a 20′ field.

NGC6992

H14⁵	20:56.4	+31 43	SNR
70.0′ × 20.0′			

This is the eastern component of the Veil Nebula, and is the brightest and most complex part, particularly the southern portion. Like NGC6960, it is primarily composed of hundreds of intertwined, thin, narrow streamers of nebulosity, which are well seen in photographs. This segment of the Veil Nebula is predominantly oriented NW–SE. Herschel discovered the nebula on 5 September 1784, describing it as: 'Branching nebulosity, extending in right ascension near 1.5 degrees and in polar distance 52 arc minutes. The following part divides into several streams uniting again towards the south.' It is unclear how much of the portion now designated NGC6995 Herschel may have seen; certainly it is not difficult and visually shows an enormous amount of structure. It is well seen without an OIII filter, though a filter greatly enhances the view. NGC6992 and NGC6995 taken together are larger, brighter and more detailed than NGC6960. NGC6992 is broad and to the N and E a small, stubby sub-branch breaks away from the nebula. At the centre of the field, a brighter condensation is visible and high magnification reveals a faint star here. Indeed, much of the charm of this magnificent object is the rich star field in which the nebula complex lies. NGC6995 to the S exhibits much structural detail, including many individually resolved filaments and bright and dark zones.

NGC6997

H58⁸	20:56.5	+44 38	OC
III2m n	8.0′ × 8.0′	10.0 V	

The DSS image shows a triangular grouping of bright stars involved in nebulosity. The cluster is situated within the boundaries of the North American Nebula on the western side. Small telescopes show this as a faint, scattered, irregularly round cluster of stars of uniform lustre. About 20 stars are resolved and are well isolated from the surrounding field. The cluster is coarse and quite scattered. Moderate apertures show the cluster is made up of bright, scattered stars bathed in the field nebulosity. Forty stars are members of the cluster. Distance: 1600 ly.

NGC7000

H37⁵	20:58.8	+44 20	EN
160.0′ × 140.0′			

This very large nebula (the North American Nebula) is well seen at low magnification; it appears as a complex cloud displaying striations as well as bright and dark zones. It is much brighter to the S and E; this is the portion known as 'Mexico'. Also prominent is the 'Florida' region, which is illuminated by a bright, magnitude 7 field star. The 'Gulf of Mexico' is dark and contrasts nicely with 'Mexico'. The spine of 'Mexico' is very bright and adorned with stars. It is widest at its southern tip. To the N the nebula fades gradually into a bright Milky Way star field. To the W across a dark gulf is the portion known as the Pelican Nebula, a large, formless glow visually, elongated N–S. The illuminating star is presumed to be HD199579, magnitude 5.96, spectral type O6V. Herschel recorded the nebula on 24 October 1786, describing it as: 'Very large diffused

nebulosity. Brighter middle. 7′ or 8′ long, 6′ broad and losing itself very gradually and imperceptibly.'

NGC7008

H192¹	21:00.6	+54 33	PN
3	1.8′ × 1.6′	12.8B	10.7V

Photographs show this moderately bright planetary nebula has a complex structure. The brightest portion of the shell is very slightly elongated N–S, relatively bright and opaque to the N and W. Very faint filamentary structure is visible to the SE. Dark cavities are visible E of the central star and to the SW. As well as the central star, at least ten stars can be seen in the nebula. Two brighter field stars border the nebula to the SSE. Visually, this is a spectacular planetary nebula, situated in a very rich Milky Way field. Very bright at medium and high magnification, a broken, almost annular ring is visible around the magnitude 13.2 central star. The nebula is brightest to the NW, where it is most sharply defined. It fades somewhat uncertainly to the SE. One other star is visible in the shell to the ENE. Herschel discovered the object on 14 October 1787 and classed it as a bright nebulae, his description reading: 'Considerably bright, irregular form, 3′ long, 2.5′ broad. Nebulosity.' Curiously, he did not mention any involved stars. The spectral type of the central star is O7 and the expansion velocity of the gas shell is 11 km/s. Radial velocity in approach: 76 km/s. Distance: about 2600 ly.

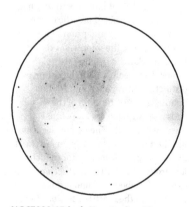

NGC7000 15-inch Newtonian 48×

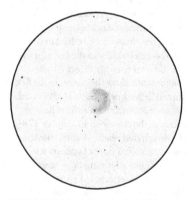

NGC7008 15-inch Newtonian 272×

NGC7008

NGC7013

H203[2]	21:03.6	+29 54	GLX
SA(r)0/a	4.5′ × 1.4′	12.36B	

Photographs show an inclined galaxy, probably a barred spiral, with a peculiar morphology. A bright core with an apparently foreshortened bar elongated to the E is connected to a bright ring. The ring is surrounded by a fainter inner envelope and beyond this a very faint outer envelope with evidence of large dust patches to the NW, E and SE. The outer envelope is tilted to the plane of the brighter inner galaxy. Visually, only the inner, bright portion of the galaxy is visible; it is gradually brighter to the middle, oval in shape and oriented N–S with somewhat hazy edges. Higher magnification broadens the galaxy and reveals hints of the fainter outer envelope. Radial velocity: 1000 km/s. Distance: 45 mly. Diameter: 58 000 ly.

NGC7024

H57[8]	21:06.0	+41 30	OC
10.0′ × 5.0′			

Discovered on 17 October 1786, Herschel described this object as: 'A cluster of coarse scattered pretty small stars of several sizes, not rich.' The DSS photograph shows a scattered group of slightly brighter stars in a rich field. Visually, the cluster is fairly obvious; it is a collection of magnitude 10 or fainter stars, well separated from the sky background and dominated at the centre by a brighter chain of four stars elongated NE–SW. Otherwise the cluster is extended SE–NW with about 25 stars resolved although only 14 are recognized as members.

NGC7031

H74[8]	21:07.3	+50 50	OC
III2m	15.0′ × 15.0′	10.0B	9.1V

The DSS image shows a small, coarse grouping of fairly bright stars in an area of 3.5′ × 3.0′, with a brighter star at the centre. Because of the brightness of its members, the cluster stands out well against the Milky Way field. In *Star Clusters*, Archinal and Hynes (2003) indicate that the extent of the cluster is much greater than it appears, on the order of 15′ × 15′ with a total of 62 stars involved, saying: 'Group of 10 brighter stars of this cluster all on NE side.' Visually, the cluster appears small and rather coarse, standing out well from the sky background. About a dozen magnitude 11–13 stars are involved, arrayed around the brightest star (magnitude 11.31). Spectral type: B5.

Age: about 56 million years. Distance: 2900 ly.

NGC7044

H24[6]	21:12.9	+42 29	OC
I1r	7.0′ × 5.0′	12.0B	

In a rich Milky Way field, this is a rich and compressed cluster of stars with a narrow range of brightness. Photographically, the cluster appears elongated in an E–W direction and a little more compressed to the centre. A prominent straight line of stars is visible in the northern half, oriented ENE–WSW. In small telescopes, this is a very dim and small cluster; it is a faint nebulous haze with a few stars shining through at low magnification. At medium magnification, a diamond-shaped asterism of four very faint stars defines the cluster's extremities. As the brightest members are about magnitude 15, this is a difficult cluster to resolve. Discovered by Herschel on 17 October 1786, he described it as: 'A very

NGC7044

NGC7062

compressed and very rich cluster of extremely small stars about 6′ long and 4′ broad nearly parallel.'

NGC7062

H51[7]	21:23.2	+46 23	OC
II2m	5.0′ × 5.0′	9.0B	8.3V

In photographs this is a moderately rich cluster of faint stars with a narrow range of brightness. The cluster is elongated in an E–W direction. The stellar background is rich, but the cluster members are slightly brighter, helping to separate the cluster from the background. In a small telescope, this is a 'suspicious' small grouping of stars at low magnification. High magnification brings out about 15 very faint stars dominated by the four brightest stars, which form a diamond-shaped asterism elongated ESE–WNW. Resolution is better at high magnification in a moderate-aperture telescope. Herschel discovered the cluster on 19 October 1788 and found it to be: 'A pretty compressed cluster of pretty small stars, considerably rich, irregularly round, 5′ or 6′ diameter.' About 85 stars are considered to be members. Spectral type: A1. Age: about 100 million years. Distance: 5500 ly.

NGC7067

H50[7]	21:24.2	+48 01	OC
I1p	3.5′ × 1.75′	10.6B	9.7V

Photographs reveal a small cluster of moderately bright stars, which is fairly

compressed and elongated along a N–S axis. The range of brightness of the cluster members is fairly narrow. The cluster is 3.0′ W of a bright field star. Visually, the cluster is small, faint and fairly weak in a moderately rich field. It is dominated by a narrow triangle of stars with the apex pointing S, several very faint members are involved, and there is an unresolved, weak haze in the N, extending towards the NW. The brightest star is magnitude 11.17 and almost 50 stars are accepted to be members. Spectral type: B0. Age: 13 million years. Distance: 12 000 ly.

NGC7082

H52[7]	21:29.4	+47 05	OC
IV2p	24.0′ × 24.0′	7.7B	7.2V

Photographs show a large, coarse cluster of scattered stars of moderate brightness range in a rich Milky Way field with the cluster poorly separated from the sky background. In a small telescope, the brightest 20 stars dominate and are contained in an area of about 15′ × 15′. The main feature is a loose chain of bright stars running ENE–WSW near the centre of the field. Discovered on 19 October 1788, Herschel described the object as: 'An extensive cluster of large stars, considerably rich, above 20′ diameter.' This is a moderately aged cluster, estimated to be 1.6 billion years old. About 180 stars are considered to be members. Distance: 4300 ly.

NGC7086

H32[6]	21:30.5	+51 35	OC
II2m	12.0′ × 12.0′	9.2B	8.4V

In photographs this is a small, scattered cluster of stars displaying a moderate brightness range. The cluster is moderately condensed and because the cluster members are generally brighter than the background star field, it stands out well. Herschel recorded the cluster on 21 September 1788, and called it: 'A beautiful cluster of pretty compressed

NGC7128

stars 8′ or 9′ diameter, nearly round, considerably rich.' In a small telescope it is a faint cluster, featuring a string of stars oriented E–W at the cluster centre. In total about 18 stars are visible, all fainter than magnitude 10; these make a more attractive group at lower magnification. The total cluster membership is about 80 stars. Spectral type: A5. Age: about 80 million years. Distance: 3850 ly.

NGC7128

H40[7]	21:44.0	+53 43	OC
I3m	4.0′ × 4.0′	10.5B	9.7V

The photographic and visual appearances of this cluster are quite similar. In a moderate-aperture telescope, this is a small, compressed and rich cluster of stars, with a moderate range of brightness with a brighter field star at the edge of the cluster to the ESE. A prominent, curving chain of stars cuts across the centre of the cluster, running N–S, but otherwise the cluster is elongated ESE–WNW. The cluster stands out well against the background star field. First noted by Herschel on 14 October 1787, he described it as: 'A cluster of small stars of several sizes, 3′ or 4′ diameter, pretty rich like a forming one.' About 70 stars are members. Spectral type: B2. Age: relatively young at 10 million years. Distance: 7500 ly.

Delphinus

The five Herschel discoveries in this diminutive constellation were all recorded in 1784 and 1785. At least three of the five are compelling objects: the globular clusters NGC6934 and NGC7006, and the planetary nebula NGC6905.

NGC6905

H16[4]	20:22.4	+20 07	PN
3+3	1.4′ × 0.6′	11.9B	11.1V

Photographs show this bright planetary nebula has a fainter secondary shell, oriented NNW–SSE, with the shell tapering to points. Visually, it is moderately bright and fairly well defined and is situated within a triangle of field stars. The nebula appears fairly round, about 40″ in diameter. The nebulosity is a little brighter to the E and the magnitude 13.5 central star, spectral type WC6, is intermittently visible. Distance: about 4200 ly.

NGC6934

H103[1]	20:34.2	+7 24	GC
CC8	7.1′ × 7.1′	10.5B	9.7V

In photographs this is a bright, rich and compressed globular cluster with a symmetrical form. Telescopically, it is a well-condensed globular cluster, and is intensely bright towards the middle. A magnitude 9 field star precedes the cluster in the W. At high magnification the cluster is quite mottled towards the centre with an extensive secondary halo of unresolved stars; many stars are resolved in the halo as well as in the core, including two bright ones SE of the cluster centre. The brightest stars in the cluster are magnitude 13.8 and the horizontal branch visual magnitude is 17.1. Spectral type: F7. Radial velocity in approach: 379 km/s. Distance: 49 000 ly.

NGC6950

H23[8]	20:41.2	+16 38	OC
14.0′ × 10′			

The DSS image shows a scattered grouping of brighter stars in a rich star field. Visually, this is a large and conspicuous grouping of stars of moderate brightness range, well separated from the sky background. It is roughly oval in shape and elongated N–S, with at least 35 stars resolved. This may not be a cluster but merely a condensed Milky Way patch. Total membership in the group is about 60 stars.

NGC6956

H219[3]	20:44.0	+12 31	GLX
SBb	1.9′ × 1.6′	13.26B	

The DSS image reveals a face-on, barred spiral galaxy with a bright core in a fainter, slightly tapered bar. Two strong, S-shaped spiral arms emerge from the bar, with three more fainter, broader arms emerging from the main spiral

NGC6905

©STScI/AURA/CALTECH/ROE

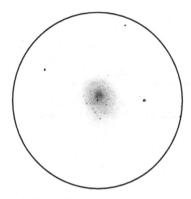

NGC6934 15-inch Newtonian 466x

arms. The arms and the bar contain many knotty condensations. A field star is 0.5′ E of the core and a smaller field galaxy is 6.75′ to the SE. At high magnification, this faint and ethereal galaxy is visible as a ghostly glow involved with the embedded field star which interferes with the view. The envelope of the galaxy is oval and elongated E–W. Discovered on 19 October 1784, Herschel called the galaxy: 'Extremely faint, very small, stellar. Verified 240 (magnification) with difficulty.' Radial velocity: 4839 km/s. Distance: 216 mly. Diameter: 119 000 ly.

NGC7006

H52[1]	21:01.5	+16 11	GC
CC1	3.6′ × 3.6′	11.4V	

The DSS photograph shows that this distant globular cluster is round and fairly compressed to the centre, but with a relatively extensive halo of outlying stars. A faint group of small galaxies is centred about 5′ SW of the cluster. Visually, although this globular cluster is smaller and dimmer than most

examples, it is still a fairly bright object, well seen at high magnification as a round patch of light with irregular edges, brighter to the middle with some evidence of graininess but no hint of resolution. The brighter central region makes up about a quarter of the visible disc. Herschel, who recorded the cluster on 21 August 1784, described it as: 'Very bright, small, round, gradually much brighter to the middle, resolvable.' This cluster and NGC2419 in Lynx are the most distant Milky Way globular clusters listed in the *New General Catalogue*. The brightest member stars are magnitude 15.6 and the horizontal branch visual magnitude is

18.8. Spectral type: F4. Radial velocity in approach: 385 km/s. Distance: 128 000 ly.

Draco

The north circumpolar constellation of Draco is one of the largest in the sky and extends over more than ten hours of right ascension. For this reason the constellation is well placed in the sky from late winter until early autumn. Herschel recorded 86 objects here, all galaxies except for the bright planetary nebula NGC6543 and NGC3210, a pair of field stars. Most of the objects were recorded in the last years of Herschel's survey, when he concentrated most of his observing efforts on north circumpolar constellations, an area he had tended to ignore in his early observing career. Many large, sometimes bright and nearby galaxies can be found here, including NGC4236, NGC5907 and NGC5866, but most of the objects are faint and often remote, requiring great care by the observer to identify correctly.

NGC2908

H977[3]	9:43.5	+79 41	GLX
0.8′ × 0.6′	14.46B		

Photographs show a small, faint, face-on galaxy, probably a spiral, with a slightly brighter core and one distinct spiral arm emerging from the S and curving N. Telescopically, the galaxy is rather faint but is well seen at both medium and high magnification as a round patch of light, very slightly brighter to the middle with well defined edges. Radial velocity: 6179 km/s. Distance: 276 mly. Diameter: 64 000 ly.

NGC2938

H963[3]	9:38.3	+76 19	GLX
SB(rs)cd	1.6′ × 1.1′	14.37B	

This is a very dim, slightly inclined barred spiral galaxy, elongated E–W. Images reveal a slightly brighter, narrow bar which is brighter to the E than to the W, with faint spiral arms attached. Visually, this is a very dim galaxy, only confirmed at high magnification as a very diffuse, unconcentrated patch of light. The surface brightness is even and the edges are very hazy and ill defined. Radial velocity: 2416 km/s. Distance: 108 mly. Diameter: 50 000 ly.

NGC2963

H315[3]	9:47.8	+72 58	GLX
SBab	1.2′ × 0.6′	14.36B	

Photographs show an inclined barred spiral galaxy, with a small, bright core and slightly fainter bar. To the S is a half ring attached to the bar while to the N is a faint, opaque envelope with a spiral arm emerging towards the N. NGC2957 is 2.75′ to the NW. These two galaxies are part of a apparent group of four galaxies oriented on a line 5.75′ long running ESE–WNW. In a moderate-aperture telescope only NGC2963 is visible; it is small and quite faint, located 2.0′ N of a pair of faint field stars. It is relatively well defined and very slightly elongated ESE–WNW; the surface brightness is

NGC2963

even across the disc. Radial velocity: 6658 km/s. Distance: 297 mly. Diameter: 104 000 ly.

NGC2977

H282[1]	9:43.8	+74 52	GLX
Sb:	1.8′ × 0.8′	13.28B	

This is an inclined spiral galaxy with a small, brighter core. Four spiral arms can be traced with several knotty condensations involved. Visually, this fairly bright galaxy is best seen at medium magnification as a well defined, oval galaxy oriented SE–NW. The disc is even in surface brightness but somewhat grainy. A small, brighter core is embedded and a faint field star is 1.4′ ESE of the core. Radial velocity: 3170 km/s. Distance: 142 mly. Diameter: 74 000 ly.

NGC3057

H978[3]	10:05.6	+80 17	GLX
2.2′ × 1.3′	13.59B		

Photographs show a very dim barred spiral galaxy with a narrow, short, slightly brighter bar embedded in a grainy, very faint outer envelope, elongated N–S. Two very weak spiral arms emerge, one from the N curving abruptly to the S, and one from the S abruptly extended to the NNE. Visually, it is rather faint but is seen as a diffuse,

©STScI/AURA/CALTECH/ROE

round patch of light, a little brighter to the middle and fairly well defined. Radial velocity: 1669 km/s. Distance: 75 mly. Diameter: 46 000 ly.

NGC3061

H903[2]	9:56.2	+75 52	GLX
(R')SB(rs)c	2.0' × 1.5'	13.48B	

Images show a dim, face-on barred spiral galaxy with a small, slightly brighter core and a weak bar with two patchy spiral arms attached. The telescopic appearance is very faint, and it is best seen at medium magnification as a dim, roundish patch of diffuse light, fairly even in surface brightness with no concentration to the middle. The edges are poorly defined. Radial velocity: 2587 km/s. Distance: 115 mly. Diameter: 67 000 ly.

NGC3144

H964[3]	10:15.5	+74 13	GLX
1.3' × 0.6'	14.31B		

In photographs this is a small, inclined spiral galaxy with a small, brighter core. Two spiral arms emerge from the core and the galaxy is elongated N–S. A field star is 30" ENE of the core. Visually, this galaxy is very faint and is seen as a round, blurry patch of light with ill defined edges and even surface brightness. Radial velocity: 6572 km/s. Distance: 293 mly. Diameter: 111 000 ly.

NGC3147

H79[1]	10:16.9	+73 24	GLX
SA(rs)bc	4.3' × 3.7'	11.26B	10.6V

This is a large, bright, face-on spiral galaxy with a brighter core. Images show bright, narrow, internal spiral arms expanding to fainter, broader outer arms and many dust patches and dust lanes mark the spiral arms. Visible even at low magnification, this is a very bright galaxy visually, well seen at high magnification. It appears almost round with a bright, grainy disc surrounded by a very faint, hazy outer envelope. A bright, nonstellar core is embedded. Radial velocity: 2943 km/s. Distance: 131 mly. Diameter: 164 000 ly.

NGC3155

H965[3]	10:17.6	+74 21	GLX
S?	1.0' × 0.7'	14.10B	

The DSS image shows a slightly inclined, possible barred spiral galaxy with a N–S bar in a fainter disc oriented NE–SW. The galaxy is quite faint visually and is seen as an oval, unconcentrated glow, evenly bright and fairly well defined at the edges. A faint triangular triplet about 3.0' to the NE points towards the galaxy. Radial velocity: 3070 km/s. Distance: 137 mly. Diameter: 40 000 ly.

NGC3183

H283[1]	10:21.8	+74 11	GLX
SB(s)bc:	2.3' × 1.4'	12.68B	

This is a face-on barred spiral galaxy. Images reveal a bright, oval bar with dust lanes involved and a slightly brighter, round core. Two thin, principal spiral arms emerge from the bar. Two smaller, possibly satellite, galaxies are situated 1.0' and 2.25' NNE of the core.

Telescopically, this is a large and fairly bright galaxy; the bar is well seen as a bright, elongated streak, oriented ESE–WNW, embedded in a fainter, oval glow oriented in the same position angle. The outer glow is diffuse and rather ill defined and several faint field stars are near or within the confines of the galaxy. Herschel observed this object on 2 April 1801 and described it as: 'Considerably bright, considerably large, easily resolvable.' The last comment is probably due to the faint field stars nearby. Radial velocity: 3214 km/s. Distance: 132 mly. Diameter: 96 000 ly.

NGC3197

H966[3]	10:14.4	+77 49	GLX
1.3' × 0.8'	14.39B		

Images show a faint, face-on spiral galaxy, slightly elongated N–S, with a small, round, brighter core and faint spiral arms. It is a fairly dim object visually, but is reasonably well seen at high magnification as a fairly well

NGC3147

©STScI/AURA/CALTECH/ROE

defined, round patch of light, evenly bright across its envelope. It is located midway along a line oriented NE–SW joining two field stars. Radial velocity: 8224 km/s. Distance: 367 mly. Diameter: 139 000 ly.

NGC3210

H979[3]	10:28	+79 50	DS

This pair of stars is the first of three Herschel objects visible together in a telescopic field. The other two are the galaxies NGC3212 and NGC3215. Herschel recorded the triplet on 26 September 1802, describing them as: 'Three, the place is that of the last. Two last, very faint, very small, preceding one stellar; all in a line, about 1.0′ distant from each other, preceding one most northerly, about 2.0′ more than the last.' This entry never appeared in the original *Philosophical Transactions* publication but was added by John Herschel in his *Cape Observations* (Herschel, 1847) and subsequently included in *The Scientific Papers* as part of Herschel's final catalogue of 500 discoveries. Visually, this object is clearly seen as an E–W aligned pair of stars, with a separation of about 0.5′, with the preceding star a little fainter.

NGC3212

H980[3]	10:28.1	+79 47	GLX
S?	1.9′ × 1.1′	14.50B	

This galaxy forms a probable interacting pair with NGC3215 and is also catalogued as Arp 181. In the DSS image it is a slightly inclined barred spiral galaxy with a round, brighter core and a slightly curving bar, brighter to the W. Two faint spiral arms emerge from the bar. The eastern arm, which curves to the SW has a small condensation at its southernmost point. Deep images show a third spiral plume, possibly connected with a small condensation ESE of the principal galaxy and extending to the W below the pair of stars designated NGC3210. NGC3215 is 1.25′ to the ESE and almost certainly physically related. The minimum core-to-core separation

between the two galaxies would be about 156 000 ly at the presumed distance. Note that NGC3210 = H979[3] is a pair of faint stars 1.0′ NW from NGC3212. Telescopically, NGC3212 appears fainter and smaller than its neighbour; it is a difficult object with only the core seen with certainty as a round and well defined patch of light. Radial velocity: 9858 km/s. Distance: 440 mly. Diameter: 243 000 ly.

NGC3215

H981[3]	10:29.0	+79 46	GLX
SB?	1.3′ × 1.2′	14.23B	

This barred spiral companion of NGC3212 has a round, brighter core and a slightly curving bar photographically. Two faint spiral arms emerge from the bar, with a faint star involved in the spiral arm to the E. Telescopically, it is larger and brighter than its neighbour; it is a moderately bright, very slightly oval patch of light, just a little brighter to the middle in a fairly well defined disc. Radial velocity: 9612 km/s. Distance: 430 mly. Diameter: 162 000 ly.

NGC3252

H316[3]	10:34.4	+73 46	GLX
SBd? sp	2.0′ × 0.4′	14.03B	

Photographs show an almost edge-on galaxy, possibly a spiral, with an elongated, slightly brighter, bar-like core. There is some evidence of dust patches in the outer envelope, which is oriented NE–SW. A bright double star is 4.25′ N. This is a fairly faint galaxy visually, best seen at medium magnification as a hazy, slightly elongated patch of light, oriented NE–SW and evenly bright along its major axis. Radial velocity: 1282 km/s. Distance: 57 mly. Diameter: 33 000 ly.

NGC3329

H284[1]	10:44.7	+76 49	GLX
(R)SA(r)b:	2.0′ × 1.2′	12.96B	

This is a slightly inclined spiral galaxy. Photographs show a bright, round core in a slightly fainter inner envelope. The

inner envelope consists of strong spiral arms laced with dust which expand to fainter, broader outer spiral structure. The barred spiral galaxy UGC5841 is 7.25′ S. NGC3329 is quite bright visually; it is a fairly well defined oval patch of light with a grainy texture and is very slightly brighter to the middle. The disc is oriented SE–NW and there is a very faint and hazy secondary envelope. Radial velocity: 2084 km/s. Distance: 93 mly. Diameter: 54 000 ly.

NGC3343

H317[3]	10:46.1	+73 21	GLX
E	1.3′ × 0.9′	14.58B	

Photographs show an elliptical galaxy with an elongated, brighter core and a fainter outer envelope, oriented NE–SW. Well seen at high magnification, this is a very small, condensed galaxy, almost round with well defined edges and even brightness across its disc. Radial velocity: 6421 km/s. Distance: 287 mly. Diameter: 108 000 ly.

NGC3403

H335[2]	10:53.9	+73 41	GLX
SAbc:	3.0′ × 1.2′	12.94B	

This highly inclined spiral galaxy is of the multi-arm type, featuring a bright inner disc and arms and much fainter outer spiral structure which extends to about 4.0′ on the DSS image. Although rather dim visually, this galaxy is well seen at high magnification as an elongated patch of light, oriented ENE–WSW. The edges are fairly well defined and the disc is even in surface brightness. Radial velocity: 1389 km/s. Distance: 62 mly. Diameter: 72 000 ly.

NGC3465

H967[3]	10:59.3	+75 12	GLX
Sab	1.2′ × 1.2′	14.71B	

Photographs show a face-on spiral galaxy with a bright core with two principal spiral arms attached. At least two subarms emerge from the principal arms within half a revolution around the core. Spiral galaxy NGC3500 (UGC6090) is 9.0′ to the E, while

NGC3465/3500/3523

©STScI/AURA/CALTECH/ROE

NGC3523 is 14.25′ ESE. Visually, NGC3465 is the westernmost object in a medium-magnification field; it is a hazy, roundish patch of light, a little concentrated towards the middle, situated 3.5′ WSW of a pair of prominent field stars. Radial velocity: 7353 km/s. Distance: 328 mly. Diameter: 115 000 ly.

NGC3500

H968[3]	11:01.7	+75 13	GLX
Sab	1.3′ × 0.6′	14.99B	

In photographs this highly inclined, possibly barred, spiral galaxy has a bright, elongated core and broad, sweeping spiral arms, oriented NE–SW. It was not plotted in the first edition of *Uranometria 2000.0* and is also catalogued as UGC6090. It is the middle of three NGC galaxies with NGC3465 9.0′ to the W and NGC3523 6.75′ SE. Visually, it appears slightly fainter than NGC3465 with only the core visible as a hazy, roundish patch of light. Radial velocity: 3600 km/s. Distance: 161 mly. Diameter: 61 000 ly.

NGC3523

H904[2]	11:02.8	+75 08	GLX
Sbc	1.5′ × 1.5′	13.93B	

This is a face-on spiral galaxy with a small, round, brighter core. Photographs reveal at least five distinct spiral arms. A nebulous bridge joins two of the spiral arms in the SSW and the spiral galaxy NGC3500 is 6.75′ to the NW. Best seen at medium magnification, NGC3523 is a round, somewhat diffuse patch of light, broadly brighter to the middle with fairly well defined edges. It is very slightly the largest and brightest of the three galaxies in the field. Radial velocity: 7298 km/s. Distance: 326 mly. Diameter: 142 000 ly.

NGC3562

H337[2]	11:12.9	+72 53	GLX
E	1.7′ × 1.3′	13.17B	

Photographs show an elliptical galaxy with a bright core and extensive outer envelope, elongated NNW–SSE. There are three much fainter galaxies within a 7.0′ radius to the N and ENE. The galaxy

is fairly well seen at high magnification; it is moderately bright and round with a bright, somewhat grainy disc and a star-like brighter core. It is quite well defined at the edges with a faint field star 1.0′ to the NNE. Radial velocity: 6875 km/s. Distance: 307 mly. Diameter: 152 000 ly.

NGC3682

H262[1]	11:27.7	+66 35	GLX
SA(s)0/a:?	2.1′ × 1.5′	13.24B	

While the classification is somewhat uncertain, the DSS shows an inclined spiral galaxy with a bright, elongated core. The core appears blunted to the E, certainly fainter, perhaps caused by a large dust patch. In the fainter outer envelope there is some evidence of spiral structure, particularly a dust lane which borders the brighter core to the S. A second dust lane in the S is further from the core. The outer disc is otherwise smooth textured, resembling a lenticular envelope. Visually, the object is fairly bright and takes magnification well; it is an elongated patch of light, oriented almost due E–W with a fairly bright inner disc which brightens to a condensed core. A very faint and ill defined secondary envelope is intermittently visible. Radial velocity: 1624 km/s. Distance: 72 mly. Diameter: 44 000 ly.

NGC3735

H287[1]	11:36.0	+70 32	GLX
SAc: sp	4.2′ × 0.8′	12.32B	

Photographs show a large, almost edge-on spiral galaxy with a bright, elongated core and dusty outer spiral arms. Telescopically, this is a bright galaxy, a fairly large, elongated streak of light, best seen at medium magnification but rewarding at high magnification as well. It is fairly flat and is well defined at the edges, the main envelope is a little mottled and uneven in brightness. It is brighter to an elongated core and the galaxy is oriented ESE–WNW. Radial velocity: 2819 km/s. Distance: 126 mly. Diameter: 154 000 ly.

NGC3747

H969[3]	11:32.5	+74 23	GLX
0.5′ × 0.15′	14.0B?		

This galaxy is not plotted in the first edition of *Uranometria 2000.0*. The DSS image shows a very small, flat galaxy, oriented E–W and relatively bright along its major axis, which curves northward slightly at its western tip. Visually, the galaxy is extremely faint and small and is best seen at high magnification as an intermittent blur; it is very slightly elongated E–W. Radial velocity: 5737 km/s. Distance: 256 mly. Diameter: 37 000 ly.

NGC3752

H905[2]	11:32.5	+74 38	GLX
Sab	1.7′ × 0.7′	13.74B	

This galaxy is misplotted in the first edition of *Uranometria 2000.0*; the correct coordinates are given here. Photographs show an inclined spiral galaxy with a somewhat peculiar morphology. The small, brighter core has three long, curved spiral arms: one

emerges from the E, curving SE, and two very close together emerge from the W and curve to the NW. This bright portion of the galaxy is offset to the S, with an extensive, grainy, fainter envelope to the N. Telescopically, the galaxy is best seen at high magnification as an elongated, moderately bright object, oriented SE–NW. It is bright along its major axis with a brighter core. The ends taper slightly and the envelope is a little grainy while the edges are somewhat diffuse. Radial velocity: 2047 km/s. Distance: 29 mly. Diameter: 45 000 ly.

NGC3879

H881[2]	11:46.9	+69 24	GLX
Sdm:	2.4′ × 0.5′	13.71B	

Images reveal an almost edge-on galaxy oriented SE–NW, probably a spiral. It is brighter to the NW where bright, knotty patches are visible in the main envelope. There is some evidence of dust,

particularly near the knotty patches and towards the SE in the envelope. Visually, this is a very faint and diffuse patch of light; it is roundish and only very slightly brighter to the middle. The faint extensions may be visible in excellent conditions and field stars are 1.2′ SW and 2.5′ WNW. Radial velocity: 1552 km/s. Distance: 69 mly. Diameter: 48 000 ly.

NGC3890

H940[3]/H971[3]	11:49.3	+74 18	GLX
S?	0.9′ × 0.8′	14.21B	

This galaxy was recorded twice by Herschel: first on 12 December 1797 and again on 2 April 1801. Though the classification is uncertain, the DSS photograph shows a face-on spiral galaxy with a small, round core and two main spiral arms in a fainter outer envelope. Though fairly faint visually, this galaxy is well seen at both medium and high magnification as a round

NGC3752

patch of light which is broadly brighter to the middle with hazy edges. Radial velocity: 6945 km/s. Distance: 310 mly. Diameter: 81 000 ly.

NGC3961

H905[3]	11:55.0	+69 20	GLX
(R)SB(r)a:	1.3′ × 1.3′	14.58B	

This is an almost face-on barred galaxy with an elongated core and a smaller bar. The bar is attached to a possible ring which brightens around the points of attachment. In the S, a single spiral arm emerges from the bar and ring, circling clockwise to the N where it fades. A very faint object visually, this galaxy is somewhat difficult with direct vision. It is a hazy, round patch of light, fairly even in surface brightness and moderately well defined at the edges. Radial velocity: 6842 km/s. Distance: 305 mly. Diameter: 115 000 ly.

NGC4034

H903[3]	12:01.5	+69 19	GLX
Scd:	1.9′ × 1.2′	14.60B	

The DSS photograph shows an almost face-on spiral galaxy, possibly a dwarf, with a small core in a diffuse spiral disc. This is an extremely dim and difficult object visually; it is a very slightly elongated blur, oriented N–S and is seen only intermittently with averted vision. The disc is even in surface brightness and the edges are hazy. Radial velocity: 2490 km/s. Distance: 111 mly. Diameter: 61 000 ly.

NGC4120

H904[3]	12:08.5	+69 33	GLX
SA(rs)cd:	1.8′ × 0.4′	14.31B	

Photographs show an almost edge-on galaxy, probably a spiral, with a slightly brighter major axis which curves through the fainter envelope. There is some evidence of dust patches and a faint field star is visible in the envelope 0.25′ S of the centre of the galaxy which is oriented NNW–SSE. This is a dim, irregular patch of light visually; it is very slightly oval and even in surface brightness. The edges are poorly defined

and it is a little easier to see at medium magnification. Radial velocity: 2372 km/s. Distance: 106 mly. Diameter: 56 000 ly.

NGC4128

H263[1]	12:08.5	+68 46	GLX
SA0: sp	2.6′ × 0.9′	12.93B	

This is an almost edge-on lenticular galaxy. Photographs show a bright, elongated core in a fainter outer envelope, oriented ENE–WSW. A fainter, likely companion galaxy, NGC4128A, is 2.0′ to the NW. Telescopically, the galaxy is small but quite bright with only the elongated central region visible. It is bright, condensed and opaque and is well defined at the edges. Radial velocity: 2449 km/s. Distance: 109 mly. Diameter: 83 000 ly.

NGC4133

H278[1]	12:08.6	+74 56	GLX
SABb:	1.8′ × 1.5′	13.13B	

Photographs show an almost face-on spiral galaxy, probably barred, with a bright, elongated central region and two main spiral arms which branch out to four arms away from the core. It is a fairly bright galaxy visually and is well seen at high magnification. It is situated along the S side of a triangle of field stars; the somewhat ill defined envelope is grainy and broadly brighter to the middle where there is a sizable bright core. The disc is elongated SSE–NNW. Radial velocity: 1498 km/s. Distance: 67 mly. Diameter: 35 000 ly.

NGC4159

H941[3]	12:10.7	+76 09	GLX
Sdm	1.3′ × 0.5′	14.28B	

Images reveal an almost edge-on galaxy, perhaps barred, with a thin, slightly brighter, elongated central region in a fainter outer envelope, elongated NNE–SSW. Visually, the galaxy is seen as a roundish, hazy patch of light, very broadly brighter to the middle with ill defined edges. Field stars are 0.5′ E and 1.0′ NE of core. Radial

velocity: 1879 km/s. Distance: 84 mly. Diameter: 32 000 ly.

NGC4210

H850[3]	12:15.3	+65 59	GLX
SB(r)b	1.9′ × 1.4′	13.5B	

The DSS photograph shows an inclined barred spiral galaxy with a small, bright core and a slightly fainter bar, lying SE of a magnitude 7 field star. Two bright spiral arms emerge from the bar and can be traced for one and a half revolutions around the core. A large anonymous galaxy is 9.5′ to the NW, immediately SE of the bright field star. Telescopically, NGC4210 is fairly large and moderately bright; it is a fat oval of light elongated E–W, very broadly brighter to the middle with a slightly grainy disc and fairly well defined edges. The galaxy is probably a member of the NGC4256 group of galaxies. Radial velocity: 2848 km/s. Distance: 127 mly. Diameter: 70 000 ly.

NGC4236

H51[5]	12:16.7	+69 28	GLX	
SB(s)dm	21.9′ × 7.2′	10.05B	9.6V

This faint, highly inclined, barred spiral galaxy is probably an outlying member of the M81/NGC2403 galaxy group. It features a long, narrow central bar and low-surface-brightness spiral arms, marked by small knots of nebulosity. In a moderate-aperture telescope, the galaxy is best seen at low magnification as a very faint, diffuse and very poorly defined object, elongated SSE–NNW.

NGC4210

NGC4236

©STScI/AURA/CALTECH/ROE

At medium magnification, a very faint star near the core is intermittently visible, mimicking a stellar nucleus. The galaxy is a little brighter along its major axis but the texture appears fairly smooth. Herschel discovered this galaxy on 6 April 1793, describing it as: 'Very faint, much extended 70° north preceding south following. About 25′ long and losing itself imperceptibly, about 6′ or 7′ broad.' Radial velocity: 126 km/s. Distance: 7.2 mly. Diameter: 47 000 ly.

NGC4238

H851³	12:16.9	+63 25	GLX
Sd	1.8′ × 0.5′	14.25B	

This is a highly inclined spiral galaxy with a brighter core. Photographs show dust patches and knotty condensations along the spiral arms. This galaxy may be an outlying member of a group of galaxies centred on NGC4521. NGC4238 is a very dim galaxy visually, but with patience the elongated disc, oriented NE–SW, can be made out at medium magnification. The surface brightness is even along the major axis and the edges are fairly well defined. Radial velocity: 2871 km/s. Distance: 128 mly. Diameter: 67 000 ly.

NGC4250

H264¹	12:17.4	+70 48	GLX
SAB(r)0⁺	2.7′ × 2.1′	13.20B	

Images show an elongated theta-shaped barred galaxy with a bright, round core.

NGC4250

The faint bar is connected to a faint, elongated ring, oriented SSE–NNW. Surrounding this is a very faint, suspected outer ring. Only the core of this galaxy is visible at high magnification. It is a round, faint glow with a bright, stellar core. The edges of the disc are hazy and ill defined. Radial velocity: 2149 km/s. Distance: 96 mly. Diameter: 78 000 ly.

NGC4256

H846[2]	12:18.7	+65 54	GLX
SA(s)b: sp	4.5′ × 0.8′	12.75B	

This galaxy is the principal member of the NGC4256 group of galaxies, a loose association of seven objects including the Herschel objects NGC4210 and NGC4332. The DSS image shows an almost edge-on spiral galaxy with a large, bright, elongated core. The outer envelope is defined by a thin, distinct dust lane which runs along much of the length of the galaxy. This is a bright,

NGC4256

large and well defined galaxy visually and averted vision bring out the extensions well. It is well seen at both medium and high magnification and features a sharp stellar core with a slight bulge in the middle. The extensions taper to points and the disc is oriented NE–SW. Radial velocity: 2645 km/s. Distance: 118 mly. Diameter: 155 000 ly.

NGC4291

H275[1]	12:20.3	+75 22	GLX
E3	1.9′ × 1.6′	12.38B	

This bright, elliptical galaxy, which has a bright, elongated core and a fainter outer envelope, is the principal galaxy of the NGC4291 group, which comprises a total of eleven galaxies, including NGC4319 and NGC4386 which are described below. Visually, the galaxy is a bright, round, well condensed object with some mottling visible in its well defined envelope. A bright central core is prominent. The galaxy is located W of

a bright triangle of field stars and two faint field stars border the outer envelope to the S and W. Photographs show it elongated E–W and the bright companion galaxy NGC4319 is 6.0′ ESE. Radial velocity: 1899 km/s. Distance: 85 mly. Diameter: 47 000 ly.

NGC4319

H276[1]	12:21.7	+75 19	GLX
SB(r)ab	3.0′ × 2.3′	12.93B	

The DSS image shows a bright, slightly inclined, theta-shaped, barred spiral galaxy with a peculiar morphology. It has a bright, slightly elongated core with a slightly fainter bar, which is attached to an inner ring. A faint, broad spiral arm emerges from the E of the ring, curves S, then abruptly NW. A faint spiral arm in the NW curves N, then E and has a faint condensation at its tip and the galaxy is elongated NNW–SSE. The quasar Markarian 205 is 0.75′ S of the core, while the bright

NGC4291/4319

companion galaxy NGC4291 is 6.0′ WNW. In moderate-aperture telescopes, NGC4319 appears a little smaller and fainter than NGC4291, oval in shape and oriented N–S and a little brighter to the middle. Immediately S is the quasar Markarian 205, which is stellar in appearance and usually visible with direct vision. Herschel discovered NGC4291, NGC4319 and NGC4386 on 10 December 1797. There is no mention of any of the field 'stars' in his description but it is quite possible that Herschel was the first person ever to see the image of a quasar if he spied the faint 'star' immediately S of NGC4319. Radial velocity: 1499 km/s. Distance: 67 mly. Diameter: 59 000 ly.

NGC4331

H942[3]	12:22.4	+76 11	GLX
Im?	2.2′ × 0.4′	14.78B	

This irregular galaxy appears as a faint, elongated ray, oriented N–S, which grows broader towards the S with a brighter condensation N of centre and a smaller, faint patch in the eastern part of the envelope, S of centre. Visually, only the bright irregular patch to the N is visible at high magnification; it is very dim and is seen as a hazy, very slightly oval patch of light immediately S of a faint field star. A faint double star is 0.75′ W. Radial velocity: 1713 km/s. Distance: 77 mly. Diameter: 49 000 ly.

NGC4332

H847[2]	12:22.8	+65 51	GLX
SB(s)a	2.2′ × 1.6′	13.29B	

The DSS photograph shows a slightly inclined, S-shaped, barred spiral galaxy. The bright, elongated core has a curved, slightly fainter bar which is attached to very faint outer spiral arms. Probable dust patches are evident, particularly in the SE bar and immediately NNW of the core. A very faint field star lies near the tip of the SE bar. Another, brighter, field star is 2.0′ ENE. NGC4332 is a moderately bright object telescopically, and is a little more distinct at medium magnification. It is an elongated,

slightly oval disc, oriented SE–NW, very slightly brighter to the middle and well defined at the edges. Radial velocity: 2886 km/s. Distance: 129 mly. Diameter: 83 000 ly.

NGC4363

H938[3]	12:23.5	+74 57	GLX
SAb:	0.8′ × 0.6′	14.68B	

This is very likely a dwarf galaxy. Photographs show a faint, oval disc, oriented N–S, featureless except for a brighter core An anonymous, brighter and more compact galaxy is 8.2′ SSW. Visually, NGC4363 is very faint and is best seen at medium magnification as a round, even-surface-brightness patch of light with well defined edges. Radial velocity: 1568 km/s. Distance: 70 mly. Diameter: 16 000 ly.

NGC4386

H277[1]	12:24.5	+75 32	GLX
SAB0°:	2.5′ × 1.3′	12.66B	

This is a member of the NGC4291 group of galaxies. Photographs show this lenticular galaxy has a bright, inner envelope surrounding the bright core as well as a fainter, extensive outer envelope. Visually, it appears brighter than the companion galaxies NGC4291 and NGC4319, which are located just outside a high-magnification field. It is a bright, well-condensed oval of light oriented NW–SE; the central region is quite bright but nonstellar and the envelope is well defined. Radial velocity: 1820 km/s. Distance: 81 mly. Diameter: 59 000 ly.

NGC4391

H852[3]	12:25.3	+64 56	GLX
SA0⁻:	1.1′ × 1.1′	13.73B	

Photographs show a face-on lenticular galaxy with a bright, round core and a fainter outer envelope. This galaxy can be seen in the same medium-magnification field as NGC4441 and appears quite small but moderately bright, quite round and even in surface brightness across the disc which is well defined and sharp-edged. A distinctive

three-star triangular asterism is 2.5′ to the SW. Radial velocity: 1440 km/s. Distance: 64 mly. Diameter: 21 000 ly.

NGC4441

H848[2]	12:27.3	+64 48	GLX
SAB0⁺ pec	3.9′ × 3′	13.34B	

Although this is classified as a lenticular galaxy, the DSS image reveals a face-on spiral galaxy with a peculiar morphology. The core of the galaxy is bright and slightly irregular in outline and is embedded in a fainter inner envelope. Small dust patches border the core to the SW and S. Three very faint and asymmetrical spiral arms are attached: two emerge from the SW and curve S, while one broad arm emerges from the N and continues in that direction. The galaxy is elongated N–S. It can be seen in the same medium-magnification field as NGC4391, though they are probably not physically associated. Visually, NGC4441 is a little larger but slightly fainter than NGC4391; it is an irregularly round patch of light, very slightly brighter towards the middle with no evidence of the faint outer spiral pattern. Radial velocity: 2838 km/s. Distance: 127 mly. Diameter: 143 000 ly.

NGC4521

H849[2]	12:32.8	+63 57	GLX
S0/a	2.5′ × 0.5′	12.94B	

Photographs show an edge-on lenticular galaxy with a large, bright, elongated core and fainter extensions. A star or bright condensation borders the core to the SSE. This is the principal galaxy of a small group which includes the faint, barred spiral galaxy UGC7700 4.0′ to the SSW and the Herschel objects NGC4441 and NGC4545. Although small visually, NGC4521 is a moderately bright, edge-on galaxy with an extended, bright and well defined disc encasing a bright core that bulges slightly. The galaxy is oriented SSE–NNW and a field star is 2.0′ NNW. Radial velocity: 2651 km/s. Distance: 118 mly. Diameter: 86 000 ly.

NGC4521

©STScI/AURA/CALTECH/ROE

elongated ENE–WSW with a bright, elongated central region encompassing a sharp, almost stellar core. A fainter secondary envelope surrounds the central region and has hazy edges. There is some graininess in the disc. Companion galaxy NGC4572 is 7.5′ to the WNW. Radial velocity: 2122 km/s. Distance: 95 mly. Diameter: 83 000 ly.

NGC4648

H274[1]	12:41.8	+74 25	GLX
E3	1.7′ × 1.3′	12.95B	

In photographs this elliptical galaxy has a bright, concentrated oval core and an extensive, fainter outer disc. Although small, this is quite a bright object visually; a moderately bright, slightly elongated envelope, oriented E–W, surrounds a bright, star-like core. The edges are a little hazy and ill defined and the galaxy is well seen at high magnification. This galaxy may be a member of the NGC4291 group of galaxies, though its radial velocity is a little lower than most of the members of that group. Radial velocity: 1557 km/s. Distance: 70 mly. Diameter: 34 000 ly.

NGC4545

H850[2]	12:34.6	+63 31	GLX
SB(s)cd:	2.7′ × 1.5′	13.19B	

This is a slightly inclined barred spiral galaxy with a very small, round core. Photographs show a narrow bar which connects to two brighter spiral arms. These arms quickly fragment to a fainter, multi-arm pattern which exhibits many knots and condensations; the galaxy is elongated N–S. Visually, the galaxy appears quite dim and is seen as a roundish, diffuse glow, fairly even in surface brightness with fairly well defined edges, and a faint field star is just beyond the disc to the ENE. Radial velocity: 2844 km/s. Distance: 127 mly. Diameter: 100 000 ly.

NGC4572

H939[3]	12:35.9	+74 15	GLX
S	1.5′ × 0.5′	14.73B	

This galaxy and NGC4589 are possible outlying members of the NGC4291

group of galaxies. Photographs show a small, low-surface-brightness, elongated, S-shaped spiral with a small and very slightly brighter core. Two long, thin curved spiral arms, oriented N–S, emerge from the core. Telescopically, this galaxy and NGC4589 can be seen in the same high-magnification field, though NGC4572 is far fainter and smaller. It appears as a very hazy and ill defined glow, quite dim and irregular in outline. Companion galaxy NGC4589 is 7.5′ to the ESE. Radial velocity: 2346 km/s. Distance: 105 mly. Diameter: 46 000 ly.

NGC4589

H273[1]	12:37.4	+74 12	GLX
E2	3.0′ × 2.7′	11.71B	

This is probably a member of the NGC4291 group of galaxies and photographs show an elliptical galaxy with a bright, elongated core in a fainter outer envelope. Telescopically, the galaxy is large, bright and slightly

NGC4693

H906[3]	12:47.1	+71 12	GLX
Sd	2.5′ × 0.5′	14.34B	

Images reveal a highly inclined spiral galaxy with a small, bright, round core. The outer envelope shows some evidence of a spiral pattern and the envelope is dusty, particularly along the SE flank. The galaxy is moderately faint visually but is fairly well seen at high magnification as an elongated streak of light oriented NE–SW, fairly even in surface brightness and well defined along the edges. Radial velocity: 1806 km/s. Distance: 81 mly. Diameter: 59 000 ly.

NGC4749

H907[3]	12:51.2	+71 38	GLX
Sb?sp	1.7′ × 0.3′	14.31B	

This faint, edge-on spiral galaxy presents a similar aspect both visually and photographically. It is a well defined streak, oriented SSE–NNW with a very slight bulge in the middle and is very broadly brighter along its major axis, though no core is visible. Along with NGC4693 this galaxy is a probable member of the NGC4750 group of galaxies. Radial velocity: 1827 km/s. Distance: 81 mly. Diameter: 40 000 ly.

NGC4750

H78[4]	12:50.1	+72 52	GLX
(R)SA(rs)ab	2.1′ × 2.0′	11.94B	

This is the principal member of a loose collection of eight galaxies. The DSS image shows a slightly inclined galaxy, probably spiral, with a brighter, round core embedded in a bright inner ring. A blunt spur emerges from the ESE of the ring. The ring is surrounded by a fainter outer envelope, which is grainy and elongated ESE–WNW. The galaxy CGCG335-024 is 7.1′ NNW. From just a cursory glance at the DSS image, it is easy to mistake this galaxy for a planetary nebula and Herschel placed the object in his Class IV. Discovered on 8 November 1798, his description reads: 'Considerably bright, round, about 1.5′ in diameter. Somewhat approaching to a planetary nebula with a strong hazy border.' Telescopically, this is a very bright and rather large galaxy, visible even at low magnification. A fat oval of light, the envelope is opaque and grainy with a small, but nonstellar, brighter core which seems slightly offset to the N in the envelope. The edges are moderately well defined but a very faint outer haze is suspected. Radial velocity: 1764 km/s. Distance: 79 mly. Diameter: 48 000 ly.

NGC4857

H908[3]	12:57.2	+70 12	GLX
SABb	1.3′ × 0.7′	14.60B	

This is a slightly inclined spiral galaxy, which photographs show may be barred. The brighter, slightly elongated core has two faint, narrow spiral arms emerging from the central region. A faint galaxy is 4.6′ NNE. NGC4857 is quite a faint object visually, but is fairly well seen at high magnification. It is a

slightly oval patch of light, fairly even in surface brightness, somewhat ill defined at the edges and elongated ESE–WNW. Radial velocity: 8732 km/s. Distance: 390 mly. Diameter: 148 000 ly.

NGC4954

H937[3]	13:02.3	+75 24	GLX
S0?	0.8′ × 0.5′	14.32B	

This is the brightest of four spiral galaxies in a 3.0′ chain running SSE–WNW; a very large telescope would be required to see all four. Photographs show an inclined lenticular, or perhaps spiral, galaxy with a bright, elongated core and a fainter outer envelope, which suggest possible spiral structure; the galaxy is elongated ENE–WSW. It is small and quite faint visually, a roundish, unconcentrated patch of light, evenly bright and fairly well defined. Radial velocity: 9430 km/s. Distance: 421 mly. Diameter: 98 000 ly.

NGC5667

H807[2]	14:30.4	+59 29	GLX
Scd: pec	2.0′ × 1.1′	12.95B	

Photographs show a bright irregular galaxy featuring a bright bar with several knotty condensations along its major axis, including two especially bright knots at the bar's southern extremity. The bar is surrounded by a fainter outer envelope which displays condensations, especially to the E. Weak spiral arms are visible in the W. A faint field star borders the galaxy to the N. Visually, the galaxy is moderately

bright, opaque and elongated N–S with a small slightly brighter core. The envelope is fairly bright and quite well defined. Radial velocity: 2083 km/s. Distance: 93 mly. Diameter: 54 000 ly.

NGC5678

H237[1]	14:32.1	+57 55	GLX
SAB(rs)b	3.3′ × 1.7′	12.08B	

In photographs, this inclined spiral galaxy has a bright inner envelope, marked by many knotty condensations, and a brighter core. Two bright spiral arms expand to fainter, broader arms at the galaxy's extremities and the pattern is asymmetrical. There are large dust patches N and S of the central region. A very small, compact galaxy is immediately N, while a bright field star is 2.75′ NNW. This is a large and fairly bright galaxy in moderate apertures; it is elongated N–S, with a bright, inner envelope and a much fainter secondary envelope which broadens into a fat oval when using averted vision. It is a little brighter to a broadly concentrated middle and holds magnification well. Radial velocity: 2059 km/s. Distance: 92 mly. Diameter: 88 000 ly.

NGC5777

H806[3]	14:51.3	+58 58	GLX
Sbc	3.1′ × 0.4′	14.01B	

Photographs show an almost edge-on spiral galaxy, oriented SE–NW, with a bright, elongated core and a dust lane along the galaxy's major axis; the dust lane is best seen where it crosses the

NGC5667

NGC5678

©STScI/AURA/CALTECH/ROE

©STScI/AURA/CALTECH/ROE

NGC5777

core. The dim face-on galaxy UGC9570 is located 2.75′ ESE. Visually, good conditions are required to see the full extent of the galaxy. It is quite faint but visible as an extended patch of light of even surface brightness with a small, star-like core and is fairly well defined at the edges. Radial velocity: 2291 km/s. Distance: 102 mly. Diameter: 92 000 ly.

NGC5866

H215[1]	15:06.5	+55 46	GLX
S03	6.3′ × 3.0′	10.73B	9.9V

This galaxy was once thought to be Messier's missing M102, but a letter written by his assistant Pierre Méchain revealed that M102 was actually a duplicate observation of M101. It is the principal galaxy of a group of four which includes NGC5907 and photographs show a large, bright, edge-on lenticular galaxy with an elongated core and bright, tapered extensions. There is a broad, faint outer envelope which is squarish in shape. Though not seen in the DSS image here, the galaxy features a razor thin dust lane along its major axis, tilted very slightly from the plane of the galaxy. Visually, this is a superb, bright object, well defined, with an even surface brightness, lenticular in shape and elongated ESE–WNW. The dust lane is not visible in medium apertures, though it might be a good challenge for

large amateur instruments. Radial velocity: 819 km/s. Distance: 37 mly. Diameter: 67 000 ly.

NGC5879

H757[2]	15:09.8	+57 00	GLX
SA(rs)bc:?	5.0′ × 1.6′	12.12B	

Photographs show this highly-inclined spiral galaxy is of the multi-arm type with an elongated, bright core and thin, but somewhat fragmentary, spiral arms. Telescopically, the galaxy is small but quite bright and is well seen at high magnification. The bright core is embedded in a small, bright, opaque envelope, is quite well defined and is elongated N/S. A bright field star is 7.3′ to the NNW. Radial velocity: 922 km/s. Distance: 41 mly. Diameter: 60 000 ly.

NGC5881

H818[2]	15:06.3	+62 59	GLX
1.0′ × 0.7′	15.90B		

This galaxy is unclassified but the DSS image shows a face-on spiral galaxy, possible barred, with a bright core and two spiral arms, which start out narrow and bright. The arm emerging NE of the core and curving SW is short, but the arm emerging SW of the core fades and broadens in the E before winding completely around the galaxy. NGC5881 is moderately bright and well seen at high magnification as a round, mottled patch of light that is fairly well defined and brighter to the middle. It lies between two field stars oriented NNE–SSW. Field star 0.75′ SW. Radial velocity: 6741 km/s. Distance: 301 mly. Diameter: 88 000 ly.

NGC5894

H763[2]	15:11.7	+59 49	GLX
SBdm?	3.6′ × 0.6′	13.25B	

The DSS photograph reveals an almost edge-on spiral galaxy with a brighter, elongated core. The highly inclined spiral structure is defined by small dust patches with a thin dust lane running along most of the major axis on the W side. Visually, the galaxy is somewhat dim but is well seen at high magnification as an elongated sliver of light which is well

NGC5866

defined and fairly even in surface brightness, but with a brighter core. The galaxy is oriented NNE–SSW. Radial velocity: 2620 km/s. Distance: 117 mly. Diameter: 123 000 ly.

NGC5905

H758[2]	15:15.4	+55 31	GLX
SB(r)b	4.0′ × 2.6′	13.21B	

This face-on barred spiral galaxy has a brighter, round core. Photographs show a faint bar is connected to three, thin spiral arms. One spiral arm emerges from the NNE bar, while the other two come from the SSW of the bar. The arms have several knotty condensations along their length and fainter spiral fragments make up much of the disc. Field stars border the galaxy on the E, S and SW. Visually, the galaxy is moderately faint and quite diffuse, a roundish patch of light with a slightly brighter, star-like core. An attractive double star is located 4.0′ to the SSE. The galaxy is a probable member of the NGC5908 group of galaxies. Radial velocity: 3540 km/s. Distance: 158 mly. Diameter: 184 000 ly.

NGC5907

H759[2]	15:15.9	+56 19	GLX
SA(s)c: sp	11.3′ × 1.4′	11.06B	

This is one of the more spectacular galaxies in Herschel's catalogue. He came upon it on 5 May 1788, but his description was rather terse, commenting only: 'Pretty bright. Faint nucleus in the middle. 8′ or 10′ long, 2′ broad.' Photographs reveal a very thin, almost edge-on spiral galaxy, with a very small, nonbulging core. The spiral extensions are slightly fainter and a complex dust lane extends almost the whole length of the major axis which is elongated SSE–NNW. The galaxy is a rewarding sight in moderate-aperture telescopes, very flat, edge-on, with a brighter core. The core is surrounded by a lozenge-shaped bright zone oriented along the major axis. With direct vision the ends taper to points but averted vision reveals flares of faint light, more so in the northern extremity than in the S.

NGC5907

©STScI/AURA/CALTECH/ROE

The dust lane is not seen and the galaxy is surprisingly bright along the major axis despite the presence of the lane of absorbing material. Radial velocity: 818 km/s. Distance: 37 mly. Diameter: 119 000 ly.

NGC5908

H760[2]	15:16.7	+55 25	GLX
Sb	2.7′ × 1.7′	12.80B	

This is the principal galaxy in a small group of seven. Photographs show an

NGC5908

©STScI/AURA/CALTECH/ROE

almost edge-on spiral galaxy with a bright, very large, elongated core. The bright extensions are grainy textured with a very distinct, well defined dust lane along the major axis. Visually, the galaxy is bright and appears as an oval patch elongated SSE–NNW. It is broadly brighter to the middle and the edges are fairly well defined. Radial velocity: 3456 km/s. Distance: 154 mly. Diameter: 121 000 ly.

NGC5949

H906[2]	15:28.0	+64 46	GLX
SA(r)bc?	2.3′ × 1.1′	12.91B	

The DSS photograph shows an inclined galaxy, probably a spiral with a high-surface-brightness disc and a slightly brighter core. Knotty condensations are seen throughout, defined by dust patches and short dust lanes; higher-resolution photographs show a very fragmented spiral pattern surrounding a tiny nucleus. Visually, this is a moderately bright galaxy; it is a well defined oval oriented SE–NW with an opaque envelope that is broadly brighter along its major axis. Radial velocity: 600 km/s. Distance: 27 mly. Diameter: 18 000 ly.

NGC5963

H761[2]	15:33.5	+56 35	GLX
S pec	3.3′ × 2.6′	13.02B	

Images show an inclined galaxy which is probably a spiral. It features a large, even-surface-brightness disc and extremely faint, thin spiral arms, barely visible against the sky background. Visually, though fairly bright, the galaxy is a small, well defined oval oriented NE–SW and is slightly brighter to the middle with no trace of the outer spiral structure. The brighter edge-on galaxy NGC5965 is 8.5′ NNE but is not a physical companion. Radial velocity: 813 km/s. Distance: 36 mly. Diameter: 35 000 ly.

NGC5965

NGC5965

H762[2]	15:34.0	+56 42	GLX
Sb	5.6′ × 0.8′	12.66B	

Photographically, this is an attractive galaxy and is morphologically similar to NGC4565. It is a large, bright, almost edge-on galaxy with a bright, elongated central region and well defined spiral extensions, oriented NE–SW. A prominent dust lane along the major axis crosses in front of the core. In moderate apertures the galaxy is quite bright; it is a flat, much extended galaxy with a bright core and well defined extremities. It is visible in the same medium-magnification field as NGC5963. Radial velocity: 3570 km/s. Distance: 159 mly. Diameter: 260 000 ly.

NGC5982

H764[2]	15:38.7	+59 21	GLX
E3	3.0′ × 2.1′	11.98B	

Photographs reveal a large, bright, elliptical galaxy with a bright, elongated core and a fainter outer envelope, oriented ESE–WNW. Visually, at low magnification it is a round, bright, condensed glow with a much brighter middle. High power brings out the faint extensions. NGC5982 is the middle of three galaxies, the others being NGC5981 and NGC5985, which form a remarkable ESE–WNW chain of galaxies with different morphologies, though the spread in radial velocities indicates that the galaxies may not be physically related. Although NGC5981 is not entered as a separate entry in Herschel's catalogue, there is no doubt that he observed it. In the entry for NGC5982, recorded on 25 May 1788, Herschel writes: 'Pretty bright, small, irregularly round. One preceding suspected, very faint, a little extended.' This is a perfect visual description of NGC5981, which is a thin, edge-on galaxy, similar to NGC5907 in morphology. Radial velocity: 3180 km/s. Distance: 142 mly. Diameter: 124 000 ly.

NGC5985

H766[2]	15:39.6	+59 20	GLX
SAB(r)b	5.5′ × 3.0′	11.95B	

Photographically, this is a large, inclined, barred spiral galaxy with a small, bright, round core and short bright extensions. The inner envelope is faint with a slightly brighter ring of matter surrounding the core. Three thin spiral arms emerge with many knotty condensations along their length, especially in the SSE. The arms are well defined by dust patches and dust lanes and the disc is oriented NNE–SSW. In moderate-aperture telescopes, this is a rewarding object. It is the most easterly of a trio, the others being NGC5981 and NGC5982, and appears as a large well defined glow, decidedly oval in shape and even in surface brightness with a small, slightly brighter core. Herschel called it: 'Pretty bright, considerably large, irregularly extended, resolvable.' Radial velocity: 2680 km/s. Distance: 120 mly. Diameter: 192 000 ly.

©STScI/AURA/CALTECH/ROE

NGC5981/5982/5985

©STScI/AURA/CALTECH/ROE

NGC5987

H765[2]	15:39.9	+58 05	GLX
Sb	5.3′ × 1.4′	12.72B	

Images of this interesting object show a large, highly inclined spiral galaxy with a bright core and a bright, very elongated inner envelope. The fainter outer spiral arms feature a prominent double dust lane, with the inner lane bordering the core and the outer lane in the envelope. The structure of the lanes is complex with extensive branching. A

NGC5987

©STScI/AURA/CALTECH/ROE

dense, broad dust patch is suspected along the the ESE flank of the galaxy E of the core and the galaxy is elongated ENE–WSW. A bright field star is 1.2′ NW. In moderate-aperture telescopes, this is an attractive galaxy, though small, and even at high magnification only the core and brighter inner region are visible. The galaxy is well condensed, displaying a concentrated, bright core with faint extensions. It would probably require a large aperture to see the full extent of the galaxy well. Radial velocity: 3172 km/s. Distance: 142 mly. Diameter: 260 000 ly.

NGC5989

H738[3]	15:41.5	+59 45	GLX
Scd?	0.9′ × 0.9′	13.53B	

This is a small, face-on, four-branch spiral galaxy with a small, brighter core. The spiral arms are narrow and bright in a slightly fainter envelope. There are dust patches along the spiral arms and a large dust patch in the envelope W of the core. Visually, this is a fairly nondescript

object, being a small, oval glow oriented N–S. Medium magnification brings out a slightly brighter core. Radial velocity: 3042 km/s. Distance: 136 mly. Diameter: 36 000 ly.

NGC6015

H739[3]	15:51.4	+62 19	GLX
SA(s)cd	6.0′ × 2.1′	11.60B	

The DSS photograph shows an inclined, multi-arm spiral galaxy with a bright core. Several bright, spiral arms emerge from the core and are well defined by dust patches and dust lanes. There are knotty condensations along the spiral arms and out into the fainter outer envelope. A faint field star is 1.9′ S, a brighter field star is 2.1′ W. In a moderate-aperture telescope, this is a large galaxy and, though bright, it is quite nebulous in appearance; it is broadly brighter along its major axis with hazy, ill defined edges and a very faint secondary envelope, elongated NNE–SSW. Radial velocity: 1005 km/s. Distance: 45 mly. Diameter: 79 000 ly.

NGC6015

©STScI/AURA/CALTECH/ROE

NGC6088

H812³	16:10.7	+57 28		GLX
E/Sb	0.4′ × 0.3′	0.6′ × 0.25′	14.7B/14.7B:	

Photographs reveal that this is a double system featuring a galaxy classified as elliptical to the N, and an inclined spiral, elongated E–W, to the S. Herschel discovered this object on 24 April 1789, describing it as: 'Very faint, very small, a little extended.' This indicates either that he saw both galaxies as an unresolved, elongated object or that he saw only the spiral galaxy, even though it is slightly fainter than the other component. Visually, this object is very faint and difficult to see at any magnification, but is best with medium power. It is a round, formless glow about 30″ in diameter and is not resolved into two components. Though there is a slight difference in their radial velocities, the two galaxies probably form a physical pair. N component: Radial velocity: 9983 km/s. Distance: 446 mly.

Diameter: 52 000 ly. S component: Radial velocity: 9596 km/s. Distance: 428 mly. Diameter: 75 000 ly.

NGC6127

H810²	16:19.2	+57 59	GLX
E	1.4′ × 1.4′	13.00B	

Photographs show an elliptical galaxy with a bright core and an extensive envelope; a condensation S of the core may be a satellite galaxy. The galaxy is moderately bright in a moderate-aperture telescope; it is a small, round, well-condensed glow which displays a brighter core. Radial velocity: 5006 km/s. Distance: 224 mly. Diameter: 91 000 ly.

NGC6140

H740³	16:20.9	+65 23	GLX
SB(s)cd pec	6.3′ × 4.6′	12.25B	

Photographs show a large, inclined, barred spiral galaxy. The bright bar,

oriented NE–SW, has two large, knotty S-shaped spiral arms attached and is located SE of a bright field star. Telescopically, the galaxy is quite bright, fairly large and well seen at high magnification. An elongated oval, the envelope is grainy and moderately well defined with a brighter core embedded. Radial velocity: 1090 km/s. Distance: 49 mly. Diameter: 89 000 ly.

NGC6143

H811²	16:21.7	+55 05	GLX
SAB(rs)bc:	1.0′ × 0.9′	14.49B	

Images show a round, face-on spiral galaxy with a smaller, brighter core. The spiral arms are fragmentary and knotty. At high magnification, this is a small, though moderately bright galaxy. The envelope is grainy and the edges are somewhat ill defined. A brighter, star-like core is intermittently visible and the galaxy is located between two bright field stars. Radial velocity: 5484 km/s. Distance: 245 mly. Diameter: 71 000 ly.

NGC6182

H813³	16:29.5	+55 31	GLX
Sa	1.7′ × 0.4′	14.53B	

In photographs, this is an almost edge-on spiral galaxy, oriented NW–SE, with a small, bright core, elongated with faint extensions and very faint secondary extensions. Two faint companion galaxies are located to the SW and a bright field star is 3.2′ NW. Visually, only the central portion of this galaxy is visible; it has a bright, star-like core surrounded by a hazy, ill defined envelope. Radial velocity: 5315 km/s. Distance: 237 mly. Diameter: 117 000 ly.

NGC6338

H812²	17:15.3	+57 25	GLX
cD;S0	1.7′ × 1.1′	13.47B	

This is the brightest member of a galaxy group which contains at least six members; the DSS image shows them well. NGC6338 is a face-on lenticular galaxy with a bright, elongated core and an extensive, fainter envelope. A faint

NGC6338

©STScI/AURA/CALTECH/ROE

resolvable, about 3.0′ in diameter. Very gradually brighter to the middle. I suppose it to be a cluster of stars extremely compressed. 300 [magnification] confirms the supposition, and shews a few of the stars; it must be immensely rich.' The DSS image reveals a face-on barred spiral galaxy with a bright core and a very short, fainter bar. Knotty, dusty spiral arms emerge from the bar, broadening as they spiral outward. A large HII region is N of the core and there are two fainter ones to the ENE. The galaxy is also catalogued as Arp 38. Field stars are at 1.1′ SW, and 2.0′ SE. The visual appearance reveals a fairly bright but quite diffuse and ethereal galaxy. It is brighter and more readily visible at medium magnification, but high power brings out a faint star-like core. The surface brightness is quite even across the envelope and the galaxy is round with ill defined edges. Radial velocity: 1504 km/s. Distance: 67 mly. Diameter: 49 000 ly.

NGC6434

H741[3]	17:36.8	+72 05	GLX
SBbc:	2.3′ × 1.0′	13.02B	

Images show a highly inclined, barred spiral galaxy with a bright, slightly elongated core and slightly fainter bar. The bar is attached to an inner pseudo-ring which brightens slightly at the contact point in the SW. Fainter, knotty spiral arms emerge from the ring. In moderate-aperture telescopes, the galaxy is a difficult galaxy to view, owing to the presence of a bright field star 2.1′ S. Best seen at high magnification, it is fairly condensed at the core, well defined at the edges and elongated E–W. Radial velocity: 2678 km/s. Distance: 120 mly. Diameter: 80 000 ly.

NGC6543

H37[4]	17:58.6	+66 38	PN
3a+2	6.75′ × 5.25′	9.78B	

In the DSS image, the core of this bright planetary nebula is overexposed, but the filamentary outer shell is well seen. The inner shell is opaque and featureless and

---

star or a compact object is in the outer envelope to the N. At least a dozen other galaxies are in a 15′ field, including NGC6345 3.75′ S and NGC6346 5.25′ S. Telescopically, NGC6338 is moderately faint but readily seen at medium magnification, appearing almost round with a condensed envelope surrounding a brighter core. High magnification brings out both NGC companions to the S as small, hazy patches of light. Radial velocity: 8416 km/s. Distance: 376 mly. Diameter: 186 000 ly.

NGC6340

H767[2]	17:10.4	+72 18	GLX
SA(s)0/a	3.2′ × 3.0′	11.97B	11.0V

Photographs show an almost face-on spiral galaxy with a large, brighter core and faint spiral arms defined by narrow dust lanes. A wide double star is 1.75′ NNW. Two possible companion galaxies are in the field:

IC1251 is 6.5′ almost due N and IC1254 is 7.75′ NE. In addition, there are about a half dozen very small and faint field galaxies. Visually, NGC6340 is a fairly bright, extensive galaxy located SSE of a bright, uneven pair of field stars. The halo is bright and condensed and fades very gradually into the sky background. It brightens considerably to the centre to a star-like core which is intermittently seen with direct vision. Radial velocity: 1388 km/s. Distance: 62 mly. Diameter: 58 000 ly.

NGC6412

H41[6]	17:29.6	+75 42	GLX
SA(s)c	2.5′ × 2.5′	12.38B	

This is one of the more curious entries in Herschel's catalogue. Discovered on 12 December 1797, Herschel placed this galaxy in his Class VI: very compressed and rich clusters of stars, and provided a detailed description, which reads: 'Round,

just slightly out-of-round. It is surrounded by a narrow, faint envelope. Two very short, sharp spikes, or ansae, emerge from the inner shell, one to the N and one to the S. Surrounding the inner shell is a very tenuous outer shell of gas, round but very faint and broken up into discrete small patches of gas. The inner shell is about $0.7' \times 0.7'$ in size. Visually, the nebula is small but extremely bright, featuring sharply defined edges, and is slightly elongated N–S. No structure is visible, even at high magnification, but the envelope appears quite mottled and the central star is well seen. Catalogued by Herschel on 15 February 1786, he called

it: 'A planetary nebula, very bright. Has a disc of about 35″ diameter but very ill defined edge. With long attention a very bright well defined round centre becomes visible.' The central star, HD164963, has a visual magnitude of 10.9 and the spectral type is O7+WR. Radial velocity in approach: 66 km/s. Expansion velocity: 19 km/s. Distance: 3950 ly.

NGC6742

H742³	18:59.3	+48 28	PN
2c	0.5′ × 0.45′	15.00B	13.4V

Images show this planetary nebula is almost round, but slightly elongated

E–W. It is fairly opaque with a slightly darker patch in an arc from S to W, bordering the central star. A condensation or very faint field star is just within the border of the shell of gas to the W, while a bright field star is 3.1′ SW. Telescopically, although fairly faint, the planetary nebula is well seen at medium magnification and takes higher magnification well, appearing as a perfectly round, well defined disc with an even surface brightness. The central star is magnitude 19.4 and not visible. Radial velocity in approach: 159 km/s. Distance: 13 000 ly.

Equuleus

This is the smallest constellation in the northern hemisphere (only Crux in the south is smaller), and Equuleus is home to one Herschel object, described below.

There are a handful of fainter galaxies in the constellation but all escaped detection by Herschel.

NGC7046

H858³	21:14.9	+2 50	GLX
SB(rs)cd	1.9′ × 1.3′	14.17B	

Photographs reveal a slightly-inclined multi-armed spiral galaxy, probably barred, with a slightly brighter, elongated core. The principal spiral arms emerge from the bar and branch quickly into multiple fragments. The

arms are very knotty and dusty and one spiral arm extends well away from the galaxy in the SE. Telescopically, this is a fairly faint galaxy, appearing as a hazy patch of light with ill defined edges. The galaxy is broadly brighter to the middle and is elongated E–W. Radial velocity: 4328 km/s. Distance: 193 mly. Diameter: 107 000 ly.

NGC7046

©STScI/AURA/CALTECH/ROE

Eridanus

Herschel swept this large, meandering constellation throughout his observing career, recording 80 objects in total. They are all galaxies except for the triple star NGC1498, the bright planetary nebula NGC1535 and the extensive nebula complex IC2118. There are some bright and interesting galaxies here, including NGC1187, NGC1232 and NGC1332, and though many of the galaxies can be challenging visually, they are frequently organized into groups, presenting many interesting fields to explore.

NGC1084

H64[1]	2:46.0	−7 35	GLX
SA(s)c	3.2′ × 1.8′	11.19B	10.7V

On the DSS image this spiral galaxy appears overexposed, the bright spiral arms and central region being burned out. High-resolution images reveal a smaller, brighter core embedded in a massive, high-surface-brightness, three-arm spiral pattern featuring knotty spiral arms emerging from the central region into a fainter, outer envelope. Dust lanes define the spiral arms along their inner edges. The galaxy is very bright and large visually, oval in shape and oriented NNE–SSW. It is quite well defined at the edges and opaque with brighter mottlings suspected with averted vision over much of the disc. The surface brightness is fairly even, however, and no brighter core is seen. This may be an outlying member of the NGC1052 galaxy group. Radial velocity: 1391 km/s. Distance: 62 mly. Diameter: 58 000 ly.

NGC1114

H449[3]	2:49.2	−17 00	GLX
SA(r)c	1.9′ × 0.8′	13.15B	

The DSS photograph shows an inclined spiral galaxy with knotty spiral structure and a faint field star immediately E of the core. A curious knotty-textured straight spike emerges from the core and extends to the N. Visually, the galaxy is quite bright and grainy textured, but fairly even in surface brightness across a well defined disc. It is elongated N–S and a brighter spur is visible in the envelope on the eastern flank N of the core. Radial velocity: 3416 km/s. Distance: 153 mly. Diameter: 84 000 ly.

NGC1125

H450[3]	2:51.8	−16 38	GLX
(R′)SAB(rl:)0+	1.8′ × 0.9′	13.45B	13.1V

This is a peculiar case of two galaxies at very different radial velocities which appear to be interacting. The DSS image shows NGC1125 is a highly inclined spiral galaxy, probably barred, with a very bright and much elongated central region which tapers to points. Faint, tightly wound spiral arms surround the bright, central disc. There is a disturbance in the bright central region to the SW and the spiral arm which emerges here has a distinct condensation which points to a smaller galaxy. This smaller galaxy is 0.75′ SW of the core of NGC1125 and appears to be interacting with the larger galaxy: a faint plume of material emerges from the outer regions of the small galaxy, pointing at the condensation in NGC1125's spiral arm. The radial velocities of the two galaxies are very different, however; that of NGC1125 is 3230 km/s, while the magnitude 15 smaller galaxy has a radial velocity of 9247 km/s. If both galaxies are at the presumed distance of NGC1125, the minimum core-to-core separation between the two would be about 32 000 ly. Visually, only NGC1125 is visible in a moderate-aperture telescope; it is fairly bright and well defined and its surface brightness is fairly even along the major axis. The main envelope is elongated ENE–WSW and tapers to points. Radial velocity: 3230 km/s. Distance: 144 mly. Diameter: 76 000 ly.

NGC1140

H470[2]	2:54.6	−10 02	GLX
IBm pec:	1.7′ × 0.9′	12.78B	12.5V

Photographs show an elongated galaxy with a peculiar morphology. The disc of the galaxy is lenticular or irregular with a bright elongated core and a fainter outer envelope, including a faint plume extending to the NNW. Two knotty condensations are in the outer envelope to the N, while one to the S has a peculiar large nebulous condensation to the SW. Telescopically, this is an oval, moderately well defined galaxy with a bright envelope oriented N–S which brightens broadly towards the middle. Radial velocity: 1452 km/s. Distance: 66 mly. Diameter: 33 000 ly.

NGC1162

H469[3]	2:58.9	−12 24	GLX
E:	1.6′ × 1.5′	13.81B	

In photographs this elliptical galaxy has a bright core, slightly elongated ENE–WSW, with a fainter outer envelope. Moderately bright visually, it appears almost round, with a bright, well defined envelope which is broadly brighter to the middle. Radial velocity: 3900 km/s. Distance: 174 mly. Diameter: 81 000 ly.

NGC1172

H502[2]	3:01.6	−14 50	GLX
E⁺:	2.3′ × 1.7′	12.76B	

The classification of this galaxy is somewhat uncertain and photographs suggest it is lenticular, owing to the extensive, faint outer envelope which surrounds the round and very bright core. Elongated NNE–SSW, the galaxy is located within the northern boundary of a triangle of bright field stars. The DSS image suggests the long axis of the galaxy extends at least 3.2′. Visually, the galaxy appears moderately bright and decidedly elliptical with a round, opaque and evenly bright envelope and well defined edges. The faint outer envelope is not seen visually. Observed on 30 December 1785, Herschel's description reads: 'Faint, extremely small, stellar, preceding a pretty bright star.' Radial velocity: 1621 km/s. Distance: 72 mly. Diameter: 49 000 ly.

NGC1187

H245[3]	3:02.6	−22 52	GLX
SB(r)c	5.5′ × 4.1′	11.36B	

Photographs show a bright, slightly inclined, barred spiral galaxy with a bright, slightly elongated core and knotty spiral arms which fade abruptly and transition to an extensive pattern of fainter outer spiral arms. The faint galaxy ESO480-20 is 4.25′ NNW, next to a bright field star which is 4.5′ NW of NGC1187's core. Telescopically, this is a large and fairly bright galaxy which is somewhat diffuse and grainy in texture. Broadly brighter to the middle, the galaxy is irregularly oval and oriented SE–NW towards a bright field star. The edges are ragged and ill defined. Recorded on 9 December 1784, Herschel described this galaxy as: 'Very faint, considerably large, irregularly extended, resolvable, unequally bright.' Radial velocity: 1319 km/s. Distance: 59 mly. Diameter: 95 000 ly.

NGC1187

NGC1199

H503[2]	3:03.6	−15 37	GLX
E3:	2.4′ × 1.9′	12.39B	

This elliptical galaxy is the brightest of a compact group of five galaxies which includes NGC1189, NGC1190, NGC1191, and NGC1192. The DSS image shows that all five galaxies appear in a 6.0′ × 6.0′ area and the field is probably very interesting in a large-aperture telescope, though moderate apertures will likely only show NGC1199. In photographs it has a bright, elongated core and a fainter outer envelope with a nebulous, compact object in the outer envelope to the SE. There are a number of very small compact objects or background galaxies arrayed around the galaxy. If all five galaxies are at the same distance, the 6′ × 6′ area corresponds to an area about 196 000 ly across. Telescopically, NGC1199 is large and quite bright, very slightly oval and oriented NE–SW with a bright, moderately well defined envelope in which a bright, round, nonstellar core is embedded. Herschel recorded the galaxy on 30 December 1785, calling it: 'Pretty bright, small, irregular form, much brighter to the middle.' He was not able to detect any of the fainter galaxies in the field. Radial velocity: 2570 km/s. Distance: 112 mly. Diameter: 78 000 ly.

NGC1199

NGC1200

H475[2]	3:03.8	−11 59	GLX
SA(s)0⁻	2.9′ × 1.8′	13.61B	

The DSS photograph reveals a lenticular galaxy with a large, bright core and a faint, extensive outer envelope, very slightly elongated ENE–WSW. It is the brightest of a small group of galaxies including NGC1195 (6.0′ WSW) and NGC1196 (6.75′ SW). An anonymous, edge-on galaxy is 3.1′ ESE and a faint, E–W oriented chain of galaxies is located 7.0′ to the SSE. NGC1200 is moderately faint visually, appearing as an opaque, roundish, well defined patch of light with a small, star-like, brighter core. Radial velocity: 4007 km/s. Distance: 179 mly. Diameter: 151 000 ly.

NGC1208

H285[2]	3:06.2	−9 32	GLX
SA(r)0/a?	2.0′ × 1.1′	13.32B	

Images reveal an inclined lenticular galaxy with a large, bright, elongated core and a faint outer envelope. Visually, the galaxy is moderately bright, oval and elongated ENE–WSW. The envelope is opaque and quite well defined and only very slightly brighter to the middle. Radial velocity: 4320 km/s. Distance: 193 mly. Diameter: 112 000 ly.

NGC1209

H504[2]	3:06.0	−15 37	GLX
E6:	2.4′ × 1.1′	12.35B	

Photographs show a highly flattened elliptical galaxy with a large, bright elongated central region and a faint, extensive outer envelope which is decidedly squarish in outline. The long axis of the galaxy appears to extend at least 3.3′ in the DSS image. Telescopically, this is a large and fairly bright galaxy; it is an elongated oval oriented E–W and broadly brighter to the middle. The outer envelope is fairly bright and fades slowly to a hazy, ill defined outer envelope. The smaller companion galaxy NGC1231 is 6.75′ ENE. Radial velocity: 2547 km/s. Distance: 114 mly. Diameter: 79 000 ly.

NGC1232

H258[2]	3:09.8	−20 35	GLX
SAB(rs)c	7.4′ × 6.5′	10.56B	9.9V

This bright, face-on, multi-arm spiral galaxy has a small, brighter nucleus and a possible very short bar. The DSS photograph shows three principal spiral arms emerging from the inner disc with subsidiary spiral arms branching off after about half a revolution. High-resolution photographs show the arms well resolved into HII regions, clusters and associations. The smaller, Magellanic barred galaxy NGC1232A is 4.0′ to the ESE and together both galaxies have been catalogued as Arp 41. There is some controversy surrounding this pair based on whether they form a physically related pair or not. Radial velocity measurements are inconclusive: some measurements show the smaller galaxy has a radial velocity similar to that of NGC1232, while others suggest it has a much larger radial velocity in the region of 6500 km/s. However, both galaxies have a similar degree of resolution of presumed HII regions, suggesting that they may be related. At the presumed distance of NGC1232, the minimum separation between the two galaxies would be about 80 000 ly. Telescopically, NGC1232 is moderately

NGC1200

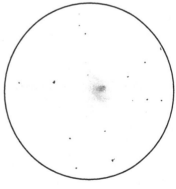

NGC1232 15-inch Newtonian 94x

NGC1232

©STScI/AURA/CALTECH/ROE

Telescopically, this pair of galaxies is well seen, despite the presence of a bright field star 3.0′ N of NGC1241. The galaxy is large and bright; it is an irregularly oval object, elongated SE–NW with a slightly brighter nebulous bar running through the middle. The outer envelope is patchy with ill defined edges. Radial velocity: 4015 km/s. Distance: 179 mly. Diameter: 146 000 ly.

NGC1242

H591[3]	3:11.3	−8 54	GLX
SB(rs)c:	1.2′ × 0.7′	14.32B	13.7V

Photographically, this companion of NGC1241 is a small, slightly inclined spiral galaxy with a brighter core and two stubby, narrow spiral arms. Visually, it is smaller and slightly fainter than NGC1241, though it appears better defined. Elongated SE–NW, its surface brightness is fairly even. Herschel discovered NGC1241 on 10 January 1785 but did not pick up the fainter NGC 1242 until a subsequent sweep on 15 December 1786, calling it: 'Extremely faint, stellar, about 1′ north following H286[2].' Radial velocity: 3971 km/s. Distance: 177 mly. Diameter: 62 000 ly.

NGC1247

H900[2]	3:12.2	−10 29	GLX
Sbc	3.9′ × 0.4′	13.46B	

Photographs show an almost edge-on spiral galaxy with a slightly brighter core and bright extensions. A small compact galaxy is 2.75′ SE. This is a fairly faint galaxy visually; it is an elongated patch oriented ENE–WSW of even surface brightness and is moderately well defined. Radial velocity: 3900 km/s. Distance: 173 mly. Diameter: 198 000 ly.

NGC1248

H443[3]	3:12.8	−5 13	GLX
SA(s)0°:	1.2′ × 1.0′	13.53B	

Situated 5.5′ S of a magnitude 8 field star, photographs show this lenticular galaxy has a bright, slightly elongated

bright and fairly large, brightening to a large central region. Large apertures begin to show signs of spiral structure; one broad arm curves N, then W, another S, then E. The edges are moderately well defined and a faint condensation is suspected W of the core. NGC1232A is not seen. Radial velocity: 1533 km/s. Distance: 68 mly. Diameter: 148 000 ly.

NGC1239

H262[3]	3:11.0	−2 31	GLX
SA0° pec:	1.2′ × 0.7′	14.60B	

Photographs show a lenticular galaxy with a bright, round core and a fainter outer envelope extended in a NE–SW direction. A similarly bright, edge-on anonymous galaxy is 8.0′ NNE. Visually, the galaxy appears moderately bright and well defined and is well seen at high magnification as a very slightly oval, opaque and condensed patch of light of even surface brightness. Radial

velocity: 8625 km/s. Distance: 385 mly. Diameter: 135 000 ly.

NGC1241

H286[2]	3:11.3	−8 55	GLX
SB(rs)b	2.8′ × 1.7′	12.97B	12.0V

This inclined, barred spiral galaxy forms a close pair with NGC1242, 1.6′ to the NE and they have also been catalogued as Arp 304. Their radial velocities are similar and the galaxies probably form a physical pair though there does not seem to be evidence of tidal interaction between the two galaxies. At the presumed distance, the minimum core-to-core separation between the pair would be about 84 000 ly. Photographs show NGC1241 is a barred spiral with a brighter, elongated core and a short, fainter bar. Two principal spiral arms are attached to the bar and subsidiary, fragmentary arms emerge from the main arms within a quarter of a revolution around the core.

NGC1253

core in a fainter envelope. The outer envelope is not smooth as the halo is a little irregular in outline and there is evidence of dust patches. Telescopically, the galaxy is small but moderately bright and is well seen at high magnification. It appears slightly oval, oriented roughly E–W, and the envelope is quite grainy and even in surface brightness with a tiny star-like and brighter core embedded. The edges are fairly well defined. Radial velocity: 2193 km/s. Distance: 98 mly. Diameter: 34 000 ly.

NGC1253

H17⁴	3:14.1	−2 49	GLX
SAB(rs)cd	5.2′ × 2.3′	12.25B	

This is a highly inclined multi-arm spiral with a small, bright, elongated core. The galaxy is also catalogued as Arp 279, owing to the presence of a small, irregular companion galaxy, NGC1253A, 3.9′ to the ENE. Photographs show the grainy inner envelope expands into loosely structured spiral arms with many knots and condensations; the companion shows the same degree of resolution and is very likely physically related. The minimum core-to-core separation between the galaxies at the presumed distance would be about 85 000 ly. There is a field star in the spiral arm 0.8′ WSW of the core and a field star 3.0′

ENE in line with the faint companion. Telescopically, the galaxy is seen to be involved with a triangle of faint field stars. It is large, and somewhat bright but very diffuse. It is best seen at medium magnification; the western star of the triangle is embedded in the outer envelope and hinders observation. Oval and elongated E–W, the galaxy is very slightly and broadly brighter to the middle with diffuse edges. The faint companion is not seen in moderate-aperture telescopes. Radial velocity: 1692 km/s. Distance: 76 mly. Diameter: 114 000 ly.

NGC1266

H194[3]	3:16.0	−2 24	GLX
(R')SB(rs)0 pec	1.5' × 1.0'	14.33B	

The DSS image shows an inclined, possibly spiral galaxy with a brighter core and faint outer envelope. A dust patch is suspected W of the core. A dim barred galaxy is 1.75' W and a field star is 1.5' WSW. Visually, the galaxy is a quite faint, diffuse, oval patch of light with a faint field star preceding it to the WSW. The edges are hazy and the envelope is broadly brighter to the middle. Radial velocity: 2173 km/s. Distance: 97 mly. Diameter: 42 000 ly.

NGC1284

H956[3]	3:17.6	−10 18	GLX
SA0°	1.8' × 1.4'	14.36B	

Photographs show a lenticular galaxy with a round, bright core and a fainter, elongated outer envelope, oriented ENE–WSW. A small, compact galaxy is 0.9' N. Visually, this object is somewhat faint, though is well seen at high magnification as a slightly elongated patch of light, fairly well defined in the middle with very faint N and S extensions. Radial velocity: 8940 km/s. Distance: 400 mly. Diameter: 209 000 ly.

NGC1287

H195[3]	3:18.6	−2 45	GLX
(R')SB(r)a pec	0.8' × 0.7'	14.81B	

This is a small, faint galaxy with a peculiar morphology. Photographs

show a round, brighter core with a bright, narrow and broken irregular ring surrounding the core, the whole embedded in a faint envelope. Visually, the galaxy is very faint and diffuse and is best at high magnification as a roundish patch of light with moderately well defined edges and a suspected, tiny, star-like core. Radial velocity: 8561 km/s. Distance: 382 mly. Diameter: 89 000 ly.

NGC1299

H287[2]	3:20.1	−6 17	GLX
SB(rs)b?	1.1' × 0.6'	13.81B	

The morphology of this galaxy is difficult to ascertain on the DSS image, but it appears as a small, inclined galaxy, probably a spiral, with a brighter, elongated central region in a fainter outer envelope which is more extensive to the SW. Telescopically, the galaxy appears moderately bright; it is an opaque and well defined elongated oval, oriented NE–SW with a condensed, slightly brighter core. Radial velocity: 2301 km/s. Distance: 103 mly. Diameter: 33 000 ly.

NGC1304

H444[3]	3:21.0	−4 35	GLX
S0⁻ pec	1.3' × 0.7'	14.54B	

This galaxy presents fairly similar appearances visually and photographically, though only the bright elongated central region is detected at high magnification. It is moderately faint, elongated NE–SW with an even surface brightness and well defined edges. Radial velocity: 3940 km/s. Distance: 176 mly. Diameter: 67 000 ly.

NGC1308

H568[3]	3:22.5	−2 45	GLX
SB(r)0/a	1.1' × 0.8'	14.73B	

This theta-shaped, barred galaxy has a bright, round core and a fainter bar. Photographs show that the bar is connected to a faint, elongated ring, oriented NNE–SSW. Field stars are 1.4' E and 1.9' NNW, while an edge-on galaxy is 4.9' NNE. Telescopically, this is

a very faint and difficult galaxy, which is only confirmed at high magnification. Situated within a four-star asterism, the galaxy is very small and somewhat diffuse, almost round and slightly brighter to the middle. Radial velocity: 6323 km/s. Distance: 282 mly. Diameter: 90 000 ly.

NGC1309

H106[1]	3:22.1	−15 24	GLX
SA(s)bc	2.3' × 2.2'	11.96B	

Photographs show a face-on spiral galaxy with a bright core and bright, knotty spiral arms. The arms are tightly wound, somewhat fragmentary and brighter to the W and N. Visually, the galaxy is quite bright and is well seen at high magnification as an irregularly round, but fairly well defined object. The envelope is bright but clumpy in appearance, with a brighter, round and nonstellar core. A magnitude 7 field star is 4.0' SSW. Recorded on 3 October 1785, Herschel's description reads: 'Considerably bright, considerably large, irregularly round, brighter to the middle. 3' diameter.' Radial velocity: 2073 km/s. Distance: 93 mly. Diameter: 62 000 ly.

NGC1320

H197[3]	3:24.8	−3 01	GLX
Sa: sp	1.9' × 0.6'	13.34B	

This lenticular galaxy is part of a chain of four physically related galaxies which includes NGC1321 1.75' N, NGC1322 and NGC1323. The DSS image reveals a bright, elongated core and a fainter outer envelope, oriented SE–NW. Telescopically, the brightest three of the group can be seen in a high-magnification field. NGC1320 is the largest, an elongated, moderately bright but small patch of light. It is a little brighter to the middle and well defined, and the extensions are tapered. NGC1322 to the NNE is the faintest of the three and was overlooked by Herschel; it appears round with a tiny star-like core and moderately well defined edges. NGC1323 is very faint though probably detectable in large-aperture amateur telescopes. Radial

velocity: 2634 km/s. Distance: 118 mly. Diameter: 65 000 ly.

NGC1321

H196³	3:24.9	−3 00		GLX
E?	0.9′ × 0.6′	14.20B		

The classification of this galaxy is uncertain but photographs show a lenticular galaxy with a bright, elongated core and a fainter outer envelope, elongated E–W. Situated only 1.75′ N of NGC1320, the minimum core-to-core separation between the two galaxies would be about 64 000 ly at the apparent distance. Visually, it is smaller but brighter than NGC1320, an E–W oriented oval of light which is condensed, opaque and well defined. Radial velocity: 2819 km/s. Distance: 126 mly. Diameter: 33 000 ly.

NGC1324

H445³	3:25.0	−5 45		GLX
Sb:	2.2′ × 0.7′	13.51B		

Photographs show an almost edge-on spiral galaxy with a brighter, elongated core and fainter spiral extensions. There are several small dust patches in the spiral arms. At high magnification, this moderately bright galaxy is narrow and much extended SE–NW; it is well defined at the edges with a grainy, mottled envelope of uneven surface brightness. A small, faint, flat triangle of stars is 2.5′ S. Radial velocity: 5631 km/s. Distance: 251 mly. Diameter: 161 000 ly.

NGC1325

H77⁴	3:24.4	−21 33	GLX	
SA(s)bc	4.7′ × 1.7′	12.25B	11.6V	

This is a highly inclined multi-arm spiral galaxy, with a very small, brighter core and grainy spiral arms. Photographs reveal an asymmetrical structure to the fainter outer spiral arms, which are more extensive to the SW. The lenticular galaxy NGC1319 is 6.75′ WNW. Visually, this is a large and quite bright galaxy; it is an elongated, well defined oval of light, oriented NE–SW and broadly brighter to the

middle along its major axis. A prominent field star is embedded in the outer envelope, 1.0′ ENE of the core and a fainter field star is ESE. The fainter galaxy NGC1319 is in the field to the W; it is an oval patch of light, brighter to the middle and oriented NNE–SSW. Herschel discovered NGC1325 on 19 December 1798, and because of the presence of the bright star in the envelope, placed it in his Class IV. He described it as: 'A star about 9 or 10 magnitude with a nebulous ray to the south-preceding side. The ray is about 1.5′ long. The star may not be connected with it.' Radial velocity: 1510 km/s. Distance: 67 mly. Diameter: 92 000 ly.

NGC1331

H959³	3:26.5	−21 21		GLX
E2:	0.9′ × 0.8′	14.30B		

Photographs show an elliptical galaxy with a bright round core and a fainter

outer envelope elongated N–S. In *The Scientific Papers* this galaxy is identified as IC324 with the note 'NGC1331 should be struck out.' It is part of a loose group of bright galaxies and its companion NGC1332 is 3.0′ WNW; the minimum core-to-core separation between the two galaxies would be about 56 000 ly at the presumed distance. Herschel recorded this galaxy on 19 December 1799, 15 years after discovering NGC1332. Telescopically, NGC1331 is a faint, roundish and slightly ill defined patch of light which is broadly brighter to the middle. Radial velocity: 1128 km/s. Distance: 50 mly. Diameter: 13 000 ly.

NGC1332

H60¹	3:26.3	−21 20		GLX
S(s)0⁻: sp	4.7′ × 1.4′	11.20B		

This is a prototypical bright lenticular galaxy seen almost edge-on.

NGC1325

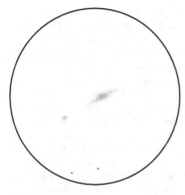

NGC1331, NGC1332
15-inch Newtonian 272x

Photographs reveal a very bright, round core and bright extensions with an almost rectangular outer envelope. This is the brightest galaxy of a loose group which includes NGC1315, NGC1319, NGC1325, NGC1325A, and NGC1331. Companion galaxy NGC1331 is 3.0′ ESE, while the fainter, peculiar spiral

ESO548-16 is 3.5′ W. Radial velocity: 1442 km/s. Distance: 65 mly. Diameter: 88 000 ly.

NGC1353

H246[3]	3:32.1	−20 49	GLX	
SA(rs)bc	3.5′ × 1.4′	12.36B	11.3V	

Photographs show a highly inclined spiral galaxy with a brighter core and a bright pair of tightly wound inner spiral arms which very nearly form a pseudo-ring. The bright arms lead into fainter outer spiral structure. Several dust patches and one major dust lane are visible in the outer spiral arms. At high magnification, this is a largish and only moderately bright galaxy, elongated SE–NW and oriented towards a field star 2.9′ to the SE. The core is a little brighter and irregularly round; the extensions are slightly fainter, fairly well defined and taper to blunt ends. Radial velocity: 1441 km/s. Distance: 64 mly. Diameter: 66 000 ly.

NGC1354

H487[3]	3:32.4	−15 13	GLX
SAB(rs)0/a:	2.4′ × 0.9′	13.30B	

This is a highly inclined lenticular galaxy with a bright, elongated central region and very faint extensions which increase the length of the major axis on the DSS image to at least 3.0′. Photographs show some evidence of dust in the outer portions of the central region to the NW and SE. A field star is visible within the central region to the NW. Visually, this is a large and fairly bright galaxy, even in surface brightness along its major axis and elongated SE–NW. It is a quite well defined oval which tapers to fairly sharp points. Radial velocity: 1733 km/s. Distance: 77 mly. Diameter: 54 000 ly.

NGC1357

H290[2]	3:33.2	−13 40	GLX
SA(s)ab	4.0′ × 3.5′	12.40B	

The DSS image shows an almost face-on spiral galaxy with a very large and bright central region. Two bright, thin spiral arms emerge with dust lanes defining the inner edges of the arms. The arms expand into a fainter, smooth-textured spiral outer envelope and the galaxy is slightly elongated E–W. The physical dimensions are at least 4.0′ × 3.5′. High-resolution photographs show considerable structure in the inner central region, which is burned out in the DSS image. A small, barred spiral galaxy is 6.4′ SE. Visually, the galaxy is bright, well defined and almost round with slightly soft edges, a bright and opaque envelope and a brighter, nonstellar core. Two bright field stars are nearby, one 3.9′ WNW, the other 3.75′ NNE. Radial velocity: 1930 km/s. Distance: 86 mly. Diameter: 100 000 ly.

NGC1358

H446[3]	3:33.7	−5 05	GLX
SAB(r)0/a	2.6′ × 2.0′	13.20B	

The DSS image shows a face-on, barred spiral galaxy with a bright, elongated

NGC1332

NGC1357

©STScI/AURA/CALTECH/ROE

a small compact object, or companion galaxy is within the halo to the S. Telescopically, the galaxy is situated between two faint field stars oriented SE–NW. Moderately faint, it is well seen at high magnification as a slightly oval patch of light, oriented NE–SW and fairly even in surface brightness. The edges are hazy and somewhat ill defined. Radial velocity: 978 km/s. Distance: 44 mly. Diameter: 18 000 ly.

NGC1376

H288[2]	3:37.1	−5 03	GLX
SA(s)cd	2.0′ × 2.0′	12.89B	

The DSS image shows a face-on spiral galaxy with a small, brighter core and brighter inner spiral arms defined by knotty condensations. The fainter, fragmentary outer spiral structure is studded with condensed patches. A lenticular galaxy is 5.5′ to the NE. Telescopically, this is a moderately large, fairly bright but diffuse object which is broadly brighter to the middle and round with edges that fade uncertainly into the sky background. Radial velocity: 4110 km/s. Distance: 184 mly. Diameter: 107 000 ly.

NGC1377

H961[3]	3:36.7	−20 54	GLX
S0°	1.7′ × 0.8′	13.53B	

Presenting a fairly similar appearance both visually and photographically, this lenticular galaxy appears small and moderately faint, but is well seen at high magnification as an elongated, well defined sliver of light, oriented E–W. It is fairly even in surface brightness with a faint field star 1.2′ NNE. Radial velocity: 1705 km/s. Distance: 76 mly. Diameter: 38 000 ly.

NGC1393

H451[3]	3:38.6	−18 26	GLX	
SA(rl)0°	1.9′ × 1.3′	13.03B	12.0V

Photographs show a lenticular galaxy with a bright core and a slightly fainter outer envelope. The outer envelope is unevenly bright, possibly due to the presence of dust. The galaxy is

core and a bright, stubby bar. High-resolution photographs show that the arms are defined by HII regions, which look like beads on a string, and a dark lane extending along the eastern flank. The arms almost form a ring around the galaxy and enclose a fainter, smooth-textured disc. Telescopically, this galaxy is paired with the edge-on lenticular galaxy NGC1355 in a high-magnification field. NGC1358 is quite bright, preceding a pair of moderately bright field stars. The envelope is round, fairly condensed, grainy in texture and moderately well defined. The central region is very bright and small and is extended ESE–WNW; this is the core and bar of the galaxy. Neighbouring galaxy NGC1355 is 6.75′ NW; it is a moderately bright, well defined E–W streak of light, evenly bright along its major axis and was evidently overlooked by Herschel. Radial velocity: 3987 km/s. Distance: 178 mly. Diameter: 135 000 ly.

NGC1362

H960[3]	3:33.8	−20 17	GLX
S0° pec:	1.2′ × 1.0′	13.84B	

This lenticular galaxy has a round core and a fainter outer envelope and is probably a dwarf galaxy. Photographs show that a small compact object or companion galaxy is within the halo to the S; this object is not seen in moderate-aperture telescopes. Visually, this is a very small, round and fairly faint galaxy. The main envelope is opaque and even in surface brightness; the edges are fairly well defined. Radial velocity: 1149 km/s. Distance: 51 mly. Diameter: 18 000 ly.

NGC1370

H559[3]	3:35.2	−20 22	GLX
E+:	1.4′ × 0.9′	13.59B	

This is probably a dwarf elliptical galaxy with an elongated core and a fainter outer envelope. Photographs show that elongated SSE–NNW and a faint compact object or star borders the outer envelope to the SW. There are eight NGC galaxies in a 1° field, including the Herschel objects NGC1400 and NGC1407, but a fairly wide spread in their radial velocities indicates that they may not all be part of the same group. The lenticular galaxy NGC1391 is 5.6' NE. Visually, NGC1393 is the brightest of a chain of three galaxies visible in a high-magnification field; the others are NGC1391 and NGC1394. NGC1393 is fairly bright; it is a slightly oval galaxy with a brighter core and moderately well defined edges. The chain is elongated NE–SW with NGC1391, which is small and slightly elongated, though well defined, in the middle. NGC1394 is the northernmost of the three galaxies; it appears as a well defined sliver of light with a field star to the NNW. Radial velocity: 2045 km/s. Distance: 91 mly. Diameter: 51 000 ly.

NGC1395

H58[1]	3:38.5	−23 02	GLX
E2	5.6' × 4.7'	10.60B	

The DSS image shows an elliptical galaxy with a slightly elongated core within a fainter outer envelope elongated E–W. A small compact object is within the halo 0.75' to the W and a faint field star is 1.0' immediately N. Telescopically, this is a bright, almost round galaxy with a bright, grainy envelope surrounding a much brighter, nonstellar core. A quite diffuse, faint secondary outer envelope forms a hazy halo around the main envelope, with a bright field star 3.6' S. Radial velocity: 1623 km/s. Distance: 72 mly. Diameter: 118 000 ly.

NGC1397

H569[3]	3:39.6	−4 40	GLX
(R)SB(r)0/a	1.3' × 1.0'	14.66B	

This is a face-on, theta-shaped, barred galaxy with a large, round, bright core and a fainter bar. The bar is attached to a faint ring. Two compact objects, or faint field stars, border the ring, the brighter one to the NNE, the other to the NNW.

The galaxy is elongated N–S. The DSS image shows that an extremely faint, thin ring of matter, measuring approximately 2.0' × 1.75' and elongated E–W, surrounds the main galaxy. A faint edge-on galaxy is 7.1' NNE. NGC1397 is extremely faint and difficult to see visually, appearing as a small, round, hazy patch of light which is very slightly brighter to the middle, with ill defined edges. Radial velocity: 5463 km/s. Distance: 244 mly. Diameter: 92 000 ly.

NGC1400

H593[2]	3:39.5	−18 41	GLX
SA0⁻	2.3' × 2.0'	11.94B	11.0V

The DSS image shows a lenticular galaxy with a slightly elongated core within an extensive fainter outer envelope. This is one of the principal galaxies of the NGC1400 group, a loose collection of galaxies which includes the bright elliptical NGC1407. NGC1400 has a low radial velocity compared with most of the surrounding galaxies, though this may be attributable to orbital motion with respect to the other galaxies. The distance and actual diameter quoted below may therefore be spuriously low. The DSS image shows several small and very faint galaxies in the immediate vicinity, including an extremely faint, diffuse object 3.0' to the ENE, the physical size of which is at least 4.0' × 3.9'. Telescopically, NGC1400 can be seen in a medium-magnification field with two other galaxies which form a roughly equilateral triangle, the other two being NGC1402 and NGC1407. Though it appears smaller than NGC1407, NGC1400 seems a little brighter and more intense, is very well defined at the edges and is quite opaque with a bright, even-surface-brightness envelope. Radial velocity: 475 km/s. Distance: 21 mly. Diameter: 14 000 ly.

NGC1401

H247[3]	3:39.4	−22 44	GLX
SB(s)0°: sp	1.9' × 0.6'	13.14B	

The visual and photographic appearances of this edge-on lenticular

galaxy are quite similar. Visually, the galaxy is small and moderately faint, but nevertheless it is well seen at high magnification as an elongated, well defined sliver of light, tapering to sharp tips. A small, round core is embedded, bulging out very slightly from the main envelope. The galaxy is elongated SE–NW and a faint field star is 0.6' N of the core. Radial velocity: 1401 km/s. Distance: 63 mly. Diameter: 35 000 ly.

NGC1407

H107[1]	3:40.2	−18 35	GLX
E0	4.6' × 4.3'	10.70B	9.7V

This large, elliptical galaxy has a round, bright core within a very broad, fainter outer envelope. The DSS image shows it surrounded by several very small, faint galaxies and globular clusters and its physical dimensions extend to about 7.5' × 7.5'. It is a member of the NGC1400 group of galaxies and visually it is the largest galaxy in a medium-magnification field which includes NGC1400 and NGC1402. It is a large, round, opaque patch of light with a bright, round, nonstellar core. The edges are a little hazy and NGC1402 can be seen to the WNW as a faint, well defined and almost stellar patch of light. Radial velocity: 1696 km/s. Distance: 76 mly. Diameter: 102 000 ly.

NGC1415

H267[2]	3:41.0	−22 34	GLX
(R)SAB(s)0/a	3.5' × 1.8'	12.66B	

The DSS image shows an inclined galaxy, probably barred, with a large,

NGC1407

bright, elongated central region and a faint, suspected bar running ESE–WNW connecting to a fainter outer envelope. There are dark patches in the outer envelope to the SSE and NNW; high-resolution photographs show that the smooth spiral arms are defined by thin dust lanes. The DSS image shows an extremely faint outer envelope which extends the apparent size of the galaxy to about 5.0′ × 2.0′. The faint galaxy ESO482-31 is 6.0′ SW. NGC1415 is moderately bright visually; it is a moderately well defined, elongated streak, oriented SE–NW. The core is brighter and nonstellar and a bright field star is 2.6′ to the NNW. Radial velocity: 1491 km/s. Distance: 66 mly. Diameter: 68 000 ly.

NGC1417

H455[2]	3:42.0	–4 42	GLX
SAB(rs)b	2.8′ × 1.6′	12.81B	

Photographs show an inclined spiral galaxy, possibly barred, with a large, bright, elongated central region and a grainy, multi-arm spiral pattern emerging. A bright field star is 1.3′ SW of the core. Companion galaxy NGC1418 is 5.0′ ESE, while the fainter galaxy IC344 is 7.2′ WNW. All three may form a physically related system. In a high-magnification field NGC1417 and NGC1418 can be seen together. NGC1417 is much larger and brighter; it is a broad, bright N–S streak of light which is fairly bright along its major axis and hazy along the extremities. Radial velocity: 4068 km/s. Distance: 182 mly. Diameter: 148 000 ly.

NGC1418

H456[2]	3:42.3	–4 44	GLX
SB(s)b	1.3′ × 0.9′	14.44B	

Images reveal an inclined spiral galaxy with a slightly brighter central region and two spiral arms attached. It forms a probable physical pair with NGC1417 5.0′ to the WNW. Visually, it appears as an oval, well defined, fairly condensed patch of light elongated N–S. Radial

velocity: 4158 km/s. Distance: 186 mly. Diameter: 70 000 ly.

NGC1421

H291[2]	3:42.5	–13 29	GLX
SAB(rs)bc	3.5′ × 0.9′	11.98B	

Photographs show a bright, highly inclined spiral galaxy which is bright along its major axis with a narrow, conspicuous spiral arm on the western flank. There is some evidence of dust, especially to the S and a bright, nebulous knot SSE of the centre, though no core or bulge is visible. Faint extensions increase the apparent length of the major axis to about 4.8′. Visually, this is a large, quite bright and well defined galaxy, much extended N–S with fairly even surface brightness along its major axis. The galaxy bulges very slightly at the middle and a faint field star lies close by to the ESE. Radial velocity: 2016 km/s. Distance: 90 mly. Diameter: 92 000 ly.

NGC1426

H248[3]	3:42.8	–22 07	GLX
E4	2.6′ × 1.7′	12.24B	11.4V

Situated 9.0′ SSE of a magnitude 8 field star, photographs show a bright, elongated elliptical galaxy with a bright core and a fainter outer envelope.

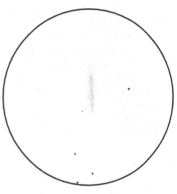

NGC1421 15-inch Newtonian 272x

NGC1417/1418/IC344

©STScI/AURA/CALTECH/ROE

Visually, the galaxy is moderately bright; it is an opaque object of even surface brightness, oval and elongated ESE–WNW. Radial velocity: 1349 km/s. Distance: 60 mly. Diameter: 46 000 ly.

NGC1439

H249[3]	3:44.8	−21 55	GLX
E1	2.6′ × 2.5′	12.29B	

Photographs show an elliptical galaxy with a bright, round core which is blunted slightly to the ESE. An extensive, fainter envelope surrounds the core. Telescopically, the galaxy is small, round and well defined with a moderately bright envelope in which the small, stellar, brighter core is embedded. Radial velocity: 1575 km/s. Distance: 71 mly. Diameter: 53 000 ly.

NGC1440

H458[2]/H594[2]	3:45.0	−18 16	GLX
(L)SB(rs)0+	2.2′ × 1.5′	12.68B	

This galaxy is a member of the NGC1400 group. Photographs show a slightly inclined barred galaxy with a bright, elongated core and a slightly fainter, short bar. A fainter envelope surrounds the inner region with slight indentations of the envelope to the ESE and W. The galaxy is elongated NE–SW. Visually, only the core and bar are well seen; the galaxy appears small but is well defined and moderately bright. The lens-shaped envelope is elongated NE–SW with a small, star-like, brighter core. This object was observed twice by Herschel, but the position given for the second observation (H594[2], 20 September 1786) was erroneous and this was undoubtedly a reobservation of H458[2]. Radial velocity: 1511 km/s. Distance: 68 mly. Diameter: 43 000 ly.

NGC1441

H597[2]	3:45.7	−4 06	GLX
SB(s)b	1.8′ × 0.7′	13.90B	

This is a highly inclined, lenticular galaxy with a bright, elongated core. Photographs show evidence of possible dust patches at the E and W extensions, as well as a suspected very narrow dust

lane which crosses the core from E to W. The very faint outer envelope displays a weak spiral pattern. This galaxy is the westernmost of a group of four NGC galaxies, including NGC1449, NGC1451 and NGC1453 that probably form a physical system. All four are well seen visually in a medium-magnification field with NGC1441 being the second brightest. A well defined E–W streak of light, it is evenly bright along its major axis. Radial velocity: 4250 km/s. Distance: 190 mly. Diameter: 99 000 ly.

NGC1452

H459[2]	3:45.4	−18 38	GLX
SB(r)a	2.3′ × 1.7′	12.82B	

The DSS image shows a slightly inclined, theta-shaped, barred galaxy with a bright, round core and a bright, narrow bar. The bar flares slightly in brightness where it attaches to a faint ring. To the ESE the ring is double while

very faint outer spiral arms emerge from the ring. The galaxy is elongated ESE–WNW and appears at least 3.0′ × 2.0′ in size, with the very small and faint galaxy NGC1455 2.9′ SSW. Visually, only the core and bar of NGC1452 are visible; the galaxy appears small and moderately bright and is elongated NNE–SSW with a small bright core embedded and well defined edges. A wide pair of field stars is about 3.0′ to the NNE. The galaxy is a probably a member of the NGC1400 group. Radial velocity: 1650 km/s. Distance: 74 mly. Diameter: 50 000 ly.

NGC1453

H155[1]	3:46.4	−3 58	GLX
E2-3	2.4′ × 2.0′	12.59B	

This is the dominant galaxy of a loose group of about ten galaxies which includes NGC1441, NGC1449 and NGC1451. Photographs show a bright galaxy with a broad, bright, slightly

NGC1452

©STScI/AURA/CALTECH/ROE

NGC1441/1449/1451/1453

Visually, NGC1482 is moderately faint but is well seen as an oval, well defined patch of light, elongated E–W with a small, sharp, brighter stellar core. Radial velocity: 1818 km/s. Distance: 81 mly. Diameter: 59 000 ly.

NGC1498

H3[7]	4:00.3	−12 01	TS

Herschel recorded seeing a star cluster on 8 February 1784, describing it as: 'A small cluster of compressed stars, some pretty large.' Dreyer noted in *The Scientific Papers* that there was 'no very pronounced cluster at the place' but very near Herschel's position there is a triangular triple star, with the members faint but of equal brightness, around magnitude 13 in an otherwise star-poor field. The DSS image shows some extremely faint stars and even fainter suspected galaxies involved with the three stars but it is unlikely that Herschel would have been able to detect them. It can be presumed that Herschel's object was the three stars.

NGC1507

H279[2]	4:04.5	−2 11	GLX
SB(s)m pec?	3.6′ × 0.9′	12.82B	

This edge-on spiral or irregular galaxy is brighter along its major axis, with short, fainter extensions. Photographs show nebulous condensations along the major axis including a bright knot in the envelope to the S. There are several small dust patches in the envelope to the N and a large one NNE of the galaxy

elongated core in an extensive outer envelope. A small condensation is visible in the ENE of the outer envelope just beyond the bright core. The galaxy is elongated NNE–SSW. Small, edge-on galaxies are 4.1′ W and 2.6′ N. Visually, it is the brightest of the NGC galaxies in a medium-magnification field; it is a bright, round, well-condensed galaxy which is moderately well defined at the edges with a very bright core. Radial velocity: 3840 km/s. Distance: 172 mly. Diameter: 120 000 ly.

NGC1461

H460[2]	3:48.5	−16 24	GLX
SA(r)0°	3.0′ × 0.9′	12.82B	

This is a bright lenticular galaxy in the DSS image with a bright, extended central region and a faint outer envelope. The galaxy is elongated NNW–SSE and field stars are 4.5′ NW and 6.0′ SSW. Visually, the galaxy has a

large, bright, roundish core, embedded in an extended, fainter outer envelope which is well defined and tapers to blunt points. Radial velocity: 1384 km/s. Distance: 62 mly. Diameter: 54 000 ly.

NGC1482

H962[3]	3:54.7	−20 30	GLX
SA0⁺ pec sp	2.5′ × 1.4′	13.10B	12.1V

Photographs show a small, bright lenticular galaxy with a bright, extended central region in an extensive, broader outer envelope. A prominent, slightly curved dust lane, running ESE–WNW, extends along the length of the galaxy bordering the central region. The galaxy is oriented ESE–WNW and bright field stars are 2.25′ ENE and 2.5′ NNW. The galaxy NGC1481 is 5.1′ to the NNW and may be a physical companion. If so, the minimum core-to-core separation would be about 121 000 ly at the presumed distance.

NGC1482

NGC1507

©STScI/AURA/CALTECH/ROE

centre. Visually, this large and moderately bright galaxy is best at medium magnification but is well seen at high magnification. It is a narrow and well defined sliver of light, oriented NNE–SSW and slightly uneven in surface brightness along its major axis. There is no central bulge but it is slightly brighter at the middle. The extremities end in blunt points and a field star is 1.25′ W of centre. Radial velocity: 810 km/s. Distance: 36 mly. Diameter: 38 000 ly.

NGC1516

H499³	4:08.1	−8 53	GLX	SB(s)bc pec
Scd? pec	0.5′ × 0.5′	0.6′ × 0.6′	15.0B	

The DSS photograph shows two face-on galaxies in contact; these are designated NGC 1516A and NGC1516B. The galaxy to the SE (NGC1516A) has a bright, irregular central region and a fainter outer envelope, while the galaxy to the NW (NGC1516B) is slightly

elongated ENE–WSW with disturbed plumes to the W and NW. The galaxies have very similar radial velocities and probably form a physical pair. Visually, the galaxies are not resolved as separate objects even at high magnification; they are seen as an elongated, hazy smudge of light, quite faint and very diffuse, with even surface brightness and fairly well defined at the edges. The image is elongated SSE–NNW. Herschel recorded this object on 30 January 1786 and described it as: 'Very faint, small, extended, easily resolvable.' The last comment suggests that Herschel may have been able to see two separate objects. Radial velocity: 9947 km/s. Distance: 441 mly. Diameter: 64 000 ly.

NGC1535

H26⁴	4:14.2	−12 44	PN	
4+2c	0.75′ × 0.7′	11.6B	10.6V	

Discovered on 1 February 1785, Herschel initially considered this a

planetary nebula but later thought it might be a very distant cluster of stars. His description reads: 'Very bright, perfectly round or very little elliptical planetary but ill defined disc. Second observation, resolvable on the borders, and is probably a very compressed cluster of stars at an immense distance.' The nebula is too bright to be well seen on the DSS image, appearing only as an opaque, oval disc. Even in a small telescope, this is an impressive sight; it is one of the dozen brightest planetary nebulas in the sky. A strong blue colour is noted at high magnification; the texture is extremely grainy with the central star (magnitude 11.59) shining through. Moderate apertures reveal a fairly bright object, steadily brighter to the middle with ragged, ill defined edges. The expansion velocity is 19 km/s and the nebula shows a very slight radial velocity of 3 km/s in approach. The central star has been catalogued as HD 26847, spectral type O5.

NGC1552

H490³	4:20.3	−0 40	GLX
SAB(r)0⁺?	1.8′ × 1.3′	14.24B	

Images show that this lenticular galaxy has a brighter, slightly elongated central region in a fainter outer envelope, oriented E–W. The galaxy appears small visually, but is moderately bright and well defined and is best seen at high magnification, which brings out the outer envelope. Very slightly oval in

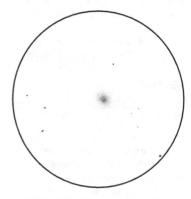

NGC1535 15-inch Newtonian 293x

shape, the envelope is somewhat mottled. A small, brighter core is embedded and is best seen at medium magnification. Radial velocity: 4909 km/s. Distance: 219 mly. Diameter: 115 000 ly.

NGC1576

H587[3]	4:26.2	−3 37	GLX
S0⁻:	1.4′ × 0.9′	14.34B	

The DSS image shows a lenticular galaxy with a brighter, elongated central region in a fainter outer envelope. A faint plume emerges from the envelope to the ESE and two field stars bracket the galaxy, one 1.1′ ESE, the other 1.4′ WNW. In a medium-aperture telescope, the galaxy is moderately faint but stands out well at medium magnification. It is a condensed patch of light, even surface brightness and slightly elongated E–W with prominent bracketing stars. Radial velocity: 4686 km/s. Distance: 209 mly. Diameter: 85 000 ly.

NGC1600

H158[1]	4:31.7	−5 05	GLX
E3	2.5′ × 1.7′	11.94B	

Photographically and visually, this is a bright galaxy with a large, bright core and an extensive outer envelope. The disc of this galaxy is much mottled and it

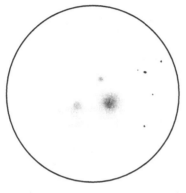

NGC1600, NGC1601
15-inch Newtonian 313x

is slightly elongated almost due N–S. It is the largest and brightest member of a compact group which includes NGC1601, NGC1603 and NGC1606. Herschel observed this galaxy at least twice. Although they are immediately adjacent to the N and E, respectively, and not particularly difficult to see, neither NGC1601 nor NGC1603 was noted by Herschel. Radial velocity: 4603 km/s. Distance: 206 mly. Diameter: 150 000 ly.

NGC1609

H585[3]	4:32.6	−4 22	GLX
(R')S0⁺ pec:	1.1′ × 0.7′	14.57B	

The DSS image shows a slightly inclined, elongated lenticular galaxy with a bright, elongated core and an extensive outer envelope, oriented E–W. A small galaxy is 0.75′ SW, while a field star is 0.75′ NW. NGC1609 is part of a loose assemblage of NGC galaxies which includes NGC1607, NGC1611,

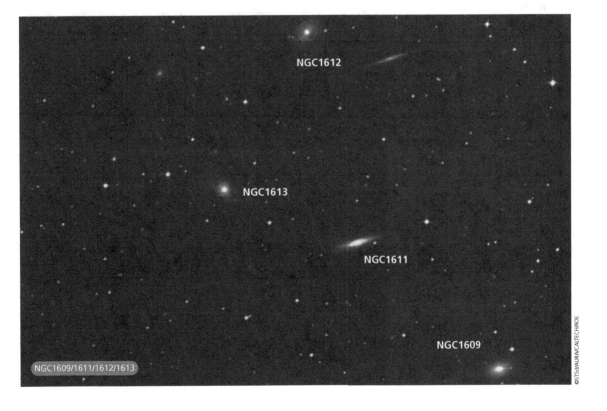

NGC1612

NGC1613

NGC1611

NGC1609

NGC1609/1611/1612/1613

NGC1612 and NGC1613. At medium magnification this galaxy can be seen with NGC1611 and NGC1613 in the field. NGC1609 is a faint, very small, condensed patch of light, with the field star visible to the NW. Radial velocity: 4229 km/s. Distance: 189 mly. Diameter: 60 000 ly.

NGC1611

H586[3]	4:33.0	−4 17	GLX
(R')SB(rs)0⁺ pec?	1.9′ × 0.5′	14.40B	

Photographs show an almost edge-on, elongated lenticular galaxy with a bright, elongated core and an extensive outer envelope. There is evidence of a dust patch in the western part of the bright central region. A very faint field star is 0.5′ W and NGC1609 is 6.75′ SW. Visually, the galaxy is quite faint and elongated ESE–WNW with a slight central bulge and fairly well defined edges. Radial velocity: 4229 km/s. Distance: 189 mly. Diameter: 104 000 ly.

NGC1618

H524[2]	4:36.1	−3 09	GLX
SB(r)b:	2.2′ × 0.8′	13.51B	

This is a highly inclined spiral galaxy, probably barred, with an elongated, brighter central region oriented N–S in a mottled spiral pattern oriented NNE–SSW. Visually, it is the first of three galaxies visible in a low-power field immediately N of the bright star Nu

Eridani. The other galaxies are NGC1622 and NGC1625. Curiously, though they are all of similar brightness, Herschel only noted NGC1618. The core of this galaxy is bright and well condensed, embedded in a bright, oval outer envelope. Nu Eridani should be kept out of the field to view to see these three galaxies to advantage. Radial velocity of NGC1618 is 4817 km/s. Distance: 215 mly. Diameter: 138 000 ly.

NGC1620

H514[2]	4:36.6	−0 09	GLX
SAB(rs)bc	3.2′ × 1.0′	12.87B	

This is a highly inclined spiral galaxy. Photographs show a small, elongated, brighter core which may be a bar. The fainter outer envelope is not well resolved but seems to indicate a multi-arm spiral structure defined by dust patches, with very faint outer structure. Visually, the galaxy is best seen at

medium magnification; it is a moderately large and diffuse galaxy and is fairly even in surface brightness, though just marginally brighter along its major axis. The galaxy is elongated NNE–SSW, with a prominent field star 4.75′ to the ENE. Radial velocity: 3445 km/s. Distance: 154 mly. Diameter: 143 000 ly.

NGC1635

H515[2]	4:40.1	−0 33	GLX
(R)SB(r)0/a	1.5′ × 1.4′	13.38B	

This is an almost face-on, theta-shaped, barred galaxy. Photographs showing a bright, round core and a slightly fainter bar, both ends of which curve to the N. The bar is attached to a ring which is a little brighter to the SW and has faint condensations in the W. The ring is embedded in a large, faint outer envelope. The inner ring and bar are elongated E–W, while the outer

**NGC1618 (right) and NGC 1622 (left)
15-inch Newtonian 169x**

NGC1635

©STScI/AURA/CALTECH/ROE

envelope is elongated N–S. The galaxy is moderately bright visually, and holds magnification fairly well. It is a fairly round and well defined galaxy with a small, prominent and brighter core. The envelope itself is fairly even in surface brightness and a faint pair of field stars is visible about 1.0′ to the NNW. Radial velocity: 3208 km/s. Distance: 143 mly. Diameter: 63 000 ly.

NGC1636

H522²	4:40.7	−8 36	GLX
(R')SB(rs)ab:	1.2′ × 0.9′	13.93B	

Photographs reveal an almost face-on, barred spiral galaxy with a small, bright, round core and two bright spiral arms in a slightly fainter envelope. It is slightly elongated N–S and a faint edge-on galaxy is 6.75′ SE. Well seen at high magnification, NGC1636 is a fairly bright galaxy, oval and extended N–S; it is well defined at the edges with a bright envelope that brightens gradually to the middle. A

faint field star is visible to the NNE. Radial velocity: 4069 km/s. Distance: 182 mly. Diameter: 63 000 ly.

NGC1637

H122¹	4:41.5	−2 51	GLX
SA(s)cd	3.2′ × 2.8′	11.56B	

Photographically, this is an almost face-on spiral galaxy with a small, round,

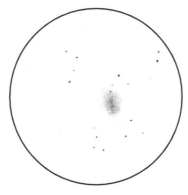

NGC1637 15-inch Newtonian 95x

brighter core and a very faint, short bar. The spiral arms are knotty and form a fragmentary, multi-arm pattern with the core offset to the SSW. A large dust lane emerges E of the core and broadens as it curves to the N. Deep photographs begin to resolve some stars which are brighter than blue magnitude 22. Visually, it is a large, very bright galaxy, easily visible at low magnification. It is a fat oval of light, the extremities of which fade gradually into the sky background. Elongated NNE–SSW, high magnification reveals a mottled core with brighter condensations. Radial velocity: 639 km/s. Distance 29 mly. Diameter: 26 000 ly.

NGC1638

H525²	4:41.6	−1 49	GLX
SAB(rs)0°?	2.2′ × 1.5′	13.11B	

Images show a slightly inclined spiral galaxy with a bright, slightly elongated core in a very faint spiral pattern. The arms are broad and defined by faint dust lanes. Visually, the galaxy is fairly bright, small and rather condensed. The core is very bright and the ENE–WSW elongation of the galaxy is noted, as is the faint secondary halo delineated by the spiral arms. Radial velocity: 3241 km/s. Distance: 145 mly. Diameter: 93 000 ly.

NGC1643

H588³	4:43.6	−5 17	GLX
SB(r)bc pec?	1.3′ × 0.9′	14.08B	

The DSS image reveals an almost face-on spiral galaxy with a very peculiar morphology. The large, bright, slightly elongated core may conceal a bar, but it appears overexposed in the image. Two narrow, high-surface-brightness spiral arms emerge. The spiral arm to the S is detached in the SE and a separated arm with four brighter condensations is visible immediately E. The N spiral arm has a bright condensation in it and appears attached to either an irregular companion galaxy or a detached spiral arm with four condensations involved. The galaxy is elongated N–S and a small galaxy, perhaps barred, is visible 0.5′

NGC1637

©STScI/AURA/CALTECH/ROE

SE. NGC1643 is quite a faint object visually, however; it is small and somewhat squarish, moderately well defined at the edges and fairly even in surface brightness with a slight apparent elongation N–S. Herschel discovered this object on 28 November 1786 and called it: 'Very faint, small.' Because of the varying descriptions of the galaxy down through the years, it was once suspected of variability: it has been described as 'Extremely faint' (John Herschel), 'Faint or very faint' (d'Arrest) 'Faint' (Dreyer, 1877), 'Bright, pretty large' (Roberts, 1903). Two supernovae have been recorded in the galaxy recently, one in 1995 and another in 1999. The morphology of the galaxy suggests a high rate of star formation and the varying brightness descriptions may be attributable to the presence or absence of supernovae activity at the time of observation. Radial velocity: 4791 km/s. Distance: 214 mly. Diameter: 81 000 ly.

NGC1646

H523[2]	4:44.3	−8 33	GLX
Pec	2.3′ × 1.2′	13.64B	

The DSS image shows a small galaxy with a peculiar morphology. The primary galaxy appears to be lenticular with a bright round core and a fainter outer envelope. The envelope flares, however, particularly to the S, where it appears quite broad and surrounds the image of a faint foreground star. A small, condensed, round object appears in silhouette immediately SSE of the core of the principal galaxy and may be a satellite galaxy. A second, much smaller, condensed object (perhaps a faint star) is in the envelope immediately ESE of the principal galaxy's core and a small, edge-on galaxy can be seen 0.75′ to the ESE. Another galaxy, NGC1648, is 4.1′ NE. The bright field star 56 Eridani is 4.9′ WNW. Visually, this moderately bright galaxy is a small, irregularly round patch of light, even in surface brightness and fairly well defined. A faint field star is seen immediately to the NNE, while a brighter one follows 2.0′ to the E. Radial velocity: 4721 km/s. Distance: 211 mly. Diameter: 141 000 ly.

NGC1653

H526[2]	4:45.8	−2 25	GLX
E[+]:	1.8′ × 1.7′	12.92B	

Presenting a similar appearance both visually and photographically, this elliptical galaxy appears small, fairly bright and round; it is quite bright to the middle with a well-condensed core. The outer regions of the galaxy are diffuse. Radial velocity: 4252 km/s. Distance: 190 mly. Diameter: 99 000 ly.

NGC1659

H589[3]	4:46.5	−4 47	GLX
SA(r)bc pec	1.5′ × 1.0′	13.20B	

Photographically, this is a slightly inclined spiral galaxy, elongated NE–SW, with a bright core and grainy spiral structure. The form is asymmetrical, with the core offset towards the SW. The spiral arms are brighter to the S than to the N, where the arms appear filamentary and less massive. A fainter and much smaller galaxy is 8.0′ WNW. At the eyepiece, NGC1659 is best seen at medium magnification as an oval, well defined patch of light which is brighter along its major axis with a tiny, bright core. Radial velocity: 4497 km/s. Distance: 201 mly. Diameter: 88 000 ly.

NGC1665

H457[2]	4:48.2	−5 25	GLX
SA(s)0[+] pec?	1.9′ × 1.1′	13.87B	

This lenticular galaxy is seen almost face-on in photographs; the slightly oval, brighter core is surrounded by an elongated inner disc which is slightly brighter along a NE–SW axis, suggesting a possible bar. A much fainter, ring-like outer envelope hints at possible spiral structure. A fainter edge-on galaxy is 4.9′ NNE. Visually, the galaxy is a faint, oval patch of light

NGC1646

©STScI/AURA/CALTECH/ROE

oriented NE–SW. The edges are moderately well defined and the galaxy appears a little brighter to the middle. Radial velocity: 2673 km/s. Distance: 119 mly. Diameter: 66 000 ly.

NGC1700

H32[4]	4:56.9	−4 52	GLX
E4	3.0′ × 1.9′	12.03B	

Herschel recorded this object on 5 October 1785, placing it in his Class IV and describing it as: 'Very bright, very small, much brighter to the middle, like a star affected with irregular burs.' The DSS photograph shows an elongated elliptical galaxy with a bright core in a fainter outer envelope which is irregular at the edges and appears elongated SE–NW, while the core is elongated E–W. The bright spiral galaxy NGC1699 is 6.3′ N, while a bright field star is 2.5′ SW. Telescopically, NGC1700 is well seen with direct vision; the sharp, stellar nucleus is surrounded by a very smooth, well defined outer envelope. It is oval in shape and oriented E–W. High magnification also shows NGC1699 as a condensed oval object, oriented SE–NW. Radial velocity: 3796 km/s.

Distance: 170 mly. Diameter: 148 000 ly.

NGC1779

H500[3]	5:05.3	−9 09	GLX
(R′)SAB(r)0/a?	2.7′ × 1.4′	13.03B	

Images show an elongated lenticular or spiral galaxy with a bright, oval core, which may contain a bar, and a faint outer envelope. This envelope is elongated ESE–WNW and is smooth-textured; there may be a pseudo-ring surrounding the central region. Visually, only the central region of the galaxy is seen; it is small and fairly dim, roundish and broadly brighter to the middle with moderately well defined edges. Radial velocity: 3203 km/s. Distance: 143 mly. Diameter: 113 000 ly.

IC2118

H38[5]	5:06.9	−7 13	SNR-RN
180.0′ × 60.0′			

This object was originally identified by Dreyer as NGC1909 in *The Scientific Papers* but was later thought to be nonexistent as nothing could be found at Herschel's position (he provided two positions to indicate an object of great extent).

However, he listed the extensive nebula as following Rigel by 11 minutes 9 seconds and 11 minutes 35 seconds instead of preceding. Making the correction brings the observer to the position of IC2118, also known as the Witch's Head Nebula, a suspected SNR remnant or possible reflection nebula. Herschel's 20 December 1786 discovery description is certainly apt: 'Strongly suspected nebulosity of very great extent. Not less than 2 degrees 11 minutes of polar distance and 26 seconds of right ascension in time.' On the DSS image the nebula is quite dim; it is a large intricate mass of faintly glowing gas greatly elongated N–S, the brightest portion being a wedge-shaped mass with the apex pointing E, about 70′ E of 65 Eridani. Visually, the object is somewhat difficult and a good transparent night at a dark site is necessary to get a good view. About 20′ E of 65 Eridani is a very faint and extremely hazy nebula displaying no structure or brightness variations and measuring about 1° E–W and about 2° N–S in extent. It is just a general brightening of the sky background, especially compared to the sky immediately W of 65 Eridani. Rigel may be the illuminating star.

Fornax

Despite its far southerly location, Herschel managed to record nine of the brighter galaxies in this constellation, several of which were never more than a few degrees above Herschel's southern horizon. The most spectacular object here is the barred spiral NGC 1097, which may show extensive detail for observers who can observe it from southern sites.

NGC686

H459[3]	1:49.0	−23 48	GLX
SA0⁻:	1.5′ × 1.2′	13.57B	

The visual and photographic appearances of this face-on lenticular galaxy are quite similar. Telescopically, it is bright, very slightly oval in shape and elongated N–S. The brightness increases first slowly and then more rapidly to the centre of a bright core. The edges are hazy and ill defined and the object forms a triangle with two bright field stars, one to the WNW, the other to the SSW. Radial velocity: 4587 km/s. Distance: 205 mly. Diameter: 90 000 ly.

NGC922

H239[3]	2:25.1	−24 47	GLX
SB(s)cd pec	1.9′ × 1.9′	12.40B	12.2V

The DSS photograph reveals a face-on galaxy, possibly a barred galaxy or an irregular, with chaotic, asymmetrical structure. The bar is brighter to the NE and fragmentary spiral structure extends away to the S and SE, ending abruptly at a partial ring which is defined by a chain of several HII knots. A bright, partial ring structure defined by many nebulous knots forms a half circle from the S to the E and then to the N. Here it fades to a broader, more chaotic arc which runs S, almost completing the ring. Visually, the galaxy appears moderately large but it is somewhat faint and diffuse, irregularly round with an envelope which varies in brightness. It is slightly elongated N–S and a small, elongated brighter core is embedded. The edges are moderately

well defined and a field star is located 2.1′ NNW. Radial velocity: 3029 km/s. Distance: 135 mly. Diameter: 75 000 ly.

NGC1097

H48[5]	2:46.3	−30 17	GLX
(R'1:)SB(r'l)b	9.3′ × 6.3′	10.14B	9.5V

Together with the small galaxy NGC1097A located 3.5′ to the NW, this galaxy has also been catalogued as Arp 77. This is one of the more striking of the barred spiral galaxies and is well seen in the DSS image. It is a very large and bright object with a large, elongated core and a strong bar. There is a great deal of dust in the bar, particularly to the SE and NW where it attaches to the spiral arms. Initially, the arms are tightly wound around the core but then

NGC922

they expand outward. The outer arms appear relatively faint in comparison to the bar but there are a number of brighter knots and condensations and the galaxy is elongated SE–NW. If the elliptical galaxy NGC1097A is a physical companion, and it appears within the confines of the spiral arms in the NW, then the minimum core-to-core separation at the presumed distance would be about 54 000 ly. In a moderate-aperture telescope only the bright core and bar are well seen, but the galaxy is exceptional in a large-aperture telescope. It features a large, bright and irregularly round core embedded in a slightly fainter, elongated bar. The bar on the NNW flank is divided by a dark lane running along the major axis and a thin, moderately long spiral arm is attached here, curving to the E. The SSE bar is broad and hazy with a short, blunt spiral arm attached and curving W. NGC1097A is visible as a bright, condensed, well defined object. Herschel discovered NGC1097 on 9 October 1790 and described it as: 'Very bright, extended 75 degrees north preceding, south following, 8' long. A very bright nucleus, confined to a small part, or about 1' in diameter.' Radial velocity: 1190 km/s. Distance: 53 mly. Diameter: 144 000 ly.

NGC1097

©STScI/AURA/CALTECH/ROE

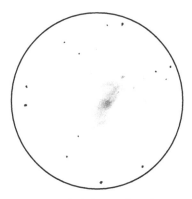

NGC1097 25-inch Newtonian 195x

NGC1201

H109[1]	3:04.1	−26 04	GLX
SA(r)0°:	3.2' × 1.6'	11.72B	10.6V

In photographs this is a prototypical lenticular galaxy featuring three distinct brightness zones; a bright core, a slightly fainter, elongated disc and a faint outer envelope. Visually, the galaxy is fairly bright and well condensed. Elongated NNE–SSW, it appears broadly brighter to the middle and the edges are quite well defined, indicating that the faint outer envelope is not visible. Field stars are located 3.8' NNE and 2.8' NNW. Radial velocity: 1599 km/s. Distance: 71 mly. Diameter: 66 000 ly.

NGC1344

H257[1]	3:28.3	−31 04	GLX
E5	5.6' × 3.5'	11.22B	10.3V

Photographs show a bright, elongated elliptical galaxy with a brighter core and a slightly fainter outer envelope. The galaxy is elongated NNW–SSE and bright field stars are located 6.0' E and 5.7' N. Visually, this galaxy is quite bright and is well seen at high magnification as a large, elongated galaxy with a bright, extensive envelope which brightens to the middle, though no core is visible. The edges are hazy and fade slowly to a very faint outer envelope. Radial velocity: 1061 km/s. Distance: 47 mly. Diameter: 77 000 ly.

NGC1366

H857[3]	3:33.9	−31 12	GLX
S0°	2.0' × 0.8'	12.82B	

Images reveal an edge-on, lenticular galaxy with a bright core and extensions in a slightly fainter outer envelope. The galaxy is elongated N–S and a bright field star is located 6.9' N. Visually, the galaxy is small but it has fairly high surface brightness and is well seen at high magnification as a round, bright, nonstellar core embedded in a faint, tapered envelope, elongated N–S. The edges are moderately well defined and the bright field star does not interfere with observation. Radial velocity: 1186 km/s. Distance: 53 mly. Diameter: 31 000 ly.

NGC1371

H262[2]	3:35.0	−24 56	GLX
(R')SAB(r'l)a	5.8′ × 3.9′	11.59B	

The DSS image shows a slightly inclined, barred spiral galaxy with a bright, slightly elongated core and a very short bar. The bar is connected to spiral arms which are broad and smooth textured. The galaxy is elongated SE–NW and a faint field star is involved in the outer envelope 1.4′ SSE. High-resolution blue-sensitive plates show considerable evidence of star formation and many HII regions in the outer spiral arms. Visually, only the central region of this large spiral galaxy is visible, but it is fairly bright and takes magnification well. Slightly oval and oriented roughly E–W, the galaxy brightens towards the middle, but no bright core is visible. The outer envelope is hazy, faint and ill defined. Radial velocity: 1366 km/s. Distance: 61 mly. Diameter: 103 000 ly.

NGC1385

H263[2]	3:37.5	−24 30	GLX	
SB(s)cd	3.4′ × 2.0′	11.46B	10.9V	

This slightly inclined barred spiral galaxy displays chaotic structure in photographs. The bar is offset to the W in the envelope, is extended E–W and is attached to an irregular, knotty patch to the N and a smoother patch to the S. The bar is connected to a knotty spiral arm in the W, which extends to the S. The northern spiral arm is connected to the bar in two places on the eastern side and a fainter, grainy envelope with dust

NGC1385

©STScI/AURA/CALTECH/ROE

patches envelopes the galaxy. On the DSS image, the outer envelope can be traced to about 4.5′ × 2.5′. Overall it is elongated N–S. Visually, the galaxy is brightish, but somewhat irregular with a bright, slightly elongated core in an irregularly bright and lumpy outer envelope. The galaxy is oriented E–W with ill defined edges and the faint outer envelope and spiral arms are not seen. Radial velocity: 1402 km/s. Distance: 63 mly. Diameter: 62 000 ly.

NGC1425

H852[2]	3:42.2	−29 54	GLX
SA(rs)b	6.0′ × 2.8′	11.32B	

Images show a highly inclined spiral galaxy with a brighter core embedded in an elongated, bright central region. The multiple spiral arms are grainy with a few knots, dust patches and lanes in the spiral structure. Telescopically, this galaxy is large and fairly bright; it is an elongated oval of light, steadily increasing in brightness towards the middle, though no core is visible. The galaxy is oriented SE–NW and the edges are hazy and fade slowly into the sky background. Three bright field stars to the N are on a line which matches the galaxy's orientation. They are 2.5′ NNW, 1.4′ NE and 1.9′ NE of the core. Radial velocity: 1397 km/s. Distance: 62 mly. Diameter: 109 000 ly.

Gemini

Herschel recorded a total of 23 objects here, a mix of open clusters, asterisms, planetary nebulae and galaxies. Interesting objects include NGC2158, a remote but very rich star cluster, and NGC2392, well worth examining carefully with high magnification on transparent nights. The galaxies are faint and nondescript for the most part but many of the star clusters are interesting if not spectacular. Herschel made frequent forays into this constellation, beginning in 1783 and extending into the later years of the survey.

NGC2129

H26[8]	6:01.0	+23 18	OC
I3m	6.0′ × 6.0′	7.3B	6.7V

Images reveal a coarse cluster of stars which stands out well in a fairly poor field of stars. The cluster is dominated by two bright stars and cluster members display a wide range of brightnesses. There is some concentration of stars, particularly into a broken ring, but the centre of the ring is largely empty. The cluster is resolved in small-aperture telescopes with about 20 stars visible and the concentration of faint stars around the two brightest members is a distinct feature. Total membership of the cluster is probably more than 70 stars. Spectral type: B1. Age: about 16 million years. Radial velocity in recession: 18 km/s. Distance: 6500 ly.

NGC2158

H17[6]	6:07.5	+24 06	OC
II3r	5.0′ × 5.0′	9.5B	8.6V

This very rich and remote open cluster was first observed by Herschel on 16 November 1784 and concisely described as: 'A very rich cluster of very compressed and extremely small stars 4′ or 5′ diameter. A miniature of the 35 cluster of the Connaissance des Temps which it precedes 1′18″ and is 2′ north.' The '35 cluster' is a reference to Messier's M35. Photographs show a very rich and compressed cluster of stars with a square outline in a rich field. In small telescopes, at medium magnification, it is broadly concentrated to the middle and is slightly extended NW–SE, which reflects the distribution of the brightest stars in the cluster. A magnitude 9 star bordering the cluster to the SE is conspicuous. Moderate-aperture telescopes resolve the cluster well at high magnification and it is well separated from the starry field. The brightness range is relatively narrow. Total cluster membership is almost 1000 stars. The cluster is relatively old at an age of about 3.2 billion years. Spectral type: F0. Distance: 12 630 ly.

NGC2224

H35[7]	6:27.6	+12 38	OC?
20.0′ × 20.0′			

Recorded on 24 December 1786, Herschel called this object: 'A cluster of small, pretty much compressed stars

NGC2158

with suspected milky nebulosity.'
Visually, at the presumed position is a
coarse N–S collection of scattered stars
with a narrow brightness range. About
11 stars are resolved but there is no
unresolved glow; the brightest member
is the star HD257536, magnitude 10.4.
About 20 stars may be members.

NGC2234

H9[8]	6:29.5	+16 43	MWP
35.0′ × 35.0′			

Near the position given by Herschel is a
scattered grouping of stars situated S of a
magnitude 8 star. The brightness range
is moderate and the grouping is not well
separated from the field, which is fairly
rich in stars. About 35 stars are resolved
but this is probably nothing more than a
slight concentration of stars in a Milky
Way field. Recorded on 19 February
1784, Herschel's description reads:
'A cluster of very much scattered stars of
various magnitudes, near a half degree,
not rich.'

NGC2266

NGC2266

H21[6]	6:43.2	+26 58	OC
II2m	5.0′ × 5.0′	9.5V	

Similar photographically and visually,
this is a well concentrated and relatively
rich cluster of stars displaying a well
defined, triangular shape. The cluster is
well separated from the sky background
with the cluster members all of a similar
brightness. A brighter field star borders
the cluster on the SW. Resolution of the
cluster is a challenge with smaller
apertures but a conspicuous row of stars
on the SE flank is visible and was
remarked upon by Herschel. Medium
apertures at high magnification do a
good job of resolving the cluster. About
50 stars are considered members; the
cluster is fairly remote at a distance of
about 11 000 ly.

NGC2274

H615[2]	6:47.2	+33 34	GLX
E	1.9′ × 1.9′	13.51B	

This galaxy forms a pair with NGC2275
1.9′ N and the galaxies are probably

physically related: the radial velocities
are fairly similar. At the presumed
distance, the core-to-core minimum
separation between the two galaxies
would be about 125 000 ly. The DSS
image shows an elliptical galaxy with a
round, very bright core and a faint and
extensive outer envelope. A very faint
field star borders the core of the galaxy
to the WSW. Another much fainter
spiral galaxy is 7.4′ WNW.
Telescopically, both galaxies are
moderately faint but can be seen
together in a high-magnification field.
NGC2274 is a little brighter, a round
patch of light, brighter to the middle
with ill defined extremities. Radial
velocity: 5058 km/s. Distance: 226 mly.
Diameter: 125 000 ly.

NGC2275

H614[2]	6:47.2	+33 36	GLX
S?	1.5′ × 1.1′	13.67B	

Paired with NGC2274, this galaxy
shows something of a peculiar

morphology in photographs. Very
probably a spiral, the galaxy has a very
bright, irregular-shaped core with a field
star located immediately to the SW. A
short, stubby bar emerges from the core
in the S, while a broad, faint plume
emerges from the core in the N and
curves to the W. This is broader at its
extremity and slightly brighter along its
SW limb. A much fainter band
continues to the S, curves to the E and
sweeps by the bar to the S. The galaxy is
elongated N–S. Visually, the galaxy is
slightly fainter than NGC2274 and only
the bright core is visible as a hazy,
roundish patch of light. Radial velocity:
4821 km/s. Distance: 215 mly.
Diameter: 94 000 ly.

NGC2289

H898[3]	6:50.7	+33 29	GLX
S0	1.0′ × 0.7′	14.24B	

This is part of a compact, curving chain
of galaxies which includes NGC2288,

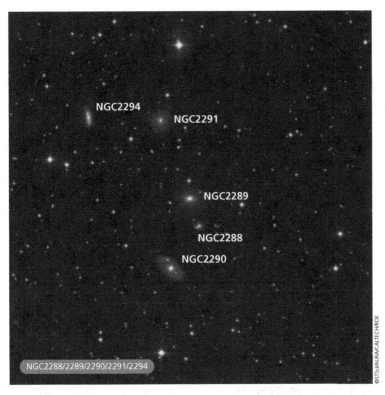

NGC2288/2289/2290/2291/2294

©STScI/AURA/CALTECH/ROE

Telescopically, this is a small, faint, but well-resolved cluster located in a fairly rich field. It is visible as a nebulous haze at low power and at medium magnification resolution is good with only faint traces of unresolved stars. High magnification shows the cluster well; the stars are about magnitude 13 or a little fainter. The NW flank is very flat and the overall outline is triangular. This cluster was one of Herschel's earliest discoveries: he recorded it on 30 December 1783 as: 'A very compressed cluster of extremely small stars. Irregular form, 5' or 6' in diameter.' Thirty stars are considered members of this cluster.

NGC2331

H40[8]	7:07.2	+27 21	OC
IV2m	19.0' × 19.0'		

In photographs and at the eyepiece, this is a coarse, scattered group of about 30 stars in a poor star field. The brightest members are magnitude 9 and form a ragged group oriented N–S; a small oval group of half a dozen fainter stars is included in the SE.

NGC2333

H899[3]	7:08.4	+35 11	GLX
Sa	1.0' × 0.7'	14.01B	

Photographs show a slightly inclined spiral galaxy with a bright core and a fainter outer envelope. Visually, the galaxy is a quite faint, oval patch of light with moderately well defined edges and a little brighter along its major axis. The galaxy is elongated NE–SW. Radial velocity: 4716 km/s. Distance: 211 mly. Diameter: 61 000 ly.

NGC2339

H769[2]	7:08.3	+18 47	GLX
SAB(rs)bc	2.7' × 2.0'	12.47B	

Images show that this face-on, barred spiral galaxy has a small, slightly brighter core and a weak bar. The spiral pattern fragments quickly into a multi-armed configuration with knotty condensations and dust patches. Several faint field stars are involved and the

NGC2290, NGC2291 and NGC2294. Photographs show a lenticular galaxy with a brighter, elongated core in a faint outer envelope, elongated E–W. In the first edition of *Uranometria 2000.0*, the labels for NGC2288 and NGC2289 should be reversed. Visually, only this galaxy and NGC2290 to the S can be detected. It appears as a roundish, well defined patch of light 0.6' S of a faint field star. This group of galaxies may be associated with NGC2274 and NGC2275 to the W, as they have very similar radial velocities. Radial velocity: 4921 km/s. Distance: 220 mly. Diameter: 64 000 ly.

NGC2290

H897[3]	6:51.0	+33 26	GLX
(R)SAa:	1.3' × 0.7'	14.01B	

Located 2.5' SSE of NGC2289, images show an inclined, theta-shaped, barred galaxy with a bright core and a very short, stubby bar. The inner region is surrounded by a faint envelope. A ring surrounds the inner galaxy; the ring is double at its NE extremity, divided by a thin dark zone. Two very faint field stars are involved, though not seen visually: one to the SW, the other to the NNE. Telescopically, the galaxy is very faint but slightly brighter and larger than NGC2289. It is fairly well defined, an oval patch of light, elongated NE–SW. Radial velocity: 5027 km/s. Distance: 225 mly. Diameter: 85 000 ly.

NGC2304

H2[6]	6:55.0	+18 01	OC
II1m	5.0' × 3.0'	10.0V	

Photographs show a small, crescent-shaped and not particularly rich cluster which is compressed towards the middle and elongated in an E–W direction. The brightness range is moderate and, because the stars are a little brighter than the surrounding field, it is well separated from the background.

galaxy is elongated N–S. A small galaxy is located 6.75′ ESE and may be visible in very large apertures. Visually, this is a bright galaxy, though the main envelope is rather diffuse and the edges fade unevenly into the sky background. It is gradually brighter to the middle with a small, faint, stellar nucleus which is visible at high magnification. A very faint field star borders the galaxy to the ENE. Herschel recorded the galaxy on 22 February 1789, and thought it easily resolvable. Radial velocity: 2130 km/s. Distance: 95 mly. Diameter: 74 000 ly.

NGC2355

H6[6]/H6[7]	7:16.9	+13 47	OC
II2m	8.0′ × 8.0′	9.7V	

The DSS photograph shows a fairly rich cluster of stars, well separated from the background field as many of the stars are bright. The cluster is elongated NW–SE and has a crab-like appearance with a rectangular centre and short chains of stars at each corner. Telescopically, this is a bright and fairly compressed cluster which is well seen at medium magnification as a roundish, well defined object, well isolated from the background with the stars all in a fairly narrow brightness range. One star to the E is about a magnitude brighter than the rest and may not be part of the cluster. Forty stars are recognized as members of the cluster. It was recorded twice by Herschel: on 8 March and 16 March 1784.

NGC2371/NGC2372

H316[2]/H317[2]	7:25.6	+29 29	PN
3a+6	2.2′ × 1.0′	13.0B	11.2V

Although this planetary nebula is a single object, Herschel recorded it as two separate objects of his Class II: faint nebulae. NGC2371 is the SW portion and NGC2372 is the portion to the NE. Photographically, they are two bright, irregular patches bracketing the very faint (magnitude less than 15.5) central star. With much fainter portions to the ESE and WNW, the nebulosity forms a continuous shell around the central star. In addition, there are faint,

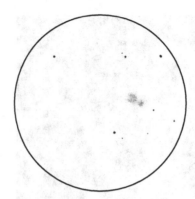

NGC2371 NGC2372
15-inch Newtonian 313x

detached portions of the nebula located to the ESE and WNW. Visually, this is a moderately bright planetary nebula that holds magnification well. The two brightest patches are resolved at high magnification, and appear elongated almost due E–W. The western half is

brighter and more condensed, the eastern half has ill defined and ragged extremities. The central star is not visible. Herschel's 12 March 1785 description reads: 'Two, south preceding, north following, distance 1′, chevelure mixed. Both faint, small, equal, having a nucleus.' This indicates that Herschel may have seen the central star. Radial velocity in recession: 21 km/s. Distance: 3900 ly.

NGC2385

H900[3]	7:28.4	+33 50	GLX
Sb	0.6′ × 0.25′	15.01B	

This galaxy is part of a group of seven very faint NGC galaxies which form an E–W chain a little less than 1° in length. Photographs show a highly inclined galaxy, probably a spiral, with a bright core and a fainter outer envelope. Herschel objects NGC2388 and NGC2389 are 5.3′ ESE and 7.5′ ENE, respectively. Visually, all

NGC2371/2372

NGC2385/2388/2389

©STScI/AURA/CALTECH/ROE

the knotty structure as unresolved stars. Radial velocity: 3932 km/s. Distance: 176 mly. Diameter: 102 000 ly.

NGC2392

H45[4]	7:29.2	+20 55	PN
3b+3b	0.75′ × 0.75′	9.9B	9.1V

This planetary nebula is overexposed on the DSS image, appearing as a round, opaque object with no detail visible. Popularly known as the Eskimo or Clown Face Nebula, this is a sensational object at high magnification in a moderate-aperture telescope. The central star is magnitude 10.47 and is immediately surrounded by a dark zone. Surrounding this zone is the bright inner ring: a doughnut-shaped object 13″ across. The outer envelope is just a little fainter, almost perfectly round, very mottled and a little darker at the eastern edge. The nebula is immediately S of a magnitude 8 field star. The spectral type of the central star (HD59088) is O7. Distance: 2930 ly (estimated).

three galaxies can be seen in the same medium-magnification field, but they are very faint. NGC2385 seems slightly brighter than the others; it is a little brighter to the middle with poorly defined edges and is elongated ENE–WSW. Radial velocity: 4134 km/s. Distance: 185 mly. Diameter: 32 000 ly.

NGC2388

H901[3]	7:28.8	+33 49	GLX
S?	1.0′ × 0.6′	14.63B	

Photographs show a small, slightly inclined spiral galaxy, with a bright core and strong, S-shaped spiral arms embedded in a fainter envelope. Visually, this is the middle galaxy of the trio which also includes NGC2385 and NGC2389. It is difficult and best seen with averted vision as a hazy, elongated patch situated between two faint field stars and is elongated ENE–WSW.

Radial velocity: 4110 km/s. Distance: 184 mly. Diameter: 53 000 ly.

NGC2389

H703[3]	7:29.1	+33 51	GLX
SAB(rs)c	2.0′ × 1.4′	13.39B	

Herschel discovered this faint galaxy on 5 February 1788, five years before he recorded NGC2385 and NGC2388. Photographs show a small, slightly inclined, barred spiral galaxy with a small, bright core and a short, bright bar. A narrow, bright partial ring is attached to the bar in the S and curves E to N. A broad, open pattern of knotty, fragmented spiral arms expands from a dusty central envelope. Visually, the galaxy is extremely faint; it is a hazy slightly elongated patch of light with poorly defined edges. Herschel described the galaxy as: 'Very faint, very small, perhaps a patch of stars.' It is possible that he interpreted

NGC2395

H11[8]	7:27.1	+13 35	OC
IV2m	13.0′ × 6.0′	8.8B	8.0V

Similar visually and photographically, this is a rather weak, scattered cluster, approximately 15′ in diameter. The brightest stars are magnitude 10 and about 20–25 cluster members are

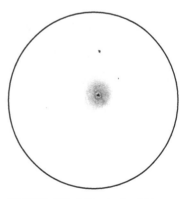

NGC2392 15-inch Newtonian 466x

visible. The cluster is elongated SE–NW with most of the members being about magnitude 11 or fainter. The total membership of the group is more than 50 stars. Spectral type B4. Age: about 50 million years. Distance: 3850 ly.

NGC2420

H1[6]	7:38.5	+21 34	OC
I1r	6.0′ × 6.0′	9.0B	8.3V

Images show a rich and compressed cluster of faint stars with a moderate brightness range. The group is well separated from the field and roughly triangular in shape. The cluster is difficult to resolve in small apertures, but the brightest members are seen to define the outer edges of the cluster and form a V shape with its apex pointed towards the brighter field star to the SSW. Moderate apertures at high magnification resolve the group well. Total membership of the cluster is probably more than 300 stars. Spectral type: F. Age: about 4 billion years. Radial velocity in recession: 115 km/s. Distance 6340 ly.

NGC2420

©STSCI/AURA/CALTECH/ROE

NGC2435

H616[2]	7:44.2	+31 38	GLX
Sa(f)	2.1′ × 0.5′	13.66B	

This is a highly inclined spiral galaxy with a large, bright, elongated core and faint spiral extensions. Photographs show evidence of dust patches in the spiral arms to the NE and SW. An extremely faint field star is immediately E of the core and a brighter field star is located just off the NE tip of the galaxy 1.1′ from the core. Visually, this is a moderately bright galaxy which takes magnification well. Only the bright,

slightly elongated core is well seen; it is well defined, is slightly brighter to the middle and is oriented NE–SW. Radial velocity: 4153 km/s. Distance: 186 mly. Diameter: 113 000 ly.

NGC2481

H302[2]	7:57.3	+23 46	GLX
S?	1.4′ × 0.4′	13.63B	

Although the published classification suggests this might be a spiral, the DSS image seems to indicate it is a highly inclined lenticular galaxy with a bright, elongated core and bright extensions in

a narrow, fainter outer envelope. Very faint field stars or possible compact companions are located immediately S and NNW of the core, while the possible companion galaxy NGC2480 is located 1.0′ NNW. Visually, the galaxy is small but bright and well defined with a prominent core. It is elongated NNE–SSW with sharply defined edges and a bright, edge-on envelope. The companion galaxy is only suspected and is a morely likely target for large-aperture telescopes. Radial velocity: 2263 km/s. Distance: 101 mly. Diameter: 41 000 ly.

Hercules

The 33 objects recorded here by Herschel are mostly galaxies, only NGC6058 (a planetary nebula) and NGC6229 (a globular cluster) being 'local' objects. Although most of the galaxies are faint, several of them, including NGC6160 and NGC6166, are markers for the remote galaxy clusters Abell 2197 and Abell 2199 and the patient observer with a moderate- or large-aperture telescope can spend many enjoyable evenings tracking down the fainter members of these rich galaxy clusters. Herschel explored this constellation frequently, particularly in the later stages of the survey.

NGC6028

H33[3]	16:01.3	+19 22	GLX
(R)SA0[+]:	1.3' × 1.2'	14.35B	

The DSS image shows a possibly barred galaxy with a bright, lens-shaped core elongated E–W in a faint envelope. An extremely faint large ring surrounds the core and a compact object immediately S of the core may be a field star or a satellite galaxy. Visually, the galaxy is quite small and faint, appearing as a roundish patch of light which is a little brighter to the middle and moderately well defined. Radial velocity: 4574 km/s. Distance: 204 mly. Diameter: 77 000 ly.

NGC6052

H140[3]	16:05.2	+20 33	Sc
0.9' × 0.7'	13.45B		

This peculiar object is very likely an interacting pair of galaxies and it has also been catalogued as Arp 209. The DSS image shows a high-surface-brightness object of chaotic form, brighter to the middle with at least four short plumes radiating from the core. Visually, the object is fairly bright, opaque and quite small, but is well seen at high magnification. A bright stellar core is intermittently visible, with a boxy, well defined outer envelope. Radial velocity: 4843 km/s. Distance: 216 mly. Diameter: 57 000 ly.

NGC6058

H637[3]	16:04.4	+40 41	PN
3+2	0.4' × 0.4'	13.30B	

In photographs, this is a small, moderately bright planetary nebula of annular form. The central star is magnitude 13.6 and of spectral type O9 and the surrounding ring of gas is brighter to the NNW and SSE. A faint field star is involved in the ring immediately E. Fainter gas is visible outside the ring to the S and NNW. Herschel discovered the object on 18 March 1787, classing it as a very faint nebula. His description states: 'Very faint, extremely small. 300 (magnification) shewed 2 very small stars with nebulosity.' Visually, this is a small planetary nebula with a moderately bright envelope which takes magnification well. The envelope is grainy in texture and almost round, though faint N–S extensions are suspected. The central star is not seen with any certainty. Radial velocity in recession: 3 km/s. Expansion velocity of the gas shell: 33 km/s. Distance: about 8500 ly.

NGC6073

H74[3]	16:10.1	+16 42	GLX
Sc:	1.3' × 0.7'	14.30B	

This is a small, inclined spiral galaxy. Images show a small, brighter core and three brighter spiral arms emerging into a slightly fainter, broad spiral pattern. At high magnification this is a small, fairly diffuse, oval galaxy which brightens broadly from the edges to a bright, star-like core, which is intermittently visible. The galaxy is elongated ESE–WNW. Radial velocity: 4742 km/s. Distance: 212 mly. Diameter: 80 000 ly.

NGC6106

H151[2]	16:18.8	+7 25	GLX
SA(s)c	2.5' × 1.4'	12.79B	

Photographs show an inclined, spiral galaxy with a bright central envelope and a small, slightly brighter core. A large bright knot is visible in the envelope to the S and a spiral arm extends westward from this knot. The

NGC6058

spiral arms expand outward into a loose pattern with several knots and brighter condensations. Although moderately large visually, this is a fairly faint galaxy; it is an oval patch elongated SE–NW, broadly brighter along its major axis and poorly defined at the edges. A brighter core is intermittently visible. Radial velocity: 1525 km/s. Distance: 68 mly. Diameter: 49 000 ly.

NGC6146

H638[3]	16:25.2	+40 53	GLX
E/S0	1.3′ × 1.0′	13.51B	

Photographs show a bright, elliptical galaxy elongated E–W with a faint outer envelope. This galaxy is one of the principal members of the Abell 2197 galaxy cluster, an 'L' system in the Rood–Sastry classification scheme, owing to the dominating chain of galaxies in the cluster. Three companion galaxies, including NGC6145, are nearby to the NNW. At high magnification, NGC6146 is a small but moderately bright, almost round galaxy with a bright, well defined envelope surrounding a brighter core. A faint field star is immediately ENE and NGC6145 is visible in the field as an oval, even-surface-brightness blur. Radial velocity: 8977 km/s. Distance: 401 mly. Diameter: 152 000 ly.

NGC6150

H639[3]	16:25.8	+40 30	GLX
E?	1.3′ × 0.7′	14.54B	

Though this is classified as a possible elliptical galaxy, photographs show an almost edge-on galaxy, probably a spiral. Very elongated NNE–SSW with a large, bright elongated core and very faint outer envelope, the galaxy is a member of the Abell 2197 galaxy cluster. The edge-on companion galaxy NGC6150B is located to the WSW and, at magnitude 15.3, is a target for large amateur telescopes. Visually, NGC6150 is a small, faint patch of light which is almost round, well defined and brighter to the middle. Radial velocity: 8876 km/s. Distance: 396 mly. Diameter: 150 000 ly.

NGC6154

H680[3]	16:25.6	+49 50	GLX
SB(r)a	2.3′ × 2.3′	14.07B	

Photographs show a theta-shaped, barred galaxy with a bright, round core and a fainter bar, joined to a pseudo-ring made up of overlapping spiral arms. Two much fainter and somewhat broader spiral arms emerge, encircling the bright inner structure. Visually, the galaxy is a moderately faint, slightly oval patch of light with even surface brightness and moderately well defined edges. Radial velocity: 6185 km/s. Distance: 276 mly. Diameter: 185 000 ly.

NGC6155

H690[2]	16:26.1	+48 22	GLX
S?	1.5′ × 1.0′	13.35B	

This is a spiral galaxy with a peculiar morphology. It is slightly inclined with a very small, brighter core and a bright secondary envelope. Images reveal that the spiral structure is very knotty, with a broad outer disc surrounding the galaxy. A bright field star is located 3.2′ to the WSW. In a moderate-aperture telescope the galaxy is a fairly bright, slightly elongated oval with a bright envelope and fairly well defined edges. The galaxy brightens to a broad core and is elongated SSE–NNW. Radial velocity: 2587 km/s. Distance: 115 mly. Diameter: 50 000 ly.

NGC6158

H647[2]	16:27.7	+39 23	GLX
E/S0	0.9′ × 0.6′	14.72B	

Photographs show a bright, elliptical galaxy elongated E–W with a fainter outer envelope. This galaxy is a member of the Abell 2199 galaxy cluster and several companions are nearby, particularly to the S. In moderate-aperture telescopes, this is a small, faint galaxy and is well seen at high magnification. It is well condensed and quite round, though the extremities are only moderately well defined. Radial velocity: 9136 km/s. Distance: 408 mly. Diameter: 107 000 ly.

NGC6160

H652[2]	16:27.7	+40 55	GLX
cD;E	1.6′ × 1.1′	14.18B	

Images show a bright, elliptical galaxy with an extensive, fainter outer envelope. This is one of the dominant galaxies of the Abell 2197 galaxy cluster and is a large, massive object. Several large and small galaxies are nearby, forming a swarming halo around the dominant galaxy. NGC6160 is readily visible at medium magnification but is a little more difficult at higher power. It is an elongated blur with a brighter core, oriented NNE–SSW with moderately well defined edges. Herschel considered this galaxy resolvable. Radial velocity: 9682 km/s. Distance: 432 mly. Diameter: 201 000 ly.

NGC6166

H875[2]	16:28.6	+39 33	GLX	
cD;E	2.3′ × 1.7′	12.90B	11.8V	

Herschel discovered this dominant member of the Abell 2199 galaxy cluster on 30 May 1791, describing it as: 'Pretty bright, small, little extended, very gradually much brighter middle.' Photographically, this peculiar, bright elliptical galaxy with a fainter outer envelope is of the rare type known as a cD galaxy. Usually elliptical, these are massive galaxies often surrounded by several companions which the main galaxy may be in the process of annexing. NGC6166 is quite typical: two distinct galaxies are visible bordering the core immediately E and several companions nearby swarm around the galaxy. Visually, the galaxy can be seen with relatively small-aperture telescopes, appearing as a well defined, rather smooth, nebulous spot with direct vision. Averted vision brings out a more extensive outer envelope, elongated NE–SW. In moderate apertures, the galaxy is brighter to the middle and a nebulous extension can be seen in the NE: this is the image of two unresolved companion galaxies. The galaxy MCG+7-34-54 is visible to the SW and immediately S of NGC6166 is MCG+7-34-66. Radial velocity:

9257 km/s. Distance: 413 mly. Diameter: 277 000 ly.

NGC6173

H640[3]	16:29.8	+40 49	GLX
cD;E	2.0′ × 1.5′	13.14B	

Images show a massive elliptical galaxy with a fainter outer envelope. This is one of three dominant galaxies in the Abell 2197 galaxy cluster, the others being NGC6160 and NGC6146. Several faint companions are in the field, including NGC6174, which is due N. A bright field star is 7.0′ SE. Visually, this galaxy and NGC6175 can be seen together in a medium-magnification field. NGC6173 is the brighter and larger of the two galaxies. It is a well defined oval galaxy with even surface brightness, elongated SE–NW. Radial velocity: 8944 km/s. Distance: 399 mly. Diameter: 232 000 ly.

NGC6175

H641[3]	16:29.9	+40 39	GLX
1.3′ × 0.7′	14.82B		

This galaxy is unclassified but photographs show a highly inclined spiral with a small, bright core. It is a member of the Abell 2197 galaxy cluster. Its knotty spiral arms are accompanied by dusty patches and a prominent dust lane borders the outer disc to the S. A companion galaxy, magnitude 14.8, is immediately SSE of the core and may be tidally disrupted by the larger galaxy. It has a bright core and the halo is asymmetrically extended towards the S. Several companion galaxies are nearby, especially to the E. Visually, NGC6175 is a faint blur, elongated E–W, just N of a line joining two faint field stars. The interacting companion galaxy is not visible in a moderate-aperture telescope. Radial velocity: 9145 km/s. Distance: 408 mly. Diameter: 152 000 ly.

NGC6177

H890[3]	16:30.6	+35 04	GLX
(R′)SB(s)b	1.7′ × 1.2′	14.74B	

Photographs show a face-on, barred spiral galaxy with a bright core, a slightly fainter, dusty bar and two dusty spiral arms. Telescopically, the surface brightness is fairly high and the galaxy is visible as a hazy extension of a magnitude 12 field star. It is a greatly elongated oval haze which lacks a brighter core and is oriented NNE–SSW. Radial velocity: 9458 km/s. Distance: 423 mly. Diameter: 209 000 ly.

NGC6181

H753[2]	16:32.3	+19 50	GLX
SA(rs)c	2.7′ × 1.2′	12.23B	

Photographs show a bright, face-on spiral with a dense, large, boxy core and three thin, knotty spiral arms. The galaxy is quite bright visually, appearing as a bright, oval, grainy-textured disc which is broadly brighter to the middle. A faint, curved plume, one of the spiral arms, extends towards the N along the eastern flank, giving the object the appearance of a fat comma. Radial velocity: 2493 km/s. Distance: 112 mly. Diameter: 88 000 ly.

NGC6186

H730[3]	16:34.4	+21 32	GLX
(R′)SB(s)a	1.7′ × 1.4′	14.09B	

In photographs this barred spiral galaxy features a large, irregularly round core at the centre of a prominent, curved bar. Two faint spiral arms emerge from the bar; they are smooth textured but a large, irregular brighter patch is in the arm immediately S of the core. Visually, the galaxy is moderately faint, an elongated oval disc oriented ENE–WSW with ill defined edges and a bright core. Radial velocity: 3061 km/s. Distance: 137 mly. Diameter: 68 000 ly.

NGC6195

H893[3]	16:36.5	+39 02	GLX
Sb	1.5′ × 0.9′	13.97B	

Images show an inclined spiral galaxy with a medium-sized, bright core, elongated NE–SW. Two thin spiral arms extend a long distance from the core. To the SW, the spiral arm is detached and a thin band is visible. Visually, this galaxy is a faint, hazy patch of light just N of a line joining two magnitude 13 field stars. It appears almost round and is very ill defined with even surface brightness. Radial velocity: 9182 km/s. Distance: 410 mly. Diameter: 179 000 ly.

NGC6207

H701[2]	16:43.1	+36 50	GLX
SA(s)c	2.5′ × 1.0′	11.85B	11.6V

This inclined spiral galaxy is situated NE of the great globular cluster M13. Photographs reveal a bright, irregular central region with knotty, fragmentary spiral structure emerging into a smooth-textured, fainter outer envelope. This galaxy is fairly bright visually and appears as an extended oval, oriented NNE–SSW, with a bright, grainy disc which is fairly well defined and brighter to an apparently double core, with a faint field star immediately N of the actual core. Radial velocity: 1012 km/s. Distance: 45 mly. Diameter: 33 000 ly.

NGC6229

H50[4]	16:47.0	+47 32	GC
CC4	4.0′ × 4.0′	9.4B	

Photographs show a small but fairly bright, much compressed globular star cluster with a bright core. The cluster is well resolved around the edges and symmetrical in structure. Visually, this is a bright, much compressed cluster with a large, condensed and well defined brighter core. The boundary between the core and the outer envelope is sharply defined and the outer envelope is smooth in texture fading gradually into the sky background. There is no hint of resolution and a bright field star is 6.1′ WNW, while another is located 6.6′ SW. The horizontal branch visual magnitude is 18.0 and the brightest stars are magnitude 15.5. Distance: 103 000 ly.

NGC6239

H727[3]	16:50.1	+42 44	GLX
SB(s)b pec?	3.0′ × 1.0′	12.89B	12.4V

Images show a highly inclined galaxy, probably barred, with a bright, elongated core or bar. This bar displays

NGC6239

and no core is visible. Faint spiral arms emerge from the bar, with suspected dust patches N and S of the bar. A very bright knot, or perhaps a small companion galaxy, is embedded in the spiral arm to the E and the galaxy is elongated E–W. Visually, the galaxy is very dim; it is an ethereal glow which is slightly brighter along its major axis with edges which fade uncertainly into the sky background. Radial velocity: 1084 km/s. Distance: 48 mly. Diameter: 51 000 ly.

NGC6267

H123³	16:58.2	+22 59	GLX
SB(r)bc	1.3′ × 1.0′	13.82B	

Images reveal a small, face-on, barred spiral galaxy with a very small, round core and a thin, fainter bar. The bright spiral arms are attached to the bar at each end; two emerge from the bar in the NNW. The arms have many brighter, knotty condensations and they are straight and make sharp, 90° turns. In moderate apertures, this galaxy is fairly faint, though easily visible at medium magnification. The envelope is very smooth and diffuse with ill defined edges and only a little brighter to the middle. It appears oval in shape and is oriented NNE–SSW. Radial velocity: 3120 km/s. Distance: 139 mly. Diameter: 53 000 ly.

NGC6278

H124³	17: 00.9	+23 01	GLX
S0	1.9′ × 1.0′	13.74B	

Photographs show a slightly inclined lenticular galaxy with a bright,

dusty patches along its length. To the WNW the bar breaks up into three condensations, which are attached to a faint spiral arm that curves back to the ESE. A short spiral arm is visible just S of the core and appears detached from the bar. High-resolution images show a number of HII regions dotting the central bar and very faint plumes extending from the galaxy to the WNW. Visually, this moderately bright galaxy exhibits a fairly condensed main envelope. This envelope is oriented ESE–WNW with only a very gradual brightening to the middle. The brightness drops off quite suddenly to the sky background. Radial velocity: 1095 km/s. Distance: 49 mly. Diameter: 43 000 ly.

NGC6241

H735³	16:50.1	+45 24	GLX
Sbc:	0.9′ × 0.8′	15.65B	

The DSS image shows a small, face-on spiral galaxy with a large, bright core.

Three distinct spiral arms emerge from the core. The galaxy is slightly elongated E–W. The field has many faint galaxies nearby, including Zwicky's Triplet 8.0′ WNW and UGC10586 6.75′ to the ESE. In a moderate-aperture telescope, the galaxy is small, faint and almost round, appearing at medium magnification as a hazy, out-of-focus star. High power brightens it somewhat and averted vision brings out a brighter core. The edges are fairly well defined. Discovered on 29 April 1788, Herschel's description reads: 'Extremely faint, pretty small, with 300 (magnification) irregular form.' Radial velocity: 9193 km/s. Distance: 410 mly. Diameter: 108 000 ly.

NGC6255

H689³	16:54.8	+36 30	GLX
SBcd:	3.6′ × 1.5′	13.57B	

This is a faint, highly inclined, barred spiral galaxy. The DSS image shows that a long, thin, brighter bar is at the centre

NGC6255

NGC6267

©STScI/AURA/CALTECH/ROE

the galaxy in his Class IV; discovered on 11 June 1788, Herschel described the object as: 'A very small, faint star involved in extremely faint nebulosity.' Radial velocity: 8523 km/s. Distance: 381 mly. Diameter: 244 000 ly.

NGC6372

H137[3]	17:27.5	+26 28	GLX
Sb? pec	1.7′ × 1.1′	13.67B	

Images reveal a slightly inclined spiral galaxy with a brighter, slightly elongated core. There are two principal spiral arms: the arm to the NE has several condensations, including a bright one which may be a foreground star. The galaxy NGC6371 is located 3.0′ NW. Visually, the galaxy is a dim oval smudge elongated E–W, slightly brighter to the middle and poorly defined along the edges. NGC6371 to the NW is visible only intermittently as a roundish, dim patch of light. Radial velocity: 4913 km/s. Distance: 219 mly. Diameter: 108 000 ly.

NGC6389

H901[2]	17:32.7	+16 24	GLX	
Sbc	3.3′ × 1.9′	12.84B	12.4V	

Photographically, this slightly inclined spiral galaxy has a small, round, brighter core. Grainy spiral arms with dust patches and dust lanes are evident and a very faint and narrow outer spiral arm emerges from the SE, curving counterclockwise to the N. At the eyepiece, this galaxy is best at medium magnification: it is a faint, moderately large galaxy, oval in shape and elongated ESE–WNW. Faint condensations are visible near the core; these are faint field stars involved in the envelope to the SW. The edges fade gradually into the sky background and the envelope is broadly concentrated to the centre. Radial velocity: 3262 km/s. Distance: 146 mly. Diameter: 140 000 ly.

NGC6500

H957[3]	17:56.0	+18 20	GLX
SAab:	2.2′ × 1.6′	13.02B	

In photographs, this object forms a pair with NGC6501, located 2.25′ NNE. The

---

elongated core embedded in a fainter outer envelope which is oriented ESE–WNW. The galaxy NGC6277 is located 2.25′ NW. Visually, this is a moderately bright galaxy which is very much brighter to a sharp, stellar core. The envelope is round and fairly well defined. Radial velocity: 2932 km/s. Distance: 131 mly. Diameter: 72 000 ly.

NGC6283

H728[3]	16:59.4	+49 56	GLX
S	1.2′ × 1.0′	13.88B	

This is a small, possibly dwarfish, face-on spiral galaxy with a large, bright core. Photographs show knotty, ill defined spiral arms emerging from the core; the whole is embedded in a fairly bright envelope and the galaxy is slightly elongated E–W. At high magnification, it is moderately bright, almost round with a grainy envelope brightening to a small, round core. The edges are slightly ragged and somewhat

poorly defined. The galaxy is preceded by an easy double star situated 3.5′ to the WSW. Radial velocity: 1276 km/s. Distance: 57 mly. Diameter: 20 000 ly.

NGC6301

H57[4]	17:08.6	+42 20	GLX
Scd:	2.2′ × 1.0′	14.31B	

Photographs show a slightly inclined spiral galaxy with a very small, brighter core. Faint, fairly smooth spiral arms emerge from the core and the galaxy is elongated ESE–WNW. A field star is embedded in the outer spiral arms 0.25′ SW of the core. This field star hinders visual observation, making the galaxy very difficult to see. The star forms a triangle with two brighter stars to the W. The galaxy is suspected at medium magnification; higher magnification does not improve visibility much. The disc is a small, formless glow best seen with averted vision. The presence of the star in the envelope prompted Herschel to place

NGC6500/6501

outer halo. Visually, the galaxy is moderately bright, fairly round and embedded in a hazy outer envelope. It brightens suddenly to the core, which appears nonstellar. The edge-on galaxy NGC6549 is 3.8′ to the SW but is probably not physically related. Radial velocity: 2338 km/s. Distance: 105 mly. Diameter: 87 000 ly.

NGC6548

galaxies are very likely physically related and the minimum core-to-core separation would be about 93 000 ly at the presumed distance. NGC6500 is a slightly inclined spiral galaxy with a round, brighter core; the inner envelope is bright with thin spiral arms emerging, expanding to broad, very faint, spiral arms particularly to the SE, and the galaxy is elongated NNE–SSW. Visually, though, the galaxy is faint, located 5.8′ W of a magnitude 6 field star. A small, bright core is embedded in a round, well defined envelope; the outer envelope is a little too faint to see visually. This is the brightest of a small group of six galaxies. Radial velocity: 3162 km/s. Distance: 141 mly. Diameter: 90 000 ly.

NGC6501

H958[3]	17:56.1	+18 22	GLX
SA0+:	1.8′ × 1.6′	13.11B	

Photographically, this lenticular galaxy has a large, bright core and a broad, fainter envelope, elongated NNE–SSW. The brightness of the outer envelope is a little uneven and dust may be involved. At the eyepiece, this galaxy is a little fainter and smaller than its companion NGC6500; it is a small, round, well defined galaxy with a brighter, almost stellar core. Radial velocity: 3227 km/s. Distance: 144 mly. Diameter: 75 000 ly.

NGC6548

H555[3]	18:06.0	+18 35	GLX
SB0	2.8′ × 2.5′	12.74B	

This galaxy was originally identified as NGC6550 by J. L. E. Dreyer. Images show a face-on, theta-shaped, barred galaxy with a large, bright core and a narrow, bright bar, elongated SSE–NNW. A very faint, broad ring is attached to the bar. When discovered by Herschel on 22 June 1786, he thought it resolvable; there are several extremely faint stars involved with the

NGC6555

H902[2]	18:07.8	+17 36	GLX
SAB(rs)c	2.0′ × 1.5′	12.99B	

In photographs this spiral galaxy has a very small core which is only slightly brighter than the envelope of the galaxy. Several knotty, dusty spiral arms emerge from the core, broadening as they expand outward. Visually, it is best at medium magnification; it is a fairly large, though faint galaxy, irregularly round and only slightly brighter to the core. The extremities fade very gradually into the sky background and the galaxy is very slightly elongated SE–NW. Radial velocity: 2388 km/s. Distance: 107 mly. Diameter: 62 000 ly.

Hydra

Of the 52 Herschel objects located in this southern constellation, only four are not galaxies: the open cluster NGC2548, the planetary nebulae NGC2610 and NGC3242 and the remote globular cluster NGC5694. Herschel recorded all the objects here between the years 1785 and 1793. NGC2548 was first seen by Caroline Herschel in 1783, though it was discovered by Charles Messier many years before. This meandering constellation, the longest in the sky, slopes gradually to the southeast, so the Herschel objects in this region tend to be at more southerly declinations the further E one goes. Indeed, the most southerly Herschel object of all, the galaxy NGC3621, is located here. From Herschel's observing location the galaxy would have culminated less than 6° above the southern horizon. While most of the galaxies are not particularly well known, many of them are fairly bright and show well in medium- and large-aperture telescopes, especially for observers at southern locations.

NGC2548

H22[6]	8:13.8	−5 48	OC
I3r	30.0′ × 30.0′	6.1B	5.8V

This cluster was originally one of Messier's missing objects; Dr T. F. Morris identified it in July 1959 as the object that was listed as No. 48 in Messier's famous catalogue. Herschel first observed the cluster on 1 February 1786 but his note published in the *Philosophical Transactions* (Herschel, 1789) indicates that his sister Caroline first observed the cluster on 8 March 1783. Photographs show the cluster situated in a rich star field and visually this is a large, bright cluster well seen at low magnification with at least 70 stars visible, predominantly of magnitude 8–12. The most obvious feature is a bright, crescent-shaped subgroup of six stars near the centre, with a chain of about 20 stars curving first E, then SE from this group. A total of 80 stars are probable members; the spectral type is A0 and the age of the cluster is about 300 million years. The radial velocity is 7 km/s in approach; the distance is about 2050 ly.

NGC2555

H256[3]	8:18.0	+0 44	GLX
SB(rs)ab	1.9′ × 1.4′	13.33B	

Images show a face-on barred spiral galaxy with a large, bright, round core and a short, stubby bar. The bar is attached to two narrow, bright spiral arms which are embedded in a fainter envelope. The galaxy is slightly elongated ESE–WNW and field stars are involved in the outer envelope to the NW. A bright field star is 1.1′ to the SE, while another is 1.75′ NE. The galaxy is fairly large and moderately bright visually, appearing almost round and broadly brighter to the middle. The edges are poorly defined though the inner envelope is a little more distinct. Radial velocity: 4261 km/s. Distance: 190 mly. Diameter: 105 000 ly.

NGC2610

H35[4]	8:33.4	−16 09	PN
4+2	1.0′ × 0.9′	12.7V	

Photographs reveal an annular planetary nebula with a fairly prominent central star. The bright inner ring is circular while faint extensions are visible to the SE, SW and NW. Telescopically, the magnitude 15.9 central star is not visible but the nebula is a moderately bright object, situated SE of a bright triangle of field stars. It is a fairly well defined, hazy, roundish patch of light, with a faint field star on the edge of the nebula to the NNE. Herschel recorded this object on 31 December 1785 and described it as: 'A small star with a brush south preceding. Faint and small it resembles fig. 7 Phil. Trans. Vol. LXXIV. Tab. 17. (Plate VII.)' This is a reference to a rough sketch that Herschel made of NGC2261, Hubble's Variable Nebula, and its associated star R Monocerotis. Though visually somewhat similar, NGC2610 and NGC2261 are two entirely different kinds of nebula. Distance: 5500 ly.

NGC2618

H257[3]	8:36.2	+0 42	GLX
(R')SA(rs)ab	2.7′ × 2.3′	13.96B	

This is a very slightly-inclined spiral galaxy which images show has a bright core surrounded by a faint, grainy envelope with spiral arms showing faintly near the edges. The faint edge-on galaxy UGC4493 is 5.3′ SSE. The galaxy is faint and diffuse visually, a round unconcentrated patch of light which is fairly even in surface brightness. Radial velocity: 3878 km/s. Distance: 173 mly. Diameter: 136 000 ly.

NGC2695

H280[2]	8:54.5	−3 05	GLX
SAB(s)0°?	1.7′ × 1.4′	12.84B	

Photographs show a lenticular galaxy with a large, bright core and a fainter outer envelope, elongated N–S. Three field stars are involved in the outer envelope to the E and one field star is at the edge of the envelope to the W. Telescopically, the galaxy is a moderately bright, fairly high-surface-brightness object with a faint, stellar core embedded in a grainy oval envelope. The edges are poorly defined and there is very faint evidence of a hazy outer envelope. Also in the field to the NNE is NGC2697, a smaller and slightly

fainter, slightly oval patch of light. Radial velocity: 1671 km/s. Distance: 75 mly. Diameter: 37 000 ly.

NGC2708

H281[2]	8:56.1	−3 22	GLX
SAB(s)b pec?	3.0′ × 1.2′	12.81B	

Images show a highly inclined spiral galaxy with a large, bright, extended inner region and a fainter, extended outer envelope. A thick, dark dust lane borders the core to the NE and is slightly above the rotational plane of the galaxy. The galaxy is elongated NNE–SSW and the very faint, disturbed outer envelope increases the apparent size to about 4.3′ × 2.5′. Field stars are 0.9′ NNE and 1.1′ NW. It is part of a loose group of six NGC galaxies: the nearest is NGC2709, which is 7.2′ almost due N. Visually, this is a moderately bright, but somewhat diffuse patch of light, very much extended with a fairly even-surface-brightness envelope which is moderately well defined. A stellar core is at the centre

and a faint field star borders the tip of the envelope in the NNE. In a medium-magnification field the companion galaxies NGC2698 and NGC2699, both moderately bright, are easily visible. Radial velocity: 1842 km/s. Distance: 82 mly. Diameter: 103 000 ly.

NGC2718

H557[2]	8:58.9	+6 18	GLX
(R′)SAB(s)ab	2.1′ × 2.1′	13.48B	

The DSS image shows a face-on barred spiral galaxy with a small, round, bright core. The bar is long and the spiral arms are initially bright, curving off the bar, but fade considerably thereafter to a broad and faint spiral pattern. The galaxy is moderately bright visually; it is a quite diffuse, round disc of even surface brightness, though it is somewhat grainy textured and fairly ill defined at the edges. Radial velocity: 3707 km/s. Distance: 166 mly. Diameter: 101 000 ly.

NGC2721

H529[2]	8:58.9	−4 46	GLX
SB(rs)bc pec	2.1′ × 1.5′	13.13B	

This is a slightly inclined spiral galaxy, probably barred. Photographs show a bright, elongated core and thin, bright, narrow spiral arms wound tightly around the core. The faint, outer spiral arm is very long and grainy and winds around the galaxy in a clockwise direction, while the inner arms wind counterclockwise. The galaxy is elongated SSE–NNW and several very faint field stars surround it. In a moderate-aperture telescope this object is somewhat faint; it is a roundish patch of light, brighter to the middle, irregular in shape and somewhat ill defined at the edges. Radial velocity: 3542 km/s. Distance: 158 mly. Diameter: 97 000 ly.

NGC2722

H264[3]	8:58.8	−3 43	GLX
SA(rs)bc pec:	2.0′ × 1.1′	13.47B	

Photographs show an inclined spiral galaxy with a large, bright core and very faint, grainy spiral arms. A detached condensation is 1.25′ E and a field star is 2.0′ WNW. Telescopically, only the

NGC2708

©STScI/AURA/CALTECH/OE

NGC2763

©STScI/AURA/CALTECH/ROE

region in a bright ring probably created by a pair of tightly wound spiral arms. The outer spiral pattern is much fainter, broad and smooth textured. Telescopically, only the core and inner ring are visible as a small but well defined object which is quite bright, oval and oriented E–W with a bright, star-like core. Radial velocity: 1859 km/s. Distance: 83 mly. Diameter: 96 000 ly.

NGC2784

H59[1]	9:12.3	−24 10	GLX
SA(s)0°:	5.5′ × 2.2′	11.24B	10.2V

This is a prototypical lenticular galaxy; the DSS image shows the faint, outer envelope particularly well. This envelope has the third brightness gradient in the galaxy: the DSS image shows the central region as a large, elongated core, but high-resolution images show an even brighter, round core embedded in the central region. Telescopically, the galaxy is situated within a bright, four-star, diamond-shaped asterism at medium magnification. It is a very bright and large galaxy; the main envelope is oval and fairly diffuse, while the sizable core is bright and elongated along the major axis of the galaxy, which is oriented ENE–WSW. Radial velocity: 479 km/s. Distance: 21 mly. Diameter: 35 000 ly.

NGC2811

H505[2]	9:16.2	−16 19	GLX
SB(rs)a	2.4′ × 0.8′	12.24B	

The DSS shows a bright, highly inclined galaxy with a very large, elongated core and a fainter outer envelope. High-resolution photographs show a smooth spiral pattern with evidence of dust lanes but little apparent star formation. Visually, this is a quite bright and well defined object, appearing as an elongated sliver of light, oriented NNE–SSW. It has high surface brightness, a tiny, brighter core and extensions which taper to blunt points. Radial velocity: 2171 km/s. Distance: 97 mly. Diameter: 68 000 ly.

bright central region is seen; it is a fairly hazy object, somewhat faint but visible as a slightly oval disc, elongated ENE–WSW, a little brighter to the middle and fairly well defined at the edges. Radial velocity: 2593 km/s. Distance: 116 mly. Diameter: 67 000 ly.

NGC2763

H275[3]	9:06.8	−15 30	GLX
SB(r)cd pec	2.3′ × 1.9′	12.65B	

This face-on spiral galaxy has a brighter core and narrow, bright spiral arms near the core. Photographs show bright knots in the spiral arm E of the core and the spiral arms broaden and fade further from the core, particularly to the W, giving the galaxy an asymmetrical appearance. The arms are grainy and there is evidence of dust, particularly a broad dust lane to the SW. Telescopically, the galaxy is fairly bright though somewhat diffuse; it is a slightly oval patch of light, oriented N–S with a slightly elongated central region and a

hazy, ill defined envelope. A field star is 1.0′ N of the core. Radial velocity: 1696 km/s. Distance: 76 mly. Diameter: 51 000 ly.

NGC2765

H520[2]	9:07.6	+3 23	GLX
S0	2.0′ × 1.1′	13.23B	

In photographs this is an almost edge-on lenticular galaxy with a brighter core and bright extensions embedded in a fainter outer envelope. The visual appearance is quite similar; it is a faint, thin galaxy, elongated almost due E–W with an intermittently visible, small and well defined bright core. Radial velocity: 3617 km/s. Distance: 161 mly. Diameter: 94 000 ly.

NGC2781

H66[1]	9:11.5	−14 49	GLX
SAB(r)0+	4.0′ × 1.8′	12.51B	

Images reveal a highly inclined spiral galaxy with a large and bright central

NGC2815

H242[3]	9:16.3	−23 38	GLX
(R')SB(r)b	3.4' × 1.2'	12.77B	11.8V

Images show a large, highly inclined, barred spiral galaxy with a bright core and a greatly foreshortened bar. Two spiral arms emerge from the bar: the arm emerging from the E of the bar is particularly narrow and bright. The arms overlap after about half a revolution. Telescopically, the galaxy is moderately bright; it is an extended oval of light oriented NNE–SSW. The galaxy is broadly brighter to the middle and its edges are diffuse. Radial velocity: 2329 km/s. Distance: 104 mly. Diameter: 103 000 ly.

NGC2848

H488[3]	9:20.2	−16 32	GLX
SAB(s)c:	2.6' × 1.8'	12.45B	

The DSS image shows an inclined, spiral galaxy, probably barred, with a brighter, elongated core and knotty spiral arms emerging from the core. The galaxy is

elongated NE–SW and is bordered by three faint field stars to the NNE. The lenticular galaxy NGC2851 is located 5.3' ENE. NGC2848 is a fairly large and moderately bright object visually, appearing as a slightly oval patch of light, oriented N–S, which is slowly brighter to the middle. The envelope is a little mottled and hazy at the edges. NGC2851 can be seen to the ENE; a field star separates the two galaxies with NGC2851 appearing as a smaller, oval patch oriented N–S with a brighter core. Radial velocity: 1838 km/s. Distance: 82 mly. Diameter: 62 000 ly.

NGC2855

H132[1]	9:21.5	−11 55	GLX
(R)SA(rs)0/a	3.6' × 2.9'	12.55B	

Photographs show a bright, slightly inclined spiral galaxy with a large, bright core and tightly wound, grainy spiral arms defined by very thin dust lanes. The galaxy is elongated SE–NW and a bright, uneven double star is 4.1'

NNE. The galaxy is bright visually, even in smaller apertures, and is well defined with a brighter core but no stellar nucleus; the faint outer structure is not visible. Radial velocity: 1710 km/s. Distance: 76 mly. Diameter: 80 000 ly.

NGC2863

H520[3]	9:23.6	−10 24	GLX
Sm	0.8' × 0.8'	13.0B	

The DSS photograph shows a small, bright, face-on, barred spiral galaxy with a bright core and a bright, curved bar expanding into a dense, knotty and fragmented spiral structure. A faint field star borders the galaxy in the NNW, another is 0.6' to the S and the faint companion galaxy NGC2868 is 2.4' W. Telescopically, the galaxy is moderately faint and only the bar is well seen as a N–S elongated thin ray of light which is well defined at the edges with a small, sharply brighter core. Herschel described the galaxy as extended. Radial velocity: 1604 km/s. Distance: 72 mly. Diameter: 17 000 ly.

NGC2889

H555[2]	9:27.2	−11 38	GLX
SAB(rs)c	2.2' × 1.9'	12.50B	

This is a face-on, barred spiral galaxy. Deep photographs reveal that the galaxy has one bright, thick, knotty spiral arm. A second arm is fainter and branches at its root into three broadly parallel arms. The galaxy is well seen visually and is moderately bright with a fairly extensive outer halo. Oval in form, it is very slightly elongated N–S. Medium magnification reveals a galaxy broadly concentrated to the core. High magnification shows a bright, nonstellar nucleus. The outer halo fades uncertainly into the sky background. Bright field stars are 1.5' SSE and 3.1' SSW. Radial velocity: 3151 km/s. Distance: 141 mly. Diameter: 90 000 ly.

NGC2902

H276[3]	9:30.9	−14 44	GLX
SA(s)0°:	1.7' × 1.4'	13.24B	

This is a prototypical lenticular galaxy with a large, bright core. The DSS image

NGC2848

NGC2889

shows a faint envelope around the core but high-resolution photographs show an even fainter outer envelope. Small, compact galaxies are located 2.75′ ENE and 3.9′ SW. Visually, the galaxy is moderately faint and the envelope is somewhat diffuse, brightening to the middle. It is moderately well defined, there are hints of a star-like nucleus and the galaxy is slightly elongated N–S. Radial velocity: 1802 km/s. Distance: 81 mly. Diameter: 40 000 ly.

NGC2907

H506[2]	9:31.6	−16 44	GLX
SA(s)a?	2.4′ × 1.7′	12.70B	

The DSS image shows an almost edge-on spiral galaxy with a very large, bright, elongated core and bright extensions in a fainter outer envelope. This is a 'Sombrero' galaxy with a prominent dust lane along the major axis but high-resolution photographs show three more thin dust lanes stacked perpendicular to the main dust lane along the NNE flank.

Visually, the galaxy is a misty, oval patch of light, fairly even in surface brightness and oriented ESE–WNW. Radial velocity: 1893 km/s. Distance: 84 mly. Diameter: 59 000 ly.

NGC2921

H597[3]	9:34.7	−20 57	GLX
SAB(r)b	3.0′ × 1.1′	13.03B	

Photographs show a highly inclined spiral galaxy with a bright core and a

NGC2907

slightly fainter inner spiral arm in a very faint, grainy outer envelope. Visually, the galaxy is paired with the smaller and fainter NGC2920, 6.0′ to the NW. NGC2921 is a moderately dim object, an oval envelope with a brighter core. The envelope is diffuse and elongated E–W. A threshold star is suspected near the tip of the western envelope: photographs show that this star is 0.9′ WNW of the core. Radial velocity: 2756 km/s. Distance: 123 mly. Diameter: 107 000 ly.

NGC2935

H556[2]	9:36.7	−21 08	GLX
(R′2)SAB(s)b	4.8′ × 3.5′	12.26B	10.0V

Photographs show a barred spiral galaxy with a small core and a slightly oval bar, laced with thin dust lanes. The two principal spiral arms are narrow and relaxed but overlap after half a revolution to form a pseudo-ring. They are studded with star-forming regions. Visually, only the core and central bar are visible. The galaxy is moderately bright and features a large, bright core region oriented SE–NW, oval in shape, with a star immediately to the SE. This central region appears surrounded by a very faint, diffuse envelope, which is the vestige of the outer spiral structure. Radial velocity: 2066 km/s. Distance: 92 mly. Diameter: 129 000 ly.

NGC2983

H289[3]	9:43.7	−20 29	GLX
(RL)SB(s)0+	2.5′ × 1.5′	12.74B	11.7V

The DSS photograph shows an inclined, barred galaxy with a very bright core. A bright, slightly curved bar is embedded in a fainter outer envelope. Blunt spiral arcs are at the end of the bar, transitioning into the smooth-textured disc which is elongated E/W. A small, compact galaxy is 5.75′ SSW, while a bright field star is 7.1′ NNW. Visually, the galaxy is moderately bright but fairly small and has a round, diffuse envelope which is very poorly defined at the edges. The core, however, is round, sizable and bright. Radial velocity: 1830 km/s. Distance: 82 mly. Diameter: 60 000 ly.

NGC2983

NGC2986

H311[2]	9:44.3	−21 17		GLX
E2	3.2′ × 2.8′	11.69B		

Photographs show a typical E2 elliptical galaxy with a large, bright core and an extensive outer envelope, elongated NNE–SSW. Telescopically, this is quite a bright galaxy with a bright and distinct envelope, almost round and fairly well defined. The envelope is mottled and increases in brightness to a bright central region. The faint field galaxy ESO566-4 is 2.25′ to the WSW and appears as a very small and faint, slightly oval patch of light. Radial velocity: 2098 km/s. Distance: 94 mly. Diameter: 87 000 ly.

NGC2992

H277[3]	9:45.7	−14 20		GLX
Sa pec	4.0′ × 1.2′	13.06B	12.2V	

This galaxy, together with NGC2993 which lies 3.1′ to the SSE, has also been catalogued as Arp 245. They were discovered on 8 February 1785, when Herschel described them as: 'Two 3′ or 4′ distant. The most north very faint, small. The south very faint, very small. Both stellar'. In photographs, NGC2992 is a bright, elongated lens galaxy with a peculiar morphology. The core is bright, elongated and bisected by a thin dust lane. There is a short, faint extension to the S, and a much larger one to the N which brightens at its extremity. A faint, broad bridge of material connects the two galaxies along a SE–NW axis. On deep photographs, this broad bridge continues along the length of NGC2992's eastern flank to the broad, northern plume. According to Sandage and Bedke (1994), there is evidence of recent star formation near the core of the galaxy. The radial velocities of both galaxies are similar and if they are a pair, the minimum projected separation between NGC2992 and NGC2993 would be about 86 000 ly. Visually, NGC2992 is larger than its companion, elongated NNE–SSW and bright along its major axis. The galaxy is well defined at the edges. Large telescopes may detect a very faint edge-on galaxy in the field, 5.0′ SSW. Radial velocity: 2121 km/s. Distance: 95 mly. Diameter: 110 000 ly.

NGC2993

H278[3]	9:45.8	−14 22		GLX
Sa pec	3.3′ × 1.8′	13.11B	12.6V	

Paired with NGC2992, this peculiar galaxy has a bright, irregularly-shaped core in a faint outer envelope. A plume of material emerges from the outer envelope in the N and extends to the ESE where it curves, broadens and fades. Visually, NGC2993 is star-like at low magnification and is suspected of being elongated at medium magnification. The core is bright and the edges are moderately well defined. Radial velocity: 2240 km/s. Distance: 100 mly. Diameter: 96 000 ly.

NGC3052

H272[3]	9:54.5	−18 38	GLX
SAB(rs)c	2.0′ × 1.3′	12.81B	

This is a slightly inclined, three-branch spiral galaxy with a small, round core. In photographs, the spiral arms are of high surface brightness; the western arm broadens and fades slightly as it spirals outward, the eastern one splits into two distinct arms fairly near its root. Visually, NGC3052 is a fairly faint object, an oval glow, oriented E–W. The brightness increases gradually to the middle but the galaxy lacks a bright core. Radial velocity: 3580 km/s. Distance: 160 mly. Diameter: 93 000 ly.

NGC2992/2993

NGC3072

H273[3]	9:57.6	−19 21	GLX
S0/a? sp	2.0′ × 0.6′	13.83B	

Images show an almost edge-on lenticular galaxy with a slightly brighter core and bright extensions. Visually, the galaxy is quite dim and not easily seen, appearing as a diffuse oval of light oriented ENE–WSW which is a little brighter along its major axis and hazy at the edges. Radial velocity: 3232 km/s. Distance: 144 mly. Diameter: 84 000 ly.

NGC3078

H268[2]	9:58.4	−26 56	GLX
E2-3	2.9′ × 2.3′	12.07B	11.1V

Photographs show a bright elliptical galaxy with an elongated core and a fainter outer envelope, elongated N–S. The galaxy is moderately bright visually; it is an opaque and slightly oval disc, which is a little brighter to the middle and fairly well defined at the edges. Radial velocity: 2283 km/s. Distance: 102 mly. Diameter: 86 000 ly.

NGC3081

H596[3]	9:59.5	−22 50	GLX
(R1)SAB(r)0/a	2.2′ × 1.5′	12.89B	11.8V

The DSS image shows a bright, slightly inclined, theta-shaped, barred galaxy with a very bright core. A faint, broad bar extends from the core to the ring which is much brighter, narrow, well defined and oriented ENE–WSW. Deep photographs show the ring is made up of two tightly wound spiral arms which after half a revolution each fade suddenly and form a dim, ring-like outer spiral structure. This outer structure is elongated SSE–NNW and brings the apparent size of the galaxy to about 3.2′ × 2.5′. Only the inner ring and core are seen telescopically, and appear at medium magnification as an oval, well defined patch of light which is broadly brighter to the middle. Several prominent field stars are situated to the NE and NNW. Radial velocity: 2186 km/s. Distance: 98 mly. Diameter: 91 000 ly.

NGC3081

©STScI/AURA/CALTECH/DOE

NGC3091

H293[2]	10:00.2	−19 38	GLX
E3:	3.0′ × 1.9′	12.13B	10.6V

Photographs show a bright, elongated elliptical galaxy with a bright core and an extensive outer envelope. NGC3091 is the principal member of a small group of six galaxies (Sandage and Bedke (1994) number at least ten, mostly dwarf ellipticals). Visually, this elliptical galaxy is small, though moderately bright; it appears round, well defined and well concentrated to the centre. It is surprisingly easy to pick up at low magnification and precedes two magnitude 9 field stars. The surface brightness is constant across the disc and no stellar nucleus is visible. Large-aperture telescopes may detect one or more fainter field galaxies nearby: one is 1.25′ NW, a second 2.2′ S, a third 4.1′ SE and the fourth, NGC3096, a bright barred lenticular, is 4.8′ ESE. Radial velocity: 3765 km/s. Distance: 168 mly. Diameter: 147 000 ly.

NGC3145

H518[3]	10:10.2	−12 26	GLX
SB(rs)bc	2.8′ × 1.3′	12.40B	

Well seen on the DSS image, this is a bright, highly inclined, barred spiral galaxy with a bright, elongated central region which includes the bar and is oriented E–W. Three thin, bright spiral arms emerge from the bar, with a fourth, subsidiary arm seen along the eastern

NGC3091

©STScI/AURA/CALTECH/DOE

flank. The arms expand into a fainter, mottled outer envelope. A likely physical companion galaxy, NGC3143, is much fainter and situated 8.75′ to the S. Telescopically, NGC3145 is a difficult object, as it is located only 7.5′ SW from Lambda Hydrae which floods a low-magnification field with light. The galaxy appears as a faint, oval glow oriented NNE–SSW, gradually brighter to a bright core. This core appears elongated along the major axis. Radial velocity: 3471 km/s. Distance: 155 mly. Diameter: 126 000 ly.

NGC3242

H27[4]	10:24.8	−18 38	PN
4+3b	1.0′ × 0.75′	8.60B	7.0V

This is the Ghost of Jupiter, a bright planetary nebula which is poorly seen on the DSS image, appearing as an opaque disc. Higher-resolution images show a bright central star surrounded by an elliptical inner ring, embedded in a fainter outer disc. In a small telescope at low magnification, it is a bright, nebulous disc of light, almost circular, with sharply defined edges. Medium magnification reveals that the brightness fades rather slowly at the edges, with the disc appearing opaque and evenly illuminated. The magnitude 12.1 central star is not detected. Moderate apertures bring out the bluish tint well and though no star is visible the inner third of the nebula is much brighter and opaque. The disc is mottled, particularly at the edges. Recorded on 7 February 1785, Herschel described it as: 'Beautiful, brilliant, planetary disc ill defined, but uniformly bright, the light of the colour of Jupiter. 40′ diameter. Second observation near 1′ in diameter by estimation.' Distance: 2600 ly.

NGC3411

H522[3]	10:50.3	−12 51	GLX
E+0/SA0−	1.9′ × 1.9′	12.93B	

Photographs show an elliptical galaxy with a bright central region and a very faint outer envelope. Three small, compact galaxies are in the immediate vicinity and may be detectable in large-aperture telescopes. The first is 0.8′ WSW, the second 2.0′ ESE and the third 3.75′ SE. The galaxy is quite small visually, and at medium magnification it is round and brighter to the middle with a faint stellar core and quite well defined edges. Radial velocity: 4407 km/s. Distance: 197 mly. Diameter: 109,000 ly.

NGC3585

H269[2]	11:13.3	−26 45	GLX
E7/S0	5.9′ × 3.3′	10.82B	

Photographs show this galaxy displays characteristics of both elliptical and lenticular objects, in particular a probable edge-on disc component typical of lenticular galaxies. The galaxy is very bright with an elongated core in an extensive, fainter outer envelope. Visually, the galaxy is fairly bright and is well seen at high magnification as an ESE–WNW oriented oval disc, opaque and brighter to the middle, embedded in a slightly fainter secondary envelope which fades slowly into the sky background. Radial velocity: 1207 km/s. Distance: 54 mly. Diameter: 92 000 ly.

NGC3621

H241[1]	11:18.3	−32 49	GLX
SA(s)d	12.3′ × 7.1′	10.10B	9.6V

High-resolution photographs show a large, slightly inclined spiral galaxy with a very small, brighter core and poorly defined, knotty spiral arms emerging from the centre. The brightest stars resolve at about blue magnitude 20 and the spiral structure is better defined on the E side, where a large number of dust patches are visible. The DSS image shows very faint but extensive outer spiral structure, particularly to the S, where an extensive dust lane is suspected. Visually, the galaxy is large but quite dim and is best seen at medium magnification as an

NGC3621

©STScI/AURA/CALTECH/ROE

oval disc oriented N–S, broadly brighter to the middle with hazy edges. The disc is fairly smooth textured and the galaxy is situated within a triangle of field stars, the brightest two of which are 3.8′ SSE and 2.5′ SSW of the core. Radial velocity: 530 km/s. Distance: 24 mly. Diameter: 84 000 ly.

NGC3885

H828[3]	11:46.8	−27 55	GLX
SAB(r:)0/a:	2.8′ × 1.0′	12.82B	11.9V

Photographs reveal an inclined spiral galaxy with a large, bright, elongated core embedded in a fainter outer envelope. There is some evidence of dust in the bright, central region, particularly a dark patch W of the core and a broad, fainter patch SE of the core. High-resolution images show extensive dust lanes defining the spiral structure in the outer disc. Visually, only the bright inner disc is visible, appearing as a small and fairly faint oval patch of light oriented ESE–WNW, well defined and even in surface brightness. A field star is 1.4′ ESE of the core. Radial velocity: 1770 km/s. Distance: 79 mly. Diameter: 65 000 ly.

NGC3904

H864[2]	11:49.2	−29 17	GLX
E2-3:	3.0′ × 2.0′	11.80B	11.0V

In photographs this slightly inclined elliptical galaxy has a broad, bright core and an extensive faint outer envelope. The outer envelope, elongated NNE–SSW, is quite grainy and may have small patches of dust involved. The galaxy is quite bright visually and is seen as a slightly oval disc, fairly opaque and a little brighter to the middle with slightly hazy edges. Radial velocity: 1393 km/s. Distance: 62 mly. Diameter: 55 000 ly.

NGC3923

H259[1]	11:51.0	−28 48	GLX
E4-5	5.9′ × 3.9′	10.77B	9.8V

Photographs show a slightly inclined elliptical galaxy with a broad, brighter core and an extensive faint outer envelope. The outer envelope is quite grainy and there are small patches of dust and at least one very narrow dust lane involved. A small, brighter condensation appears on the edge of the bright core almost due W. The galaxy may form a widely separated physical pair with NGC3904. Telescopically, NGC3923 is a large and fairly bright galaxy, well seen at high magnification as a grainy, oval disc, well defined at the edges and broadly brighter along its major axis, which is oriented NE–SW. Radial velocity: 1558 km/s. Distance: 70 mly. Diameter: 119 000 ly.

NGC4087

H754[3]	12:05.6	−26 33	GLX
SA0⁻:	2.0′ × 1.7′	13.15B	

Images show a lenticular galaxy with a brighter core and a faint outer envelope, slightly elongated NE–SW. This is a small and quite faint galaxy visually, brighter to a small core in an irregularly round, somewhat ill defined envelope. Radial velocity: 3162 km/s. Distance: 141 mly. Diameter: 82 000 ly.

NGC4105

H865[2]	12:06.7	−29 46	GLX
E3	2.7′ × 2.0′	11.57B	10.9V

This galaxy forms a probable interacting pair with NGC4106 which is situated 1.2′ to the ESE. At the presumed distance, the minimum core-to-core separation would be about 28 000 ly. Images show an elliptical galaxy with a large, bright core and an extensive outer envelope which exhibits little if any evidence of the ongoing encounter. A small, bright condensation or field star is in the outer envelope to the WNW and the galaxy is slightly elongated SSE–NNW. The two galaxies form a pair in a high-magnification field. NGC4105 is larger, but both galaxies are evenly bright. NGC4105 appears round and broadly brighter to the middle with a small, bright core. A field star is 2.5′ S. Radial velocity: 1760 km/s. Distance: 79 mly. Diameter: 62 000 ly.

NGC4106

H866[2]	12:06.8	−29 46	GLX
SB(s)0⁺	2.3′ × 1.3′	12.28B	11.3V

Photographs show this barred lenticular galaxy has a bright core, slightly elongated E–W, and a large outer envelope. Evidence of interaction with NGC4105 is compelling, with a broad plume of material extending to the E, while a southern plume appears to be drawn back towards the larger galaxy in the WNW. Visually, the galaxy appears a little more concentrated than its larger neighbour; it is a round, opaque patch, fairly well defined and brighter to a

NGC4105/4106

©STScI/AURA/CALTECHROE

star-like core. Radial velocity: 1974 km/s. Distance: 88 mly. Diameter: 59 000 ly.

NGC4970

H765[3]	13:07.5	−24 02	GLX
S0°:	1.8′ × 1.0′	13.33B	

This is an inclined, lenticular galaxy. Photographs show a brighter, bar-shaped core and a very faint outer envelope. Visually, the galaxy appears moderately bright, well defined and elongated SE–NW, following a magnitude 10 field star by 2.5′. Radial velocity: 3120 km/s. Distance: 139 mly. Diameter: 73 000 ly.

NGC4993

H766[3]	13:09.7	−23 24	GLX
(R′)SAB(rs)0⁻:	1.4′ × 1.3′	13.45B	

This lenticular galaxy presents similar appearances visually and photographically. In a moderate-aperture telescope, the galaxy is a small, moderately faint, roundish patch of light, poorly defined and brighter to the middle, which appears slightly elongated N–S. Field stars are 1.5′ W and 2.75′ SW. Radial velocity: 2784 km/s. Distance: 124 mly. Diameter: 51 000 ly.

NGC5061

H138[1]	13:18.1	−26 50	GLX	
E0/SA0⁻	4.0′ × 3.3′	11.21B	10.4V	

The classification of this galaxy is somewhat uncertain: it is described variously as elliptical or lenticular. Photographs show an elongated, bright core and a faint outer envelope which is quite extensive. The galaxy is oriented ESE–WNW and there are faint field stars in the outer envelope to the NE and SW. A bright field star is 2.2′ E. Along with NGC5101, this is one of the principal members of a loose galaxy group comprising about ten members. Visually, this is quite a bright galaxy with a very bright, nonstellar core. The envelope is somewhat diffuse but fairly well defined, though it appears mottled. The faint field star to the NE is detectable, and is seen to border the

halo. Radial velocity: 1930 km/s. Distance: 86 mly. Diameter: 100 000 ly.

NGC5078

H566[2]	13:19.8	−27 24	GLX
SA(s)a: sp	4.8′ × 2.6′	12.04B	

This galaxy is probably a member of the NGC5061 galaxy group. The DSS image shows an almost edge-on spiral galaxy of the 'Sombrero' type, with a large, bright, elongated core and a thick, dark dust lane bisecting the core along the major axis. The galaxy is tilted slightly so that the ENE flank is brighter than the WSW side. The galaxy is asymmetrical with both the dust lane and outer envelope more extensive to the NNW than to the SSE. IC879 is a probable companion galaxy 2.3′ to the WSW; at the presumed distance the minimum separation between the two galaxies would be 61 000 ly. Telescopically, NGC5078 is a bright object; it is an elongated sliver oriented NNW–SSE, well defined and bright along its major axis and very slightly brighter to the middle. The ends taper to points and IC879 is faintly visible immediately to the WSW as an elongated, hazy patch of light. Radial velocity: 2033 km/s. Distance: 91 mly. Diameter: 127 000 ly.

NGC5085

H780[2]	13:20.3	−24 26	GLX	
SA(s)c	3.5′ × 2.8′	13.47B	11.1V	

The DSS image reveals an almost face-on spiral galaxy with a small, brighter core and grainy spiral arms with several brighter condensations involved. A bright field star is 4.1′ S. There are two principal arms; the rest of the spiral pattern is fragmentary and defined by dust patches and HII regions. Visually, the galaxy is a rewarding object, appearing as a large oval, oriented NE–SW, with a diffuse but well defined envelope. It brightens to a sizable, well defined core and the envelope is somewhat mottled with suspected dust lanes surrounding the core. Radial velocity: 1828 km/s. Distance: 81 mly. Diameter: 83 000 ly.

NGC5101

H567[2]	13:21.8	−27 26	GLX	
(R′1R′2)SB(rl)0/a	5.4′ × 5.4′	11.58B	10.5V	

This is quite a remarkable galaxy, well seen in the DSS image as a bright, theta-shaped, barred galaxy with a bright, slightly elongated core and a tapered bar oriented ESE–WNW. The bar brightens and expands where it is attached to the fainter, elongated pseudo-ring, which is actually made up of two tightly wound inner spiral arms. There are two

NGC5078

NGC5101

NGC5328

broadly brighter to the middle and situated W of a small triangle of field stars. A linear asterism of four stars is seen 5.5′ to the WSW and it is curious that Herschel did not note it as it is moderately bright and appears slightly nebulous at medium magnification. Radial velocity: 4210 km/s. Distance: 188 mly. Diameter: 82 000 ly.

NGC5694

H196[2]	14:39.6	−26 32	GC
CC7	4.3′ × 4.3′	11.6B	

Images reveal a small, compact globular cluster with a grainy, brighter core and a compressed halo of resolved stars. There are bright field stars 1.3′ SW and 2.5′ SSW. Visually, the cluster is quite small and round, with moderately high surface brightness and ragged extremities. It brightens to a sizable, condensed core but there is no evidence of resolution; the brightest stars are magnitude 15.5 and the spectral type is F3. Herschel, however, thought this object was resolvable. Radial velocity is 184 km/s in approach and the cluster is one of the more remote in the *New General Catalogue* at a distance of 107 000 ly.

extremely faint, broad and large outer spiral arms expanding outward from the inner structure, which appears to form an outer pseudo-ring. Although attached to the inner structure, there appears to be no specific point where the outer arms are attached. The outer arms appear very smooth textured in the DSS image but high-contrast photographs show some clumpiness and evidence of star formation. A bright field star is 3.25′ NNW, a fainter one is 1.75′ W. Not surprisingly, only the bright inner galaxy is visible visually. The galaxy is a quite bright and large oval object with a bright, oval and nonstellar core with dark patches to the N and S. The envelope is a little diffuse but well defined with a faint star bordering the envelope to the west. Radial velocity: 1734 km/s. Distance: 77 mly. Diameter: 121 000 ly.

NGC5328

H923[3]	13:52.9	−28 29	GLX
E1:	2.2′ × 1.5′	12.71B	

Photographs show an elliptical galaxy with a bright, elongated core and a fainter outer envelope. It is the brightest member of a compact group of galaxies, including NGC5330, located 1.75′ NE. Eight galaxies are within a 5.0′ radius and the field is an interesting one for large-aperture telescopes. Visually, NGC5328 is moderately bright and is seen as an E–W elongated oval with a brighter core and fairly well defined edges. The disc is quite grainy and is steadily brighter to the core. None of the field galaxies is seen, probably due to the group's proximity to the horizon from mid-northern latitudes. Radial velocity: 4621 km/s. Distance: 206 mly. Diameter: 132 000 ly.

NGC5592

H924[3]	14:23.9	−28 41	GLX
SB(s)bc?	1.5′ × 0.9′	13.50B	12.7V

Photographs show an inclined, barred spiral galaxy with a bright, narrow spiral arm to the N and a broad, diffuse and ill defined arm S of the core. The almost edge-on galaxy ESO446-59 is 7.0′ to the ENE. Visually, NGC5592 is not very bright and is seen as an irregular, elongated haze oriented E–W,

NGC5694

Lacerta

All 10 of the Herschel objects in this constellation, a mix of star clusters and galaxies, were recorded between the years 1786 and 1790. While none of the objects is particularly noteworthy, all are situated in fairly rich fields of stars, which make for interesting views, especially for the galaxies.

NGC7197

H599[2]	22:02.9	+41 03	GLX
Sa	1.6′ × 0.8′	14.35B	

Photographs show a highly inclined spiral galaxy with a broad, brighter core and a fainter outer envelope. The galaxy is oriented ESE–WNW with a coarse grouping of field stars to the N and a bright field star 4.1′ ESE. In a moderate-aperture telescope, the galaxy is faint but readily visible as an extended ethereal glow in a rich starfield. It is broadly brighter to the middle and the brightness extends along the major axis. The edges fade uncertainly into the sky background and the envelope is very smooth textured. Radial velocity: 4604 km/s. Distance: 206 mly. Diameter: 96 000 ly.

NGC7209

H53[7]	22:05.2	+46 30	OC	
III1m	15.0′ × 15.0′	8.2B	7.7V	

This cluster is a coarse gathering of brighter stars in a busy Milky Way field to the S of the bright red variable star HT Lacertae. Telescopically, 40–50 members are visible in a 15′ × 15′ area; the stars exhibit a narrow brightness range and overall the group shows little concentration to the centre. About 100 stars are members of this cluster. Spectral type: B9. Age: approximately 300 million years. Distance: 2900 ly.

NGC7223

H862[3]	22:10.2	+41 00	GLX
SB(rs)bc	1.7′ × 1.2′	12.58B	

Images show a face-on, barred spiral galaxy with a very small, brighter, elongated core which stretches into a short bar. The spiral arms emerge from the bar and wind tightly around the central region, forming a pseudo-ring, before expanding outward into thin, knotty spiral arms. A faint field star is involved in the spiral arm to the SE. Visually, the galaxy is a very faint, oval glow best seen at medium magnification. The galaxy is situated between two faint field stars oriented almost N–S and the faint field star involved in the envelope is visible as well. The envelope is smooth textured and broadly brighter to the middle and the edges are very poorly defined. Radial

NGC7223
©STScI/AURA/CALTECH/OE

velocity: 4908 km/s. Distance: 219 mly. Diameter: 108 000 ly.

NGC7231

H606[2]	22:12.4	+45 19	GLX
SBa	1.9′ × 0.6′	13.85B	

Photographs show a highly inclined spiral galaxy with a brighter, elongated central region and a faint outer envelope. A dust patch is prominent in the outer envelope to the E of the central region and the galaxy is elongated E–W. At high magnification, the galaxy is fairly faint with a brighter core and a faint, extended envelope. The edges are well defined. Radial velocity: 1304 km/s. Distance: 58 mly. Diameter: 32,000 ly.

NGC7243

H75[8]	22:15.3	+49 53	OC	
II2m	30.0′ × 30.0′	6.5B	6.4V	

Similar photographically and visually, this is a bright, coarse cluster of stars located in a rich Milky Way field. Telescopically, the brighter stars are elongated in an E–W direction and stand out well against the background and there is a spur of stars trending toward the S as well. A conspicuous double star, with both members about magnitude 9, lies near the heart of the cluster. About 40 stars are visible. There is a wide brightness range amongst the cluster members. The cluster exhibits the very slight radial velocity of 3 km/s in recession. Spectral type: B7. Age: 110 million years. Distance: 2700 ly.

NGC7245

H29[6]	22:15.3	+54 20	OC
II2m	5.0′ × 5.0′	9.8B	9.2V

The DSS image shows a small, rich cluster of stars in a very rich Milky Way starfield. The cluster appears triangular in outline and is compressed and richer in the E with a prominent starless zone near the centre. There is a very narrow range of brightness amongst cluster members, and a small, rich open cluster, King 9, is visible 5.0′ to the NNE. Telescopically, the cluster is compressed and rich, dominated by very faint stars with a bright field star following to the E. The cluster is elongated E–W and broader on the eastern side, tapering finally to three faint field stars. King 9 is visible as an unresolved haze with a couple of faint stars shining through; the group is elongated N–S. The brightest members of NGC7245 are magnitude 12.75. About 170 stars are members. Distance: 6000 ly.

NGC7245

©STScI/AURA/CALTECH/ROE

NGC7248

H863[3]	22:16.9	+40 30	GLX
SA0⁻:	1.7′ × 0.9′	13.59B	12.4V

Images reveal a lenticular galaxy with a brighter, elongated core and an extensive outer envelope. The galaxy is elongated ESE–WNW and there is a very faint and small galaxy 0.6′ to the NNW. In a moderate-aperture telescope NGC7248 is small and moderately bright and holds magnification well. It is oval in shape, the edges are fairly well defined and there is a small, brighter core. Three attractive double stars bracket the galaxy, one pair to the E, two to the W. Radial velocity: 4608 km/s. Distance: 206 mly. Diameter: 102 000 ly.

NGC7250

H864[3]	22:18.3	+40 35	GLX
Sdm?	1.7′ × 0.8′	13.19B	12.6V

The DSS image shows a possible spiral galaxy with an elongated core, offset towards the W in a fainter outer envelope. The envelope is asymmetrical and a little more extensive to the SE and the galaxy is oriented N–S. A small, elongated companion galaxy appears embedded in the envelope 0.4′ NNW of the core and a field star is 0.9′ to the SSE. Visually, the galaxy is a moderately bright sliver of light that holds magnification well. It is fairly well defined with even surface brightness. Radial velocity: 1389 km/s. Distance: 62 mly. Diameter: 31 000 ly.

NGC7296

H41[7]	22:28.2	+52 17	OC
II2p	3.0′ × 3.0′	9.7V	

Photographs show a small, scattered group of brighter stars in a rich starfield. In a small-aperture telescope this small cluster is only partially resolved at low magnification. A brighter field star precedes the group to the W and at medium magnification about a dozen magnitude 12 or fainter stars follow, forming two chains running parallel in an E–W direction. Twenty stars are accepted as members.

NGC7426

H576[3]	22:56.1	+36 21	GLX
E	1.5′ × 1.1′	13.56B	

This elliptical galaxy has a brighter, elongated core and a faint outer envelope. The galaxy is elongated ENE–WSW and images show it should be classified as an E4 elliptical. A bright, wide double star is 3.75′ to the W. These stars hinder visibility of the galaxy visually: the galaxy is faint and only the core is seen. The galaxy appears almost round, opaque and condensed with fairly well defined extremities and a brighter, star-like core. Radial velocity: 5536 km/s. Distance: 247 mly. Diameter: 108 000 ly.

Leo

Not surprisingly, Herschel returned to this constellation almost every year of his 19-year survey, each time finding something new. In the end, he recorded 158 objects here, almost all of them galaxies. The variety of galaxies is overwhelming. There are magnificent, large, bright and nearby objects, such as NGC2903, NGC 3521 and NGC3628. There are mid-range galaxies involved in small groups, such as the galaxies surrounding NGC3190 or NGC3607. And then there are the many Herschel objects involved in the Abell 1367 galaxy cluster, a small patch of sky that can be explored over many nights by the patient observer in search of ever smaller and fainter prey. Many objects are extremely faint, however, and care is need to observe galaxies such as NGC3080 or NGC3196. One could conceivably speed through Leo in a single observing season, but why? There is so much to explore here that patient observers who take their time are richly rewarded by the wonders to be found within Leo's borders.

NGC2872

H57²/H546²	9:25.7	+11 26	GLX
E2	2.0′ × 1.9′	12.85B	

This elliptical galaxy forms a visual pair with NGC2874, which is situated 1.2′ to the ESE, but there is a considerable spread in their radial velocities. Photographs show the galaxy has a bright, slightly elongated core in a fainter outer envelope, oriented SE–NW. A third, much fainter galaxy, NGC2873, is 2.0′ NE. Telescopically, NGC2872 and NGC2874 form a moderately bright pair, with NGC2872 appearing the larger and brighter of the two; it is an opaque, almost round disc of fairly even surface brightness with well defined edges. Radial velocity: 3079 km/s. Distance: 138 mly. Diameter: 80 000 ly.

NGC2874

H58²/H547²	9:25.8	+11 26	GLX
SB(r)bc	2.9′ × 0.9′	13.37B	

Photographs show a highly inclined, barred spiral galaxy with an elongated core and a very short bar. A possible ring is attached to the bar, but it is more likely a tightly wound spiral arm, which is narrow and brighter in the S and to the NE of the core. The spiral structure is asymmetrical, being broader and fainter

to the SW. The galaxy is elongated NE–SW and may form a pair with NGC2872; the two galaxies are also catalogued as Arp 307. The minimum actual separation between the two galaxies would be 48 000 ly at the presumed distance, though there is no visual evidence of interaction. Both

NGC2872/2874

©STScI/AURA/CALTECHROE

galaxies were discovered on 15 March 1784 and subsequently recorded on 3 March 1786. Herschel used a different offset star on this occasion and thought he had recorded two new nebulae. Visually, only the brighter core of NGC2874 is visible, appearing as a roundish, concentrated patch of light with well defined edges. Radial velocity: 3666 km/s. Distance: 164 mly. Diameter: 138 000 ly.

NGC2893

H297[3]	9:30.3	+29 32	GLX
(R)SB0/a	1.1′ × 1.1′	13.88B	13.1V

Images show a face-on, theta-shaped, barred galaxy with a brighter, elongated core and a bar attached to an extremely faint, thin ring. The galaxy is moderately bright visually; it is a roundish disc with a brighter core and well defined edges. Radial velocity: 1650 km/s. Distance: 74 mly. Diameter: 24 000 ly.

NGC2894

H8[3]	9:29.4	+7 43	GLX
Sa	2.0′ × 1.0′	13.23B	

Images show a highly inclined spiral galaxy, oriented NNE–SSW, with a large, bright, elongated core in a fainter, tightly wound spiral pattern. The spiral arms are dusty, with an extensive dust lane seen well away from the core along the western flank. Telescopically, the galaxy is a moderately bright object in an interesting field and is situated amongst three field stars. At moderate magnification, this gives the illusion of an E–W oriented galaxy with faint sparkles along its envelope, but high magnification resolves the three stars, the two fainter ones to the W and a brighter one to the E. Photographs show an even fainter star on the W side bordering the core. The galaxy itself appears round, small and well condensed with a sharp stellar core. Discovered on 23 January 1784, Herschel called the object: 'Extended, easily resolvable, 3 of the stars visible.' Radial velocity: 2016 km/s. Distance: 90 mly. Diameter: 52 000 ly.

NGC2903

H56[1]	9:32.2	+21 30	GLX
SB(s)d	12.6′ × 6.0′	9.53B	9.0V

This bright, barred spiral galaxy is well seen on the DSS image, which reveals a slightly inclined spiral galaxy with an elongated core and bright, dense spiral arms displaying complex structure including many knots, dust patches and dust lanes. The spiral arms are tightly wound with faint, thin extensions at the extremities extending to the S and to the N. The bar is best seen on short-exposure images, which suppress the spiral pattern. Something of the spiral nature of the galaxy can be seen in a moderate aperture, where the elongated disc, oriented N–S, can be seen to curve slightly at both extremities. The oval core is large and intensely bright, and is embedded in the disc which is very mottled in texture. A brighter condensation N of the core is probably NGC2905, described below. Radial velocity: 476 km/s. Distance: 21 mly. Diameter: 79 000 ly.

NGC2905

H57[1]	9:32.2	+21 31	HII?
0.5′ × 0.2′	10.0B		

Herschel noted this condensation N of the core of NGC2903 on the same night as he discovered the galaxy, 16 November 1784. His description reads: 'Two, at 1′ distance. Both considerably

bright, considerably large, appear like one much extended.' Images show a very small knot NNE of the core and another corresponding to the SSW of the core. Visually, both condensations are intermittently visible; the one to the N of the core is a small, roundish knot about 1' in diameter; the one S of the core appears fainter and elongated N–S. This agrees with what is seen on short-exposure images. Dark skies and good transparency are required to see both condensations well.

NGC2906

H495[2]	9:32.2	+8 27	GLX
Scd:	1.4' × 0.9'	12.97B	

Photographs show an inclined spiral galaxy with a small, brighter, elongated core and fainter spiral arms. The galaxy is moderately faint visually and is best seen at medium magnification as an oval glow oriented ENE–WSW with fairly sharply defined edges. The disc is even in brightness with a very small, brighter core. Radial velocity: 2013 km/s. Distance: 90 mly. Diameter: 37 000 ly.

NGC2911

H40[2]	9:33.8	+10 09	GLX
SA(s)0: pec	4.1' × 3.2'	12.71B	

In photographs, this is a slightly inclined, possible spiral galaxy with a brighter core and a very faint outer envelope. Dust patches are visible S of the core and a large, elongated dust patch is immediately W of the core, extending to the NW. This dust patch, in particular, appears to be silhouetted against the faint outer envelope and is not necessarily integral to it. This galaxy has also been catalogued as Arp 232. Visually, the galaxy is moderately bright and is extended NW–SE with ragged edges and a grainy outer envelope. It grows brighter along the major axis but no stellar core is visible. The fainter neighbouring galaxies NGC2914 and UGC5093 are 4.8' ESE and 8.1' SSE, respectively. NGC2911 is the principal galaxy of a small group which also includes NGC2912,

NGC2913 and NGC2919. Radial velocity: 3062 km/s. Distance: 137 mly. Diameter: 163 000 ly.

NGC2914

H513[3]	9:34.0	+10 07	GLX
SB(s)ab	1.0' × 0.6'	14.07B	

This is a slightly inclined spiral galaxy. Images show a bright, slightly elongated core and two faint spiral arms. The northern spiral arm is narrow and has a faint condensation embedded in the E. The southern spiral arm is broader and short. The galaxy has also been catalogued as Arp 137 and its companion galaxy NGC2911 is 4.8' WNW. Another companion, UGC5093, is 4.8' S. Visually, NGC2914 is a round patch of light; only the core is visible and it is even in surface brightness with well defined edges. Radial velocity: 3038 km/s. Distance: 136 mly. Diameter: 40 000 ly.

NGC2916

H260[2]	9:35.0	+21 42	GLX
SA(rs)b?	2.5' × 1.7'	12.74B	12.0V

In photographs, this is a slightly inclined spiral galaxy with a small, brighter core and knotty, dusty spiral arms. A very small, brighter condensation, possibly a star, is visible immediately N of the core. At the eyepiece, this galaxy is moderately bright, an oval patch of light oriented almost due N–S. High magnification brings out a faint stellar core, which is intermittently visible at lower powers. Though the outer envelope is bright, its edges are poorly defined. Radial velocity: 3651 km/s. Distance: 163 mly. Diameter: 118 000 ly.

NGC2918

H298[3]	9:35.8	+31 42	GLX
E	1.5' × 1.0'	13.67B	

Presenting a similar appearance both visually and photographically, this

NGC2911/2914

galaxy is moderately bright and very slightly oval in shape, oriented ENE–WSW with a brighter core and well defined edges. It lies halfway between two field stars which are oriented N–S. Radial velocity: 6770 km/s. Distance: 302 mly. Diameter: 132 000 ly.

NGC2939

H4[3]	9:38.1	+9 31	GLX
Sb-c	2.5′ × 0.9′	13.61B	

This is one of Herschel's earliest Class III discoveries. Photographs show a highly inclined spiral galaxy with an elongated core and dusty, grainy spiral arms. A faint star or condensation is immediately E of the core. The galaxy NGC 2940 is 5.75′ almost due N. In a medium-magnification field NGC2939 is a moderately faint, elongated streak oriented SSE–NNW. It is bright along the central two thirds of its major axis and has fairly diffuse edges. NGC2940 is quite a bit fainter and difficult with direct vision. It is a round, weakly concentrated glow with poorly defined edges. Radial velocity: 3216 km/s. Distance: 144 mly. Diameter: 105 000 ly.

NGC2948

H519[3]	9:38.9	+6 57	GLX
SBbc	1.9′ × 1.1′	13.78B	

The DSS image reveals a small, face-on barred spiral galaxy with a very small, brighter core and a thin bar. Its spiral arms are bright where they emerge from the bar, and fade as they expand away. Two thin, knotty arms extend to the N, and two similar arms extend to the S. In moderate apertures, this galaxy can be seen at high magnification as a moderately bright galaxy, oval in shape and oriented N–S. There is a hint of a faint stellar nucleus. The edges are diffuse and ill defined and a faint field star lies just outside the envelope to the SE. Radial velocity: 4852 km/s. Distance: 217 mly. Diameter: 120 000 ly.

NGC2964

H114[1]	9:42.9	+31 51	GLX
SAB(r)bc	3.2′ × 1.8′	12.04B	

This galaxy forms a probable linear triple system with NGC2968 situated 6.3′ to the NE and NGC2970, 10.9′ NE. On the DSS image NGC2964 is an inclined spiral galaxy with a small, elongated core and two principal, dusty spiral arms. A third spiral arm seems to consist of individual, aligned fragments W of the core. The minimum separation between NGC2964 and NGC2968 is about 106 000 ly at the presumed distance, while the distance from NGC2968 to NGC2970 would be 183 000 ly. Telescopically, all three galaxies can be seen together in a medium-magnification field, with NGC2964 the largest and brightest; it is an irregularly oval glow with a fairly opaque and well defined disc. The core is large and bright and the disc is elongated E–W. Radial velocity: 1289 km/s. Distance: 58 mly. Diameter: 54 000 ly.

NGC2968

H491[2]	9:43.2	+31 56	GLX
I0	3.0′ × 2.0′	12.74B	

This is a peculiar lenticular galaxy. Photographs show a bright core, elongated E–W and a fainter outer envelope which is elongated NNE–SSW. There are prominent, chaotic dust patches in the outer envelope N and S of the core and a very faint, possibly barred galaxy is 0.75′ WSW, though it may be a background object. NGC2964 is 6.3′ SW, while the likely companion galaxy NGC2970 is 5.0′ NE. The core-to-core separation between NGC2968 and NGC2970 would be about 100 000 ly at the presumed distance. Visually NGC2968 is a quite opaque, oval disc elongated NNE–SSW, and is broadly brighter to the middle. A hazy secondary envelope is suspected. NGC2970 appears as a very small, concentrated spot E of a bright field star. Radial velocity: 1528 km/s. Distance: 68 mly. Diameter: 59 000 ly.

NGC2964/2968

NGC2984

H34[3]	9:43.6	+11 04	GLX
S0?	0.7′ × 0.7′	14.31B	

This quite faint galaxy presents a similar appearance both visually and photographically. It is best seen at medium magnification as a round, diffuse patch of light, even in surface brightness with well defined edges. Radial velocity: 6050 km/s. Distance: 270 mly. Diameter: 55 000 ly.

NGC3020

H51[3]	9:50.1	+12 49	GLX
SB(r)cd:	3.2′ × 2.6′	13.33B	

This is the largest and brightest member of a compact group of galaxies which includes NGC3016 8.0′ SSW, NGC3019 4.0′ S and NGC3024 5.75′ ESE. Photographs show an inclined, barred spiral galaxy with a thin, bright bar and knotty spiral arms. The spiral arms are bright near the bar but fade as they expand outward. Visually, it is a much elongated patch of light, oriented E–W and a little brighter to the middle though no nucleus is visible. NGC3019 is difficult to see in moderate apertures, though NGC3016 stands out as a small, round and moderately concentrated patch of light that Herschel appears to have missed. On their discovery on 19 March 1784, Herschel described NGC3020 and NGC3024 as: 'Two about 20° north preceding, south following, 6′ or 7′ distant. Both extremely faint. Preceding is the largest.' Radial velocity: 1330 km/s. Distance: 59 mly. Diameter: 55 000 ly.

NGC3024

H52[3]	9:50.5	+12 46	GLX
Sc: sp	2.0′ × 0.4′	13.82B	

This almost edge-on galaxy, possibly a spiral or barred spiral, is a companion to NGC3020. It is bright along its major axis and has knotty condensations in its extremities. Visually, the galaxy is elongated ESE–WNW; it is well defined and quite thin with a brighter core. A magnitude 14 field star is located off the SE tip. Radial velocity: 1305 km/s. Distance: 58 mly. Diameter: 34 000 ly.

NGC3041

H98[2]	9:53.1	+16 41	GLX
SAB(rs)c	3.7′ × 2.4′	12.30B	

The DSS image shows this slightly inclined spiral galaxy has a small, bright core and faint, grainy, multiple spiral arms. Visually, this is a moderately bright, large and diffuse galaxy bordered on the SW by a fairly bright field star which hinders observation. It forms a triangle with two fainter stars to the WNW. The galaxy is very gradually brighter to the middle and the outer envelope is irregular, very poorly defined and elongated E–W. Radial velocity: 1313 km/s. Distance: 59 mly. Diameter: 60 000 ly.

NGC3053

H600[3]	9:55.6	+16 26	GLX
S(r). . .	1.3′ × 0.8′	13.97B	

Photographs show a slightly inclined spiral galaxy with a small, bright, elongated core, surrounded by dark patches. The inner spiral pattern appears to form a pseudo-ring with a bright arm emerging in the N and a fainter, broader arm in the S. In a medium-magnification eyepiece, the galaxy is located amongst a loose group of magnitude 10 and fainter stars. It is faint but readily visible with direct vision. It is a small, elongated and moderately well defined streak of light oriented SE–NW with a suddenly brighter, almost stellar core. Radial velocity: 3637 km/s. Distance: 162 mly. Diameter: 62 000 ly.

NGC3060

H601[3]	9:56.4	+16 51	GLX
Sb	2.2′ × 0.4′	13.62B	

This galaxy was discovered with NGC3053 on 14 January 1787. Images show a highly inclined, probably barred,

NGC3016/3019/3020/3024

©STScI/AURA/CALTECH/ROE

spiral galaxy with an elongated, brighter core and a foreshortened bar attached to a bright inner spiral arm or ring with faint, broadening spiral arms emerging. Visually, the galaxy is faint but easily visible; it is an elongated well defined glow oriented ENE–WSW. The surface brightness is uniform along its major axis. Radial velocity: 3590 km/s. Distance: 160 mly. Diameter: 102 000 ly.

NGC3067

H492[2]	9:58.4	+32 22	GLX
SAB(s)ab?	2.2′ × 0.8′	12.73B	

This is a highly inclined galaxy, probably a spiral, with a peculiar structure. Photographs show a faint star-like core involved in a bright, knotty, extended inner region, which is brighter to the ESE. A bright knot is E of the core and the fainter outer envelope has prominent dust patches to the E and W. In a moderate-aperture telescope, the galaxy is very bright, moderately large and elongated ESE–WNW. The bright envelope is mottled with sharply defined edges and blunt extensions, and overall brightens slightly to the centre. Radial velocity: 1441 km/s. Distance: 64 mly. Diameter: 41 000 ly.

NGC3068

H293[3]	9:58.5	+28 53	GLX	
S0[-]: pec	0.9′ × 0.8′	15.19B	14.1V	

The DSS image shows a very small lenticular galaxy with a brighter core and faint extensions, elongated ESE–WNW. A faint companion is 0.5′ SW and the pair are also catalogued as Arp 174 due to an extremely faint plume of matter which is being expelled by the smaller galaxy. This is not visible in the DSS image, though it is apparent in high-resolution photographs. Visually, NGC3068 is a very faint galaxy, a small nebulous patch. Its surface brightness is even and its edges are poorly defined. Discovered on 12 March 1785, this was very much a threshold object for Herschel. His description reads: 'Extremely faint, extremely small, verified 240

(magnification) doubtful.' In his explanatory notes, he stated: 'Suspected, extremely faint, extremely small, stellar, 240 left it doubtful but shew'd the same suspicious nebulous appearance which other stars of equal size were free from.' The French observer Guillaume Bigourdan was unable to find it with the 12-inch Paris refractor. Radial velocity: 6274 km/s. Distance: 280 mly. Diameter: 73 000 ly.

NGC3070

H59[2]	9:58.0	+10 22	GLX
E	1.4′ × 1.4′	13.26B	

Similar visually and photographically, this elliptical galaxy is small, moderately bright and round and is quite compressed to a bright, star-like nucleus. At high magnification, a faint secondary envelope is suspected; the inner envelope's edges are quite sharply defined. Also in the field 5.0′ NNW is NGC3069, which is faintly visible as a diffuse patch of light SSE of a faint pair of stars. Radial velocity: 5251 km/s. Distance: 234 mly. Diameter: 96 000 ly.

NGC3080

H934[3]	9:59.9	+13 03	GLX
Sa	0.8′ × 0.8′	14.97B	

Photographs show this remote galaxy as a very small, face-on spiral with a small, round, bright core and very faint spiral arms. The slightly fainter galaxy IC585 is 4.25′ SW. NGC3080 was discovered on 1 April 1794, when Herschel used that evening's position of the planet Uranus as his reference for the location of the galaxy. Visually, this is a very faint object, located with certainty only at high magnification. It is small, round and just a little brighter to the core with ill defined edges. Radial velocity: 10 514 km/s. Distance: 470 mly. Diameter: 109 000 ly.

NGC3088

H24[3]	10:01.1	+22 24	GLX
S0	0.9′ × 0.8′	14.40B	

This is the brighter component of a physical pair which shows definite

evidence of interaction in the DSS image: a plume of matter emerges from the core in the direction of the fainter, edge-on companion NGC3088B, located immediately SE. Another slightly fainter field galaxy, UGC5381, is 6.75′ SW. At medium magnification, NGC3088 and NGC3088B are seen to be in contact; they form an unresolved smudge elongated ESE–WNW and a little broader to the W. The smudge is gradually brighter along the major axis and the edges are well defined. It is uncertain whether Herschel saw both galaxies. Recorded on 12 March 1784, the object was described as: 'Very small. 240 (magnification) left some doubt.' Radial velocity: 6903 km/s. Distance: 308 mly. Diameter: 81 000 ly.

NGC3107

H898[2]	10:03.3	+13 38	GLX
Sbc:	0.7′ × 0.6′	13.55B	

Photographs show a slightly inclined galaxy, probably a barred spiral as the central region features a bright, N–S wedge attached to bright, knotty spiral arms. Telescopically, this galaxy is difficult to see at low magnification because of a bright field star located 1.8′ SSE. High magnification reveals a moderately well defined oval glow oriented NW–SE with a faint star-like core intermittently visible. This is one of Herschel's later Class II discoveries and the only one he made in 1794 (on 22 March) and, as with NGC3080 above, he used the planet Uranus as his reference 'star' to describe the location of this galaxy. Radial velocity: 2711 km/s. Distance: 121 mly. Diameter: 25 000 ly.

NGC3129

H65[3]	10:08.4	+18 25	DS

Recorded by Herschel on 21 March 1784 and described as: 'Very small, extended, resolvable, better with 240 (magnification)', this is a close, equally bright pair of stars, about magnitude 13, oriented NE–SW and easily resolved at medium magnification. The DSS image reveals no nearby galaxies.

NGC3153

H53[3]	10:12.9	+12 40	GLX
Scd:	2.1′ × 0.9′	13.93B	

Photographs show a slightly inclined spiral galaxy with a thin, brighter bar at the core and four knotty, low-surface-brightness spiral arms emerging. The arms expand into a fainter, grainy envelope and a small companion galaxy is 5.25′ E. At the eyepiece, this galaxy is best seen at a medium magnification as it is moderately faint with only a slight brightening to its core. It is oval, much elongated and oriented N–S, the outer envelope is rather poorly defined and a faint pair of field stars lies due N from the galaxy. Curiously, Herschel's description reads: 'Extremely faint, small, little extended, resolvable, 3 or 4 stars in it.' In photographs, three tiny condensations are visible in the spiral arm on the eastern flank, but it would be remarkable if Herschel was able to see these. Radial velocity: 2702 km/s. Distance: 121 mly. Diameter: 74 000 ly.

NGC3162

H43[2]	10:13.5	+22 44	GLX
SAB(rs)bc	3.0′ × 2.5′	12.29B	

Images reveal an almost face-on, barred spiral galaxy with a small, bright, round core and a very short bar. Bright spiral arms emerge from the bar and widen quickly into broad, faint outer spiral arms. Visually, this is a moderately large galaxy which is gradually brighter to the middle and slightly extended NE–SW. Only the bright inner region of the galaxy is detected. Radial velocity: 1231 km/s. Distance: 55 mly. Diameter: 48 000 ly.

NGC3177

H25[3]	10:16.6	+21 07	GLX
SA(rs)b	1.4′ × 1.2′	13.01B	

The DSS image shows an almost face-on spiral galaxy with a bright spiral arm which emerges from the NNE and curves E to S and a somewhat fainter arm which emerges from the S, splits quickly and curves W then N. The outer arms are very faint and clumpy.

Herschel classed this as a very faint nebulae, but visually, though small, it appears fairly bright. It is round and well condensed with a brighter core and well defined edges. Radial velocity: 1228 km/s. Distance: 55 mly. Diameter: 23 000 ly.

NGC3190

H44[2]	10:18.1	+21 50	GLX
SA(s)a pec sp	4.4′ × 1.5′	11.88B	11.1V

This is the principal galaxy of the NGC3190 group, which involves three other bright nearby galaxies, NGC3190, NGC3185 and NGC3187, as well as NGC3221, NGC3177, NGC3162 and possibly the galaxies surrounding NGC3227 2° to the SSE. Photographs show a highly inclined spiral galaxy with a large, bright, extended core in a smooth outer envelope. The principal dust lane is thick and somewhat inclined to the core of the galaxy; it is disturbed to the WNW, where a tidal interaction with NGC3187 may have occurred. Two very thin dust lanes are also visible immediately E of the galaxy's central region. Unfortunately, the principal dust lane is not seen in moderate-aperture telescopes, though large-aperture ones should show it well. The galaxy is bright and appears as a large, elongated streak oriented ESE–WNW. The envelope is well defined and even in surface brightness. Radial velocity: 1200 km/s. Distance: 54 mly. Diameter: 69 000 ly.

NGC3193

H45[2]	10:18.4	+21 54	GLX
E2	3.0′ × 2.7′	11.99B	10.9V

With NGC3187 and NGC3190, this galaxy is also known as Arp 316. Visually and photographically, the appearance of this galaxy is quite similar. Visually, the galaxy appears

NGC3190/3193

quite round with a bright centre in a fainter, well defined envelope and is adjacent to a magnitude 9 field star located 1.25′ N. Radial velocity: 1328 km/s. Distance: 59 mly. Diameter: 52 000 ly.

NGC3196

H348[3]	10:18.8	+27 40	GLX
0.3′ × 0.1′		16.19B	

Recorded on 11 April 1785, Herschel described this object as: 'Extremely faint, a little extended, a little doubtful.' On the DSS photograph this is a very small and compact lenticular galaxy with a brighter, elongated core and faint extensions, elongated ESE–WNW. Visually, this is one of the most difficult objects in the entire Herschel catalogue. It required repeated attempts before a confirmed observation was achieved. It appears as a very tiny, very faint, hazy patch of light at medium and high magnification, preceding a faint field star by 0.5′. The present author is somewhat sceptical that such a faint object would be detectable in Herschel's telescope. There are no alternative candidate objects in the immediately vicinity, however. Cautiously, this can be accepted as Herschel's object. Radial velocity: 15 131 km/s. Distance: 676 mly. Diameter: 59 000 ly.

NGC3216

H330[3]	10:21.7	+23 54	GLX
E:	1.3′ × 1.0′	14.65B	

The visual and photographic appearances of this elliptical galaxy are quite similar. Although fairly faint, this galaxy is visible at medium magnification as a small, nebulous patch which is brighter to the middle, round and fairly well defined at the edges. The DSS image shows two fainter, edge-on galaxies in the field which may be visible in large-aperture instruments: one is 4.5′ SE, the other is located 2.5′ NNW. Radial velocity: 11 704 km/s. Distance: 523 mly. Diameter: 198 000 ly.

NGC3226

H28[2]	10:23.4	+19 54	GLX
E2: pec	3.2′ × 2.8′	12.34B	

This galaxy forms an interacting pair with NGC3227 and together they are designated as Arp 94. The classification of this galaxy is somewhat uncertain and it could possibly be an S0. Deep photographs show a bright core and an extensive, ill defined outer envelope in contact with the larger companion galaxy. Visually, the galaxy is bright and round; it has a bright, well-condensed core surrounded by a faint envelope extending back to NGC3227. In a low-power field, two fainter galaxies, NGC3222 and NGC3213, are visible but were overlooked by Herschel, possibly because of the presence of Gamma Leonis to the W. Radial velocity: 1073 km/s. Distance: 48 mly. Diameter: 44 000 ly.

NGC3227

H29[2]	10:23.5	+19 52	GLX
SAB(s) pec	5.4′ × 3.6′	11.55B	

Photographs show a slightly inclined, large, bright, barred spiral galaxy, with a small core and a short dusty bar. The spiral arms are massive and broad near the core and grainy in appearance with many dust patches scattered throughout. The outer parts of the arms are fainter and smooth textured. A large dust cloud is in the outer spiral arms W of the core. The galaxy is interacting with its neighbour NGC3226 2.1′ NNW. The western spiral arm, which emerges from the galaxy to the S, seems to connect with and pass through NGC3226 and is faintly visible in the NE as a broad glow. Deep photographs show an extensive tidal plume emerging from the eastern spiral arm, curving back towards and connecting with NGC3226. NGC3227 is a bright galaxy in moderate-aperture telescopes; the core is bright and well

NGC3226/3227

©STScI/AURA/CALTECH/ROE

condensed and a broad, bright, extended outer envelope, oriented SSE–NNW, extends back to the smaller companion. Herschel's description of the two galaxies reads: 'Two. About 2′ asunder. Both faint, considerably large, round.' Radial velocity: 1079 km/s. Distance: 48 mly. Diameter: 76 000 ly.

NGC3239

H10[4]	10:25.1	+17 10	GLX
IB(s)m pec	5.0′ × 3.3′	11.67B	

Herschel classified this Magellanic-type galaxy as a Class IV object, owing to the presence of a bright field star silhouetted in front of it. His description reads: 'A pretty considerable star with a very faint brush north following. With 240 [magnification] two very small stars visible in it, but not connected.' In photographs, this is an irregular galaxy with a chaotic structure. There is no central core, but the galaxy is brightest along its major axis, which is elongated E–W. There are several bright condensations, including one to the N of the core and a large complex one to the SE which involves at least four condensations. Two faint plumes of matter extend towards the S: one comes from the central region, the other from the western end. A bright field star is immediately S of the galaxy. There are several very faint galaxies in the field, including one 3.2′ ESE, and four others about 6.0′ to the N and NNE. The galaxy has also been catalogued as Arp 263. It is a difficult object visually owing to the magnitude 8 field star, which greatly hinders visibility. Best seen at medium magnification, the galaxy is a diffuse, formless glow, brightest to the N and E of the field star. There is no central brightening. The field star forms a triangle with two fainter field stars located W of the galaxy. Radial velocity: 666 km/s. Distance: 30 mly. Diameter: 43 000 ly.

NGC3248

H347[2]	10:27.8	+22 50	GLX
S0	2.5′ × 1.1′	13.91B	

Images show a lenticular galaxy with a bright core, a slightly fainter inner disc

and an extensive and very faint outer envelope, oriented SE–NW. Visually, only the bright core of this galaxy is seen at medium magnification; it is small but moderately bright. The core is round and brighter to the middle, with a grainy envelope and fairly well defined edges. Herschel thought this object resolvable. Radial velocity: 1458 km/s. Distance: 65 mly. Diameter: 47 000 ly.

NGC3270

H331[3]	10:31.5	+24 51	GLX
SAB(r)b:	3.0′ × 0.8′	13.91B	

Photographs show a highly inclined barred spiral galaxy with a brighter, elongated core expanding into a fainter, extended S-shaped spiral pattern seen obliquely. At medium magnification the galaxy is moderately bright with a brighter, star-like core in a slightly fainter, grainy and well defined envelope. The galaxy is elongated almost due N–S. Radial velocity:

6207 km/s. Distance: 277 mly. Diameter: 242 000 ly.

NGC3274

H358[2]	10:32.3	+27 40	GLX
SABd?	2.1′ × 1.0′	13.21B	

The DSS image reveals a highly inclined, barred spiral galaxy or possibly an irregular, with a brighter, elongated core. The outer envelope is asymmetrical, fainter and broader to the W with a few very faint condensations. Several knotty condensations are involved in the envelope to the E and there is a prominent dust patch SSW of the core. The galaxy is curiously blunted along its NNE flank. It is a relatively nearby galaxy and stars begin to resolve at about blue magnitude 19 on high-resolution photographs. Visually, the galaxy is bracketed by a faint triangle of field stars; the easternmost star is an unequal pair. NGC3274 is moderately bright, broadly concentrated to the

NGC3239

centre and fairly smooth textured. It appears irregularly round with poorly defined edges. Radial velocity: 491 km/s. Distance: 22 mly. Diameter: 14 000 ly.

NGC3299

H54[3]	10:36.4	+12 42	GLX
SAB(s)dm	2.2' × 1.7'	13.81B	

This is a very dim, almost face-on galaxy that photographs show has a slightly brighter central region in a low-surface-brightness envelope. Almost certainly a dwarf, the galaxy is slightly elongated N–S. Telescopically, the galaxy is moderately large but very dim; it is a diffuse patch of light, almost round and only very slightly brighter to the middle. The edges are hazy and ill defined. The smaller but brighter galaxy NGC3306 can be seen in the same medium-magnification field to the ESE. Radial velocity: 541 km/s. Distance: 24 mly. Diameter: 16 000 ly.

NGC3300

H55[3]	10:36.6	+14 10	GLX
SAB(r)0°:?	1.5' × 0.8'	13.37B	

Photographically, this is a slightly inclined barred lenticular galaxy with a very bright core and a bright curved bar embedded in a fainter, smooth-textured elongated disc. Visually, the galaxy is small but moderately bright, elongated N–S with well defined edges and fairly even in surface brightness. Herschel considered the galaxy might be resolvable and thought he saw some stars in it. Radial velocity: 2980 km/s. Distance: 133 mly. Diameter: 58 000 ly.

NGC3301

H46[2]	10:36.9	+21 53	GLX
(R')SB(rs)0/a	3.5' × 1.0'	12.29B	

This is a highly inclined spiral galaxy, possibly barred, with a bright core, a slightly fainter inner envelope or ring and a faint outer envelope. At the eyepiece, this is a small, though moderately bright galaxy, lenticular in form with a bright, well defined core, elongated NE–SW. The core seems slightly displaced to the W and a faint

star or condensation is suspected to the NE of the core. Along the major axis to the SE the galaxy appears sharply bounded. Radial velocity: 1254 km/s. Distance: 56 mly. Diameter: 57 000 ly.

NGC3303

H66[3]	10:37.1	+18 08	GLX
S0?/Sb	3.0' × 2.1'	14.44B	

The DSS image shows a pair of interacting galaxies with a peculiar morphology. The principal galaxy is situated to the E: it is a bright lenticular with an elongated core and extremely faint plume extending to the SSE. The companion galaxy has a small, bright core offset to the S in a partial ring-shaped outer envelope. It is situated immediately NW of NGC3303 and its outer disc blends into the image of the principal galaxy. High-resolution images show broad, faint plumes of material to the E. The bright portions of the interacting galaxies combine to form a 0.6' × 0.4' image and the pair are also catalogued as Arp 192. In a moderate-aperture telescope, this is a faint galaxy, best seen with averted vision. It is a nebulous, fairly well condensed patch of light with even surface brightness across the envelope. The edges are ragged and poorly defined and the system is unresolved as separate images. Herschel discovered this object on 21 March 1784, commenting: 'Very faint, small, extended, resolvable. The same with 240 [magnification].' Herschel may have seen the two galaxies as separate objects. The following data are for the galaxy pair. Radial velocity: 6201 km/s. Distance: 277 mly. Diameter: 242 000 ly.

NGC3332

H5[3]/H272[1]	10:40.4	+9 11	GLX
(R)SA0⁻	1.4' × 1.3'	13.71B	

This galaxy was observed three times by Herschel: first on 18 January 1784 and then on 4 and 9 March 1796. The observing conditions caused the galaxy to appear very faint in 1784 and Herschel lost track of the object because he had 'no person at the clock' and was

not able to recover the object after exposing his vision to candlelight. However, his description of the field indicates that the 1784 and 1796 observations were of the same object. Photographs show a very small lenticular galaxy with a brighter core and a faint outer envelope. There is an extremely faint ring of matter surrounding the galaxy at a distance, which is brightest in the NE. It is asymmetrically situated towards the N and appears to contact the outer envelope of the galaxy in the S. This ring extends the size of the galaxy to about 3.3' × 3.2'. Visually, this is a faint galaxy, located about 1.25' NW of a faint field star. The envelope is oval, well defined and suspected of being elongated NNW–SSE. It is brighter to the middle to a stellar core comparable in brightness to the field star previously noted. Radial velocity: 5647 km/s. Distance: 252 mly. Diameter: 242 000 ly (including outer ring).

NGC3338

H77[2]	10:42.1	+13 45	GLX
SA(s)c	4.7' × 2.6'	11.44B	

Photographs reveal a large, inclined spiral galaxy with a small, slightly elongated core in a broad spiral pattern. The spiral arms are grainy with many small patches of dust defining them. A field star is located 2.75' W. On the DSS image, the galaxy measures at least 5.0' × 3.0', which is larger than its previously published size. At medium magnification the galaxy is large, prominent and fairly bright but rather diffuse with a brighter oval core embedded in a mottled outer halo. The halo fades gradually outward to ill defined edges and is elongated E–W. Radial velocity: 1207 km/s. Distance: 54 mly. Diameter: 78 000 ly.

NGC3345

H26[1]	10:43.6	+11 59	DS
nonexistent?			

At the position given by Herschel is a moderately wide, equal pair of

NGC3338

broadening to a grainy, multi-arm spiral pattern after less than half a revolution. Visually, the galaxy is moderately large but very diffuse and slightly elongated E–W. The envelope is fairly even in brightness and the edges are ill defined; the overall surface brightness is low. Radial velocity: 1170 km/s. Distance: 52 mly. Diameter: 37 000 ly.

NGC3356

H107[3]	10:44.2	+6 45	GLX
Sbc	1.7' × 0.8'	13.78B	

Photographs show a highly inclined spiral galaxy, possibly barred, with a bright core and narrow, grainy spiral arms emerging. There are a number of condensations in the arms and the galaxy is elongated E–W. The similarly bright NGC3362 is about 13.0' to the SE, while NGC3349 is 5.5' W. At medium magnification, NGC3356 is a fairly faint object in a star-poor field. It is a very small, elongated patch of light oriented E–W, slightly brighter along its major axis and moderately well defined at the edges. A field star is 2.8' due S. Radial velocity: 6056 km/s. Distance: 271 mly. Diameter: 134 000 ly.

NGC3367

H78[2]	10:46.6	+13 45	GLX	
SB(rs)c	2.2' × 2.1'	12.02B	11.5V	

Images reveal a face-on, barred spiral galaxy with a small, bright, elongated core and a bright bar. Narrow, grainy spiral arms emerge from the bar, expanding into broader outer spiral

magnitude 11 stars oriented ESE–WNW and the designation NGC3345 is now associated with these stars. However, Herschel's discovery description of 19 March 1784 calls the object: 'Considerably bright, pretty large, not round, much brighter to the middle.' He observed it only once and J. L. E. Dreyer noted that there was no nebula at the position but that M95 is about 15.0' to the SSE. Since the description is fairly consistent with the appearance of M95 and not consistent with the appearance of the pair of stars, H26[1] may be M95. However, Herschel had knowingly observed the Messier object only eight days earlier, on 11 March 1784, calling the bright galaxy: 'A fine, bright nebula, much brighter in the middle than at the extremes, of a pretty considerable extent, perhaps 3 or 4' or more. The middle seems to be of the magnitude of 3 or 4 stars joined together, but not exactly round; from the brightest part of it there is a sudden transition to the

nebulous part, so that I should call it cometic.' It seems unlikely that Herschel would not recognize the Messier object eight days later and mistake what he saw for a new object. However, he was still relatively new to deep-sky observing at the time and this may be a case of inexperience on his part. There is a remote possibility that Herschel happened upon a telescopic comet on 19 March but no way to verify this. In this case H26[1] would be nonexistent. It seems certain that the double star is not Herschel's object. Guardedly, Dreyer may be right: H26[1] = M95.

NGC3346

H7[5]	10:43.7	+14 52	GLX
SB(rs)cd	2.4' × 2.3'	12.45B	

This object is a face-on, barred spiral galaxy with a very small core embedded in a brighter bar. Images show a partial ring structure attached to the bar with two spiral arms emerging and

NGC3346

arms. Deep photographs show complex structure in these outer regions. Visually, this is a fairly bright, round galaxy which is brighter to the middle. A pair of faint field stars are nearby to the SW. The galaxy's edges are fairly well defined, fading quickly into the sky background. Although in the same field as the Leo Group of galaxies, it is presumed to be a background object. Radial velocity: 2947 km/s. Distance: 132 mly. Diameter: 81 000 ly.

NGC3370

H81[2]	10:47.1	+17 16	GLX
SA(s)c	3.2′ × 1.8′	12.19B	

The DSS image shows a bright, slightly inclined multi-arm spiral galaxy with a brighter, elongated core. Two main spiral arms emerge from the core and expand outward to form a broader spiral pattern; there are two minor arms emerging from the core as well. The arms are very knotty with many brighter condensations throughout, as well as patches of dust. The galaxy is elongated SE–NW and a very dim companion galaxy is 3.9′ W. In moderate apertures, this is a moderately large galaxy, broadly brighter to the middle to a brighter core. The envelope is well condensed and moderately well defined along the edges. Radial velocity: 1198 km/s. Distance: 53 mly. Diameter: 50 000 ly.

NGC3377

H99[2]	10:47.7	+13 59	GLX	
E5	5.2′ × 3.0′	11.13B	10.4V	

Photographs show an elliptical galaxy with an extensive, elongated, brighter core in a broad outer envelope. Visually, this elliptical galaxy is elongated NE–SW and is very bright and intense, holding magnification well. The edges are crisply defined and the major axis is very bright. The core is bright and star-like and another very bright condensation near the core makes the galaxy appear bi-nuclear. A very faint companion galaxy, NGC3377A, is 7.1′ NW. Deep photographs show several possible globular clusters in the

outer halo of this Leo Group galaxy. Radial velocity: 573 km/s. Distance: 26 mly. Diameter: 40 000 ly.

NGC3379

H17[1]	10:47.8	+12 35	GLX
E1	5.1′ × 4.7′	10.23B	

This galaxy forms a triplet with NGC 3384 7.25′ to the ENE and NGC3389 10.0′ ESE. Together they are the brightest members of the Leo Group of galaxies. There is a considerable spread in the radial velocity of these three galaxies which presumably can be explained by gravitational interaction. However, the spread leads to widely different distances for the individual members; a compromise distance for the group would be about 35 mly at the presumed distance. Photographs show NGC3379 is a large, bright elliptical galaxy with a very bright core and an extensive, fainter outer envelope. It is also catalogued as M105 and was originally discovered by Méchain on 24

March 1781. Visually, the triplet can all be seen in the same high-magnification field, forming a very attractive group with NGC3379 appearing large, bright and round with a grainy and somewhat well defined envelope. It brightens to a large central region, with a bright nonstellar core embedded. Radial velocity: 814 km/s. Distance: 37 mly. Diameter: 54 000 ly.

NGC3384

H18[1]	10:48.3	+12 38	GLX
SB(s)0⁻:	5.4′ × 2.8′	10.89B	

In the DSS image, this Leo Group member has a large, bright, round core and an extensive, elongated outer envelope. Visually, the galaxy is the largest and brightest of the triplet which also includes NGC3379 and NGC3389, with a roundish disc and a very bright, non-stellar core. The disc is surrounded by a hazy secondary envelope oriented NE–SW. Radial velocity: 607 km/s. Distance: 27 mly. Diameter: 43 000 ly.

NGC3370

©STScI/AURA/CALTECH/ROE

NGC3379/3384/3389

NGC3389

H41[2]	10:48.5	+12 32	GLX
SA(s)c	2.8′ × 1.3′	12.51B	

Images show a large, bright, inclined spiral galaxy with a brighter, elongated core. Two main spiral arms emerge from the core; they are grainy with many brighter condensations and dust patches. Although it is the faintest of the triplet of Leo Group galaxies, the others being NGC3379 and NGC3384, NGC3389 is a rewarding object visually, best seen at medium magnification as an elongated grainy disc with a bright spine extended along its major axis and oriented roughly E–W. Radial velocity: 1211 km/s. Distance: 54 mly. Diameter: 44 000 ly.

NGC3412

H27[1]	10:50.9	+13 25	GLX	
SB(s)0°	3.6′ × 2.0′	11.44B	10.5V	

Images reveal a bright lenticular galaxy with an oval central region in a bright and extensive outer disc. Deep photographs show a bar inclined about 35° to the disc. Telescopically, this is a bright, easy galaxy with a bright, well-concentrated and star-like core, surrounded by a well defined, slightly elongated secondary halo. The fainter outer envelope can be seen as extensions oriented SSE–NNW. The galaxy is probably a member of the Leo Group. Radial velocity: 748 km/s. Distance: 33 mly. Diameter: 35 000 ly.

NGC3425

H108[3]	10:51.4	+8 34	GLX
S0	1.2′ × 1.1′	14.49B	

In photographs this lenticular galaxy has a round core and a very faint outer envelope. A small compact galaxy is 0.6′ SSW, while the inclined barred spiral NGC3417 is 8.1′ SW. Visually, NGC3425 is a small, though moderately well defined galaxy which forms a triangle with two bright field stars to the S. The envelope is round and small, and the core is faint, though well condensed and stellar. Herschel considered the galaxy resolvable. Radial velocity:

6518 km/s. Distance: 291 mly. Diameter: 102 000 ly.

NGC3433

H20[3]	10:52.1	+10 09	GLX
SA(s)c	3.5′ × 3.2′	13.09B	

Photographs show a face-on spiral galaxy with a very small, round and bright core. Two major spiral arms of the grand-design type emerge from the

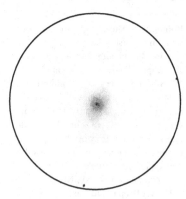

NGC3412 15-inch Newtonian 313x

core and after about half a revolution each, broaden, fade and branch into outer spiral structure. The main arms are grainy but there are no knotty condensations visible. Small galaxies are 2.0′ ESE, 2.75′ S, 3.9′ SW and 5.0′ WSW and are part of a chain of five galaxies that are probably in the background. In moderate-aperture telescopes this is a faint, though sizable galaxy, gradually brighter than the sky background. The envelope is ill defined and almost round and the galaxy is only gradually brighter to the middle. Radial velocity: 2616 km/s. Distance: 117 mly. Diameter: 119 000 ly.

NGC3434

H497[3]	10:52.0	+3 48	GLX
SA(r)b	2.1′ × 1.7′	13.85B	

The DSS image shows a face-on spiral galaxy with a very small, round, bright core surrounded by a bright inner ring. Faint, narrow spiral arms emerge from the ring and are grainy in texture. Small, faint galaxies are 2.75′ NNE (CGCG038-037), and 2.25′ WSW (CGCG038-035). NGC3434 is a fairly faint galaxy visually; it is an oval, slightly elongated patch of light, oriented N–S and fairly well defined. At its centre is a condensed, bright spot which is best seen at medium magnification. Radial velocity: 3511 km/s. Distance: 157 mly. Diameter: 96 000 ly.

NGC3437

H47[2]	10:52.6	+22 56	GLX	
SAB(rs)c	2.6′ × 0.8′	12.59B	12.0V	

This is a bright, highly inclined spiral galaxy with a large, broad, bright central region. Photographs show high-surface-brightness, knotty spiral arms emerging from the central region with some evidence of dust, particularly bordering the central region to the SSW. Visually, this is a large and bright galaxy which takes magnification well. It is an elongated streak oriented ESE–WNW with a bright, tapered envelope, fairly even in surface brightness with a small, condensed core.

The edges are fairly well defined and there is some graininess to the envelope. Radial velocity: 1225 km/s. Distance: 55 mly. Diameter: 41 000 ly.

NGC3455

H82[2]	10:54.5	+17 17	GLX
(R′)SAB(rs)b	2.5′ × 1.5′	14.32B	

Images reveal an inclined spiral galaxy with a small, brighter core in a slightly fainter, grainy central region. Knotty spiral arms emerge from the central region and the western spiral arm winds back towards the E and fades but this arm completely encircles the galaxy. The galaxy is elongated ENE–WSW. A field star is 2.0′ N and the edge-on galaxy NGC3454 is 3.5′ N. Visually, this is a faint, though fairly large galaxy and is broadly brighter to the middle, though no core is visible. The envelope has poorly defined edges. NGC3454 may be visible in large-aperture telescopes. Radial velocity: 1024 km/s. Distance: 46 mly. Diameter: 33 000 ly.

NGC3462

H16[2]	10:55.3	+7 42	GLX
S0:	1.8′ × 1.2′	13.44B	

This small, lenticular galaxy has very similar visual and photographic appearances. Although small, the galaxy is visually bright, well condensed and very slightly elongated ENE–WSW. The main envelope is quite opaque with fairly even surface brightness and the edges are sharply defined. Photographs show an extensive outer halo. Radial velocity: 6331 km/s. Distance: 283 mly. Diameter: 148 000 ly.

NGC3473

H67[3]	10:58.0	+17 07	GLX
SBb:	1.1′ × 1.0′	14.71B	

Photographs show a slightly inclined spiral galaxy, possibly barred, with a very small core. The inner envelope is slightly fainter than the core and possibly a ring. Thin spiral arms emerge from the central region and a field star borders the outer envelope to the N.

Another field star is 0.9′ S. The lenticular galaxy NGC3474 is 1.9′ SSE, while a faint edge-on galaxy is 1.2′ NNW. Visually, NGC3473 is an extremely faint galaxy; it is a hazy patch of light with a faint field star involved in the N. It is poorly defined and almost round. Radial velocity: 9098 km/s. Distance: 406 mly. Diameter: 130 000 ly.

NGC3475

H332[3]	10:58.3	+24 14	GLX
Sa	1.7′ × 1.1′	14.56B	

Photographs show a slightly inclined spiral galaxy, with a very bright, elongated core and thin, asymmetrical and very faint spiral arms, oriented ENE–WSW. It is the brightest of four galaxies in a compact group, the others are 2.1′ NW, 1.8′ SSE and 2.4′ SSW. In moderate-aperture telescopes, this is a very dim and small object, suspected at medium magnification and requiring high power to confirm. It is located on a line of three faint field stars oriented N–S and appears oval and gradually elongated ENE–WSW. Though faint, it is fairly well defined and slightly brighter to the middle. Radial velocity: 6379 km/s. Distance: 285 mly. Diameter: 141 000 ly.

NGC3485

H100[2]	11:00.0	+14 50	GLX
SB(r)b:	2.6′ × 2.6′	12.67B	

This is a face-on, barred spiral galaxy with a small, brighter core. Images show that the bar is surrounded by a tightly wound inner ring which transitions into an open spiral structure. The spiral arms are grainy with a few brighter condensations. Telescopically, this moderately bright galaxy is best at medium magnification and is seen as a diffuse, quite large circular patch with a small, prominent core. The outer envelope is a little grainy, the edges are well defined and a field star is 1.6′ W. Radial velocity: 1351 km/s. Distance: 60 mly. Diameter: 46 000 ly.

NGC3489

H101[2]	11:00.3	+13 54	GLX
SAB(rs)0⁺	3.5′ × 2.0′	11.06B	10.3V

The DSS image shows a bright, lenticular galaxy with a large, brighter core and possible ring structure. High-resolution images show an intricate pattern of dust in the disc, particularly to the E. The galaxy is visually fairly bright and quite large; it has a slightly oval, grainy envelope with a brighter, round inner disc and a very bright nonstellar core embedded. The edges are quite hazy and ill defined. Radial velocity: 589 km/s. Distance: 26 mly. Diameter: 27 000 ly.

NGC3491

H21[3]	11:00.6	+12 10	GLX
S0⁻:	0.9′ × 0.7′	14.26B	

Visually and photographically similar, this is a small, circular galaxy of relatively high surface brightness and is quite distinct at high magnification. The surface brightness is even across the disc and the edges are sharply defined. Photographs show the galaxy is elongated ESE–WNW with a fainter outer disc. Radial velocity: 6292 km/s. Distance: 281 mly. Diameter: 74 000 ly.

NGC3495

H498[3]	11:01.3	+3 38	GLX
Scd(f)	4.9′ × 1.2′	12.43B	

Photographs show a large, highly inclined spiral galaxy in which the spiral structure contains many knotty condensations and dust patches. The core is extremely small with no bulge. At the eyepiece, though the galaxy has low surface brightness, it stands out well. Situated within a triangle of magnitude 11 or fainter field stars, it is much elongated with a mottled major axis and no bright core and is oriented NNE–SSW. The bright field star 58 Leonis, located almost 10′ to the W, hinders observation somewhat. Radial velocity: 1008 km/s. Distance: 45 mly. Diameter: 64 000 ly.

NGC3495

NGC3498

H75[3]	11:01.7	+14 21	TS

Herschel recorded this object on 8 April 1784 and called it: 'Extremely faint, not small, doubtful.' It was observed only once and Herschel noted that the lack of stars in the field made it impossible for him to focus an eyepiece giving a

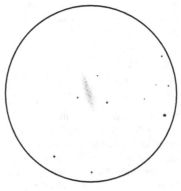

NGC3495 15-inch Newtonian 146x

magnification of 240, so he could not confirm the observation. The object is now associated with a very faint triple star which is near the presumed position, but the grouping is small, about 0.5′ across in an E–W direction. There are no nebulae in the 15′ field surrounding the triple star on the DSS image, but there is a sufficiently bright star about 5.0′ to the NW which Herschel could have certainly used to focus his eyepiece. J. L. E. Dreyer notes that the object was looked for repeatedly by Heinrich d'Arrest but not found. Telescopically, the field is found by locating an ENE–WSW oriented chain of moderately bright stars, the easternmost star of which is a wide, unequal pair. The triple star is S of the two stars to the W, forming an equilateral triangle. The triple star system is extremely faint and not resolvable at high magnification, which shows only two members and these only intermittently. At medium and high magnification, the group appears consistently hazy and nebulous. Associating NGC3498 with H75[3] is decidedly uncertain.

NGC3506

H22[3]	11:03.2	+11 05	GLX
Sc:	1.2′ × 1.1′	13.21B	

This is a face-on spiral galaxy with a small, bright core and two bright main spiral arms emerging from the core; the arms start to fade after half a revolution. High-resolution photographs show four arm fragments emerging from the N spiral arm. Telescopically, the galaxy is moderately bright, though a little less distinct at high magnification. It is a round, diffuse patch, slightly concentrated to the middle with moderately well defined edges. Radial velocity: 6306 km/s. Distance: 282 mly. Diameter: 98 000 ly.

NGC3507

H7[4]	11:03.5	+18 08	GLX
SB(s)b	3.4′ × 2.9′	12.07B	

This is a beautiful example of an S-shaped, face-on, barred spiral galaxy. Photographs show a small, bright core and a slightly fainter bar which curves slightly into two main spiral arms. The spiral arms are grainy, particularly in the outer spiral structure. A field star is 0.25′ ENE of the core and it greatly hinders observation of the galaxy. At high magnification the bright core is small and round and the bar is seen faintly. Medium magnification brings out the outer envelope, which is very faint and diffuse and somewhat elongated E–W. The edges are poorly defined. Radial velocity: 907 km/s. Distance: 40 mly. Diameter: 40 000 ly.

NGC3509

H598[3]	11:04.4	+4 50	GLX
SA(s)bc pec	2.2′ × 1.1′	14.22B	

This highly inclined spiral galaxy with a peculiar morphology has also been catalogued as Arp 335. It is well seen in the DSS image and may possibly be two galaxies interacting since the two

NGC3507

bright, elongated condensations seen on the image could be the cores of separate objects. The spiral structure is asymmetrical and clumpy with two dark patches SW of the smaller, more central core. A single, large spiral arm emerges from the NE and curves sharply towards the SW. The galaxy is elongated NE–SW. Telescopically, this is a moderately faint, elongated galaxy, somewhat bright along its major axis, which is oriented NNE–SSW. A narrow, hazy secondary envelope surrounds the bright spine and is ill defined. Radial velocity: 7587 km/s. Distance: 339 mly. Diameter: 207 000 ly.

NGC3521

H13[1]	11:05.8	−0 02	GLX
SAB(rs)bc	11.0′ × 5.1′	9.90B	9.0V

One of Herschel's early discoveries, located on 22 February 1784, this is a classic, highly inclined spiral galaxy with a bright, elongated core.

Photographs show a complex, multi-arm spiral with many dust patches in the inner spiral pattern and two strong dust lanes in the outer spiral arms on the W side. Faintly visible on the E side is a broad column rising well above the plane of the galaxy. Herschel was not much of an artist (his son John, by contrast, created some highly evocative

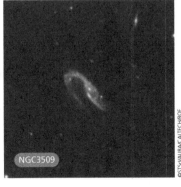

NGC3509

NGC3521

©STScI/AURA/CALTECH/ROE

drawings of deep-sky objects) but he did a rough sketch of this galaxy for his classic paper on the construction of the heavens, published in 1784. Visually, this is a superb object for moderate-aperture telescopes. It is a large, bright, much elongated galaxy, oriented NNW–SSE; the outer envelope is a well defined oval with a bright, well-developed and star-like core. A dark lane, very thin and razor sharp, is suspected just W of the core and the outer envelope is fairly mottled. Radial velocity: 669 km/s. Distance: 30 mly. Diameter: 95 000 ly.

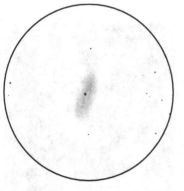

NGC3521 15-inch Newtonian 146x

NGC3524

H23³	11:06.5	+11 24	GLX
S0/a	1.5′ × 0.5′	13.34B	

Presenting very similar appearances visually and photographically, this galaxy is small but has high surface brightness and holds magnification well. Bright and opaque along its major axis, the galaxy is elongated NNE–SSW and is sharply defined with tapered ends and a slightly brighter core. It is situated SSE of a bright pair of field stars, the nearest of which is 0.9′ from the core. Radial velocity: 1270 km/s. Distance: 57 mly. Diameter: 25 000 ly.

NGC3535

H111³	11:08.5	+4 50	GLX
SA(s)a pec:	1.2′ × 0.5′	14.41B	

Photographs reveal a small, highly inclined spiral galaxy with a bright, elongated core in a faint extended envelope. A prominent dust lane borders the core on the W side and the galaxy is elongated N–S. Visually, the galaxy is a small, faint and very diffuse patch of light, very even in surface brightness and ill defined at the edges. Radial velocity: 6847 km/s. Distance: 306 mly. Diameter: 107 000 ly.

NGC3547

H42²	11:09.9	+10 43	GLX
Sb:	2.0′ × 0.9′	13.20B	

Images show a very small, inclined galaxy, probably spiral, with a very small core and a bright, knotty inner envelope. An extremely faint outer envelope is visible, which is a little more extensive in the S. Visually, the galaxy is moderately bright and a little more distinct at medium magnification. Roughly oval in shape and elongated N–S, the disc is somewhat grainy. No core is visible and the edges are a little irregular in outline. Radial velocity: 1484 km/s. Distance: 66 mly. Diameter: 39 000 ly.

NGC3559

H79³	11:10.7	+12 01	GLX
S pec	1.5′ × 1.0′	13.88B	

The DSS photograph reveals an inclined spiral galaxy, with a very small core and knotty spiral arms that are brighter in the SE. The spiral structure is asymmetrical and somewhat chaotic; it is thin and faint towards the NW where there may be large dust patches involved and brighter though fragmentary along the SE flank. Telescopically, the galaxy is fairly faint, best seen at medium magnification as a diffuse oval patch, elongated NE–SW with a grainy, irregular texture to the envelope and diffuse edges. No core is visible. Radial velocity: 3159 km/s. Distance: 141 mly. Diameter: 62 000 ly.

NGC3567

H89[3]	11:11.3	+5 50	GLX
S0 pec	1.0′ × 0.6′	14.59B	

The DSS image shows a small, lenticular galaxy with a companion galaxy, MCG +1-29-12, 0.6′ to the SE which may be interacting with it. The galaxy has a bright, elongated core oriented ESE–WNW and a very faint outer envelope which is offset towards the N and NW. The companion, meanwhile, has a broad and very faint plume emerging towards the S before curving E. Assuming that they form a physical pair, the minimum core-to-core separation of the galaxies would be about 50 000 ly at the presumed distance. Telescopically, NGC3567 is a very faint, blurry smudge of light, almost round, and a little brighter to the middle. The companion is extremely faint and intermittently visible as a star-like spot, and is best seen at medium magnification. Radial velocity: 6219 km/s. Distance: 278 mly. Diameter: 81 000 ly.

NGC3593

H29[1]	11:14.6	+12 49	GLX
SA(s)0/a	5.3′ × 2.3′	11.85B	

This is a highly inclined spiral galaxy with a bright, elongated core embedded in a fainter outer envelope. Photographs show that a prominent dust lane extends along the major axis, just N of the core; this would seem to be the near side of the galaxy, but the spiral arms are difficult to trace. Telescopically, the galaxy is quite bright and best seen at high magnification as a grainy, elongated disc oriented E–W. The central disc has high surface brightness, is embedded in a fainter outer envelope and is very broadly brighter to the middle, with a grainy-textured disc elongated E–W. A sharp, stellar core is intermittently visible. Radial velocity: 541 km/s. Distance: 24 mly. Diameter: 37 000 ly.

NGC3596

H102[2]	11:15.1	+14 47	GLX
SAB(rs)c	4.1′ × 4.1′	11.80B	

This face-on spiral galaxy has a small, round and bright core. The DSS image

NGC3593

shows that two bright, main spiral arms emerge, expanding into broad, symmetrical, flocculent spiral structure. There are several brighter nebulous knots and a sudden transition to a much fainter outer spiral pattern. Although fairly bright and large visually, this galaxy is somewhat diffuse and is more distinct at medium magnification. It appears as a very slightly oval, fairly well defined disc with a tiny, star-like core and patchy brightenings are suspected in the disc. Radial velocity: 1113 km/s. Distance: 50 mly. Diameter: 59 000 ly.

NGC3599

H49[2]	11:15.4	+18 07	GLX
SA0:	2.7′ × 2.1′	12.88B	

This is the first of five fairly bright Herschel objects visible in a 1° field, the others being NGC3605, NGC3607, NGC3608 and NGC3626. In photographs, the bright, slightly elongated core is surrounded by an oval inner disc and an extensive and very faint outer envelope. Telescopically, it is a small bright galaxy, round and very much brighter to a very bright core. The outer envelope is moderately well defined and grainy in appearance. The object was discovered on 14 March 1784 and the observed graininess led Herschel to believe that this lenticular

NGC3596

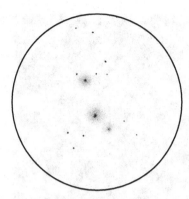

NGC3605, NGC3607, NGC3608
15-inch Newtonian 146x

galaxy might be resolvable. Radial velocity: 764 km/s. Distance: 34 mly. Diameter: 27 000 ly.

NGC3605

H27[3]	11:16.8	+18 01	GLX
E4-5	1.1′ × 0.6′	13.16B	

Similar in appearance visually and photographically, this elliptical galaxy is part of a very attractive group of three galaxies visible together in a medium- to high-magnification field. It has a bright core and is elongated NNE–SSW. It is the smallest of three Herschel objects in the field: NGC3607 is 2.9′ NE and NGC3608 is 8.4′ NNE. NGC3605, NGC3599 and NGC3626 were all discovered on 14 March 1784. Radial velocity: 600 km/s. Distance: 27 mly. Diameter: 8600 ly.

NGC3607

H50[2]	11:16.9	+18 03	GLX
SA(s)0°:	4.9′ × 2.5′	10.93B	

This lenticular galaxy is the brightest and largest in a field of three. Deep photographs show a thin dust lane bordering the bright central bulge and separating it from the outer, fainter disc. Like most lenticular galaxies, it has three distinct brightness zones: a large, bright and oval central region, a fainter inner disc and an extensive and very faint outer envelope. Visually, the galaxy appears almost round, with a

broad, mottled outer envelope and a very bright core. Photographs show the galaxy is elongated ESE–WNW. Radial velocity: 892 km/s. Distance: 40 mly. Diameter: 57 000 ly.

NGC3608

H51[2]	11:17.0	+18 09	GLX
E2	3.2′ × 2.6′	11.57B	

This is an elliptical galaxy, the most northerly of three in the field, and is quite bright with a very bright core and an extensive, mottled outer envelope. It appears round though photographs show it to be very slightly elongated E–W. Presumably this galaxy, NGC3607 and NGC3605 form a group and are all at approximately the same distance. The wide range of radial velocity for the three galaxies may indicate orbital motion and therefore the derived sizes may be in error as well. Radial velocity: 1186 km/s. Distance: 53 mly. Diameter: 50 000 ly.

NGC3611

H521[2]/H626[2]	11:17.5	+4 33	GLX
SA(s)a pec	2.1′ × 1.7′	12.85B	

This is an inclined spiral galaxy with a large, elongated core and very faint outer spiral structure. Photographs show the arms are smooth textured but faint; however, small, round clumps are evident, particularly to the N and W. High-resolution images reveal the likelihood of star formation near the core. A bright field star 2.2′ NNW is immediately W of a very faint field galaxy, UGC6306. Telescopically, NGC3611 is a moderately bright galaxy, visible as a small, oval patch of light which is broadly brighter to the middle and elongated NNE–SSW. The edges fade slowly into the sky background and suggest a faint secondary envelope. This object was recorded twice by Herschel: it was discovered on 27 January 1786 and observed again on 30 December 1786 when it was thought to be a separate object. It subsequently received the designation NGC3604. Radial velocity: 1472 km/s. Distance: 66 mly. Diameter: 40 000 ly.

NGC3615

H333[3]	11:18.0	+23 24	GLX
E	1.2′ × 0.8′	14.01B	

The DSS image shows a small lenticular galaxy with a bright, elongated core in a faint outer envelope. The classification should be about E4 or perhaps S0, due to the extensive envelope. The galaxy is elongated NE–SW and the companion galaxy NGC3618, a Herschel object, is 7.4′ NE. There are many very small, faint galaxies in the field. The two Herschel galaxies can be seen together in a medium-magnification field, with NGC3615 being the brighter of the two; it is almost round with a bright star-like core and poorly defined edges. Radial velocity: 6636 km/s. Distance: 296 mly. Diameter: 104 000 ly.

NGC3616

H76[3]	11:18.2	+14 44	nonexistent

NGC3618

H334[3]	11:18.5	+23 29	GLX
SABb:	1.1′ × 0.8′	14.49B	

Photographs show a small face-on galaxy with a tiny, star-like nucleus in a brighter core. Grainy spiral structure emerges from the core with dust patches evident in the E and S and the arms themselves have fairly high surface brightness. Companion galaxy NGC3615 is 7.4′ SW and probably forms a physical pair. At the presumed distance, the core-to-core separation between the two galaxies would be about 650 000 ly. Visually, this galaxy is smaller and fainter than its neighbour; it is an ethereal glow which is a little brighter to the middle and very slightly elongated N–S. Radial velocity: 6759 km/s. Distance: 302 mly. Diameter: 97 000 ly.

NGC3626

H30[2]/H52[2]	11:20.1	+18 21	GLX
(R)SA(rs)0+	2.7′ × 1.9′	11.81B	

This object was recorded twice by Herschel, first on 15 February 1784 and then on 14 March 1784. Herschel made a substantial error when recording the

NGC3626

position of the object at its first observation, placing the galaxy 12' N of its true location. It is the fifth of five bright Herschel objects in a 1° field; the others are NGC3599, NGC3605, NGC3607 and NGC3608. Photographs show this bright galaxy to be an inclined ring galaxy with a very bright, elongated core which has a prominent dust lane bordering the core to the W. Some images show this dark ring completely encircling the bright core. The spiral arms are smooth with dust bands in the outer spiral structure. At the eyepiece, the galaxy is bright with a well-condensed core. The outer envelope appears elongated NNW–SSE at high magnification; this envelope is very faint and best seen at medium magnification. Radial velocity: 1428 km/s. Distance: 64 mly. Diameter: 50 000 ly.

NGC3628

H8[5]	11:20.3	+13 36	GLX	
SAb pec sp	14.8' × 3.0'	9.97B	9.5V

The DSS image shows this galaxy very well, revealing a very large and bright spiral galaxy seen edge on. It has also been designated Arp 317 due to its complex and unusual structure. The bright core is almost completely obscured by a thick, complex, dust lane which extends along the major axis. The dust lane broadens to the E and in the W splits into two components: a short, stubby spur passes below the main dust lane which broadens as it extends to the E. The outer arms of the galaxy appear disturbed in the E and in the W and flare

above and below the galactic plane. It is the northernmost galaxy of the 'Leo Trio', the others being the bright galaxies M65 and M66. Visually, this galaxy is well seen even in small-aperture telescopes, despite the presence of the obscuring dust lane. It appears very flat and edge-on, with a smooth texture and is very slightly brighter to the core. The N flank is fairly sharply defined, while the S flank fades indefinitely into the sky background. Moderate-aperture telescopes show the dust lane well and under the best conditions the galaxy is seen extending ESE–WNW across almost half the field of a medium-magnification eyepiece. Radial velocity: 761 km/s. Distance: 34 mly. Diameter: 148 000 ly.

NGC3629

H338[2]	11:20.5	+26 58	GLX
SA(s)cd:	2.0' × 1.4'	13.12B	

Photographs show a slightly inclined spiral galaxy with a small, elongated,

brighter core embedded in dusty, low-surface-brightness spiral arms, reminiscent of M33. Several knotty condensations are involved and the galaxy is elongated NNE–SSW. Visually, the galaxy is fairly faint and is best seen at medium magnification as a diffuse patch of light which is gradually brighter to the middle, though no core is visible. The edges fade gradually into the sky background and the galaxy is very slightly elongated. A faint field star is 1.0' to the ESE. Radial velocity: 1474 km/s. Distance: 66 mly. Diameter: 38 000 ly.

NGC3630

H32[2]	11:20.3	+2 58	GLX
S0	1.7' × 0.7'	12.90B	

This Herschel object was designated NGC3645 by Dreyer in *The Scientific Papers* but the actual NGC3645 is part of a small group of remote galaxies and a very faint object, not matching Herschel's description, which reads:

NGC3628

'Pretty bright, very small, brighter to the middle.' Moreover, Herschel's position measured from the star Tau Leonis falls very close to the position for NGC3630. In photographs the galaxy is a high-surface-brightness, edge-on lenticular galaxy that Herschel is not likely to have missed. Although small visually, this is a bright and very well defined galaxy with a bright, slightly bulging core. It is an opaque, elongated object oriented NE–SW, and the slightly fainter extensions taper to points. The galaxy is probably physically associated with the bright elliptical galaxy NGC3640, situated to the NE. Radial velocity 1368 km/s. Distance: 61 mly. Diameter: 30 000 ly.

NGC3640

H33[2]	11:21.1	+3 14	GLX
E3	4.6′ × 4.1′	11.33B	10.4V

Photographically, this is a bright elliptical galaxy with a large, slightly elongated core offset to the E in a very faint outer envelope. The envelope is irregular in outline with a dark zone in the western half and the galaxy is elongated E–W. Telescopically, this is quite a bright galaxy; it is a slightly oval, well defined patch of light, with even surface brightness that falls slightly at the edges. The envelope is grainy and a small bright core is in the middle. Also visible is companion galaxy NGC3641, 2.5′ to the SSE, which is seen as a small, condensed patch of light. Radial velocity: 1135 km/s. Distance: 51 mly. Diameter: 68 000 ly.

NGC3646

H15[3]		11:21.7	+20 10	GLX
SA:(r)bc pec (ring)	3.0′ × 1.8′	11.77B	11.1V	

The DSS image shows a large, bright, inclined spiral galaxy with a small, elongated core embedded in a fainter inner envelope. Knotty spiral arms surround the inner envelope with several brighter condensations and large dust patches N, E and SW of the core. High-resolution photographs seem to show that the inner spiral pattern is not connected to the bright, massive outer

NGC3646

arms, which form a ring around the inner galaxy. The galaxy is larger and shows considerably more detail than typical galaxies at similar radial velocities. The spiral galaxy NGC3649 is 7.8′ ENE and is not probably physically related as it has a higher radial velocity. Visually, NGC3646 is a very large and moderately bright galaxy; it is a large oval which is broadly brighter to the middle with a small, brighter core. Elongated NE–SW, the envelope is quite mottled and the edges are poorly defined. Radial velocity 4190 km/s. Distance: 187 mly. Diameter: 163 000 ly.

NGC3649

H16[3]	11:22.2	+20 13	GLX
SB(s)a	1.2′ × 0.6′	14.87B	

Photographs reveal an inclined spiral galaxy with a bright core and two very faint spiral arms. Faint field stars are 0.4′ S and 0.4′ SW (this one may actually be a compact galaxy as a flare in the DSS image emerges to the

SSW). Visually, this galaxy can be seen in the same medium-magnification field as NGC3646. It is smaller than NGC3646 but moderately bright, elongated SE–NW with a bright, small core and the faint field star to the S is visible. Photographically, the cores of both galaxies are similar in brightness but the difference in radial velocities is large, 731 km/s, so they may not form a physical pair. Radial velocity: 4921 km/s. Distance: 220 mly. Diameter: 77 000 ly.

NGC3651

H335[3]	11:22.4	+24 18	GLX
E	1.0′ × 0.9′	14.17B	

The DSS image shows a small elliptical galaxy with a round, bright core and a faint outer envelope. An S0 companion galaxy is involved immediately S though the resolution is not good enough to reveal any disruption in either galaxy. NGC3651 is part of a

NGC3651/3653

©STScI/AURA/CALTECH/ROE

compact group of at least eight galaxies, including NGC3651 which is 1.4′ SE. In a moderate-aperture telescope both NGC3651 and NGC3653 are readily visible and form a close pair, oriented SE–NW. NGC3651 is the larger and slightly brighter, a condensed galaxy which is brighter to the middle, almost round and moderately well defined at the edges. There is no sign of the S0 companion or any other galaxy in the field but the field is interesting and challenging for large amateur telescopes, Herschel recorded both galaxies on 10 April 1785 and described them as: 'Two. 2′ or 3′ distant. Both very faint, very small, the most southerly faintest.' Radial velocity: 7583 km/s. Distance: 339 mly. Diameter: 99 400 ly.

NGC3653

H336[3]	11:22.5	+24 16	GLX
S0	0.9′ × 0.6′	14.59B	

Located 1.4′ SE of NGC3651, this is a small lenticular galaxy with a bright core and a faint outer envelope, elongated E–W. Visually, only the core of the galaxy is visible; it is small, round and brighter to the middle. Although NGC3653 would seem to be physically associated with the large elliptical NGC3651 there is a large difference in their radial velocities: 1276 km/s. Radial velocity: 8859 km/s. Distance: 396 mly. Diameter: 104 000 ly.

NGC3655

H5[1]	11:22.9	+16 35	GLX	
SA(s)c:	1.5′ × 1.0′	12.32B	11.9V	

Images show a small, slightly inclined spiral galaxy with a large, slightly brighter central region and short, high-surface-brightness spiral arms, probably of the multi-arm type. Visually, the galaxy is bright and elongated NE–SW with a small, bright core and a condensed, grainy outer envelope and fairly well defined edges. Radial velocity: 1402 km/s. Distance: 63 mly. Diameter: 27 000 ly.

NGC3659

H53[2]	11:23.8	+17 49	GLX
SB(s)m?	1.5′ × 0.8′	12.92B	

In photographs this is a small, highly inclined spiral galaxy, probably barred, with a brighter, narrow, elongated central region. The outer envelope is asymmetrically bright with the ENE region much brighter. Visually, the galaxy is moderately faint and is best seen at medium magnification. It is bright along its major axis but lacks a brighter core. Elongated ENE–WSW, it is well defined at its extremities. Discovered on 14 March 1784, Herschel thought it resolvable. Radial velocity: 1219 km/s. Distance: 54 mly. Diameter: 24 000 ly.

NGC3662

H4[4]	11:23.8	−1 06	GLX
SAB(r)bc pec	1.4′ × 0.9′	13.77B	

Images show a small, faint almost face-on spiral galaxy elongated N–S, with a very small, brighter core in a faint inner envelope. Two grainy spiral arms emerge and a field star is immediately ENE of the core. Visually, this is a faint and very small galaxy; it is a round patch of light with an apparent bright core, though this is actually the foreground star. Herschel recorded the object on 22 February 1784, placing it in his Class IV with the description: 'Extremely faint and small like a star with a very faint brush south preceding. 240 [magnification] shews the star.' Radial velocity: 5450 km/s. Distance: 244 mly. Diameter: 99 000 ly.

NGC3666

H20[1]	11:24.4	+11 21	GLX
SA(rs)c:	4.5′ × 1.3′	12.69B	

The DSS image shows a large, highly inclined spiral galaxy with an elongated, bright, knotty central region and weak, grainy spiral structure. The galaxy is moderately bright visually, though there is some interference from the bright field star 8.5′ to the ENE in a medium-magnification field. It is an extended oval of light oriented E–W which is broadly brighter along the major axis with a hazy outer envelope. A faint field star is 1.5' to the NNE. Radial velocity: 972 km/s. Distance: 43 mly. Diameter: 57 000 ly.

NGC3670

H337[3]	11:24.7	+23 57	GLX
SB0/a	1.1' × 0.7'	14.42B	

Photographs reveal a small, slightly inclined spiral or lenticular galaxy, with a large, bright, oval core and a faint outer envelope, elongated NNE–SSW. Dark patches N and S of the core may be dust. A small, possibly spiral, galaxy is 3.75' NNW. Visually, only the core of NGC3670 is visible. It is small, moderately bright, condensed and almost round with a brighter middle. The edges are fairly well defined and the envelope has a granular texture. Radial velocity: 7008 km/s. Distance: 313 mly. Diameter: 100 000 ly.

NGC3679

H112[3] (MCG–01-29-021)	11:26.2	−5 35	GLX
Im? pec	1.1' × 0.8'	14.81B	

Herschel's 24 April 1784 entry for this object reads: 'Extremely faint, considerably large, round, resolvable, near a very bright star. Moonlight.' Herschel further remarked that there was so much moonlight that he was uncertain of the actual existence of the nebula, which he said preceded the star, almost touching it. He subsequently searched for the object on 29 December 1786 and on 20 March 1789 but never saw it again, though he did come across a nebula '. . . making a trapezium with 3 small stars, very faint, extended, very small.' Herschel's offset measurements from Psi Leonis (which he conceded might be inaccurate) bring us very close to the position noted above and an object identified as NGC3679 in the first edition of *Uranometria 2000.0*, an irregular Magellanic galaxy. Visually, it is a very faint and slightly extended patch of light, oriented SE–NW, even in surface brightness and moderately well defined. There are three faint stars nearby to the

W and SW which may be Herschel's trapezium candidates. This object is identified as MCG–01-29-021 = NGC3679 on the SIMBAD database. The DSS identifies NGC3679 as a small elliptical galaxy at right ascension 11:25.8, declination −5 45, 2.2' S of and very slightly following a bright field star. This elliptical galaxy is identified on the NASA/IPAC Extragalactic Database as MCG–01-29-012 but the database coordinates place the galaxy at right ascension 11:21.8. It is unlikely that Herschel observed this galaxy and the present author feels that this is not the object that he thought he saw on 24 April 1784. That object probably does not exist.

NGC3681

H159[2]	11:26.5	+16 52	GLX
SAB(r)bc	2.5' × 2.0'	12.42B	

This is part of a loose grouping of four prominent galaxies visible in a low-magnification field, the others being NGC3684, NGC3686 and NGC3691. A number of fainter, non-NGC and dwarf galaxies are in the field, suggesting this is a physical group or cloud of galaxies. Curiously, Herschel missed one of the brighter galaxies, NGC3684. Photographically, this ring galaxy is almost face-on with a very bright, elongated core encircled by thin, grainy spiral arms and a diffuse outer envelope, elongated NNW–SSE. At the eyepiece, it is by far the smallest of the four galaxies

NGC3681 NGC3684 NGC3686 NGC3691
15-inch Newtonian 95x

in the field but stands out well owing to its bright, intense core. It is well condensed and appears circular. Radial velocity: 1171 km/s. Distance: 52 mly. Diameter: 38 000 ly.

NGC3686

H28[3]/H160[2]	11:27.7	+17 13	GLX
SB(s)bc	3.2' × 2.5'	12.00B	

In photographs this is a classic, large, bright, face-on spiral galaxy with a brighter core and a very short bar. Two bright, very knotty, spiral arms emerge from the bar, broadening and fading as they spiral outward in a classic grand-design pattern. Many small dust patches are involved and a faint field star is in the spiral arm 1.1' S of the core. A brighter field star is 2.2' N. Visually, the galaxy is bright; it is a large, smooth oval of light which is a little brighter to the centre. Herschel recorded this galaxy twice, on 14 March and 17 April 1784, but the position of his first observation, H28[3], is more accurate. Radial velocity: 1090 km/s. Distance: 49 mly. Diameter: 45 000 ly.

NGC3689

H339[2]	11:28.2	+25 40	GLX
SAB(rs)c	1.7' × 1.1'	13.01B	

Photographs show a small, slightly inclined spiral galaxy, possibly barred, with a brighter, oval core and two grainy, high-surface-brightness, spiral arms emerging. A third arm fragment emerges from the E. Visually, the galaxy is small, condensed and moderately

NGC3686
©STScI/AURA/CALTECH/ROE

bright. It is fairly even in surface brightness across the envelope, oriented E–W with a grainy texture and well defined extremities. Radial velocity: 2703 km/s. Distance: 121 mly. Diameter: 60 000 ly.

NGC3691

H54[2]	11:28.2	+16 55	GLX
SBb?	1.3′ × 1.0′	12.64B	

Images show a probable dwarf spiral or irregular galaxy, with a very small core surrounded by a grainy outer envelope, elongated NNE–SSW. It is the faintest member of a loose group that includes NGC3681, NGC 3684 and NGC3686. Visually, the galaxy is oval and gradually elongated. Its texture is smooth, diffuse and there is no brightening to the middle. Radial velocity: 1018 km/s. Distance: 45 mly. Diameter: 17 000 ly.

NGC3692

H152[2]	11:28.4	+9 24	GLX
Sb	3.2′ × 0.7′	13.02B	

This is a highly inclined spiral galaxy, which the DSS image suggests may be barred with a bright inner ring and fainter, grainy spiral arms. Visually, this is a very elongated, bright galaxy, oriented E–W with a field star located 2.1′ NE of the core. The envelope is fairly even in surface brightness and the tips taper to soft points. The disc is distinct and well defined. The galaxy forms part of a loose group with NGC3705 and IC2887. Radial velocity: 1633 km/s. Distance: 73 mly. Diameter: 68 000 ly.

NGC3701

H349[2]	11:29.4	+24 05	GLX
Sbc	1.7′ × 0.8′	13.50B	

Images reveal a small, inclined spiral galaxy with a brighter core and faint, grainy spiral arms. At the eyepiece this is a moderately faint galaxy with fairly even surface brightness across the envelope. Poorly defined at the edges, the galaxy is elongated NW–SE. Radial velocity: 2755 km/s. Distance: 123 mly. Diameter: 61 000 ly.

NGC3705

H13[2]	11:30.1	+9 17	GLX
SAB(r)ab	4.9′ × 2.0′	11.76B	11.1V

This is an inclined spiral galaxy with a bright, almost complete inner ring. Photographs show the narrow, grainy spiral arms emerging from the ring. They are relaxed with knotty condensations evident. Visually, this galaxy is very bright and large. The envelope is oval, elongated ESE–WNW and very grainy in texture with moderately well defined edges. The large, condensed and slightly elongated core is bright and quite distinct. The galaxy is part of a loose group with NGC3692 and IC2887. Radial velocity: 926 km/s. Distance: 41 mly. Diameter: 59 000 ly.

NGC3710

H350[2]	11:31.0	+22 46	GLX
E	1.0′ × 0.8′	14.09B	

The photographic and visual appearances of this very small elliptical galaxy are quite similar. The surface brightness is high and the galaxy is almost round with a small, brighter core in a condensed, well defined envelope. Photographs show the galaxy slightly elongated E–W with a faint, diffuse outer envelope. A magnitude 8 field star is 4.25′ NE and a widely separated, slightly fainter pair is about 6.0′ SE. Radial velocity: 6445 km/s. Distance: 288 mly. Diameter: 84 000 ly.

NGC3713

H367[2]	11:31.6	+28 09	GLX
S0⁻:	1.2′ × 0.8′	14.24B	

Images show a small, elongated, lenticular galaxy with a bright core and faint extensions. A faint, compact galaxy is 5.4′ WNW, immediately N of a magnitude 11 field star. Visually, NGC3713 can be seen in the same medium-magnification field as NGC3714, which is just over the border in Ursa Major and described in that section. NGC3713 is a small oval galaxy, gradually brighter to the middle, fairly well defined and elongated NW–SE.

Radial velocity: 6963 km/s. Distance: 311 mly. Diameter: 108 000 ly.

NGC3728

H351[2]	11:33.2	+24 27	GLX
Sb	1.8′ × 1.1′	14.57B	

In photographs this is an inclined spiral galaxy with a bright core embedded in a fainter inner envelope, surrounded by extremely faint, fragmentary spiral structure which is elongated NNE–SSW. Only the core of this galaxy is detected visually; it is small but moderately bright and fairly well seen at high magnification as a condensed, slightly elongated patch of light. Even in surface brightness, it is oriented N–S with sharply defined edges. Radial velocity: 6931 km/s. Distance: 310 mly. Diameter: 162 000 ly.

NGC3731

H80[3]	11:34.2	+12 32	GLX
E	0.8′ × 0.7′	14.42B	

This is a small, faint elliptical galaxy with an anonymous, very faint galaxy 0.5′ S. Visually, although faint, NGC3731 stands out well at medium magnification as a round, condensed patch, brighter to the middle and well defined at the edges. The radial velocity of the fainter galaxy is 26 465 km/s, far higher than that of NGC3731. The magnitude is about 17 and it may be visible in very large-aperture telescopes. Radial velocity: 3044 km/s. Distance: 136 mly. Diameter: 32 000 ly.

NGC3768

H29[3]	11:37.2	+17 51	GLX
S0	1.5′ × 0.9′	13.80B	

In photographs, this is a slightly inclined lenticular galaxy with a very bright core and faint outer envelope, elongated NW–SE. Telescopically, the galaxy is very small, but moderately bright; it is a round and well defined object with even surface brightness. NGC3764 is visible 5.4′ WNW; photographs show it forms a double system with a larger but much dimmer spiral galaxy immediately W.

Radial velocity: 3408 km/s. Distance: 152 mly. Diameter: 66 000 ly.

NGC3772

H352[2]	11:37.8	+22 42	GLX
SBa	1.3′ × 0.7′	14.19B	

Images show a highly inclined, barred spiral galaxy with a bright, round core and a narrow, foreshortened bar connected to diffuse, smooth-textured spiral arms. Visually, the galaxy is small and quite faint, and is best seen at medium magnification as an elongated patch of light oriented NNE–SSW. The surface brightness of the galaxy is fairly even and its edges are moderately well defined. Radial velocity: 3517 km/s. Distance: 157 mly. Diameter: 60 000 ly.

NGC3773

H81[3]	11:38.2	+12 07	GLX
SA0:	1.2′ × 1.0′	13.51B	

Photographs show a very small, face-on galaxy, probably lenticular and possibly a dwarf, with a bright, round core and a very faint outer envelope. In a moderate-aperture telescope the galaxy is a moderately bright, round, well defined patch of light with a brighter, star-like core. Radial velocity: 907 km/s. Distance: 40 mly. Diameter: 14 000 ly.

NGC3790

H109[3]	11:39.8	+17 43	GLX
S0/a	0.8′ × 0.3′	14.66B	

The DSS photograph shows this galaxy is part of a group which includes NGC3801, NGC3802, NGC3803 and NGC3806. Visually, it is a small, inclined, lenticular galaxy, with a bright, elongated core and fainter extensions, oriented NNW–SSE. Radial velocity: 3365 km/s. Distance: 150 mly. Diameter: 35 000 ly.

NGC3798

H340[2]	11:40.2	+24 43	GLX
SB0	2.5′ × 1.8′	13.69B	

Photographs reveal an inclined, theta-shaped, barred galaxy with a bright, elongated core. A short, slightly fainter

NGC3790/3801/3802

bar is attached to the inner ring. Two very faint, broad and smooth spiral arms emerge from the ring and the galaxy is elongated NE–SW. At the eyepiece, only the bright inner region of this galaxy is detected. At high magnification, it is a small but very high-surface-brightness patch of light, oval and elongated slightly E–W. It is opaque and even in surface brightness; no core is visible and the edges are very sharply defined. A magnitude 9 field star is 4.6′ almost due S. Radial velocity: 3517 km/s. Distance: 157 mly. Diameter: 114 000 ly.

NGC3800

H103[2]	11:40.2	+15 21	GLX
SAB(rs)b: pec	2.0′ × 0.6′	13.93B	

Together with NGC3799 1.2′ SW, this is also known as the interacting pair Arp 83. Photographs show a faint plume of material extending from NGC3799's eastern spiral arm towards the larger NGC3800. NGC3800 itself is a highly inclined spiral galaxy, probably barred, with a brighter, elongated core and grainy spiral arms, elongated NE–SW. It was discovered on the very productive night of 8 April 1784 (when Herschel recorded no less than 30 new nebulae), Herschel did not see the small companion galaxy but described NGC3800 as: 'Faint. Small. Elongated. Resolvable. 2 or 3 stars visible in it.' However, there are no field stars involved. Visually, NGC3800 is a

moderately bright, much extended galaxy, a little brighter to the middle but no core is visible. Just off its SW tip is NGC3799, which is slightly fainter and ill defined. NGC3799 is round and gradually brighter to the middle. Radial velocity: 3244 km/s. Distance: 144 mly. Diameter: 84 000 ly.

NGC3801

H161[2]	11:40.3	+17 44	GLX
S0/a	3.5′ × 2.1′	13.18B	

This is the dominant galaxy of a compact group which includes NGC3790, NGC3802, NGC3803 and NGC3806. Photographs show an inclined galaxy, probably lenticular, with a brighter core and fainter extensions which are irregular in outline. Two prominent dust lanes emerge from the core, one extends to the E, the other emerges from the N before bending

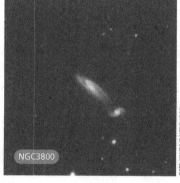

NGC3800

sharply W; together they are somewhat reminiscent of a pair of eyebrows. The galaxy is elongated ESE–WNW. Telescopically, it is the brightest and largest of four galaxies visible in a medium-magnification field. It is a mottled, elongated patch, broadly brighter to the centre with somewhat diffuse edges. Radial velocity: 3257 km/s. Distance: 145 mly. Diameter: 148 000 ly.

NGC3802

H30³	11:40.3	+17 46	GLX
S0/a	1.6′ × 0.4′	14.47B	

This edge-on galaxy, probably a spiral, is located 2.25′ N of NGC3801 and photographs show it is bright along its major axis with a thin dust lane running the length of the major axis. A very faint field star is involved immediately NNE of the core. Visually, NGC3802 is a little fainter than NGC3801 but distinct; it is an elongated galaxy with a brighter core and is well defined and oriented E–W. A faint field star is 1.0′ to the E. Radial velocity: 3262 km/s. Distance: 146 mly. Diameter: 68 000 ly.

NGC3805

H375³	11:40.8	+20 22	GLX
S0⁻:	1.2′ × 0.8′	13.72B	

This is a lenticular galaxy with a bright, elongated core and a fainter outer envelope. Images show that it is elongated ENE–WSW. It is an outlying and somewhat isolated member of the Abell 1367 galaxy cluster, located WNW of the main cluster concentration. At high magnification, the galaxy appears small and moderately bright, with a hazy, round envelope with a brighter core embedded. The edges of the disc are fairly well defined. Radial velocity: 6544 km/s. Distance: 292 mly. Diameter: 102 000 ly.

NGC3808

H338³	11:40.7	+22 27	GLX
SAB(rs)c: pec	1.7′ × 0.9′	14.23B	

This is the brighter member of an interacting pair of galaxies which has also been catalogued as Arp 87. In the

DSS image it is a face-on spiral galaxy with a bright, elongated central region and two broad spiral arms. The northern arm sweeps toward the small companion galaxy, sometimes called NGC3808A and sometimes NGC3808B, depending on the source, 1.0′ to the N. The southern arm is blunter and fades vaguely into the sky background. The companion is somewhat disrupted and is classified as I0? pec. The radial velocities of both galaxies are similar and the core-to-core separation would be at least 91 000 ly at the presumed distance. Telescopically, both galaxies can be seen dimly at high magnification, about 3.75′ NNE from a magnitude 9 field star. NGC3808 is a diffuse, round, nebulous spot, brighter to the middle with hazy edges, while the companion galaxy is a faint, extremely small, stellar patch immediately to the N. Radial velocity: 7033 km/s. Distance: 314 mly. Diameter: 155 000 ly.

NGC3810

H21¹	11:41.0	+11 28	GLX
SA(rs)c	3.9′ × 2.7′	11.4B	

This is a large, bright, multi-arm spiral galaxy, viewed almost face-on, with a brighter, elongated core. Photographs show that six spiral arms emerge from the bright central region, broadening as they expand outward and studded with brighter condensations. The galaxy is elongated NNE–SSW. Visually, this is a bright and large galaxy but it does not hold magnification well. The central region is quite bright but the outer envelope is diffuse and poorly defined. It is gradually brighter to the middle but

NGC3810

©STScI/AURA/CALTECH/ROE

no nucleus is visible. The central region appears quite mottled and the overall form is very slightly oval. Radial velocity: 912 km/s. Distance: 41 mly. Diameter: 47 000 ly.

NGC3812

H320³	11:41.1	+24 51	GLX
E	1.7′ × 1.6′	13.94B	

The visual and photographic appearances of this elliptical galaxy are similar. At high magnification it is a very small, but very high-surface-brightness object and is round, with a bright, well defined envelope and a brighter, stellar core. A bright field star is 1.75′ ESE. It is part of a small group of galaxies which includes NGC3814 and NGC3815; the latter galaxy can also be seen in the field, 7.0′ to the ESE. Radial velocity: 3567 km/s. Distance: 159 mly. Diameter: 79 000 ly.

NGC3815

H339³	11:41.7	+24 49	GLX
Sab	1.7′ × 0.9′	14.05B	

Photographs show a highly inclined spiral galaxy with a bright, elongated core and two spiral arms; the north arm is more extensive but fainter and seems to curve above the plane of the galaxy. Large-aperture telescopes may reveal NGC3814, 2.6′ almost due W. Visually, NGC3815 is well seen as an even-surface-brightness oval, elongated ENE–WSW and a little fainter than NGC3812, which can be seen in the same field. Radial velocity: 3677 km/s. Distance: 164 mly. Diameter: 81 000 ly.

NGC3821

H376³	11:42.2	+20 20	GLX	
(R)SAB(s)ab	1.4′ × 1.3′	13.67B	12.9V	

The DSS photograph shows a face-on barred spiral with a bright round core embedded in an elongated inner envelope. Two faint, smooth spiral arms encircle the galaxy, forming a pseudo-ring. A faint field star is 0.25′ SW of the core, embedded in the inner envelope. The galaxy is an outlying member of the Abell 1367 galaxy cluster. Telescopically, the galaxy is small but stands out well owing to its perceived 'double core', which is actually a bright core and faint field star located immediately SW. Both appear in a common envelope which seems elongated NE–SW and are best seen at high magnification. Herschel's discovery description of 26 April 1785 did not note the star, calling the galaxy only: 'Very faint, very small.' Radial velocity: 5739 km/s. Distance: 256 mly. Diameter: 104 000 ly.

NGC3826

H341[2]	11:42.4	+26 30	GLX
E	0.9′ × 0.7′	14.40B	

Photographs show an elliptical galaxy with a bright elongated core and a faint outer envelope, oriented ENE–WNW. The edge-on galaxy UGC6677 is 5.1′ NE. Telescopically, the galaxy is small and moderately bright; the envelope is round and faint, but fairly well defined with a bright, star-like core embedded. Radial velocity: 9078 km/s. Distance: 406 mly. Diameter: 106 000 ly.

NGC3832

H340[3]	11:43.5	+22 44	GLX
SB(rs)bc	2.1′ × 2.1′	13.69B	

Images show a face-on barred spiral with a very small, round core embedded in a faint bar which brightens where it attaches to two narrow, open spiral arms. Telescopically, the galaxy is small and quite faint; it is a hazy patch of diffuse light, slightly elongated ESE–WNW. The surface brightness is even and the edges are moderately well defined. Radial velocity: 6869 km/s. Distance: 307 mly. Diameter: 187 000 ly.

NGC3842

H377[3]	11:44.0	+19 57	GLX	
E	1.4′ × 1.0′	12.80B	11.8V

This elliptical galaxy is the principal member of the Abell 1367 galaxy cluster which is classed as a type 'F' system in the Rood–Sastry classification scheme, owing to the flattened distribution of some of the brightest cluster members. The galaxy features a bright, elongated core in an extensive, fainter outer envelope and is elongated

N–S. The DSS image shows a compact swarm of galaxies around NGC3842, a scene which is repeated visually in moderate-aperture telescopes. At medium magnification at least 12 galaxies can be seen in the field, with NGC3842 appearing round, moderately bright and condensed to the core with moderately well defined edges. The Herschel object NGC3851 can be seen in the field and is somewhat isolated, about 4.3′ to the ENE. Radial velocity: 6266 km/s. Distance: 280 mly. Diameter: 114 000 ly.

NGC3851

H378[3]	11:44.3	+19 59	GLX
E	0.3′ × 0.2′	15.31B	

This is a very small, faint, elliptical galaxy, which is similar in appearance visually and photographically, with an elongated core oriented E–W. A faint field star is 0.6′ W, while a brighter field star 2.1′ WSW. This galaxy is a member of the Abell 1367 galaxy cluster. Radial

velocity: 6355 km/s. Distance: 284 mly. Diameter: 25 000 ly.

NGC3860

H386[3]	11:44.6	+19 49	GLX
Sa	1.0′ × 0.5′	14.25B	

The DSS image shows an inclined spiral galaxy with a bright core and two spiral arms. A member of the Abell 1367 galaxy cluster, it is located ESE of NGC3842 and several compact galaxies are nearby, particularly to the S. Visually, NGC3860 is moderately bright, slightly oval in shape and oriented NE–SW. It is slightly brighter to the middle and moderately well defined. Radial velocity: 5545 km/s. Distance: 248 mly. Diameter: 72 000 ly.

NGC3862

H385[3]	11:45.1	+19 36	GLX	
E	1.2′ × 1.2′	14.0B	13.71V

Visually and photographically, the appearance of this Abell 1367 elliptical galaxy is quite similar. At medium

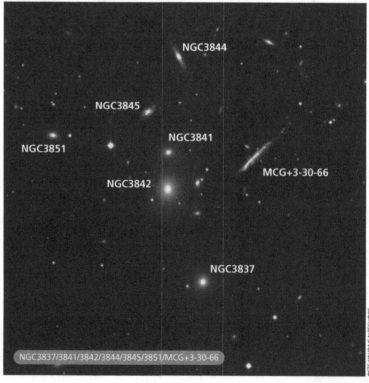

NGC3837/3841/3842/3844/3845/3851/MCG+3-30-66

©STScI/AURA/CALTECH/ROE

magnification it is bright, round, and condensed with sharply defined edges and a brighter, star-like core. It can be seen in the field with five other galaxies: NGC3857, NGC3859, NGC3864, NGC3867 and NGC3868. A slightly fainter, compact object seen on the DSS image 0.8′ to the NNW may be visible in large-aperture telescopes. Radial velocity: 6460 km/s. Distance: 289 mly. Diameter: 101 000 ly.

NGC3872

H104[2]	11:45.8	+13 46	GLX
E5	1.7′ × 1.2′	12.72B	

This elliptical galaxy is elongated NNE–SSW and photographs show an extensive, fainter envelope around the bright elongated core. The galaxy is moderately bright visually and is seen as a roundish, somewhat condensed patch of light which is broadly brighter to the middle with moderately well defined edges. Radial velocity: 3115 km/s. Distance: 139 mly. Diameter: 69 000 ly.

NGC3875

H387[3]	11:45.8	+19 46	GLX
S0/a	1.4′ × 0.5′	14.50B	

Images show a small, lenticular galaxy with a bright core and well defined fainter extensions, elongated E–W. It forms a double system with the brighter galaxy NGC3873, 0.75′ to the WNW and is a member of the Abell 1367 galaxy cluster. Visually, only the core of this galaxy is visible as a tiny, round, nebulous patch, paired with NGC3873, which appears brighter and larger in the field. Herschel evidently missed NGC3873; his discovery description of NGC3875 on 27 April 1785 states only: 'Very faint, very small, resolvable.' Radial velocity: 6908 km/s. Distance: 308 mly. Diameter: 126 000 ly.

NGC3883

H372[3]	11:46.8	+20 42	GLX
SA(rs)b	3.0′ × 2.4′	13.82B	

The DSS photograph shows a face-on, multi-arm spiral galaxy with a very small, elongated core, possibly barred, embedded in a small, faint inner envelope. A pseudo-ring surrounds the core and several very faint, narrow, knotty spiral arms emerge from the ring. The structure is very symmetrical and a bright field star is 5.1′ SSW. NGC3883 is a possible outlying member of the Abell 1367 galaxy cluster. Visually, only the core of this dim galaxy is visible and it appears as a very faint and small, roundish patch with ill defined edges. Radial velocity: 6979 km/s. Distance: 312 mly. Diameter: 272 000 ly.

NGC3884

H388[3]	11:46.2	+20 25	GLX
SA(r)0/a	2.1′ × 1.3′	13.55B	

This is a slightly inclined spiral galaxy with a small, brighter core. Photographs show a partial pseudo-ring made up of almost overlapping spiral arms which expand to very faint and broad outer structure. NGC3884 is a possible outlying member of the Abell 1367 galaxy cluster. This galaxy appears a little larger and more distinct than NGC3883; it is a moderately bright but hazy patch of light with ragged edges, which is almost round and a little brighter to the middle. Radial velocity: 6898 km/s. Distance: 308 mly. Diameter: 188 000 ly.

NGC3900

H82[1]	11:49.2	+27 01	GLX
SA(r)0+	3.7′ × 1.6′	12.29B	

The DSS photograph reveals a highly inclined spiral galaxy with a large, brighter core and a bright inner spiral arm or partial ring. The outer spiral structure is grainy and much fainter. Telescopically, this is a bright, moderately large galaxy, situated on the western flank of a bright triangle of field stars. The bright, oval envelope is quite mottled and well defined and is elongated N–S with a bright, sizable core embedded. The core is slightly oval and elongated along the major axis. Radial velocity: 1776 km/s. Distance: 79 mly. Diameter: 85 000 ly.

NGC3902

H321[3]	11:49.3	+26 07	GLX
SAB(s)bc:	1.6′ × 1.3′	13.98B	

Photographs show an inclined spiral galaxy of the grand-design type, with a small, round core and two principal spiral arms which are knotty and bright. A fainter spiral spur emerges from the eastern spiral arm. Visually, the galaxy is quite dim and is best seen at medium magnification as a slightly elongated, oval patch of light, oriented E–W, with a very faint but well defined disc. Radial velocity: 3575 km/s. Distance: 160 mly. Diameter: 74 000 ly.

NGC3912

H342[2]	11:50.1	+26 29	GLX
SAB(s)b? pec	2.0′ × 0.9′	13.34B	

Photographs show a highly inclined galaxy, probably a spiral, with a large, bright, elongated central region. A small, conspicuous dark patch is located in the outer envelope immediately to the SSW. Visually, the galaxy is situated in a star-poor field; it is a moderately bright oval of light, elongated N–S and fairly well defined. Tiny, star-like nebulous clumps appear on the E and N flanks. Radial velocity: 1755 km/s. Distance: 78 mly. Diameter: 46 000 ly.

NGC3920

H341[3]	11:50.1	+24 57	GLX
S?	1.1′ × 0.9′	14.11B	

Photographs show a round, probably spiral, galaxy with a brighter core in a fainter, envelope. Visually, this is quite a dim galaxy, best seen at medium magnification as a round, uncondensed patch of light. It is small with even surface brightness across the disc. The edges are well defined. Initially, the galaxy NGC3911 was identified as Herschel object H341[3] by J. L. E. Dreyer; it lies immediately W but is considerably dimmer. Radial velocity: 3605 km/s. Distance: 161 mly. Diameter: 51 000 ly.

NGC3926

H379[3]	11:51.5	+22 03	GLX
E	0.4′ × 0.4′	14.7B	

The DSS image shows a very small, round, elliptical galaxy which forms a double system with a smaller galaxy immediately to the W. There is a difference of about 800 km/s in their radial velocities, perhaps due to gravitational interaction. The cores are possibly separated by as little as 34 000 ly. The pair are the brightest in a group of several galaxies in the field. Visually, this object is very faint and best seen as a slightly oval patch of light; two brighter cores are visible intermittently, the one to the E is the brighter. The edges are moderately well defined and the galaxy is best seen at medium magnification. Herschel discovered this object on 26 April 1785 and described it as: 'Very faint, very small, a little extended, easily resolvable or a small patch of stars.' Radial velocity: 7656 km/s. Distance: 342 mly. Diameter: 40 000 ly.

NGC3937

H389[3]	11:52.7	+20 39	GLX
S0⁻:	1.8′ × 1.6′	13.44B	

The DSS image shows a lenticular galaxy with an oval core and an extensive disc, elongated NNE–SSW in a field with several fainter, non-NGC galaxies, including one 2.9′ to the W. It is part of a loose swarm of galaxies which includes Herschel objects NGC3940, NGC3947 and NGC3954. Visually, the galaxy is moderately faint and small but stands out well as a well defined, round, condensed patch, with a brighter, star-like core. In a medium-magnification field the non-Herschel galaxy NGC3943 is visible as a very small, slightly elongated patch of light, located NE of a bright field star. Radial velocity: 6624 km/s. Distance: 296 mly. Diameter: 155 000 ly.

NGC3940

H380[3]	11:52.7	+21 01	GLX
E	1.3′ × 1 2′	13.82B	

The visual and photographic appearances of this elliptical galaxy are quite similar. Small and fairly faint, this galaxy is a condensed, well defined object at medium magnification, very slightly oval and elongated E–W. The edges are well defined and the brightest star in the field is 5.1′ ENE. The galaxy MCG+4-28-78 is 9.0′ to the NW. Radial velocity: 6377 km/s. Distance: 285 mly. Diameter: 108 000 ly.

NGC3944

H322[3]	11:53.1	+26 14	GLX
S0⁻:	1.4′ × 1.1′	14.23B	

Photographs show a lenticular galaxy with a bright elongated core and a faint outer envelope, elongated NE–SW. Visually, the galaxy is very small, but moderately bright, featuring a small, bright core in a much fainter, diffuse outer envelope. Its edges are not well defined and it is best seen at medium magnification. Radial velocity: 3614 km/s. Distance: 161 mly. Diameter: 66 000 ly.

NGC3947

H403[2]	11:53.4	+20 45	GLX
(R)SB(rs)b	1.4′ × 1.2′	13.93B	

Photographs show a barred spiral galaxy with a small, round core and a fainter bar. The bar brightens where it is attached to grainy spiral arms. There are several faint galaxies in the field and NGC3947 is part of a loose swarm which includes Herschel objects NGC3937, NGC3940 and NGC3954. Visually, the galaxy can be seen in a medium-magnification field with NGC3954. It is a small, nebulous patch, oval in shape and elongated E–W. Its surface brightness is even and its edges are hazy. Radial velocity: 6154 km/s. Distance: 275 mly. Diameter: 112 000 ly.

NGC3951

H342[3]	11:53.7	+23 24	GLX
Sa	1.0′ × 0.3′	14.30B	

The appearance of this galaxy is quite similar visually and photographically. It is moderately bright and best seen at medium magnification as a condensed, opaque and well defined oval, elongated N–S with even surface brightness across the disc and a faint field star following to the ESE. The field galaxy MCG+4-28-87 is 7.25′ NW. Radial velocity: 6424 km/s. Distance: 287 mly. Diameter: 84 000 ly.

NGC3954

H381[3]	11:53.7	+20 54	GLX
E?	1.4′ × 1.1′	14.39B	

Photographs show an elliptical galaxy with a bright core in a fainter inner disc surrounded by an extremely faint outer envelope. Visually, it is a very small, round, condensed patch of light with well defined edges. It is a little brighter than NGC3947, which can be seen in the same medium-magnification field. Radial velocity: 6976 km/s. Distance: 312 mly. Diameter: 127 000 ly.

NGC3968

H162[2]	11:55.5	+11 58	GLX
SAB(rs)bc	3.3′ × 1.8′	13.36B	

On the DSS image this galaxy has a small, round core and a very short bar. Two spiral arms emerge from the bar but branch quickly into a loosely-wound, multi-arm pattern, elongated N–S. Field galaxies are 2.8′ SSE and 2.5′ NE. Visually, the galaxy is moderately bright, appearing almost round, with a bright, star-like core, best seen at medium magnification. The edges of the disc are diffuse and the main envelope is fairly even in surface brightness with a bright field star 2.5′ ENE. Radial velocity: 6316 km/s. Distance: 282 mly. Diameter: 270 000 ly.

NGC3983

H343[3]	11:56.4	+23 53	GLX
S0/a	1.1′ × 0.3′	14.57B	

Photographs show a small, edge-on lenticular galaxy with a round, bright, bulging core and bright extensions, elongated ESE–WNW. Telescopically though, only the bright core is visible; the galaxy is very faint and difficult, a very small, round patch of light, and is fairly well defined. Radial velocity: 4232 km/s. Distance: 189 mly. Diameter: 61 000 ly.

NGC3987

H323[3]	11:57.3	+25 12	GLX
Sb	2.3′ × 0.4′	13.92B	

This galaxy is a member of the NGC4005 group of galaxies, which straddles the Coma–Leo border. The DSS photograph shows an ENE–WSW chain of four galaxies which include NGC3989, NGC3993 (a Herschel object) and NGC3997. NGC3987 is an edge-on spiral galaxy with a bright central region and a distinct dust lane is traceable the entire length of the galaxy. Visually, three of the galaxies in the chain are visible in a medium-aperture telescope (NGC3989 is not). NGC3987 is the largest and westernmost member of the chain; it is a well defined, elongated streak of light which is a little brighter along its major axis and is situated 2.5′ S of a field star. Radial velocity: 4477 km/s. Distance: 200 mly. Diameter: 134 000 ly.

NGC3993

H324[3]	11:57.6	+25 15	GLX
Sb: sp	1.9′ × 0.4′	14.6B	

This spindle-shaped galaxy is the third in the four-galaxy chain described under NGC3987. Photographs show an elongated galaxy with a very small, brighter core in an extensive, grainy disc, oriented SE–NW. Visually, NGC3989 is the faintest of the three galaxies seen in a high-magnification field, appearing as a roundish, hazy patch of light, 1.5′ SSE of a pair of field stars. Herschel discovered NGC3987 and NGC3993 on 6 April 1785 and described them as: 'Very faint, a little extended. Suspected another north following, extremely faint, 5 or 6′ distance, pretty sure.' This north-following galaxy is almost certainly NGC3997, which is well seen, situated between two field stars to the ENE of NGC3983. Radial velocity: 4802 km/s. Distance: 215 mly. Diameter: 119 000 ly.

NGC4002

H344[3]	11:58.0	+23 13	GLX
S0⁻a	1.0′ × 0.5′	14.47B	

Photographs show a small, inclined lenticular galaxy with a large bright core and bright extensions, elongated ESE–WNW. A field star is 4.0′ ESE. Companion galaxy NGC4003 is 4.75′ S. Visually, both galaxies can be seen in the same field: NGC4002 is a little brighter, a well-condensed, slightly oval patch of light with a very faint field star 0.6′ ESE of the core. Radial velocity: 6551 km/s. Distance: 292 mly. Diameter: 85 000 ly.

NGC4003

H345[3]	11:58.0	+23 08	GLX
SB0	1.5′ × 1.2′	14.53B	

This galaxy forms a likely physical pair with NGC4002; the minimum core-to-core separation between the two galaxies would be about 404 000 ly at the presumed distance. Images show a small, inclined, barred spiral galaxy with a round core and a faint bar attached to very faint, narrow spiral arms. Elongated SSE–NNW, the galaxy is very faint visually and only the core is detected; it is a very small, round and condensed spot. Radial velocity: 6476 km/s. Distance: 289 mly. Diameter: 126 000 ly.

NGC4004

H354[3]	11:58.1	+27 53	GLX
Pec	2.4′ × 0.7′	14.06B	

Images show a small, inclined, spiral galaxy, oriented N–S, with a peculiar morphology. The core is oval and bright with two principal spiral arms. To the N the spiral structure is blunted and broadens, defined by three E–W oriented nebulous knots, while the southern spiral arm is straight with knotty condensations. The galaxy is a likely member of the NGC4008 group of galaxies and companion galaxy IC2982 (also identified as NGC4004B) is 3.1′ W,

while NGC3988 is 9.0′ W. Though moderately faint, NGC4004 is well seen visually at medium magnification as a well defined, elongated patch of light, with a field star 1.0′ to the ESE. Oriented N–S, the envelope is even in surface brightness and though IC2982 is not seen, NGC3988 is seen in the same field to the W as a small, round, well defined patch of light. Radial velocity: 3352 km/s. Distance: 150 mly. Diameter: 105 000 ly.

NGC4005

H325[3]	11:58.2	+25 07	GLX
Sb	1.0′ × 0.6′	13.90B	

This galaxy is the principal galaxy of the NGC4005 group of galaxies, which straddles the border between Leo and Coma Berenices. Photographs show an inclined spiral galaxy, probably barred, with a large, elongated central region, perpendicular to the fainter outer disc. Visually, the galaxy is situated 1.5′ ESE of a bright field star. It is a small, elongated oval oriented E–W, well defined and broadly brighter to the middle. The medium-magnification field also shows NGC4015, NGC4021 and NGC4022 and there are a total of 15 NGC galaxies within a 15.0′ radius. Radial velocity: 4443 km/s. Distance: 199 mly. Diameter: 58 000 ly.

NGC4008

H368[2]	11:58.3	+28 12	GLX
S0/E5?	2.5′ × 1.3′	12.97B	

This is the principal galaxy of a group of 14 which includes the Herschel objects NGC4017 and NGC4004. Photographs show an elongated lenticular galaxy with a bright core and an extensive, fainter outer envelope elongated NNW–SSE. The faint companion galaxy UGC6968 is 8.25′ NE. Visually, NGC4008 is a bright oval galaxy with high surface brightness, a well defined envelope and a tiny, brighter, star-like core. Radial velocity: 3605 km/s. Distance: 161 mly. Diameter: 117 000 ly.

Leo Minor

Herschel recorded 36 objects in this constellation between the years 1785 and 1788. All of them are galaxies and there are several bright and interesting examples including NGC2859, NGC3003 and NGC3245. NGC3158 is located in an interesting field of faint galaxies which will be of interest to patient observers with moderate- or large-aperture instruments.

NGC2859

H137[1]	9:24.3	+34 31	GLX
(R)SB(r)0+	4.3′ × 3.8′	11.86B	

The DSS image shows a slightly inclined, theta-shaped, barred galaxy with a bright core and a fainter short bar which brightens where it is in contact with the inner ring. A second, much fainter ring surrounds the galaxy. A bright field star is 6.1′ WNW. Telescopically, the galaxy is fairly bright with a bright, circular core embedded in a fainter outer envelope which is round and fades uncertainly into the sky background. Recorded on 28 March 1786, Herschel described the object as: 'Very bright, round, very suddenly much brighter to the middle. Chevelure 3′ in diameter.' Radial velocity: 1657 km/s. Distance: 74 mly. Diameter: 93 000 ly.

NGC2955

H541[3]	9:41.3	+35 53	GLX
(R′)SA(r)b	1.7′ × 0.9′	13.58B	

Photographs reveal a slightly inclined spiral galaxy with a brighter core and two fairly thin, knotty spiral arms. Visually, this is a faint and moderately difficult galaxy located 2.25′ N of a magnitude 11 field star. The extremities are poorly defined and the galaxy is slightly elongated NNW–SSE. It appears oval with a faint stellar core suspected with averted vision. Curiously, Herschel thought this galaxy resolvable. He observed it at least three times: the discovery date was 28 March 1786.

Radial velocity: 6990 km/s. Distance: 312 mly. Diameter: 154 000 ly.

NGC2965

H751[3]	9:43.2	+36 14	GLX
S0	1.1′ × 1.0′	14.44B	

The DSS image shows a lenticular galaxy with a bright core and a faint, extensive, outer envelope, elongated ENE–WSW. This is the brightest galaxy of a small group which includes the barred spiral galaxy NGC2971, which is 6.75′ SE. Three faint possible satellites form a line to the S: all of them are within 1.5′ of NGC2965 and may be detected in large-aperture telescopes. Visually, NGC2965 is fairly bright and appears as a round, fairly opaque disc, very slightly brighter to the middle with well defined edges. The much fainter galaxy NGC2971 is visible to the SE as a

NGC2859

diffuse, roundish patch of light, broadly brighter to the middle. Radial velocity: 6712 km/s. Distance: 300 mly. Diameter: 96 000 ly.

NGC3003

H26[5]	9:48.6	+33 25	GLX
SBbc	5.8′ × 1.3′	12.25B	

Photographs show a large, highly inclined spiral galaxy, possibly barred, with a very thin elongated core. The outer envelope is grainy with a few knots and traces of spiral structure. In a moderate-aperture telescope the galaxy is large but somewhat diffuse and much extended ENE–WSW. The envelope is very mottled; its surface brightness is uneven and it has hazy, poorly defined edges and extensions that fade uncertainly into the sky background. Radial velocity: 1446 km/s. Distance: 65 mly. Diameter: 109 000 ly.

NGC3021

H115[1]	9:51.0	+33 33	GLX
SA(rs)bc:	1.5′ × 0.8′	12.54B	

This is an inclined spiral galaxy. Photographs show a slightly brighter central region and bright spiral arms. A very faint field star borders the galaxy to the NE. High-resolution photographs reveal a multi-arm spiral pattern of high surface brightness. Visually, the galaxy is small but moderately bright, elongated almost due E–W with a bright, grainy-textured envelope. It is a little brighter to the middle with well defined extremities which exhibit a sharp fall-off in brightness at the edges. The faint field star mentioned above is just visible. A bright field star is 1.0′ SE. Radial velocity: 1510 km/s. Distance: 68 mly. Diameter: 29 000 ly.

NGC3074

H542[3]	9:59.7	+35 24	GLX
SAB(rs)c	2.3′ × 2.1′	13.46B	

This is a face-on spiral galaxy, probably of the grand-design type. Photographs show a small, round core and two principal spiral arms, each of which

quickly branches into two. The galaxy is quite faint visually and is seen as a round, somewhat diffuse patch of light, quite even in surface brightness and well defined at the edges. Radial velocity: 5122 km/s. Distance: 229 mly. Diameter: 153 000 ly.

NGC3099

H478[3]	10:02.7	+32 42	GLX
0.9′ × 0.5′	15.21B		

The DSS image shows a very small, lenticular galaxy with a bright core and a faint, extensive outer envelope, elongated ESE–WNW. It is unclassified but may be a cD elliptical. This is the brightest galaxy of a small group that includes companion galaxy NGC3099B, which is 1.2′ WNW. About two dozen faint galaxies are within a 7.0′ radius of NGC3099. In a moderate-aperture telescope the galaxy is extremely faint and is seen only intermittently as a small, almost

round nebulous patch, showing little condensation to the centre and poorly defined edges. Discovered on 7 December 1785, Herschel's terse comment reads: 'Extremely faint, small, left doubtful.' This is a very remote object. Radial velocity: 15 141 km/s. Distance: 676 mly. Diameter: 177 000 ly.

NGC3104

H48[4]	10:03.9	+40 45	GLX
IAB(s)m	3.3′ × 2.2′	13.97B	

This large, faint irregular galaxy with a diffuse, slightly brighter central region has also been catalogued as Arp 264. Photographs reveal many knotty condensations, including a large one to the S, and the galaxy is elongated NE–SW. The galaxy and the field stars 8.5′ SE and 7.25′ SSW form an equilateral triangle. In a moderate-aperture telescope this is a very faint and difficult object, owing to the

NGC3003

©STScI/AURA/CALTECH/ROE

presence of a magnitude 13 field star which hinders visibility somewhat. It is visible at medium magnification as a faint, irregular glow located for the most part NE of the field star. There is no concentration to the middle and the extremities are very poorly defined. The galaxy was discovered on 18 March 1787, when the faint nebulosity around the field star suggested a planetary nebula to Herschel, who described the object as: 'A very faint star affected with very faint nebulosity. Extended south preceding north following, 1' long. 300 (magnification).' Radial velocity: 599 km/s. Distance: 27 mly. Diameter: 26 000 ly.

NGC3106

H320[2]	10:03.9	+31 11	GLX
S0	2.1' × 1.9'	13.32B	

The DSS image shows a face-on spiral galaxy with a small, round core and very faint and narrow spiral arms which extend out to about 3.0' × 2.7'. Based on the available image, the published classification and size would appear to be in error. Visually, the galaxy is fairly bright, located mid-way between two field stars on a line oriented ESE–WNW. It is quite condensed at the centre with hazy extremities. Radial velocity: 6167 km/s. Distance: 275 mly. Diameter: 240 000 ly.

NGC3158

H639[2]	10:13.8	+38 46	GLX
E3:	1.8' × 1.5'	12.90B	

This elliptical galaxy is the brightest member of a fairly compact group of about 30 galaxies: eight of them are relatively bright, the rest quite faint and small and most of them are contained within a 15.0' field. The photographic and visual appearances of NGC3158 are similar. It is fairly bright, almost round, with grainy texture and a fainter outer envelope. The core is bright and the edges are fairly well defined. Images show the outer envelope is elongated NNW–SSE. In a high-magnification field, five other galaxies are readily visible, including the Herschel object

NGC3163, which is the westernmost member of a chain of three E–W oriented galaxies, 6.75' to the SSE. Radial velocity: 6982 km/s. Distance: 312 mly. Diameter: 163 000 ly.

NGC3163

H640[2]	10:14.1	+38 39	GLX
SA0⁻:	0.9' × 0.8'	14.36B	

The DSS image reveals a lenticular galaxy with a broad, bright core in a fainter outer envelope. A possible dwarf companion galaxy is in the outer envelope immediately E, though this may just be a very faint field star. NGC3163 is the third of three galaxies in a row, NGC3159 and NGC3161 being the others. Visually, NGC3163 is visible as a small, concentrated nebulous spot, a little brighter to the middle with moderately well defined edges. Its brightness is very similar to that of its two companions to the W. Discovered on 17 March 1786, Herschel considered both this galaxy and NGC3158 resolvable. Radial velocity: 6251 km/s. Distance: 279 mly. Diameter: 73 000 ly.

NGC3245

H86[1]	10:27.3	+28 30	GLX
SA(r)0°:?	3.3' × 2.0'	11.66B	

This lenticular galaxy is the principal member of a small group of galaxies which includes the Herschel objects NGC3254, NGC3265, NGC3277. Photographs show NGC3245 has a bright elongated core in an extensive outer envelope and the fainter companion galaxy NGC3245A is situated 9.0' to the NNW. Photographs show a bright, elongated core in an extensive outer envelope. Visually, this is a bright and large galaxy and is well seen at high magnification as an oval disc, oriented N–S with a large, elongated and bright central region surrounding a star-like core. The disc is slightly fainter but opaque and mottled with well defined edges. NGC3245A may be visible in large-aperture instruments. Radial velocity:

1270 km/s. Distance: 57 mly. Diameter: 54 000 ly.

NGC3254

H72[1]	10:29.3	+29 30	GLX
SA(s)bc	4.8' × 1.3'	12.29B	

Photographs show a highly inclined spiral galaxy with a small but brighter elongated core and grainy, fainter spiral arms. In a moderate-aperture telescope, the galaxy is moderately bright and is elongated NE–SW with a brighter, stellar core. The envelope is mottled and ill defined and the extensions fade gradually into the sky background. A widely separated pair of magnitude 9 field stars is 5.5' E. Radial velocity: 1315 km/s. Distance: 59 mly. Diameter: 82 000 ly.

NGC3265

H349[3]	10:31.1	+28 48	GLX
E:	0.9' × 0.6'	14.04B	13.5V

Based on its photographic appearance, the classification of this elliptical galaxy should probably be about E3. Its visual and photographic appearances are quite similar: in a moderate-aperture telescope the galaxy is very faint with a star-like core and is very slightly extended ENE–WSW. A very faint field star is immediately SE. Radial velocity: 1277 km/s. Distance: 57 mly. Diameter: 15 000 ly.

NGC3277

H359[2]	10:32.9	+28 31	GLX
SA(r)ab	2.8' × 2.1'	12.48B	11.7V

A cursory glance at the DSS image indicates a face-on lenticular galaxy, but higher-resolution images show an almost face-on spiral galaxy with a large, bright core and very faint, thin spiral arms in a very faint and extensive outer envelope. Visually, this is a fairly bright, almost round galaxy. The envelope is bright and brightens to a sizable core. The envelope is mottled and though it is not well defined, the light at the edges drops off quite suddenly. Radial velocity: 1365 km/s. Distance: 61 mly. Diameter: 50 000 ly.

NGC3294

H164[1]	10:36.3	+37 20	GLX
SA(s)c	3.5′ × 1.8′	12.15B	

This is a bright, inclined spiral galaxy. Photographs show a very small, bright core and a bright, knotty, multi–arm, spiral pattern. This is a large and bright object visually, an elongated oval oriented ESE–WNW, brighter along its major axis with a disc that is fairly well defined at the edges. The grainy disc is best seen at medium magnification and a field star is 5.25′ ENE. Radial velocity: 1578 km/s. Distance: 70 mly. Diameter: 72 000 ly.

NGC3304

H615[3]	10:37.6	+37 27	GLX
SB(s)a?	1.7′ × 0.5′	14.53B	

This highly inclined galaxy is probably a spiral. Photographs show a bright core and fainter grainy extensions which show evidence of dust. Telescopically, this galaxy is moderately faint and is seen as an extended disc, elongated NNW–SSE with a well defined envelope and a brighter, small core. Radial velocity: 6889 km/s. Distance: 308 mly. Diameter: 152 000 ly.

NGC3327

H348[2]	10:40.0	+24 05	GLX
SA(r)b:	1.2′ × 0.9′	14.18B	

Images show a slightly inclined spiral galaxy with a brighter core and a grainy outer envelope. A faint field star is involved immediately W. Visually, the galaxy is fairly pale, though moderately well defined and appears elongated due E–W. It is brighter to a small, faint, stellar core. The field star involved is very faint though it may contribute to the elongated appearance. Radial velocity: 6246 km/s. Distance: 279 mly. Diameter: 97 000 ly.

NGC3334

H641[2]	10:41.4	+37 18	GLX
S0?	1.1′ × 0.9′	13.85B	

The visual and photographic appearances of this lenticular galaxy are quite similar. At high magnification, the galaxy is small but fairly bright with a round, even-surface-brightness disc that is fairly well defined at the edges. Three field stars follow closely in the NE. Radial velocity: 7195 km/s. Distance: 321 mly. Diameter: 103 000 ly.

NGC3344

H81[1]	10:43.5	+24 55	GLX
(R)SAB(r)bc	7.1′ × 6.5′	10.5B	9.7V

This is a face-on, multi-arm spiral galaxy with a small, round, bright core and a very thin, faint bar. The spiral arms are grainy with many knots and thin dust lanes. Four field stars are involved in the disc E of the core. In moderate-aperture telescopes, this is a bright and interesting galaxy. Herschel noted the two brightest stars but the third one, the closest to and SE of the core, can be seen well at high magnification. The galaxy's outer envelope is circular and diffuse and, as it approaches the bright pair, it becomes fainter. The edges of the envelope are poorly defined. The core is bright, round and definitely nonstellar. Radial velocity: 533 km/s. Distance: 24 mly. Diameter: 49 000 ly.

NGC3380

H360[2]	10:48.2	+28 36	GLX
(R′)SBa?	1.7′ × 1.6′	13.48B	

The DSS photograph shows a possible inclined spiral galaxy or barred galaxy with a peculiar morphology. The large, bright core has curved extensions emerging into the slightly fainter disc. A very faint ring surrounds the galaxy, connected to the inner galaxy to the N. The ring is almost round, but the inner galaxy is elongated NNE–SSW. Not surprisingly, the appearance in a moderate-aperture telescope is restricted to the inner disc, which appears as a small, faint patch of light, well defined along its edges and fairly

NGC3344

even in surface brightness. It is oval and slightly elongated. Radial velocity: 1568 km/s. Distance: 35 000 ly.

NGC3381

H565[2]	10:48.4	+34 42	GLX
SB pec	2.0′ × 1.9′	12.62B	

The DSS image shows an almost face-on, barred spiral galaxy with a bright central bar but with no core attached to two bright, grainy spiral arms. The arms broaden to asymmetrical outer structure, which is more extensive to the WSW. Visually, the galaxy is moderately bright and fairly large. It is almost round and gradually brighter to the middle. The texture of the envelope is grainy and the edges are moderately well defined. Radial velocity: 1614 km/s. Distance: 72 mly. Diameter: 42 000 ly.

NGC3395

H116[1]	10:49.8	+32 59	GLX
SAB(rs)cd pec:	2.1′ × 1.2′	12.40B	12.1V

This galaxy is paired with NGC3396 1.1′ ENE and together they have also been catalogued as Arp 270. The DSS image shows an inclined spiral galaxy with disturbed spiral structure involving interaction with neighbour NGC3396; a faint plume of material connects the two galaxies. The core is somewhat chaotic and displaced towards the N. High-resolution images show several nebulous knots in the spiral structure. Visually, NGC3395 is the brighter and larger of the two galaxies; it is elongated N–S with a bright, grainy and fairly well defined envelope and a brighter, star-like core embedded. At their discovery on 7 December 1785, Herschel combined the descriptions of the two galaxies to read: 'Two; the 1st, considerably bright, considerably large, irregularly extended; the 2nd, pretty bright, pretty large, irregularly extended. Distance 1′ at the vertex at the north ends.' Radial velocity: 1604 km/s. Distance: 72 mly. Diameter: 44 000 ly.

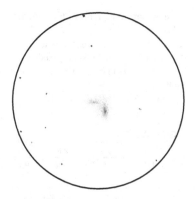

NGC3395 NGC3396 15-inch Newtonian 146x

NGC3396

H117[1]	10:49.9	+32 59	GLX	
IBm pec	2.5′ × 0.9′	12.43B	12.5V	

Although this is classified as an irregular galaxy, there is some evidence of rudimentary spiral structure which curves back towards companion galaxy NGC3395, and the bright, elongated central region suggests a possible bar. High-resolution photographs indicate probable intense star formation, particularly in the bright central region, as well as an extensive and very faint outer envelope of stars. Visually, the galaxy has a grainy envelope with a star-like core embedded and the poorly defined envelope is elongated E–W. Radial velocity: 1604 km/s. Distance: 72 mly. Diameter: 52 000 ly.

NGC3400

H361[2]	10:50.8	+28 28	GLX
SB(s)a:	1.3′ × 0.8′	14.06B	

Photographs show a highly inclined, barred galaxy with a large, bright core and fainter bar which brightens where it is attached to the outer disc, which is elongated E–W. At medium magnification, this galaxy is small and

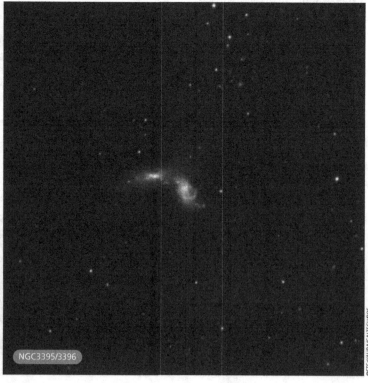

NGC3395/3396

moderately bright, slightly oval in shape and gradually elongated N–S, indicating that only the core and bar are visible. It is well condensed with fairly well defined edges. Radial velocity: 1370 km/s. Distance: 61 mly. Diameter: 23 000 ly.

NGC3413

H493[2]	10:51.3	+32 46	GLX
S0	2.2′ × 0.7′	13.15B	

Although it can be seen in the same medium-magnification field as the galaxies NGC3424 and NGC3430, its lower radial velocity indicates this is a foreground object. The photographic and visual appearances of this highly inclined lenticular-type galaxy are quite similar. The galaxy is a N–S extended oval with a large, round core, fainter extensions and well defined edges. Based on the presumed distance, this is probably a dwarf galaxy. Radial velocity: 623 km/s. Distance: 28 mly. Diameter: 17 000 ly.

NGC3414

H362[2]	10:51.3	+27 59	GLX
S0 pec	3.5′ × 2.6′	12.06B	

The DSS image reveals an edge-on lenticular galaxy with peculiar structure. The large, bright, round core has fainter, narrow extensions emerging along a NNE–SSW axis and is embedded in a broad, faint outer

**NGC3413 NGC3424 NGC3430
15-inch Newtonian 272x**

envelope. A small, compact galaxy is 2.0′ NNW. The edge-on galaxy UGC5958 is 7.0′ S and the Herschel object NGC3418 is 8.0′ NNE. In a moderate-aperture telescope NGC3414 and NGC3418 can be seen together in a high-magnification field. NGC3414 is very bright, with a bright, round core and a grainy envelope. It appears round with ragged ill defined edges; the faint extensions are not visible. The galaxy has also been catalogued as Arp 162. Radial velocity: 1374 km/s. Distance: 61 mly. Diameter: 62 000 ly.

NGC3418

H363[2]	10:51.3	+28 07	GLX
SAB(s)0/a:	1.4′ × 1.1′	14.19B	

Photographs show an inclined spiral galaxy with a large, bright core and a possible bar embedded in two broad, smooth-textured spiral arms, elongated E–W. Visually, the galaxy is small but

moderately bright, slightly elongated, with a fairly even surface brightness and well defined edges. Radial velocity: 1252 km/s. Distance: 56 mly. Diameter: 23 000 ly.

NGC3424

H494[2]	10:51.8	+32 54	GLX
SB(s)b:?	2.8′ × 0.8′	13.13B	

Photographs show an almost edge-on spiral galaxy, possibly barred, with an elongated, brighter central region in a grainy envelope. There are condensations in the western extensions and a faint field star is involved E of the core. The oblique spiral galaxy NGC 3430 is 6.0′ NE. Visually, this galaxy is a well defined sliver of light with fairly even surface brightness along its major axis, which is elongated SSE–NNW. A faint field star is involved in the outer envelope to the E. Both this galaxy and NGC3430 may be physically related to the

NGC3414

NGC3424/3430

©STScI/AURA/CALTECH/ROE

A condensation in the NE is intermittently visible. Radial velocity: 609 km/s. Distance: 27 mly. Diameter: 53 000 ly.

NGC3451

H364[2]	10:54.4	+27 14	GLX
Sd	1.7′ × 0.8′	13.35B	

Photographs show a small, highly inclined spiral galaxy with a small, round core and grainy, dusty spiral arms. In a moderate-aperture telescope, this is a somewhat faint, oval galaxy which is a little brighter at medium magnification. It is elongated NE–SW and is broadly brighter to the middle, though no core is visible. The edges fade gradually into the sky background. Radial velocity: 1292 km/s. Distance: 58 mly. Diameter: 29 000 ly.

NGC3486

H87[1]	11:00.4	+28 58	GLX	
SAB(r)c	7.1′ × 5.2′	11.07B	10.5V	

Photographs show a large, very slightly inclined, multi-arm spiral galaxy with a small, round core and a very faint bar connected to a brighter inner ring. Thin, low-surface-brightness spiral arms emerge from the ring and many small nebulous knots are involved with the spiral arms, which are intact for only short arcs around the galaxy core. In a moderate-aperture telescope, the galaxy is large and fairly bright, but ethereal with a small, brighter core and a grainy envelope and is almost round with ill defined edges. Radial velocity: 648 km/s. Distance: 29 mly. Diameter: 60 000 ly.

NGC3504

H88[1]	11:03.2	+27 58	GLX	
(R)SAB(s)ab	2.7′ × 2.7′	12.62B	11.8V	

This slightly inclined, barred galaxy has a large, bright core and a fainter, thin bar which is attached to a dusty, bright, oval, inner envelope. Two faint outer arms emerge from the envelope at the bar and form an almost complete

NGC3395, NGC3396 pair, as they all have similar radial velocities. Radial velocity: 1473 km/s. Distance: 66 mly. Diameter: 54 000 ly.

NGC3430

H118[1]	10:52.2	+32 57	GLX
SAB(rs)c	4.0′ × 2.2′	12.19B	

This is an inclined spiral galaxy with a small, round core and grainy, dusty spiral arms. Photographs show that the spiral structure is most likely multi-arm, with fairly high surface brightness near the core, but fading and broadening at the outskirts. In a moderate-aperture telescope, the galaxy is a large oval patch of light, with grainy texture and broadly concentrated to the middle. The extremities are fairly well defined and the galaxy is elongated NNE–SSW. Radial velocity: 1565 km/s. Distance: 70 mly. Diameter: 81 000 ly.

NGC3432

H172[1]	10:52.5	+36 37	GLX	
SB(s)m	6.8′ × 1.5′	11.65B	11.7V	

This is a large edge-on galaxy, with an elongated, slightly distorted central region and fainter extensions. There is some evidence of dust and several knotty condensations along the bright major axis of the galaxy; high-resolution photographs begin to resolve individual stars at about blue magnitude 21. The galaxy has also been catalogued as Arp 204. The DSS image also shows a low-surface-brightness, dwarf irregular galaxy 3.4′ WSW of the core. Visually, NGC3432 is a very bright galaxy; it is a very elongated sliver of light with three stars involved, two SE of the core and the other immediately E of the core. The galaxy is well defined along its major axis and averted vision extends its length. No core is visible and the galaxy is brighter in its NE half.

NGC3432

©STScI/AURA/CALTECH/ROE

NGC3510

H365[2]	11:03.7	+28 53	GLX
SB(s)m	3.7′ × 0.6′	13.55B	

The DSS image shows a low-surface-brightness edge-on galaxy with a bar-like, brighter central region and fainter, slightly curved extensions. The galaxy is relatively nearby and stars begin to resolve at about blue magnitude 22. In moderate apertures, this is a faint galaxy, though fairly large, especially when seen with averted vision which brings out the fainter extensions. The surface brightness is quite even, with no brighter core visible. The galaxy is elongated NNW–SSE with a magnitude 7 field star 7.25′ to the WNW. Radial velocity: 672 km/s. Distance: 30 mly. Diameter: 33 000 ly.

NGC3512

H366[2]	11:04.0	+28 02	GLX	
SAB(rs)c	1.6′ × 1.5′	13.02B	12.9V	

Photographs show a small, face-on spiral galaxy with a small, round core and bright, knotty spiral arms. Visually, the galaxy is small but fairly bright, well condensed and a little brighter to the middle. The envelope is round with moderately well defined edges. The galaxy probably forms a pair with NGC3504. Radial velocity: 1340 km/s. Distance: 60 mly. Diameter: 28 000 ly.

pseudo-ring. Visually, the galaxy is well seen at high magnification; only the faint, smooth-textured outer spiral arms are not seen. The galaxy is moderately large and quite bright with a bright core and a bright but ill defined outer envelope, which corresponds to the oval inner region. The galaxy is elongated NNW–SSE. A very faint field star is just beyond the envelope to the ESE. Radial velocity: 1498 km/s. Distance: 67 mly. Diameter: 52 000 ly.

Lepus

The 13 Herschel galaxies in Lepus were all recorded between the years 1784 and 1786. A few of the objects, such as NGC1964, are fairly bright but the most interesting object visually is probably the galaxy pair NGC1888 and NGC1889.

NGC1781

H268[3]	5:07.9	−18 11	GLX
(R)SB(s)0° pec:	1.3′ × 1.1′	13.74B	

Images reveal a lenticular galaxy with a large, bright, elongated core which probably includes a bar and a slightly fainter outer envelope. Visually, this is a moderately faint galaxy, which is irregularly round, brightens slightly to the middle and has moderately well defined edges. Radial velocity: 4855 km/s. Distance: 217 mly. Diameter: 82 000 ly.

NGC1832

H292[2]	5:12.1	−15 41	GLX
SB(r)bc	2.6′ × 1.7′	11.64B	

Photographs show a slightly inclined spiral galaxy with a large, bright, elongated core and bright, knotty spiral arms. High-contrast images reveal the central bar and resolve the inner ring: the outer spiral structure emerges from several points along the ring. Visually, the galaxy precedes a bright field star 1.1′ E. At medium magnification the galaxy is a moderately bright, diffuse object with hazy extremities. The main body is extended N–S. At high magnification the central region is bright and large, extending to about 1.0′ × 0.45′. Occasionally, a faint stellar core is visible. The core is well condensed and well defined. Radial velocity: 1808 km/s. Distance: 81 mly. Diameter: 61 000 ly.

NGC1888

H289[2]	5:22.5	−11 30	GLX
SB(s)c pec	3.4′ × 1.1′	12.77B	

This galaxy is paired with the elliptical system NGC1889, which lies immediately E, and together they are also catalogued as Arp 123. The DSS image shows a highly inclined spiral galaxy with a bright elongated envelope, a prominent dust lane and a disturbed outer spiral arm to the SSE. In the image NGC1889 appears to blend into the spiral and seems to be behind the larger NGC1888. Telescopically, NGC1888 is moderately bright with a bright central region and faint extensions, oriented NNW–SSE. It appears a little broader to the NW, tapering to a narrow point in the SE. In poor conditions NGC1889 appears as a slight bulge to the NE of the core but good skies and high magnification reveal it as a separate object ENE of NGC1888's core; it is seen as a small, round, well defined patch of light. It would seem that Herschel was aware of NGC1889, but did not record it as a separate object; he called NGC1888 'triangular', implying that he saw the unresolved bulge. Observed on 31 January 1785, his description reads: 'Faint, pretty large, irregularly triangular. Faint, resolvable.' Radial velocity: 2306 km/s. Distance: 103 mly. Diameter: 102 000 ly.

NGC1954

H590[3]	5:32.8	−14 04	GLX
SA(rs)bc pec:	4.0′ × 2.0′	12.56B	

Images show a face-on spiral galaxy with a small, round, core embedded in a fainter envelope. The loosely structured spiral arms are defined by nebulous knots in an extended, loosely curved pattern

NGC1888

NGC1954

that forms part of a pseudo-ring. Two faint field stars are involved to the NNW, while fainter anonymous galaxies are 4.5′ SSE and 4.5′ ESE. Telescopically, NGC1954 is a very faint galaxy, best seen at medium magnification as an oval haze, oriented SSE–NNW with a faint pair of stars to the NNW. It is an ill defined glow, very broadly brighter to the middle; high magnification occasionally shows a small, brighter core. Radial velocity: 2992 km/s. Distance: 134 mly. Diameter: 156 000 ly.

NGC1964

H21[4]	5:33.4	−21 57	GLX
SAB(s)b	5.6′ × 2.1′	11.53B	10.8V

This is a large, highly inclined spiral galaxy with a large, elongated, bright core and thin, grainy spiral arms. Dust is evident along the inner edge of the spiral arm on the WNW flank as well as in the inner disc. Four faint stars are involved with the spiral arms N, W and SW of the central region. Telescopically, although the galaxy is bright, only its inner portion is well seen and the three prominent field stars to the NNW hinder the view somewhat. The core area is bright and elongated along the major axis. The secondary envelope is much fainter and diffuse and the edges are poorly defined. Only one of the faint field stars, the one W of the core, is seen. and the galaxy is elongated NNE–SSW. Herschel discovered the galaxy on 20 November 1784 and described it as: 'Very small, stellar, very bright nucleus and very faint chevelure not quite central.' Radial velocity: 1503 km/s. Distance: 67 mly. Diameter: 110 000 ly.

NGC1979

H240[3]	5:33.8	−23 19	GLX
SA	2.0′ × 1.8′	12.91B	

This lenticular galaxy has a slightly elongated core and a faint outer envelope, elongated N–S. Photographs show a small cluster of dim galaxies about 7.0′ to the SE. Telescopically, though somewhat faint, this galaxy is nevertheless well seen as a well defined circular patch of light with a prominent star-like core embedded. A field star

NGC1964

©STScI/AURA/CALTECH/ROE

follows the galaxy 1.75′ to the ESE. Radial velocity: 1543 km/s. Distance: 69 mly. Diameter: 40 000 ly.

NGC1993

H269[3]	5:35.5	−17 49	GLX
SA(rs)0⁻:	1.4′ × 1.2′	13.62B	

In photographs this lenticular galaxy has a round core and a faint outer envelope and there is a very faint galaxy 5.9′ ENE. Telescopically, the galaxy is small but moderately bright and is well seen as a condensed, opaque patch of light, even in surface brightness and well defined at the edges. Radial velocity: 2987 km/s. Distance: 133 mly. Diameter: 54 000 ly.

NGC2073

H241[3]	5:45.9	−21 59	GLX
SA(rs)0⁻:	1.4′ × 1.3′	13.45B	

Photographs show a lenticular galaxy with a round core and a faint outer envelope. A fainter anonymous galaxy is 6.75′ NW. The galaxy is faint visually

but is well seen at both medium and high magnification as a round patch of light, moderately well defined and a little brighter to the middle. Radial velocity: 2821 km/s. Distance: 126 mly. Diameter: 51 000 ly.

NGC2076

H267[3]	5:46.8	−16 46	GLX
S0⁺: sp	2.4′ × 1.4′	13.93B	

Photographs show this edge-on galaxy is a 'Sombrero' type with a large, bright, elongated central region bisected by a prominent dust lane running the length of the major axis. Several very faint field stars are seen in silhouette against the disc of the galaxy, which is elongated NE–SW. Visually, the galaxy appears largish but quite dim, and is visible as a diffuse and irregularly-shaped patch of light with poorly defined edges. A brightening in the envelope to the SW may be a very faint field star and the galaxy forms a triangle with two bright field stars to the NW. Radial velocity:

NGC2139

©STScI/AURA/CALTECH/IOE

1991 km/s. Distance: 89 mly. Diameter: 62 000 ly.

NGC2089

H270[3]	5:47.8	−17 36	GLX
SAB0⁻:	2.0′ × 1.1′	13.04B	

Images reveal a lenticular galaxy with a slightly elongated core and a faint elongated outer envelope, oriented NE–SW. At high magnification, only the bright central region is seen; it is a well defined, roundish, opaque patch of light, with even surface brightness. Several field stars are nearby, particularly to the SE and a faint one is visible to the NNE. Herschel recorded the galaxy on 6 February 1785, calling it: 'Very faint, extremely small, stellar. Verified 240 (magnification) with difficulty.' Radial velocity: 2832 km/s. Distance: 126 mly. Diameter: 74 000 ly.

NGC2124

H225[3]	5:57.9	−20 04	GLX
SA(s)b?	2.7′ × 0.8′	13.43B	

This is a highly inclined spiral galaxy with a bright, elongated core and faint extensions, but the DSS image shows little detail. A half dozen very faint field stars are scattered across the disc and the field is fairly rich. Herschel may have seen some of these stars as he considered this galaxy resolvable. Visually, the galaxy is moderately faint but large and elongated N–S. Well seen at high magnification, the surface brightness is even across the major axis and the edges are fairly well defined. The faint extensions taper gradually to points. Radial velocity: 2818 km/s. Distance: 126 mly. Diameter: 99 000 ly.

NGC2139

H264[2]	6:01.1	−23 40	GLX
SAB(rs)cd	2.9′ × 2.2′	11.98B	11.7V

The DSS photograph shows a slightly inclined, barred spiral galaxy with a brighter, elongated bar and grainy spiral arms with several bright condensations. High-resolution photographs show that the bright condensation S of the bar is actually a satellite galaxy which may be in the process of interacting with the larger galaxy; a very faint, straight plume of material emerges from the main galaxy and points to the SSE. Visually, this is a faint, diffuse and ill defined patch of light which is broadly brighter to the middle. It is very slightly oval, but irregular in outline and oriented N–S with a faint field star near the border to the N. In *The Scientific Papers* H264[2] is identified as IC2154, which was discovered by Lewis Swift. The NASA/IPAC Extragalactic Database equates NGC2139 with IC2154. Radial velocity: 1664 km/s. Distance: 74 mly. Diameter: 63 000 ly.

NGC2196

H265[2]	6:12.2	−21 48	GLX
(R':)SA(rs)ab	2.8′ × 2.2′	11.96B	

Images show a slightly inclined spiral galaxy with a large, oval central region, which is bright and fairly smooth textured, and thin, grainy spiral arms, elongated NE–SW. There is a prominent chain of HII regions in the outer spiral arm on the NW flank. Visually, the galaxy is bright and moderately large and reveals something of its spiral character in large-aperture instruments. The core is large, bright, and oval in shape; the spiral structure reveals itself as a faint ring surrounding the core and slightly offset towards the SW. There are several moderately bright field stars nearby. Radial velocity: 2148 km/s. Distance: 96 mly. Diameter: 78 000 ly.

Libra

Herschel recorded the 17 objects he found in this constellation between the years 1784 and 1788. All are galaxies except for the loosely structured globular cluster NGC5897 and while there are no spectacular objects, a few of the galaxies form interesting pairs.

NGC5595

H121[3]	14:24.2	−16 43	GLX
SAB(rs)c	2.2′ × 1.2′	12.58B	

Images show an inclined spiral galaxy with a brighter central region and bright, distorted spiral arms, including knotty condensations to the SW. The neighbouring galaxy NGC5597 is located 4.0′ to the ESE and is almost certainly a physical companion. The minimum core-to-core separation at the presumed distance would be 137 000 ly. Telescopically, the two galaxies appear as a moderately bright pair, but NGC5595 is far easier to see. It appears as an elongated, bright bar in a very faint envelope, elongated NE–SW. Radial velocity: 2641 km/s. Distance: 118 mly. Diameter: 75 000 ly.

NGC5597

H122[3]	14:24.5	−16 46	GLX
2.3′ × 1.7′	12.61B		

This companion galaxy to NGC5595 is a barred spiral galaxy viewed face-on with a bright central bar and two broad, grainy spiral arms emerging; there are a few large HII regions embedded. Telescopically, the galaxy is difficult to see, especially at high magnification. It is quite diffuse and appears as a slightly oval, ill defined patch, elongated ESE–WNW. Radial velocity: 2613 km/s. Distance: 117 mly. Diameter: 78 000 ly.

NGC5605

H120[3]	14:25.1	−13 10	GLX
(R′)SAB(rs)c	1.6′ × 1.3′	13.13B	

The DSS image shows a face-on spiral galaxy with a small, round core and asymmetrical and somewhat fragmentary, spiral arms. A field star is 3.0′ W. The galaxy is most distinct at medium magnification, though it is fairly faint with a slightly brighter middle. The edges are hazy and the disc appears round. Radial velocity: 3335 km/s. Distance: 149 mly. Diameter: 69 000 ly.

NGC5716

H671[3]	14:41.1	−17 29	GLX
SB(rs)c?	1.8′ × 1.3′	13.70B	

Photographs reveal a slightly inclined, barred spiral galaxy. It is elongated E–W and has a bright bar and bright, knotty spiral arms tightly wound around the central bar. There are fainter, asymmetrical arm fragments to the N that extend almost to a wide pair of bright field stars 1.0′ NE. Visually, the galaxy is fairly faint and quite diffuse, appearing as an irregularly round object which is a little brighter to the middle with hazy, poorly defined edges. Radial

NGC5595/5597

NGC5728

©STScI/AURA/CALTECH/ROE

broad, grainy spiral pattern. A faint edge-on galaxy is 3.7′ SSE. Telescopically, NGC5757 is fairly bright and is well seen at medium magnification when the core is prominent and at high magnification when the bar stands out in an otherwise ill defined haze. Radial velocity: 2616 km/s. Distance: 117 mly. Diameter: 61 000 ly.

NGC5768

H373[3]	14:52.1	−2 32	GLX
SA(rs)c:	2.0′ × 1.5′	13.47B	

Photographs reveal an almost face-on barred spiral galaxy with a small core and a very faint bar attached to a grainy spiral pattern, elongated E–W. A faint field star is 0.5′ S. Telescopically, the galaxy is well seen, but is quite diffuse; it is a little more distinct at medium magnification. It is a roundish patch of light, very broadly brighter to the middle, somewhat grainy and moderately well defined at the edges. Radial velocity: 1952 km/s. Distance: 87 mly. Diameter: 51 000 ly.

NGC5791

H691[3]	14:58.8	−19 16	GLX
E4/S0⁻:	2.5′ × 1.4′	12.70B	

The classification of this galaxy is a little uncertain and photographs show a lenticular galaxy with a large, elongated core and fainter, elongated extensions, oriented NNW–SSE. A field star is 4.75′ SSE and the spiral galaxy IC1081 is 2.75′ NE. Telescopically, NGC5791 is fairly bright and is seen as a slightly oval disc with a grainy envelope and a brighter nonstellar core. The edges are somewhat ill defined and the galaxy is well seen at high magnification. Radial velocity: 3293 km/s. Distance: 147 mly. Diameter: 103 000 ly.

---

velocity: 4094 km/s. Distance: 183 mly. Diameter: 96 000 ly.

NGC5728

H184[1]	14:42.4	−17 15	GLX
(R1)SAB(r)a	3.6′ × 3.6′	12.23B	

The DSS image portrays an inclined, barred galaxy with a large, bright core and a thin bar attached to a thin pseudo-ring which is made up of two tightly wound and almost overlapping spiral arms. A second set of very faint spiral arms surrounds the galaxy and brings its apparent size to 3.6′ × 3.6′. A faint field star is immediately NE of the core; another field star is involved immediately E of the southern bar and a bright field star is 3.8′ S. Visually, this galaxy is quite bright and well defined, featuring a large, bright and extended core embedded in fainter extensions, elongated NE–SW. Radial velocity: 2744 km/s. Distance: 123 mly. Diameter: 128 000 ly.

NGC5729

H508[3]	14:42.0	−9 03	GLX
Sb pec:	2.6′ × 0.7′	12.98B	

Photographs show a highly inclined spiral galaxy with a bright, elongated central region, possibly a bar. The spiral structure along the western flank appears dusty. Two faint field stars are immediately E of the core. Well seen at medium magnification, the galaxy is moderately bright and fairly well defined with even surface brightness along its major axis, which is oriented almost due N–S. Radial velocity: 1783 km/s. Distance: 80 mly. Diameter: 60 000 ly.

NGC5757

H690[3]	14:47.8	−19 05	GLX
(R)SB(rs)b: pec	1.8′ × 1.8′	13.27B	

Images reveal a face-on, barred spiral galaxy with a small, round core and a curved narrow bar that expands to a

NGC5792

H683[2]	14:58.4	−1 05	GLX
SB(rs)b	8.3′ × 1.5′	12.12B	

The DSS photograph shows a large, highly inclined, barred spiral galaxy with a small, brighter core and bar

NGC5792

embedded in an almost complete inner ring. Two grand-design spiral arms emerge from the ring; they are narrow and ribbon-like. Several dust patches are involved in the inner spiral pattern. Telescopically, the galaxy is fairly bright, but the presence of a bright field star 1.0′ NW of the core hinders the view somewhat. The galaxy is a large but quite diffuse, oval patch of light which is broadly brighter to the middle and elongated E–W, with ill defined edges. Radial velocity: 1921 km/s. Distance: 86 mly. Diameter: 208 000 ly.

NGC5812

H71[1]	15:01.0	−7 27	GLX
E0	2.9′ × 2.9′	12.19V	

This elliptical galaxy has a large round core and a fainter extensive outer envelope, which on the DSS image extends to 2.9′ × 2.9′. Companion galaxy IC1084 is 5.0′ to the ESE. Telescopically, the galaxy is quite large and bright with a high-surface-

brightness envelope that is grainy and fairly well defined. The core is fairly large and broadly brighter to the middle. Radial velocity: 1951 km/s. Distance: 87 mly. Diameter: 74 000 ly.

NGC5861

H192[2]	15:09.3	−11 19	GLX
SAB(rs)c	3.0′ × 2.0′	12.34B	

This is an inclined, grand-design spiral galaxy with a small, bright core and narrow, bright, knotty spiral arms. While the pattern to the S is regular, the pattern to the N becomes somewhat chaotic with knots and spiral fragments evident. A bright field star is 2.5′ SSW, while the lenticular galaxy NGC5858 is 9.3′ NW. The galaxy is large and moderately bright visually, but the envelope is diffuse and very broadly brighter to the middle. The edges are quite diffuse and not well defined and the broad oval is elongated NNW–SSE. NGC5858 is visible but is much smaller and round and is well defined with high

surface brightness. Radial velocity: 1826 km/s. Distance: 81 mly. Diameter: 71 000 ly.

NGC5878

H736[3]	15:13.8	−14 16	GLX
SA(s)b	4.3′ × 1.4′	12.33B	

Images reveal an inclined spiral galaxy with a large, oval central region and grainy, multi-arm spiral structure. A faint field star is involved in the outer envelope 0.75′ S of the core. A small, anonymous, barred spiral galaxy is 7.75′ SE. Telescopically, this galaxy is moderately bright, oval and elongated N–S. It is broadly brighter to the middle and is moderately well defined. Radial velocity: 1960 km/s. Distance: 88 mly. Diameter: 109 000 ly.

NGC5885

H116[3]	15:15.1	−10 05	GLX
SAB(r)c	3.6′ × 2.9′	12.34B	

This is an almost face-on spiral galaxy with a small round core and a very faint, short bar. Photographs show that two principal arms spring from the ends of the bar and branch out to a multi-arm pattern after a quarter of a revolution around the core. The arms are knotty and broaden as they curve outward. Telescopically, this is a large but somewhat faint and very diffuse galaxy; it is almost round and very gradually brighter to the middle. The edges are poorly defined and a bright field star is 1.5′ NE of the core. Radial velocity: 1983 km/s. Distance: 89 mly. Diameter: 93 000 ly.

NGC5885

NGC5897

©STScI/AURA/CALTECH/ROE

6.0′ or 7.0′ in diameter, irregularly round, faint red colour.' The brightest cluster members are magnitude 13.3 and the radial velocity is 10 km/s in recession. Distance: 41 000 ly.

NGC5898

H138³	15:18.2	−24 06	GLX
E0	2.7′ × 2.3′	12.44B	11.5V

In photographs this elliptical galaxy has a bright round core and an extensive fainter envelope. An extremely faint galaxy is involved in the outer envelope immediately N. The likely companion galaxy NGC5903 is 5.2′ ENE, while the lenticular galaxy MCG−04-36-07 is 5.0′ ESE. In a moderate-aperture telescope the three galaxies form a compelling triplet, although the MCG galaxy is very faint and only seen at medium magnification. NGC5898 is bright and very slightly oval in form with a bright, grainy disc and a condensed, bright core. The edges of the disc are hazy. Radial velocity: 2066 km/s. Distance: 92 mly. Diameter: 72 000 ly.

NGC5897

H19⁶/H8?⁶	15:17.4	−21 01	GC
CC11	11.0′ × 11.0′	10.2B	9.5V

The DSS image shows a large, weakly concentrated, almost round globular cluster. The cluster members exhibit a fairly wide range in brightness and there are outlying cluster members scattered across the field. Visually, this large cluster is quite faint but readily visible at low magnification as a gradual brightening in the sky background. This brightness increases slowly to a brighter central region. High magnification resolves some of the brighter members, particularly in the eastern half of the cluster, against the background nebulous glow. Herschel may have seen this cluster on 25 April 1784, and catalogued it as H8⁶. However, conditions on that evening were poor: there were frequent interruptions by cloud and a bright moon was present in the sky. The identification is far from certain and H8⁶ may be lost. The object was definitely recorded on 10 March 1785 and described as: 'A beautiful large cluster of the most minute and most compressed stars of different sizes.

NGC5903

H139³	15:18.6	−24 04	GLX
E2	3.0′ × 2.3′	12.1B	11.5V

Photographs show this elliptical galaxy has a bright, elongated core and an extensive fainter envelope, elongated N–S. Visually, it appears a little more concentrated than the neighbouring galaxy NGC5898. It is a round, grainy disc, hazy at the edges, with a star-like, brighter core. Radial velocity: 2509 km/s. Distance: 112 mly. Diameter: 98 000 ly.

Lynx

The 42 Herschel objects in this constellation are all galaxies, except for the remote globular cluster NGC2419. Herschel recorded these objects between the years 1785 and 1790 and while there are some bright and/or nearby galaxies in this area (for instance, NGC2537 – the Bear Paw galaxy) owners of large-aperture telescopes may be more interested in the marker galaxies of the Abell 569 galaxy cluster (NGC2320) and the Abell 779 galaxy cluster (NGC2832), which may prove rewarding to explore.

NGC2320

H861[2]	7:05.6	+50 36	GLX
E	1.5' × 1.0'	12.92B	

The classification of this galaxy may be in error; the DSS image suggests an inclined spiral galaxy with a large, bright, elongated core and a slightly fainter inner ring in an extensive, fainter outer envelope, oriented SE–NW. A possible dust lane borders the core to the SE. A bright field star is 1.5' ENE and the probable companion galaxy NGC2322 is 5.0' SE. In a high-power field, both galaxies can be seen but they are fairly faint. NGC2320 is seen WSW of the bright field star. It is a somewhat diffuse, hazy oval with a bright core that is intermittently visible. This is one of several galaxies over several square degrees of the sky which all have similar radial velocities and are probably members of the Abell 569 galaxy cluster, a relatively nearby, but not particularly rich, group of galaxies. Radial velocity: 5989 km/s. Distance: 267 mly. Diameter: 117 000 ly.

NGC2322

H874[3]	7:05.9	+50 31	GLX
SBa:	1.4' × 0.5'	14.50B	

This is a highly inclined spiral galaxy, probably barred, with a bright, elongated core. Photographs show two thin, condensed, spiral arms emerging from the core into a fainter outer envelope, elongated SE–NW. It is probably a member of the galaxy cluster Abell 569. Visually, the galaxy is a little fainter than NGC2320 to the NW; it is a SE–NW oriented, moderately well defined patch of light with a brighter core that is frequently visible. Radial velocity: 6367 km/s. Distance: 284 mly. Diameter: 116 000 ly.

NGC2326

H734[2]	7:08.4	+50 44	GLX
SB(rs)b	2.7' × 2.5'	13.77B	

Photographs show a face-on, barred spiral galaxy with a small, round core and fainter, straight bar. The bar is attached to tightly wound, narrow spiral arms, which wind for about two turns around the centre of the galaxy. Companion galaxy NGC2326A, an asymmetrical spiral, is 4.75' SE and is probably physically related. Though well seen at high magnification, NGC2326 is a quite diffuse and ghostly galaxy; the brightest portion is slightly oval and oriented E/W and is embedded in a very faint, roundish disc with poorly defined edges. Radial velocity: 6030 km/s. Distance: 269 mly. Diameter: 211 000 ly.

NGC2329

H735[2]/H875[3]	7:09.2	+48 37	GLX
S0⁻:	1.2' × 1.0'	13.54B	

Images show a lenticular galaxy with a bright core and an extensive, fainter outer envelope. A very faint star or condensation borders the core to the N. Several faint galaxies are in the field including UGC3696, which is 2.75' ENE, and MCG+8-13-72, which is 2.75' SSW. This is one of the dominant galaxies in the cluster Abell 569, a relatively nearby and not particularly rich group of galaxies. Telescopically, this is a moderately bright though small galaxy and is seen as a round, well-condensed patch of light, well defined at the edges and quite opaque. There are several brightish stars in the field. This object was recorded twice by Herschel: first on 9 February 1788 and then on 28 December 1790. Radial velocity: 5832 km/s. Distance: 260 mly. Diameter: 91 000 ly.

NGC2332

H862[2]	7:09.5	+50 11	GLX
S0:	1.2' × 0.9'	13.88B	

Photographs show a lenticular galaxy with a bright, elongated core in a

NGC2326

fainter outer envelope, elongated ENE–WSW. The fainter galaxy IC457 is 2.0′ SSW. Both galaxies are probably members of the Abell 569 galaxy cluster. Visually, NGC2332 is moderately faint but well defined; it is a slightly oval, even-surface-brightness patch of light. Quite small, it holds magnification well. A field star is 6.25′ N. Radial velocity: 5879 km/s. Distance: 262 mly. Diameter: 92 000 ly.

NGC2340

H736[2]	7:11.0	+50 10	GLX
E	2.6′ × 1.3′	12.72B	

Photographs reveal an elliptical galaxy with a bright elongated core in a fainter, extensive outer envelope, elongated ENE–WSW. It is probably a member of the Abell 569 galaxy cluster and there are several fainter galaxies in the field, including IC458 at 6.5′ WSW, IC461 at 7.0′ SW, IC464 at 2.75′ SSW, and IC465 at 5.75′ NE. These galaxies may be visible in large-aperture telescopes. In a moderate-aperture telescope, NGC2340 is slightly diffuse, but moderately bright; it is visible as an oval patch of light with hazy edges, broadly brighter to the middle. At high magnification, IC464 is visible as a star-like spot to the SSW. Radial velocity: 5968 km/s. Distance: 266 mly. Diameter: 201 000 ly.

NGC2415

H821[2]	7:36.9	+35 15	GLX
Im?	0.9′ × 0.9′	12.55B	

The DSS photograph shows an irregular galaxy with a very bright central region and short disturbed spurs or spiral arms emerging towards the NW and SE. The structure is very peculiar and the classification is uncertain. In a moderate-aperture telescope, the galaxy is moderately bright and well condensed at high magnification. Round with a brighter core, the envelope is fairly grainy and there is a sharp drop-off in brightness at the edges to a narrow, ragged

secondary envelope. A field star is 1.75′ ENE. Herschel discovered the galaxy on 8 March 1790 and thought it resolvable. Radial velocity: 3763 km/s. Distance: 168 mly. Diameter: 44 000 ly.

NGC2419

H218[1]	7:38.1	+38 53	GC
CC2	4.6′ × 4.6′	10.4V	

Discovered on 31 December 1788, Herschel classified this remote globular cluster as a bright nebulae. It is small, round and very compressed, and photographs show this globular cluster has an extensive halo of faint outlying stars. In a moderate-aperture telescope it is brightest and best seen at medium magnification as a round cluster which is broadly brighter to the middle. The disc is quite smooth and there is no hint of resolution. At high magnification, the cluster is quite large and the edges are ill defined and fade gradually into the sky background. A

bright field star is 4.0′ W. The brightest cluster members are magnitude 17.3, while the horizontal branch visual magnitude is 20.2. The absolute visual magnitude is one of the highest for globular clusters: −9.57. The spectral type is F5 and the radial velocity is 20 km/s in approach. This is one of the more distant of the Milky Way globular clusters: 272 000 ly.

NGC2426

H822[2]	7:43.2	+52 49	GLX
E	1.1′ × 1.1′	14.36B	

In photographs this is an elliptical galaxy with a bright round core and a fainter outer envelope. A pair of anonymous galaxies lie 4.9′ NE, just N of a field star. Visually, the galaxy is extremely faint and small, almost star-like and is seen as a round, condensed patch, well defined at edges and 2.75′ NNE of a bright field star. Radial velocity: 5722 km/s. Distance: 256 mly. Diameter: 82 000 ly.

NGC2419

©STScI/AURA/CALTECH/DOE

NGC2431

H829[3]	7:45.3	+53 04	GLX
(R')SB(s)a:	0.9' × 0.9'	14.56B	

Images show a face-on, barred spiral galaxy with a round core and a fainter bar in an oval envelope attached to two spiral arms. Visually, it is a very small and faint object, visible as a tiny round patch of light with even surface brightness and well defined edges. A faint field star is 1.6' SSE. Radial velocity: 5727 km/s. Distance: 256 mly. Diameter: 67 000 ly.

NGC2469

H836[3]	7:58.1	+56 41	GLX
Sbc pec:	1.0' × 0.7'	13.13B	

The DSS shows this galaxy is part of an E–W string of six NGC galaxies which may all be visible in large-aperture telescopes. It is an inclined spiral galaxy, elongated N–S, with a bright core attached to three bright spiral arms. There are several condensations in the spiral structure. Telescopically, the galaxy is quite faint; it is a small, oval glow, broadly brighter towards the middle and elongated almost due N–S. A faint field star is 0.8' WNW, while a bright field star, which hinders the view, is 2.25' NNE. Radial velocity: 3550 km/s. Distance: 158 mly. Diameter: 46 000 ly.

NGC2474

H830[3]	7:57.9	+52 51	GLX
E0	0.6' × 0.6'	14.09B	

Photographs show an elliptical galaxy forming a double system with NGC2475 0.4' to the NE. The galaxies share a common envelope and there is a bright field star 2.2' NE of NGC2474. Visually, both objects are seen at high magnification, with NGC2475 the slightly larger and brighter. The galaxies are just resolved at high magnification as two round condensed glows, oriented NE–SW with well defined edges. The core-to-core separation between the two galaxies at the presumed distance would be a minimum of 30 000 ly. Herschel

discovered this object on 17 March 1790, describing it as: 'Considerably faint, pretty small, brighter middle.' Evidently, Herschel saw only one object. J. L. E. Dreyer identified this object with NGC2474, but it is possible that Herschel saw the combined image of NGC2474 and NGC2475. Radial velocity: 5675 km/s. Distance: 253 mly. Diameter: 44 000 ly.

NGC2488

H837[3]	8:01.8	+56 34	GLX
S0⁻:	1.4' × 0.8'	13.44B	

This is a lenticular galaxy with a bright, oval core and an extensive outer envelope, elongated E–W. Photographs show there are several very faint galaxies in the field, including UGC4164 5.8' NNE. Visually, the galaxy is moderately faint but brighter than NGC2497, which is nearby to the N. Well seen at high magnification, NGC2488 appears round and broadly concentrated with moderately well defined edges. Radial velocity: 8786 km/s. Distance: 393 mly. Diameter: 160 000 ly.

NGC2493

H750[3]	8:00.4	+39 50	GLX
SB0	2.0' × 2.0'	13.09B	

Images reveal a lenticular galaxy with a bright, round core and a bar embedded in a very faint, round outer envelope. The galaxy is moderately bright visually, though only the core is visible as a round, concentrated patch of light, even in brightness and well defined. Radial velocity: 3904 km/s. Distance: 174 mly. Diameter: 101 000 ly.

NGC2497

H838[3]	8:02.3	+56 57	GLX
E?	1.4' × 1.2'	14.29B	

The DSS image shows an elliptical galaxy with a bright, round core. There are several very faint galaxies in the field which may be accessible in very large amateur instruments. Visually, the galaxy is a moderately faint, round, condensed patch of light which is a little

brighter to the middle and fairly well defined at the edges. Radial velocity: 8234 km/s. Distance: 368 mly. Diameter: 150 000 ly.

NGC2500

H709[3]	8:01.9	+50 44	GLX	
SB(rs)d	2.9' × 2.6'	12.23B	11.6V

This galaxy is part of a physical system involving the galaxies NGC2537, NGC2541 and NGC2552, which are widely separated in the sky. The DSS photograph shows a large, barred spiral galaxy with an extremely small core in a faint bar. The spiral pattern is quite fragmentary and several large, bright knots are involved. Visually, this is a large, moderately bright, but very diffuse galaxy, situated within a diamond-shaped asterism of faint field stars. The disc is grainy textured, broadly brighter to the middle and fairly round, with very hazy, ill defined edges. Radial velocity: 552 km/s. Distance: 25 mly. Diameter: 21 000 ly.

NGC2505

H839[3]	8:04.0	+53 34	GLX
SBa	1.5' × 0.9'	14.42B	

Situated 8.25' SSW of a magnitude 7 field star, photographs show an inclined, barred spiral galaxy with a bright core and two faint spiral arms, elongated N–S. Telescopically, the bright field star interferes with observation somewhat; the galaxy is a moderately faint and hazy, slightly oval, diffuse patch of light which is broadly brighter to the middle with very ill

NGC2500

defined edges. Radial velocity:
9797 km/s. Distance: 438 mly.
Diameter: 191 000 ly.

NGC2532

H726[2]	8:10.2	+33 57	GLX
SAB(rs)c	1.9′ × 1.6′	13.06B	

This is a face-on spiral galaxy with a
very small bright core. Photographs
show bright inner spiral structure
expanding to fainter dusty, knotty spiral
arms. Visually, the galaxy is moderately
bright and is best seen at medium
magnification. The inner region is a
little brighter to the middle and
elongated E–W, and is embedded in a
diffuse and ill defined secondary
envelope which is almost round. Radial
velocity: 5229 km/s. Distance: 233 mly.
Diameter: 129 000 ly.

NGC2534

H840[3]	8:12.9	+55 40	GLX
E1? pec	1.3′ × 1.2′	13.70B	

Situated 2.8′ NNW of a magnitude 8
field star, photographs show a small
elliptical galaxy with a brighter core and
faint outer envelope. A narrow dust lane
runs N–S to the E of the core. At high
magnification, this is a small but fairly
high-surface-brightness galaxy, almost
round with a star-like brighter core. The
disc is grainy and the edges moderately
well defined. Radial velocity:
3503 km/s. Distance: 157 mly.
Diameter: 59 000 ly.

NGC2537

H55[4]	8:13.2	+46 00	GLX	
SB(s)m pec	2.2′ × 1.9′	12.27B	11.7V	

This unusual galaxy is very likely a
dwarf and is sometimes known as the
Bear Paw galaxy due to its peculiar
morphology. It has also been catalogued
as Arp 6. Photographs show three
bright, knotty bars oriented N–S in a
bright, round envelope which is
surrounded by a very faint outer halo.
The westernmost bar is the brightest
and has at least three HII regions
involved. The unusual structure may be
the result of an ongoing collision: the

NGC2537

central bar looks very much like a small,
irregular galaxy merging with a larger
galaxy. A bright field star is 6.5′ W. Also
in the field is the faint face-on spiral
galaxy NGC2537A, 4.6′ to the E.
Because of its high surface brightness,
many of the unusual structural features
of NGC2537 are visible in a moderate-
aperture telescope, including the bright
arc to the N which is quite well defined
and the brighter N–S oriented core. The
southern flank is flat and ends abruptly.
This galaxy is a member of a nearby
group that includes NGC2500,
NGC2541 and NGC2552. Radial
velocity: 448 km/s. Distance: 20 mly.
Diameter: 13 000 ly.

NGC2541

H710[3]	8:14.7	+49 04	GLX	
SA(s)cd	6.3′ × 3.2′	12.25B	11.8V	

This is the largest of the four galaxies in
the group which includes NGC2500,
NGC2537 and NGC2552. The DSS
photograph shows a large, inclined
spiral galaxy with an irregular, brighter

core. Anaemic, fragmentary spiral arms
emerge with several very small,
condensed knots involved.
Telescopically, the galaxy is well seen at
medium magnification but it is a quite
diffuse, low-surface-brightness object. It
is oval in shape, the edges are somewhat
ill defined and it is elongated N–S. It is
very broadly brighter to the middle,
where a very faint star-like core is
intermittently visible. Radial velocity:
577 km/s. Distance: 26 mly. Diameter:
47 000 ly.

NGC2543

H719[2]	8:13.0	+36 15	GLX
SB(s)b	2.4′ × 1.1′	13.01B	

This is an inclined, S-shaped, barred
spiral galaxy with a small, brighter core
and two grainy spiral arms emerging
from the bar. In the DSS image, the
northern spiral arm has a faint, broad
extension that increases the size to
about 3.0′ × 1.1′. Telescopically, the
galaxy is fairly bright, oval in form and

NGC2543

©STScI/AURA/CALTECH/ROE

shows a large, highly inclined spiral galaxy with a very bright, elongated central region. There are extensive, fainter outer spiral arms with several dust patches and lanes involved in the NW flank. The spiral pattern is presumed to be multi-arm, defined principally by the dust but the outer structure is surprisingly smooth and unresolved, despite the low radial velocity. Telescopically, the galaxy is magnificent; it is a very bright and large galaxy, almost edge-on and elongated NE–SW. The centre is bright and quite elongated, embedded in fainter extensions which taper to soft points. The SE flank is slightly convex and sharply defined compared to the NW flank, which is more diffuse. Discovered on 5 February 1788, Herschel described the galaxy as: 'Very brilliant, much extended south preceding, north following. 8′ long, 3′ broad. Beautiful.' Radial velocity: 376 km/s. Distance: 16.8 mly. Diameter: 43 000 ly.

NGC2691

H658[2]	8:54.8	+39 32	GLX
Sa?	0.9′ × 0.5′	13.93B	

The classification of this galaxy is uncertain and in photographs it appears as a lenticular or possibly spiral galaxy with a large, elongated core and extensive outer disc, elongated N–S. This galaxy is not particularly bright visually and appears initially as a fuzzy star 0.9′ S of a faint field star. High magnification brings out a well defined, roundish, grainy patch of light with a star-like core.

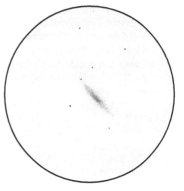

NGC2683 15-inch Newtonian 272x

elongated NE–SW. The core is a little brighter and the envelope is more sharply defined along the SE flank. Radial velocity: 2449 km/s. Distance: 109 mly. Diameter: 96 000 ly.

NGC2552

H711[3]	8:19.3	+50 01	GLX	
SA(s)m?	3.5′ × 2.3′	12.69B	12.1V	

This is the faintest of a group of relatively nearby galaxies including NGC2500, NGC2537 and NGC2541 that span 5° of sky and presumably form a small, gravitationally bound group. At the presumed distance, the 5° circle would be about 2.2 mly across. In photographs NGC2552 appears to be an irregular galaxy, possibly a spiral, with a slightly brighter central region and a very faint outer envelope with several small knots. Resolution into stars begins at about blue magnitude 21. Visually, it is a very faint, large and very diffuse object. It is a dim, oval patch of light, elongated ENE–WSW that is only very slightly brighter to the middle and the edges are quite hazy.

Radial velocity: 557 km/s. Distance: 25 mly. Diameter: 25 000 ly.

NGC2649

H727[2]	8:44.1	+34 43	GLX
SAB(rs)bc:	1.8′ × 1.5′	13.33B	

The DSS image shows a face-on spiral galaxy of the grand-design type, with a small, round, bright core and bright, symmetrical, spiral arms, elongated E–W. A faint field star is at the tip of the northern spiral arm 0.5′ from the core. Telescopically, the galaxy is quite faint and round, with even surface brightness across the disc. The edges are fairly sharply defined and the faint field star is seen to border the disc immediately to the N. Radial velocity: 4214 km/s. Distance: 188 mly. Diameter: 99 000 ly.

NGC2683

H200[1]	8:52.7	+33 25	GLX
SA(rs)b	8.7′ × 2.5′	9.99B	

This is a spectacular example of an almost edge-on galaxy. The DSS image

NGC2683

©STScI/AURA/CALTECH/ROE

Radial velocity: 3970 km/s. Distance: 177 mly. Diameter: 47 000 ly.

NGC2704

H625[3]	8:56.9	+39 22	GLX
SB(r)ab	1.1′ × 1.1′	14.36B	

Photographs show an inclined, theta-shaped, barred galaxy with an elongated core. The ring surrounding the bar is probably a pair of tightly wound, overlapping, inner spiral arms. The outer spiral structure is broad, faint and smooth textured. Visually, the galaxy is a moderately faint, slightly elongated patch of light, even in brightness and well defined, oriented ESE–WNW. Radial velocity: 7105 km/s. Distance: 317 mly. Diameter: 102 000 ly.

NGC2719

H540[3]	9:00.3	+35 44	GLX
Im pec?	1.3′ × 0.3′	14.03B	

The pair comprising this galaxy and the peculiar companion galaxy NGC2719A located 0.3′ immediately S has also been

catalogued as Arp 202. The DSS image shows an edge-on galaxy with a bright, clumpy major axis, brighter to the SE, and a faint, disturbed extension off the SE tip. The companion galaxy is clumpy and irregularly shaped, with high surface brightness. Evidently a physical pair, the minimum core-to-core separation of the two galaxies would be about 12 000 ly at the presumed distance. Visually, only the primary galaxy is visible; it is a blurry, indistinct narrow sliver of light with even surface

NGC2719

©STScI/AURA/CALTECH/ROE

brightness and well defined edges, elongated SE–NW. In a medium-magnification field the non-Herschel object NGC2724 is also visible as an indistinct, small patch of light located NE of a moderately bright field star. Radial velocity: 3051 km/s. Distance: 136 mly. Diameter: 52 000 ly.

NGC2746

H825[3]	9:06.0	+35 22	GLX
SB(rs)a	1.6′ × 1.5′	14.17B	

Images show an almost face-on barred spiral galaxy with a small, round core. The faint bar is attached to grainy, faint outer spiral arms. This galaxy is moderately faint telescopically, but is seen as a round, hazy patch of light which is very slightly brighter to the middle and fairly well defined at the edges. It is situated 0.8′ S of a field star, the middle one of three in a curving line oriented ENE–WSW. Radial velocity: 7038 km/s. Distance: 314 mly. Diameter: 146 000 ly.

NGC2755

H626[3]	9:07.9	+41 43	GLX
S?	1.3′ × 0.6′	14.33B	

The DSS photograph shows an inclined spiral galaxy with a bright core and bright inner spiral arms which expand to much fainter, relaxed and narrow outer spiral arms. The galaxy is quite faint visually; it is a difficult object best seen at medium magnification as an oval patch of light, even in surface brightness and oriented ESE–WNW. Radial velocity: 7542 km/s. Distance: 337 mly. Diameter: 127 000 ly.

NGC2759

H647[3]	9:08.6	+37 37	GLX
S0⁻:	1.0′ × 0.7′	14.01B	

Photographs show an inclined, lenticular galaxy with an oval core and faint outer envelope, elongated ENE–WSW. Faint galaxies are at 5.75′ SW, 7.2′ SSE and 3.1′ N. This galaxy is fairly faint visually and is seen as a round, uncondensed glow, even in brightness and well defined at the edges. Radial velocity: 6926 km/s. Distance: 309 mly. Diameter: 90 000 ly.

NGC2770

H490[2]	9:09.6	+33 07	GLX
SA(s)c:	3.7′ × 1.0′	12.76B	

Images reveal a highly inclined spiral galaxy with a very small core and grainy spiral structure. Visually, the galaxy is moderately bright and much extended NW–SE with a brighter core visible with averted vision. The SW flank is more sharply defined than the NE one, which appears quite diffuse, and photographs show a probable broad dust lane bordering the SW flank. Two faint field stars are N and NE of the galaxy. Companion NGC2770B is extremely faint and located 2.9′ W. Herschel thought NGC2770 resolvable. Radial velocity: 1911 km/s. Distance: 85 mly. Diameter: 92 000 ly.

NGC2778

H564[2]	9:12.3	+35 01	GLX
E	1.1′ × 0.8′	13.35B	

This elliptical galaxy is the brightest of a small galaxy group which includes NGC2779 1.6′ NNE, NGC2780 7.2′ SSE, and fainter galaxies 4.4′ E and 7.5′ WSW. Photographs show an elongated core in an extensive fainter envelope and visually the galaxy appears as a moderately bright, well defined patch of light, somewhat grainy but even in surface brightness. NGC2780 can be seen in the same medium-magnification field. Radial velocity: 2021 km/s. Distance: 90 mly. Diameter: 29 000 ly.

NGC2780

H826[3]	9:12.7	+34 55	GLX
Sab	0.9′ × 0.6′	14.32B	

The DSS image shows a barred spiral galaxy seen at a slightly oblique angle; the principal arms form two bright arcs perpendicular to the central bar in a somewhat diffuse outer envelope. The galaxy is probably a companion of NGC2778 with a minimum separation of 189 000 ly at the presumed distance. Telescopically, NGC2780 is a diffuse patch of light which is fairly well defined and even in surface brightness and is seen following a pair of field stars by

1.75′. Radial velocity: 1950 km/s. Distance: 87 mly. Diameter: 23 000 ly.

NGC2782

H167[1]	9:14.1	+40 07	GLX	
SAB(rs)a pec	3.5′ × 2.6′	12.32B	11.6V

This face-on spiral galaxy has also been catalogued as Arp 215 because of its unusual morphology. The DSS image shows a large, round, bright core embedded in a somewhat chaotic spiral structure which consists of a ring-like inner arm which broadens to a faint, asymmetric outer arm. A second broad, faint arm, which is tangent to the inner structure, unwinds in the opposite direction from the NW before curving S and then E, extending for a considerable distance away from the galaxy. Visually, this is quite a bright galaxy, appearing large and round, a little diffuse at the edges with a disc of uneven brightness that brightens to a large, condensed, nonstellar core. Radial velocity: 2535 km/s. Distance: 113 mly. Diameter: 115 000 ly.

NGC2798

H708[2]	9:17.4	+42 00	GLX	
SB(s)a pec	2.6′ × 1.0′	13.02B	12.3V

This highly inclined spiral galaxy forms an interacting pair with the edge-on spiral galaxy NGC2799 1.5′ to the E and is also catalogued as Arp 283. Also in the field is the faint spiral galaxy UGC4904, 5.25′ S. The DSS image shows a very bright, elongated core with a bright curved bar attached to very smooth but faint spiral arms. The northern spiral arm is connected to the companion galaxy

NGC2782

whose western spiral extension is warped slightly. The core-to-core separation between the two is about 34 000 ly at the presumed distance. Both galaxies are visible in a high-magnification field and although it is moderately bright, only the core of NGC2798 is visible. It appears irregularly round, a little brighter towards the middle and fairly well defined at the edges. NGC2799 appears as a very small, hazy and unconcentrated spot. Radial velocity: 1726 km/s. Distance: 77 mly. Diameter: 59 000 ly.

NGC2832/NGC2831

H113[1]	9:19.8	+33 45	GLX	
E⁺2:; cD	2.1′ × 1.4′	12.79B	13.2V

J. L. E. Dreyer has associated this Herschel object with the faint, edge-on galaxy NGC2830, but there is a much stronger argument to link the object with the combined images of NGC2832 and NGC2831. Herschel discovered this object on 7 December 1785, and classified it as a bright nebula, describing it as: 'Considerably bright, considerably large, a little extended, irregular form, much brighter on the following side.' This is an excellent visual description of the double system NGC2832–NGC2831 and there can be no doubt that these are what Herschel saw. The DSS image shows NGC2832 as an elliptical galaxy with a large, bright core and extensive fainter outer envelope oriented SSE–NNW. NGC2831 is a round, compact object 0.4′ to the SW and within the larger galaxy's halo, though its radial velocity is considerably lower at 5147 km/s. The two galaxies

NGC2798

NGC2832
NGC2831
NGC2830
NGC2825
NGC2829
ABELL779 galaxy cluster
NGC2825/2826/2829/2830/2831/2832
NGC2826

©STScI/AURA/CALTECH/ROE

NGC2844

H628[3]	9:21.8	+40 09	GLX
SA(r)a:	1.7′ × 0.8′	13.69B	

This is a highly inclined spiral galaxy. Images show a large, bright, elongated core and fainter spiral arms, elongated N–S. It may be physically related to the close pair NGC2852 and NGC2853 situated about 17.0′ to the E. The galaxy is moderately bright visually, but only the core is seen as a well defined, roundish patch of light which is evenly bright across the disc. Radial velocity: 1479 km/s. Distance: 66 mly. Diameter: 33 000 ly.

NGC2852

H629[3]	9:23.2	+40 10	GLX
SB0:	0.9′ × 0.8′	14.14B	

On the DSS image this galaxy has a bright, elongated core in the slightly fainter disc with an outer envelope that is only suspected. The galaxy forms a close pair with NGC2853 2.2′ NNE; the minimum core-to-core separation between the two would be about 50 000 ly at the presumed distance. Published sources seem to have switched the classifications of the two galaxies; the correct classifications appear here. Both galaxies are fairly faint visually, though NGC2852 is a little more distinct; it is a round, fairly concentrated patch of light of even surface brightness and well defined at the edges. Radial velocity: 1766 km/s. Distance: 79 mly. Diameter: 21 000 ly.

NGC2853

H630[3]	9:23.3	+40 12	GLX
SAB(r)a?	1.9′ × 1.1′	14.43B	

This is an inclined spiral galaxy. Photographs show a bright, elongated core and very faint, narrow spiral arms, elongated NNE–SSW. A faint field star is involved in the spiral arm immediately S of the core and the companion galaxy NGC2852 is 2.2′ SSW. Telescopically, only the core of NGC2853 is visible, seen as a broad patch of light, even in brightness and moderately well defined at the edges. Radial velocity: 1758 km/s. Distance: 79 mly. Diameter: 43 000 ly.

---

are also catalogued as Arp 315. NGC2830 is an edge-on galaxy, probably a spiral, 1.2′ to the WSW. These galaxies are at the core of the galaxy cluster Abell 779. Telescopically, NGC2832 is large and fairly bright but NGC2831 is quite difficult; it is a small, slightly concentrated patch in the halo of NGC2832 to the SW. NGC2830 is even more difficult, a hazy, slightly elongated streak oriented ESE–WNW, fairly well defined and even in surface brightness. NGC2832 appears as a bright, concentrated patch, broadly brighter to the middle and somewhat ill defined at the edges. Radial velocity: 6915 km/s. Distance: 309 mly. Diameter: 189 000 ly.

NGC2838

H627[3]	9:20.7	+39 19	GLX
0.8′ × 0.7′	14.50B		

This galaxy is unclassified but on the DSS image it appears to be a lenticular galaxy with a bright round core. A fainter anonymous galaxy is 1.2′ to the WSW. Though moderately faint visually, this galaxy is well seen as a small, circular spot of even brightness and well defined edges. Radial velocity: 8255 km/s. Distance: 369 mly. Diameter: 86 000 ly.

NGC2840

H827[3]	9:20.8	+35 23	GLX
SB(rs)bc	0.9′ × 0.7′	14.57B	

Photographs show a small, inclined spiral galaxy with a small, round core and faint spiral arms. Visually, the galaxy is quite faint, but its location 0.75′ ESE of a field star helps to pinpoint it. It is a somewhat diffuse, slightly oval glow of even brightness, fairly well defined at the edges and oriented E–W. Radial velocity: 7473 km/s. Distance: 334 mly. Diameter: 87 000 ly.

Lyra

The single Herschel object discovered in this small constellation was recorded near the end of Herschel's sky sweeps, in June 1802. There are a handful of similar galaxies within the confines of the constellation but they were all overlooked, as was the large but dim open cluster NGC6791.

NGC6646

H907[2]	18:29.6	+39 52	GLX
Sa	1.7′ × 1.4′	13.59B	

Photographs show an inclined spiral galaxy with a large, bright, elongated core and faint, narrow spiral arms in a faint envelope. Elongated NE–SW, it is situated in a triangle of bright field stars. In moderate apertures this is a moderately bright galaxy in a star-rich field. The outer envelope is rather diffuse but a bright star-like core dominates the centre. The envelope appears textured while the extremities fade gradually into the sky background. This was the penultimate entry in Herschel's Class II: faint nebulae, and was recorded on 26 June 1802 with the description as: 'Faint, small, irregular form.' Radial velocity: 5971 km/s. Distance: 267 mly. Diameter: 130 000 ly.

NGC6646

©STScI/AURA/CALTECH/ROE

Monoceros

Much of the area of this faint constellation is within the confines of the winter Milky Way and the 34 Herschel objects recorded here are all star clusters or nebulae. They were discovered between the years 1783 and 1790 and though some of the clusters are bright and attractive, especially NGC2301, there are a small number of objects that may only be asterisms or brighter patches in the Milky Way. Surprisingly, Herschel missed the Rosette Nebula, no doubt dazzled by the bright star cluster NGC2244. Together, the clusters and nebulae form an attractive target for the modern amateur astronomer with a good telescope.

NGC2170

H19⁴/H44⁴	6:07.5	−6 24	RN
3.75′ × 3.5′			

The DSS image of this reflection nebula shows a bright, irregularly shaped core and faint outer extensions which stream back towards the SW. Faint field stars bracket the nebula to the N and S while very faint detached patches of nebulosity are visible to the ESE and ENE. Visually, at medium magnification the nebula is small and faint and surrounds a magnitude 9.5 star (spectral type B1). The field stars to the N and S are visible as well. The nebula is well condensed near the central star and roundish, and fades gradually into the sky background. Discovered on 16 October 1784, Herschel called the nebula: 'A star of the 9 magnitude, with milky chevelure, irregularly elliptical.' It was subsequently observed by Herschel on 28 November 1786 and erroneously recorded as a new nebula.

NGC2182

H38⁴	6:09.5	−6 20	RN
3.0′ × 1.8′	9.0B		

Images show this nebula is involved with a bright field star. The star is HD42261, magnitude 9 and spectral type B4. The brightest portion of the nebula is crescent-shaped and arcs from NNW to ESE. There are faint extensions mostly to the SW and a faint, detached portion surrounding a field star to the WNW. In a moderate-aperture telescope, this is a very faint nebula surrounding the magnitude 9 star; it is very small and delicate and best seen with averted vision. The edges are poorly defined and the star interferes with visibility. Herschel discovered this nebula more than a year after NGC2170 and NGC2185, on 24 February 1786, and described it as: 'A considerable star affected with very faint milky chevelure.'

NGC2185

H20⁴	6:11.1	−6 13	RN
2.75′ × 2.75′	12.9B		

Photographs show this object is the brightest portion of a nebula complex

NGC2170

which also involves a small group of stars to the SW and NGC2183 to the W. It is involved with a magnitude 12 star; the nebula streams towards the SSE with dark dust lanes involved. Visually, both NGC2183 and NGC2185 are visible in the same field, so it is surprising that Herschel did not record both objects, especially considering NGC2183 appears marginally brighter. NGC2185 is an irregular patch of light surrounding the magnitude 12 star, but the star hinders visibility somewhat. It was discovered on the same night as NGC2170, when Herschel commented: 'A star of the 11 or 12 magnitude affected like the foregoing (NGC2170), but very faint.'

NGC2215

H20[7]	6:21.0	−7 17	OC
II2m	8.0′ × 8.0′	8.8B	8.4V

This is a scattered group of mostly bright stars in a rich Milky Way field, somewhat elongated NNW–SSE. Small apertures resolve about 20 magnitude 11 on fainter stars and show it as a well defined object; at high magnification moderate apertures pick up many of the fainter members. Forty stars are probable members. Spectral type: B9. Age: about 350 million years. Distance: 3200 ly.

NGC2225

H26[7]	6:26.6	−9 39	OC
5.0′ × 4.5′			

The DSS image shows a U-shaped grouping of stars with the open end towards the N and the brightest cluster members along the eastern flank. In a moderate-aperture telescope, however, the field is very similar to Herschel's 30 January 1786 description, which reads: 'A cluster of extremely small and pretty compressed stars with a few large but not rich. In the shape of a hook.' At high magnification the 'hook' outline is obvious and involves about ten stars including a very tight pair to the N and three stars S of this pair involved in an unresolved, nebulous haze slightly elongated ENE–WSW. This hazy patch

has been separately catalogued as NGC2226, though it is not a separate object and in photographs involves at least 15 stars. Total cluster membership is about 40 stars.

NGC2232

H25[8]	6:26.6	−4 45	OC
III2p	50.0′ × 50.0′	3.9V	

Photographs show a very large, coarse and scattered group of bright stars, including HD45321 to the N and 10 Monocerotis to the E. Visually, a very-low-magnification eyepiece is needed to take in the full extent of this cluster which comprises 20 bright stars with a moderate brightness range. Most of the presumed members are individually plotted in the first edition of *Uranometria 2000.0*, though the cluster symbol should be larger. About 20 stars are considered members, the brightest of which is magnitude 5.03. Spectral type: B3. Age: 22 million years. Distance: 1200 ly.

NGC2236

H5[7]	6:29.7	+6 50	OC
II2m	8.0′ × 8.0′	9.1B	8.5V

Images show a well-resolved cluster of stars with a moderate range in brightness. In moderate apertures, this is a small, well-condensed cluster grouped around a magnitude 9 central star, which is probably not a cluster member. The central region is quite round and a conspicuous chain of stars curves away from the central region in the N, sweeping W then S, reminding one of a galaxy's spiral arm. Most cluster members are magnitude 12 or fainter. More than 200 stars are thought to be members. Spectral type: A0. Distance: 10 850 ly.

NGC2244

H2[7]	6:32.4	+4 52	OC
II3r n	30.0′ × 30.0′	5.3B	4.8V

This bright, coarse and scattered cluster is involved with the Rosette Nebula (NGC2237). Interestingly, though Herschel observed this cluster on at least

three occasions, he never saw the accompanying nebula. Visible in binoculars, telescopically the cluster is bright and well seen at low magnification and is dominated by four magnitude 7 stars in a rectangular shape. At least 50 cluster members are visible scattered across the field (total membership in the cluster is believed to be about 100 stars); higher magnification reveals more visually. This is a very young star cluster. Spectral type: O5. Age (estimated): about 3 million years. Radial velocity in recession: 34 km/s. Distance: about 4900 ly.

NGC2245

H3[4]	6:32.7	+10 10	RN
7.0′ × 5.0′	11.0B		

Photographs reveal an irregularly shaped nebula with a bright inner region and faint, crab-like plumes that extend from the core and are oriented NNE–SSW with a dark patch to the NE. Visually, medium magnification shows an obviously nebulous, faint star followed by a brighter, magnitude 8 star 2.0′ ENE. The nebulosity fans away to the S and the involved star is magnitude 10.8 and spectral type B1p. The nebula is ill defined and averted vision increases its extent somewhat. Discovered an 16 January 1784, Herschel's description reads: 'Pretty bright, milky, like a star with an electrical brush.'

NGC2251

H3[8]	6:34.7	+8 22	OC
III2m	10.0′ × 6.0′	7.7B	7.3V

Well seen both photographically and visually, this is an attractive, greatly elongated cluster oriented SE–NW. The two brightest stars are about magnitude 9, while the rest are magnitude 11–13. The cluster members are arrayed in four distinct groups and the cluster stands out well, despite its location in a rich Milky Way field, with at least 30 cluster members visible. More than 90 stars are accepted as members. Spectral type: B3. Age: 300 million years. Radial velocity in recession: 8 km/s. Distance: 5075 ly.

NGC2252

H50[8]	6:35.0	+5 23	OC
III2m n	18.0' × 4.0'	7.7V	

The visual and photographic appearances of this large cluster are quite similar. Visually, this is a scattered cluster of bright and faint stars arranged in a long, N–S chain which hooks towards the W at its northern extremity. The range in brightness is moderate and the group is well separated from the sky background. About 30 stars are probable members, the brightest being about magnitude 9, and the cluster is spectral type A. Herschel recorded the cluster on 27 January 1786, describing it as: 'A cluster of stars arranged in a broad row. 25' long, 6' or 8' broad. Not very compressed but pretty rich.'

NGC2254

H22[7]	6:36.0	+7 40	OC
I1m	4.0' × 4.0'	9.7B	9.1V

This cluster is a compressed, well-resolved object in photographs but visually appears as a small and much condensed cluster and is seen as a hazy spot at low power. Medium or high magnification does little to improve resolution. At least five magnitude 12 or fainter stars are resolved, surrounded by a crescent-shaped haze of unresolved members at high magnification. At least 90 stars are thought to be members of the cluster. Spectral type is B9. Distance: 6875 ly.

NGC2259

H28[6]	6:38.6	+10 53	OC
II1p n	3.5' × 3.5'	10.8V	

Photographs show a compressed grouping of faint stars with a narrow brightness range. An E–W chain of stars near the centre distinguishes the cluster, which is well separated from the sky background. Visually, the cluster follows a magnitude 9 field star and is a very faint, hazy patch of light, broadly concentrated to the centre, with ill defined edges. It appears slightly crescent-shaped, concave towards the W. Resolution into individual stars

is uncertain at medium magnification; the brightest member stars are magnitude 14. Total membership of the cluster is about 25 stars. Discovered on 11 January 1787, Herschel's description reads: 'A cluster of extremely compressed and extremely small stars, considerably rich, irregular form. The following and most compressed part round.'

NGC2260

H48[8]	6:38.1	−1 28	OC?
20.0' × 20.0'			

Photographs show a coarse grouping of bright stars with a moderate brightness range, well separated from the sky background. It may include a coarse clump of stars to the NE of the brightest cluster star. Telescopically, this is a coarse and rather poor cluster of stars in a medium-magnification field with a wide scatter in brightness. Although perhaps not part of the grouping, a prominent question-mark-shaped asterism precedes in the W with a widely separated pair of stars providing connection to the scattered grouping to the E. Including this asterism, about 25 stars are involved.

NGC2261

H2[4]	6:39.2	+8 44	E+RN
2.0' × 1.0'			

Popularly known as Hubble's Variable Nebula, this object is easily swept up at low power and is a striking object at medium magnification. Herschel placed this object in his Class IV: planetary nebulae, but it is now known to be a combination emission and reflection nebula. Although Herschel observed this object at least four times, he never noted any changes in the appearance of the nebula. It is a fan-shaped object which broadens towards the N with its apex at the illuminating star R Monocerotis. At the time of the author's observation (13 November 1993) R Monocerotis was bright; the western portion of the nebula was brighter than the E, though the eastern limit was a little more sharply defined. Visual

observers may note subtle changes in the appearance of this nebula over time spans of years.

NGC2262

H37[7]	6:39.6	+1 08	OC
II2r	4.0' × 3.5'	11.3V	

Discovered on 27 December 1786, Herschel described this rich cluster as: 'A cluster of very compressed extremely small stars, considerably rich. 3 or 4' in diameter, most compressed to the middle.' Images show a compressed grouping of faint stars with a narrow brightness range. Most members are arranged along a N–S axis and there is an inner compressed group surrounded by more scattered outliers. Visually, the cluster is quite faint and poorly resolved, and is best seen at medium magnification as a hazy, slightly elongated patch of light which is broadly brighter to the middle with a few faint sparkles shining through. Averted vision at high magnification resolves the group a little better, particularly a chain of stars in the northern part. About 180 stars are probable members.

NGC2264

H27[5]/H5[8]	6:41.1	+9 53	OC+EN
III3m n	40.0' × 40.0'	3.9V	

Herschel catalogued this complex object twice: on 18 January 1784 he recorded it as a star cluster calling it: 'Double and

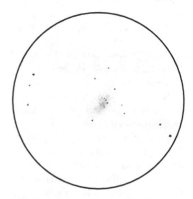

NGC2262 15-inch Newtonian 272x

attended by more than 30 considerably large stars'; on 26 December 1785 he entered it as a Class V: very large nebula: 'Some pretty bright stars 7′ or 8′ south preceding 15 Monocerotis are involved in extremely faint milky nebulosity which loses itself imperceptibly'. Visually, this is a large, scattered cluster, attractive only at low magnification. It is dominated by the bright variable star S Monocerotis around which a pale, nebulous glow is visible. A faint glow is also visible to the SW around a loose clump of five stars. Most of the stars are magnitude 8–10 and about 40 cluster members are visible at low power. This star cluster is popularly known as the Christmas Tree. Attached to the S of the cluster is the Cone Nebula, which is very difficult to see visually. Forty stars are probable cluster members. As might be expected, this is quite a young object. Spectral type: 07. Age (estimated): 20 million years. Radial velocity in recession: 21 km/s. Distance: about 2600 ly.

NGC2269

| H3[6] | 6:43.3 | +4 37 | OC |
| I1p | 3.5′ × 2.0′ | 10.4B | 10.0V |

In photographs, this is an oval group of faint stars with a moderate range of brightnesses, oriented SSE–NNW and moderately well separated from the sky background. Visually, it is a sparse, elongated, but fairly obvious cluster of stars, well separated from the sky background and fully resolved at high magnification. A dozen stars are

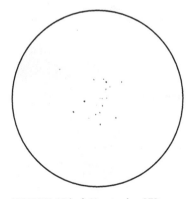

NGC2269 15-inch Newtonian 272x

involved in an elongated row, with half a dozen other stars flanking the row on the E and W sides. The faintest members are threshold objects. About a dozen stars are probable members. Distance: about 4700 ly.

NGC2270

| H36[7] | 6:43.9 | +3 26 | OC? |

The DSS image shows a row of bright stars oriented N–S and the row of stars is on the W side of a poor, coarse scattering of stars with a moderate brightness range, though most of them are quite faint. Visually, this is a delicate grouping of stars which stands out well in a medium-magnification field. A triangle of bright stars is on the W side, with most of the cluster E and NE of the triangle. The cluster is well resolved with most of the stars organized into pairs and triplets. About 30 stars appear to be involved. Recorded on 26 December 1786, Herschel described the object as: 'A cluster of very scattered stars, considerably rich and of great extent.'

NGC2286

| H31[8] | 6:47.6 | −3 10 | OC |
| I2m | 15.0′ × 15.0′ | 8.2B | 7.5V |

The visual and photographic appearances of this coarse cluster are quite similar. Visually, the cluster appears as a large and scattered group of stars of moderate brightness range, poorly separated from the sky background. About 80 stars are probable members, the brightest of which are about magnitude 9. Spectral type: B4. Distance: 4140 ly.

NGC2301

| H27[6] | 6:51.8 | +0 28 | OC |
| I3r | 12.0′ × 8.0′ | 6.3B | 6.0V |

Discovered on 27 December 1785, Herschel described this attractive open cluster as: 'A very beautiful cluster of much compressed small and large stars, above 20′ diameter'. Visually, this is a bright, striking cluster, displaying a prominent N–S orientation, dominated by half a dozen magnitude 8–9 stars. A

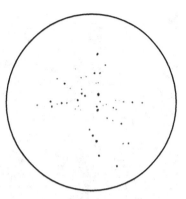

NGC2301 15-inch Newtonian 95x

prominent spur of stars, magnitude 10–12, shoots off towards the E from the heart of the cluster. There is a roundish concentration of bright and faint stars near the core and a faint spur trending towards the W. At least 50 stars are visible in an area approximately 25′ in diameter and about 80 stars are probable members. Spectral type: B8. Age: 110 million years. Distance: 2500 ly.

NGC2302

| H39[8] | 6:51.9 | −7 04 | OC |
| III2m | 5.0′ × 5.0′ | 9.1B | 8.9V |

Photographs show a coarse cluster of bright stars with a moderate brightness range in a rich field. The cluster is elongated roughly E–W and well separated from the sky background. In small apertures, the cluster is easy to overlook but its most conspicuous feature is a tightly grouped curve of three magnitude 9 field stars on the W side, with several fainter cluster members immediately to the E. About 30 stars are probable members of the cluster. Distance: about 3500 ly.

NGC2306

| H51[8] | 6:54.6 | −7 11 | OC? |
| 20.0′ × 10.0′ | | | |

Immediately W of the obvious cluster NGC2309 is a scattering of stars with a large range in brightness that is extended roughly E–W. It is little more than a

NGC2301

©STScI/AURA/CALTECH/ROE

gathering of field stars, the brightest of which are magnitude 8 and about two dozen stars are visible. Herschel's 23 February 1786 entry describes it only as: 'A cluster of very scattered stars.'

NGC2309

H18[6]	6:56.2	−7 12	OC
I2m	5.0′ × 5.0′	10.5V	

Photographically and visually, this is a moderately rich cluster of stars with a narrow range in brightness. Roughly triangular in shape with the apex of the cluster to the N, the group is moderately well separated from the sky background and fairly well resolved with medium-aperture telescopes. Small-aperture telescopes show a small and faint cluster that is difficult to resolve, with about a dozen members visible on a hazy background. A bright field star is visible to the NNE and about 40 stars are probable members of this cluster.

NGC2311

H60[8]	6:57.8	−4 35	OC
III2m	7.0′ × 7.0′	9.6V	

Images reveal a scattered cluster of faint stars with a moderate range in brightness, elongated SE–NW and moderately well separated from the background. In small telescopes this is a poor cluster with about a dozen faint stars resolved. In his description, Herschel noted: '... may be a projecting point of the Milky Way.' Fifty stars are considered members.

NGC2316

H304[2]	6:59.7	−7 46	RN
4.0′ × 3.0′			

The DSS image shows a small nebula with a roughly oval, brighter core. Fainter, irregular extensions emerge and in the S it seems involved with a small cluster of faint stars. Telescopically, this is a small but moderately bright nebula, quite irregular in form and slightly brighter to the middle. Even at high magnification, the involved star or stars are difficult to detect; a broad, flat triangle of field stars is immediately S.

NGC2319

H1b[8]	7:01.1	+3 04	AST?
14.0′ × 3.0′			

This object was not originally in the catalogue printed in the *Philosophical Transactions* but was included by Caroline Herschel in her Zone Catalogue. Recorded on 18 December 1783, the description included in *The Scientific Papers* reads: 'A cluster of very small stars, not rich.' The DSS photograph shows a chain of moderately bright stars running E–W and well separated from the sky background. Telescopically, the group is well seen at medium magnification as a long, almost straight chain of stars displaying a moderate range in brightness. A couple of faint pairs are involved in the chain, which is oriented ESE–WNW. To the N of the chain is a scattering of stars. About 25 well-resolved stars are present and are moderately well separated from the sky background.

NGC2324

H38[7]	7:04.2	+1 03	OC
II2r	8.0′ × 5.0′	8.8B	8.4V

Well seen in the DSS image, telescopically this is a faint, though attractive cluster located in a rich Milky

NGC2316

Way field. Partial resolution begins at low magnification and is more complete at medium powers. The main body of the cluster is a nebulous haze, almost 'dumb-bell' in form, with about half a dozen magnitude 12 stars sprinkled over the haze. The resolved members are fainter than magnitude 12; perhaps 40 are visible in good transparency. Discovered on 27 December 1786, Herschel described the cluster as: 'A beautiful cluster of very small stars of several sizes. Considerably compressed and rich middle, 10′ or 12′ diameter'. This group is quite remote and more than 130 stars are probable members. Spectral type: B9. Age: about 660 million years. Distance: about 10 400 ly.

NGC2335

H32[8]	7:06.6	−10 05	OC
III2m n	7.0′ × 7.0′	7.8B	7.2V

This cluster does not stand out well on the DSS image and visually it is a fairly weak cluster in a small-aperture

telescope; about 20 stars are resolved, centred on a diamond-shaped asterism of brighter stars. A close pair lies to the N. A few more stars are resolved with moderate apertures but this is not an impressive group. More than 50 stars are considered members and the cluster is part of the Canis Major OB1 association. Spectral type: B9. Age: 160 million years. Distance: 3300 ly.

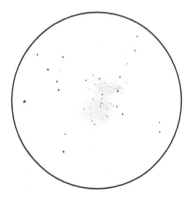

NGC2324 15-inch Newtonian 146x

NGC2324

is not hot enough to provide the necessary excitation. The true illuminating source is a spectroscopic binary with an orbital period of almost 16 days. The edge of the almost round shell is diffuse and fades gradually into the sky background. The expansion velocity of the gas shell is uncertain, but believed to be 40 km/s. Radial velocity in recession: 22 km/s. Distance: 3900 ly.

NGC2349

H27[7]	7:10.0	−8 37	OC?
10.0′ × 10.0′			

This object was first discovered by Caroline Herschel in 1783 and was recorded on 24 February 1786 by Herschel as: 'An irregular cluster of extremely small stars, considerably compressed 9 or 10′ long, 4 or 5′ broad with an extending branch towards south preceding. Caroline Herschel discovery 1783.' Visually, fairly near the presumed position is a rich collection of about 35 faint stars of moderate brightness range, with a slightly curving branch of stars that are a little brighter to the SW. This grouping is probably no more than a Milky Way condensation situated ENE of an E–W scattering of bright stars.

NGC2343

H33[8]	7:08.3	−10 39	OC
II2p n	6.0′ × 6.0′	7.0B	6.7V

Photographs show a coarse cluster of bright stars with a moderate range of brightness, the brightest being a close pair in the E. In small telescopes, the cluster is well differentiated from the background, with about 20 fairly bright stars resolved; the outline is roughly triangular in shape and the cluster appears quite compressed. This cluster is also a possible member of the Canis Major OB1 association. More than 50 stars are members of the cluster. Spectral type: A0. Age: about 100 million years. Distance is 2850 ly.

NGC2346

H65[4]	7:09.4	−0 48	PN
3b+6	2.5′ × 1.3′	10.8B	11.3V

The DSS image shows a bipolar-type planetary nebula with a bright inner

envelope elongated E–W and fainter crab-like extensions emerging to the SSE and NNW. Visually, only the central portion of the planetary nebula is visible and it is best seen at high magnification with averted vision as a grainy shell surrounding a bright central star. This star is evidently not the illuminating source, however, as it

NGC2346

NGC2353

H34[8]	7:14.6	−10 18	OC
III3p	18.0′ × 18.0′	7.3B	7.1V

Photographs reveal a very large, showy cluster of brilliant little jewels located to the N of a magnitude 6 field star. In a small telescope, at least 40 stars are resolved in a 20′ × 20′ area; many of the stars arranged in tight pairs and the range of brightness of the members is moderate. The cluster is well separated from the sky background, and total membership of the cluster is at least 100 stars. The cluster is a member of the Canis Major OB1 association. Spectral type: B0. Age: 13 million years. Radial velocity in recession 33 km/s. Distance: about 3400 ly.

NGC2506

H37[6]	8:00.2	−10 47	OC
I2r	9.0′ × 8.0′	8.3B	7.6V

The DSS image shows a very rich and compressed cluster in a rich field. Visually, this is a very attractive cluster: at low magnification it is mostly unresolved and quite compressed. It initially appears annular in form, and higher magnification reveals a triangle of stars to the S which help to fill in the ring. At the eyepiece, the cluster is more properly crescent shaped with two magnitude 11 stars resolved on its western cusp. Many more magnitude 12 or fainter stars are suspended over the unresolved, nebulous haze. The main mass of the cluster appears elongated E–W. This is an extremely rich cluster, numbering more than 800 members. The cluster is moderately old and quite remote. Spectral type: A. Age: 4 billion years (at least). Distance: 10 500 ly.

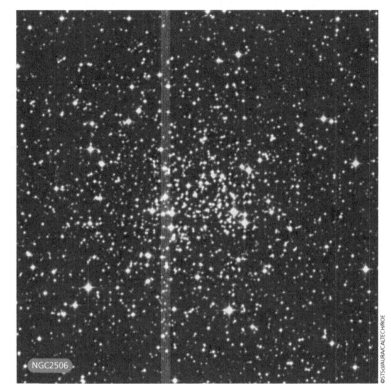

NGC2506

©STScI/AURA/CALTECH/ROE

Ophiuchus

Of the 15 Herschel objects in this sprawling constellation, 13 are globular clusters and all of the objects were recorded between the years 1784 and 1793. One object, the large open cluster NGC6633, was observed by Caroline Herschel in the summer of 1783, though it was probably originally discovered by Philippe De Chéseaux, while the globular cluster NGC6171 was first recorded by Pierre Méchain.

NGC6171

H40[6]	16:32.5	−13 03	GC
CC10	13.0′ × 13.0′	10.0B	8.9V

The discovery of this large globular cluster has been attributed to Messier's collaborator Pierre Méchain in April 1782 and it is now catalogued as M107. Herschel himself did not record the cluster until 12 May 1793, describing it as: 'A very beautiful extremely compressed cluster of stars, extremely rich, 5.0′ or 6.0′ in diameter. Gradually more compressed towards the centre.' In photographs the cluster is large with a small, bright, compressed core of stars. The outer halo is extensive, well resolved and symmetrical. Telescopically, the cluster is moderately large at high magnification, quite grainy in texture and uneven in brightness across its face. The main mass is distinctly irregular in outline with the brightest portion visible in the NW. A very faint haze of unresolved stars forms a secondary halo around the central region and individual stars and mottled clumps are visible particularly in the SE. The brightest members are magnitude 13 and the horizontal branch visual magnitude is 15.6. Spectral type: G0. Radial velocity in approach: 60 km/s. Distance: 21 000 ly.

NGC6235

H584[2]	16:53.4	−22 11	GC
CC10	5.0′ × 5.0′	12.0B	11.0V

Photographs show a small globular cluster with a small, slightly compressed core of stars elongated N–S. The outer halo is small, poorly resolved and irregular in outline. Visually, this cluster is only moderately bright; it is quite diffuse around its extremities with a fairly smooth texture and a concentration in brightness towards the NW. Resolution is difficult even at high magnification; the brightest member stars are magnitude 14 and the horizontal branch visual magnitude is 16.7. Herschel recorded the cluster on 26 May 1786 and called it: 'Pretty bright, considerably large, gradually brighter to the middle. Easily resolved. Undoubtedly stars.' Radial velocity in recession: 85 km/s. Distance: 31 000 ly.

NGC6284

H11[6]	17:04.5	−24 46	GC
CC9	3.5′ × 3.5′	10.7B	9.7V

Photographs show a compressed globular cluster with a bright, irregularly round core and an asymmetric outer halo, well differentiated from the sky background. Telescopically, the cluster is fairly bright and is seen as a bright core, steadily brighter towards the middle, with an extensive, symmetrical outer envelope extending out to a diamond-shaped asterism on the E side. There is no resolution into individual stars but the outer halo is somewhat grainy. The horizontal branch visual magnitude is 16.6. Distance: 39 000 ly.

NGC6287

H195[2]	17:05.2	−22 42	GC
CC7	4.8′ × 3.8′	11.5B	10.3V

This is a small globular cluster with a small, compressed core of stars, slightly oblate and extended NNE–SSW. Photographs reveal an extensive outer halo, well resolved and asymmetrical, extending towards the WSW and the

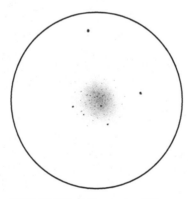

NGC6171 15-inch Newtonian 272x

NGC6284

©STScI/AURA/CALTECH/ROE

NGC6287

cluster is well differentiated from the sky background. Visually, this is a moderately bright globular cluster; it is very irregular in outline and quite mottled across its face, suggesting stars at the edge of resolution. It is moderately brighter to the middle, and the brightest members stars are magnitude 14.5, with the horizontal branch visual magnitude at 17.1. Herschel found the cluster resolvable. Radial velocity in approach: 211 km/s. Distance: about 22 000 ly.

NGC6293

H12[6]	17:10.2	−26 35	GC
CC4	3.5′ × 3.5′	10.0B	9.0V

Photographs show this globular cluster has a large, bright core with a fairly well resolved outer halo that blends into a rich star field. Visually, the cluster is bright and has a triangular-shaped, bright core which is quite mottled and uneven in brightness. The extensive

outer halo is definitely grainy, fading slowly into the sky background. Faint sparkles are seen consistently in the outer envelope. The brightest members are magnitude 14.3 and the horizontal branch visual magnitude is 16.5. Distance: 28 000 ly.

NGC6304

H147[1]	17:14.5	−29 28	GC
CC6	8.0′ × 8.0′	10.3B	9.0V

The DSS photograph shows a bright, almost round globular cluster with a very compressed core of stars in a densely populated Milky Way field. The outer halo is extensive, well resolved and symmetrical but poorly differentiated from the sky background. Visually, although it resists resolution, this is a moderately large and quite bright globular cluster with a bright, central region which intensifies to a fairly bright core. An extensive, fainter outer envelope surrounds the central

region. This outer envelope is very slightly oval in shape and elongated E–W. The brightest stars are magnitude 14.5 but the horizontal branch visual magnitude is 16.5, so only large amateur telescopes are likely to show any resolution into stars. Spectral type: G4. Radial velocity in approach: 98 km/s. Distance: 19 500 ly.

NGC6316

H45[1]	17:16.6	−28 08	GC
CC3	5.4′ × 5.4′	10.3B	9.0V

Situated in a rich Milky Way field, photographs show this globular cluster has a very compressed core of stars, elongated N–S. The outer halo is extensive, fairly rich and well resolved and it is poorly differentiated from the sky background. Telescopically, the cluster appears small even at high magnification. It is fairly bright and compact, not resolved, but round and opaque; it brightens quickly to a prominent, small core and is fairly well defined at the edges. The brightest stars are magnitude 15 but the horizontal branch visual magnitude is 17.8. Radial velocity in recession: 68 km/s. Distance: about 42 000 ly.

NGC6342

H149[1]	17:21.2	−19 35	GC
CC4	4.4′ × 3.0′	11.2B	11.0V

Photographs reveal a small, irregularly round globular cluster with a compressed core of stars. The outer halo is small, resolved and slightly elongated

NGC6304

NE–SW. In a moderate-aperture telescope, this is a faint, tenuous globular cluster, distinctly elongated NE–SW with a faint field star just beyond its SSW edge. The main envelope is tenuous and poorly defined at the edges and gradually brightens to the middle. There is no resolution, even at high magnification, which is not surprising since the horizontal branch visual magnitude is 16.9. Herschel recorded the object on 28 May 1786, calling it: 'Considerably bright, pretty small, little extended, easily resolvable.' The brightest cluster members are magnitude 15 and the cluster is receding at a radial velocity of 83 km/s. Distance: 39 000 ly.

NGC6355

H46[1]	17:24.0	−26 21	GC
5.0′ × 5.0′	12.5B	11.1V	

This small globular cluster appears irregularly round and has a somewhat compressed core of stars. Photographs show a fairly well resolved but irregular outer halo, quite rich and fairly well differentiated from the rich Milky Way field. Telescopically, this globular is a little faint and diffuse, roughly circular in shape and fairly well defined. The surface brightness is fairly even with a slightly brighter bar extending N–S. There is no resolution into individual stars. The horizontal branch visual magnitude is 17.2. Distance: 23 000 ly.

NGC6356

H48[1]	17:23.6	−17 49	GC
CC2	10.0′ × 10.0′	10.0B	8.9V

Images show a globular cluster with a very compressed round core of stars. The outer halo is rich and well resolved, but does not extend much more than 5′ before merging with the rich star field. Visually, it is a fairly bright globular cluster; the main mass is well defined with a sharp drop off in brightness to a faint, narrow secondary halo. There is no resolution even at high magnification; the envelope is quite smooth in texture and brightens suddenly to a small bright core.

Herschel discovered the cluster on 17 June 1784 and his description states: 'Bright, large, round, gradually brighter to the middle. Easily resolvable.' The brightest member stars are magnitude 15.1, while the horizontal branch visual magnitude is 17.7. Spectral type: G5. Radial velocity in recession: 32 km/s. Distance: 54 500 ly.

NGC6369

H11[4]	17:29.3	−23 46	PN
4+2	1.1′ × 1.1′	13.67B	

This bright planetary nebula is somewhat overexposed on the DSS image, but this serves to show the faint outer structure of faint gas loops to the E, N and W surrounding the bright annulus. The magnitude 15.5 central star (HD158269, spectral type WC) is visible within the ring. Telescopically, the ring structure is visible even in small-aperture telescopes at high magnification. In a moderate aperture, the annular ring appears well defined at the edges and very faintly bluish in colour. The ring is unevenly bright; the northern arc is brighter with a still brighter patch in the NW. The central star is not seen, nor is any of the faint outer structure. Distance: 3900 ly.

NGC6401

H44[1]	17:38.6	−23 55	GC
CC8	5.0′ × 5.0′	11.3B	10.7V

Images reveal a small globular cluster with a very compressed core of stars and an outer halo blending into a rich Milky

NGC6369

Way field. Visually, this globular cluster is a little faint and is seen as a diffuse haze which is poorly defined at the edges with a condensed, brighter core. A field star is immediately E of the core in the outer halo and there is no resolution into stars. The brightest individual stars are magnitude 15.5 and the horizontal branch visual magnitude is 18.0. Distance: 21 000 ly.

NGC6426

H587[2]	17:44.9	+3 00	GC
CC9	3.0′ × 2.75′	10.9B	

On photographs, this is a small globular cluster with a weakly compressed core, dominated by a N–S chain of faint stars just E of centre. It is well resolved throughout, oblate in form and elongated ENE–WSW. Visually, the cluster is a faint, broadly concentrated patch of light which is a little brighter to the core. The cluster is fairly round but with brighter patches visible to the WNW and NNE. The edges are poorly defined and, though not resolvable, high magnification does bring out a handful of very faint stars that may be cluster members. The brightest members are magnitude 14. Radial velocity in approach: 155 km/s. Distance: 57 000 ly.

NGC6517

H199[2]	18:01.8	−8 58	GC
CC4	4.0′ × 3.8′	11.1B	

This is a rich and compressed globular cluster photographically; the core is irregularly round and a small halo of

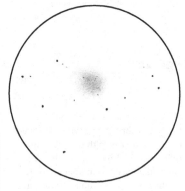

NGC6426 15-inch Newtonian 272x

NGC6426

©STScI/AURA/CALTECH/ROE

than magnitude 16, and the horizontal branch visual magnitude is 18.0. Radial velocity in approach: 37 km/s. Distance: about 20 000 ly.

NGC6633

H72[8]	18:27.7	+6 34	OC
III2m	25.0′ × 10.0′	5.0B	4.6V

This is one of Caroline Herschel's discoveries, recorded on 31 July 1783 but not included in William Herschel's catalogue until 30 July 1788. Photographs show a large, coarse, scattered cluster of bright stars elongated NE–SW, located in a rich field but well separated from the sky background. Visually, the cluster is best seen at low magnification. It is very bright and quite rich, located in a fairly dense Milky Way field. There are many bright stars of about magnitude 9, while the rest are between magnitude 9 and magnitude 13. The cluster is very large, at least 30.0′ in diameter, including suspected outliers which may not be part of the actual cluster. More than 50 stars are visible and membership of the cluster is about 160 stars, the brightest of which are magnitude 8. Spectral type: B6. Age: about 660 million years. Radial velocity in approach: 23 km/s. Distance: about 1000 ly.

fainter resolved stars appears elongated NNE–SSW. At the eyepiece, this small globular cluster is not resolved even at high magnification but the central region is fairly well defined. It is brighter to the middle, the core appears irregular and very slightly mottled and there is a short nebulous spike protruding towards the S. A hazy outer envelope surrounds the cluster. Herschel thought the cluster resolvable. The brightest member stars are fainter

Orion

The 25 objects recorded in Orion were all discovered between the years 1783 and 1786. There is a good mix of objects here: the galaxies are mostly faint and nondescript but the nebulae are for the most part remarkable and probably form extensions of the Orion Nebula complex. There are open clusters and asterisms as well, the most interesting of which is probably NGC2169.

NGC1662

H1[7]	4:48.5	+10 56	OC
II3m	12.0' × 4.0'	7.0B	6.4V

Photographs show a coarse, scattered cluster of stars with a wide range in brightness, arranged in a long chain oriented NNW–SSE with a spur of six stars emerging from the centre towards the SW. In a medium-magnification eyepiece, the cluster appears coarse and is well detached from the sky background. It is dominated by stars of magnitude 9–11, arranged around a bright diamond-shaped asterism. There are 22 members visible, with two of the members in the asterism being close pairs of stars. About 60 stars are suspected of being cluster members. Spectral type: A0. Age: about 300 million years. Distance: 1250 ly.

NGC1663

H7[8]	4:49.5	+13 10	OC?
IV2p	9.0' × 7.0'		

The visual and photographic appearances of this cluster are quite similar. At medium magnification this is a loose and modest cluster, well separated from the sky background and exhibiting a narrow brightness range of the members. Altogether, about 15 stars are resolved, all of which are fainter than magnitude 11, and the cluster is slightly elongated E–W. About 30 stars may be members of the cluster.

NGC1670

H501[3]	4:49.9	−2 44	GLX
SA0°:	1.8' × 0.9'	14.12B	

Photographs show a lenticular galaxy with a bright elongated core in a very faint disc. Visually, the galaxy can be found just N of a roughly equilateral triangle of faint field stars. It is small with diffuse extremities and a well-condensed, small and fairly bright core. High magnification enhances the view. The faint extensions, elongated ENE–WNW, are not detected visually. In photographs, a very faint field star is superimposed on the outer envelope immediately E of the core. Radial velocity: 4522 km/s. Distance: 202 mly. Diameter: 106 000 ly.

NGC1678

H502[3]	4:51.7	−2 38	GLX
SA0° pec?	1.1' × 0.8'	14.27B	

Photographs show that this lenticular galaxy is elongated ENE–WSW with a bright oval central region and fainter extensions. Smaller galaxies are in the field 2.4' NE and 3.8' N. Telescopically, this galaxy immediately follows a magnitude 10 field star by 1'. It is brighter and more conspicuous than NGC1670. It is round, small and well condensed, averted vision brings out a brighter core and the edges are well defined. Radial velocity: 4683 km/s. Distance: 209 mly. Diameter: 67 000 ly.

NGC1682

H527[2]	4:52.3	−3 06	GLX
SA?0': pec	1.0' × 1.0'	14.52B	

Photographs show a face-on lenticular galaxy with a bright core situated N of a bright field star. Telescopically, NGC1682 forms a pair in a high-magnification field with NGC1684, which is 2.9' to the E. It is small and condensed with a bright core. The fainter, edge-on lenticular galaxy NGC1683 is 4.8' N. Radial velocity: 4309 km/s. Distance: 193 mly. Diameter: 56 000 ly.

NGC1684

H528[2]	4:52.5	−3 06	GLX
E+ pec:	2.8' × 1.8'	13.00B	

In photographs, this bright elliptical galaxy has an oval core and an extensive outer disc. It forms a physical pair with NGC1682 to the W and the

NGC1682/1684

minimum core-to-core separation between the two galaxies would be about 163 000 ly at the presumed distance. Visually, this galaxy is elongated E–W with a bright disc, which is brighter to the middle, and a hazy outer envelope. Radial velocity: 4341 km/s. Distance: 194 mly. Diameter: 158 000 ly.

NGC1713

H516[2]	4:58.9	−0 30	GLX
cD;E+:	1.5′ × 1.3′	13.89B	

This is a lenticular or elliptical galaxy with a slightly elongated core and a faint and extensive outer envelope, elongated NNE–SSW. The DSS photograph shows several small, faint galaxies in the field, including NGC1709 2.6′ to the WNW. Telescopically, NGC1713 is a moderately faint galaxy; it is round and very broadly brighter to the middle with well defined edges. Radial velocity: 4350 km/s. Distance: 194 mly. Diameter: 85 000 ly.

NGC1729

H503[3]	5:00.1	−3 21	GLX
SA(s)c	1.7′ × 1.4′	13.0B	

In photographs, this is an almost face-on spiral galaxy with a bright core and bright, massive spiral arms. Several knotty condensations are in the spiral arms and field stars are 1.0′ E and 1.1′ N. Visually, the galaxy is a diffuse, uncertain patch of light which is poorly defined and a little brighter to the middle. It appears to be very slightly extended almost due N–S. Radial velocity: 3541 km/s. Distance: 158 mly. Diameter: 78 000 ly.

NGC1762

H453[3]	5:03.6	+1 34	GLX
SA(rs)c	1.7′ × 1.1′	13.35B	

Photographs show a slightly inclined spiral galaxy with a small, round core and two prominent spiral arms in a nebulous envelope, elongated N–S. A faint field star is immediately E of the core. Visually, the galaxy is well seen at high magnification and features an envelope of fairly high surface

brightness that is almost round and moderately well defined, with a prominent core or star embedded. Radial velocity: 4675 km/s. Distance: 209 mly. Diameter: 103 000 ly.

NGC1788

H32[5]	5:06.9	−3 21	RN
8.0′ × 5.0′			

The DSS image shows that the brightest portion of this nebula forms a fat crescent elongated roughly E–W with a few faint stars involved. The nominal published size is increased to 21.0′ × 10.5′ when taking the very faint outer reaches of the nebula into account. The dark nebula LDN1616 is on its SW flank and extremely faint loops of nebulosity extend towards the S and the N. Telescopically, although relatively small, this is a fairly bright and obvious nebula, even at low magnification. It consists of a small, bright, condensed but not opaque core that is almost round, surrounded by a

much fainter, diffuse envelope. The outer envelope is more extensive to the N and involves a bright field star 1.8′ to the NNW. Recorded on 1 February 1786, Herschel called this object: 'Considerably bright, very large, much diffused and vanishing near, and south following, a bright star.' The star HD293815 illuminates the nebula; it is magnitude 10.12, spectral type B9V.

NGC1908

H33[5]	5:26.0	−2 32	nonexistent

On 1 February 1786 Herschel recorded an object that he described as: 'Diffused extremely faint milky nebulosity. The means of verifying this phaenomenon are difficult.' There is nothing at the position indicated.

NGC1924

H447[3]	5:27.9	−5 19	GLX
SB(r)bc	1.6′ × 1.2′	13.26B	

Images reveal an almost face-on, barred spiral galaxy with a large, bright central

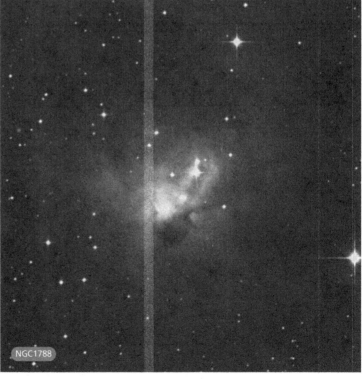

NGC1788

©STScI/AURA/CALTECH/DOE

NGC1924

©STScI/AURA/CALTECH/ROE

region and bright, narrow spiral arms forming a broken ring-like structure in a grainy outer envelope. It is elongated SE–NW with a faint star involved in the outer envelope immediately E of the core. In moderate apertures, this is a round, pale and moderately faint galaxy. Though small and diffuse with poorly defined edges, it takes magnification well and is a little brighter to the middle though no core is visible. A prominent field star is 6.5′ WNW. Herschel's 5 October 1785 discovery description reads: 'Very faint, considerably large, irregularly round, near a hook of very small stars.' Radial velocity: 2421 km/s. Distance: 108 mly. Diameter: 53 000 ly.

NGC1977

H30[5]	5:35.5	−4 52	EN+RN
20.0′ × 10.0′			

With NGC1973 and NGC1975, this object forms the northern, detached extension of the Orion Nebula. It is well seen on the DSS image as a complex network of bright and dark nebulae with a lace-like pattern, brightest around 42 and 45 Orionis and fading towards the N as it approaches the open cluster NGC1981. Visually, this is a pale, though obvious nebula. It is not concentrated and is ill defined, but is seen elongated due E–W and highlighted by a row of three bright

stars including 42 and 45 Orionis. 42 Orionis (spectral type: B1V) is believed to be the source of illumination. The nebula is most intense around these stars and more sharply defined to the N. Northwest of NGC1977 is the nebula NGC1973, which is visible as a small, separate, glowing patch of light surrounding the magnitude 7.36 star HD36958.

NGC1980

H31[5]	5:35.4	−5 54	EN
14.0′ × 14.0′			

The DSS image shows that this is the southern extension of the Orion Nebula. Several bright stars are involved, including the likely illuminating star Iota Orionis. The object appears as a lace-like pattern of nebulosity generally elongated NE–SW. At low magnification care is needed to pick up the nebulosity around the scattered group of bright and faint stars situated around Iota Orionis. About two dozen stars are involved, with a wide range in brightness and the nebulosity is most pronounced around Iota Orionis and the widely separated, bright pair of stars 8.5′ to the SW. The nebulosity is almost featureless, fading gradually into the sky background.

NGC1982

H1[3]	5:35.6	−5 16	EN+RN
20.0′ × 15.0′			

The field of this object is poorly seen on the DSS image, as it involves the bright

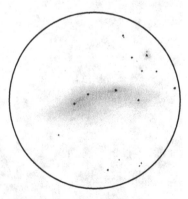

NGC1977 15-inch Newtonian 95x

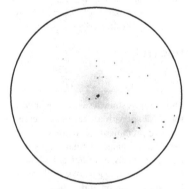

NGC1980 15-inch Newtonian 83x

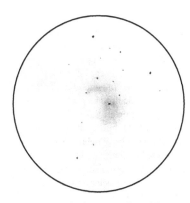

NGC1982 15-inch Newtonian 146x

nebulae M42 and M43. NGC1982 is part of M43, a detached portion of the Orion Nebula, which appears as a comma-shaped object seemingly separated from M42 by dark, obscuring material in high-resolution images. NGC1982 has been equated with M43, which Charles Messier entered in his catalogue on 4 March 1769, calling it a star encircled by nebulosity above the Orion Nebula. Messier's drawing of M42 and M43 in 1771 clearly shows the star circled by a faint haze. The object Herschel recorded on 3 November 1783, however, is described as: 'Very faint, small, much extended. In the large nebula.' This can hardly be interpreted as a description of M43, which would have certainly appeared large and bright in Herschel's reflector. In *The Scientific Papers*, J. L. E. Dreyer notes: 'H1³ is an appendage to the north of M43', so the 'large nebula' Herschel mentions is very likely M43 itself, and strictly speaking NGC1982 = H1³ should actually be interpreted as the faint, curving portion of the nebula N of the circular mass surrounding the illuminating star (HD37061, magnitude 6.85). Visually, M43 is quite bright; it is a detached portion of the Orion Nebula and appears as a large, 'comma'-shaped object, brightest in the area immediately surrounding the illuminating star. Subtle dark bands cross the nebula, particularly NNE of the star, defining the fainter nebulosity that Herschel recorded as distinct from M43.

NGC1990

H34⁵	5:36.2	−1 12	nonexistent

Herschel's 1 February 1786 discovery description for this object reads: 'I am pretty certain Epsilon Orionis is involved in unequally diffused nebulosity.' Popular star atlases show an irregularly round symbol indicating that a large nebula surrounds the bright star but examination of the DSS image shows a glow around the star that looks suspiciously like light scatter from the optics. There is extremely faint nebulosity about 15' to the NNW, WSW and ESE, but nowhere does it appear to approach or connect with the glow surrounding Epsilon Orionis and it is highly unlikely that nebulosity this faint could be detected visually with the kind of telescopes Herschel used. Several visual observations by the author have not shown anything that could not be characterized as light scatter around Epsilon Orionis and it is unlikely that any visually detectable nebulosity exists around the star. The conclusion is that NGC1990 is nonexistent.

NGC1999

H33⁴	5:36.5	−6 42	EN+RN
16.0' × 12.0'	9.5B		

Herschel's 5 October 1785 description reads: 'A star with milky chevelure or a very bright nucleus with much nebulosity.' Visually, this is a small, bright nebula that responds well to high magnification. The star is offset to the SW of the nebulosity, which is surrounded by a small, bright patch of light. The outer envelope is bright and well defined to the SE, NE and NW. A dark zone borders the star to the SW. Photographically, this dark zone is chevron shaped and a faint, milky nebulosity surrounds the bright nebula to a radius of at least 10.0'. The brightest portion of the nebula is about 2.3' in diameter. The spectral type of the central star is B8.

NGC1999

NGC2022

H34[4]	5:42.1	+9 05	PN
4+2	0.4' × 0.4'	14.9B	10.1V

This object is not well seen on the DSS image, appearing as a slightly oval, opaque disc. Discovered on 28 December 1785, Herschel correctly surmised that this was a planetary nebula, describing it as 'Considerably bright, small, nearly round like a star with a large diameter. With 240 [magnification] like an ill defined planetary nebula.' Visually, it is a bright and well defined object. The surface appears quite opaque and fairly uniform in brightness. The edges are sharply defined and no trace of the magnitude 14.9 central star (HD37882) is visible. The spectrum is continuous. Distance: about 6850 ly.

NGC2023

H24[4]	5:41.6	−2 14	E+RN
10.0' × 10.0'			

This is a bright portion of the extensive nebula complex SE of the bright star Zeta Orionis, which includes NGC2024, IC431, IC432, IC434 and the dark nebula B33, popularly known as the Horsehead Nebula. NGC2023 is NE of the Horsehead, and is a bright, irregularly round patch with streamers extending towards the S and E where they merge with the faint nebula IC434. Visually, it is visible as a tenuous, diffuse glow surrounding a 7.82 magnitude star (HD37903), which is the

illuminating source. The nebula is irregularly round and appears a little brighter and more extensive to the W. The immediate vicinity of the star is quite dark; the nebulosity begins to brighten about 20″ from the illuminating star.

NGC2024

H28[5]	5:41.9	−1 51	EN
30.0' × 30.0'			

Also known as the Flame Nebula, this is a large, bright and complex nebula immediately ENE of the bright star Zeta Orionis, which is the illuminating source for this nebula. It is part of an extensive nebula complex that includes NGC2023, IC431, IC432, IC434 and the dark nebula B33. Visually, the nebula is roughly horseshoe shaped, opening up towards the NNW. There are three distinct parts, separated by dark zones with a dark bay seen along the NW flank of the nebula. The bright portions appear diffuse and little structure is visible within them. The portion to the SE is the brightest and most sharply defined. Recorded by Herschel on 1 January 1786, his vivid description reads: 'Wonderful black space included in remarkable milky nebulosity, divided in 3 or 4 large patches; cannot take up less than a half degree, but I suppose it to be much more extensive.'

On 1 February 1786, Herschel documented an entry for a large nebula which he designated H35[5]. Using the stars 36 Orionis and 56 Orionis as the

reference points, his description reads: 'Diffused milky nebulosity, extending over no less than 10 degrees in polar distance and many degrees of right ascension. It is of very different brightness, and in general extremely faint and difficult to be perceived. Most probably the nebulosities of the 28th, 30, 31, 33, 34, and 38th of this class (Class V: very large nebulae) are connected together, and form an immense stratum of far distant stars, to which must also belong the nebula in Orion.' Although the description is somewhat vague and references objects that are probably nonexistent (NGC1908 – H33[5] and NGC1990 – H34[5]), there is a possibility that Herschel observed portions of the nebula IC434 and/or Barnard's Loop. Both objects are extremely difficult visually with moderate-aperture telescopes. The present author has attempted to observe Barnard's Loop without filters and has never met with any success. He has successfully observed IC434, the reflection nebula associated with the Horsehead Nebula, only once, while using a hydrogen beta filter, and he has never seen it without a filter. H35[5] was omitted from John Herschel's *General Catalogue* and was not assigned a number by J. L. E. Dreyer in the *New General Catalogue*. The identity of H35[5] is therefore uncertain and its observation by William Herschel unsubstantiated.

NGC2063

H2[8]	5:46.8	+8 48	AST
9.0' × 9.0'			

This is an E–W scattering of faint stars with a fairly narrow brightness range. About 15 stars are in the field and 10 are presumed to be members, though this may just be a chance alignment of stars. Herschel's 26 December 1784 descriptions reads: 'A small cluster of very small scattered stars.'

NGC2071

H36[4]	5:47.2	+0 18	RN
12.0' × 12.0'			

This is a large, complex reflection nebula with two stars involved at its

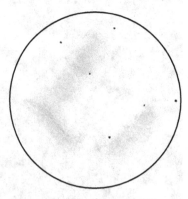

NGC2023 15-inch Newtonian 95x   NGC2024 15-inch Newtonian 177x

NGC2071

©STScI/AURA/CALTECH/ROE

nondescript and scattered cluster, well isolated from the sky background. About ten cluster members dominate, these being arranged in a 'tuning fork' pattern. Several cluster members can be resolved along its length, but generally a hazy unresolved background is evident. The cluster is much extended WNW–ESE. Herschel's 1 January 1786 discovery description reads: 'A cluster of pretty compressed, pretty small, scattered stars with many extremely small suspected between them. 7′ or 8′ diameter.' About 50 stars are likely members, with the brightest being magnitude 10. Distance: 2600 ly.

NGC2169

H24[8]	6:08.4	+13 57	OC
III3m	6.0′ × 6.0′	6.0B	5.9V

Photographically and visually this cluster presents a similar appearance. In small telescopes, this is a well resolved, somewhat compressed cluster of bright stars organized into two distinct groupings: the SE grouping is larger and has more stars than the NW grouping. The arrangement of the stars suggests the number '37'. The brightness range of the member stars is moderate and the cluster is well separated from the sky background. Thirty stars are probable members. Spectral type: B1. Age: 50 million years. Distance: about 3000 ly.

NGC2180

H6[8]	6:09.6	+4 43	OC/AST
8.0′ × 5.0′	9.0B		

Images show a scattered grouping of stars of moderate brightness range, probably an asterism, arrayed around a brighter field star, with most of the stars situated to the NW. Visually, this object is a coarse cluster of stars of moderate brightness range with about 15 stars in an area about 10.0′ in diameter. Four of the stars are bright and form a W-pointing triangular asterism; the rest of the stars are fainter, well resolved and well separated from the sky background.

core. Photographically, the brightest portion is very roughly triangular in shape and extends E–W. Dark channels define the nebula to the N and SW. Faint streamers extend the size of the nebula to about 12.0′ in diameter. The main illuminating star is HD290861, magnitude 9.5, spectral type F8. Visually, the entire nebula complex including M78 and the faint patch NGC2064 can be seen together in a medium-magnification field. NGC2071 is an ill defined glow around the illuminating star. It is almost circular and quite diffuse. Herschel recorded the nebula on 1 January 1786, calling it: 'A star affected with very faint, extensive, milky chevelure. The star not quite central.'

NGC2110

H448[3]/H510[3]	5:52.2	−7 27	GLX
SAB0⁻	1.8′ × 1.4′	13.50B	

Photographs show a lenticular galaxy with a bright, elongated core and a faint

outer envelope, elongated NNW–SSE in a fairly rich star field. Visually, the galaxy is small, fairly faint and almost round with hazy edges. The envelope is broadly brighter to the middle. Herschel recorded this galaxy twice: on 5 October 1785 he called it: 'Very faint, small, round, resolvable, a little brighter to the middle'; almost six months later, on 24 February 1786 he observed it again, this time describing it as: 'Extremely faint, extended, easily resolvable, probably a patch of stars.' Radial velocity: 2205 km/s. Distance: 98 mly. Diameter: 51 000 ly.

NGC2112

H24[7]	5:53.9	+0 24	OC
II2m n	18.0′ × 10.0′	9.1V	

In photographs, this is a scattered grouping of stars with a moderate brightness range, roughly elongated E–W with a brighter star in its NW quadrant. Visually, it is a large, though

NGC2169

©STScI/AURA/CALTECH/ROE

centred around a four-star chain oriented NNE–SSW. About a dozen stars are resolved; the grouping is elongated E–W and well separated from the sky background. About 30 stars are likely members, the brightest of which shine at magnitude 12. Distance: 6000 ly.

NGC2194

H5[6]	6:13.8	+12 48	OC
II2r	9.0′ × 7.0′	9.0B	8.5V

Photographs reveal a rich cluster of faint stars with a moderate brightness range. Two chains of stars, one running N–S and the other running E–W and joining in the SW, are the dominant feature. The group is in a rich field and is only moderately well differentiated from the sky background. Visually, this is a small, though bright, fairly compressed cluster that stands out well in a rich Milky Way field. The stars are fairly uniform in brightness with magnitudes ranging from 13 to 15 and about 40 members are visible over a faint, nebulous background. The cluster appears 'wasp-waisted', being concave in outline to the S and N and extended E–W. Total membership of the cluster is about 200 stars. Distance: 8700 ly.

NGC2186

H25[7]	6:12.2	+5 27	OC
II2m	4.0′ × 3.0′	9.2B	8.7V

Photographs show a small grouping of stars of moderate brightness range.

The brightest members form a wedge running E–W. Some fainter stars below the wedge may be involved with the cluster. Telescopically, it is a small but moderately bright cluster

Pegasus

Most of the 68 Herschel objects in this large autumn constellation were recorded between the years 1783 and 1785. A gap of five years followed before the last five objects were discovered in the autumn of 1790. All are galaxies except for one object, the asterism NGC7186. Not surprisingly, most of the galaxies here are faint, but there are a few bright and compelling objects such as NGC7331,

NGC7479 and NGC7814 as well as a host of moderately prominent objects like NGC7332, NGC7339 and NGC7741 which are visually interesting. There are some faint groups here as well, the most interesting of which is the Pegasus I Cloud, dominated by the marker galaxies NGC7619 and NGC7626. Amateurs with large-aperture telescopes will find this an interesting field to explore.

NGC14

H591[2]	0:08.8	+15 49	GLX
(R)IB(s)m pec	2.5′ × 1.8′	13.02B	

Photographically, this galaxy has a complex and somewhat peculiar structure. It has also been catalogued as Arp 235. The core is bright and irregular in form, surrounded by a slightly fainter inner envelope featuring many lumpy condensations and dust patches. The whole is surrounded by an extensive, faint and almost round outer envelope. In a moderate-aperture telescope, the galaxy is moderately faint but stands out well against the sky background. An oval spot of light, it is slightly brighter to the middle, milky in texture with soft edges and oriented very slightly NNE–SSW. Herschel recorded the galaxy on 18 September 1786 and described it as: 'Faint, pretty large, irregular form, unequally bright.' Radial velocity: 1012 km/s. Distance: 45 mly. Diameter: 33 000 ly.

NGC16

H15[4]	0:09.1	+27 44	GLX	
SAB0⁻	1.9′ × 1.2′	13.00B	12.1V

This fairly bright, moderately condensed galaxy is suddenly brighter to a bright, nonstellar core. High magnification brings out a faint secondary halo and this lenticular galaxy is elongated NNE–SSW. In photographs, the central

bar is oriented perpendicular to the outer envelope. Herschel placed the galaxy in his Class IV, recording it on 8 September 1784 and describing it as: 'A faint star with small chevelure and 2 burs.' Radial velocity: 3272 km/s. Distance: 146 mly. Diameter: 81 000 ly.

NGC23

H147[3]	0:09.9	+25 55	GLX
SB(s)a	3.1′ × 2.0′	12.84B	

Photographically, this spiral galaxy has a peculiar morphology. The five-sided core is bisected by a narrow, straight dust lane. A bright spur curves southward, terminating at a field star, condensation or companion galaxy and there are two extremely faint spiral arms. Visually, this is a fairly bright small galaxy, featuring a very bright nucleus. The magnitude 12 star (condensation?) lies SE of the nucleus by less than 30″, giving the impression of two galaxies in contact. The object is visible against the galaxy's outer envelope. NGC26 is visible in the field 9.0′ SE. Radial velocity: 4734 km/s. Distance: 212 mly. Diameter: 190 000 ly.

NGC52

H183[3]	0:14.6	+18 33	GLX
S?	1.9′ × 0.4′	14.29B	

The classification of this galaxy is somewhat uncertain; the DSS image

shows an edge-on, probably spiral galaxy with a thin dust lane along its major axis. Visually, the galaxy is quite faint and somewhat diffuse with fairly low surface brightness. It is a boxy streak that is fairly well defined, elongated SE–NW and evenly bright along its major axis. Radial velocity: 5542 km/s. Distance: 247 mly. Diameter: 137 000 ly.

NGC7042

H209[3]	21:13.8	+13 33	GLX
Sb	2.0′ × 1.8′	12.84B	

Photographs show this spiral galaxy is almost face-on with a small, round core and narrow spiral arms, which are generally stable for less than half a revolution before interacting with other spiral fragments. The smaller spiral galaxy NGC7043 is 5.3′ NE. Both galaxies can be seen in a high-magnification field in a moderate-aperture telescope. NGC7042 is round and moderately well defined; there is little or no brightening to the middle. NGC7043, though smaller and fainter, is a well-condensed patch of light. Radial velocity: 5274 km/s. Distance: 235 mly. Diameter: 137 000 ly.

NGC7137

H261[2]	21:48.2	+22 10	GLX
SAB(rs)c	1.8′ × 1.5′	13.05B	

The DSS image reveals a face-on spiral galaxy with a bright core and high-surface-brightness, knotty, spiral arms. High-resolution images reveal that two spiral arms emerge from the core but one of them splits into five arm fragments while the other remains intact. Visually, this galaxy is fairly large and moderately bright, though quite diffuse. The main envelope is round with a sharp fall in brightness at the edges to a very faint secondary halo. The galaxy is only a little brighter to the middle. Radial velocity: 1892 km/s. Distance: 84 mly. Diameter: 44 000 ly.

NGC7156

H452³	21:54.6	+2 57	GLX
SAB(rs)cd:	1.6′ × 1.4′	13.31B	

Images show a face-on, three-branch spiral with a small, brighter core encircled by an inner ring from which the spiral arms emerge. The arms are faint, knotty and symmetrical. In a moderate-aperture telescope, this galaxy is small and moderately faint, though the surface brightness of the main envelope is fairly high. The edges are well defined and the form is quite round and well concentrated to the middle. Herschel thought the galaxy resolvable. Radial velocity: 4145 km/s. Distance: 185 mly. Diameter: 86 000 ly.

NGC7177

H247²	22:00.7	+17 44	GLX
SAB(r)b	3.1′ × 2.0′	11.86B	

In photographs this inclined, barred spiral galaxy with a bright inner ring and bar is theta shaped and surrounded by a faint, broad spiral pattern. A brightening or foreground star is visible at the southern junction of the bar and ring. Deeper photographs reveal that the outer arms are thin and dotted with small condensations, though only moderate recent star formation is suspected for this galaxy. In a small-aperture telescope, the galaxy is readily visible as a small, roundish smudge of light. The core is moderately bright and well condensed, though nonstellar. A secondary envelope is faintly visible with averted vision. The galaxy is

bright and well seen in a moderate-aperture telescope as a mottled, oval halo which is brighter to the middle with a sharp, stellar core. Radial velocity: 1345 km/s. Distance: 60 mly. Diameter: 54 000 ly.

NGC7186

H165³	22:01.0	+35 06	AST
2.0′ × 0.5′			

The DSS image shows a small, NE–SW oriented chain of ten very faint field stars, well separated from the sky background. Herschel recorded this object as a very faint nebula; his 13 September 1784 entry reads: '5 or 6 stars forming a parallelogram with mixed nebulosity. Verified 240 [magnification].' In a moderate-aperture telescope four stars of this grouping are resolved, the stars at the NE and SW ends; between them is a suspected nebulous haze, made up of faint, unresolved members. It is unlikely that this group is anything more than a chance alignment of stars.

NGC7217

H207²	22:07.9	+31 22	GLX
(R)SA(r)ab	3.9′ × 3.2′	11.03B	10.4V

This is an almost face-on multi-arm spiral galaxy. The DSS image reveals a large, bright core with tightly wound arms emerging. The outer spiral pattern is dusty with many nebulous knots involved and the arms overlap, creating the impression of a ring. A very faint and hazy outer disc is suspected. This is quite a bright galaxy visually. A brighter central region is embedded in a slightly fainter inner disc. This is surrounded by a much fainter, diffuse outer envelope, which is poorly defined. Radial velocity: 1168 km/s. Distance: 52 mly. Diameter: 60 000 ly.

NGC7280

H248²	22:26.5	+16 09	GLX
SAB(r)0⁺	2.2′ × 1.5′	13.09B	

In photographs, this is an inclined galaxy, possibly barred, with a bright elongated core and a much fainter,

smooth-textured outer envelope. A possible companion galaxy, probably a Magellanic dwarf, is 4.6′ ENE and may be detectable in large-aperture telescopes. At high magnification, this faint galaxy forms a diamond-shaped asterism with three magnitude 13 field stars, the westernmost of which is a faint double. The galaxy has poorly defined edges, is marginally brighter to the middle and is elongated WSW–ENE. Recorded on 15 October 1784, Herschel called this object: 'Faint, pretty small, a quartile with three small stars.' Radial velocity: 2030 km/s. Distance: 91 mly. Diameter: 58 000 ly.

NGC7311

H428²	22:34.2	+5 35	GLX
Sab	1.6′ × 0.8′	13.53B	

Photographs show a highly inclined galaxy, possibly barred, with a bright elongated core leading into diffuse, grainy spiral arms. In moderate apertures, this is a rather faint galaxy which is a little brighter to the middle to a moderately condensed core. The envelope is elongated N–S with diffuse edges and overall the galaxy is more distinct at high magnification. Herschel thought the galaxy might be resolvable. Radial velocity: 4691 km/s. Distance: 210 mly. Diameter: 98 000 ly.

NGC7316

H180³	22:35.9	+20 19	GLX
SBc	1.1′ × 0.9′	13.61B	

On the DSS image it is difficult to determine the morphology of the galaxy but it appears to be a face-on, barred spiral galaxy with two bright arms emerging from the bar; after half a revolution the arms suddenly fade to faint, narrow arms. Visually, this is a fairly faint object; visibility is hindered by the magnitude 7 field star only 3.2′ SSW. It is slightly oval and a little brighter to the middle; the outer envelope is well defined and a magnitude 14 field star is visible immediately SSW. Radial velocity:

5745 km/s. Distance: 257 mly. Diameter: 82 000 ly.

NGC7321

H237[3]	22:36.4	+21 38	GLX
SB(r)b	3.0′ × 1.3′	13.88B	

Images show a slightly inclined spiral galaxy, probably barred. The core is small and faint, with a possible bar connected to a brighter inner ring, with four spiral arms emerging. Visually, this galaxy is moderately faint and best seen at medium magnification. It is elongated N–S, slightly brighter to the middle with extremities that are moderately well defined. Radial velocity: 7340 km/s. Distance: 328 mly. Diameter: 190 000 ly.

NGC7331

H53[1]	22:37.1	+34 25	GLX	
SA(s)b	10.5′ × 3.7′	10.20B	9.5V	

This is a well-known and frequently observed galaxy: it is a highly inclined spiral of the multi-arm type. It features a large, bright core, though no bulge is detected, perhaps due to the angle of inclination. The spiral structure is well defined with prominent dust lanes visible against the core and in the near spiral arm. The outer spiral structure is asymmetrical, being more extensive on the S side. Several small, bright galaxies are in the field to the E, but it is presumed that they are in the background as the radial velocities of these galaxies are quite high. In moderate-aperture telescopes, NGC7331 is a very bright spiral galaxy, quite elongated and oriented N–S with a prominent, bright and well-condensed core which is slightly offset towards the N. The extensions are quite bright with a distinctly grainy texture. It is surrounded by a very faint, secondary envelope, which is ragged and ill defined, particularly to the S. Discovered 5 September 1784, Herschel called it: 'Very bright, considerably large, much extended. Much brighter to the middle. Resolvable.' Radial velocity: 1030 km/s. Distance: 46 mly. Diameter: 140 000 ly.

NGC7331

©STScI/AURA/CALTECH/ROE

NGC7332

H233[2]	22:37.4	+23 48	GLX
S0 pec sp	3.3′ × 0.7′	11.97B	

Images show that this edge-on lenticular galaxy has a prominent rectangular core and bright extensions. In a moderate-aperture telescope, this galaxy forms a superb, bright contrasting pair with NGC7339 5.25′ to the ESE. NGC7332 is very bright and sharply defined, much elongated and oriented NNW–SSE with an intensely bright core. The extensions of the galaxy taper to well defined points. The two galaxies probably form a physical pair: the core-to-core separation between the two would be about 52 000 ly at the presumed distance. Radial velocity: 1370 km/s. Distance: 61 mly. Diameter: 59 000 ly.

NGC7335

H166[3]	22:37.3	+34 27	GLX	
SA(rs)0+	1.3′ × 0.6′	14.48B	13.3V	

Photographs show this lenticular galaxy as the brightest of four field

galaxies immediately E of NGC7331. It features a brighter, elongated core with a broad, prominent secondary envelope. In moderate-aperture telescopes this galaxy is faint, though readily visible in the field with NGC7331 as a well defined oval smudge oriented NNW–SSE. With care, the other three field galaxies can be detected in medium-aperture telescopes. Herschel came upon this galaxy on 13 September 1784, eight days after his discovery of NGC7331. His description reads: 'Extremely faint, very small. Elongated north following and 4′ or 5′ distance from I. 53 [NGC7331].' Radial velocity: 6529 km/s. Distance: 291 mly. Diameter: 110 000 ly.

NGC7339

H234[2]	22:37.8	+23 47	GLX
SAB(s)bc:?	3.0′ × 0.7′	13.10B	

This is a bright, highly tilted galaxy, very likely a spiral. Photographs show that it has an elongated, though dusty core, with asymmetrical extensions that

NGC7332/7339

are brighter though shorter to the W, faint but longer to the E. Dust patches are seen throughout the disc and a prominent dust lane is visible along the northern flank. Visually, though it is fainter than its companion NGC7332, it is still an intriguing galaxy, Well defined and elongated E–W, the envelope is fairly smooth and a little brighter along its major axis. Radial velocity: 1511 km/s. Distance: 67 mly. Diameter: 59 000 ly.

NGC7385

H216³	22:49.9	+11 36	GLX
cD;E pec:	1.7′ × 1.3′	13.16B	

Photographs show this galaxy and NGC7386 as the brightest of eight moderately bright NGC galaxies in a 15.0′ field. Several smaller fainter galaxies are also involved so that the field is an interesting one to explore with a large-aperture telescope. NGC7385 is a slightly inclined elliptical galaxy with a large, bright core and a suspected thin dust cloud bordering the core to the S. The galaxy is moderately bright in medium apertures; it is a slightly elongated oval galaxy with a core gradually brightening to the middle. It is fairly well defined and oriented ENE–WSW. A magnitude 10 field star is visible to the NW. Many of the brighter members of this group may be detected with large amateur instruments. Radial velocity: 8016 km/s. Distance: 358 mly. Diameter: 177 000 ly.

NGC7386

H217³	22:50.0	+11 42	GLX
cD;SA0:	2.0′ × 1.4′	13.52B	

Photographs show an elongated lenticular galaxy with a large secondary envelope. It is the northernmost bright member of a group of eight which includes NGC7385. Telescopically, it is a round, well-condensed glow which is very smooth-textured and forms a galaxy triangle with NGC7385 5.5′ to the SSW and NGC7387, a non-Herschel object, 5.3′ to the SE. Radial velocity: 7487 km/s. Distance: 334 mly. Diameter: 194 000 ly.

NGC7432

H465³	22:58.0	+13 08	GLX
E	1.2′ × 1.0′	14.46B	

The DSS image shows an elliptical galaxy with a faint, secondary envelope and a condensation or faint companion bordering the bright core to the E. In a moderate-aperture telescope the galaxy is a quite faint, roundish object which is brighter to the middle and fairly well defined. Radial velocity: 7785 km/s. Distance: 348 mly. Diameter: 121 000 ly.

NGC7436

H243³	22:58.0	+26 09	GLX
E	1.0′ × 0.8′	13.77B	

On the DSS image, this elliptical galaxy displays an extensive outer envelope and may, in fact, be an S0 galaxy. It is the brightest of a compact group of six galaxies covering an area 3.5′ × 3.5′ in extent, including NGC7431, NGC7433

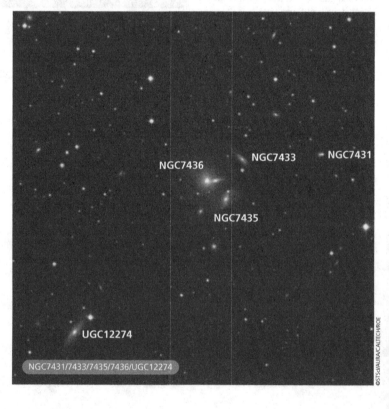

NGC7436 NGC7433 NGC7431

NGC7435

UGC12274

NGC7431/7433/7435/7436/UGC12274

and NGC7435, and there are several more in the immediate field. In a moderate-aperture telescope, the image is quite faint and poorly defined; it is an elongated oval oriented E–W with rather smooth surface brightness. This is the combined glow of NGC7436 and an anonymous edge-on spiral galaxy, separated by only 0.3′. The almost edge-on spiral NGC7433 is faint, small and round, and precedes to the NW by 1.5′ and the barred spiral NGC7435 is 1.0′ to the SW. Herschel discovered NGC7436 on 2 December 1784 and his description reads: 'Very faint, small, easily resolvable.' This comment may indicate that he was able to separate NGC7436 and the edge-on spiral and see them as separate objects. It is also possible that this is an indication that he was able to see one or more of the nearby galaxies as separate objects as well. Radial velocity: 7574 km/s. Distance: 338 mly. Diameter: 98 000 ly.

NGC7448

| H251[2] | 23:00.1 | +15 59 | GLX |
| SA(rs)bc | 2.5′ × 1.0′ | 12.10B | |

The structure of this slightly inclined spiral galaxy is somewhat unusual and it has also been catalogued as Arp 13. High-resolution photographs show a galaxy with a bright core embedded in a bright but fragmentary inner spiral structure, some of which is defined by dust patches. The outer spiral structure is a little more regular but the brightness falls off abruptly with several large, nebulous knots involved, particularly to the N. The outer spiral pattern is much more extensive to the S. The galaxy is quite bright visually; it is a large and fairly well defined oval, oriented N–S with a brighter core in a fairly opaque envelope that is slightly grainy and fades slowly to the edges. Radial velocity: 2370 km/s. Distance: 106 mly. Diameter: 77 000 ly.

NGC7454

| H249[2] | 23:01.1 | +16 23 | GLX |
| E4 | 2.5′ × 1.5′ | 12.82B | |

The visual and photographic appearances of this elliptical galaxy are quite similar. Telescopically, this is a fairly bright galaxy; it is a well defined oval object oriented SSE–NNW with an envelope that is broadly brighter to the middle with a star-like core embedded. A widely separated pair of field stars lies just to the NW. Radial velocity: 2198 km/s. Distance: 98 mly. Diameter: 71 000 ly.

NGC7457

| H212[2] | 23:01.0 | +30 09 | GLX |
| SA(rs)0⁻? | 4.1′ × 2.5′ | 11.87B | 11.9V |

The DSS photograph shows a lenticular galaxy with the characteristic, three-brightness-zone profile of the class. The fainter, spindle-shaped galaxy UGC12311 is 7.75′ to the NE. NGC7457 is fairly bright visually, oval in shape and elongated ESE–WNW with a bright, elongated central region, reminiscent of a bar, in a slightly fainter disc which fades uncertainly into the sky background. Radial velocity: 1014 km/s. Distance: 45 mly. Diameter: 54 000 ly.

NGC7463

| H210[3] | 23:01.9 | +15 59 | GLX |
| SABb: pec | 2.9′ × 0.7′ | 13.78B | |

Photographs show a highly inclined barred spiral, featuring a bright bar, a short, bright inner arm and very faint and broad outer spiral arms. It is the first of three galaxies (the others being NGC7464 and NGC7465) in the field. In moderate apertures, NGC7463 is the brightest of the group and also the largest. It is a well defined object that is

NGC7463/7465

quite bright throughout; it is slightly brighter to the middle, and is oriented E–W. NGC7464, located 0.9′ SE of the core, may be visible in larger instruments. Radial velocity: 2516 km/s. Distance: 112 mly. Diameter: 95 000 ly.

NGC7465

| H211[3] | 23:02.0 | +15 58 | GLX |
| (R′)SB(s)0°: | 1.2′ × 0.8′ | 13.36B | |

Images show an inclined 'Sombrero' galaxy with a large, bright central region and a thin dust lane E of the core. Moderate-aperture telescopes show only the central region of this galaxy as a round, well-condensed spot with a much brighter core. It can be seen in the same high-power field as NGC7463 2.5′ to the WNW and it is probably a companion of the larger galaxy. Radial velocity: 2143 km/s. Distance: 96 mly. Diameter: 33 000 ly.

NGC7468

| H202[3] | 23:03.0 | +16 36 | GLX |
| E3: pec | 0.9′ × 0.6′ | 14.08B | |

Photographs show the peculiar morphology of this presumed E3 elliptical galaxy. It may be two bright galaxies in contact, as it appears to be bi-nuclear with bright cores and two very faint wisps of material extending away from the cores, one N, one S. Visually, it is a rather faint galaxy; it is a small, diffuse spot with even surface brightness, elongated N–S. The edges are moderately well defined. Herschel discovered this object on 15 October 1784, his description reading: 'Very faint, very small. Stellar. Verified 240.' Radial velocity: 2257 km/s. Distance: 101 mly. Diameter: 26 000 ly.

NGC7469

| H230[3] | 23:03.3 | +8 52 | GLX |
| (R′)SAB(rs)a | 1.5′ × 1.1′ | 12.90B | 12.3V |

Together with the fainter, disturbed galaxy IC5283 (magnitude 15.2) located 1.4′ NNE, this inclined ring galaxy is also catalogued as Arp 298. In photographs, the bright core is attached to a moderately bright inner ring. The

NGC7469

outer ring is very faint and detached. Visually, it is a difficult galaxy to confirm, owing to the brightness of its core, which overwhelms the inner envelope. This envelope is best detected at medium magnification. It is very diffuse and is oriented ESE–WNW. The core is small and intensely bright, and is easily mistaken for a star. A faint star is suspected WNW of the core. This star exists, as does another immediately E of the galaxy and a little further out. Herschel discovered the galaxy on 12 November 1784 and had some difficulty with it, describing it thus: 'Extremely faint. Extremely small. 240 (magnification) left some doubt.' Radial velocity: 5049 km/s. Distance: 226 mly. Diameter: 98 000 ly.

NGC7479

H55[1]	23:04.9	+12 19	GLX	
SB(s)c	4.0′ × 3.1′	11.73B	10.8V

This is a beautiful, face-on, two-arm, barred spiral galaxy. The DSS image reveals a slightly brighter core and the bar displays thin dust lanes, which extend far into the bright inner spiral arms. The arms are of the grand-design type with many knots and condensations. There are fainter, fragmentary arms branching off the main arms along their inner edges. Telescopically, this galaxy is moderately bright and appears as an elongated oval glow, oriented N–S with a bright bar defining the major axis. There is a slight, brighter bulge in the middle and the envelope is hazy and diffuse, best seen at medium magnification and a little

brighter along the eastern flank. A field star borders the galaxy to the N and a very faint star is in the envelope, 0.75′ WSW of the core. Radial velocity: 2546 km/s. Distance: 114 mly. Diameter: 132 000 ly.

NGC7497

H203[3]	23:09.1	+18 11	GLX
SB(s)d	4.9′ × 1.1′	13.00B	

In photographs, this is a highly inclined spiral galaxy with a very small core and dusty spiral arms. Telescopically, the galaxy is somewhat dim but large and is best seen at medium magnification as a fairly well defined streak of light which is a little brighter along its major axis. The profile is fairly flat and elongated NE–SW. Radial velocity: 1885 km/s. Distance: 84 mly. Diameter: 120 000 ly.

NGC7515

H220[3]	23:12.9	+12 41	GLX
S?	1.7′ × 1.6′	13.22B	

Images show a face-on spiral galaxy with a small, bright core offset to the N

in an asymmetrical, faint spiral structure. Visually, the object is well seen at high magnification as a moderately bright galaxy with a mottled envelope, oval with diffuse edges and oriented N–S. A double star about 7.0′ S points directly at it. Radial velocity: 4639 km/s. Distance: 207 mly. Diameter: 102 000 ly.

NGC7550

H181[3]	23:15.3	+18 57	GLX
SA0⁻	1.4′ × 1.2′	13.16B	

The trio comprising this galaxy, NGC7547 and NGC7549 is also catalogued as Arp 99. There are two other galaxies in the field as well, NGC7553 and NGC7558, and the group of five is also known as the Hickson 93 galaxy group. The photographic and visual appearances of NGC7550 are quite similar: it is a round, well-condensed object with a brighter core. With care all five galaxies can be seen in a medium-magnification field in moderate-aperture telescopes and it is

NGC7479

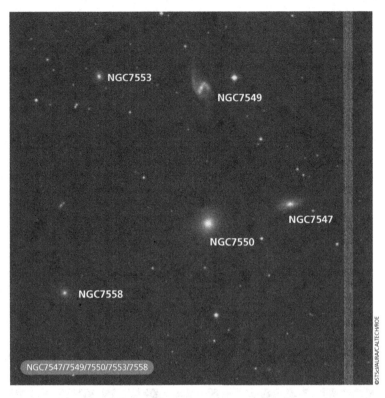

NGC7547/7549/7550/7553/7558

second observation he called it very faint, which led his son John to suspect that the nebula was variable in brightness. Radial velocity: 4338 km/s. Distance: 194 mly. Diameter: 101 000 ly.

NGC7570

H238[3]	23:16.8	+13 29	GLX
SBa	2.0′ × 2.0′	13.96B	

Photographs show an almost face-on, barred spiral galaxy. The very bright and round core is embedded in a faint bar with two tightly wound spiral arms forming a pseudo-ring. After half a revolution the spiral arms fade and broaden to a faint, smooth-textured outer spiral pattern. Visually, the galaxy is small and fairly faint; it is a hazy patch of light at medium magnification. At high magnification, it is extended NE–SW, though ill defined, and the light is a little more intense along the major axis with a brighter core in the middle. Radial velocity: 4862 km/s. Distance: 217 mly. Diameter: 126 000 ly.

NGC7578

H182[3]	23:17.2	+18 41	GLX
A = S0° pec	1.4′ × 1.4′	14.34B	
B = E1:	1.0′ × 1.0′	14.96B	

This pair of galaxies has been catalogued as Arp 170 and they are also the central galaxies in the Abell 2572 galaxy cluster. In the DSS image, both galaxies are very small, round and condensed. These two galaxies share a common, fainter and irregularly round envelope, which is more extensive to the S of the pair but there is no indication of tidal disturbance between the galaxies, even though the measured radial velocities are almost identical. Several cluster members are nearby, including NGC7578C, an S0 galaxy immediately to the ENE. Telescopically, the pair is best seen at medium magnification as a hazy patch of light, ill defined and roundish. With averted vision, two star-like points are visible, oriented NE–SW. These are the cores of the individual galaxies: component A is the brighter image and appears immediately NE of

somewhat surprising that Herschel did not discover NGC7547 and NGC7549 as well. NGC7547 is 3.0′ WNW, NGC7549 is 4.75′ N, NGC7553 6.6′ NNE and NGC7558 5.75′ ESE. Herschel recorded NGC7550 on 18 September 1784, describing it as: 'Very faint, very small, round. Verified 240 (magnification).' Radial velocity: 5249 km/s. Distance: 234 mly. Diameter: 96 000 ly.

NGC7559

H221[3]	23:15.9	+13 17	GLX
S0⁻:	1.3′ × 1.2′	14.7B	

Photographs show this lenticular galaxy has a smaller, fainter galaxy embedded in its outer halo NNW of the core. It is unlikely that this companion would be easily visible in any but the largest amateur telescopes. Both galaxies have identical radial velocities so they are probably companions. The barred spiral galaxy NGC7563 is 6.1′ to

the SSE and is a likely companion galaxy. In moderate apertures, NGC7559 is a fairly faint, round, well defined patch of light with even surface brightness and poorly defined edges. Radial velocity: 4761 km/s. Distance: 212 mly. Diameter: 80 000 ly.

NGC7563

H222[3]	23:16.0	+13 11	GLX	
SBa	1.8′ × 0.8′	13.88B	12.8V

This is an inclined, theta-shaped, barred galaxy with a bright, elongated core and a bright bar attached to a faint, smooth-textured ring which is elongated SSE–NNW. In moderate-sized telescopes, the core is bright and a brighter star-like nucleus is visible in the well defined envelope. The bar of the galaxy appears as a faint extension to the core and is oriented E–W. The galaxy precedes a prominent five-star asterism. Herschel initially described the galaxy as extremely faint, but then at a

component B. Herschel came upon this object on 18 September 1784, calling it: '4 or 5 small stars with nebulosity. 240 [magnification] doubtful.' Radial velocity: 12 243 km/s. Distance: 547 mly. Diameter: component A: 224 000 ly; component B: 160 000 ly.

NGC7619

H439²	23:20.2	+8 12	GLX
E	2.5′ × 2.3′	12.11B	

Along with NGC7626, this galaxy was discovered on 10 September 1785. It is a bright elliptical galaxy, the brightest of a rich group known as the Pegasus I Cloud and sometimes referred to as the NGC7619 galaxy group. The determination of the redshift of these two galaxies by Milton Humason in 1929 led Edwin Hubble to formulate his theory of the redshift–distance relation. The distance for the cloud as a whole is about 160 mly. Visually, the core of NGC7619 is bright and moderately large. The outer envelope is bright, well condensed and well defined, and is very slightly elongated NNE–SSW. A small companion galaxy, NGC7617, is 2.6′ SSW. Radial velocity: 3911 km/s. Distance: 175 mly. Diameter: 127 000 ly.

NGC7623

H435³	23:20.5	+8 24	GLX
SA0°:	1.4′ × 1.0′	13.85B	

This galaxy is located about 11.0′ N of NGC7626 and is a member of the Pegasus I Cloud. Photographs show a bright, elongated core in an irregularly shaped, asymmetrical envelope which is broader S of the core. Best seen at high magnification, it is a small, very compact galaxy, well condensed with a brighter star-like core. The envelope is oval in shape and elongated almost due N–S and the edges are sharply defined. Observers with very-large-aperture telescopes may detect the three faint galaxies nearby, one to the E, the other two to the SW. Radial velocity: 3889 km/s. Distance: 174 mly. Diameter: 71 000 ly.

NGC7625

H250²	23:20.5	+17 14	GLX
SA(rs)a pec	1.6′ × 1.4′	12.94B	

This peculiar galaxy, probably a spiral, has also been catalogued as Arp 212. The DSS image shows a galaxy with a bright core and three large dust lanes, two below the core, one above, in a fainter outer envelope. Deep, high-contrast images show the southern dust lanes are a curiously 'trifid' structure with a large, bright condensation in the dust lane to the SSW. Visually, this is a bright and fairly large galaxy that holds magnification well. The bright core is a little more prominent at medium magnification. The main envelope is grainy and of high surface brightness. The edges drop off suddenly in brightness to a faint, diffuse and ill defined outer halo. A magnitude 7 field star is 6.75′ ENE. Radial velocity: 1804 km/s. Distance: 81 mly. Diameter: 38 000 ly.

NGC7626

H440²	23:20.7	+8 13	GLX
E pec:	2.6′ × 2.3′	12.18B	

The galaxy and NGC7619 are the dominant members of the Pegasus I Cloud. Photographs show an elliptical galaxy with a bright, round core and an extensive outer envelope elongated NNE–SSW. Visually, it appears slightly fainter than NGC7619 but larger. The envelope is bright with the brightness increasing to the middle but the core is not as intense as that of NGC7619. The edges are a little more diffuse and the envelope is round. Radial velocity: 3554 km/s. Distance: 159 mly. Diameter: 120 000 ly.

NGC7634

H441²	23:21.6	+8 52	GLX
SB0	1.2′ × 0.9′	13.61B	

This inclined, barred galaxy has a bright, elongated core and a much

NGC7625

fainter outer envelope and is a possible outlying member of the Pegasus I Cloud. Visually, it is a faint and difficult object. At medium magnification it is a nebulous spot but at high magnification the nebulosity is overwhelmed by the bright core and a magnitude 13 star located to the SSE. With averted vision the outer envelope is just visible as a diffuse, poorly defined circular glow which extends to the magnitude 13 field star. The galaxy core is bright and star-like and appears offset to the NW in the outer envelope. Radial velocity: 3375 km/s. Distance: 151 mly. Diameter: 53 000 ly.

NGC7647

H473[3]	23:24.1	+16 47	GLX
cD;E	1.3′ × 0.7′	14.66B	

This elliptical galaxy has a bright, round core and a large, very faint secondary halo and is the brightest member of the galaxy cluster Abell 2589. The DSS image shows a swarm of faint galaxies surrounding the giant elliptical galaxy; most of the prominent galaxies appear to be spirals. Visually, this is a small and very faint galaxy, best seen with averted vision. It is very broadly concentrated to the centre and is poorly defined. The galaxy was discovered by Herschel on 29 November 1785 when he described it as: 'Extremely faint, considerably large, some doubt. Preceding a row of stars.' Radial velocity: 12 496 km/s. Distance: 558 mly. Diameter: 211 000 ly.

NGC7648

H218[3]	23:23.8	+9 39	GLX
S0	1.6′ × 1.0′	13.75B	

Despite its classification as an S0 galaxy, the DSS image shows a slightly inclined spiral galaxy with a large, bright core and narrow spiral arms defined by dust lanes. The correct classification is probably Sb. The galaxy is moderately bright at high magnification; it is an oval patch of light which is broadly brighter to the middle, the edges are fairly well defined and the galaxy is elongated E–W. A very faint star follows

immediately to the E and another is a little further to the SE. Radial velocity: 3711 km/s. Distance: 166 mly. Diameter: 77 000 ly.

NGC7659

H212[3]	23:26.0	+14 12	GLX
S0/a	0.8′ × 0.2′	14.69B	

The DSS image shows an edge-on lenticular galaxy with a faint star or condensation in its eastern extension. Visually, the galaxy is quite faint and small, though it is fairly well defined; it is a small sliver of light with even brightness along its major axis, which is extended ESE–WNW. Radial velocity: 4129 km/s. Distance: 185 mly. Diameter: 43 000 ly.

NGC7671

H226[3]	23:27.3	+12 28	GLX
SA0:	1.4′ × 0.8′	13.79B	

In photographs this lenticular galaxy has a bright core and an extensive outer envelope and is elongated SE–NW. The spiral galaxy NGC7672 is 5.6′ SSE. Telescopically, both galaxies are visible in a high-magnification field, with NGC7671 being much brighter; it is a round, opaque and condensed object which brightens quickly to the middle. A bright field star precedes it by 2.0′. Radial velocity: 4285 km/s. Distance: 191 mly. Diameter: 78 000 ly.

NGC7678

H226[2]	23:28.5	+22 25	GLX
SAB(rs)c	2.5′ × 1.7′	12.50B	

This barred spiral galaxy is a starburst galaxy which has also been catalogued as Arp 28. Photographs show a bright, face-on, barred galaxy with spiral arms of the grand-design type. The inner arms form a pseudo-ring where they emerge from the bar and the southern arm, though markedly shorter, is far brighter due to the presence of a string of massive HII regions. The other spiral arm is not continuous, fracturing in two places along its length. It, too, has its complement of HII regions, but they are much fainter. In moderate-aperture

telescopes, this is a moderately bright, though diffuse galaxy which is broadly concentrated to the middle with poorly defined extremities. It is bracketed by three magnitude 10–11 field stars. At high magnification, a faint stellar core is occasionally glimpsed as is a faint mottling in the envelope. Radial velocity: 3668 km/s. Distance: 164 mly. Diameter: 119 000 ly.

NGC7680

H860[3]	23:28.5	+32 24	GLX
S0⁻:	1.7′ × 1.7′	13.67B	

Photographs show a lenticular galaxy, the brightest of several in the field, including MCG+5-55-22, which is 6.0′ to the WNW. None of the fainter galaxies is noted in a moderate-aperture telescope. Though small, NGC7680 is a moderately bright galaxy, well seen at medium and high magnification. It is round and well defined with a bright, opaque envelope, even in surface brightness. Radial velocity: 5132 km/s. Distance: 238 mly. Diameter: 118 000 ly.

NGC7681

H242[2]	23:28.6	+17 16	GLX
S0°: sp	1.8′ × 1.5′	15.61B	

In photographs, this is a peculiar galaxy, perhaps with a double core, and features a bright central region and a very faint, asymmetrical outer envelope. A faint field star, or perhaps a condensation, is visible in the outer envelope NE of the core. Two anonymous galaxies are in the field, one 4.75′ WSW, the other 6.0′ WSW. They may be visible in large-aperture telescopes. Visually, although very small, this high-surface-brightness galaxy is well condensed with a sharp, stellar nucleus. At medium magnification, the core is very bright and overwhelms the surrounding envelope, which is faint but easily visible. High magnification broadens and enhances the envelope, diminishing the core. Discovered by Herschel on 11 October 1784, his description reads: 'Faint, small, irregularly round, near and preceding two or three stars.' Radial velocity: 7017 km/s. Distance: 313 mly. Diameter: 164 000 ly.

NGC7691

H213[3]	23:32.6	+15 50	GLX
SAB(rs)bc	2.3′ × 1.7′	14.06B	

Photographs show a face-on, multi-arm spiral galaxy with a small core, dusty arms and condensations in the spiral arm to the S. The galaxy is fairly faint visually; it is a round, nebulous patch of light broadly and very slightly brighter to the middle. The edges are poorly defined and bright field stars are immediately S and E of the galaxy. Radial velocity: 4204 km/s. Distance: 188 mly. Diameter: 126 000 ly.

NGC7711

H244[2]	23:35.7	+15 18	GLX
S0	4.0′ × 1.3′	13.09B	

The DSS image shows a probably spiral galaxy of peculiar morphology. The core of the galaxy is bright and elongated and the faint outer envelope appears disturbed on the W side. On the E is an extensive looping spiral arm, which gives the galaxy a decidedly asymmetrical appearance. Telescopically, the galaxy is moderately bright; it is an oval patch of light which is well defined and well condensed. The surface brightness is fairly even across the envelope and the galaxy is elongated E–W. Radial velocity: 4218 km/s. Distance: 188 mly. Diameter: 219 000 ly.

NGC7720

H146[3]	23:38.5	+27 02	GLX	
cD;E[+] pec:	2.4′ × 1.3′	13.43B	13.3V	

The DSS image shows that this large elliptical galaxy has a smaller companion immediately N of its core. The radial velocity of this companion is considerably lower than that of the primary, however, at 8288 km/s, but nevertheless, the two galaxies probably form a physical pair. NGC7720 is the brightest member of the Abell 2634 galaxy cluster and a compact swarm of

NGC7720

©STScI/AURA/CALTECHROE

elliptical and lenticular galaxies surrounds it. Visually, NGC7720 is a small and faint, round object with moderately well defined extremities. Its brightness rises towards the middle to a bright core and the attached companion, which shines at magnitude 15.0, is not detected. None of the other members of the galaxy cluster is visible. However, Herschel, who discovered the galaxy on 10 September 1784, wrote: 'Very faint, extended. Some small stars with nebulosity.' This implies not only that he probably saw both galaxies as an unresolved nebulous object, but, more interestingly, that he may have seen some of the fainter companion galaxies as nebulous stars. Radial velocity: 9336 km/s. Distance: 417 mly. Diameter: 288 000 ly.

NGC7741

H208[2]	23:43.9	+26 05	GLX	
SB(s)cd	4.0′ × 2.7′	11.82B	11.3V	

Photographs show that this face-on barred spiral galaxy has a large, bright and well-developed bar with only a very slight bulge identifying the core. The two spiral arms broaden quickly into the outer envelope and there are many HII regions. High-contrast, deep images begin to resolve stars at about blue magnitude 21.5. Visually, this is a moderately bright, large and diffuse object with poorly defined extremities. The envelope is oval and oriented E–W. A faint hint of a brighter central bar is visible, also oriented E–W. This bar is best seen with averted vision. The outer envelope is marginally fainter N of the bar. Two bright field stars are seen 2.0′ NNW of the core. Radial velocity: 931 km/s. Distance: 41 mly. Diameter: 49 000 ly.

NGC7742

H255[2]	23:44.3	+10 46	GLX
SA(r)b	1.7′ × 1.7′	12.29B	

On the DSS image much of the internal structure of this ring spiral galaxy is burned out, though something of the

NGC7741

©STScI/AURA/CALTECH/ROE

very faint outer spiral structure can be seen. High-resolution photographs show a round, bright core with a knotty, detached, bright ring, and the outer spiral structure emerging from this ring. Telescopically, the galaxy is quite bright; it is a round, well defined object with an opaque disc and a sizable, nonstellar brighter core. The disc is a little mottled in texture and a field star is 1.2′ ESE of the core. The likely companion galaxy NGC7743 is 50.0′ to the S. Radial velocity: 1809 km/s. Distance: 81 mly. Diameter: 40 000 ly.

NGC7743

H256[2]	23:44.4	+9 56	GLX
(R)SB(s)0+	2.3′ × 1.9′	12.40B	

Photographs show a face-on spiral galaxy, very probably barred, with a bright, elongated core and two smooth-textured, tightly wound spiral arms. The galaxy is moderately bright visually and has an irregularly round, poorly defined

disc that is broadly brighter to the middle to a bright, star-like core. Radial velocity: 1854 km/s. Distance: 83 mly. Diameter: 55 000 ly.

NGC7753

H213[2]	23:47.1	+29 29	GLX
SAB(rs)bc	3.3′ × 2.1′	13.36B	

Together with NGC7752, situated 1.9′ SW, this galaxy has also been catalogued as Arp 86. Photographs show a nearly face-on spiral with a very faint, short bar emerging from the brighter core. The faint spiral arms have some condensations and the northern spiral arm extends towards NGC7752, which appears slightly disturbed. Visually, NGC7753 is the larger and brighter of the pair; it is a roundish patch of light with a brighter core which is somewhat offset towards the N. The envelope is uneven in brightness and there appears to be a darkish patch to the SW, where a faint field star borders

the envelope. NGC7752 is a small, faint oval with an opaque core. Radial velocity: 5352 km/s. Distance: 239 mly. Diameter: 229 000 ly.

NGC7760

H854[3]	23:49.3	+30 58	GLX
E?	1.4′ × 1.1′	14.59B	

Images show an elliptical galaxy with a bright core and an extremely faint, asymmetrical outer envelope which is more extensive to the SSE. A bright star or compact companion is involved to the SW, and a very faint star is involved to the NNW. In a medium-aperture telescope, the galaxy is virtually stellar at medium magnification with just a trace of an outer envelope. High magnification shows a bright, well-condensed object with an outer envelope oriented NNE–SSW. The core is sharp and stellar. The faint star or bright condensation in the envelope is visible immediately to the SW. Herschel discovered this object on 9 October 1790 and described it as: 'Two very small close stars with nebulosity between.' Radial velocity: 5434 km/s. Distance: 243 mly. Diameter: 99 000 ly.

NGC7769

H230[2]	23:51.1	+20 09	GLX
(R)SA(rs)b	3.2′ × 2.6′	12.50B	

Images reveal this face-on spiral galaxy has a very complex, dusty central region, surrounded by two very faint, smooth-textured spiral arms. These arms can be traced out to about 3.2′. With NGC7770 and NGC7771 this galaxy forms an attractive triplet in a medium-magnification field. Photographs show at least three other faint, dwarfish galaxies which may be involved as well. Visually, NGC7769 is very slightly larger than and about equal in brightness to NGC7771. It is round with a fairly opaque envelope and is brighter to a condensed, though nonstellar, nucleus. The edges are moderately well defined. NGC7770 and NGC7771 are in the field 5.3′ ESE. Herschel discovered both NGC7769 and NGC7771 on 18 September 1784, and

NGC7769/7771

©STScI/AURA/CALTECH/ROE

the presumed distance. Radial velocity: 4442 km/s. Distance: 198 mly. Diameter: 144 000 ly.

NGC7773

H851[2]	23:52.3	+31 16	GLX
SBbc	1.2′ × 1.2′	14.44B	

Images show this face-on spiral has an elongated core and a suspected bar. There is a four-branch spiral structure with a brighter condensation E of the core. The galaxy is quite faint at high magnification and is seen as a slightly oval, nebulous patch 0.6′ SW from a faint field star. The galaxy is broadly brighter to the middle and the edges are hazy and ill defined. Radial velocity: 8671 km/s. Distance: 387 mly. Diameter: 135 000 ly.

NGC7794

H466[3]	23:58.6	+10 42	GLX
S?	1.2′ × 1.0′	13.85B	

The DSS image shows a face-on spiral galaxy which may be barred. The core is bright and two arms emerge, broadening and fading, but there are also conspicuous condensations visible both to the N and the SSW. In a medium-magnification eyepiece, this is a faint, ethereal galaxy located NNE of a magnitude 12 field star. The envelope is quite smooth, milky in appearance and much elongated N–S. No core is visible. Radial velocity: 5419 km/s. Distance: 242 mly. Diameter: 85 000 ly.

NGC7798

H232[2]	23:59.4	+20 45	GLX
SBc	1.4′ × 1.3′	12.95B	12.4V

Images reveal a face-on spiral, probably barred, with two spiral arms of the grand-design type. A large condensation is seen in the N spiral arm just at the point where it fades as it expands outward. The S spiral arm is bright for a considerable distance before broadening and fading. At high magnification, this is a fairly bright galaxy, large and gradually brighter to a well defined core. The outer edges are rather diffuse and the galaxy is located

called NGC7769: 'Faint, pretty large, round, bright middle. Resolvable.' Radial velocity: 4376 km/s. Distance: 195 mly. Diameter: 182 000 ly.

NGC7771

H231[2]	23:51.4	+20 07	GLX
SB(s)a	2.5′ × 1.0′	12.94B	12.3V

Photographs show a highly inclined, two-armed spiral galaxy. The core is bright and a very bright, narrow spiral arm crosses in front of it. This arm has many brighter condensations along its length which are probably massive HII regions. The outer spiral arms are fainter and the southern arm may be disturbed and in contact with companion galaxy NGC7770; it broadens as it passes the smaller galaxy and high contrast shows plumes of luminous matter exchanging between the two galaxies. A broad plume of material extends towards the WSW for

about 1.0′ and another broad plume extends to the NNE from the northern flank of the galaxy about 1.3′. At moderate magnification, NGC7771 is an elongated, well defined system oriented ENE–WSW; it is a little brighter to the centre and quite bright along its major axis. It is curious that Herschel did not note NGC7770 immediately SSW, which is a well defined, circular patch of light, evenly bright across its envelope. Herschel's description for NGC7771 reads: 'Faint, pretty large, extended along the parallel, contains a stellar nebula or star.' This may be a reference to NGC7770 as an unresolved nebula, connected to NGC7771. The separation, centre-to-centre, between NGC7770 and NGC7771 is only 1.1′. At the presumed distance of NGC7771, this would result in a core-to-core separation of about 64 000 ly. The core-to-core separation between NGC7769 and NGC7771 would be 306 000 ly at

NNE from a magnitude 10 field star.
Radial velocity: 2566 km/s. Distance:
114 mly. Diameter: 47 000 ly.

NGC7800

H10[2]	23:59.6	+14 49	GLX
Im?	2.3′ × 1.0′	13.56B	

The DSS photograph shows an inclined,
irregular galaxy with a brighter bar and
a chaotic, faint, outer structure.
Telescopically, the galaxy is faint and
best seen at medium magnification. It is
a diffuse, small and elongated blur
oriented NE–SW with moderately well
defined edges and is fairly even in
surface brightness along its major axis.
Herschel discovered the galaxy on 24
December 1783 and thought it
resolvable. Radial velocity: 1903 km/s.
Distance: 85 mly. Diameter: 57 000 ly.

NGC7805

H855[3]	0:01.4	+31 26	GLX
SAB0°: pec	0.7′ × 0.6′	14.20B	

This lenticular galaxy is part of a three-
galaxy system which has also been
catalogued as Arp 112. The other
principal member is NGC7806 and the
third member is a faint, curved
companion, 1.0′ E of NGC7806.
Visually, only NGC7805 and NGC7806
are visible in a high-magnification field.
NGC7805 is the brighter and larger of
the two and only the bright, condensed
core of the galaxy is seen, appearing
round and condensed with sharply
defined edges. The core-to-core
separation of NGC7805 and NGC7806
is 0.75′, or about 49 000 ly at the
presumed distance. Herschel discovered
the pair on 9 October 1790, describing
them as: 'Two nebulae, both extremely
faint. Stellar. Distance 1′ from 30° south
preceding to north following.' Radial
velocity: 4993 km/s. Distance: 223 mly.
Diameter: 46 000 ly.

NGC7806

H856[3]	0:01.5	+31 27	GLX
SA(rs)bc? pec	0.9′ × 0.5′	14.32B	

The DSS image shows a peculiar spiral
with an elongated core and a dust patch

NGC7805/7806

©STScI/AURA/CALTECH/ROE

to the N. There appears to be just one
principal spiral arm which winds for a
full revolution around the galaxy, with
some evidence of interaction with
NGC7805 to the SW. The spiral arm
extends well to the N, increasing the
length of the major axis to about 1.5′ on
the DSS photograph. Visually, only the
core is visible; it is irregularly round and
well condensed with edges that are a
little ragged. Radial velocity: 4949 km/s.
Distance: 221 mly. Diameter: 97 000 ly.

NGC7810

H984[3]	0:02.4	+12 57	GLX
S0	1.2′ × 0.9′	14.3B	

Photographs show that this may be a
spiral galaxy: a large, bright central
region is surrounded by a fainter outer
envelope which has suspected
condensations involved, particularly to
the E. Visually, the galaxy is quite faint
and is seen as a hazy streak of light with
a very faint field star 0.5′ WNW of the

core. Four moderately bright field stars
are located to the NNW. The galaxy has
a faint stellar core embedded in the
envelope, which is elongated E–W.
Radial velocity: 5675 km/s. Distance:
253 mly. Diameter: 89 000 ly.

NGC7814

H240[2]	0:03.3	+16 09	GLX
SA(S)ab:	5.5′ × 2.6′	12.0B	

This classic lenticular galaxy has a
large, flattened core surrounded by an
extensive, flattened halo. Seen almost
precisely edge-on, the galaxy features a
very thin, well defined dust lane that
extends all along a very thin disc. High
resolution photographs show a second,
fainter dust lane along the NE flank. In
moderate-aperture telescopes, the
galaxy is very bright and fairly large
with a bright core and fainter extensions
elongated NW–SE. The extremities are
fairly well defined, but are not as distinct
at the NW and SE points. The dust lane

NGC7814

is not detected, but may be visible in very large amateur telescopes. Discovered on 8 October 1784, Herschel thought the galaxy easily resolvable. Radial velocity: 1200 km/s. Distance: 53 mly. Diameter: 86 000 ly.

NGC7817

H227[2]	0:04.0	+20 45	GLX
SAbc: sp	3.0' × 0.8'	12.7B	

Photographs show a highly inclined spiral galaxy with grainy spiral arms and a faint star superimposed S of the core. At high magnification, it is a moderately bright galaxy, quite flat, with a slightly brighter core. It is oriented NE–SW and there is a pair of magnitude 13 stars located S of the southern extension. The galaxy is large and well defined along its major axis and a magnitude 8 field star lies 9.0' WSW. Herschel thought this galaxy was resolvable. Radial velocity: 2469 km/s. Distance: 110 mly. Diameter: 96 000 ly.

Perseus

The 32 Herschel objects in Perseus represent a good mix of star clusters, nebulae and galaxies. Showpieces include the Double Cluster (known since antiquity), NGC650–NGC651 (discovered by Pierre Méchain) and the emission nebula NGC1491. While some of the galaxies are bright, the interested observer might like to use the faint galaxy NGC1278 as a marker for the rich and compelling galaxy cluster Abell 426. All the objects in this constellation were recorded between the years 1784 and 1790, except for the star cluster NGC1342, which was discovered in 1799.

NGC650/NGC651

H193[1]	1:42.3	+51 34	PN	
3+6	4.0′ × 2.2′	12.20B	10.1V	

As is frequently the case with bright planetary nebula, the DSS image of this object is quite overexposed, burning out any detail in the brightest portion of the nebula while showing some of the faint outer structure well. The central elongated bar is oriented NE–SW and two fainter lobes of looping, tenuous gas emerge in the ESE and WNW. This object was originally discovered by Pierre Méchain on 5 September 1780 and was entered in the Messier catalogue as M76. Méchain thought the object was a nebula but when Messier himself observed it on 21 October 1780 he thought it was composed of very small stars surrounded by nebulosity.

Herschel's only published observation of this object was recorded on 12 November 1787, and, as he was able to resolve the planetary nebula into two separate components, he probably thought that this was worthy of note, saying in his description: 'Two close together. Both very bright, distance 2′ south preceding, north following. One is 76 of the Conn. (Connaissance des Temps).' Visually, this is a very bright, large and well defined nebula; it is rectangular in shape and fairly opaque except for a slight dimming near the centre. It is brightest at the SW end where it is also most opaque. The central star, which is magnitude 15.9, is not visible. The expansion velocity of the

NGC650/651

NGC650 NGC651
15-inch Newtonian 293x

©STScI/AURA/CALTECH/ROE

NGC869

shell is estimated at 42 km/s. Radial velocity in approach: 19 km/s. Distance: 3600 ly.

NGC869

H33[6]	2:19.0	+57 09	OC
I3r	18.0′ × 18.0′	4.3B	3.7V

Though this cluster and its companion NGC884 have been known since antiquity and were first recorded by Hipparchus in 130 BC, Herschel did not catalogue the pair until 1 November 1787, describing NGC869 as: 'A very beautiful and brilliant cluster of large stars. Very rich. The middle contains a vacancy.' This cluster is the western portion of the famous Double Cluster in Perseus, a large and very rich cluster of stars with a wide brightness range. It is very compressed towards the middle with a bright elongated group of stars oriented ESE–WNW. Even in a small-aperture telescope it is superb; the field is absolutely peppered with stars, quite compressed to the centre with many threshold stars popping into view. The cluster is remarkable for the number of subgroups of 3–5 stars, scattered over the field, the most compelling of which is a crescent-shaped group of five stars just ESE of centre. Cluster membership is more than 300 stars. The cluster is part

NGC884

©STScI/AURA/CALTECH/ROE

of the Perseus OB1 association. Spectral type: B0. Age: 5.6 million years. Radial velocity in approach: 22 km/s. Distance: 7275 ly.

NGC884

H34⁶	2:22.4	+57 07	OC
I3r	18.0′ × 18.0′	4.4B	3.8V

Herschel described the eastern portion of the famous Double Cluster in Perseus as: 'A very beautiful brilliant cluster of large stars, irregularly round. Very rich, near half a degree in diameter.' While not quite as rich photographically as NGC869, there is a rich concentration of stars just W of the centre with a chain of stars running SSE–NNW. A compelling object even in small telescopes, it is not quite as compressed as its neighbour, nonetheless up to 100 stars are readily seen. Small, tight subgroups, though visible, are not as numerous as in NGC869. The clusters form a true pair in space, separated by only 32 ly, with NGC884 the closer of the two clusters. Cluster membership is probably more than 300 stars, and the cluster seems to be slightly younger than its neighbour. Spectral type: B0. Age: 3.2 million years (estimate). Radial velocity in approach: 21 km/s. Distance: 7243 ly.

NGC1003

H238²/H198³	2:39.3	+40 52	GLX
SA(s)cd	5.5′ × 1.9′	12.09B	

This galaxy was catalogued twice by Herschel on 6 October 1784 during a night characterized by haze and unsettled weather. In photographs, it is a highly inclined spiral galaxy with a brighter, irregular core and knotty, faint spiral arms. Visually, it is a faint though moderately large galaxy; it is a smooth, glowing patch which is a little brighter to the core. It is elongated E–W with a magnitude 10 star to the SW and a fainter magnitude 12 star just N of the core. Radial velocity: 747 km/s. Distance: 33 mly. Diameter: 54 000 ly.

NGC1003 15-inch Newtonian 177x

NGC1023

H156¹	2:40.4	+39 04	GLX	
SB(rs)0⁻	8.7′ × 3.0′	9.56B	9.4V	

This is the principal member of a relatively nearby group of galaxies which includes NGC1003, NGC1058 and NGC891 in Andromeda. The DSS image shows an inclined lenticular galaxy with a bright, large, elongated core in an extensive, fainter outer envelope. A faint, Im-type dwarf companion galaxy can be seen in silhouette against the outer envelope in the E and the two galaxies have also been catalogued as Arp 135. Telescopically, NGC1023 is a fairly large

and bright galaxy, featuring a large, extended, cigar-shaped envelope which tapers to rounded ends. The envelope is grainy in texture, with a brighter oval core embedded and an even brighter star-like nucleus. Oriented almost due E–W, the envelope is fairly well defined. The dwarf companion is not visible in a

NGC1023 15-inch Newtonian 293x

moderate-aperture telescope. Radial velocity: 754 km/s. Distance: 34 mly. Diameter: 85 000 ly.

NGC1058

H633²	2:43.5	+37 21	GLX
SA(rs)c	3.8′ × 3.7′	11.96B	

This bright, face-on spiral galaxy is a member of the nearby NGC1023 group of galaxies. The DSS image shows a small, round core and narrow, knotty spiral arms, well defined by dust patches and lanes, especially to the SW. The spiral structure is fairly symmetrical with a very faint field star involved in a spiral arm WNW of the core. High-resolution photographs begin to resolve individual stars at visual magnitude 22. Telescopically, the galaxy, though moderately bright, is quite diffuse and ill defined, and is almost round with a grainy envelope which is slightly brighter to the middle. Radial velocity:

NGC1023

NGC1058

629 km/s. Distance: 28 mly. Diameter: 31 000 ly.

NGC1122

H601[2]	2:52.8	+42 12	GLX
SABb	1.7′ × 1.3′	13.35B	

This galaxy was originally given the designation NGC1123 and later thought to be nonexistent, but it has been positively identified with NGC1122 and was plotted as such in the first edition of *Uranometria 2000.0*. The DSS image shows an inclined spiral galaxy, possibly with a bar, as the central region is a bright and elongated oval with two strong spiral arms emerging. Several very faint field stars are seen in silhouette in the spiral arms or just beyond the galaxy's apparent limits. Visually, it is a diffuse and faint object; it is slightly oval and slowly brighter towards the middle, though no core is visible. Radial velocity: 3716 km/s. Distance: 166 mly. Diameter: 82 000 ly.

NGC1129

H602[2]	2:54.5	+41 35	GLX	
E	4.0′ × 3.1′	13.40B	13.6V	

The DSS photograph shows an elliptical galaxy with a bright, round core and extensive, fainter outer envelope. The outer envelope is quite asymmetrical, being more extensive towards the E. A very small, compact companion is within the halo immediately WSW of

NGC1129

the core. NGC1129 appears to be a giant elliptical galaxy and is the principal galaxy of a swarm of tiny, fainter galaxies seen in the immediate field, including NGC1130 1.6′ NNW and NGC1131 1.75′ SE. Telescopically, both NGC1129 and NGC1130 can be seen in a moderate-aperture telescope. NGC1129 is fairly large, moderately bright, almost round and broadly brighter to the middle. The envelope is quite diffuse and fades slowly into the sky background. NGC1130 can be seen immediately NNW; it is a very small, concentrated patch of light which is round and well defined. Radial velocity: 5309 km/s. Distance: 237 mly. Diameter: 276 000 ly.

NGC1138

H580[3]	2:56.5	+43 03	GLX
SB0	1.6′ × 1.4′	13.99B	

Photographs show a barred lenticular galaxy with a brighter, elongated core and faint bar embedded in a faint, round outer envelope. The bar and core are elongated E–W while the outer envelope is slightly extended N–S. Visually, the galaxy forms an equilateral triangle with two field stars to the S, with the easternmost component being a double star. The galaxy is small and quite faint; it is best seen at medium magnification as a round, concentrated patch of light, that is almost star-like, well defined and even in surface brightness. Radial velocity: 2452 km/s. Distance: 110 mly. Diameter: 51 000 ly.

NGC1160

H199[3]	3:01.2	+44 58	GLX
SBc	1.8′ × 0.8′	13.50B	

This is a highly inclined spiral galaxy with a slightly brighter central region and grainy spiral arms, elongated NE–SW. NGC1161, a lenticular galaxy, is 3.6′ S. Visually, NGC1160 appears smaller and fainter than its neighbour; it is a diffuse patch slightly elongated NE–SW and fairly even in surface brightness, with a small, triple asterism 1.75′ N. Radial velocity: 2647 km/s. Distance: 118 mly. Diameter: 62 000 ly.

NGC1161

H239[2]	3:01.2	+44 55	GLX
S0	3.0' × 2.0'	12.13B	

The DSS image shows an elongated lenticular galaxy with a large, bright core and a fainter outer envelope, elongated NNE–SSW. Bright field stars are located 0.75' W, 1.25' WSW and 1.4' E, while the spiral galaxy NGC1160 is 3.6' N. The radial velocity of NGC1161 is quite a bit lower than its apparent neighbour and the two galaxies may not form a physical system. Telescopically, the two galaxies form an interesting pair and the fainter galaxy, NGC1160, is noticed first, since the bright pair of stars W of NGC1161 overwhelm the galaxy's light. However, at high magnification NGC1161 stands out well as a bright, well-condensed galaxy which is quickly brighter to the middle, grainy textured and ill defined at the edges. Radial velocity: 2073 km/s. Distance: 93 mly. Diameter: 81 000 ly.

NGC1167

H178[3]	3:01.7	+35 12	GLX
SA0⁻	2.8' × 2.3'	13.45B	13.0V

This galaxy is classified as lenticular, but examination of the DSS image shows hints of a possible spiral pattern, both within the bright inner envelope and in the faint outer disc. There are subtle suggestions of possible dust lanes, especially ENE of the core, which is bright and round. The galaxy is slightly elongated ENE–WSW. The galaxy is moderately bright though diffuse visually; it is round and broadly brighter towards the middle. The edges are quite hazy and indistinct and an unequal pair of stars is visible 4.0' to the S. Radial velocity: 5041 km/s. Distance: 225 mly. Diameter: 183 000 ly.

NGC1169

H620[2]	3:03.6	+46 23	GLX
SAB(r)b	4.2' × 2.5'	12.34B	

This is an inclined, barred spiral galaxy with a bright, elongated core and a short, fainter bar attached to an inner ring. The DSS image reveals thin, delicate, spiral arms emerging from the ring; they are of the multi-arm type and high-resolution photographs resolve discrete patches at blue magnitude 22 which are probably HII regions. A faint field star is involved in the core immediately SSW of centre and the galaxy is elongated NNE–SSW. Telescopically, only the inner core of this galaxy is visible as a round, fairly concentrated patch of light with diffuse edges. It is moderately bright and the faint field star is to the SSW almost touching the core. Radial velocity: 2508 km/s. Distance: 112 mly. Diameter: 137 000 ly.

NGC1175

H607[2]	3:04.5	+42 20	GLX
SA(r)0⁺	1.9' × 0.7'	13.84B	

This is a highly inclined galaxy, possibly spiral, with a large, elongated core, a bright inner envelope and fainter outer extensions, elongated NNW–SSE. Photographs show the companion galaxy NGC1177 1.6' to the NNE. If they form a physical pair the minimum separation between the two galaxies would be about 118 000 ly at the presumed distance. NGC1175 is fairly faint and small visually, an elongated patch of light of even surface brightness with moderately well defined edges. It is a little more distinct at medium magnification. Radial velocity: 5635 km/s. Distance: 252 mly. Diameter: 139 000 ly.

NGC1186

H43[4]	3:05.5	+42 50	GLX
SB(r)bc:	3.2' × 1.2'	12.25B	

Herschel listed this galaxy on 24 October 1786 though he had seen it in an earlier sweep. Placing it in his Class IV, he described it as: 'A pretty bright star with 2 faint branches.' His description from the earlier sweep stated: 'A pretty bright star with a very faint nebulosity to the north following side of very little extent.' Photographs show a highly inclined, barred spiral galaxy, oriented SE–NW, with an elongated core and a fainter bar attached to a bright inner ring. The ring is surrounded by fainter outer spiral arms. A field star is in the ring immediately SW of the core. Visually, the galaxy appears bright but very small at medium magnification, while increased power dims the galaxy but brings out a grainy envelope with a slightly brighter core. The field star is faint and offset to the SW. Radial velocity: 2852 km/s. Distance: 128 mly. Diameter: 118 000 ly.

NGC1193

H608[2]	3:05.8	+44 23	OC
I2m	3.0' × 2.8'	12.6V	

Photographs show a rich and compressed cluster of faint stars with a very narrow range in brightness, the brightest being magnitude 14.0. The most concentrated part is oval in shape and elongated N–S. The cluster is well differentiated from the sky background and a bright field star is 4.0' WNW. Visually, the cluster is best at high magnification and is seen in a rich field of stars. It appears as an unresolved, hazy glow bordered on the E and W by faint field stars. It is broadly brighter to the middle and, with attention, a few faint stars shimmer into view in a moderate-aperture telescope. Herschel classified the object as a faint nebula, but his 24 October 1786 discovery description reads: 'Faint cluster, easily resolvable, some stars visible.' About 40 stars are probable members of this cluster.

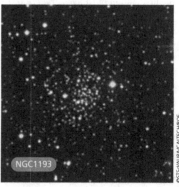

NGC1193

NGC1207

H578[3]	3:08.2	+38 23	GLX
SA(rs)b	2.0′ × 1.4′	13.66B	

Images reveal a slightly inclined spiral galaxy, possibly barred with a small round core embedded in a broken inner ring. Narrow, grainy spiral arms emerge from the ring and the galaxy is elongated SSE–NNW. A faint field star is embedded in the ring immediately NW of the core. The fainter field galaxy CGCG524-054 is 5.7′ WSW. Visually, NGC1207 is situated in a rich star field; it is fairly bright, round and diffuse and the edges are very ill defined. The brightening seen near the core is actually the faint field star. Radial velocity: 4900 km/s. Distance: 219 mly. Diameter: 127 000 ly.

NGC1245

H25[6]	3:14.7	+47 15	OC
II2r	10.0′ × 7.0′	9.2B	8.4V

In photographs this is quite a rich cluster of stars with a narrow brightness range. Many of the member stars are grouped into long, meandering chains which define otherwise starless zones. Visually, this is a beautiful, rich cluster of stars, well resolved, well defined and easily separated from the sky background. Members are magnitude 12 or fainter; the overall form of the cluster is rather box-like, elongated E–W, and there are prominent starless zones near the middle, particularly ESE of the centre. The stars are sprinkled like diamond dust and a striking T-shaped asterism is visible on the NNW edge. Herschel recorded the cluster on 11 December 1786 and his description reads: 'A beautiful compressed and rich cluster of small and large stars, 7 or 8′ in diameter. The large stars arranged in lines like interwoven letters.' Indeed, in photographs one can make out a C-shaped asterism involving seven stars in the SW and a broad V-shaped grouping of stars SE of the aforementioned T-shaped asterism. About 200 stars are probable members, the brightest being magnitude 12 and

the B9 spectral type indicates an age of 1.1 billion years. Distance: 7300 ly.

NGC1278

H603[2]	3:19.9	+41 33	GLX
E pec:	1.4′ × 1.2′	13.60B	14.1V

On the night of 17 October 1786 Herschel unknowingly came across the field of the rich galaxy cluster Abell 426, recording three objects there: NGC1278, NGC1293 and NGC1294. Curiously, though, he missed two of the brightest members, NGC1272 and NGC1275, and a host of small galaxies similar in brightness to the ones he recorded. However, the field is rich in faint stars which may have helped to mask the tiny members of this galaxy cluster. Moderate-aperture telescopes should readily pick up two dozen members of the cluster in dark skies, more if the observer is careful and determined. NGC1278 is an elliptical galaxy with a bright, elongated core and fainter outer

envelope, elongated E–W in photographs. Telescopically, it appears as a small, concentrated bright patch. Herschel's description reads: 'Pretty bright stellar or pretty compressed star with small, very faint chevelure.' There is a distinct possibility that Herschel actually observed NGC1275 and not NGC1278, as it is brighter and Herschel's position is actually closer to NGC1275. Guardedly, the present author accepts NGC1278 = H603[2]. Radial velocity: 6192 km/s. Distance: 276 mly. Diameter: 113 000 ly.

NGC1293

H574[3]	3:21.6	+41 24	GLX
E0	0.9′ × 0.9′	14.50B	

This elliptical galaxy is paired with NGC1294 and can be seen in the same high-magnification field. Visually, the galaxy is round and gradually brighter towards the middle with fairly well defined edges. Several galaxies are nearby, including NGC1294, which is

NGC1281

NGC1276

NGC1277

NGC1278

NGC1274

NGC1273

NGC1275

NGC1272

ABELL426 galaxy cluster

NGC1272/1273/1274/1275/1276/1277/1278/1281

©STScI/AURA/CALTECH/ROE

2.0′ SSE. The radial velocity of NGC1293 is considerably lower than its presumed companion and it may not be a member of the Abell 426 cluster. Radial velocity: 4274 km/s. Distance: 191 mly. Diameter: 50 000 ly.

NGC1294

H575[3]	3:21.7	+41 22	GLX
SA0⁻?	1.3′ × 1.1′	14.33B	

This Abell 426 member has a bright core and a fainter outer envelope and is paired with NGC1293 2.0′ to the NNW. If they form a physical pair, the core-to-core separation between the two galaxies would be at least 175 000 ly. The telescopic appearance of this galaxy is similar to its companion, though it does appear very slightly brighter and larger. Herschel grouped the two galaxies together in his description, which reads: 'Two. Both very faint, stellar. Very little brighter to the middle but the south is the brightest and largest.' Radial velocity: 6685 km/s. Distance: 299 mly. Diameter: 113 000 ly.

NGC1342

H88[8]	3:31.6	+37 20	OC
III2m	17.0′ × 17.0′	7.3B	6.7V

This scattered cluster of magnitude 8–12 stars has many of its members arranged in a curving band running E–W and dipping at the centre towards the S, as well as a string of stars curving from N to S at the western extreme. The string includes one very close pair at the W edge. This was Herschel's last entry in his Class VIII: coarsely scattered clusters of stars, and was recorded on 28 December 1799. About 100 stars are considered members. Spectral type: A2. Age: about 300 million years. Distance: 1745 ly.

NGC1348

H84[8]	3:33.8	+51 26	OC
III2m	6.0′ × 6.0′		

Photographs show a small cluster of faint stars of moderate brightness range with half a dozen brighter stars involved. Most of the stars are around

the periphery with a starless zone near the centre. Visually, this is a small and poor cluster, well separated from the sky background. The stars are of moderate brightness range and are best resolved at high magnification. The cluster appears as a small nebulous patch at low magnification and there is the suspicion of a nebulous, unresolved haze at medium magnification but threshold stars are resolved at high magnification. About ten stars are visible. The total membership of the cluster is about 30 stars.

NGC1444

H80[8]	3:49.4	+52 40	OC
IV1p	4.0′ × 4.0′	7.0B	6.6V

This is a small, coarse cluster of stars of moderate brightness range, arrayed around a bright field star (ADS 2783, which contains four components). The brightest members are arranged in a meandering chain of stars running N–S immediately W of the bright field star. In a small telescope, this is a poor object; apart from the multiple star only seven extremely faint stars are visible. As many as 60 stars may be members of this cluster which may be part of the Camelopardalis OB1 association. Spectral type: B0. Age: 160 million years. Radial velocity in recession: 2 km/s. Distance: 3060 ly.

NGC1491

H258[1]	4:03.4	+51 19	EN
20.0′ × 20.0′	11.3B		

The DSS image reveals a bright nebula with complex structure, including several bright, narrow tendrils. The brightest portion of the nebula is irregular in form, elongated N–S and measures 4.5′ × 3.0′. Fainter wisps of nebulosity with several dark channels involved are visible, particularly to the N, E and S. Visually, only the crescent-shaped inner nebulosity surrounding the magnitude 11 central star is visible. It is fairly bright and well seen at medium magnification without a nebula filter, and is brightest to the W, with filamentary structure N and S of

the illuminating star. An ultra-high-contrast filter improves the view and the edges of the nebula are more sharply defined. The spectral type of the illuminating star is B0. Herschel discovered the object on 28 December 1790 and described it as: 'Very bright, irregular form, resolvable, with a brighter middle. 5′ long, 4′ broad. A pretty large star in it towards the following side, but unconnected.'

NGC1513

H60[7]	4:10.0	+49 31	OC
II1m	9.0′ × 9.0′	8.7B	8.4V

Photographs show a scattered cluster of stars, irregularly round and well separated from the background with a brighter star to the NNE. The visual impression is of a fairly large, scattered but moderately rich cluster of stars, with a narrow brightness range. Best at medium magnification, a prominent circle of stars is SE of the main concentration of scattered, fainter stars. About 30 stars are involved. As many as 50 stars may be members with the brightest at magnitude 11. Spectral type: B8. Age: about 430 million years. Distance: 2550 ly.

NGC1528

H61[7]	4:15.4	+51 14	OC
II2m	18.0′ × 18.0′	6.8B	6.4V

This is a large, bright, very attractive cluster of stars of moderate brightness range, which is oval in form with a prominent, elongated chain of stars running from the brightest member in the W and oriented ESE–WNW. In a small telescope at least 45 members are visible, distributed fairly evenly across the field and well separated from the sky background. A total of 165 stars are probably members of this cluster. Spectral type: A0. Age: about 270 million years. Distance: 2430 ly.

NGC1545

H85[8]	4:20.9	+50 15	OC
IV2p	12.0′ × 12.0′	7.2B	6.2V

Photographs show a bright but ill defined cluster of stars and visually it is a

NGC1491

©STScI/AURA/CALTECH/ROE

poor, scattered cluster of stars, dominated by a bright triangle of magnitude 8–9 stars. The rest of the cluster members are magnitude 11 or fainter, and altogether about two dozen stars are involved. Herschel, however, considered it: 'A coarsely scattered cluster of large stars, pretty rich.' He observed it on 28 December 1790. About 65 stars are considered to be members. Spectral type: B8. Age: about 10 million years. Distance: 2475 ly.

NGC1579

H217[1]	4:30.2	+35 16	RN
12.0′ × 8.0′			

The constituent parts of this reflection nebula exhibit a wide brightness range. The DSS image shows that the brightest portion of the nebula is small and roughly triangular in shape. A large dark channel is visible immediately SE and most of the fainter nebulosity is located to the W and S of the brightest

part; the illuminating star is magnitude 12. Telescopically, this bright nebula is obvious even at low magnification and is well seen at higher powers. Situated within a group of bright field stars, the brightest portion is slightly elongated

E–W and brightens evenly towards the middle. The edge of the bright portion is fairly well defined but there is a suspicion of fainter nebulosity surrounding this. Herschel recorded the object on 3 December 1788, his

NGC1579

©STScI/AURA/CALTECH/ROE

comments reading: 'Considerably bright, considerably large, much brighter to the middle. Stands nearly in the centre of a trapezium.' Indeed, the placement of four of the brighter field stars resembles a much larger version of the Orion Nebula's famous Trapezium.

NGC1582

H70[8]	4:32.0	+43 51	OC
IV2p	24.0′ × 24.0′	7.0V	

Photographs show a poor, scattered cluster of stars with about half a dozen bright ones involved; it is not well separated from the sky background. Visually, this is a large and quite bright cluster of stars of moderate brightness range in an area about 35′ in diameter and is best seen at low magnification. It is fairly coarse and scattered, with a number of stars surrounding a relatively starless zone to the W. About 40 stars are visible, though probably only 20 are cluster members. *Sky Catalogue 2000.0* (Hirshfeld and Sinnott, 1985) quotes a diameter of about 37′ for the cluster. The brightest stars are magnitude 9. Herschel discovered the cluster on 3 February 1788 and his description of it reads: 'A cluster of coarsely scattered large stars, pretty rich above 20′ in diameter.' Spectral type: B8.

NGC1605

H26[6]	4:35.0	+45 15	OC
III1m	5.0′ × 5.0′	11.6B	10.7V

The DSS shows a moderately rich cluster of faint stars with a narrow brightness range, which is almost round and is well detached from the sky background. Visually, the cluster is very faint and is not well seen. About eight equally bright stars are resolved; four of them appear

NGC1624

©STScI/AURA/CALTECH/ROE

nebulous. Herschel recorded the cluster on 11 December 1786 and described it as: 'A very faint and very compressed cluster of extremely small stars near 4′ in diameter.' About 40 stars are probable members, the brightest being magnitude 12.52. Spectral type: B3. Distance: 8300 ly.

NGC1624

H49[5]	4:40.4	+50 27	EN/OC
F-2-R C-AF	II1p n	5.0′ × 5.0′	11.8V

This combination star cluster–emission nebula is well seen on the DSS image as a compressed cluster of faint stars involved in a round, well defined nebula, which is broadly brighter to the middle with faint dark channels

involved, especially to the E. Telescopically, this nebula is quite bright and best seen at medium magnification. High magnification resolves a small 'Sagitta'-like asterism. The nebula is brightest around the triangle of stars at the centre, while the two stars forming the 'arrow head' emerge from the nebula in the W. The edges of the nebula are hazy but this is a striking object. Discovered on 28 December 1790, Herschel placed this object in his Class V, describing it as: '6 or 7 small stars, with faint nebulosity between them, of considerable extent, and of an irregular figure.' About a dozen stars are probable members of the cluster, the brightest being magnitude 11.77. Spectral type: O.

Pisces

Of the 74 Herschel objects recorded in this constellation, 73 are galaxies. The exception, NGC552, is today associated with a faint single star though it is far from certain that Herschel observed this object. All the objects were observed by Herschel in the years between 1784 and 1790. While there are no galaxies that can be considered spectacular, there are many bright examples, including NGC488, NGC474 and NGC660. Patient observers with moderate- or large-aperture telescopes will find some interesting details in the peculiar galaxy. NGC520 and many of the fainter galaxies are located in small groups or clusters which may prove to be compelling fields to explore. Notable clusters include the Pisces Cloud (with NGC383 acting as a marker) and the group of galaxies surrounding NGC507.

NGC12

H868³	0:08.7	+4 37	GLX
SAB(rs)c:	1.6′ × 1.3′	13.84B	

Photographs show a face-on spiral galaxy with a small, round core and grainy spiral arms. Telescopically, the galaxy is an almost round, diffuse disc, the envelope is faint and the edges are ill defined. A faint, star-like core is seen intermittently at medium magnification. Radial velocity: 4058 km/s. Distance: 181 mly. Diameter: 85 000 ly.

NGC36

H456³	0:11.4	+6 23	GLX
SAB(rs)b	2.8′ × 1.7′	14.31B	

Images reveal an inclined barred spiral galaxy with a small, round core and a very faint bar attached to an inner pseudo-ring. Two thin spiral arms broaden to fainter outer spiral structure, elongated NNE–SSW. A possible companion galaxy is 1.0′ E and a faint, broad plume of material appears to connect the two galaxies in the N. The minimum core-to-core separation between the two galaxies at the presumed distance would be 80 000 ly. At high magnification, NGC36 is a roundish, moderately faint object with a faint stellar core embedded in a hazy envelope, with a faint field star 1.75′ to the NE. Radial velocity: 6151 km/s. Distance: 275 mly. Diameter: 224 000 ly.

NGC57

H241²/H243²	0:15.4	+17 18	GLX
E	2.2′ × 1.9′	12.70B	

In the DSS image, this elliptical galaxy has a bright, slightly elongated core and a fainter outer envelope. A peculiar filament emerges from the core in the NE but this may be a plate defect. Visually, this is a moderately bright object, small, but with an intense, though nonstellar, core. It is quite condensed at the middle; the outer envelope is round but very ragged, fading irregularly into the sky background. Herschel catalogued this galaxy twice: on 8 October and 11 October 1784. Using different reference stars to establish the location, he erroneously thought the sightings were of separate objects. Radial velocity: 5587 km/s. Distance: 249 mly. Diameter: 160 000 ly.

NGC61

H428³	0:16.5	−6 14	GLX
1.2′ × 0.7′	0.9′ × 0.5′	15.0B	

This is a double galaxy, designated NGC61A and NGC61B. The DSS photograph shows two lenticular galaxies, which have large, bright cores and share a common envelope. NGC61B, the southern component, has an asymmetrical outer envelope extending towards the E. Telescopically, medium magnification shows the pair as a N–S elongated, fairly well defined patch of light. The object is resolvable as two distinct galaxies at high magnification: the S component is brighter and slightly broader than the N component. Herschel evidently did not note the dual nature of this object; recording it on 10 September 1785; he described it as: 'Very faint, small, irregular form, a little brighter to the middle.' Radial velocity: 8150 km/s. Distance: 364 mly. Diameter: component A 95 000 ly; component B 127 000 ly.

NGC95

H257²	0:22.2	+10 30	GLX
SAB(rs)c pec	1.9′ × 1.1′	13.28B	

This multi-arm spiral galaxy has a unique and somewhat unusual structure that is well seen on the DSS image. At least five spiral-arm segments emerge from the bright core, with those on the E flank being relatively short and straight. An arm which emerges from the N is interrupted at the edge of the disc before continuing to wind around the galaxy for another half of a revolution. There are several brighter knots in the arms, especially along the rim of the galaxy, which appears to end abruptly. However, the image shows fainter, outer spiral structure particularly to the N and W. Telescopically, the galaxy is best seen at medium magnification. A round, somewhat mottled, hazy envelope surrounds a brighter core. The edges are fairly well defined and at high magnification the field is star-poor except for a triangle of stars N of the galaxy. Radial velocity: 5507 km/s. Distance: 249 mly. Diameter: 136 000 ly.

NGC125

H869[3]	0:28.8	+2 50	GLX
(R)SA0+ pec	1.6' × 1.4'	13.31B	

This spiral galaxy may be a part of a small, compact group of galaxies which includes NGC126, NGC127, NGC128 and NGC130, although its radial velocity is considerably higher than that of NGC128, the dominant galaxy of the group. On the DSS photograph, NGC125 has a large, bright core and very faint, asymmetrical spiral structure. Two faint field stars are 0.75' and 1.0' S. The separation between NGC125 and NGC128 is 6.25'. At the

distance of NGC128 the minimum separation, core-to-core, would be about 350 000 ly. In a moderate-aperture telescope, only the central part of NGC125 is visible. It is quite faint, though its core is bright, well defined and round and it is seen together with NGC128 in a high-magnification field. Discovered on 25 December 1790, Herschel's comments were: 'Very faint, very small, brighter middle, preceding and in the field with II 854 [NGC128], north following two small stars.' Radial velocity: 5408 km/s. Distance: 242 mly. Diameter: 112 000 ly.

NGC128

H854[2]	0:29.2	+2 52	GLX
S0 pec sp	3.0' × 0.9'	12.65B	

This galaxy is the dominant member of a compact group which certainly includes NGC126, NGC127, NGC130 and quite possibly NGC125, which is farther away to the WSW. In photographs this edge-on lenticular

galaxy exhibits a peculiar structure. The central region is bright and peanut-shaped with fainter extensions. The southern extension curves slightly towards the SSW, while the northern extension broadens to a roughly circular glow. The small spiral galaxy NGC127 is immediately to the W and a faint plume of matter extends eastward to NGC128. Visually, NGC128 is quite bright and much elongated N–S. The core is bright and well condensed and appears a little broader than the main envelope of the galaxy. The extensions are well defined and quite bright. Herschel recorded the galaxy on 25 December 1790, calling it: 'Pretty bright, very small, round, very gradually much brighter to the middle. Pretty well defined on the margin.' The companion galaxies are all extremely close and NGC127, in particular, seems to be interacting with the larger galaxy: plumes of matter and the southern spiral extension seem to be drawn towards NGC128. NGC127 is 0.75' NW of the core of NGC128, NGC126 is 3.5' SSW and NGC130 is 0.9' ENE. The core-to-core separations would then be about 42 000 ly, 197 000 ly and 51 000 ly, respectively. Radial velocity: 4342 km/s. Distance: 194 mly. Diameter: 169 000 ly.

NGC137

H471[2]	0:31.1	+10 11	GLX
S0	1.3' × 1.3'	13.75B	

This lenticular galaxy presents similar appearances visually and photographically. It is a quite faint and difficult object initially, and is seen as a hazy spot at medium magnification. High magnification shows it as a round, well defined nebulous patch with a tiny, brighter core. Radial velocity: 5397 km/s. Distance: 245 mly. Diameter: 91 000 ly.

NGC180

H876[3]	0:38.0	+8 38	GLX
SB(rs)bc	2.3' × 1.7'	13.85B	

Photographs show a barred spiral galaxy with a small, round core and a

NGC130
NGC127
NGC128
NGC125
NGC126

NGC125/126/127/128/130

©STScI/AURA/CalTech/ROE

faint bar attached to a broken inner ring. Narrow, knotty, spiral arms emerge from the ring, broadening as they expand outward; the disc is elongated NNW–SSE. A field star is involved in the spiral structure 0.75′ NW of the core. Visually, the galaxy appears 'attached' to the faint field star, which hinders observation somewhat. The bar is the brightest part of the galaxy and is seen at medium magnification as a ghostly flare emerging from the star and extending to the SE. Surrounding the bar is a very faint, poorly defined outer halo. Radial velocity: 5394 km/s. Distance: 241 mly. Diameter: 161 000 ly.

NGC182

H870[3]	0:38.2	+2 44	GLX
(R′)SAB(rs)a pec:	2.0′ × 1.7′	13.42B	

This is a slightly inclined galaxy, elongated E–W. Photographs show that it is probably barred with a small core, a faint, oval, inner envelope and narrow, outer spiral arms. A bright field star is 3.5′ NW and the galaxy is part of an extended chain of twelve NGC galaxies, including the Herschel objects NGC193, NGC194, NGC198 and NGC200. Telescopically, NGC182 is moderately faint and a little more distinct at medium magnification. It is seen as a round, moderately well defined glow with a slightly brighter, star-like core embedded. The bright field star 3.5′ to the NW hinders observation. Radial velocity: 5357 km/s. Distance: 239 mly. Diameter: 139 000 ly.

NGC193

H595[3]	0:39.3	+3 20	GLX
SAB(s)0⁻:	1.9′ × 1.5′	13.34B	

This lenticular galaxy is the brightest of a small group visible in a medium magnification field. Both photographically and visually, it has a bright, slightly elongated core and a faint outer envelope and is elongated ENE–WSW. A faint field star is 0.5′ WSW, while a brighter field star lies 2.5′ ESE between this galaxy and NGC204. Several faint galaxies are in the field

including NGC204 6.5′ ESE and NGC203 8.25′ NE. The very faint NGC202 is 5.3′ due N of NGC203. Radial velocity: 4491 km/s. Distance: 200 mly. Diameter: 111 000 ly.

NGC194

H856[2]	0:39.3	+3 02	GLX
E	1.3′ × 1.2′	13.11B	

Visually and photographically, this elliptical galaxy presents similar appearances. In moderate apertures, the galaxy is bright and almost round with a small, prominent bright core and sharply defined edges. It is located 5.8′ SSE of a magnitude 7 field star. Photographs show that it is very slightly elongated NE–SW. Also visible in a medium-magnification field is NGC199 7.0′ to the NNE and almost due E of the bright field star. Radial velocity: 5324 km/s. Distance: 238 mly. Diameter: 90 000 ly.

NGC198

H857[2]	0:39.4	+2 48	GLX
SA(r)c	1.2′ × 1.2′	13.10B	

In photographs this is a face-on spiral galaxy with a very small, round core and a strong, symmetrical, multiple-spiral-arm pattern. Several faint galaxies are in the field and the brighter companion NGC200 is 6.25′ NNE. This is part of the galaxy group mentioned under NGC182. Visually, the galaxy is moderately bright, well defined and perfectly round with a bright, opaque central region. NGC200 is readily visible to the NNE and if the galaxies are at the same distance, the minimum separation would be about 436 000 ly. Radial velocity: 5371 km/s. Distance: 240 mly. Diameter: 84 000 ly.

NGC200

H858[2]	0:39.6	+2 53	GLX
SB(s)bc	1.7′ × 1.0′	13.49B	

Photographs reveal this inclined, barred spiral galaxy has a small core and a faint bar attached to narrow, knotty spiral arms, which broaden as they expand outward. In a moderate-aperture

telescope the galaxy can be seen in the field with NGC198 and initially appears round but high magnification brings out faint extensions oriented NNW–SSE and a bright core. Radial velocity: 5265 km/s. Distance: 235 mly. Diameter: 116 000 ly.

NGC213

H200[3]	0:41.2	+16 28	GLX
SB(rs)a	1.6′ × 1.2′	14.2B	

This slightly inclined, barred spiral galaxy has an elongated core and a faint bar. Photographs show that the bar is attached to an inner ring with faint spiral arms emerging. Visually, this faint and diffuse object is best seen at medium magnification, appearing as a double star involved in faint nebulosity, which is elongated E–W. The westernmost star is actually the bright core of the galaxy and is very slightly fainter than the field star which follows 0.4′ ESE of the core. Photographs show a very faint field star in the spiral arm immediately N of the galaxy core. Radial velocity: 5580 km/s. Distance: 249 million years. Diameter: 116 000 ly.

NGC234

H245[2]	0:43.4	+14 20	GLX
SAB(rs)c	1.6′ × 1.6′	13.55B	

Photographs show a face-on multi-arm spiral galaxy with a small core and bright, knotty spiral arms. Four major spiral arms are seen and the structure is symmetrical. Telescopically, the galaxy is fairly bright and appears as a round, somewhat ill defined nebulous patch of light that is broadly brighter to the middle. Radial velocity: 4577 km/s. Distance: 204 mly. Diameter: 95 000 ly.

NGC251

H204[3]	0:47.8	+19 34	GLX
Sc	2.4′ × 1.9′	14.30B	

Photographs show an inclined spiral galaxy, elongated E–W with a small, slightly elongated core in a fainter inner envelope. Thin spiral arms emerge from the envelope; they take on a very fragmentary multi-arm structure at the extremities. A faint double star in the outer spiral arm is 0.75′ E of the core, while a slightly brighter field star is 1.0′ NE. It is faint and difficult to see visually as the two field stars to the E interfere somewhat. The galaxy is quite a hazy, ill defined, round patch of light that is broadly brighter to the middle. It is a little easier to see at medium magnification. Radial velocity: 4688 km/s. Distance: 209 mly. Diameter: 146 000 ly.

NGC257

H863²	0:48.1	+8 19	GLX
Scd:	1.9′ × 1.3′	13.30B	

Photographs show that this inclined spiral galaxy has a small, slightly elongated core. Bright, knotty spiral arms emerge from the core; the three major spiral arms all terminate on the northern side of the galaxy. At medium magnification, this galaxy is moderately bright, smooth textured and only a little brighter to the middle. It is elongated E–W and the extremities are well defined. Radial velocity: 5380 km/s. Distance: 240 mly. Diameter: 133 000 ly.

NGC266

H153³	0:49.8	+32 16	GLX	
SB(rs)ab	3.5′ × 3.5′	12.64B	11.6V	

This is a face-on, barred spiral galaxy with a large, bright core and a fainter bar. Photographs show that two major spiral arms emerge from the bar; they

are narrow and bright and each completes almost a full revolution around the bar. A compact galaxy borders the spiral arm 1.2′ SW and a bright field star is 3.75′ S. At high magnification this is a fairly bright galaxy, a well defined oval elongated E–W with a bright, star-like core. The envelope has a fairly even surface brightness; visually only the bar and core are visible, the spiral structure is not seen. Radial velocity: 4822 km/s. Distance: 215 mly. Diameter: 219 000 ly.

NGC296

H214²	0:55.1	+31 33	GLX
SBb:	2.2′ × 1.0′	13.09B	

Photographically, this is a highly inclined spiral galaxy, possibly barred, with a large, bright core and bright, tightly wound spiral arms, speckled with dust patches. A bright field star is 1.0′ ESE of the core. It is misplotted as NGC295 in the first edition of *Uranometria 2000.0*. Telescopically, the galaxy is bright and is seen immediately WNW of the bright field star which hinders observation. It is a ghostly, though prominent, patch of light, a little extended NNW–SSE. In a medium-magnification field, a second very faint galaxy is visible to the N as a roundish patch of light that is a little brighter to the middle. This may be NGC295: it was misplotted in the first edition of *Uranometria 2000.0* as NGC296. Radial velocity: 5613 km/s. Distance: 250 mly. Diameter: 160 000 ly.

NGC315

H210²	0:57.8	+30 21	GLX
E⁺:	2.3′ × 1.8′	12.2B	

The DSS image shows an elliptical galaxy with a large, elongated core and an extensive outer envelope, elongated NE–SW. A bright field star is 3.5′ SE. The lenticular galaxy NGC311 is 5.5′ SW while NGC318 is 5.75′ NE. Visually, all three galaxies are visible in a high-magnification field; NGC315 is by far the largest and brightest of the three. It

is moderately large and fairly well defined and the envelope is almost round and grainy in appearance; the galaxy is opaque with a brighter core. Radial velocity: 5095 km/s. Distance: 227 mly. Diameter: 152 000 ly.

NGC379

H215²	1:07.3	+32 31	GLX
S0	1.1′ × 0.5′	13.81B	

Photographically and visually, this lenticular galaxy presents a similar appearance. It is moderately large and fairly well defined; the high-brightness envelope is slightly oval and grainy in appearance with a brighter core. It is the northernmost member of the Pisces Cloud group of galaxies, the principal members of which have also been catalogued as Arp 331. Sixteen NGC objects are visible here in less than 1° × 1°. Herschel object NGC380 is 2.4′. Radial velocity: 5730 km/s. Distance: 255 mly. Diameter: 82 000 ly.

NGC380

H216²	1:07.3	+32 29	GLX
E2	1.0′ × 1.0′	13.54B	

Photographs show an elliptical galaxy with a bright, round core and a fainter outer envelope. Visually, this Pisces Cloud member is a well defined circular glow. NGC379 is 2.4′ N while the principal galaxy, NGC383, is 4.4′ SSE. Radial velocity: 4579 km/s. Distance: 205 mly. Diameter: 60 000 ly.

NGC383

H217²	1:07.4	+32 25	GLX
SA0⁻:	1.8′ × 1.6′	13.04B	

Images show that this is a lenticular galaxy with a bright, round core and an extensive outer envelope. This is the principal galaxy of the Pisces Cloud galaxy group. Companion galaxy NGC382 is 0.5′ SSW while NGC379 is 6.8′ N and NGC380 is 4.4′ NNW. There is a large spread in the radial velocities of NGC379, NGC380 and NGC383 (about 1200 km/s), leading to a wide spread in

NGC266
©STScI/AURA/CALTECH/ROE

NGC379/380/382/383/385/386/387

©STScI/AURA/CALTECH/ROE

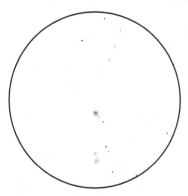

NGC379, NGC380, NGC383, NGC382, NGC385, NGC384 15-inch Newtonian 177x

distance estimates. Presumably, this is a physically associated group and a distance of about 235 mly can be cautiously adopted for the cluster as a whole. Radial velocity: 5250 km/s. Distance: 234 mly. Diameter: 123 000 ly.

NGC392

H218[2]	1:08.4	+33 07	GLX
S0⁻:	1.2' × 0.9'	13.69B	

This is the brightest of a group of three galaxies in the field; likely companion galaxy NGC394 is 1.0' NE, while NGC397 is 2.1' SE. If they form a physical group the minimum separation between NGC392 and NGC394 would be 63 000 ly while the distance between NGC392 and NGC397 would be about 132 000 ly. Photographs show a lenticular galaxy with a bright, elongated core and faint outer envelope, elongated NE–SW. Visually, NGC392 is the brightest of the trio, and is best seen at high magnification as a moderately bright, round and condensed galaxy, grainy in texture and brighter to the middle. The edges are diffuse and fade gradually. NGC394, a very small, faint and condensed round spot, is NNE. NGC397, a roundish, moderately well defined, diffuse patch of light, is SE and faintest of the three. Radial velocity:

4845 km/s. Distance: 217 mly. Diameter: 76 000 ly.

NGC407

H219[2]	1:10.6	+33 07	GLX
S0/a: sp	2.0' × 0.4'	14.31B	

Part of a quintet of galaxies which includes NGC408 3.0' ESE, NGC410 4.8' ENE, NGC414 8.4' E and CGCG501-119 8.4' NE. The DSS image shows this edge-on lenticular galaxy has a bright, bulging core and two thin, slightly curved dust lanes, one emerging N of the core, the other S of the core. The faint extensions curve slightly and the galaxy is elongated N–S. Visually, only the core of this spindle-shaped galaxy is visible; it is a small, round, well defined patch of light. NGC410 and NGC414 are also visible in the high-magnification field. Radial velocity: 5725 km/s. Distance: 256 mly. Diameter: 149 000 ly.

NGC410

H220[2]	1:11.0	+33 09	GLX
E⁺:	2.4' × 1.3'	12.52B	

This galaxy is the brightest of the quintet which includes NGC407, NGC408, NGC414 and CGCG501-119. Photographs show an elliptical galaxy with a bright, elongated core and an extensive outer envelope, elongated NE–SW. Visually, it is the brightest of three galaxies visible in a high-magnification field; it is a quite bright and large galaxy which is round and broadly brighter to the middle with diffuse, hazy edges. NGC414 located to the SE is only a little fainter and is decidedly oval and oriented N–S. It is actually a double system: the DSS image shows the cores of the two galaxies are almost touching. Radial velocity: 5446 km/s. Distance: 243 mly. Diameter: 170 000 ly.

NGC420

H154[3]	1:12.1	+32 06	GLX
S0:	2.0' × 2.0'	13.49B	

Photographs reveal a lenticular galaxy with a bright, round core and extensive, very faint, outer envelope. A bright field star is 6.5′ NE. Visually, this galaxy is a round, moderately bright object with a small, bright core embedded in a fairly bright envelope with moderately well defined edges. Radial velocity: 5159 km/s. Distance: 230 mly. Diameter: 134 000 ly.

NGC421

H155³	1:12.1	+32 09	SS
nonexistent			

NGC459

H205³	1:18.0	+17 33	GLX
Sbc	0.7′ × 0.6′	15.85B	

The DSS image shows a face-on galaxy with a small, round core in a fainter inner disc surrounded by a patchy spiral pattern. This is an extremely faint object visually and is seen only intermittently as a vague, roundish patch of light 0.9′ NNW of a faint field star which interferes with the view. It is best seen at high magnification and dark skies are required to confirm it with a moderate-aperture telescope. Radial velocity: 12 818 km/s. Distance: 573 mly. Diameter: 117 000 ly.

NGC467

H108¹	1:19.2	+3 18	GLX
SA(s)0° pec?	3.3′ × 1.7′	13.15B	11.9V

Photographs show a lenticular galaxy with a bright, round core and a fainter outer envelope. There is some evidence of dust in the outer halo and a faint plume of material extends to the S for a considerable distance. A faint edge-on galaxy is 1.6′ ESE. Visually, the galaxy is moderately bright, round and well condensed to a brighter core with a mottled envelope. The galaxy precedes a magnitude 8 field star which is 3.5′ ESE. Radial velocity: 5541 km/s. Distance: 248 mly. Diameter: 238 000 ly.

NGC470

H250³	1:19.7	+3 25	GLX
SA(rs)b	2.8′ × 1.7′	12.55B	11.8V

This galaxy forms a pair with NGC474, located 5.4′ to the E and they have also

been catalogued as Arp 227. Images reveal a slightly inclined spiral galaxy of the grand-design type. A small, brighter core is embedded in a bright, complex inner spiral pattern, with several fragmentary dust patches involved. On high-resolution photographs, the spiral arms are broad and dotted with HII regions. Very faint spiral arms emerge from the inner spiral pattern. In a moderate-aperture telescope, both galaxies can be seen in a high-magnification field. NGC470 is the larger of the two; it is a large oval patch of light, elongated SE–NW. The edges are somewhat ill defined and the core is not visible. Radial velocity: 2448 km/s. Distance: 109 mly. Diameter: 89 000 ly.

NGC473

H206³	1:19.9	+16 33	GLX
SAB(r)0/a:	1.7′ × 1.1′	13.32B	

Images show a slightly inclined lenticular galaxy with an extensive, brighter central region embedded in a

fainter outer envelope. Telescopically, the galaxy is moderately bright with a somewhat ill defined disc, oriented SSE–NNW, broadly brighter to the middle to a small bright core. Radial velocity: 2244 km/s. Distance: 100 mly. Diameter: 50 000 ly.

NGC474

H251³	1:20.1	+3 25	GLX
(R′)SA(s)0°	7.1′ × 6.3′	12.38B	11.5V

The DSS image shows a face-on spiral galaxy of peculiar morphology. The large, brilliant, round core is embedded in a fainter, round inner envelope which displays evidence of a possible N–S bar. Smooth, very faint spiral arms emerge from the inner envelope extending well to the S and W. An extensive dust patch is visible bordering the inner envelope to the WSW. High-resolution photographs show an extensive, broad, faint band of matter to the S and a thin, peculiar arc of matter curving E–W and intersecting the core. It is very likely

NGC470/474

©STScI/AURA/CALTECH/ROE

that this morphology is due to tidal interaction, possibly with NGC470, or a merger with a third galaxy. However, there is little evidence of disturbed structure in NGC470 and absorption of a third galaxy must have happened long ago as there is evidently nothing left of this possible companion. Visually, only the core and inner envelope of NGC474 are visible. The bright core is embedded in a hazy, ill defined, envelope. At the apparent distance, the core-to-core separation between NGC470 and NGC474 would be about 168 000 ly. Radial velocity: 2398 km/s. Distance: 107 mly. Diameter: 221 000 ly.

NGC488

H252[3]	1:21.8	+5 15	GLX
SA(r)b	5.8′ × 4.3′	11.13B	

Photographs reveal this galaxy as a classic example of the multi-arm spiral type. It is large, almost face-on and features a large, bright core and multiple, thin, tightly wound spiral arms. Many dust patches and lanes are involved and there are nebulous condensations in the arms throughout, suggesting areas of star formation. A bright field star is involved in a spiral arm 1.7′ SSE of the core. Another bright star is 3.0′ SSW. Visually, the galaxy is large and bright and is well seen at high magnification. The overall outline is oval and oriented N–S. The core is quite bright and, especially at medium magnification, a faint secondary halo is quite evident. This secondary halo is not well defined and is very patchy. The entire galaxy is quite mottled. Curiously, Herschel included this galaxy in his Class III: very faint nebulae. It was discovered on 13 December 1784, along with NGC 470 and NGC474, which are also described as faint. Conditions might have been poor for observing on this occasion. Radial velocity: 2350 km/s. Distance: 105 mly. Diameter: 177 000 ly.

NGC495

H156[3]	1:22.9	+33 28	GLX
(R′)SB(s)0/a pec	1.3′ × 0.8′	13.69B	

This is a face-on barred spiral galaxy. Photographs show that it has a large, bright core, a faint bar and two tightly wound, smooth-textured spiral arms with the disc elongated N–S. Several companion galaxies are nearby, including the Herschel objects NGC499 3.25′ to the E and NGC496, which is 4.5′ NE. All are members of the NGC507 galaxy group. Visually, only the core is visible as a small, round patch of light W of NGC499. Radial velocity: 4260 km/s. Distance: 190 mly. Diameter: 72 000 ly.

NGC496

H157[3]	1:23.3	+33 33	GLX
Sbc	1.6′ × 0.9′	14.31B	

Images reveal an inclined spiral galaxy with a very small core embedded in a brighter inner envelope. Dusty spiral arms emerge from the inner envelope and appear to be of the multi-arm type. Elongated NE–SW, this is the largest but dimmest of the trio which includes NGC495 and NGC499. Best seen at high magnification, this object is very faint and diffuse; it is an oval, ill defined patch of light. Although probably a member of the NGC507 group of galaxies, NGC496 has a considerably higher radial velocity than the group average. Herschel discovered this group on the night of 12 September 1784 and described NGC495, NGC496 and NGC499 together, calling them: 'Three forming a rectangular triangle. In the legs extremely faint, very small; at the rectangle very faint, pretty large.' Radial velocity: 6152 km/s. Distance: 275 mly. Diameter: 128 000 ly.

NGC499

H158[3]	1:23.2	+33 28	GLX	
S0⁻	1.6′ × 1.3′	13.32B	12.3V	

In photographs this is an inclined lenticular galaxy with a large, bright core and an extensive outer envelope, elongated ENE–WSW. It forms a trio with the Herschel objects NGC495 and NGC496 and is part of the NGC507 galaxy group. Visually, it is the brightest of the trio and appears as a small, slightly oval, opaque, well defined patch of light. Radial velocity: 4545 km/s. Distance: 203 mly. Diameter: 95 000 ly.

NGC507

H159[3]	1:23.7	+33 15	GLX
SA(r)0°	3.1′ × 3.1′	12.57B	

This galaxy is the principal member of the NGC507 group of galaxies, a rich group which is thought to be part of a chain of galaxy clusters in Andromeda and Pisces which includes the Abell 262 galaxy cluster and the Pisces Cloud, whose principal member is NGC393. A 2° × 2° field in the DSS image contains at least 20 bright galaxies around NGC507 and dozens of fainter ones. There is quite a high dispersal in the radial velocities of the brightest galaxies in the region but they are all evidently members of one cluster and the distance to the group would seem to be about 225 mly. NGC507 is a face-on lenticular galaxy with a large, bright core and an extensive faint outer envelope which extends to NGC508, 1.4′ to the N. The two galaxies have also been catalogued as Arp 229. Visually, it is the brightest galaxy in a medium-magnification field which also shows NGC494, NGC495, NGC496, NGC499, NGC504 and NGC508. It is round and brighter to the middle with a hazy secondary envelope. Herschel's description of the galaxy grouped it with NGC508 and he called them: 'Two. Both extremely faint, small but unequal.' Radial velocity: 5079 km/s. Distance: 227 mly. Diameter: 204 000 ly.

NGC508

H160[3]	1:23.7	+33 17	GLX
E0:	1.3′ × 1.3′	14.20B	

This elliptical galaxy and NGC507 have also been catalogued as Arp 229. Photographs show it has a large, bright core and an extensive, faint outer envelope. Visually, only the bright core is seen as a small, moderately faint, round, condensed patch of light. Radial velocity: 5671 km/s. Distance: 253 mly. Diameter: 96 000 ly.

NGC514

H252[2]	1:24.1	+12 55	GLX
SAB(rs)c	3.5′ × 2.8′	12.30B	11.7V

Photographically, this is a slightly inclined spiral galaxy, elongated E–W with a very small, round core. There are two main spiral arms with bright nebulous condensations and dust patches involved, but three other spiral fragments emerge as the arms unwind. Visually, the galaxy is a moderately bright, round and fairly large nebulous patch of light with somewhat ill defined edges. There is some mottling in the envelope as well as a brighter, round core. A bright field star is 3.0′ ESE. Radial velocity: 2570 km/s. Distance: 115 mly. Diameter: 117 000 ly.

NGC515

H167[3]	1:24.6	+33 29	GLX
S0	1.0′ × 0.8′	14.46B	

Images show a lenticular galaxy with a bright, elongated core in a smooth-textured envelope, elongated SE–NW. A dust patch is suspected immediately S of the core and a faint field star or compact object is involved immediately NW of the core. NGC515 is probably a member of the NGC507 galaxy group and a suspected companion galaxy, NGC517, is 2.8′ SSE, although its radial velocity is almost 1000 km/s lower. Visually, both galaxies can be seen in the same high-magnification field. NGC515 is small, opaque and well condensed, slightly oval and well defined. Radial velocity: 5242 km/s. Distance: 234 mly. Diameter: 68 000 ly.

NGC517

H168[3]	1:24.7	+33 27	GLX
S0	1.7′ × 0.5′	13.72B	

This is an almost edge-on lenticular galaxy with a bright, elongated core and faint extensions, oriented NNE–SSW. It is a probable member of the NGC507 galaxy group and the possible companion galaxy NGC515 is 2.8′ NNW. NGC517 is well seen at high magnification as a small, opaque, well-condensed object with well defined

edges and is very slightly larger than NGC515. Radial velocity: 4349 km/s. Distance: 194 mly. Diameter: 96 000 ly.

NGC520

H253[3]	1:24.6	+3 48	GLX
Irr pec	4.5′ × 1.8′	12.20B	11.4V

The DSS image shows a bright galaxy (or two galaxies) with a very peculiar morphology; it has also been catalogued as Arp 157. One possible interpretation is that a smaller, irregular galaxy (represented by the crescent-shaped fragment S of the larger galaxy's dust lane) has side-swiped the larger galaxy. Both have very bright central regions and they meet at an angle of about 10°. The main galaxy's dust lane is highly disturbed: it is a broad oval in the eastern portion, narrowing as it crosses the core, then broadening again, defining a faint plume of material emerging towards the W. A prominent plume of material emerges from the core, broadening and curving towards the SSE. In a moderate-aperture telescope, the galaxy is bright and of a peculiar form. The main envelope is fairly even in surface brightness. The western end is very slightly broader than the middle. The eastern tip forks into two, with the main body curving slightly to the E, while a short spur juts off to the ESE. Overall, the envelope is quite well defined and is elongated ESE–WNW. Herschel's 13 December 1784 discovery comments say only that the object is: 'Extremely faint, considerably large, extended.' Radial velocity: 2353 km/s. Distance: 105 mly. Diameter: 137 000 ly.

NGC524

H151[1]	1:24.8	+9 32	GLX
SA(rs)0+	3.3′ × 3.3′	11.40B	

This lenticular galaxy has a very large, bright core and faint, extensive outer envelope. High-resolution photographs show thin dust arcs involved in the outer envelope. Telescopically, this is a large and very bright galaxy, well seen at high magnification. The round, bright disc is very grainy and is surrounded by a fainter, ill defined halo.

A large, round, bright core is embedded in the inner disc. Radial velocity: 2467 km/s. Distance: 110 mly. Diameter: 106 000 ly.

NGC532

H556[3]	1:25.3	+9 16	GLX
Sab? sp	3.3′ × 0.8′	13.95B	

This is a highly inclined spiral galaxy, elongated NE–SW with a small core and grainy spiral arms. There is evidence of dust along the eastern flank and a field star is 4.2′ ENE. Although this galaxy is quite faint, it is fairly well seen at medium magnification as a small even-surface-brightness sliver of light, fairly well defined along its major axis. Radial velocity: 2448 km/s. Distance: 109 mly. Diameter: 105 000 ly.

NGC552

H172[3]	1:26.2	+33 24	SS
nonexistent			

This is a faint field star 0.5′ W of the edge-on galaxy NGC553. This object is not nebulous and H172[3] can be considered nonexistent. See next entry.

NGC553

H173[3]	1:26.2	+33 24	GLX
S0	0.75′ × 0.2′	15.0B:	

Both this object and H172[3] were originally considered nonexistent: the French observer Guillaume Bigourdan was unable to locate either object at Herschel's position on two occasions. When Herschel discovered these objects on 13 September 1784 he made a small error, about 1° in right ascension, in measuring their positions, which meant that the locations he reported were slightly wrong. To compound the problem, both NGC553 and NGC552 are extremely faint. Herschel's own description reads: 'Two. Both very small, stellar. A little doubtful.' On the DSS image there is a faint edge-on galaxy, preceded by a faint field star 0.5′ to the W. This field star would seem to be NGC552 while the galaxy is NGC553. Visually, this is a very difficult object to view; the author used a 25-inch telescope under New Mexico skies to observe both objects successfully. At medium magnification, the field shows what looks like a nebulous double star; the galaxy itself is very hazy and ill defined but is definitely seen as a N–S elongated streak with a slight bulge in the middle. The star immediately to the W is resolvable and its stellar nature confirmed. A drawing by the author confirmed that the field was identical with the DSS image. Nevertheless, it is not certain that this was Herschel's object. Radial velocity: 5261 km/s. Distance: 235 mly. Diameter: 51 000 ly.

NGC660

H253[2]	1:43.0	+13 38	GLX
SB(s)a pec	8.3′ × 3.2′	11.94B	11.2V

Photographically, this is a remarkable spiral galaxy with a peculiar morphology, possibly due to a recent merger of two galaxies. The major axis is elongated NE–SW and is bright with a well defined dust lane. A second, more prominent dust lane is visible in the NE, aligned at an angle of approximately 45° to the first dust lane. This lane runs N–S. Two large plumes of material emerge from the main envelope: the one in the NE extends to the N, the one in the SW extends to the S. Herschel discovered this object on 16 October 1784. Visually, it is a large, bright, though somewhat diffuse galaxy; the central region oriented along the major axis is bright, though the galaxy lacks a bright core. The extremities are diffuse and fade irregularly into the sky background. Some mottling of the central region is evident to the NE. Radial velocity: 938 km/s. Distance: 42 mly. Diameter: 103 000 ly.

NGC665

H588[2]	1:44.8	+10 26	GLX
(R)S0°?	1.9′ × 1.2′	13.18B	

Photographs show a slightly inclined lenticular galaxy with a large, bright core and a possible bar embedded in an oblique ring, elongated ESE–WNW. Visually, only the core of this galaxy is visible as a small, condensed and round patch of light, which is fairly well defined and suddenly brighter to a small core. Radial velocity: 5498 km/s. Distance: 246 mly. Diameter: 136 000 ly.

NGC676

H42[4]	1:49.0	+5 54	GLX
S0/a: sp	4.0′ × 1.2′	10.71B	12.0V

The DSS photograph shows a highly inclined lenticular galaxy with a large bright core and slightly fainter extensions, elongated N–S. The star BD +04 244, magnitude 10.44, is 5.1″ SSW of the core but is completely burned out in the photograph. Visually though, the star is easy and the galaxy itself is a tough object, owing to the interference from the star. It is best seen at medium magnification as an elongated, ghostly streak of light. Herschel recorded the object on 30 September 1786 in his Class IV, and described it as: 'A star about 8 or 9 magnitude with a very faint branch in the meridian, each branch 1′ long.' Radial velocity: 1568 km/s. Distance: 70 mly. Diameter: 81 000 ly.

NGC693

H859[2]	1:50.5	+6 09	GLX
S0/a?	2.1′ × 1.0′	13.24B	12.4V

Photographs show a highly inclined lenticular galaxy with a bright, elongated inner envelope embedded in a faint outer glow. The inner envelope is brighter in the E and there is evidence of dust on the northern flank. The disc is elongated ESE–WNW and a field star is 1.4′ ENE. Telescopically, the galaxy is small but fairly bright and well defined; it is a slightly elongated streak of light with an even-surface-brightness envelope which tapers to points. A small bright core is visible and the field star to the ENE hinders visibility somewhat. Radial velocity: 1629 km/s. Distance: 73 mly. Diameter: 44 000 ly.

NGC660

NGC706

H596[2]	1:51.8	+6 18	GLX
Sbc?	1.9′ × 1.4′	13.22B	12.5V

Images show an almost face-on spiral galaxy with a small, bright, round core and bright, knotty spiral arms. An unequal pair of stars is 0.9′ N. The galaxy is quite bright visually; it is a fat, opaque oval of light, oriented N–S. The surface brightness is fairly even, though it brightens very gradually to the middle, and the edges are well defined. The brighter of two field stars is seen immediately N. Radial velocity: 5041 km/s. Distance: 225 mly. Diameter: 92 000 ly.

NGC718

H270[2]	1:53.2	+4 12	GLX
SAB(s)a	2.3′ × 2.1′	12.55B	

Photographs show an almost face-on spiral galaxy, probably barred, with a bright, round core and faint extensions embedded in a fairly tightly wound inner spiral pattern. There are possible dust patches involved and two smooth-textured spiral arms emerge from the inner spiral pattern. Visually, the galaxy is moderately bright at high magnification and is seen as a round, condensed patch of light with a tiny, brighter core. The envelope is well defined and even in surface brightness. Radial velocity: 1788 km/s. Distance: 80 mly. Diameter: 54 000 ly.

NGC741

H271[2]	1:56.4	+5 38	GLX
E0:	2.0′ × 1.7′	12.30B	

The DSS image reveals an elliptical galaxy with a bright core and a fainter outer envelope. It is the brightest member of an extremely compact group of small galaxies. At least six galaxies are within a 3.5′ radius of NGC741 (and two more are 6.0′ to the NE), including a very small, compact object embedded in the galaxy immediately E of the core. This may be the image of a star, however. NGC742 is 0.8′ E and at the presumed distance, the core-to-core separation between the two galaxies would be about 63 000 ly.

Visually, both NGC741 and NGC742 can be seen in a high-power field. NGC741 is much brighter and larger, quite well defined, round and much brighter to a well-condensed nonstellar core. Radial velocity: 5617 km/s. Distance: 251 mly. Diameter: 142 000 ly.

NGC742

H272[2]	1:56.5	+5 38	GLX
cE0:	0.2′ × 0.2′	15.03B	

This is the second brightest galaxy in the compact group dominated by NGC741. It is a compact elliptical galaxy with a bright core and a fainter outer envelope. The visual appearance is quite similar to the photographic one and despite its size and dimness, it is quite easy to see. Radial velocity: 6025 km/s. Distance: 269 mly. Diameter: 16 000 ly.

NGC7458

H590[2]	23:01.5	+1 45	GLX
E	1.4′ × 1.2′	13.62B	

Photographs show an elliptical galaxy with a bright, elongated core and a fainter outer envelope, oriented NNE–SSW. The faint galaxy MCG+0-58-22 is 4.75′ E. Visually, the galaxy is moderately bright, round and broadly brighter to a sizable central region. The edges are diffuse and fade gradually into the sky background. Radial velocity: 5114 km/s. Distance: 228 mly. Diameter: 93 000 ly.

NGC7506

H184[3]	23:11.8	−2 10	GLX
(R′)SB(r)0+:	1.7′ × 1.1′	14.34B	

In photographs this is a highly inclined, barred spiral galaxy with a brighter core and a faint bar attached to an oblique inner ring and a very faint outer ring, or smooth spiral arm. It is part of a loose grouping of 12 NGC galaxies which includes the Herschel objects NGC7556 and NGC7566. Visually, it is a fairly faint galaxy, elongated E–W and bracketed by a diamond-shaped asterism of four magnitude 10 field stars. It is small and moderately well condensed with a faint stellar nucleus

intermittently visible. Radial velocity: 4002 km/s Distance: 179 mly. Diameter: 88 000 ly.

NGC7537

H429[2]	23:14.6	+4 30	GLX
SAbc:	2.2′ × 0.6′	13.84B	

This is the smaller and fainter member of a galaxy pair which includes NGC7541. Photographs show a highly inclined spiral galaxy with a large, bright, elongated core and grainy spiral arms. Several dust patches and lanes are involved in the spiral structure. Companion galaxy NGC7541 is 3.0′ NE. Visually, NGC7537 is quite faint but is well defined, gradually brighter along its major axis and elongated almost due E–W. Radial velocity: 2815 km/s. Distance: 126 mly. Diameter: 81 000 ly.

NGC7541

H430[2]	23:14.7	+4 32	GLX
SB(rs)bc: pec	3.5′ × 1.2′	12.48B	

This galaxy is paired with NGC7537 and has a similar radial velocity, so they are probably physically related. There is no evidence of tidal distortion, but the core-to-core separation between the two galaxies at the presumed distance would be only 110 000 ly. Images reveal a highly inclined spiral galaxy with a small core embedded in a bright spiral structure. Several dust patches and lanes are involved and a field star is 2.0′ E. Companion galaxy NGC7537 is 3.0′ SW and a small, faint, possible dwarf companion galaxy is 6.0′ N. In a medium-magnification field, both galaxies are well seen and NGC7541 is bright and very elongated almost due E–W with a magnitude 12 field star at its eastern tip. It is well defined and a little brighter to the middle. Radial velocity: 2830 km/s. Distance: 126 mly. Diameter: 129 000 ly.

NGC7556

H235[2]	23:15.9	−2 20	GLX
S0⁻	1.9′ × 1.1′	14.0B	

Photographs show this bright lenticular galaxy surrounded by several very small

NGC7537/7541

©STScI/AURA/CALTECHROE

galaxy. Visually, it is moderately faint and elongated ESE–WNW with a well-condensed core and diffuse edges. Radial velocity: 8079 km/s. Distance: 361 mly. Diameter: 137 000 ly.

NGC7685

H426[3]	23:30.6	+3 54	GLX
SAB(s)c:	1.9′ × 1.4′	13.94B	

This is an almost face-on spiral galaxy. Photographs show a small core and a possible short bar. Faint, grainy spiral arms emerge from the central region and the galaxy is elongated N–S. A small, compact galaxy is 4.5′ NE. In moderate apertures, this is a faint and very diffuse galaxy, best seen at medium magnification as a moderately large, low-surface-brightness object. It is round with very ill defined edges and slightly brighter to the middle. Radial velocity: 5775 km/s. Distance: 258 mly. Diameter: 143 000 ly.

NGC7694

H187[3]	23:33.4	−2 42	GLX
Im pec:	1.4′ × 0.8′	13.98B	

Images show this irregular galaxy has a peculiar morphology. A bright oval core is embedded in an asymmetrical envelope which is more extensive to the E. Two bright condensations are in the envelope E of the core and a compact galaxy is 1.0′ S of the core. Visually, the galaxy is moderately faint, appearing very gradually elongated E–W. It brightens slowly to a compressed core which is offset to the W with an ill defined, hazy envelope. Radial velocity: 2393 km/s. Distance: 107 mly. Diameter: 44 000 ly.

NGC7701

H188[3]	23:34.6	−2 51	GLX
1.0′ × 0.6′	14.83B		

The DSS image reveals a lenticular galaxy with a bright, elongated core and

and fainter galaxies, including one compact object embedded in the envelope 0.3′ SW of the core and a second 0.8′ W of the core. The galaxy is part of a loose grouping of 12 NGC galaxies which includes Herschel objects NGC7506 and NGC7566. In a medium-magnification field, it can be seen with NGC7566 which follows it to the E. It is larger and brighter than NGC7566 and it precedes a magnitude 10 field star. Very gradually, then suddenly brighter to the middle, it is oval and elongated ESE–WNW. Radial velocity: 7628 km/s. Distance: 341 mly. Diameter: 188 000 ly.

NGC7562

H467[2]	23:16.0	+6 41	GLX
E2-3	2.2′ × 1.5′	12.61B	

Photographs show that this elliptical galaxy has a large, bright, elongated core in an extensive outer envelope, elongated E–W. The edge-on galaxy

NGC7562A is 2.25′ SSE and a small compact galaxy is 3.5′ S. NGC7557 is 4.75′ WNW. Visually, although small, NGC7562 is very bright and takes magnification well. The main body is oval and brightens quickly to a sizable bright core. Surrounding the main envelope is a faint outer envelope that fades into the sky background. NGC7557 is also visible as a very faint, round, hazy patch of light with ill defined edges. Radial velocity: 3755 km/s. Distance: 168 mly. Diameter: 107 000 ly.

NGC7566

H185[3]	23:16.6	−2 21	GLX
(R')SB(rs)a pec?	1.3′ × 0.7′	14.29B	

Images show an inclined spiral galaxy with a bright, elongated core and faint spiral arms. Herschel thought this object resolvable, owing to the presence of a 1.5′ triangle of faint field stars which bracket the

NGC7694

near the core, but they may be indicative of a ring or pseudo-ring structure. A fainter, thin secondary envelope is visible and there is an extremely faint, large and ill defined outer envelope. A very dim galaxy is 1.9′ W. Telescopically, the galaxy is moderately faint and a little more distinct at medium magnification. It is a round, diffuse patch of light which is broadly brighter to the middle with hazy ill defined edges. Radial velocity: 3381 km/s. Distance: 151 mly. Diameter: 53 000 ly.

NGC7778

H231³	23:53.3	+7 52	GLX
E	1.0′ × 1.0′	13.72B	

The visual and photographic appearances of this elliptical galaxy are quite similar and it can be seen in a medium-magnification field along with NGC7779, NGC7781 and NGC7782. It is round, sharply defined and moderately bright. Photographs show a very faint halo around the bright core. Companion galaxy NGC 7779 is 1.7′ E. NGC7781 is 6.2′ ESE and NGC7782 is 10.0′ NE. These galaxies are at similar radial velocities and presumably form a group. Radial velocity: 5352 km/s. Distance: 239 mly. Diameter: 69 000 ly.

NGC7779

H232³	23:53.4	+7 52	GLX
(R′)SA0/a:	1.4′ × 1.1′	13.66B	

The DSS image shows a face-on spiral galaxy with a small, bright core and thin, symmetrical and very faint spiral structure. Visually, this galaxy appears to be a twin of NGC7778 1.7′ to the E but it is very slightly larger and a little brighter. Only the core of this galaxy is visible, however; it is round and sharply defined. If the two galaxies form a physical pair, the core-to-core separation would be about 116 000 ly at the presumed distance. Radial velocity: 5250 km/s. Distance: 234 mly. Diameter: 95 000 ly.

---

fainter envelope, oriented N–S. The galaxy is unclassified but may be an S0. Two galaxies are in the field: NGC7699 is 2.9′ SSW, while NGC7700 is 6.0′ S. In a moderate-aperture telescope, NGC7701 has a bright, non-stellar core embedded in an oval envelope oriented N–S and well defined edges. Radial velocity: 5447 km/s. Distance: 243 mly. Diameter: 71 000 ly.

NGC7750

H427³	23:46.7	+3 47	GLX
(R′)SB(rs)c pec:	1.6′ × 0.8′	13.49B	

Photographs show an inclined spiral galaxy with a small core and grainy spiral arms, elongated N–S. At high magnification, this is a moderately large and fairly bright galaxy. The outline is oval and the envelope is grainy textured with ill defined edges. Radial velocity: 3064 km/s. Distance: 137 mly. Diameter: 64 000 ly.

NGC7751

H437³	23:47.0	+6 51	GLX
S?	1.2′ × 1.2′	13.87B	

The DSS image shows a small round galaxy of peculiar morphology. The very bright, round core is in a slightly fainter, round inner envelope. Faint dust patches are suspected in the envelope

NGC7751

NGC7782

H233[3]	23:53.9	+7 58	GLX
SA(s)b	2.4′ × 1.3′	13.08B	

This is an inclined spiral galaxy with a small, bright core and thin, tightly wound spiral arms. High-resolution images show the spiral pattern to be multi-arm. This galaxy has a similar radial velocity to NGC7778, NGC7779, NGC7780 and NGC7781 but if it is indeed a physical member of the group, it is considerably larger. Visually, it is the brightest and largest of the group, a grainy oval envelope oriented N–S and brighter along its major axis. Radial velocity: 5513 km/s. Distance: 2461 mly. Diameter: 172 000 ly.

NGC7785

H468[2]	23:55.3	+5 55	GLX
E5-6	2.0′ × 1.3′	12.64B	

Images reveal an elongated elliptical galaxy with a bright elongated core and extensive outer envelope. Suspected dust lanes in the outer envelope suggest the galaxy may be more correctly classified as an S0 lenticular galaxy. A bright field star is 4.3′ almost due W.

Visually, the galaxy is well seen at high magnification as a small but very high-surface-brightness object, quite grainy in texture. The edges are very sharply defined, suggesting that only the bright core is visible and the galaxy is elongated SE–NW. Herschel thought this galaxy was resolvable. Radial velocity: 3936 km/s. Distance: 176 mly. Diameter: 102 000 ly.

NGC7797

H867[3]	23:59.0	+3 38	GLX
Sbc	1.0′ × 0.9′	14.84B	

Photographs show a face-on spiral galaxy with a small, round core and grainy spiral arms. Visually, the galaxy is fairly faint and is best seen at medium magnification as a round patch of light with a star-like, brighter core and a diffuse outer envelope. Radial velocity: 8973 km/s. Distance: 401 mly. Diameter: 117 000 ly.

NGC7816

H436[3]	0:03.8	+7 28	GLX
Sbc	1.7′ × 1.5′	14.0B	

Images show that this is a face-on spiral galaxy with a small, round core and grainy spiral arms. The galaxy NGC7818 is located 7.5′ SE, though it is not visible in moderate-aperture telescopes. Though moderately faint, NGC7816 is well seen at medium and high magnification as a round, nebulous patch of light, somewhat diffuse but with moderately well defined edges. It is broadly brighter to the middle. Radial velocity: 5368 km/s. Distance: 240 mly. Diameter: 119 000 ly.

NGC7832

H190[3]	0:06.6	−3 42	GLX
E+	1.7′ × 0.5′	13.0B	

This galaxy appears similar both visually and photographically. It is a quite bright and moderately large galaxy, oval in shape and elongated NNE–SSW. The bright, well defined envelope surrounds a bright, small core. Photographs show a compact field galaxy 5.25′ to the E which is faintly visible at high magnification. Radial velocity: 6296 km/s. Distance: 281 mly. Diameter: 139 000 ly.

Puppis

Most of the objects recorded in this constellation are star clusters, but there are a couple of planetary nebulae and a few galaxies as well. The constellation was low in the sky for Herschel and a few of the star clusters never rose more than 10° above the southern horizon. A total of 25 objects were recorded, all between the years 1783 and 1793. Notable objects include the star cluster/nebula NGC 2467 and the the bright cluster NGC 2539.

NGC2396

H36[8]	7:28.0	−11 43	OC
IV1m	10.0′ × 8.0′	7.4V	

Photographs show this scattered cluster of stars in a rich field. Telescopically, this is a fairly attractive cluster, well separated from the sky background and situated S of a bright unequal pair of stars. The brightness range of the stars is moderate, with one magnitude 9 star in the W and the rest of the cluster members magnitude 11 or fainter. The group is elongated E–W with about 30 stars involved.

NGC2401

H65[7]	7:29.4	−13 58	OC
I1p	2.0′ × 2.0′	12.6V	

The DSS photograph shows a small, compressed, almost round cluster of faint stars with a very narrow brightness range. The cluster is very faint telescopically but distinct and visible as a roundish, nebulous patch at both medium and high magnification. Well separated from the sky background, a few very faint stars are resolved. Herschel's 8 March 1793 description reads: 'A small cluster of very small stars considerably rich and compressed.' About 20 stars are considered members.

NGC2413

H52[8]	7:33.3	−13 07	AST
10.0′ × 8.0′			

In a rich field, photographs show a poor, scattered group of stars of moderate brightness range, roughly oval in shape and oriented N–S. Visually, this group of stars stands out well at low magnification. It is a somewhat scattered collection of about 20 stars of moderate brightness range: six of the stars are fairly bright and four of the six form a diamond shape that helps to define the group, which is very likely just an asterism.

NGC2414

H37[8]	7:33.3	−15 27	OC
I3m	6.0′ × 4.0′	8.3B	7.9V

Photographs show a compressed cluster of stars arrayed around a bright star. The field is rich in faint stars but the group stands out well from the background. Although not particularly rich visually, this cluster is distinct,

NGC2401

surrounding a magnitude 8.2 star (HD60308). The brightness range is moderate with fewer than 20 stars resolved. Herschel's 4 February 1785 description reads: 'A small cluster of pretty compressed stars of various sizes, not very rich.' Thirty-five stars are considered to be members of the cluster. Distance: 13 500 ly.

NGC2421

H67[7]	7:36.3	−20 37	OC
I2r	8.0′ × 8.0′	8.7B	8.3V

In photographs this is a fairly rich cluster of mostly bright stars with a moderate brightness range in a rich field. Visually, the cluster appears fairly large, moderately rich and well separated from the sky background with the stars of a fairly narrow brightness range. At least 35 stars are visible but actual membership of the cluster is probably 70 stars. Distance: 6100 ly.

NGC2422

H38[8]	7:36.6	−14 30	OC
I3m	25.0′ × 20.0′	4.4B	4.4V

This is one of Charles Messier's 'missing objects'; it was recorded on 19 February 1771 but was subsequently considered lost for almost two centuries due to a positional error printed in the *Connaissances des Temps*. It was independently rediscovered by Caroline Herschel on 26 February 1783 and listed in William Herschel's catalogue on 4 February 1785. However, credit for the discovery should go to the little-known Italian observer Giovanni Hodierna, who observed it before 1654. Photographs show a large, scattered cluster of stars with a wide range of brightness in a rich field. The densest portion of the cluster is in the middle and is about 5.0′ × 5.0′ in size. The cluster can be seen with the naked eye from a dark-sky site and telescopically is a very attractive, large and bright object, best seen at low magnification as a moderately compressed cluster of stars of wide brightness range. The four brightest stars form a diamond-shaped asterism

with faint stars surrounding each. The cluster is well defined with at least 50 members visible; many more can be seen with higher magnification. The total membership of the cluster is more than 100 stars with the brightest star shining at magnitude 5.68. Spectral type: B3. Age: 78 million years. Radial velocity: 9 km/s. Distance: 1700 ly.

NGC2423

H28[7]	7:37.1	−13 52	OC
II2m	12.0′ × 12.0′	7.1B	6.7V

Photographs show a large scattered cluster in a rich star field. In small telescopes this is a fairly bright, attractive cluster located N following M47. The stars are primarily magnitude 10–11 or fainter and are grouped around one magnitude 9 luminary with a subgroup of eight stars located to the ENE. More than 80 stars are considered

to be members. Spectral type: B5. Age: 35 million years. Distance: 2440 ly.

NGC2425

H87[8]	7:38.3	−14 52	OC
II1m	5.0′ × 3.0′		

Photographs show a rich cluster of faint stars of moderate brightness range which is elongated in an ESE–WNW direction. Though the field is rich in faint stars, the cluster is well separated from the sky background. The cluster is quite faint visually but distinctly seen as an elongated asterism at high magnification with a clump of half a dozen stars in the WNW surrounded by a faint haze. The rest of the cluster to the E forms a broad V-shape, again with at least half a dozen stars involved in a nebulous haze. About 30 stars are considered to be members with the brightest shining at magnitude 14.

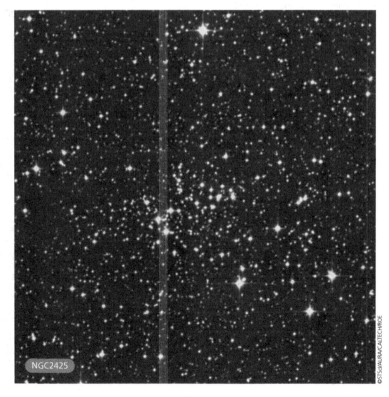

NGC2425

NGC2428

H47[8]	7:39.2	−16 31	OC/AST?
13.0′ × 13.0′			

The DSS image shows an irregular chain of moderately bright stars, oriented SE–NW in a fairly rich field of stars. Visually, this is a distinct group of about 25 stars in a 12.0′ field. The stars are of a moderate brightness range and fairly well separated from the sky background with a chain of about seven stars oriented NE–SW on the group's NW flank. Although up to 50 stars may be considered associated, this is probably an asterism and not a true cluster.

NGC2430

H46[8]	7:39.4	−16 21	MWP
8.0′ × 8.0′			

Photographs show a coarse grouping of fairly bright stars in a rich field of stars with the two brightest stars (HD61612, magnitude 10.4 and HD61553, magnitude 8.0), NW of the main concentration. Visually, this is little more than a scattering of stars of moderate brightness range, with six to eight of them brighter and not well separated from the sky background.

NGC2432

H36[6]	7:40.9	−19 05	OC
II2m	7.0′ × 6.0′	10.2V	

Images show a prominent elongated cluster of stars with a moderate range of brightness in a rich field. Telescopically, the cluster is dominated by a prominent N–S line of stars with a fairly narrow brightness range, quite compressed and well separated from the background. The stars are resolved, faint and tightly packed with a compressed V-shaped asterism in the middle. About two dozen stars are visible but the total membership of the cluster may be 50 stars.

NGC2438

H39[4]	7:41.8	−14 44	PN
4+2	1.1′ × 1.1′	11.7B	10.8V

This is the well-known planetary nebula associated with the star cluster M46, though the nebula is evidently a foreground object and not part of the cluster. Located in the northern quadrant of the cluster, photographs show a round, opaque ring with a small, darker centre and the central star showing through. Extremely faint nebulosity is seen outside the main annulus, arranged as a very thin outer ring, about 2.5′ in diameter. Visually, the nebula is a fairly large, milky spot of uniform brightness. The central star, at a magnitude of 17.5, is far too faint to be seen but a cluster member borders the nebula immediately to the ESE. Discovered by Herschel on 19 March 1786, he surmised that it was a planetary nebula, saying: 'Pretty bright, round, resolvable, within the 46th of the Connaissances des Temps almost of an equal light throughout, 2′ diameter. No connection with the cluster, which is free from nebulosity.' The central star displays a continuous spectrum, the radial velocity is 76 km/s and the distance about 2900 ly.

NGC2440

H64[4]	7:41.9	−18 13	PN
5+3	1.2′ × 1.0′	10.8B	9.4V

The DSS photograph shows a planetary nebula of complex structure located 3.0′ W of a bright field star. It appears decidedly double-lobed and dense, dark cavities in the nebulosity are seen on the E and W sides, while a fainter tuft of nebulosity emerges from the N. It is a very interesting object visually. At high magnification it is a bright object of fairly irregular form. The magnitude 18.9 central star is not visible and the bright and opaque nebulosity is faintly bluish and grainy in texture. Prominent projections emerge from the irregular disc to the NNE and SSW. Distance: 3600 ly.

NGC2438

©STScI/AURA/CALTECH/ROE

NGC2440

NGC2467

H22[4]	7:52.6	−26 23	OC/EN
I3m n	15.0′ × 15.0′	7.1V	

This is a scattered cluster of stars involved in the emission nebula Sharpless 2–311. A rich concentration of stars NNE of the brightest portion of the nebula is the cluster Haffner 19. Photographs show the field is rich in stars and the cluster is not well separated from the sky background. The nebula is complex, the brightest portion is oval and about 8′ × 5′ in size, bordered on the N by a narrow, dark channel. There is a great deal of faint nebulosity fanning out towards the E and SE and on the DSS image the nebula is at least 30.0′ × 30.0′ in extent. Visually, the nebulosity surrounding the brightest star to the W is easily seen at low magnification. It is a diffuse, roundish patch of light. The cluster is quite scattered and somewhat faint, the most prominent portion being a V-shaped group of stars elongated E–W with the

apex in the W. About 30 stars are resolved. Herschel recorded the object on 9 December 1784, describing it as: 'Large, pretty bright, round, easily resolvable. 6 or 7′ in diameter, a faint red colour visible. A star 8th magnitude not far from the centre, but not connected. Second observation 9 or 10′ in diameter.' The cluster is a part of the Puppis OB1 association with a total membership of about 50 stars. Spectral type: O5. Age: about 250 million years. Distance: 13 000 ly.

NGC2479

H58[7]	7:55.1	−17 43	OC
III1m	11.0′ × 9.0′	9.6V	

Photographs show a sparse cluster of stars, irregularly round and somewhat elongated SE–NW. At medium magnification, this is an attractive cluster of stars of narrow brightness range, well separated from the sky

background. Extending over an area of about 12.0′ × 12.0′ are about 45 magnitude 11–12 stars.

NGC2482

H10[7]	7:54.9	−24 18	OC
IV1m	10.0′ × 4.0′	7.7B	7.3V

The visual and photographic appearances of this sparse cluster of moderately bright stars are fairly similar. The cluster is dominated by a row of brighter stars oriented SE–NW and for the most part magnitude 10–11. There is an impression of a haze of unresolved stars along this chain at low magnification, associated with the rich background. At the northern edge of the chain is a triangle formed by eight stars. Resolution is best at medium magnification, when the cluster looks roughly triangular. In all about two dozen stars are resolved; the total membership of the cluster is probably 40 stars with the brightest at magnitude 10.04. Distance: 2440 ly.

NGC2489

H23[7]	7:56.2	−30 04	OC
I2m	6.0′ × 6.0′	8.6B	7.9V

A moderately rich cluster of stars on the DSS image, visually this is a fairly bright and well-resolved cluster. The stars are of a moderate brightness range and the cluster is broadly concentrated to the middle and well separated from the background. A N–S chain of five stars borders the cluster on the E side and despite being low on the horizon about 30 stars are resolved. More than 100 stars are considered members, the brightest being magnitude 11.1. Spectral type: B8. Age: about 240 million years. Distance: 8000 ly.

NGC2509

H1[8]	8:00.7	−19 04	OC
I1r	12.0′ × 12.0′	9.3V	

Images show a fairly rich cluster of stars which is quite compressed to the middle with a narrow brightness range. Telescopically, this is an attractive cluster; it is a roundish nebulous patch with a few stars shining through at low magnification. Even at high magnification resolution is not complete; a triangular portion near the middle appears particularly rich and though several stars are seen, many of them pairs, an unresolved haze persists. The cluster is elongated NE–SW and well separated from the background sky. The total membership of the cluster is about 70 stars.

NGC2525

H877³	8:05.6	−11 26	GLX
SB(s)c	3.0′ × 2.0′	12.29B	

Photographs show a barred spiral galaxy with a bright central bar and two thick, bright spiral arms emerging. Several knots and condensations are seen along the spiral arms. Deep photographs indicate resolution of individual stars occurs at a blue magnitude of 22. Visually, the galaxy is located N of a crooked row of six moderately bright stars. It is a low-surface-brightness object, elongated E–W and a little brighter to the middle with poorly defined edges. The galaxy appears fairly large at medium magnification and the texture is rather smooth. Radial velocity: 1398 km/s. Distance: 62 mly. Diameter: 55 000 ly.

NGC2527

H30⁸	8:05.3	−28 10	OC
II2m	10.0′ × 10.0′	6.8B	6.5V

Photographs show a fairly large, scattered cluster of stars with a wide range of brightness, but with many bright stars around the periphery, which help to separate the group from the sky background. The group is a scattered cluster visually, fairly bright with a moderate range in brightness. There is a well defined group of about a dozen stars in the W separated by a dark gap and then a long chain of stars oriented SSW–NNE involving another

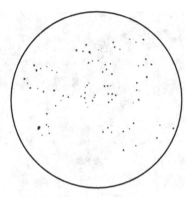

NGC2539 15-inch Newtonian 95x

dozen stars. The cluster is well defined with most of the members about magnitude 9–10 and about 30 stars are resolved. The actual membership is about 45 stars. Age: 1 billion years. Distance: 1850 ly.

NGC2539

H11⁷	8:10.7	−12 50	OC
III2m	16.0′ × 14.0′	7.0B	6.5V

Similar photographically and visually, this is a superb, well-resolved cluster that precedes the bright field star 19 Puppis, which is on the ESE edge of the cluster. The members are quite uniform in brightness, predominantly about magnitude 11–12. About 80 stars are visible in a 20′ × 15′ area, though about 60 are actual members of the cluster. There is strong visual evidence of subgrouping: most members are arranged in clumps of three to ten stars. Spectral type: A0. Age: about 660 million years. Distance: 4065 ly.

NGC2566

H288³	8:18.7	−25 29	GLX
(R')SB(r)ab	3.6′ × 1.9′	11.82B	

The DSS image reveals a barred spiral galaxy in a rich star field with a small, round core and a fainter bar. Very faint, anaemic spiral arms emerge from the bar, which are relaxed and fairly broad, and the galaxy is elongated ENE–WSW. Fainter field galaxies are at 6.25′ NW and 7.0′ NNW. Visually, only the core

NGC2525

©STScI/AURA/CALTECH/ROE

and bar are visible but the galaxy is well seen at medium magnification; it is moderately bright, fairly well defined and oriented ENE–WSW. The core is bright and almost stellar and IC2311 is seen 7.75′ to the N as a small, condensed and fairly bright object. Radial velocity: 1424 km/s. Distance: 64 mly. Diameter: 67 000 ly.

NGC2567

H64[7]	8:18.6	−30 38	OC
II2m	9.0′ × 5.0′	7.8B	7.4V

The DSS image shows a sparse cluster of stars of moderate brightness range. Telescopically, this is a compressed cluster of moderately bright, isolated stars, with a narrow to moderate brightness range. A N–S chain of six stars is just E of the main concentration, a rectangular area of brighter stars

elongated E–W. An attractive double is near the SE edge of the cluster and about 25 stars are resolved, though more than 100 are considered to be members. The brightest stars are about magnitude 10–11. Spectral type: B8. Age: approximately 70 million years. Distance: 5300 ly.

NGC2571

H39[6]	8:18.9	−29 44	OC
II3m	7.0′ × 4.0′	7.2B	7.0V

Visually and photographically this is a very sparse cluster of stars, though well separated from the sky background. The range in brightness is moderate. The three brightest stars form a flattened triangle on the E side and to the W is a circlet of stars. About 15 stars are resolved in all. The brightest is magnitude 8.8 and total membership of

the cluster is about 50 stars. Spectral type: B8. Distance: 4000 ly.

NGC2578

H902[3]	8:21.4	−13 19	GLX
SB(r)0/a pec	2.1′ × 1.1′	13.51B	

Photographs reveal an inclined spiral galaxy with a large, bright, elongated core. The spiral arms are even in brightness but a narrow spiral arm to the SW is brighter and defined by several small knots. Two bright field stars are SSE. A small spiral galaxy, MCG–2-22-3, is 3.0′ SE. Only the brighter core of NGC2578 is visible in a moderate aperture and it appears as a quite well defined, small, oval patch of light with a tiny, star-like core. Radial velocity: 4411 km/s. Distance: 197 mly. Diameter: 120 000 ly.

Pyxis

This southern constellation appeared low on Herschel's southern horizon and only two objects were recorded here: NGC2613 in 1784 and NGC2627 in 1793. Both objects are interesting visually to observers in southern locations.

NGC2613

H266[2]	8:33.4	−22 58	GLX
SA(s)b	7.1′ × 1.7′	11.11B	10.4V

The DSS image shows a large, bright and highly inclined spiral galaxy with a brighter, elongated core. The spiral arms are fairly bright, narrow and grainy with dust lanes visible in the arms along the SW flank. High-resolution images reveal an exquisitely resolved multi-arm spiral pattern. Visually, the galaxy is seen in a rich star field as a bright, elongated galaxy, oriented ESE–WNW. The envelope is moderately bright and somewhat diffuse; the edges are hazy. The central region is bright, large and elongated

along the major axis. Radial velocity: 1466 km/s. Distance: 66 mly. Diameter: 135 000 ly.

NGC2627

H63[7]	8:37.3	−29 57	OC
II2r	8.0′ × 6.0′	8.4V	

This cluster of moderately faint stars is well resolved in photographs. Visually, the members of this grouping are somewhat faint and set against a hazy, unresolved background. The cluster is oval in shape, slightly elongated E–W and the resolved stars have a narrow brightness range; there is a bright star to the E and about 20 stars resolved in all.

Cluster membership is about 60 stars with the brightest about magnitude 11.

NGC2613

©STScI/AURA/CALTECH/ROE

Sagitta

Only one object was noted by Herschel in this small constellation and most observers will have difficulty with it. The brighter, if nondescript, star cluster Harvard 20 less than a degree to the NNW was overlooked by Herschel.

NGC6839

H16[6]	19:54.5	+17 54	AST?
3.0′ × 3.0′			

Observed on 18 August 1784, Herschel described this object as: 'A very small cluster of compressed stars.' In the field are two possibilities, each near a widely separated E–W pair of magnitude 9 field stars. The first is a group of about half a dozen stars located 2.5′ SE of the eastern member of the pair of bright stars. This star is SAO105398, magnitude 8.6. This group was suggested by Brent Archinal in *The 'Non-existent' Star Clusters of the RNGC* (1993). A more convincing possibility is a group of seven resolved stars in a 'blurry' haze situated 2.8′ NNW of SAO105398. It is not likely that this is actually a cluster of stars. Photographs reveal about a dozen very faint stars of wide brightness range at this location.

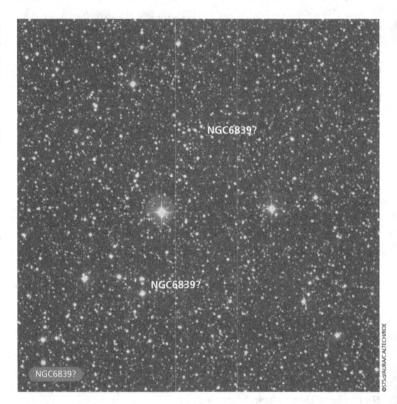

Sagittarius

The 26 objects recorded here by Herschel are for the most part a mix of open and globular clusters, with a few planetary nebulae included for good measure. Charles Messier picked up the brightest and most spectacular globular clusters here and for the northern observer the globular clusters that remain are difficult to resolve due to their southerly location. Observers with moderate- or large-aperture telescopes at lower latitudes may not have this problem. For the rest, there are some interesting objects here, including the rich open cluster NGC6520 and the planetary nebula NGC6818. The nebula/star cluster complex NGC6514 (M20) is notable for the fact that Herschel observed it on at least three occasions, and assigned it four separate catalogue numbers. All the objects were recorded between the years 1784 and 1786.

NGC6440

H150[1]	17:48.9	−20 22	GC
CC5	3.0′ × 3.0′	12.0B	10.1V

Photographs show a bright and very condensed globular cluster with an intense concentration of stars towards the middle. Very symmetrical in outline, it is round with good resolution of individual stars around the edges. It is on the border between a rich star field to the W and a field of dark nebulosity to the E. In a moderate-aperture telescope this globular is fairly bright with a much brighter core which is round and embedded in a grainy, irregularly round, outer disc. It is very slightly extended N–S and is moderately well defined at the edges. The cluster is unresolved; the brightest stars are magnitude 16.7 and the horizontal branch visual magnitude is 18.7. Distance: 23 000 ly.

NGC6445

H586[2]	17:49.2	−20 01	PN
3b+3	2.8′ × 2.8′	13.20B	

In photographs, the brightest portion of this planetary nebula has an interesting, squared off shape. It is slightly elongated SE–NW and measures about 0.9′ × 0.75′. It has a darker centre and the magnitude 19 central star is only suspected. Outside the bright inner ring are chaotic wisps of very faint nebulosity, brightest to the NE. Small apertures show a bar-shaped object with a slightly darker centre. The planetary nebula is well seen in moderate apertures at high magnification as an almost rectangular patch of very uneven brightness. Bright condensations are seen in the NW and S and a tiny knot is visible in the ESE. The centre is just slightly darker, though filled with a gauzy haze. Distance: 4600 ly.

NGC6440

NGC6445

©STScI/AURA/CALTECH/ROE

nebulae. He described it as: 'Three nebulae, faintly joined, form a triangle. In the middle is a double star, very faint and of great extent.' Curiously, however, Herschel observed the object again on 26 May 1786, this time recording it as a Class IV object, though he surely must have recognized it from his previous observations. This time he called the Trifid: 'A double star with extensive nebulosity of different intensity. About the double star is a black opening resembling the nebula in Orion in miniature.' The DSS image records the nebula well, showing it as a bright and complex object, laced with emission and reflection nebulae of different intensities as well as dozens of dark patches and channels, including the three-veined channel which gives the nebula its popular name. John Herschel was the first to call the nebula 'Trifid'. Visually, this object is very rewarding to observe and worth returning to again and again as the slightest change in observing conditions or telescope aperture changes the character of the nebula. Somewhat faint, the nebula is nevertheless easily visible and quite detailed; the 'trifid' dark channels are well seen but only the brighter portions of the nebula are readily visible. The bright star to the N is surrounded by a roundish, ill defined glow separate from the brighter nebula to the S. This glow is the brightest portion of the nebula which shines by reflected light. The double star HD 164492 is well seen where the trifid channels converge and is of spectral type O7. The star cluster is something of an afterthought, overwhelmed as it is by the emission and reflection nebulae, but 70 stars are considered members, the brightest magnitude 6.0 and the spectral type of the cluster is O5. Distance: 5200 ly.

NGC6507

H53[8]	17:59.6	−17 27	OC
IV3m	8.0′ × 7.0′	9.6V	

In a dense Milky Way field photographs show a scattering of brighter stars with a narrow brightness range which is elongated N–S and is very poorly separated from the star field. Telescopically, this is a well defined cluster, with stars of a fairly narrow range of brightness. About 25 stars are visible. The brighter field star WX Sagittari borders the cluster on the NW side. Thirty-five stars are accepted to be members, the brightest being magnitude 12.0.

NGC6514

H41[4]/H10[5]/ H11[5]/H12[5]	18:02.3	−23 02	OC/EN/RN
30.0′ × 30.0′	6.8B	6.3V	

Though the French astronomer Guillaume Bigourdan says this object

was first observed by Le Gentil before 1750, credit for the discovery of the Trifid Nebula goes to Charles Messier, who recorded it as a cluster of stars on 5 June 1764, the twentieth object in his catalogue of clusters and nebulae. On that same night he discovered the open cluster M21 and his description mentioned that both M20 and M21 were 'surrounded by nebulosity'. Herschel first described the object on 3 May 1783, several months before his sweeps for the nebulae began. He recorded it as: 'Two nebulae close together, both resolvable into stars; the preceding however leaves some doubt, though I suppose a higher power and more light would confirm the conjecture. [Telescope used] 10 feet, power 350: the instrument will not bear a higher power in this low altitude.' After Herschel began his sweeps, he formally recorded the Trifid on 12 July 1784, providing it with three separate numbers in his Class V: very large

NGC6520

H7[7]	18:03.4	−27 54	OC
I2r n	5.0′ × 2.0′	7.6V	

The DSS shows a very rich cluster of moderately bright stars, quite condensed, in an extremely rich field,

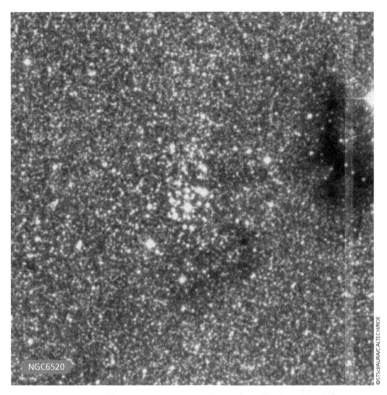

NGC6520

©STScI/AURA/CALTECH/ROE

divided into 2 parts. The most northerly about 15'. The most southerly followed by stars' is probably a visual description of M8, the Lagoon Nebula, which is immediately N of the position Herschel recorded. Herschel had observed the Lagoon Nebula seven weeks previously, on 22 May 1784, and recognized it as M8, so why he recorded the object in July as a new, large nebula is something of a mystery. The discovery of this star cluster and nebula complex has been variously credited to Flamsteed, LeGentil or Lacaille. The DSS image is decidedly overexposed and much of the subtle detail in the nebula is lost. The brightest part is about 20.0' × 20.0' in size and the nebula extends through a wide range of brightnesses, the brightest part being the tiny 'hourglass' formation in the nebula's western regions. Extensive regions of dark nebulosity extend over the whole of the nebula, with the most striking part being the broad gulf separating the brightest portion of the nebula from the embedded star cluster. Visually, the nebula and cluster fill the 30' field of a medium-magnification eyepiece. The brightest portion is to the W and is roughly circular in shape, broadly brighter towards the 'hourglass' which measures about 0.5' × 0.25' and is bordered by a prominent star. The shape can be detected at high magnification in good seeing conditions. Fainter sections of the nebula appear at the periphery in the field and are laced with subtly darker regions; a vaguely loop-like feature is in the W, while faint extensions can be seen to the N. The broad, dark obscuring region is bordered by parallel, more subtle obscuring lanes to the NW. The star cluster is bright and well resolved with more than two dozen members visible, showing a narrow brightness range. The spectral type of the cluster is O5, the age about 2 million years. Distance: 5000 ly.

but well separated from the background with the dark nebula B86 about 5.0' W. Visually, this is a bright and very attractive cluster, well resolved at medium and high magnification. The stars are of moderate brightness range, well compressed to the centre and well separated from the sky background. Many of the stars appear paired; resolution is better at high magnification but the cluster is prettier at a lower power. The cluster is elongated N–S and the dark nebula B86 is well seen to the WNW with a bright, magnitude 7 star just to the west.

members but it is difficult to separate them from the sky background. Visually, this globular cluster and NGC6528 form a contrasting pair in a medium-magnification field. NGC6522 is moderately bright and has an extensive, irregularly round outer halo, which brightens to a small, moderately condensed core. A field star is in the halo to the NE and though somewhat mottled, there is no resolution into individual stars. The brightest members are magnitude 14.1 with the horizontal branch visual magnitude 16.9. Distance: 23 500 ly.

NGC6522

| H49[1] | 18:03.6 | −30 02 | GC |
| CC6 | 3.0' × 3.0' | 10.7V | 9.5V |

The DSS image shows an extremely compressed globular cluster in a very rich Milky Way field. The central region is oblate, elongated E–W, and there is some resolution of outlying cluster

NGC6523

| H13[5] | 18:04.1 | −24 23 | EN/OC |
| 90.0' × 40.0' | 4.6V | | |

This object was originally identified as NGC6533 by Dreyer, based on an inaccurate position provided by Herschel. But Herschel's 12 July 1784 description of the object: 'Extensive milky nebulosity

NGC6526

| H9[5] | 18:04.1 | −23 35 | EN |
| 20.0' × 20.0' | | | |

Located immediately N of the Lagoon Nebula, this large, ill defined glow is readily visible, though fairly faint. Three magnitude 8 and 9 field stars mark the area, as well as many fainter ones. Probably part of the larger complex to the S, the nebula appears vaguely crescent-shaped, elongated roughly E–W with the concave portion to the N. The DSS image shows an extensive and faint nebula, laced with darker patches and channels in a rich Milky Way field.

NGC6528

H200[2]	18:04.8	−30 03	GC
CC5	2.0′ × 2.0′	12.1B	10.6V

Images show a small, very rich globular cluster with a bright and compressed core in a rich Milky Way field. The central portion of the cluster is oblate and elongated NNE–SSW; there is good resolution of stars around the edges, although the cluster is difficult to separate from the rich field. Visually, the cluster appears smaller and a little fainter than its neighbour NGC6522. It is very slightly oval in shape, fairly well defined and very broadly brighter to the middle. There is no hint of resolution; the disc is quite smooth textured. Herschel recorded both NGC6522 and NGC6528 on 24 June 1784. The brightest stars are magnitude 15.5 and the horizontal branch visual magnitude is 17.1. Distance: 21 500 ly.

NGC6540

H198[2]	18:06.3	−27 49	GC
1.5′ × 1.0′	14.6V		

Historically, this object was considered to be an open cluster but was finally proven to be a globular cluster by the astronomer Bica in 1994. The DSS image shows a very rich Milky Way field with a compressed cluster of stars, poorly resolved and elongated E–W with the line of stars on the W end curving to the S. These stars are bright compared with the background field and stand out well, though this may be just a chance alignment and the stars may not be physically related. The globular cluster itself is the small, bright clump at the centre. Herschel discovered this object on 24 May 1784 and called it: 'Pretty

faint, not large, crookedly elongated, easily resolved.' From the description, it seems quite clear that Herschel saw the combined image of the background globular cluster and the foreground grouping of faint stars as a single object. In a moderate-aperture telescope, high magnification reveals a faint, unresolved cluster easily found as an oval, elongated glow, brighter to the middle along the 'cluster' major axis which is oriented E–W. Very well defined, it stands out easily from the background star field. The horizontal branch visual magnitude of the globular cluster is 15.30. Distance: about 11 000 ly.

NGC6544

H197[2]		18:07.3	−25 00	GC
6.0′ × 5.0′	11.4V	9.9V		

The central portion of this globular cluster is quite compressed and round. Photographs show it is surrounded by a broad halo of well-resolved stars, elongated almost due N–S. The field is rich to the E of the cluster, but a large dark patch is located to the W. Visually, the cluster is situated in a fairly rich field but it is somewhat diffuse. The brighter core brightens steadily to the middle and is surrounded by a diffuse and gauzy haze. Individual stars are seen, especially towards the core but they are unevenly distributed. The brightest stars are magnitude 12.8 and the horizontal branch visual magnitude is 14.9. Distance: 8200 ly.

NGC6553

H12[4]	18:09.3	−25 54	GC
CC11	5.0′ × 5.0′	9.1B	

In a rich Milky Way field, photographs show that this is a very rich globular cluster, very symmetrical in outline. The core is quite round and there is good resolution around the edges but it is extremely difficult to separate the

NGC6540

©STScI/AURA/CALTECH/ROE

NGC6544

NGC6553

the triplet at the centre of the grouping. There are thought to be 100 stars that are members of the cluster.

NGC6568

H30[7]	18:12.8	−21 36	OC
IV1m	12.0′ × 12.0′	8.6V	

Images show a scattered cluster of stars of moderate brightness range, slightly elongated E–W and not well separated from the sky background. Visually, this rather coarse cluster is located in a fairly rich field. Most of the cluster members are fainter than magnitude 11 and form successive curving rows of stars, oriented roughly N–S. Fifty stars are accepted as members of this cluster.

NGC6569

H201[2]	18:13.6	−31 50	GC
CC8	4.0′ × 3.5′	10.8V	9.5V

Photographs reveal a rich globular cluster with a small, bright, compressed core which is almost round and surrounded by a dense halo of well resolved stars, somewhat extended E–W. The star field is rich and separation of the cluster from the background is difficult. Telescopically, this globular cluster is moderately bright and very slightly oval. It is fairly well defined at the edges but smooth-textured and only slightly and very steadily brighter to the middle. There is no resolution into individual stars, the brightest of which are magnitude 14.9. The horizontal branch visual magnitude is 17.9. Distance: 28 000 ly.

NGC6583

H31[7]	18:15.8	−22 08	OC
I2m	5.0′ × 3.0′	10.0V	'

This is a fairly rich cluster of stars of moderate brightness range in a rich field. Photographs show the brightest members form a N–S chain of stars which runs straight through the cluster. Visually, this is a very rich, though faint open cluster, appearing initially as a nebulous haze of light. High magnification brings some resolution, though the brightest members are

cluster from background star field. Telescopically, this large cluster is fairly well defined at the edges and is seen as a slightly oval glow which is very slowly brighter to the middle. At the edges there is a sudden drop in brightness and no individual stars are visible. The brightest members are magnitude 15.3

and the horizontal branch visual magnitude is 16.9. Distance: 11 500 ly.

NGC6561

H54[8]	18:10.5	−16 48	OC
15.0′ × 10.0′	8.3V		

Photographs show a large, scattered cluster of stars of moderate brightness range arrayed around a tight triplet of bright field stars. There is a prominent chain of faint stars running NNE–SSW on the NW flank. A faint patch of nebulosity is 5.5′ WSW of the central triplet. Visually, the cluster sits in a fairly rich field and is well seen as a somewhat well defined cluster of about 25–30 stars, with a very wide range in brightness and dominated by a triangle of magnitude 8 stars, seen within a 10.0′ × 10.0′ area. Discovered 27 June 1786, Herschel described the cluster as: 'A coarse scattered cluster of considerably large stars. The place is that of a small triangle.' This triangle is

©STScI/AURA/CALTECH/ROE

fainter than magnitude 13 and appear as very fine star dust. Overall, the cluster is elongated N–S and is well separated from the sky background. Thirty-five stars are recognized as members of this cluster.

NGC6596

H55[8]	18:17.5	−16 40	OC
II2m n	10.0′ × 10.0′		

The DSS image shows a scattered cluster of brighter stars involved in faint nebulosity. The cluster is richest in the N and a few members form a partial ring to the S, leaving the centre of the cluster relatively starless. Visually, this is a large, coarse cluster of stars in a fairly rich field. The brightness range of the members is moderate; about 30 stars are resolved in a 15′ field. There is a prominent starless zone in the middle, with the cluster members forming a rough circle, open to the E. The nebulosity is not seen visually. Thirty stars are accepted as members.

NGC6624

H50[1]	18:23.7	−30 22	GC
CC6	4.0′ × 4.0′	10.2B	9.1V

Photographs show a rich and compressed globular cluster in a rich star field. It is very compressed to a round core, with good resolution of stars in the outer halo. It is quite symmetrical, but is difficult to separate from the star field. Telescopically, the cluster is round, small and quite bright; the outer halo is fairly well defined and brightens quickly to a sizable bright core and a brighter, star-like nucleus. Although the disc is a little grainy, there is no resolution into individual stars, the brightest of which are magnitude 14.0. The horizontal branch visual magnitude is 16.1. Distance: 26 500 ly.

NGC6629

H204[2]	18:25.7	−23 12	PN
0.3′ × 0.3′	11.9B	9.9V	

The DSS image appears a little overexposed and in a very rich star field this is a round, bright and opaque

nebula with no central star visible. Visually, however, this planetary nebula is small and has quite high surface brightness, though even at medium magnification it can momentarily be considered just another field star. It is round and well defined with a slightly bluish, even-surface-brightness disc and the faint central star (HD169460, magnitude 12.8) is well seen at high magnification. Recorded on 24 August 1784 as a faint nebula, Herschel described it as: 'Pretty bright, small, stellar, not verified.' Distance: 6200 ly.

NGC6638

H51[1]	18:30.9	−25 30	GC
CC6	7.3′ × 9.3′	10.8B	9.7V

The DSS image shows a rich, symmetrical globular cluster with a small, compressed core and an extensive, well-resolved outer envelope. The star field is extremely rich so separation of the cluster and the background is difficult. Visually, this is a moderately bright globular cluster, though small and rather condensed. It is irregularly round with a bright core fading slowly to a poorly defined outer envelope. There is some graininess throughout but no resolution at medium magnification; the brightest stars of this cluster are magnitude 14.2 and the horizontal branch visual magnitude is 16.5. Herschel thought the cluster was easily resolvable. Spectral type: G2. Radial velocity in approach: 14 km/s. Distance: 28 000 ly.

NGC6642

H205[2]	18:31.9	−23 29	GC
CC6	4.0′ × 4.0′	11.3B	10.2V

The DSS image shows a small, rich globular cluster with a compressed, but irregularly shaped core. It displays an extensive outer envelope of well resolved stars, with two distinct chains of stars emerging, one to the WSW and the other to the SW. In a moderate-aperture telescope, this cluster is small and moderately faint and is much brighter to a well defined, condensed

core. There is some hint of graininess at the extremities at high magnification; the outer envelope appears elongated slightly N–S. There are two barely resolved cluster members here, one N and the other S of the core. A few faint stars are resolved in the halo. The horizontal branch visual magnitude is 16.3. Radial velocity in approach: 90 km/s. Distance: 26 000 ly.

NGC6645

H23[6]	18:32.6	−16 54	OC
IV1m	15.0′ × 15.0′	8.8V	

This is a rich concentration of moderately faint stars set in a brilliant Milky Way field. Discovered on 27 June 1785, Herschel described it as: 'A beautiful cluster of very small stars of various sizes. 15′ diameter very rich.' The cluster is most attractive at low magnification which helps to define its boundaries. Medium magnification aids resolution, but some of the beauty is lost. The chief defining feature is a ring of stars N of the cluster centre. To the S is a bright, widely separated triplet and NNE of this is a faint haze involving about five resolved stars. Forty stars are accepted as cluster members. Spectral type: B9.

NGC6647

H14[8]	18:31.5	−17 21	MWP
20.0′ × 8.0′	8.0V		

In Archinal and Hynes' *Star Clusters* (2003) this entry is described as a Milky Way patch. Telescopically, at the indicated position there is a coarse, scattered row of field stars, greatly elongated in a roughly N–S direction. Half a dozen bright stars and perhaps another dozen fainter ones are nearby. A small, faint, seven-star asterism is located to the SW.

NGC6698

H15[6]	18:48.2	−25 55	MWP
25.0′ × 25.0′			

Herschel recorded this object on 12 July 1784, as: 'A suspected cluster of very faint stars of considerable extent. Not verified.' In both the *New General Catalogue* and *The Scientific Papers* a typo identifies the object as NGC6678 instead of NGC6698. It is very probably a Milky Way patch. Visually, this object is a brighter patch in the Milky Way, about 25′ in diameter with about 35 stars resolved, showing a moderate range in brightness. The faint planetary nebula PK9-10.1 borders this scattered group on the E and is well seen at medium magnification in moderate-aperture telescopes.

NGC6717

H143³	18:55.1	−22 42	GC
CC8	2.6′ × 2.0′	11.3B	10.3V

This is a very small globular cluster located immediately S of the bright field star 35 Sagittari. Photographs show the centre is rich, moderately compressed and oblate, elongated ENE–WSW with an extensive halo of well-resolved stars, mostly to the S. The cluster is also catalogued as Palomar 9. Visually, this is an intriguing object; it is a small and fairly concentrated globular cluster which is very much brighter to the middle. The presence of the magnitude 6 field star 1.75′ N greatly hinders observation. The cluster is flanked immediately W and NE by magnitude 11 field stars. The envelope of the globular cluster is small, faint and difficult to see. The brightest cluster members are magnitude 14 and the horizontal branch visual magnitude is 15.6. Herschel discovered this object on 7 August 1784 and called it: 'Three very small stars with suspected nebulosity.' Radial velocity in recession: 14 km/s. Distance: 24 000 ly.

NGC6717

©STScI/AURA/CALTECH/ROE

NGC6818

H51⁴	19:44.0	−14 09	PN
4	0.5′ × 0.5′	9.90B	9.3V

The DSS image shows a very bright planetary nebula, but unfortunately the image is overexposed so that no structural detail is visible. It is a round, opaque, very dense nebula with no central star visible. At high magnification in moderate-aperture telescopes, this small planetary nebula is bright, with a dense and opaque shell. It is quite round and there is a slight drop in the brightness of the shell at the extremities, while the interior is just slightly fainter, offset to the SW. The central star, spectral type WR, is about magnitude 15 but is not visible visually. Herschel discovered this object on 8 August 1787, and described it as: 'A considerably bright, small, beautiful planetary nebula; but considerably hazy on the edges, of a uniform light; 10″ or 15″ in diameter, perfectly round. I shewed it to Monsieur de la Lande.' The expansion velocity of the nebula is 30 km/s. Radical velocity in approach: 14 km/s. Distance about 5200 ly.

Scorpius

Most of the notable objects in this constellation were too far south for Herschel to record and surprisingly, this major constellation is home to only three Herschel objects, which were recorded in the years 1784 and 1786.

NGC5998

H29[7]	15:49.4	−28 36	OC/AST?
3.5′ × 2.0′			

Recorded on 30 April 1786, Herschel called this object: 'A cluster of very small stars, pretty rich 6′ long, 4′ broad in the form of a parallelogram.' Photographs show a tight grouping of about 10 stars, broader to the S with a chain of stars curving N, then W. Visually, the asterism is quite dim and is seen as a nebulous patch of light at medium magnification. High magnification resolves the individual stars; seven are seen in all. This group may not be a physical cluster.

NGC5998

NGC6144

H10[6]	16:27.3	−26 02	GC
CC11	7.4′ × 7.4′	10.5B	9.6V

Images reveal a fairly faint loose-structured globular cluster which is poorly compressed to the middle. It is irregularly round and fairly well resolved in the core and around the edges.

Visually, the cluster is a faint and diffuse object located due E of a magnitude 10 field star. Hazy and ill defined, the globular cluster is a little brighter to the middle. It is slightly oval and elongated almost due N–S. A few of the brighter stars are resolved against a grainy, nebulous background. Herschel observed the cluster on 22 May 1784 and called it: 'A very compressed and considerably large cluster of the smallest stars imaginable. All of a dusky red colour. The next step to an easily resolved nebula.' The brightest members of the cluster are magnitude 13.4. Spectral type: G0. Radial velocity in recession: 130 km/s. Distance: 32 000 ly.

NGC6451

H13[6]	17:50.7	−30 13	OC
I2r n	8.0′ × 6.0′	8.2V	

In a very rich star field, photographs show a rich, quite compressed, moderately bright cluster of stars of moderate range of brightness. The main body is oval in shape, elongated NNE–SSW and is fairly well separated from the sky background. Telescopically, the cluster is easily seen at low magnification and is large and moderately bright. About two dozen stars are easily resolved and they are fairly similar in brightness. Most of the resolved members are parts of chains of stars which delineate the cluster boundaries. The cluster is rich with faint, unresolved members contributing to a very weak, unresolved glow. Eighty stars are considered members, the brightest being magnitude 12. Spectral type: A. Distance: about 1800 ly.

NGC6451

Sculptor

The five Herschel objects recorded in this southern constellation were discovered in the years 1783, 1785 and 1798. Each one is a notable object, in particular the spectacular nearby galaxy NGC253, which was discovered by Caroline Herschel.

NGC24

H461[3]	0:09.9	−24 58	GLX
SA(s)c	6.3′ × 1.3′	11.3	

This is a highly inclined spiral galaxy. The DSS image shows a bright core embedded in a bright inner disc and a probable multi-arm spiral pattern with many dust patches involved. Stars begin to resolve in the outer disc at about blue magnitude 21. Visually, this is a large and well defined galaxy, somewhat diffuse but bright along its major axis. It is a long cigar-shaped oval of light with extensions which taper to blunt points. The galaxy is oriented NE–SW and a field star is visible ENE of the NE tip. Radial velocity: 575 km/s. Distance: 26 mly. Diameter: 47 000 ly.

NGC253

H1[5]	0:47.6	−25 17	GLX
SAB(s)c	27.5′ × 6.8′	7.93B	7.1V

The credit for the discovery of this superb spiral galaxy goes to Caroline

Herschel, who recorded the object on 23 September 1783. A week after he formally began his sweeps of the heavens, on 30 October 1783, Herschel listed the object in his Large Nebulae class as: 'Considerably bright, much extended south preceding north following, much brighter to the middle. Above 50′ long and 7 or 8′ broad. C. H. See note.' In the note section to Herschel's Catalogue of One Thousand New Nebulae and Clusters of Stars (*Phil. Trans* **lxxix**, 1789) he states: 'This nebula was discovered September 23, 1783, by my sister Caroline Herschel,

with an excellent small Newtonian Sweeper of 27 inches focal length, and a power of 30. I have therefore marked it with the initial letters, C. H. of her name.' In the DSS image unfortunately much of the detail of this highly inclined spiral galaxy is burnt out, but something of its complex structure can be appreciated in the outer detail of its tightly wound spiral arms, which are studded with dust patches, bright HII regions, star clusters and associations. High-resolution photographs show a very small central region, which lacks a central bulge, and the spiral structure

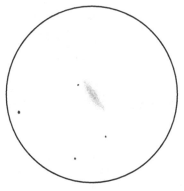

NGC24 15-inch Newtonian 272x

NGC24

©STScI/AURAC.ALTECHROE

NGC253 15-inch Newtonian 195x

can be traced directly into the core. Resolution of individual stars begins at blue magnitude 18. The galaxy is probably the principal member of the South Polar Group of galaxies which includes NGC55, NGC300 and NGC7793 as well as the Herschel object NGC247. Visually, this is one of the finest galaxies available for inspection by amateur astronomers. In a large aperture, it is superb: it is a large and very bright galaxy much extended SW–NE. The core is large, bright and oval with dust patches to the NE and a field star is situated immediately NE of the core. An elongated condensation is visible to the SW between two field stars. The disc is fairly bright in the central two thirds but is laced with dust patches; it fades slightly at the extremities but overall it is well defined, though a little fainter along the NW rim. Radial velocity: 244 km/s. Distance: 10.9 mly. Diameter: 87 000 ly.

NGC288

H20[6]	0:52.8	−26 35	GC
CC10	13.0′ × 13.0′	10.0B	9.1V

The DSS image shows a large, somewhat loosely structured globular cluster, which is a little compressed towards the middle but not dense. An extensive well-resolved halo of stars, symmetrical in form, surrounds the core and the cluster is well separated from the sky background. Telescopically, this is a large, moderately bright, quite loosely structured object with some resolution throughout. The unresolved background glow is broadly triangular in shape with the apex towards the NNW; this shape is defined by three slightly brighter stars which mark the corners of the triangle. Recorded on 27 October 1785, Herschel described the cluster as: 'Considerably bright, irregularly round, 8 or 9′ in diameter. A great many of the stars visible, so that there can remain no doubt but that it is a cluster of very small stars.' The absolute visual magnitude is −6.6 and the horizontal branch visual magnitude is 15.3, suggesting the cluster is well resolved in moderate-aperture telescopes. Radial velocity in approach: 48 km/s. Distance: 27 400 ly.

NGC613

H281[1]	1:34.3	−29 25	GLX
SB(rs)bc	5.5′ × 4.2′	10.8B	10.74V

Well seen in the DSS image, this barred spiral galaxy has a large core embedded in an oval bar. There are five principal spiral arms, which are fairly massive and well populated with large HII regions. As in many barred spirals, long dust lanes can be seen on either side of the core at the leading edge of rotation of the galaxy. Telescopically, the galaxy is bright and quite large and has a large, slightly elongated core embedded in a bright, elongated envelope. The edges of the envelope are ill defined and the galaxy is oriented ESE–WNW with a bright field star located 2.25′ NNE of the core. Radial velocity: 1444 km/s. Distance: 65 mly. Diameter: 103 000 ly.

NGC7507

H2[2]	23:12.1	−28 32	GLX
E0	3.0′ × 2.9′	11.38B	

This is a prototypical E0 system. Photographs show a large, bright and round core surrounded by a fainter secondary envelope and a very faint,

NGC253

NGC288

©STScI/AURA/CALTECH/ROE

diffuse outer envelope. The visual appearance is quite similar; it is a bright and moderately large elliptical galaxy featuring a large, bright and well-condensed core which brightens quickly to the middle and is embedded in a fainter secondary envelope, which is diffuse and ill defined. Radial velocity: 1597 km/s. Distance: 71 mly. Diameter: 62 000 ly.

Scutum

Although this Milky Way constellation is littered with nondescript star clusters and obscure planetary nebulae, Herschel overlooked them all except for three objects, a scattered star cluster (NGC6664), a probable asterism (NGC6728) and a significant globular cluster (NGC6712), all recorded in 1784.

NGC6664

H12[8]	18:36.7	−8 13	OC
III2m	12.0′ × 12.0′	8.8B	7.8V

Photographs show a scattered cluster of bright stars in a rich Milky Way field. Visually, this is a rather coarse cluster at low magnification, located due E of Alpha Scuti; medium magnification brings out several fainter members. The main body of the cluster consists of an E–W wedge of stars to the N. A spur of stars to the E is elongated N–S with the cluster members being magnitude 9 or fainter. Sixty stars are accepted as members of this cluster. Spectral type: B3. Age: about 140 million years. Radial velocity in recession: 23 km/s. Distance: about 4500 ly.

NGC6712

H47[1]	18:53.1	−8 42	GC
CC9	9.8′ × 9.8′	9.8B	8.7V

Discovered on 16 June 1784, Herschel described this globular cluster as: 'Bright, very large, irregularly faint, easily resolvable, stars visible.' Nevertheless, he assigned it to his Class I: bright nebulae. Photographs show a symmetrical, compressed cluster in a rich Milky Way field. Visually, this is a bright, attractive globular; the main envelope is somewhat triangular in outline with the apexes located to the N, SW and SE. The core is quite bright and many faint cluster members are visible across the unresolved body of the cluster, which is set in a rich Milky Way field. The brightest members are magnitude 13.3 and the horizontal branch visual magnitude is 16.3. Spectral type: G1. Radial velocity in approach: 124 km/s. Distance: about 22 000 ly.

NGC6728

H13[8]		18:57.5	−8 58	AST
18.0′ × 5.0′				

On the DSS image this is an elongated grouping of bright stars in a very rich stellar field. Visually, this is a fairly obvious grouping of stars with a moderate range in brightness, situated W of a magnitude 7 field star. The members are magnitude 10 or fainter, and the grouping is much elongated E–W and well separated from the sky background. About 20 stars are resolved, but they may not form an actual cluster.

NGC6712

©STScI/AURA/CALTECH/ROE

Serpens

This sprawling constellation is home to 22 Herschel objects, all galaxies except for the star cluster NGC6604. Though none of the galaxies is spectacular visually, most are at least moderately bright and should not pose a problem for most observers. All the objects were discovered between the years 1784 and 1787 except for NGC5928, which was recorded in 1791.

NGC5910

H400[2]	15:19.3	+20 55	GLX
0.9′ × 0.8′	14.61B		

Though described as a triple system, the DSS image shows the bright central galaxy of this group appears to have a satellite seen in silhouette immediately W of the core, making this tight group a quartet. Though the three smaller galaxies appear within NGC5910's halo there is no sign of tidal interaction. The principal galaxy is probably elliptical and features a bright, round core. There are several faint galaxies in the immediate field including UGC9813, a barred spiral located 7.75′ WSW. Telescopically, NGC5910 is quite faint and is best seen at medium magnification as a very slightly elongated patch of light, oriented E–W with a very faint star preceding the galaxy in the W. The surface brightness is even and the disc is fairly well defined. Radial velocity: 12 331 km/s. Distance: 550 mly. Diameter: 144 000 ly.

NGC5913

H374[3]	15:20.9	−2 35	GLX
SB(r)a	2.6′ × 1.0′	14.02B	

The DSS photograph shows an inclined, barred spiral galaxy with a bright core and a slightly fainter bar attached to an almost complete pseudo-ring structure. The ring is embedded in a very faint, extensive envelope, oriented NNW–SSE, and is brightest where it attaches to the bar. The broad, faint outer spiral pattern increases the size of the major axis to 2.6′. At medium magnification the galaxy is quite faint and diffuse; it is small, round and best seen at medium magnification. Radial velocity: 2013 km/s. Distance: 90 mly. Diameter: 68 000 ly.

NGC5921

H148[1]	15:21.9	+5 04	GLX
SB(r)bc	4.9′ × 4.0′	11.68B	

This is a beautiful example of an almost face-on, barred spiral galaxy and is well seen on the DSS image. The galaxy has a large, bright core and a fainter bar; high-resolution photographs show classic straight dust lanes along the bar. Two bright, narrow and tightly wound spiral arms with several knotty condensations emerge from the bar. After about half a revolution each, the arms fragment to a broader, less tightly wound spiral structure. Telescopically, this galaxy is large and quite bright with several field stars nearby, including one 1.1′ SW of the core. The galaxy has a very bright, almost stellar core, surrounded by a fairly bright central region. The outer envelope is grainy but bright, moderately well defined and elongated NNE–SSW. Radial velocity: 1513 km/s. Distance: 67 mly. Diameter: 96 000 ly.

NGC5928

H874[2]	15:25.9	+18 05	GLX
S0	1.6′ × 1.0′	13.75B	

This lenticular galaxy presents similar aspects visually and photographically. It is moderately bright at high magnification, a small oval and opaque object, elongated E–W with well defined edges, brightening steadily to the middle. A bright field star is 5.75′ to the NNE. Radial velocity: 4589 km/s. Distance: 205 mly. Diameter: 95 000 ly.

NGC5936

H130[2]	15:30.0	+12 59	GLX
SB(rs)b	1.4′ × 1.3′	13.05B	

Photographs show a face-on, barred spiral galaxy with a bright, round core surrounded by a bright, partial ring structure and knotty spiral arms. In a medium-magnification field, the galaxy can be found just S of an imaginary line connecting an optical triple in the W and a bright equal pair in the E. The galaxy is a bright and well defined, circular glow with a thin, fainter outer envelope. Although no core is visible, the central region is quite granular in texture. Radial velocity: 4065 km/s. Distance: 182 mly. Diameter: 74 000 ly.

NGC5937

H401[2]	15:30.8	−2 49	GLX
(R′)SAB(rs)b pec	1.9′ × 1.0′	13.08B	

Photographs reveal a slightly inclined spiral galaxy with a bright, elongated central region and two bright inner arms which expand to fainter, broader and asymmetrical outer arms, elongated NNE–SSW. Field stars are 2.1′ NE and 3.4′ NNE of the core. Telescopically, this galaxy is fairly bright and is seen as a round object with a bright core in an opaque, fairly well defined disc. Radial velocity: 2822 km/s. Distance: 126 mly. Diameter: 70 000 ly.

NGC5951

H654[2]	15:33.7	+15 00	GLX
SBc: sp	3.5′ × 0.8′	13.65B	

Images show a highly inclined spiral galaxy with a brighter, elongated central region and very faint, grainy spiral arms, elongated N–S. The galaxy pair NGC5953 and NGC5954 lie to the NE. In moderate apertures, this is a faint, though attractive object. A flat, much elongated streak of light, it is well defined and bright along its major axis. No core is visible. Radial velocity: 1849 km/s. Distance: 82 mly. Diameter: 84 000 ly.

NGC5953

H178[2]	15:34.5	+15 12	GLX
SAa: pec	1.6′ × 1.3′	12.99B	

The interacting pair NGC5953 and NGC5954 has also been catalogued as Arp 91. Photographs reveal that NGC5953 is a possible lenticular galaxy of peculiar morphology. The galaxy has a large, bright, squarish core and an extensive outer disc. The faint, irregular outer envelope features a very faint, triangular plume to the NNW. A faint field star or compact object is 0.75′ SW and NGC 5954 is 0.5′ NE. In medium-aperture telescopes, at high magnification, these two galaxies appear to be almost in contact. Visually, NGC5953, the brighter and larger galaxy, is round with a diffuse outer envelope. The core is bright, though mottled, and features a prominent nucleus. Herschel came upon this pair on 17 April 1784 and he combined his comments in his description: 'Two. Very close. Both small, stellar. The south (NGC5953) is largest.' Radial velocity: 2036 km/s. Distance: 91 mly. Diameter: 42 000 ly.

NGC5954

H179[2]	15:34.6	+15 12	GLX
SAB(rs)cd:	1.3′ × 0.6′	12.81B	

This inclined spiral galaxy, which is elongated N–S, has a peculiar morphology and forms an interacting pair with NGC5953. Images reveal a brighter core embedded, offset towards the E in the spiral structure. There is a bright, knotty spiral arc on the E side which fades and broadens as it curves to the W and N. A faint spiral plume in the S extends westward to the companion galaxy. Visually, this galaxy appears smaller than NGC5953 though it is, in fact, larger. Little more than the core is visible; it is a little fainter than its companion and is seen as a well defined, very slightly oval patch which is elongated N–S. Radial velocity: 2030 km/s. Distance: 91 mly. Diameter: 34 000 ly.

NGC5962

H96[2]	15:36.5	+16 37	GLX	
SA(r)c	3.0′ × 2.1′	12.09B	11.3V	

Photographs show a slightly inclined spiral galaxy with a brighter, elongated core. Multiple spiral arms emerge which are thin and weak but have several brighter condensations. When Herschel discovered this galaxy on 21 March 1784, he thought it resolvable; presumably he was detecting the knotty condensations. Visually, this galaxy is easily visible even at low magnification. It is quite bright and is elongated ESE–WNW. The outer edges fade slowly into the sky background and the central region is bright and fairly well defined, brightening quickly to the core. Radial velocity: 2033 km/s. Distance: 91 mly. Diameter: 79 000 ly.

NGC5970

H76[2]	15:38.5	+12 11	GLX
SB(r)c	2.9′ × 1.9′	11.99B	

This slightly inclined, barred spiral galaxy has an elongated, brighter core and bar attached to spiral arms which are thin but have several knotty condensations. A faint lenticular galaxy is 8.25′ SE. Visually, this bright galaxy is

NGC5953/5954

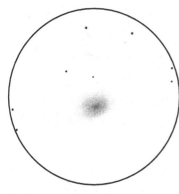

NGC5962 15-inch Newtonian 272x

located 5.4′ SW of a magnitude 8 field star. Elongated E–W, the galaxy has a bright core embedded in a bright bar which is oriented along the major axis. The outer envelope is poorly defined, but bright and more oval in shape than the central region. The bar itself is mottled and a faint condensation is visible immediately E of the core. Radial

velocity: 2021 km/s. Distance: 90 mly. Diameter: 76 000 ly.

NGC5980

H655[2]	15:41.4	+15 47	GLX
S	1.9′ × 0.7′	13.46B	

This highly inclined spiral galaxy probably involves a bar in the inner structure. Photographs show an elongated, bright, central region in a grainy spiral structure, oriented N–S. Visually, the galaxy is quite similar to NGC5951, though smaller and a little brighter. It is brighter along its major axis and is well defined but lacks a bright central region. Radial velocity: 4169 km/s. Distance: 186 mly. Diameter: 103 000 ly.

NGC5984

H656[2]	15:42.9	+14 14	GLX
SB(rs)d	2.9′ × 0.8′	13.24B	

Photographs show an almost edge-on barred spiral galaxy with a thin,

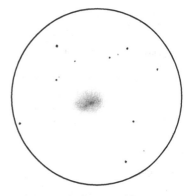

NGC5970 15-inch Newtonian 272x

NGC5970

©STScI/AURA/CALTECH/ROE

elongated, knotty, brighter central region. Visually, the galaxy is fairly faint and small; it is moderately well defined and oriented SE–NW. It is very gradually brighter along the major axis, the envelope is smooth in texture and no brighter core is visible. A faint, elongated triangle of stars is visible immediately NNE. Radial velocity: 1181 km/s. Distance: 53 mly. Diameter: 44 000 ly.

NGC5990

H425[2]	15:46.4	+2 24	GLX
(R')Sa pec?	1.5′ × 0.9′	13.17B	

Photographs show a slightly inclined spiral galaxy with a peculiar morphology. A very bright, kidney-bean-shaped core is embedded in a somewhat disrupted spiral structure with a dust patch ESE of the core. The

NGC5962

©STScI/AURA/CALTECH/ROE

diffuse, grainy envelope is elongated ESE–WNW and the very faint galaxy CGCG050-098 is 2.4′ NNW. Visually, the galaxy is fairly bright though small at medium magnification and is seen as an irregularly round patch of light which is slightly brighter to the middle, grainy textured and fairly well defined. Radial velocity: 3881 km/s. Distance: 173 mly. Diameter: 76 000 ly.

NGC5996

H97[2]	15:46.9	+17 52	GLX
SBc	1.7′ × 0.9′	13.21B	

Photographs reveal a face-on, barred spiral galaxy with a somewhat peculiar morphology; the galaxy has also been catalogued as Arp 72. There is no bright core; instead, a bright bar with flares on the ends leads to two narrow spiral arms. The arm to the S is long and straight except for a 'dog-leg' kink, which angles the outer spiral arm towards the SW. The northern spiral arm is a little more diffuse and involves a large crescent-shaped condensation. The faint field galaxy NGC5994 is immediately W of the tip of the southern spiral extension. In a moderate-aperture telescope, the galaxy is moderately bright and is gradually brighter to the middle with no core visible. The edges are diffuse and overall the galaxy is a roundish patch with an equal-brightness pair of stars immediately NNE. Herschel thought this galaxy was resolvable. Radial velocity: 3383 km/s. Distance: 151 mly. Diameter: 75 000 ly.

NGC6010

H583[2]	15:54.4	+0 32	GLX
S0/a: sp	2.2′ × 0.4′	13.14B	

Photographs show an almost edge-on lenticular galaxy with a bright, elongated core and slightly fainter, tapered extensions. The galaxy is small but fairly bright visually, appearing as a well defined streak of light, elongated ESE–WNW, with a bright star-like core. Radial velocity: 1930 km/s. Distance: 86 mly. Diameter: 55 000 ly.

NGC5996

NGC6012

H657[2]	15:54.2	+14 35	GLX	
(R)SB(r)ab:	3.5′ × 3.0′	12.87B	12.4V	

The DSS image shows a slightly inclined spiral galaxy with a peculiar morphology. A small, brighter, round core is embedded in an elongated, bright bar with a dust lane situated S of the core. There are several knotty condensations in this bar, the brightest being NNW of the core. Faint, knotty spiral arms form a pseudo-ring and very faint outer arms emerge from the pseudo-ring. These outer arms extend the apparent size to 3.5′ × 3.0′. A bright field star is 2.0′ S while a bright double star is 1.25′ NE. In a moderate-aperture telescope, the galaxy is prominent, though ill defined, and oriented SSE–NNW. It is oval in shape and the edges fade uncertainly into the sky background. The central region is only a little brighter to the middle. Radial velocity: 1935 km/s. Distance: 86 mly. Diameter: 88 000 ly.

NGC6018

H646[3]	15:57.5	+15 51	GLX
(R)SAB(s)0⁺:	1.4′ × 0.7′	14.34	

Photographs show an inclined spiral galaxy with a large, elongated, brighter core and very faint, open spiral structure, elongated ENE–WSW. A faint field star or compact object borders the core to the NE. Companion galaxy

NGC6012

NGC6021 is 5.0′ N. Visually, only the core of this galaxy is visible as a round, ill defined object which is a little brighter to the centre. Radial velocity: 5305 km/s. Distance: 237 mly. Diameter: 96 400 ly.

NGC6021

H73[3]	15:57.5	+15 56	GLX
E[+]	1.4′ × 0.8′	14.08B	

This is an elliptical galaxy with a bright, slightly elongated core in a fainter outer envelope, elongated SSE–NNW. It can be seen in the same high-magnification field as NGC6018 and appears very slightly fainter, with ill defined edges. There is a moderate difference in the radial velocities of the two galaxies and they may not be physically related. Radial velocity: 4829 km/s. Distance: 216 mly. Diameter: 88 000 ly.

NGC6070

H553[3]	16:10.0	+0 43	GLX
SA(s)cd	3.5′ × 1.9′	12.42B	

This galaxy is well seen in the DSS image as an inclined spiral galaxy with a large, slightly elongated core and bright, knotty spiral arms with many small, round condensations involved. Several very faint galaxies are in the field, including NGC6070A 4.1′ NE and NGC6070B 5.5′ NE. A bright field star is 8.0′ NNW. Telescopically, the galaxy is somewhat dim but large, appearing as a fat oval, oriented ENE–WSW with a brighter, elongated, central region in a fainter disc that fades slowly to somewhat hazy edges. Radial velocity: 2056 km/s. Distance: 92 mly. Diameter: 94 000 ly.

NGC6118

H402[2]	16:21.8	−2 17	GLX
SA(s)cd	4.7′ × 2.0′	12.44B	

Images reveal an inclined spiral galaxy with a small, elongated core embedded in a multiple-arm-spiral structure, elongated NE–SW. A few condensations are involved and a thin dust band is N of the core. Visually,

NGC6070

the galaxy is fairly large, but the edges are poorly defined. It is broadly brighter to the middle but it lacks a stellar core. Radial velocity: 1623 km/s. Distance: 72 mly. Diameter: 99 000 ly.

NGC6118

NGC6604

H15[8]	18:18.1	−12 14	OC
I3m n	6.0′ × 6.0′	7.1B	6.5V

The DSS photograph reveals a bright cluster of stars of wide brightness range in a fairly rich star field that is also bathed in light and dark nebulosity. Visually, the cluster is small and fairly coarse and surrounds a magnitude 7.48 luminary. About ten stars are resolved; except for the luminary they exhibit a narrow range in brightness and are arranged in a rough V shape with the apex pointing to the S. The cluster appears to be involved in very faint nebulosity, but this may also be very faint, unresolved cluster members. The total membership of this very young cluster is 105 stars. The cluster may be a part of the Serpens OB2 association. Spectral type: O9. Age: 4 million years. Distance: 6800 ly.

Sextans

All 14 Herschel objects recorded here are galaxies discovered between the years 1783 and 1786. Most of them are prominent objects and the spindle galaxy NGC3115 is notable, as are NGC3166 and NGC3169.

NGC2967

H275[2]	9:42.1	+0 20	GLX
SA(s)c	3.0′ × 2.8′	12.28B	

Photographs show this face-on spiral galaxy has a small, brighter core and bright, inner, spiral structure. The spiral structure is of the multi-arm type and the arms are dense with several bright knots. Beyond the bright inner structure are extensive, much fainter, loosely wound outer spiral arms. In a moderate-aperture telescope, the galaxy is bright and moderately large and is brighter to the middle to a large core. The envelope is mottled, betraying the condensations along the massive spiral arms, and the edges are poorly defined. Radial velocity: 1742 km/s. Distance: 78 mly. Diameter: 68 000 ly.

NGC2969

H527[3]	9:41.8	−8 37	GLX
SA(s)c pec	1.3′ × 1.2′	13.78B	

In photographs this is a face-on spiral galaxy with a small, bright core; two arms emerge from the core with the eastern arm branching into two after less than quarter of a revolution. Visually, the galaxy is a faint, oval patch of light, which is even in surface brightness, elongated ESE–WNW and moderately well defined at the edges. Radial velocity: 4783 km/s. Distance: 214 mly. Diameter: 81 000 ly.

NGC2974

H61[1]	9:42.6	−3 42	GLX
E4 or SA:(s:)0°	3.4′ × 2.1′	11.88B	

This galaxy is catalogued as a prototypical E4 elliptical with a large bright central region and an outer disc with at least two different brightness gradients; it is faintest at the edge. However, the DSS image reveals graininess in the inner and outer envelopes as well as suggestions of dust patches or lanes which seem to trace thin spiral patterns. The outer envelope extends well beyond the bright field star situated 0.75′ SW of the core. Visually, this is quite a bright galaxy with a bright field star involved to the SW. The inner region is bright and mottled and is brighter to a small core. There is a sharp drop off in brightness to a hazy, outer envelope, which is ill defined and extends past the field star. The envelope is elongated NE–SW. Radial velocity: 1908 km/s. Distance: 85 mly. Diameter: 85 000 ly.

NGC2979

H521[3]	9:43.1	−10 23	GLX
(R′)SA(r)a?	1.6′ × 1.1′	14.0B	

In photographs this is an inclined galaxy, possibly spiral, with a very large

NGC2967

NGC2974

NGC3044

H254[3]	9:53.7	+1 35	GLX
SB(s)c? sp	4.6′ × 0.7′	12.47B	

This is an almost exactly edge-on spiral galaxy with an extended, bright central region and fainter outer spiral structure. Photographs show the outer arms are brighter to the WNW than to the ESE and a faint spur emerges obliquely from the WNW extension. A faint, compact galaxy is 7.1′ N. Visually, this is a bright and very attractive galaxy: it is a very thin sliver of light which is much extended ESE–WNW. It is bright along the central two thirds of its length and slightly fainter at the extremities. No core is seen and the galaxy has very well defined edges. Radial velocity: 1142 km/s. Distance: 51 mly. Diameter: 68 000 ly.

NGC3044 15-inch Newtonian 272x

NGC3044

and very bright, elongated core in a fainter, unresolved outer disc. A faint field star borders the bright core to the SW. Visually, only the core is visible as a moderately faint, slightly oval and elongated disc with fairly well defined edges which is gradually brighter to the middle and oriented NNE–SSW. Radial velocity: 2539 km/s. Distance: 114 mly. Diameter: 53 000 ly.

NGC2980

H528[3]	9:43.2	−9 37	GLX
SAB(s)c?	1.6′ × 0.8′	13.60B	

This is an inclined spiral galaxy with a large, bright core. Photographs show a somewhat fragmentary spiral structure with many nebulous knots involved. Visually, the galaxy is a moderately faint, slightly oval object elongated SSE–NNW, slightly brighter to the middle and moderately well defined. The face-on spiral galaxy NGC2978, which

lies 7.75′ S, is smaller but fairly high in surface brightness. Radial velocity: 5541 km/s. Distance: 247 mly. Diameter: 115 000 ly.

NGC2990

H624[2]	9:46.3	+5 43	GLX
Sc:	1.3′ × 0.7′	13.16B	

This inclined spiral galaxy has a small bright core. The internal S-shaped spiral arms are embedded in a fainter outer envelope, elongated E–W. Deep, high-resolution photographs show that the arms have high surface brightness and are fairly massive. In medium apertures, the galaxy is a fairly faint and small patch of light with a moderately well defined envelope. It is a little brighter to the middle but no core is visible. A field star is 1.5′ NNE. Radial velocity: 2954 km/s. Distance: 132 mly. Diameter: 50 000 ly.

NGC3055

H4[6]	9:55.3	+4 16	GLX
SAB(s)c	2.1′ × 1.3′	12.65B	

In images this is an inclined, barred spiral galaxy with a narrow, bright bar. Two bright, knotty spiral arms emerge from the bar, expanding out to a broad, fainter outer spiral structure. Visually, the galaxy is fairly bright, a slightly extended oval oriented ENE–WSW. The envelope is mottled and uneven in surface brightness and is a little brighter to the middle and somewhat well defined. This visual impression may explain why Herschel classified this as a Class VI object. Recorded on 24 January 1783, he described this galaxy as: 'A cluster of very compressed small stars.' Radial velocity: 1668 km/s. Distance: 74 mly. Diameter: 46 000 ly.

NGC3110

H305[2]	10:04.0	−6 28	GLX
SB(rs)b pec	1.7′ × 0.8′	13.00B	

Dreyer correctly associated this Herschel object with NGC3110, but over time the object has mistakenly been identified as NGC3122. Both are plotted in the first edition of *Uranometria 2000.0* but the actual Herschel object is NGC3110. Photographs show a slightly inclined spiral galaxy with a peculiar morphology. The very bright and large core is elongated with two bright spiral arms attached. The eastern spiral arm is much longer than the western arm and a bright, elongated

condensation is embedded in the western arm just N of the core. A possible companion galaxy, MCG–1-26-13, is located 1.75′ SW. NGC3110 is fairly bright visually; it is an oval, opaque disc of even surface brightness which is well defined at the edges with no central brightening. A faint field star borders the disc 0.6′ to the NNW. Radial velocity: 4886 km/s. Distance: 218 mly. Diameter: 108 000 ly.

NGC3115

H163[1]	10:05.2	−7 43	GLX
S0⁻	7.2′ × 3.2′	10.08B	

Appearing similar visually and photographically, this is a very bright, well defined galaxy with an intense core which grows brighter to the nucleus. It is much extended NE–SW and spindle-shaped with sharp, tapered points. Short-exposure photographs show the thin disc with tapered ends well; the photograph below shows the extensive outer halo.

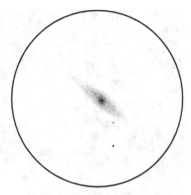

NGC3115 15-inch Newtonian 313x

An extremely faint, possibly dwarf companion galaxy is 5.6′ ESE. Curiously, Herschel did not sweep this galaxy up until 22 February 1786, though he had examined this area of sky a year earlier. Radial velocity: 493 km/s. Distance: 22 mly. Diameter: 47 000 ly.

NGC3055

NGC3115

NGC3156

H255³	10:12.7	+3 08	GLX
S0:	1.9′ × 0.9′	13.07B	

This lenticular galaxy presents similar appearances visually and photographically. A moderate-aperture telescope reveals a small oval glow with a bright, well-condensed nucleus. The major axis is extended NE–SW and the extremities taper to soft points. Bright field stars are at 2.0′ ESE, 5.4′ ENE and 8.5′ ESE. Herschel came across this galaxy on 13 December 1784, one year after discovering the brighter companions NGC3166 and NGC3169, located only 30′ to the NE. Radial velocity: 1180 km/s. Distance: 53 mly. Diameter: 29 000 ly.

NGC3166

H3¹	10:13.8	+3 26	GLX
SAB(rs)0/a	4.8′ × 2.3′	11.42B	

This is the principal galaxy of a group of five. Photographs reveal a slightly inclined spiral galaxy of peculiar morphology. It features a large, bright, almost round core and there is evidence of dust patches bordering the core to the N. There is also dust in the core on the W and S sides. The outer envelope is faint and peanut shaped, elongated E–W. The smoothness of the arms and lack of large HII regions suggest little recent star formation. At the eyepiece, the bright core and internal envelope dominate; the extensive outer envelope is too faint to see in moderate apertures. The much fainter galaxy NGC3165 is 4.5′ SW, while NGC3169 is 7.8′ ENE. The minimum projected separation between NGC3166 and NGC3169 is about 112 000 ly, while NGC3166 and NGC3165 may be separated by as little as 71 000 ly. Radial velocity: 1208 km/s. Distance: 54 mly. Diameter: 75 000 ly.

NGC3169

H4¹	10:14.2	+3 28	GLX
SA(s)a pec	5.0′ × 2.8′	11.25B	

In photographs this is a large inclined spiral galaxy of peculiar morphology. The large, bright, round core is partially obscured by a prominent, thin dust lane and extensive dust patches are visible in the inner spiral arms, S and E of the core. The spiral pattern is otherwise weak, with very faint, broad outer spiral arms, particularly to the S. The faint outer reaches of the galaxy extend to 6.0′ × 4.9′ and individual HII

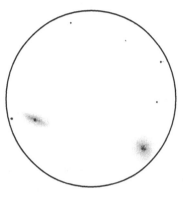

NGC3166 NGC3169
15-inch Newtonian 313x

NGC3166/3169

©STScI/AURA/CALTECH/ROE

regions are visible, especially to the S. The distortion of the spiral structure and the odd structure of NGC3166's outer envelope suggest possible tidal interaction between the two galaxies. Visually, the galaxy is bright and elongated NE–SW; the bright core stands out well but the outer envelope is faint, diffuse and poorly defined. A bright field star is 1.4′ E. The galaxy is a member of the NGC3166 galaxy group. Radial velocity: 1102 km/s. Distance: 49 mly. Diameter: 86 000 ly.

NGC3401

H88[3]	10:50.4	+5 48	nonexistent

NGC3423

H6[4]/H131[2]	10:51.2	+5 50	GLX
SA(s)cd	3.9′ × 3.2′	11.60B	

This object was catalogued twice by Herschel, first on 23 February 1784 and again on 13 April 1784. The first observation was the more interesting as he described the galaxy as: 'Faint, large, cometic. A central bright point with an extremely faint milky chevelure.' Photographs show a beautiful, face-on, multi-arm spiral with a small round core and fragmentary spiral arms studded with bright HII regions. Visually, the galaxy is large and moderately bright and is easily seen at low magnification but the envelope is quite diffuse, though broadly concentrated towards the middle with edges that fade gradually into the sky background. Radial velocity: 893 km/s. Distance: 40 mly. Diameter: 45 000 ly.

Taurus

This prominent constellation is home to 15 Herschel objects, a mix of star clusters, nebulae and galaxies. Most of the objects were recorded between the years 1783 and 1785, but the planetary nebula NGC1514 was discovered in 1790 and the galaxy pair NGC1633 and NGC1634 were discovered in 1798. The most significant of Herschel's discoveries here is NGC1514, the planetary nebula which convinced Herschel that at least some nebulae were made of luminous matter and were not simply distant, unresolved clusters of stars.

NGC1409

H263[3]	3:41.2	−1 17	GLX
SAB: pec	1.0′ × 0.8′	14.88B	

This galaxy forms a double system with NGC1410 which in photographs appears in contact. NGC1409 is possibly a lenticular galaxy with an elongated core and a faint outer envelope which is displaced for the most part toward the NW and oriented SE–NW. The core of NGC1410 is 0.25′ N. Telescopically, the two components are unresolved even at high magnification. The galaxy is a quite faint and somewhat diffuse, roundish patch of light which is broadly brighter to the middle and moderately well defined at the edges. Discovered on 6 January 1785, Herschel described this difficult object as: 'Extremely faint, stellar or a little extended, almost verified 240 (magnification).' Radial velocity: 7715 km/s. Distance: 345 mly. Diameter: 100 000 ly.

NGC1514

H69[4]	4:09.2	+30 47	PN
3+2	3.0′ × 2.9′	9.9B	9.4V

This bright planetary nebula holds the distinction of the being the object which convinced Herschel that true nebulae exist in the cosmos and are not simply remote clusters of stars. Photographically the bright inner shell is fairly complex in structure and is surrounded by an extensive secondary shell. Visually, the nebula is situated between two moderately bright field stars. The central star of the nebula is bright and easy and the surrounding nebula is quite tenuous, but bright zones are visible to the NW. The dark cavity S of the central zone is visible as well. The spectrum of the central star is A0III +sd0. Distance: 1960 ly.

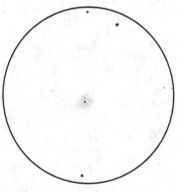

NGC1514 15-inch Newtonian 177x

NGC1514

NGC1550

H464[2]	4:19.6	+2 26	GLX
SA(s)0⁻:	2.0′ × 1.8′	13.16B	

The DSS image reveals a probable lenticular galaxy with a bright, slightly elongated core and a very faint and extensive outer envelope. There are several very faint galaxies in the field, the brightest being IC366 3.2′ to the SSE. Visually, NGC1550 is a small, moderately bright galaxy and is best at high magnification, which brings out the bright, round, well defined and opaque inner disc. It has a bright, nonstellar core embedded. Radial velocity: 3666 km/s. Distance: 164 mly. Diameter: 95 000 ly.

NGC1587

H8[2]	4:30.7	+0 40	GLX
E pec	1.7′ × 1.5′	12.72B	

Photographs reveal an elliptical galaxy with a large, bright, slightly elongated core in a fainter outer envelope. Elongated ENE–WSW, it forms a close

pair with NGC1588, 1.0′ E. The galaxies appear to share a common envelope. Telescopically, three galaxies are visible in the same medium-magnification field: NGC1587, NGC1588 and NGC1589. NGC1587 is the brightest, a round, condensed patch of light with even surface brightness and well defined edges. All three galaxies have similar radial velocities and probably form a physical group. The core-to-core separation between NGC1587 and NGC1588 would be at least 47 000 ly at the presumed distance with NGC1589 about 565 000 ly from the main pair. The three galaxies rank amongst Herschel's earliest discoveries: they were recorded on 19 December 1783. Radial velocity: 3633 km/s. Distance: 162 mly. Diameter: 80 000 ly.

NGC1588

H9[2]	4:30.8	+0 39	GLX
E pec?	1.4′ × 0.8′	13.89B	

This is a lenticular galaxy with a large, bright, slightly elongated core in a

fainter, asymmetrical outer envelope. Photographically, the outer envelope is more extensive towards the S and the galaxy is elongated N–S. It forms a close pair with NGC1587. Visually, NGC1588 is smaller than its companion and follows immediately; it is a well-condensed and sharply defined object. Radial velocity: 3447 km/s. Distance: 154 mly. Diameter: 63 000 ly.

NGC1589

H7[2]	4:30.8	+0 51	GLX
Sab	3.2′ × 1.0′	12.82B	

This is a highly inclined spiral galaxy with a large, slightly elongated core. Photographs show the core is partially obscured by a dust lane which extends along the galaxy's major axis on the western flank. The outer envelope is diffuse and appears disrupted and a field star is 1.0′ ENE. The pair NGC1587 and NGC1588 is 12.0′ S. At medium magnification the galaxy appears a little fainter than NGC1587; it is a well defined sliver of light with a brighter core, elongated NNW–SSE. Radial velocity: 3733 km/s. Distance; 167 mly. Diameter: 155 000 ly.

NGC1633

H952[3]	4:40.0	+7 21	GLX
SAB(s)ab	1.2′ × 1.0′	14.50B	

The DSS image shows an almost face-on spiral galaxy with a bright, round core and three fainter spiral arms. Visually, the galaxy is paired with NGC1634 0.7′ S. Together they are very faint and

NGC1589

difficult to distinguish and are visible as an elongated haze oriented N–S. Though the larger photographically, NGC1633 is the smaller and fainter at the eyepiece, a very diffuse and ill defined patch of light. Herschel discovered these galaxies on 9 December 1798, providing the following description for the pair: 'Two nebulae within 1.0′ of each other; in the meridian (oriented N/S). Both very faint, very small.' Radial velocity: 4943 km/s. Distance: 221 mly. Diameter: 77 000 ly.

NGC1634

H953[3]	4:40.0	+7 20	GLX
E	0.9′ × 0.6′	14.97B	

Photographs reveal an elliptical galaxy with a bright, slightly elongated core in a faint outer envelope. Paired with NGC1633 0.7′ to the N, this is the dominant galaxy in a medium-magnification eyepiece; it is a diffuse patch which is a little brighter to the middle. Two magnitude 9 field stars immediately SW hinder observation somewhat. There is a difference of almost 600 km/s in the radial velocities of the two galaxies so they may not actually form a physical pair. At the presumed distance of NGC1634, however, the core-to-core separation between the two galaxies would be about 40 000 ly. Radial velocity: 4364 km/s. Distance: 195 mly. Diameter: 51 000 ly.

NGC1647

H8[8]	4:46.0	+19 04	OC
II2r	40.0′ × 40.0′	6.8B	6.4V

Photographs show a fairly bright cluster of widely scattered stars with a moderate range of brightness. This is a large and quite bright cluster visually which is best seen at low magnification. At least 50 stars are visible, ranging from magnitude 9 to magnitude 12 in a 45′ field. The cluster is almost round, there is no central condensation and the cluster is easily separated from the sky background. Approximately 200 stars are thought to be members, with the brightest being magnitude 9. Spectral

type: B8. Age: 210 million years. Distance: 1710 ly.

NGC1750

H43[8]	5:03.9	+23 39	OC
III2p	40.0′ × 40.0′	6.1V	

William Herschel recorded both this object and NGC1758 (H21[7]) on 26 December 1785 and considered them two separate objects. His description for NGC1750 (H43[8]) reads: 'A cluster of very coarse scattered large stars joined to H21[7].' In the first edition of *Uranometria 2000.0* the cluster is identified as NGC1746. On the DSS image, we see a scattered group of stars of moderate brightness range, the brightest members in an elongated group extending ESE–WNW. The star field in the image is fairly rich and only the brightness of the principal members of the cluster separates it from the sky background. Visually, this is a large, somewhat scattered cluster which is best seen at low magnification as a grouping of several bright stars involved with many fainter ones. There is a wide brightness range and some of the stars form close pairs. The cluster is well resolved and well separated from the sky background with about 40 stars involved, though only about 20 stars may be actual members. The brightest star is magnitude 8. Spectral type: B5. Distance: 1370 ly.

NGC1758

H21[7]	5:04.4	+23 46	OC?
8.0′ × 6.0′			

This object is probably a part of the cluster designated NGC1746 and may probably be identified with a group of fainter stars on the eastern side of the larger cluster. It is not listed as a separate cluster in *Sky Catalogue 2000.0* (which recognizes NGC1746 and NGC1750), while Archinal and Hynes' *Star Clusters* (2003) assigns the tabular data for NGC1746 to NGC1758, except for slight differences in right ascension and declination. These stars are significantly fainter than those that make up NGC1746 and NGC1750 and there is no information on whether they

are part of this group or are even a separate cluster. On the DSS image there is a cross-shaped asterism at the position. Herschel's description for NGC1750 reads: 'A cluster of pretty compressed stars with many extremely small stars mixed with them.'

NGC1802

H41[8]	5:10.2	+24 06	OC/AST?
20.0′ × 20.0′			

The DSS image shows a poor cluster of brighter stars. Visually, it is best seen at medium magnification as a N–S scatter of stars, moderately well separated from the background, and featuring members with a fairly wide brightness range. About two dozen stars appear to be involved. Herschel's 7 December 1785 observation called it: 'A coarse cluster of stars or projecting point of the Milky Way.' The brightest members are about magnitude 9. This is very probably not a true cluster.

NGC1817

H4[7]	5:12.1	+16 42	OC
IV2r	20.0′ × 20.0′	8.4B	7.7V

In a small telescope, this cluster is preceded on its western boundary by a loose assemblage of four stars that are magnitude 9 or fainter. The central region is dominated by a roundish grouping of many faint 'sparkles' of magnitude 12 or fainter. In photographs, the brightest western members, including some fainter stars, form a narrow crescent-shaped grouping from the S curving to the W. Herschel discovered the cluster on 19 February 1784 and considered it quite compressed and rich. The total membership of the cluster may be nearly 300 stars. Spectral type: A0. Age: about 790 million years. Distance: 5910 ly.

NGC1996

H42[8]	5:38.2	+25 49	OC?
22.0′ × 14.0′			

Photographs show a large cluster of scattered stars with a moderate brightness range and little concentration. Visually, the cluster can be seen at medium magnification as a N–S scattering of stars which are well resolved and of moderate brightness range. About 22 stars are visible, the brightest of which form a triangle within which is a faint close pair. The cluster is not plotted in the first edition of *Uranometria 2000.0* but it is located half way between the star 125 Tauri and the star cluster DoDz4. Herschel recorded this object on 7 December 1785 and

described it as: 'A cluster of coarsely scattered stars about 15′ in diameter. The stars nearly of a size and equally scattered.'

NGC2026

H28[8]	5:43.1	+20 07	OC?
10.0′ × 10.0′			

Images show a poor, coarse cluster of scattered stars, with a moderate brightness range and little

concentration. Visually, though, this is an attractive field featuring a triangle of slightly brighter stars with fainter members clustering around two of the stars. The northernmost bright star has a chain of three fainter stars extending S. About 27 stars appear to be involved; the cluster is well resolved and well separated from the sky background. The total membership of the cluster may be 35 stars with the brightest at magnitude 9.

Triangulum

The 20 Herschel objects in this small constellation are all galaxies, except for NGC604, an HII region in the Local Group galaxy M33. All the objects were recorded between the years 1784 and 1786, although NGC598 (M33) was recorded by Charles Messier in 1764 and may very well have originally been discovered by Giovanni Hodierna.

While NGC740 is a faint and challenging object, most of the others are fairly prominent and the galaxy NGC1060 is a marker for an interesting group of fainter galaxies that will be of interest to moderate- and large-aperture telescope users.

NGC598

H17[5]	1:33.9	+30 39	GLX
SA(s)cd	70.8′ × 41.7′	6.35B	5.7V

Charles Messier catalogued this galaxy on 25 August 1764, though the Italian astronomer Giovanni Hodierna probably observed it first in 1654. Why Herschel catalogued the galaxy when he usually avoided duplicating the work of Messier is unknown. Certainly, he did not note anything that Messier had not already reported with his much smaller telescope. Recorded on 11 September 1784, Herschel's description reads: 'Milky nebulosity not less than one half degree broad. Perhaps three-quarters degree long, but not determined.' In subsequent years observing with 7-foot and 10-foot telescopes, Herschel repeatedly suspected that the 'nebula'

consisted of stars, though the ones that he may have seen were probably foreground stars in our own Milky Way. Interestingly, the brightest blue stars in the galaxy resolve in photographs at B magnitude 15, while the brightest red stars appear at V magnitude 16.5. This third largest member of the Local Group is well studied and photographs show an almost face-on spiral galaxy with a small brighter core. Two main spiral arms emerge from the core; they expand quickly into subarms and broad, faint expanses of dust, nebulae and stars. Resolution into stars is extensive; most of the dust is evident in discrete patches or short sinewy paths through the spiral arms. There are dozens of HII regions, star clusters and stellar associations, some of them quite well resolved. The Herschel object NGC604 is the brightest

of these and is located at the tip of the northern, inner spiral arm. In moderate aperture telescopes, on a good night NGC598 is a spectacular object with the two main spiral arms clearly discerned and a third, stubby and broad arm extending towards the SE. The extent of the galaxy is not much more than 25′ × 15′ however; the very faint outer regions are not detected. The core is bright, oval and elongated NNE–SSW. Several HII regions catalogued in the *New General Catalogue* are clearly visible, including NGC595 W of the core, NGC592 to the SW, and NGC588 further to the WSW. The spectacular HII region NGC604 is bright and well seen in the spiral arm to the NE. Many stars are sprinkled throughout the field. Radial velocity in approach: 44 km/s. Distance: 2.3 mly. Diameter: 65 000 ly.

NGC598

NGC604

H150[3]	1:34.5	+30 48	EN
1.93' × 1.20'	14.0		

Herschel recorded this extragalactic HII region in the northern spiral arm of NGC598 on the same night that he recorded the galaxy. In photographs it is a small, bright, opaque patch; it is irregular in outline with a very small, separate, opaque patch immediately NW. Several narrow spurs extend from the main nebula into the surrounding resolved starfields of the main galaxy's spiral arm. A foreground star is 1.0' SE. Research indicates that a supernova remnant is involved with the nebula. Visually, the nebula is a bright, irregularly elongated patch of light which holds magnification well. The texture is quite mottled, hinting at detail just beyond resolvability. Radial velocity in approach: 92 km/s. Distance: about 2.3 mly.

NGC614

H174[3]	1:35.8	+33 42	GLX
S0?	1.0' × 0.9'	13.83B	

Photographs show a lenticular or elliptical galaxy with a bright core and a fainter outer envelope elongated ESE–WNW. The likely companion galaxy NGC608 is 5.1' SSW. Visually, both galaxies can be seen in a high-magnification field. NGC614 is the larger and slightly brighter of the two; it is round and well condensed to a brighter core. A faint, diffuse secondary envelope surrounds the core. NGC608 is a little fainter, very small but well defined with a star-like core. Radial velocity: 5301 km/s. Distance: 237 mly. Diameter: 69 000 ly.

NGC661

H610[2]	1:44.2	+28 42	GLX
E[+]:	1.7' × 1.4'	13.18B	

Images show an elliptical galaxy with an elongated bright core in an extensive, fainter outer envelope, elongated NE–SW. Telescopically, it is quite bright and is well seen at high magnification as a round object with an opaque envelope which brightens to a small, condensed core. The edges are fairly well defined and the envelope is grainy in texture. Radial velocity: 3942 km/s. Distance: 176 mly. Diameter: 87 000 ly.

NGC670

H611[2]	1:47.4	+27 53	GLX
SA0	2.0' × 1.0'	13.50B	

The photographic and visual appearances of this lenticular galaxy are fairly similar. It is a moderately bright, small, oval, well defined object, oriented almost due N–S. Only a very gradual brightening to the middle is detected in a smooth-textured envelope with no core visible. A bright field star, about magnitude 9, is 7.25' ENE. Radial velocity: 3824 km/s. Distance: 171 mly. Diameter: 99 000 ly.

NGC672

H157[1]	1:47.9	+27 26	GLX
SB(s)cd	6.8' × 2.9'	11.41B	

Photographs show that this dwarfish, barred spiral galaxy forms an attractive contrasting pair with the fainter IC1727, located 8.0' SW, a galaxy which Herschel missed. The pair have very similar morphologies. Both galaxies are well resolved on high-resolution images with stars beginning to appear at about blue magnitude 22. NGC672 is highly inclined and features a long, bright bar which anchors a reverse S-shaped spiral pattern. Visually, NGC672 appears bright, large, much extended and oriented ENE–WSW. It is quite diffuse, however, and gradually brighter along its major axis though no core is visible. The extremities are only moderately well defined and taper gradually into the sky background. IC1727 is a faint low-surface-brightness patch of light with poorly defined extremities and appears slightly elongated NW–SE. The pair are also known as VV338 and have similar radial velocities: that of NGC672 is 542 km/s and that of IC1727 is 528 km/s. At the presumed distance the core-to-core separation between the two galaxies would be about 56 000 ly. Distance: 24 mly. Diameter: 48 000 ly.

NGC672

©STScI/AURA/CALTECH/ROE

NGC684

NGC738
NGC733
NGC736
NGC740
NGC733/736/738/740

NGC684

H612[2]	1:50.2	+27 39	GLX
Sb	3.2′ × 0.6′	13.33B	

Photographically, this almost edge-on spiral galaxy shows an extensive dust lane bordering the core to the S. In moderate apertures, this is a moderately bright galaxy, well seen at high magnification. It is a flat, well defined streak of light; its ends appear slightly tapered and its major axis is oriented due E–W. A brighter core is visible, a little more prominently at medium magnification, and a field star is 1.7′ ESE. Radial velocity: 3653 km/s. Distance: 163 mly. Diameter: 152 000 ly.

NGC735

H176[3]	1:56.6	+34 11	GLX
Sb	1.8′ × 0.9′	14.07B	

The DSS image shows that this is clearly an inclined, barred spiral galaxy with a bright core and a foreshortened bar. The bar is attached to an inner ring; it is brightest at the points of attachment, giving short, brighter arcs in the ring. Narrow, smooth-textured, spiral arms emerge from the ring. A faint field star is embedded at the edge of the inner ring, immediately NW of the core. The galaxy is elongated SE–NW and several field stars are nearby, including a bright one 1.4′ SW. Two small, faint, compact galaxies are located 1.4′ ENE and 1.4′ NW. Visually, the galaxy is quite faint with two field stars nearby, one immediately WNW. It is a very small,

faint streak with a bright core elongated SE–NW. The extensions are very faint, hazy and ill defined. Radial velocity: 4759 km/s. Distance: 213 mly. Diameter: 111 000 ly.

NGC736

H221?[2]	1:56.7	+33 03	GLX
E+:	1.7′ × 1.6′	13.15B	

Classified as elliptical, this may actually be an S0 galaxy. Photographs show a large, bright core and very faint outer envelope. Three field stars are immediately N of the core and a very faint condensation is in the envelope W of the core. Three NGC objects are in the field, NGC733 (a single star) is 3.6′ WNW, NGC738 is 1.3′ NE and NGC740 is 3.3′ ESE. Visually, this is a moderately bright galaxy which is well condensed and brighter to the middle with well defined edges. The brightest of the three stars to the N is visible as a magnitude 14 object just beyond the outer halo. The identification as H221[2] is a little

uncertain; see NGC740. Radial velocity: 4482 km/s. Distance: 200 mly. Diameter: 99 000 ly.

NGC740

H221?[2]	1:56.9	+33 01	GLX
SBb?	1.5′ × 0.3′	14.84B	

Photographically, this is an almost edge-on spiral galaxy with a small, brighter core encircled by a possible pseudo-ring and faint extensions. It is elongated SE–NW with a field star 1.25′ to the E. NGC733 is 5.3′ WNW, NGC738 is 3.2′ NNW and NGC736 is 3.3′ NNW. Telescopically, NGC740 is extremely faint, visible intermittently as a very small, hazy patch of light immediately W of a magnitude 10 field star which hinders the view. Only the core of this almost edge-on galaxy is visible, with NGC736 seen in the field to the NW. H221[2] was recorded 12 September 1784 and described by Herschel as: 'Faint, pretty large, much extended, resolvable, 1.5′ long.' This is a fair description of NGC740 but the description 'resolvable' might indicate that Herschel actually observed the brighter NGC736 and its three faint attendant stars. J. L. E. Dreyer equated H221[2] with NGC740 and John Herschel is credited with recording NGC736 as his h169 but did not observe NGC740. Modern sources identify H221[2] with NGC736 and credit the discovery of NGC740 to Lord Rosse in 1850. Radial velocity: 4736 km/s. Distance: 212 mly. Diameter: 92 000 ly.

NGC750

H222[2]	1:57.5	+33 13	GLX
E pec	2.6′ × 1.9′	13.11B	

The DSS image shows an interacting pair with the companion galaxy NGC751 located immediately S. The pair are also catalogued as Arp 166. Both galaxies are elliptical and share a common, faint outer envelope; in addition, a bright bridge extends to NGC751. A very faint plume of matter is being ejected from NGC750 in the N, curving to the NNW. This plume extends to a very small, faint galaxy 2.6′ NNW. A bright field star is 4.75′ SSE. Visually, the pair of galaxies is resolved as two separate objects with a faint connection between them visible at high magnification in moderate-aperture telescopes. NGC750 is the brighter and larger component. Both galaxies are relatively well defined and both are brighter to the middle to condensed cores. Herschel discovered the object on 12 September 1784; unfortunately his description reads: 'Just like the former', referring to an observation of NGC740 (actually, NGC736) which he described as: 'Faint, pretty large, much extended, resolvable, 1.5′ long.' This would seem to indicate that Herschel found NGC750 to be elongated and very possibly separated into two components. It is very likely that he had discovered NGC751 as well, though he did not catalogue it as a separate object. Radial velocity: 5298 km/s. Distance: 237 mly. Diameter: 179 000 ly.

NGC777

H223[2]	2:00.2	+31 26	GLX
E1	2.5′ × 2.0′	12.50B	11.5V

Photographs show an elliptical galaxy with a bright core and an extensive outer envelope, elongated SSE–NNW. The companion galaxy NGC778 is 7.0′ S. Visually, the galaxy is a moderately bright, small, round, well-condensed object with well defined edges. It forms a triangle with two magnitude 9 field stars to the S. NGC778 is faintly visible S of these stars and is round and well condensed with well defined edges. Radial velocity: 5137 km/s. Distance: 230 mly. Diameter: 167 000 ly.

NGC780

H583[3]	2:00.5	+28 13	GLX
1.8′ × 1.0′	14.6B		

This faint galaxy is not classified and has a peculiar morphology. Photographically, it appears likely that it is a spiral galaxy with a bright, elongated core embedded in a faint outer envelope and a dust patch or lane along the major axis on its eastern flank. There is an elongated and very faint oval disc oriented SSE–NNW, tilted about 35° to the plane of the galaxy. Visually, the galaxy is moderately faint, though well seen at medium and high magnification. It is a hazy, poorly defined patch of light, oval in form, elongated N–S and a little brighter to the middle. Two very faint field stars, one due S, the other to the NE, seem to merge with the galaxy making it look larger than it is. Herschel discovered this object on 26 October 1786 and his description reads: 'Very faint, very small, extended or three faint stars with very faint nebulosity.' Radial velocity: 5337 km/s. Distance: 238 mly. Diameter: 125 000 ly.

NGC807

H151[3]	2:04.8	+28 59	GLX
E	1.8′ × 1.2′	13.78B	

This elliptical galaxy presents similar appearances visually and photographically. At high magnification

NGC750/751

©STScI/AURA/CALTECH/ROE

the galaxy is small but fairly bright, almost round and opaque and quickly brighter to a star-like core. Photographs show three brightness gradients: a bright core, a slightly fainter inner envelope and a faint and extensive outer envelope, elongated SE–NW. One faint field star is 0.6′ N while a brighter one is 1.7′ SW. Radial velocity: 4877 km/s. Distance: 218 mly. Diameter: 114 000 ly.

NGC855

H613[2]	2:14.0	+27 53	GLX
E	2.6′ × 1.0′	13.28B	12.6V

This galaxy is relatively nearby and may be either elliptical or lenticular. Photographs show a brighter elongated core and a faint outer envelope, which is squarish and does not quite match the position angle of the inner envelope. Visually, the galaxy is moderately bright and visible as an oval patch of light, fairly even in surface brightness, well defined and elongated ENE–WSW. A bright field star is 2.8′ NNW. Radial velocity: 680 km/s. Distance: 30 mly. Diameter: 23 000 ly.

NGC890

H225[2]	2:22.0	+33 16	GLX
SAB(r)0⁻?	2.5′ × 1.7′	12.85B	11.6V

The visual and photographic appearances of this lenticular galaxy are similar. Telescopically, it is small but fairly bright, oval in shape, fairly well defined and elongated ENE–WSW. The envelope is opaque and broadly brighter to the middle with a small, round and bright core embedded. Radial velocity: 4091 km/s. Distance: 183 mly. Diameter: 133 000 ly.

NGC925

H177[3]	2:27.3	+33 35	GLX
SAB(s)d	10.5′ × 5.9′	10.59B	10.1V

The DSS image shows an inclined spiral galaxy, probably barred, with a small, extended, brighter inner core. Two low-surface-brightness spiral arms emerge from the bar as well as a spiral fragment in the E. There are several knotty condensations embedded in the

NGC925

spiral arms as well as small discrete dust patches. Several faint field stars are projected against the spiral arms. High-resolution photographs resolve many stars at blue magnitude 21. Visually, however, the galaxy is dim and best seen at low magnification. It is moderately large, elongated ESE–WNW and weakly concentrated to the centre. The extremities are very poorly defined. Herschel recorded the galaxy on 13 September 1784 and called it: 'Very faint, considerably large, irregularly round. Resolvable, 2′ or 3′ in diameter.' Evidently, he may have seen some of the faint stars projected against the galaxy's spiral arms. Radial velocity: 664 km/s. Distance: 30 mly. Diameter: 90 000 ly.

NGC949

H154[1]	2:30.8	+37 08	GLX
SA(rs)b:?	2.4′ × 1.3′	12.49B	12.0V

This is an inclined galaxy, possibly a spiral, with a peculiar morphology.

Photographs show a bright and very elongated oval inner region surrounded by a very faint outer envelope. Small bright condensations border the bright inner disc on the NW and SE and a prominent dust patch is just beyond the SE condensation. Several faint field stars are nearby. Telescopically, the galaxy is quite bright, moderately large and elongated SE–NW. The edges are fairly well defined and the main envelope has

NGC949

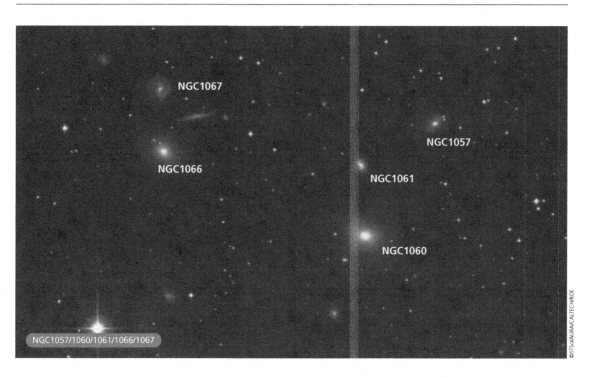

NGC1067

NGC1066

NGC1057

NGC1061

NGC1060

NGC1057/1060/1061/1066/1067

a mottled, uneven character, although it is brighter to the middle. The SE tip seems to have a 'notch' taken out of it and there is an extremely faint field star immediately S. Radial velocity: 726 km/s. Distance: 32 mly. Diameter: 23 000 ly.

NGC987

H161³	2:36.8	+33 19	GLX
SB0/a	1.5′ × 0.9′	13.51B	

Photographs show an inclined, barred spiral galaxy with a bright core and a fainter bar in a diffuse outer envelope. Two spiral arms can faintly be seen N of the core. At high magnification the galaxy is a small but moderately bright, elongated oval oriented NE–SW. Well defined along the edges, it has a bright, condensed core. Radial velocity: 4639 km/s. Distance: 207 mly. Diameter: 90 000 ly.

NGC1060

H162³	2:43.2	+32 24	GLX
S0⁻:	2.0′ × 1.7′	13.01B	11.8V

This is the dominant member of a small galaxy group: the DSS image shows about ten prominent members and several very small and faint possible members of the group. It is a lenticular galaxy with a bright, elongated core and an extensive outer envelope. Also in the field is the Herschel object NGC1066, located 7.75′ ENE, while NGC1057 is 4.7′ NNW and NGC1061 is 2.5′ north. Telescopically, NGC1060 is the brightest of five galaxies visible in a high-magnification field of view. It is a compact oval of light, opaque and brighter to the middle. The edges are fairly well defined and the disc is elongated E–W. The other galaxies seen in the field are NGC1061, NGC1057, NGC1066 and NGC1067. Radial

velocity: 5289 km/s. Distance: 236 mly. Diameter: 137 000 ly.

NGC1066

H163³	2:43.8	+32 27	GLX
E:	1.7′ × 1.5′	14.49B	

This is the second brightest galaxy in the NGC1060 galaxy group field; it is a possible member of the group although its radial velocity is over 800 km/s lower than the principal galaxy. It is an elliptical galaxy with a bright elongated core and an extensive outer envelope. Photographs show that a very faint star or compact object borders the core to the WNW and the galaxy is elongated ENE–WSW. NGC1062 is 1.7′ N, while NGC1067 is 2.3′ N. Visually, the galaxy is a high-surface-brightness object which is brighter to the middle, almost round and well defined. Radial velocity: 4445 km/s. Distance: 199 mly. Diameter: 98 000 ly.

Ursa Major

The third largest constellation in the sky is home to the second largest number of Herschel objects: 276 can be found within its confines, almost all of them galaxies. The exceptions are NGC3063 (a double star), and three nebulous knots in the spiral arms of M101 (NGC5447, NGC5461 and NGC5462). Almost all the objects were recorded between the years 1785 and 1793, though NGC3471 was discovered in 1801, while the galaxies NGC2629, NGC2641 and NGC3034 were all recorded during Herschel's last sweep of the heavens, 30 September 1802. Many of the brightest galaxies within the constellation are amongst the closest galaxies to our own Local Group and are consequently large and

detailed in moderate- and large-aperture telescopes. At the other extreme are galaxies at chillingly remote distances. There are many galaxies with radial velocities which imply distances in excess of 400 mly and the galaxy NGC4199 receives a special mention – it is evidently in excess of 800 mly away and amongst the most challenging of the Herschel objects to observe. Many of the galaxies are arranged in pairs and triplets, making for interesting fields even if the individual objects are not particularly compelling. Patience and determination will be required by the observer who accepts the challenge of observing all the Herschel objects in this constellation.

NGC2629

H982[3]	8:47.2	+72 59	GLX
SA(r)0°:	1.8′ × 1.5′	13.32B	

This is an inclined lenticular galaxy with a large, elongated core and a faint, extensive outer envelope, elongated E–W. The DSS photograph shows an edge-on, anonymous galaxy 7.3′ to the W and NGC 2641 6.2′ SSE. Visually, the two NGC galaxies can be seen in the same high-magnification field. NGC2629 is the larger and brighter; it is an oval, condensed patch of light with a bright star-like core embedded in a grainy, moderately well defined envelope. A faint field star is 0.7′ to the SE. NGC2629 and NGC2641 were the last two entries of Herschel's Class III: very faint nebulae, to be observed, and were recorded on the night of his last sweep of the heavens, 30 September 1802. Radial velocity: 3854 km/s. Distance: 172 mly. Diameter: 90 000 ly.

NGC2639

H204	−1 8:43.6	+50 12	GLX	
(R)SA(r)a:?	1.7′ × 0.9′	12.60B	11.7V	

This is an inclined multi-armed spiral galaxy with a large, elongated core and very faint, thin spiral arms. The galaxy is elongated SE–NW and the multiple arms are best seen in high-resolution, deep photographs. Visually, this is a very bright and well defined galaxy. Although small, the envelope is bright and increases in brightness to the middle to a prominent core and the extremities are sharply defined. Radial velocity: 3369 km/s. Distance: 150 mly. Diameter: 75 000 ly.

NGC2641

H983[3]	8:48.1	+72 54	GLX
S0:	1.3′ × 1.0′	14.75B	

Photographs show a lenticular galaxy with a bright, round core. A fainter envelope surrounds the core and an even fainter secondary envelope extends from this. Dark zones are suspected in

the outer envelope to the N and S. Seen in the same field as NGC2629, this is a faint, well defined, round patch of light with even surface brightness. Radial velocity: 3242 km/s. Distance: 145 mly. Diameter: 55 000 ly.

NGC2650

H908[2]	8:50.0	+70 18	GLX
SB(rs)b:	1.5′ × 1.1′	14.13B	

Situated 7.5′ W of a magnitude 7 field star, photographs show a slightly inclined, barred spiral galaxy. The core is bright and elongated and there is a slightly fainter bar, oriented E–W. Thin, relatively smooth-textured spiral arms emerge from the bar; they are tightly wound around the core. Telescopically, the galaxy is moderately faint and is best seen at medium magnification. It is a small, round, hazy patch of light which is a little brighter to the centre. A triangle of very faint stars bracket the galaxy and are best seen at high magnification. Herschel recorded the

NGC2681

©STScI/AURA/CALTECH/IOE

considerably small, resolvable, preceding some faint stars.' Radial velocity: 2888 km/s. Distance: 129 mly. Diameter: 38 000 ly.

NGC2692

H831[3]	8:57.0	+52 04	GLX
SBab:	1.3′ × 0.5′	13.96B	

This is a highly inclined, barred spiral galaxy with a bright core and bar. Photographs show that the ends of the bar curve and taper into a slightly fainter outer envelope. A faint star or compact object is involved at the southern tip of the outer envelope. A likely companion galaxy, UGC4671, is 3.3′ NW. Visually, the galaxy is moderately bright and is visible as a fairly well defined, elongated patch of light, oriented N–S, with something of a bulge at the middle. Radial velocity: 4072 km/s. Distance: 182 mly. Diameter: 69 000 ly.

NGC2693

H823[2]	8:57.0	+51 21	GLX
2.6′ × 1.8′	12.82B	11.9V	

Photographs show an elliptical galaxy, elongated NNW–SSE, with a bright, elongated core and a fainter outer envelope. The possible companion galaxy NGC2694 is 0.9′ S. It is a compact, high-surface-brightness object: the core-to-core separation between the two galaxies would be about 58 000 ly if they are at the same distance. Visually, NGC2693 is quite a bright galaxy; it is small and fairly condensed, and is visible as a grainy patch of light which is round and brighter to the middle with moderately well defined edges. Radial velocity: 4979 km/s. Distance: 222 mly. Diameter: 168 000 ly.

NGC2701

H66[4]	8:59.1	+53 46	GLX
SAB(rs)c:	2.1′ × 1.5′	12.78B	

This inclined spiral galaxy has a very small core and a grainy envelope. Photographs show a thin, bright, knotty arm emerging from the core and curving E then S. This galaxy is

galaxy on 30 September 1802, the last night of his sweeps which were begun almost 20 years before. He described the galaxy as: 'Pretty bright, pretty large, easily resolved. I believe I see some of the stars. Irregular form.' Radial velocity: 3935 km/s. Distance: 176 mly. Diameter: 77 000 ly.

NGC2681

H242[1]	8:53.5	+51 19	GLX
(R')SAB(rs)0/a	3.6′ × 3.3′	11.15B	10.3V

This galaxy is well seen on the DSS photograph as a face-on spiral galaxy with a somewhat peculiar morphology. The very bright and large core is surrounded by a thin dust lane which defines the inner edge of the inner spiral pattern and takes on the appearance of a ring as a result. The fainter, outer spiral pattern is broad, smooth textured and poorly defined. The galaxy is very bright visually; it is round, opaque and well-condensed disc which is steadily brighter

to the middle. The edges seem quite well defined, although there is a hazy, faint outer envelope fairly close to the inner disc. Radial velocity: 729 km/s. Distance: 32 mly. Diameter: 34 000 ly.

NGC2684

H712[3]	8:54.8	+49 08	GLX
S?	1.0′ × 0.9′	13.68B	

The DSS image shows a slightly inclined galaxy, probably spiral, with a small, bright core surrounded by a distorted ring. The outer envelope is grainy and asymmetric, being broader to the N and W and the galaxy is elongated NE–SW. A swarm of faint galaxies, many of them NGC objects, are within 6.0′. Almost all of them are to the E and SE. However, in moderate apertures, only NGC2684 is visible. It is gradually brighter to the middle and slightly oval in shape with moderately well defined extremities. Herschel, observing the galaxy on 9 March 1788, called it: 'Extremely faint,

moderately bright visually, though a little diffuse, and is seen as an even-surface-brightness patch of light, roughly round and fairly well defined at the edges. Immediately to the W a faint field star borders the galaxy. Recorded on 18 March 1790, this star led Herschel to enter the galaxy in his Class IV, describing it as: 'A small star with a pretty bright fan-shaped nebula. The star is on the preceding side of the diverging chevelure, and seems to be connected with it.' The galaxy is elongated NNE–SSW. Radial velocity: 2373 km/s. Distance: 106 mly. Diameter: 65 000 ly.

NGC2710

H841[3]	8:59.8	+55 41	GLX
SB(rs)b	2.0′ × 1.0′	13.65B	

In photographs this is an inclined, barred spiral galaxy with a narrow elongated bar attached to narrow, S-shaped spiral arms. Telescopically, this is quite a faint galaxy; it is a little more distinct at medium magnification and is seen as a hazy, ill defined patch of light of even surface brightness, elongated ESE–WNW. Radial velocity: 2580 km/s. Distance: 115 mly. Diameter: 67 000 ly.

NGC2726

H834[2]	9:04.9	+59 56	GLX
Sa?	1.6′ × 0.5′	13.28B	

Photographs show a highly inclined lenticular or spiral galaxy with a bright, very elongated core and fainter extensions. Visually, this small but moderately bright galaxy is somewhat extended E–W. The galaxy features a brighter oval core, while the outer envelope is hazy and fainter. A magnitude 14 star is just beyond the outer envelope, 0.3′ SSW of the core. Herschel thought the galaxy was easily resolvable. Radial velocity: 1589 km/s. Distance: 71 mly. Diameter: 33 000 ly.

NGC2742

H249[1]	9:07.6	+60 29	GLX
SA(s)c	3.0′ × 1.5′	12.08B	

Photographs reveal an inclined spiral galaxy with a very small core. The spiral pattern is multi-armed with many knots. Visually, the galaxy is a bright and very large oval, oriented E–W. Its surface brightness is fairly uniform and the galaxy is only marginally brighter to the middle. The outer portions are well defined and fade suddenly into the sky background. A magnitude 8 field star is 4.5′ NW. Herschel came upon this object on 19 March 1790, his description reading: 'Considerably bright, extended near the parallel. Easily resolvable, brighter middle, 4.0′ long, 2.0′ broad. I suppose, with a higher power and longer attention, the stars would become visible.' Radial velocity: 1362 km/s. Distance: 61 mly. Diameter: 53 000 ly.

NGC2756

H828[2]	9:09.0	+53 51	GLX
Sb	1.4′ × 0.9′	13.06B	

The DSS image reveals a slightly inclined spiral galaxy with a bright core

and bright, S-shaped inner spiral arms which appear to emerge from a bar. The arms expand to fainter, thin arms in a broad outer envelope, elongated N–S. Although somewhat faint visually, this galaxy is well seen at high magnification as an oval, well defined patch of light with a broadly bright middle which seems to bulge very slightly. Radial velocity: 4026 km/s. Distance: 180 mly. Diameter: 73 000 ly.

NGC2768

H250[1]	9:11.6	+60 02	GLX
S0/E6:	8.1′ × 4.3′	10.82B	9.9V

Images show a lenticular galaxy with a large, bright, elongated core in an extensive outer envelope. Telescopically, this is a large, bright galaxy that holds magnification well. It is very much elongated E–W with a bright sizable core. The main envelope is bright and fades gradually to an extensive secondary envelope. Radial

NGC2768

©STScI/AURA/CALTECH/ROE

velocity: 1444 km/s. Distance: 65 mly.
Diameter: 152 000 ly.

NGC2787

H216[1]	9:19.3	+69 12	GLX
SB(r)0⁺	3.2′ × 2.0′	11.60B	

This is an inclined barred lenticular
galaxy with a large, bright, elongated
core surrounded by a smooth-textured
fainter ring. Photographs show the ring
brightens immediately N and S of the
core where the bar is in contact with the
ring. Telescopically, the galaxy is bright,
located 7.5′ NE of a bright field star and
takes high magnification well. The core
is opaque and condensed with a
concentrated, bright nucleus
surrounded by a hazy outer shell,
elongated ESE–WNW. One faint field
star is just beyond the envelope to the
SSE and another is a little further to the
N. Radial velocity: 802 km/s. Distance:
36 mly. Diameter: 33 000 ly.

NGC2800

H832[3]	9:18.6	+52 31	GLX
E	1.3′ × 1.1′	13.87B	

In photographs this is an elliptical
galaxy with a brighter, elongated core
in a faint outer envelope, elongated
NNE–SSW. Telescopically, this galaxy is
quite faint and is only confirmed at high
magnification as a round, somewhat
diffuse patch of light of even surface
brightness with fairly well defined edges.
A faint field star is immediately WNW of
the core and interferes a little with the
galaxy, while a bright field star is 3.5′
SE. Radial velocity: 7664 km/s.
Distance: 342 mly. Diameter:
130 000 ly.

NGC2805

H878[3]	9:20.3	+64 06	GLX
SAB(rs)d	6.3′ × 4.8′	11.79B	

The DSS image reveals a large, face-on
spiral galaxy with a very small, brighter
core. Faint, low-mass spiral arms are
loosely wound around the core; the
arms are dotted with several knots and
condensations. One field star is 6.3′
NNW, while another is 4.25′ SSE.

NGC2805

Companion galaxies NGC2814,
NGC2820 and IC2458 lie to the NE.
They share similar radial velocities and
it is likely that the three galaxies are
physically related. Telescopically,
NGC2805 is large, round and very dim;
it is a low-surface-brightness galaxy
with ill defined edges and a very small,
star-like core. Two other galaxies
(NGC2814 and NGC2820) can be seen
in the same medium-magnification field.
Radial velocity: 1812 km/s. Distance:
81 mly. Diameter: 148 000 ly.

NGC2810

H749[3]	9:22.1	+71 50	GLX
E	1.3′ × 1.1′	13.24B	

Presenting a similar appearance
visually and photographically, this is a
very small and moderately faint galaxy,
visible at medium magnification and
well seen at higher power. It is rather
condensed, almost round and brightens
to a small core, while its edges are

fairly well defined. Radial velocity:
3687 km/s. Distance: 165 mly.
Diameter: 62 000 ly.

NGC2814

H868[2]	9:21.2	+64 15	GLX
Sb:	1.1′ × 0.3′	14.33B	

This is an edge-on galaxy, which is
shown in photographs to have a bright
major axis and a slightly curved, faint
extension in the S with a field star 1.1′ to
the SSW. Companion galaxy NGC2820
is 4.0′ E while IC2458 is 2.1′ ESE. The
minimum core-to-core separation
between NGC2814 and NGC2820
would be about 86 000 ly at the
presumed distance. Visually, the galaxy
initially appears small and round at
medium magnification, but averted
vision reveals an elongated streak
oriented N–S immediately NNE of the
aforementioned field star. Radial
velocity: 1680 km/s. Distance: 75 mly.
Diameter: 24 000 ly.

NGC2814/2820

NGC2820

H869[2]	9:21.8	+64 16	GLX
SB(s)c pec sp	4.4′ × 0.4′	13.24B	

This is an edge-on galaxy with a bright, elongated central region and fainter extensions; photographs show evidence of dust patches along the major axis and the companion galaxy IC2458 is in contact to the WSW. Visually, this is the brightest and second largest galaxy visible in a medium-magnification field; it is a well defined flat streak of light elongated ENE–WSW. Radial velocity: 1662 km/s. Distance: 74 mly. Diameter: 95 000 ly.

NGC2841

H205[1]	9:22.0	+50 58	GLX
SA(r)b	8.1′ × 3.5′	10.06B	9.2V

This is one of the finest galaxies in the Herschel catalogue. The DSS image

NGC2841

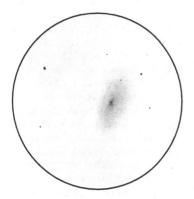

NGC2841 15-inch Newtonian 293x

shows a large, bright, highly inclined spiral galaxy with a large, elongated core. Several, thin, tightly wound spiral arms emerge from the core and many dust patches and dust lanes are involved in the somewhat fragmentary spiral structure. This is a large and very bright object visually, featuring a very bright, star-like core embedded in a bright, elongated inner disc, and is best seen at medium magnification. The outer disc is a large extended oval, oriented NNW–SSE with hazy ill defined edges. The outer disc is a little grainy and extends to the middle star of a triangle of field stars situated in the NNW. Radial velocity: 674 km/s. Distance: 30 mly. Diameter: 71 000 ly.

NGC2854

H714[3]	9:24.0	+49 13	GLX
SB(s)b	1.7′ × 0.6′	13.93B	

This galaxy is part of an attractive triple system which includes NGC2856

and NGC2857. There appears to be some interaction between NGC2854 and NGC2856 and together they are also catalogued as Arp 285. On the DSS photograph, NGC2854 is a highly inclined, barred spiral galaxy with a bright, squarish, elongated bar highlighted by a considerable dust patch on the SE flank. Two bright spiral arms emerge from the bar and after making a sharp 180° turn, both arms become long and straight, the SE arm being in line with a plume being ejected from NGC2856, which is 3.5′ to the NE. The NW arm fades suddenly and curves slightly towards the S. Visually, the three galaxies form a beautiful triplet, which can just be fit into a high-magnification field. NGC2654 is a moderately bright, well defined, elongated object oriented NE–SW and broadly brighter to the middle. Radial velocity: 2800 km/s. Distance: 125 mly. Diameter: 62 000 ly.

well defined and the galaxy seems a little more distinct along the NNE flank; no core is visible. Radial velocity: 3276 km/s. Distance: 146 mly. Diameter: 107 000 ly.

NGC2880

H260[1]	9:29.6	+62 30	GLX
SB0⁻	2.4′ × 1.5′	12.52B	

This is a lenticular galaxy with a large bright core and an extensive outer envelope, elongated SE–NW. Photographs show a compact companion galaxy 3.25′ N. At high magnification, this is a very bright galaxy featuring a small, star-like and very bright core embedded in a condensed, opaque envelope which has well defined edges. A very faint, diffuse outer envelope surrounds the inner envelope and several faint field stars accompany the galaxy, especially to the W. Radial velocity: 1690 km/s. Distance: 75 mly. Diameter: 53 000 ly.

NGC2950

H68[4]	9:42.6	+58 51	GLX
(R)SB(r)0°	2.9′ × 2.0′	11.93B	

Photographs show an inclined, barred lenticular galaxy with a large, bright core. Characteristically, the bar brightens where it touches the inner ring. This inner ring is surrounded by a very faint and broad outer ring. Four faint stars are involved in the outer ring to the W and the galaxy is elongated ESE–WNW. Visually, the galaxy is quite bright and large with a bright, slightly elongated envelope surrounding a sizable, bright core. The envelope is oriented NNW–SSE and is fairly well defined. Discovered on 19 March 1790, Herschel's description reads: ' Very bright, small, exactly round. A bright nucleus in the middle and very faint chevelure very gradually joining to the north. In a lower situation the chevelure might not be visible, and this nebula would then appear like an ill defined planetary one.' Radial velocity: 1405 km/s. Distance: 63 mly. Diameter: 53 000 ly.

NGC2854/2856

©STScI/AURA/CALTECH/ROE

NGC2856

H713[3]	9:24.2	+49 15	GLX
S?	1.1′ × 0.5′	14.01B	

Photographs show a highly inclined spiral galaxy with a very bright, elongated core and fainter spiral extensions. A thin dust lane extends along the major axis on the NE flank, while a faint, thin plume extends from the galaxy to the NE. A bright field star is 3.5′ W, while companion galaxy NGC2854 is 3.5′ SW. The face-on spiral galaxy NGC2857, which is not a Herschel object, is 7.25′ NNE. If the galaxies form a physical system the core-to-core separation between NGC2854 and NGC2856 would be about 120 000 ly at the presumed distance, while the separation between NGC2856 and NGC2857 would be 263 000 ly. Visually, NGC2856 appears marginally brighter and more distinct than NGC2854; it is a SE–NW elongated object of fairly

even surface brightness and is very well defined at the edges. NGC2857 is a very large, diffuse glow, very broadly brighter to the middle and is round and ill defined. Radial velocity: 2667 km/s. Distance: 119 mly. Diameter: 38 000 ly.

NGC2870

H846[3]	9:27.8	+57 23	GLX
Sbc	2.5′ × 0.6′	13.85B	

Photographs show an almost edge-on spiral galaxy with a small, brighter core. The outer envelope is grainy and a thin dust lane crosses S of the core, extending towards the SE. There are other dust lanes throughout the disc defining the outer spiral pattern. Although this galaxy is faint visually, its form is well seen at both medium and high magnification as an elongated patch of light of uneven surface brightness. It is oriented ESE–WNW, the edges are fairly

NGC2976

H285[1]	9:47.3	+67 55	GLX
SAc pec	5.9′ × 2.7′	10.81B	10.2V

This galaxy is a likely member of the M81/NGC2403 group of galaxies, though its radial velocity suggests it is closer than the principal members of that group. Photographs reveal a peculiar morphology. At first glance it would appear to be an irregular galaxy, but the bright, very knotty envelope has several dust patches involved and a dust lane along the SW flank, suggesting a spiral arm. There is no core visible and close examination reveals fragments of spiral structure in the envelope. Stars begin to resolve at about magnitude 18.5. Visually, this is a large and bright galaxy; the main envelope is prominent and very slightly brighter to the middle. The galaxy is elongated SE–NW with fairly well defined edges. The envelope is very mottled and a faint field star is located on the edge of the galaxy to the

NGC2976

SW. Radial velocity: 105 km/s. Distance: 4.7 mly. Diameter: 8100 ly.

NGC2985

H78[1]	9:50.4	+72 17	GLX
(R′)SA(rs)ab	4.6′ × 3.6′	11.22B	

Images show an almost face-on spiral galaxy, elongated N–S, featuring a large, brighter core in a bright inner envelope. The outer spiral pattern is very faint and somewhat asymmetrical; high-resolution photographs reveal a spiral of the multi-arm type with thin, tightly wound spiral structure. A field star is involved 1.0′ E of the core. Visually, this is a very bright galaxy. A small, intense core is surrounded by a bright envelope which fades gradually into the sky background and has ill defined edges. The envelope is grainy and quite mottled; the faint field star is prominent E of the core. Radial velocity: 1440 km/s. Distance: 64 mly. Diameter: 86 000 ly.

NGC2998

H717[2]	9:48.7	+44 05	GLX
3.3′ × 1.8′	13.11B		

This is an inclined, multi-arm spiral galaxy with a small, brighter core. Images reveal six distinct bright, narrow spiral arms with many knotty condensations. The galaxy is the brightest of a compact group of galaxies whose other members (NGC3002, NGC3005, NGC3006, NGC3008, NGC3009 and NGC3010) are arrayed to the E and NE. Visually, the galaxy is a moderately bright and fairly large, smooth-textured, oval disc oriented NE–SW with fairly well defined edges. It is very broadly brighter to the middle, though no core is seen. Radial velocity: 4795 km/s. Distance: 214 mly. Diameter: 205 000 ly.

NGC3027

H23[5]	9:55.7	+72 12	GLX
SB(rs)d:	4.1′ × 1.7′	12.20B	

The DSS image shows a very faint, barred spiral galaxy with a short, brighter bar and weak, patchy spiral arms. The arms are asymmetrical and grainy with a few brighter condensations, particularly to the SE. Telescopically, this is a large and very dim galaxy that is best seen at medium magnification. The disc has very low surface brightness, is ill defined and is elongated ESE–WNW. Radial velocity: 1176 km/s. Distance: 53 mly. Diameter: 63 000 ly.

NGC3034

H79[4]	9:55.8	+69 41	GLX
I0	11.2′ × 4.3′	8.94B	8.4V

Credit for discovery of this object, best known as M82, goes to Johan Bode who observed it along with M81 on 31 December 1774. It was subsequently observed by Messier's collaborator Pierre Méchain in August 1779 and formally recorded in Messier's catalogue on 9 February 1781. John Herschel mistakenly thought his father had discovered this object and included it in an appendix published in his *Cape*

NGC3034 15-inch Newtonian 293x

Observations in 1847, designating it H79[4]. The appendix listed eight objects discovered by his father but not included in the catalogues published in the *Philosophical Transactions*. William Herschel recorded M82 during his last sweep of the heavens (Sweep 1112) on 30 September 1802, calling it: 'A very bright, beautiful ray of light, about 8'

long, 2' or 3' broad; brightest in the middle of all its lengths.' He had previously observed it during Sweep 1100 on 8 November 1801, but in neither case did he mention that it was Bode's or Messier's object, which probably contributed to John Herschel's error. The galaxy has also been catalogued as Arp 337 and it is a member of the M81/NGC2403 group of galaxies. In photographs, it is a remarkable object, a starburst galaxy which is emitting hydrogen alpha tendrils of matter from its core, perpendicularly to the major axis of the galaxy. The structure is extremely complex with a major dust lane at its core and several dust patches radiating from the central region. The disc is very bright with a dust lane particularly prominent along the SSE flank. In a moderate-aperture telescope, this is a large and very bright galaxy, much extended ENE–WSW with a dust lane cutting the galaxy perpendicularly NNW–SSE. It is bright and more

extensive to the WSW, where it is mottled and tapers gradually to a blunt point. There is a very bright patch in the WSW extension just WSW of the dust lane and another bright patch in the ENE extension next to the dust lane: the ENE extension is a little fainter and blunter than the one to the WSW. A hazy, secondary envelope is intermittently visible. Radial velocity: 312 km/s. Distance: 14 mly. Diameter: 46 000 ly.

NGC3043

H835[2]	9:56.2	+59 18	GLX
Sb: sp	1.7' × 0.6'	8.94B	

Photographs show a highly inclined spiral galaxy with a brighter core and two bright, narrow spiral arms, possibly connected to a bar. A large dust patch is W of the core. Visually, the galaxy is fairly dim and is best seen at medium magnification. It is an elongated streak of light, oriented E–W and fairly even in surface brightness that fades quickly at the edges. Radial velocity: 3066 km/s. Distance: 137 mly. Diameter: 68 000 ly.

NGC3063

H909[2]	10:01.6	+72 08	DS

This double star, which was mistaken for a nebula, was discovered during Herschel's last sweep of the heavens on 30 September 1802. It is in the field with two galaxies, NGC3065 and NGC3066, which were both discovered 17 years earlier on 3 April 1785. The faint pair of stars forms a triangle with the two galaxies and visually the double star is best seen at high magnification as a blurry patch of light 2.3' WSW from NGC3066. The stars are difficult to resolve and appear hazy; it is easy to see how they could be mistaken for a nebula.

NGC3065

H333[2]	10:01.9	+72 10	GLX
SA(r)0°	1.7' × 1.7'	11.76B	

Photographically, this lenticular galaxy has a bright round core and an extensive, faint outer envelope. A faint double star 3.25' SSW is probably NGC3063 = H909[2], though Herschel

NGC3034

described it as: 'Faint, pretty large, round.' The galaxy forms a physical pair with NGC3066 3.0′ SSE. The minimum core-to-core separation between the two galaxies would be about 83 000 ly at the presumed distance. Visually, this is a bright pair in a high-magnification field. Only the bright core of NGC3065 is visible, so the galaxy appears smaller than it does in photographs but it is quite bright, almost round, well defined and opaque. It is located 1.5′ ESE from a moderately bright field star. Radial velocity: 2118 km/s. Distance: 95 mly. Diameter: 47 000 ly.

NGC3066

H334[2]	10:02.2	+72 07	GLX
(R′)SAB(s)bc pec	1.1′ × 0.9′	13.52B	

This is a face-on barred spiral galaxy with a small, round, brighter core. Photographs show two spiral arms emerging, which are somewhat asymmetrical in structure. The faint double star 2.25′ WSW is probably NGC3063 = H909[2]. Visually, the galaxy appears a little larger than its companion galaxy NGC3065 but is similar in brightness and is well defined with even brightness across its disc which is elongated E–W. Radial velocity: 2167 km/s. Distance: 97 mly. Diameter: 31 000 ly.

NGC3073

H853[3]	10:00.9	+55 37	GLX
SAB0−	1.3′ × 1.2′	14.11B	13.4V

In photographs this is a lenticular galaxy with a large, bright core and a very faint outer envelope. Telescopically, this galaxy is a small, very faint, unconcentrated, almost round patch of light, with diffuse edges. Large-aperture telescopes may show an oval galaxy in the field 6.75′ NNE. NGC3079 is 9.75′ ENE. Radial velocity: 1213 km/s. Distance: 54 mly. Diameter: 21 000 ly.

NGC3077

H286[1]	10:03.3	+68 44	GLX
I0 pec	5.5′ × 4.1′	10.62B	10.1V

This is a dwarf irregular galaxy of peculiar morphology. Photographs

NGC3077

show it is similar in form to an elliptical or lenticular galaxy with a brighter core and faint outer envelope. However, there are several filamentary dust lanes radiating like spokes outward from the core into the outer envelope. Another unusual feature of the galaxy is that resolution into individual stars is difficult despite the fact that the galaxy is evidently nearby and very likely a member of the M81/NGC2403 group of galaxies. Telescopically, the galaxy is situated 3.75′ SSE from a bright field star. It is a large and quite bright galaxy, elongated ENE–WSW; it is seen as a large oval broadly brighter to the middle with a mottled texture. A brighter core is intermittently visible and a hazy, faint secondary envelope fades slowly into the sky background. As with NGC2976, the radial velocity of this galaxy suggests that it is closer than the principal members of the M81/NGC2403 group. Radial velocity: 120 km/s. Distance: 5.4 mly. Diameter: 8600 ly.

NGC3079

H47[5]	10:02.0	+55 41	GLX
SB(s)c	7.9′ × 1.4′	11.41B	10.9V

The DSS image shows an almost edge-on spiral galaxy with a bright, somewhat irregular core. A bright, knotty spiral arm and a prominent dust lane pass in front of the core. A broad, fainter spiral arm is S of the core and

NGC3079

overall the faint extensions appear inclined somewhat in comparison to the bright inner region. In moderate-aperture telescopes, this is a bright, much elongated galaxy oriented SSE–NNW with a peculiar, disturbed structure, especially at high magnification. The core is offset in the disc towards the S and appears bright, well defined and elongated along the major axis. There is a dark gap just N of the core and the extension appears curved on the E side. A magnitude 13 star is visible at the N edge of the disc and a bright triangle of field stars brackets the southern extension of the galaxy. An anonymous galaxy is 6.5′ WNW and NGC3073 is 9.75′ WSW. The radial velocities of NGC3073 and NGC3079 are similar and together with the anonymous galaxy, they may form a small group. The core-to-core separation between NGC3073 and NGC3079 would be about 149 000 ly at the presumed distance and that between NGC3079 and the anonymous galaxy would be about 100 000 ly. The famous 'twin' quasar 0957+561A/B is about 13.0′ NNW. Radial velocity: 1174 km/s. Distance: 53 mly. Diameter: 121 000 ly.

NGC3102

H916[3]	10:04.6	+60 07	GLX
S0⁻:	0.9′ × 0.9′	14.27B	

Photographs show a lenticular galaxy with a large, round core and an extensive outer envelope. Visually, the galaxy is moderately bright and is best seen at high magnification as a round, condensed patch of light with a small, bright core. The disc is grainy and the edges are moderately well defined. A field star is 0.9′ SSE of the core. Radial velocity: 3140 km/s. Distance: 140 mly. Diameter: 37 000 ly.

NGC3182

H265[1]	10:19.5	+58 12	GLX
SA(r)a?	1.5′ × 1.2′	13.06B	

This lenticular galaxy has a bright core and a fainter outer envelope and is elongated NW–SE. Herschel placed it in

his bright nebula class but visually the galaxy is moderately faint and very slightly oval. The disc is hazy and somewhat ill defined with a brighter, small core embedded. Radial velocity: 2141 km/s. Distance: 96 mly. Diameter: 42 000 ly.

NGC3184

H168[1]	10:18.3	+41 25	GLX
SAB(rs)cd	7.4′ × 6.9′	10.41B	

Images show a large, bright, face-on spiral galaxy with a small, round core. The core is surrounded by a small inner disc from which two main spiral arms emerge and then expand to a broad spiral pattern. The structure of the arms is complex with several knots and condensations, including a particularly large and bright one SW from the core. A field star is involved in the spiral structure, 1.75′ N of the core. The brightest stars in the galaxy have an apparent magnitude of +22. Visually, this is a perfectly round, bright, though

very diffuse galaxy. It is a little brighter to an almost stellar core at medium magnification, though this core is faint. The envelope is large, quite diffuse and fades gradually into the sky background. The envelope extends to the bright field star. Discovered on 18 March 1787, Herschel's description reads: 'Considerably bright, round, very gradually brighter to the middle. 8′ diameter, a considerable star in it, unconnected.' Radial velocity: 597 km/s. Distance: 27 mly. Diameter: 58 000 ly.

NGC3188

H910[3]	10:19.7	+57 25	GLX
(R)SB(r)ab	0.9′ × 0.7′	15.15B	

This is a face-on, barred spiral galaxy with a small, brighter core and a N–S bar. One bright arm forms a partial ring around the galaxy before broadening and fading in the W. Photographs show a very faint galaxy, NGC3188A, is 0.6′ WSW and this may be an

NGC3184

actual physical companion. At the presumed distance, the minimum separation between the two galaxies would be about 61 000 ly. In a moderate-aperture telescope, NGC3188 is an extremely dim galaxy, only just detectable at medium magnification and not seen at higher powers. It has a very faint, roundish envelope with a very small, brighter core, which is sometimes mistaken for a star; concentrating on the core causes the outer envelope to disappear. Herschel came upon this galaxy on 8 April 1793; his description of this object is somewhat puzzling: 'Very faint, pretty large, irregular figure, resolvable, some of the stars visible.' Radial velocity: 7867 km/s. Distance: 351 mly. Diameter: 92 000 ly.

NGC3192

H704[3]	10:19.0	+46 27	GLX
SB(s)bc pec	0.8′ × 0.6′	14.28B	

The DSS image shows a very small galaxy elongated N–S, with a prominent core and two spiral arms. An almost stellar galaxy is 1.25′ W; some atlases plot two galaxies here, NGC3191 and NGC3192, but it is almost certain that only NGC3192 actually exists. The French observer Guillaume Bigourdan could not find NGC3192 with a 12-inch refractor, but Herschel came upon this object on 5 February 1788 and curiously, his description reads: 'Very small, very faint, perhaps a patch of small stars.' Visually, this object is a very small and faint, unconcentrated patch of light which is just a little brighter to the middle, fairly diffuse and almost round. Radial velocity: 9234 km/s. Distance: 412 mly. Diameter: 96 000 ly.

NGC3198

H199[1]	10:19.9	+45 33	GLX
SB(rs)c	8.5′ × 3.3′	10.92B	10.3V

Images show a highly inclined barred spiral galaxy with a small, brighter core in a short bar. The two main spiral arms

NGC3198

©STScI/AURA/CALTECH/ROE

are tangent to the bar but extend to form a pseudo-ring. Each arm branches after about half a revolution to form a broader spiral pattern with several knots and condensations. A double star is 3.6′ NNE. In a moderate-aperture telescope this is a very large and bright galaxy which is well seen even at low magnification. Much extended in a NE–SW direction, it is very well defined and bright along its major axis, though no core is visible at medium magnification. The envelope appears very smooth and the extremities are rounded. Radial velocity: 684 km/s. Distance: 31 mly. Diameter: 75 000 ly.

NGC3202

H720[2]	10:20.5	+43 01	GLX
SB(r)a	1.3′ × 0.9′	14.17B	

Photographs show an almost face-on, barred spiral galaxy with a small, brighter core and a slightly fainter bar attached to a ring. Three spiral arms emerge from the ring and the galaxy is elongated N–S. Two faint field stars border the galaxy in the E. A brighter field star is 1.2′ W. Companion galaxy NGC3205 is 4.5′ SE and NGC3207 is 5.8′ ESE. The three galaxies very probably form a physical group.

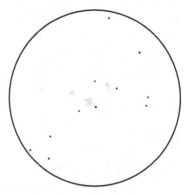

NGC3202 NGC3205 NGC3207
15-inch Newtonian 146x

NGC3202/3205/3207

slightly concave on its western flank. Radial velocity: 1215 km/s. Distance: 54 mly. Diameter: 44 000 ly.

NGC3207

H722[2]	10:21.0	+42 58	GLX
S?	0.9′ × 0.7′	14.46B	

This is a highly inclined galaxy, probably spiral, though it could be lenticular. Photographs show the galaxy has a bulging, brighter core with faint extensions elongated E–W. Visually, it is the smallest and faintest of trio NGC3207, NGC3202, NGC3205 and appears only as a round patch with poorly defined edges. Radial velocity: 7023 km/s. Distance: 314 mly. Diameter: 82 000 ly.

NGC3220

H911[3]	10:23.7	+57 02	GLX
Sb: sp	1.2′ × 0.5′	14.25B	

Photographs show an almost edge-on, possibly spiral galaxy elongated E–W, with a slightly brighter core and fainter extensions. The fainter galaxy NGC3214 is 5.0′ WNW. Visually, this probable dwarf galaxy is very faint and is best seen at medium magnification. It is an ill defined patch of light which is very slightly oval and extended E–W and is situated ESE of a faint triangle of field stars. NGC3214 may be visible in large-aperture telescopes. Radial velocity: 1236 km/s. Distance: 55 mly. Diameter: 19 000 ly.

Visually, all three galaxies fit comfortably in a high-magnification field. NGC3202 is the second largest and brightest of the group; it is an oval, moderately well defined patch of light, very slightly brighter to the middle. Radial velocity: 6726 km/s. Distance: 300 mly. Diameter: 114 000 ly.

NGC3205

H721[2]	10:20.8	+42 57	GLX
S?	1.1′ × 0.9′	14.23B	

This almost face-on spiral galaxy has a bright core, very faint spiral arms and appears elongated N–S. A faint field star is immediately NE of the core and a brighter field star is 1.0′ WSW. Examination of a DSS image reveals that the galaxy is larger than its published dimensions, appearing about 1.6′ × 1.3′. It is visible in a high-magnification field with NGC3202 and NGC3207 and appears to be the largest and brightest of the trio, an oval patch which is gradually brighter to the middle. Radial velocity: 7052 km/s. Distance: 315 mly. Diameter (based on published apparent size): 146 000 ly.

NGC3206

H266[1]	10:21.8	+56 56	GLX
SB(s)cd	2.8′ × 1.5′	13.48B	

This is an almost face-on, barred spiral galaxy with a brighter, thin bar attached to an asymmetric spiral pattern. Photographs show several knots and condensations in the broad spiral pattern with four brighter knots in the arm to the SW. Telescopically, the galaxy is moderately large and well seen at medium magnification, but it is quite a dim galaxy with relatively even surface brightness. No core is visible and the edges are very ill defined. It is elongated SSW–NNE, is somewhat crescent-shaped, and appears very

NGC3225

H882[2]	10:25.1	+58 09	GLX
Scd:	2.0′ × 1.0′	13.83B	

Images reveal an inclined spiral galaxy with a small, bright core and faint spiral arms, elongated SSE–NNW. Telescopically, the galaxy is very faint, and is best seen at medium magnification as a hazy, roundish, ill defined patch of light, with even surface brightness. A very faint field star is visible 1.0′ to the NE beyond the envelope. Radial velocity: 2204 km/s. Distance: 98 mly. Diameter: 57 000 ly.

NGC3237

H631³	10:25.8	+39 39	GLX
(R)SAB0°	1.3′ × 1.3′	14.09B	

This almost face-on, barred lenticular galaxy has a large, bright core and a very faint bar attached to a bright, elongated inner ring. Photographs show that a detached, fainter outer ring surrounds the galaxy. A faint field galaxy is 2.25′ NW. In a moderate-aperture telescope, NGC3237 is a very faint and difficult galaxy. It initially appears star-like at all magnifications, but medium powers reveal a faint outer envelope surrounding the core. It is brighter to the middle and the envelope is round and fairly well defined. Herschel discovered this galaxy on 18 March 1787, using a magnification of 300× to reveal its nonstellar nature. Radial velocity: 7078 km/s. Distance: 316 mly. Diameter: 120 000 ly.

NGC3238

H883²	10:26.7	+57 14	GLX
SA(r)0°	1.4′ × 1.3′	14.14B	

Images reveal a lenticular galaxy with a brighter core and extensive outer envelope. A fainter galaxy, MCG+10-15-79, is 5.5′ to the NW. In a moderate-aperture telescope NGC3238 is moderately faint and almost round and features a star-like core surrounded by a faint outer envelope which is quite well defined. Radial velocity: 7436 km/s. Distance: 332 mly. Diameter: 135 000 ly.

NGC3259

H870²	10:32.6	+65 03	GLX
SAB(rs)bc	1.4′ × 0.9′	13.01B	

This inclined spiral galaxy has a small, bright core in a bright inner envelope. Fainter, grainy spiral arms emerge and photographs show very faint spiral extensions which increase the length of the major axis to about 2.0′. Visually, the galaxy is small but moderately bright, very slightly oval in shape and oriented NNE–SSW. The disc is well defined and opaque and the very small brighter core stands out well. A bright field star is 3.25′ NW. Radial velocity:

1782 km/s. Distance: 80 mly. Diameter: 46 000 ly.

NGC3266

H871²	10:33.3	+64 45	GLX
SAB0°?	1.5′ × 1.3′	13.46B	

Photographs show a barred lenticular galaxy with a bright core and a N–S extension embedded in a faint outer envelope elongated E–W. At high magnification only the bright inner disc of this galaxy is seen. It appears almost round, fairly bright and small and is well condensed with well defined edges and a brighter core. Radial velocity: 1860 km/s. Distance: 83 mly. Diameter: 36 000 ly.

NGC3286

H912³/H917³	10:36.3	+58 37	GLX	
E?	0.8′ × 0.5′	14.31B	13.7V	

This object was recorded by Herschel on 8 and 9 April 1793. He used different reference stars on both nights and inadvertently thought he had recorded two different objects. NGC3286 is an elliptical galaxy which forms a probable physical pair with NGC3288, situated 3.9′ to the S. At the presumed distance, the minimum core-to-core separation between the two galaxies would be about 415 000 ly. In photographs the galaxy has a bright, elongated core in a fainter outer envelope and is elongated E–W. Visually, the pair of galaxies is quite small and faint, with NGC3286 being a little brighter; it is a round, well defined patch of light, even in surface brightness but with a brighter core suspected. Radial velocity: 8168 km/s. Distance: 365 mly. Diameter: 85 000 ly.

NGC3288

H918³	10:36.4	+58 33	GLX	
SABbc	1.1′ × 0.9′	15.03B	14.0V	

Photographs show a face-on spiral galaxy with a small, round core in fainter spiral arms. Telescopically, it can be seen in the same field as NGC3286 as a round, unconcentrated patch which is even in surface brightness and

moderately well defined. Radial velocity: 8220 km/s. Distance: 367 mly. Diameter: 118 000 ly.

NGC3298

H767³	10:37.2	+50 06	GLX
0.6′ × 0.4′	14.82B		

On the DSS photograph, this is the brightest member of a faint, scattered grouping of galaxies. The galaxy is unclassified but appears to be an elliptical, oriented SE–NW, with a brighter core in a faint outer envelope. This galaxy is very faint visually and is only seen with certainty at medium magnification when it appears as a small, unconcentrated and blurry patch of light which is almost round and fairly well defined. Radial velocity: 13 553 km/s. Distance: 605 mly. Diameter: 106 000 ly.

NGC3310

H60⁴	10:38.7	+53 30	GLX
SAB(r)bc pec	3.5′ × 3.2′	11.28B	

This is a face-on spiral galaxy with a peculiar morphology and has also been catalogued as Arp 217. The DSS image shows a bright, irregularly round core with two short spiral arms attached. Several brighter knots are N of the core and filamentary, elongated knots appear S of the core. Two faint arcs, the longer one W of the core and the shorter one SW of the core, appear well away from the main disc of the galaxy and a broad hazy plume to the N increases the physical dimensions to about 5.0′ × 4.3′. High-resolution images apparently show two nuclei separated by about 2″, suggesting that the peculiar structure may be the result of a galaxy merger. Visually, the galaxy is very bright, the central region appears irregularly round and the object is broadly brighter to the middle but with no brighter core visible. The edges are fairly well defined but appear embedded in an indistinct glow which fades uncertainly into the sky background. Recorded by Herschel on 12 April 1789, he placed this galaxy in his Class IV, describing it as: 'Very bright, round. Planetary, but very ill defined. The indistinctness on the edges is sufficiently extensive to make this a step between planetary nebula and those which are described as very suddenly much brighter to the middle.' Radial velocity: 1048 km/s. Distance: 47 mly. Diameter: 68 000 ly.

NGC3319

H700[3]	10:39.2	+41 41	GLX
SB(rs)cd	6.2′ × 3.4′	11.77B	11.1V

This inclined, barred spiral galaxy appears very faint on the DSS photograph. The brighter, narrow bar has two faint spiral arms emerging: they are thin with a few nebulous patches involved, particularly to the SW. The galaxy is large but very dim visually and is best seen at medium magnification as an ill defined elongated glow, oriented NE–SW with a thin, brighter bar in the middle. Radial velocity: 749 km/s. Distance: 33 mly. Diameter: 61 000 ly.

NGC3320

H745[2]	10:39.6	+47 24	GLX
SBc	2.1′ × 1.0′	12.79B	

Photographs show an inclined spiral galaxy inclined NNE–SSW with a slightly brighter core and grainy spiral arms. The galaxy is fairly bright visually but somewhat diffuse and is best seen at medium magnification. It is a fairly large, oval patch of light which is broadly brighter to the middle with ill defined edges and a faint star near its SSW tip. Radial velocity: 2363 km/s. Distance: 106 mly. Diameter: 65 000 ly.

NGC3348

H80[1]	10:47.2	+72 50	GLX
E0	2.0′ × 2.0′	11.94B	

This is a prototypical E0 elliptical galaxy with a bright core and fainter outer envelope. Photographs show a star immediately E of the core and a fainter one immediately N; both are within the outer halo of the galaxy. Visually, the galaxy is small but very bright and is quite round with a bright, mottled, opaque envelope which is fairly well defined at the edges. The core is sharp

and very bright and the brighter of the two stars can be seen within the envelope. Another field star is 1.6′ WNW of the galaxy core. Radial velocity: 2961 km/s. Distance: 132 mly. Diameter: 77 000 ly.

NGC3353

H842[3]	10:45.4	+55 58	GLX
BCD/Irr	1.3′ × 1.0′	13.23B	12.8V

This is an inclined spiral galaxy with peculiar morphology. Photographs show a large, bright core in a bright inner envelope, with two spurs emerging from the core to the NE and a nebulous knot in the fainter outer envelope to the SW. Telescopically, the galaxy is located 1.6′ N of a faint field star. It is moderately bright, fairly opaque and elongated NE–SW with a faint, tapered extension on the SW side. The surface brightness is even across the envelope. Radial velocity: 1009 km/s. Distance: 45 mly. Diameter: 17 000 ly.

NGC3359

H52[5]	10:46.6	+63 13	GLX
SB(rs)c	7.2′ × 4.4′	11.06B	10.6V

The DSS image reveals a beautiful, almost face-on, barred spiral galaxy. The narrow, brighter bar attaches to narrow, loosely wound spiral arms. The arms are grainy with several brighter knots and condensations and the structure is somewhat fragmentary. Telescopically, it is visible in a low-power field as a large, moderately bright, but very diffuse galaxy situated in a star-poor field. The disc is fairly even in surface brightness but quite mottled and patchy in texture with a slightly brighter central region. The edges are not well defined and the galaxy is elongated N–S. Herschel recorded the galaxy on 28 November 1793, describing it as: 'Considerably bright, extended in the meridian, very gradually brighter to the middle. About 5′ long and 3′ broad; the nebulosity seems to be of the milky kind; it loses

NGC3359

itself imperceptibly all around. The whole breadth of the sweep seems to be affected with very faint nebulosity.' Radial velocity: 1105 km/s. Distance: 49 mly. Diameter: 103 000 ly.

NGC3364

H318[3]	10:48.5	+72.25	GLX
SAB(rs)c	1.5' × 1.5'	13.47B	

Images reveal a face-on spiral galaxy with a small, round core and dim spiral arms. The arms are grainy and somewhat asymmetric. A compact galaxy is 8.5' SW. Visually, this is a very faint and extremely diffuse galaxy which is best seen at medium magnification. It is almost round with even surface brightness across the disc and is poorly defined at the edges. Radial velocity: 2854 km/s. Distance: 127 mly. Diameter: 56 000 ly.

NGC3374

H701[3]	10:47.9	+43 10	GLX
SBc	1.3' × 0.9'	14.56B	

This is a face-on barred spiral galaxy with a small, round core and a faint bar. Photographs show the bar is attached to an inner ring which probably comprises overlapping spiral structure. Three faint spiral arms emerge from the ring. CGCG212-055, a tiny edge-on galaxy, is 2.3' SSW. NGC3374 is a very faint object visually; it is a small, ill defined and hazy patch of light which is evenly bright and irregularly round. Radial velocity: 7486 km/s. Distance: 334 mly. Diameter: 126 000 ly.

NGC3392

H881[3]	10:51.0	+65 46	GLX
E?	0.9' × 0.7'	14.64B	

The photographic and visual appearances of this elliptical galaxy are quite similar and it forms a physical pair with NGC3394 4.0' to the SW. It is smaller though brighter than NGC3394 and is well seen at high magnification as a very slightly oval disc elongated ESE–WNW and brighter to the middle with well defined edges. Radial velocity:

3357 km/s. Distance: 150 mly. Diameter: 39 000 ly.

NGC3394

H872[2]	10:50.6	+65 43	GLX
SA(rs)c	1.9' × 1.6'	13.43B	

This is an almost face-on spiral galaxy with a very small core and grainy spiral arms that expand into a fainter outer envelope. It almost certainly forms a physical pair with NGC3392: the minimum core-to-core separation between the two galaxies would be about 175 000 ly. Visually, the galaxy is very diffuse and is best seen at medium magnification when it appears almost round and mottled, but with even surface brightness across its disc. The edges are poorly defined and the galaxy is situated in the middle of a pair of field stars aligned roughly N–S. Radial velocity: 3504 km/s. Distance: 157 mly. Diameter: 87 000 ly.

NGC3398

H792[3]	10:51.5	+55 23	GLX
SA:(s:)b?	1.2' × 0.5'	14.49B	

This highly inclined spiral galaxy has a small core in a grainy inner envelope. The DSS image shows extremely faint spiral arms emerging to the NE and SW, though the main disc is elongated E–W. A faint galaxy is 4.5' N while a bright field star is 4.7' ESE. This galaxy is quite dim visually and difficult to see unless conditions are excellent. It is very small and is little more than an elongated blur oriented E–W. The bright field star hinders observation and the disc is even in surface brightness. Radial velocity: 2930 km/s. Distance: 131 mly. Diameter: 46 000 ly.

NGC3407

H919[3]	10:52.3	+61 23	GLX
S0⁻:	1.2' × 0.4'	14.60B	

Images show a lenticular galaxy elongated NNE–SSW with a bright, elongated core and a fainter outer envelope. Telescopically, the galaxy is very small, though moderately bright. It is situated WSW of a triangle of faint

field stars. It is very slightly extended with a tiny, brighter core embedded in a well defined envelope. Radial velocity: 5120 km/s. Distance: 229 mly. Diameter: 80 000 ly.

NGC3408

H913[3]	10:52.2	+58 26	GLX
Sc:	1.1' × 0.8'	14.29B	

This is a face-on spiral galaxy. Photographs show a small, round core and bright, asymmetrical spiral arms. A bright spiral arm emerges from the core in the S and curves E, then N, fading abruptly after a half revolution. The galaxy is moderately faint visually and is a little more distinct at medium magnification. It appears as a round, fairly well defined object which is a little brighter to the middle. A bright field star is 4.1' SSW. Radial velocity: 9583 km/s. Distance: 428 mly. Diameter: 137 000 ly.

NGC3415

H718[2]	10:51.7	+43 43	GLX
SA0⁺:	1.9' × 1.1'	13.42B	

A casual inspection of the DSS image shows a lenticular galaxy with a bright, elongated core and a fainter outer envelope, but there is a possibility that this galaxy is, in fact, spiral. The extremely faint galaxy NGC3416 is 3.3' NNE. Visually, NGC3415 is quite bright galaxy and is best seen at high magnification as a well defined oval, grainy in texture, elongated N–S and a little brighter to the middle. A triangle of field stars is immediately SSE. Radial velocity: 3323 km/s. Distance: 148 mly. Diameter: 82 000 ly.

NGC3435

H887[2]	10:54.8	+61 17	GLX
Sb	1.7' × 1.2'	14.04B	

Photographs show an inclined, barred spiral galaxy with a small, bright core and a bar in an extended brighter inner envelope. Very faint spiral extensions emerge from the inner envelope. Visually, however, the galaxy is small, very faint and diffuse. The disc is very slightly oval and elongated NNE–SSW and has even surface brightness and fairly well defined edges. The galaxy is situated halfway between two faint field stars oriented ESE–WNW. Radial velocity: 5244 km/s. Distance: 234 mly. Diameter: 116 000 ly.

NGC3440

H914[3]	10:53.9	+57 07	GLX
SBb? sp	2.1′ × 0.5′	14.50B	

In photographs this is an almost edge-on spiral galaxy with a large, bright, irregular core. The apparent spiral structure is fragmentary and asymmetrical, oriented ENE–WSW and more extensive to the ENE, where a bar-shaped knot or compact object is embedded. Telescopically, the galaxy is quite faint and only the brighter core is visible as a very slightly elongated, dim patch of light which is a little more distinct at medium magnification. A faint field star is 1.3′ to the N. Radial velocity: 1975 km/s. Distance: 88 mly. Diameter: 54 000 ly.

NGC3445

H267[1]	10:54.6	+56 59	GLX	
SAB(s)m	1.6′ × 1.5′	12.93B	12.6V	

Together with a very faint, edge-on dwarf galaxy to the ESE, this galaxy is also catalogued as Arp 24. High-resolution photographs show a possible barred spiral galaxy with a peculiar morphology. The core is a narrow, elongated, bright patch surrounded by a knotty inner envelope with several condensations. One broad spiral arm

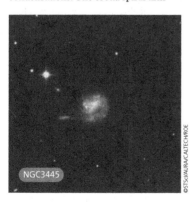

emerges and curves S to E. A plume extends from this arm to a faint companion ESE from the core. The core-to-core separation between the two galaxies would be about 35 000 ly at the presumed distance. A bright field star is 2.1′ ENE. Visually, the galaxy is fairly faint in moderate-aperture telescopes, small, round, only a little brighter than the sky background and slightly brighter to the middle. It is very slightly elongated NW–SE. Radial velocity: 2139 km/s. Distance: 96 mly. Diameter: 45 000 ly.

NGC3448

H233[1]	10:54.7	+54 19	GLX
I0	5.6′ × 1.8′	12.42B	

This galaxy and an extremely faint companion are also catalogued as Arp 205. Photographs show an irregular galaxy with a peculiar morphology. The bright, elongated inner envelope has several thin dust lanes and dust patches. The WSW extension is disturbed and broadens to an irregularly round patch. A long, broken plume of material extends

towards the E with a faint, broad patch emerging N of the galactic plane. Another faint, detached plume extends to a very faint dwarf spiral companion galaxy 4.0′ W. At the presumed distance, the core-to-core separation between the two galaxies would be about 74 000 ly. Very-large-aperture telescopes may show this companion. Visually, in moderate-aperture telescopes NGC3448 is a well defined edge-on galaxy which is evenly bright across its length. At medium magnification, a slight bulge to the centre is noted but no central brightening is visible and the galaxy is elongated ENE–WSW. Radial velocity: 1411 km/s. Distance: 63 mly. Diameter: 103 000 ly.

NGC3458

H268[1]	10:56.0	+57 07	GLX
S0?	1.4′ × 0.9′	13.01B	

This is a lenticular galaxy with a large, bright core in a slightly fainter outer envelope, elongated N–S. High-resolution photographs show a short bar at an oblique angle to the disc.

Telescopically, the galaxy is fairly bright, appearing as a roundish, diffuse patch of light with a very bright, star-like core. Radial velocity: 1889 km/s. Distance: 85 mly. Diameter: 34 000 ly.

NGC3468

H632[3]	10:57.4	+40 57	GLX
S0	1.7′ × 0.7′	14.21B	

Images show a lenticular galaxy with an elongated, bright core in a slightly fainter outer envelope. A small compact galaxy is 1.75′ WSW. Visually, NGC3468 is a small and moderately faint oval patch of light elongated N–S, moderately well defined with a grainy envelope brightening to a star-like core. Radial velocity: 7566 km/s. Distance: 338 mly. Diameter: 118 000 ly.

NGC3470

H888[2]	10:58.7	+59 31	GLX
SA(r)ab	1.6′ × 1.2′	14.23B	

Images reveal an almost face-on spiral galaxy with a bright core in a small inner envelope. Narrow spiral arms emerge; they are smooth textured and may involve dust lanes. A small, edge-on, high-surface-brightness galaxy, CGCG291-017, is 1.5′ S. NGC3470 is very diffuse visually; it is an ill defined object, irregularly round and a little brighter towards the middle. Radial velocity: 6628 km/s. Distance: 296 mly. Diameter: 138 000 ly.

NGC3471

H972[3]	10:59.1	+61 32	GLX
Sa	1.7′ × 1.0′	13.25B	

In photographs, this is a slightly inclined spiral galaxy, possibly barred, with a bright, elongated core. Two narrow spiral arms emerge, broadening into a fainter outer envelope. Telescopically, the galaxy is moderately bright, though diffuse, and is a little more distinct at medium magnification. The disc is fairly even in surface brightness and is quite well defined. The galaxy is seen as a broad, slightly extended patch of light oriented NNE–SSW. Radial velocity: 2197 km/s. Distance: 98 mly. Diameter: 49 000 ly.

NGC3478

H705[3]	10:59.5	+46 07	GLX
SB(rs)bc	2.4′ × 1.1′	13.58B	

This is an inclined barred spiral galaxy with a small, round core and a short bar. Photographs show three bright, narrow spiral arms emerging from the core area, with three additional arm fragments further out. A field star is 3.8′ S, while a very dim face-on spiral is 2.0′ S and the lenticular galaxy MCG+8-20-62 is 6.0′ ESE. Telescopically, this is a diffuse, though fairly well defined, patch of light elongated ESE–WNW with an evenly bright envelope and a slightly brighter, round, nonstellar core. Radial velocity: 6698 km/s. Distance: 300 mly. Diameter: 209 000 ly.

NGC3488

H269[1]	11:01.4	+57 41	GLX
SB(s)c	1.9′ × 1.3′	14.12B	

This is an almost face-on barred spiral galaxy. Images reveal a short, brighter bar and a fragmentary, multi-arm spiral pattern. The arms are fairly bright and grainy-textured, and the faint spiral galaxy MCG+10-16-44 is 7.0′ NNW. Visually, the galaxy is a quite faint and diffuse oval patch of light which is broadly brighter towards the middle and oriented N–S. The edges are fairly well defined and a field star is 0.8′ SSE of the core. Radial velocity: 3068 km/s. Distance: 137 mly. Diameter: 76 000 ly.

NGC3499

H793[3]	11:03.2	+56 13	GLX
S0/a	0.8′ × 0.7′	14.24B	

Situated about 14.0′ SE of Beta Ursa Majoris, this galaxy is one of the more difficult ones in Herschel's catalogue to observe. Recorded on 17 April 1789, Herschel called this object: 'Very faint, very small. Stellar. The brightness of Beta Ursae is so considerable, that it requires much attention to perceive this nebula.' Photographs show a possible dwarf lenticular galaxy. It is round and brighter to the middle with a faint outer envelope and a dust lane bordering the core on the NE flank. Visually, the galaxy is very difficult owing to the

proximity of the bright star and is most distinctly seen at medium magnification as a round, somewhat condensed patch of light, a little brighter to the middle and moderately well defined at the edges. Radial velocity: 1591 km/s. Distance: 71 mly. Diameter: 17 000 ly.

NGC3516

H336[2]	11:06.8	+72 34	GLX
(R)SB(s)0°:	1.7′ × 1.3′	12.54B	

The DSS photograph shows an almost face-on, barred lenticular galaxy with a bright core and a bar. The core is elongated ENE–WSW and is perpendicular to the fainter bar. A very faint, broad ring surrounds the core with regions of decreased luminosity to the ENE and WSW. Telescopically, the galaxy is small but quite bright and holds high magnification very well. The envelope is bright, opaque and mottled; it is fairly well defined and almost round. The very bright core appears slightly elongated ENE–WSW. Radial velocity: 2774 km/s. Distance: 124 mly. Diameter: 61 000 ly.

NGC3517

H884[2]	11:05.6	+56 31	GLX
Sb	1.0′ × 0.9′	13.90B	

This is a face-on spiral galaxy with a small, round core and grainy spiral arms. Photographs show an elongated neighbouring galaxy, possibly in contact, 0.6′ to the N. The core-to-core separation at the presumed distance would be about 75 000 ly. Telescopically, the galaxy is moderately faint but is well seen at medium magnification. It is a round, fairly well defined object, with a fairly even-surface-brightness disc. Radial velocity: 8281 km/s. Distance: 370 mly. Diameter: 108 000 ly.

NGC3527

H350[3]	11:07.3	+28 32	GLX
(R)SB(r)ab:	1.0′ × 0.9′	14.87B	

The DSS photograph shows a barred spiral galaxy with a small, bright, round core and a narrow bar attached to an almost complete inner ring. The ring is brightest where it attaches to the bar and very faint spiral structure is visible beyond the inner ring. The faint field galaxy UGC6166 is 4.3′ NNW, while MCG+5-26-58 is 6.4′ SSW. These galaxies are probably outlying members of the Abell 1185 galaxy cluster. Visually, the galaxy is quite faint and is seen situated between two faint field stars in an E–W line. It is a roundish, diffuse patch of light, with fairly even surface brightness, and is ill defined at the edges. A faint pair of stars is located 4.0′ SSW. Radial velocity: 10 060 km/s. Distance: 449 mly. Diameter: 131 000 ly.

NGC3530

H915[3]	11:08.7	+57 14	GLX
S?	0.7′ × 0.3′	14.32B	

The classification of this galaxy is uncertain but on the DSS image it appears to be a high-surface-brightness lenticular galaxy seen edge-on with a bright core and bright extensions and is elongated E–W. Visually, this is a quite bright and concentrated object; only the core is visible as a well defined, round, condensed and opaque disc. The relatively small radial velocity suggests that this may be a dwarf galaxy. Radial velocity: 2026 km/s. Distance: 91 mly. Diameter: 19 000 ly.

NGC3543

H920[3]	11:10.9	+61 21	GLX
Sc(f)	1.3′ × 0.2′	14.72B	

This dim galaxy appears fairly similar both visually and photographically. It is quite a faint object in a moderate-aperture telescope and is a little more distinct at medium magnification but its elongated form is well seen. Fairly small but well defined, the envelope is even in surface brightness and is oriented almost due N–S. Radial velocity: 1758 km/s. Distance: 79 mly. Diameter: 30 000 ly.

NGC3549

H220[1]	11:10.9	+53 23	GLX
SA(s)c:	3.2′ × 1.0′	12.79B	

Photographs show a highly inclined spiral galaxy with a small, brighter core and a multiple, knotty spiral arm pattern with many condensations and dust patches. The spiral pattern is a little more extensive to the SW. Visually, this large and fairly bright galaxy is visible as a fairly well defined, elongated oval of light oriented NE–SW. It is very slightly brighter along the major axis though no core is visible. The envelope appears smooth textured. Radial velocity: 2924 km/s. Distance: 131 mly. Diameter: 121 000 ly.

NGC3550

H351[3]	11:10.6	+28 46	GLX	
E pec	0.8′ × 0.8′	14.32B	13.2V	

The DSS image reveals that this is the brightest object in the Abell 1185 galaxy cluster and a galaxy of peculiar morphology. A small, bright, round core

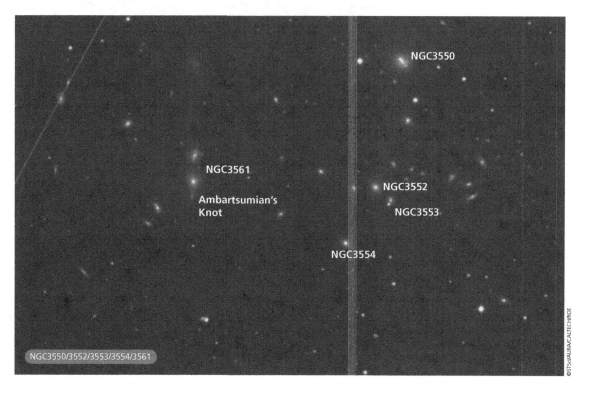

NGC3550

NGC3561

Ambartsumian's
Knot

NGC3552

NGC3553

NGC3554

NGC3550/3552/3553/3554/3561

is embedded in a fainter outer envelope. A brighter, compact object is immediately NE of the core and a slightly fainter compact object is SW of the core near the outer edge of the envelope, so that this would seem to be a triple system in a common envelope. Companion galaxy NGC3552 is 4.8′ S. There are a total of six NGC objects in the galaxy cluster, including the peculiar interacting system NGC3561 and Ambartsumian's Knot, which is a very faint object and probably visible in only the largest amateur telescopes. Telescopically, five members of the galaxy cluster can be seen in a medium-magnification field: NGC3550, NGC3552, NGC3554, NGC3558 and NGC3561. The cluster has a Rood–Sastry classification of 'C', being dominated by a core of at least three members. NGC3550 is the brightest in the cluster; it is a NE–SW elongated patch with a bright core forming a flat triangle with two field stars. The edges are fairly well defined. Radial velocity: 10 356 km/s. Distance: 463 mly. Diameter: 108 000 ly.

NGC3552

H352[3]	11:10.6	+28 42	GLX
E	0.6′ × 0.5′	15.13B	

Appearing similar visually and photographically, this is a compact elliptical system which is visible in a medium-magnification field as the brightest galaxy S of NGC3550 in the Abell 1185 galaxy cluster. It is a round, indistinct patch of light. Radial velocity: 9893 km/s. Distance: 442 mly. Diameter: 77 000 ly.

NGC3556

H46[5]	11:11.5	+55 40	GLX
SB(s)cd	8.1′ × 2.2′	10.70B	

This bright galaxy is also known as M108 but did not appear in Messier's original published catalogue. Herschel noted the galaxy on 17 April 1789, calling it: 'Very bright, much extended, resolvable, 10′ long, 2′ broad. There is an unconnected pretty bright star in the middle.' Photographs show a large,

bright, almost edge-on spiral galaxy with knotty spiral arms, several condensations and complex patterns of dust patches and lanes, particularly along the northern flank. Morphologically, the galaxy is quite similar to NGC253. This is a spectacular object in moderate-aperture telescopes. It is a large, bright, edge-on galaxy

NGC3556 15-inch Newtonian 177x

oriented E–W, with a magnitude 12 star superimposed near the centre of the nebulous envelope. Immediately W of this star a dark gap is visible and following that is a bright, elliptical concentration. To the E of the magnitude 12 star, a threshold star is visible. The E end of the disc appears wider than the W end and the entire nebulous envelope appears mottled. Radial velocity: 768 km/s. Distance: 34 mly. Diameter: 81 000 ly.

NGC3577

H723[3]	11:13.8	+48 16	GLX
SB(r)a	1.5′ × 1.5′	14.79B	

This galaxy forms a close pair visually with NGC3583, though they are not physical companions. Photographs show a face-on, barred spiral galaxy with a bright core and a bright, narrow bar attached to two, narrow, smooth-textured spiral arms. The bar flares in brightness where it touches the arms and the inner arm structure forms a

NGC3556

©STScI/AURA/CALTECH/ROE

pseudo-ring around the bar. A field star is 1.25′ SE. NGC3583 is 5.1′ to the ENE. NGC3577 is small and faint visually, and is seen as a round, somewhat concentrated patch of light of even surface brightness. Radial velocity: 5379 km/s. Distance: 240 mly. Diameter: 105 000 ly.

NGC3583

H728[2]	11:14.2	+48 19	GLX
SB(s)b	2.3′ × 1.4′	12.33B	

The DSS image shows an inclined spiral galaxy with a brighter core and a bright, S-shaped inner spiral structure. The inner spiral arms expand to fainter, broader spiral structure, oriented ESE–WNW. A tiny, possibly satellite galaxy is 0.75′ NE and the barred spiral NGC3577 is 5.1′ WSW. NGC3583 is quite bright visually, very slightly oval and elongated NW–SE. The envelope is somewhat grainy and fairly well defined and is oriented E–W with a brighter inner region betraying something of a spiral pattern and a brighter core. Radial velocity: 2179 km/s. Distance: 98 mly. Diameter: 65 000 ly.

NGC3589

H921[3]	11:15.2	+60 42	GLX
Sd:	1.5′ × 0.8′	14.55B	

In photographs this inclined spiral galaxy has a small and very faint core embedded in a difficult-to-trace spiral structure with many knotty condensations and is elongated NE–SW. Situated within a triangle of bright field stars, visually the galaxy appears mottled and diffuse; it is small but well seen at medium magnification. The surface brightness is fairly even across the disc, and the galaxy appears as a roundish, irregular patch of light. Radial velocity: 2057 km/s. Distance: 92 mly. Diameter: 40 000 ly.

NGC3594

H770[3]	11:16.1	+55 42	GLX
SB0:	1.2′ × 1.1′	14.75B	

Images show a face-on barred galaxy, possibly lenticular but more likely spiral, with a small, brighter core and a faint bar attached to the inner envelope. Faint outer spiral arms are suspected. The faint, face-on, barred spiral galaxy MCG+9-19-21 is 2.25′ SW. Though NGC3594 is faint visually, it is well seen at medium magnification. It appears as a round, slightly diffuse patch of light, somewhat grainy, but well defined with a tiny brighter core. A field star is 2.75′ NE. Radial velocity: 6327 km/s. Distance: 283 mly. Diameter: 99 000 ly.

NGC3595

H706[3]	11:15.4	+47 27	GLX
E?	1.6′ × 0.7′	12.86B	

The classification of this galaxy is uncertain, but on the DSS image it appears to be a possible SA type spiral with evident spiral arms in a fainter outer envelope elongated N–S. The core is large and oval and a bright double star is 2.0′ NNE. Although quite small visually, this is a high-surface-brightness object with a stellar, bright core in a diffuse but well defined envelope. The bright field star does not significantly affect the visibility of the galaxy. Radial velocity: 2239 km/s. Distance: 100 mly. Diameter: 47 000 ly.

NGC3600

H709[2]	11:15.8	+41 36	GLX
Sa?	3.8′ × 0.9′	13.27B	

The DSS image shows an edge-on spiral galaxy with a bright core and fainter spiral arms. Several dust patches are involved, particularly N of the core, and they are somewhat irregular in form. Telescopically, this galaxy is fairly bright, but only the inner region is visible as a moderately well defined patch of light, broadly brighter along the major axis and elongated N–S. Radial velocity: 738 km/s. Distance: 33 mly. Diameter: 37 000 ly.

NGC3610

H270[1]	11:18.4	+58 47	GLX
E5:	2.7′ × 2.6′	11.62B	

This elliptical galaxy has a large, bright, elongated core and the DSS image shows an asymmetrical, fainter outer envelope which is more extensive to the E and NE. The galaxy is elongated SE–NW and a small, compact galaxy is 3.75′ to the WSW. The galaxy is quite bright visually and is well seen at high magnification as a fairly large, diffuse patch of light that is very slightly oval in shape with very hazy edges. The core is a very bright, non-stellar oval and is quite grainy in texture, well defined and brighter to the middle. Radial velocity: 1786 km/s. Distance: 80 mly. Diameter: 63 000 ly.

NGC3613

H271[1]	11:18.6	+58 00	GLX
E6	3.6′ × 2.0′	11.74B	

This elliptical galaxy presents similar appearances visually and photographically. At high magnification it is quite a bright, fairly large object which appears as an extended, grainy disc, oriented E–W, brightening to a large, nonstellar core which is elongated along the major axis. There is a fainter secondary envelope, which is hazy and poorly defined. Radial velocity: 2113 km/s. Distance: 94 mly. Diameter: 99 000 ly.

NGC3614

H729[2]	11:18.3	+45 45	GLX
SAB(r)c	4.2′ × 2.8′	12.96B	

The DSS image shows an inclined spiral galaxy with a small core surrounded by a small ring. Two spiral arms emerge from the ring and expand to a multi-arm pattern with many condensations. The very dim galaxy NGC3614A is 2.4′ to the SW but its radial velocity of 7341 km/s suggests it is probably not a physical companion. Telescopically, NGC3614 is large but quite diffuse; it is seen as an oval patch of light, oriented E–W, which is very slowly brighter to the middle. There is no core and the edges are well defined. A bright field star is 6.75′ to the north. Radial velocity: 2368 km/s. Distance: 106 mly. Diameter: 129,000 ly.

NGC3619

H244[1]	11:19.4	+57 46	GLX
(R)SA(s)0+:	3.6′ × 2.8′	12.53B	

Photographs show a face-on galaxy, possibly spiral, with a large bright core and an extensive, fainter, inner envelope. Some evidence of dust is seen in the inner envelope to the ESE. There is an extremely faint arc well away from the galaxy in the S, curving S–W. Three compact objects are involved in the inner envelope N of the core, though two of them may just be foreground stars. Visually, the galaxy can be seen in the same high-magnification field as NGC3625, though the two galaxies may not be physically associated. The galaxy is quite bright and somewhat grainy across the disc, and is irregularly round and fairly well defined. It brightens very broadly towards the centre. Radial velocity: 1631 km/s. Distance: 73 mly. Diameter: 76 000 ly.

NGC3622

H879[2]	11:20.2	+67 15	GLX
S?	1.1′ × 0.7′	13.65B	

This is an inclined galaxy which photographs indicate may possibly be spiral. It features a brighter core in an asymmetrical outer envelope which is elongated N–S. The galaxy is moderately faint visually and is situated in a star-poor field. It appears as a roundish patch of light with a slightly brighter core. Radial velocity: 1407 km/s. Distance: 63 mly. Diameter: 20 000 ly.

NGC3625

H885[2]	11:20.5	+57 47	GLX
SAB(s)b:	2.0′ × 0.6′	13.94B	

This is a highly inclined, barred spiral galaxy, oriented SSE–NNW. Photographs show a brighter, elongated core in a slightly fainter bar. Two smooth-textured spiral arms emerge and there are possible dust patches near the bar. Telescopically, this galaxy is quite faint and appears as a very small, roundish patch of light

surrounding a star-like core. Radial velocity: 2019 km/s. Distance: 90 mly. Diameter: 52 000 ly.

NGC3631

H226[1]	11:21.0	+53 10	GLX
SA(s)c	5.0′ × 4.8′	10.71B	

This is a bright, face-on spiral galaxy of the grand-design type with two principal arms emerging from the bright core. Photographs show that the main arms quickly branch to a multiple, broad-armed spiral pattern with a particularly large and bright HII region ESE of the core involved with a large, ill defined brightening in the outer spiral arm. The galaxy has also been catalogued as Arp 27. This is a large and quite bright object visually and is seen well even at low magnification. It has a fairly large and bright core that is sharply defined and much brighter than the surrounding disc, which is quite diffuse and even in

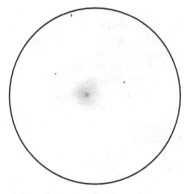

NGC3631 15-inch Newtonian 272x

surface brightness. The disc is round and fairly well defined. Herschel's 14 April 1789 description reads: 'Considerably bright, round, suddenly brighter to a nucleus and very faint chevelure, 4′ in diameter.' Radial velocity: 1218 km/s. Distance: 54 mly. Diameter: 79 000 ly.

NGC3631

NGC3642

H245[1]	11:22.3	+59 05	GLX
SA(r)bc:	5.5′ × 4.6′	11.71B	

In photographs this face-on spiral galaxy has a small bright core in a bright, multi-arm inner spiral pattern; a faint field star is 0.3′ W of the core. Very faint, thin spiral arms emerge that are loosely wound and grainy in texture. Visually, only the core of this faint spiral galaxy is visible. It is fairly bright, round and brighter to the middle with hazy, ill defined edges. Radial velocity: 1671 km/s. Distance: 75 mly. Diameter: 119 000 ly.

NGC3652

H775[2]	11:22.6	+37 46	GLX
SBc	2.3′ × 0.8′	12.40B	

The DSS photograph reveals a highly inclined spiral galaxy, probably barred, with two bright, knotty spiral arms which form a Z shape. Telescopically, the galaxy is a small but moderately bright, opaque disc with no brighter core visible. The envelope is fairly well defined, even in surface brightness and elongated SSE–NNW. Radial velocity: 2004 km/s. Distance: 89 mly. Diameter: 60 000 ly.

NGC3654

H880[2]	11:24.1	+69 26	GLX
S?	1.2′ × 0.6′	13.27B	

Photographs show an inclined spiral galaxy, probably barred, with a bright, narrow, extended inner region in a fainter outer envelope. There is evidence of a narrow dust patch immediately N of the inner bar. Telescopically, the galaxy appears as an elongated patch of light oriented NE–SW with fairly even surface brightness across the disc and moderately well defined edges. Radial velocity: 1687 km/s. Distance: 75 mly. Diameter: 26 000 ly.

NGC3656

H782[2]	11:23.6	+53 51	GLX
(R')I0: pec	1.9′ × 1.9′	13.52B	

This peculiar system has also been catalogued as Arp 155. Images reveal that the unusual morphology of

NGC3656

this galaxy is probably due to a collision between two galaxies. The principal galaxy appears to be a face-on spiral with a large core and ill defined, disturbed spiral structure. A N–S dust patch crosses in front of the core and a curving ring of matter runs N–S just E of the core. This ring interacts with the small companion galaxy, which is seen as a small, elongated, bright patch S of the principal galaxy's core. A small inclined spiral galaxy, MCG+9-19-59, is 4.6′ SW. Visually, the galaxy is fairly bright, appearing as a fat oval of light which is a little extended E–W, moderately well defined and broadly brighter to the middle. A field star is 0.75′ W of the core. Radial velocity: 2955 km/s. Distance: 132 mly. Diameter: 73 000 ly.

NGC3657

H768[3]	11:23.9	+52 55	GLX
SAB(rs)c pec	2.7′ × 2.7′	13.64B	

The DSS image shows an almost face-on, multi-arm spiral galaxy with a bright, slightly elongated core and extremely faint and thin spiral arms, hardly brighter than the sky background. Only the inner disc of this diffuse spiral is visible in a moderate-aperture telescope, appearing as a round, ill defined patch of light with a very bright, condensed core. Radial velocity: 1277 km/s. Distance: 57 mly. Diameter: 45 000 ly.

NGC3658

H59[4]	11:24.0	+38 34	GLX
SA(r)0°:	1.6′ × 1.5′	13.10B	

Photographs show a lenticular galaxy with a bright core and an extensive outer envelope, elongated NE–SW. Visually, the galaxy is small but quite bright at high magnification. Situated within an equilateral triangle of similarly bright field stars, the galaxy is round with a much brighter, star-like core and fairly well defined edges. It was first observed on 23 March 1789 when Herschel was able to see the very faint outer envelope, describing the object as: 'Considerably bright, small, round, with a bright nucleus. The nucleus is considerably well defined, and the chevelure very faint.' Radial velocity: 2055 km/s. Distance: 92 mly. Diameter: 43 000 ly.

NGC3665

H219[1]	11:24.7	+38 46	GLX
SA(s)°	2.5′ × 2.0′	11.78B	

This is the principal galaxy of a small group of seven, which also includes NGC 3658. The DSS image shows a lenticular galaxy with a bright core and an extensive outer envelope. A thin, curved dust lane crosses directly in front of the core. Visually, the galaxy is moderately large and quite bright and is broadly brighter to the middle with a quite opaque envelope embedded in a hazy, ill defined outer halo. Best seen at medium magnification, the galaxy is elongated NNE–SSW. Radial velocity: 2060 km/s. Distance: 92 mly. Diameter: 67 000 ly.

NGC3668

H845[2]	11:25.5	+63 27	GLX
Sbc	1.9′ × 1.2′	13.12B	

Images reveal an inclined spiral galaxy with a brighter core and grainy spiral arms with dust lanes in the W and SE. A star is in the spiral arm 0.5′ NW of the core. It is a moderately bright galaxy telescopically; it appears as an oval, fairly well defined patch of light oriented SE–NW and very slightly brighter to the middle. It follows two bright field stars, the brighter of which is 2.5′ SW of the core. Radial velocity: 3607 km/s. Distance: 161 mly. Diameter: 89 000 ly.

NGC3669

H829[2]	11:25.4	+57 43	GLX
SBcd: sp	2.0′ × 0.5′	13.12B	

Photographs show an edge-on galaxy, possibly a barred spiral, with a thin, elongated, brighter central region in a slightly fainter outer envelope. Visually, the galaxy is moderately faint but is well seen at medium magnification as a well defined streak, oriented SSE–NNW and fairly even in surface brightness with no

core visible. Radial velocity: 2019 km/s. Distance: 90 mly. Diameter: 52 000 ly.

NGC3671

H922[3]	11:25.8	+60 29	GLX
0.4′ × 0.4′	15.66B		

This remote galaxy is unclassified but the DSS image shows an elliptical galaxy with a brighter core and hazy ill defined patches predominantly to the W, N and E of the galaxy. Visually, it is quite faint but once located it can be seen fairly well at medium magnification as a very small, roundish, nebulous spot which is somewhat concentrated and fairly well defined with a very faint field star 0.25′ E of the core. Radial velocity: 18 087 km/s. Distance: 808 mly. Diameter: 94 000 ly.

NGC3674

H886[2]	11:26.4	+57 03	GLX
S0	1.6′ × 0.5′	12.95B	

Images show an edge-on lenticular galaxy with a bright core and slightly

fainter extensions. Telescopically, the galaxy can be seen in the same medium-magnification field as NGC3683, located 13.0′ to the SE. The pair are moderately bright and NGC3674 is smaller and appears as a short streak of light oriented NE–SW and tapering to points. It is a very well defined high-surface-brightness ray with a sharp, brighter, stellar core. Radial velocity: 2196 km/s. Distance: 98 mly. Diameter: 46 000 ly.

NGC3675

H194[1]	11:26.1	+43 35	GLX
SA(s)b	6.3′ × 3.6′	10.96B	

This is a highly inclined spiral galaxy with a large, elongated central region and grainy, tightly wound spiral arms. Photographs reveal complex dust lanes entwined with the multi-arm pattern along the eastern flank, which is the near side of the galaxy. Visually, this is a large and very bright galaxy, oriented N–S with an elongated bright and grainy envelope. The disc is fairly well

NGC3675

defined but appears more sharply defined along the E flank. The inner region is large, elongated and bright with a small, brighter core embedded. Discovered on 14 January 1788, Herschel described the galaxy as: 'Very bright, considerably large, much extended in the meridian. Bright nucleus, 6′ long, 2′ broad, chevelure.' Radial velocity: 799 km/s. Distance: 36 mly. Diameter: 65 000 ly.

NGC3683

H246[1]	11:27.5	+56 53	GLX
SB(s)c?	1.8′ × 0.7′	13.03B	

This is an almost edge-on spiral galaxy with a bright, extended, asymmetrical envelope. The DSS photograph shows a fainter extension towards the NW, as well as a faint outer envelope along the SW flank. A large, prominent dust lane begins in the SE, bordering the bright inner envelope, and extends to the NW, curving northward into the bright envelope. Visually, the galaxy is quite bright and fairly well defined; it is a large streak of light oriented SE–NW, and fairly even in surface brightness with no core visible. Radial velocity: 1793 km/s. Distance: 80 mly. Diameter: 42 000 ly.

NGC3687

H770[2]	11:28.0	+29 31	GLX	
(R′)SAB(r)bc?	1.8′ × 1.8′	13.66B	12.6V	

The DSS photograph shows a face-on barred spiral galaxy with a small, round core and a faint bar attached to an inner, partial ring. Grainy, thin spiral arms emerge from the ring forming a fragmentary spiral pattern. A very faint and small compact object is 0.75′ E, bordering the outermost spiral arm. Telescopically, the galaxy is quite bright and is best seen at medium magnification located N of a flat triangle of field stars. Almost round, it has an opaque envelope and a very small, brighter core. The envelope is moderately well defined at the edges. Radial velocity: 2485 km/s. Distance: 111 mly. Diameter: 58 000 ly.

NGC3690

H247[1]	11:28.5	+58 33	GLX
IBm pec	2.7′ × 2.0′	13.47B	

This is an interacting system involving a pair of highly disrupted galaxies separated by 0.5′ core-to-core on an E–W line. The actual separation would be about 21 000 ly at the presumed distance. The DSS image shows each galaxy has a very bright, irregularly-shaped core with asymmetrical, disrupted spiral structure. Faint companion galaxies are at 1.25′ NNW (IC694?) and 2.4′ NE. Visually, this pair of galaxies is quite bright and can be resolved at high magnification. Only the core of the W component can be seen, but it is bright and well condensed, small, round, but not stellar with well defined edges. The E component is larger and a little more diffuse but quite bright. It is somewhat wedge or fan shaped, a little mottled and moderately well defined. Herschel discovered this object

on 18 March 1790 and it is surprising that he did not remark on its dual nature, calling it only: 'Very bright, pretty large, a little extended near parallel. Much brighter middle.' Radial velocity: 3204 km/s. Distance: 143 mly. Diameter: 113 000 ly.

NGC3714

H353[3]	11:31.8	+28 22	GLX
I?	0.5′ × 0.4′	14.27B	

Photographs show a possible irregular galaxy with a high-surface-brightness, kidney-shaped core in a slightly fainter envelope. This galaxy has a similar radial velocity to NGC3713, 13.0′ SSW in Leo and they probably form a widely separated physical pair. Visually, NGC3714 is a small condensed patch of light, with even surface brightness and well defined edges. Radial velocity: 6959 km/s. Distance: 311 mly. Diameter: 45 000 ly.

NGC3690

©STScI/AURA/CALTECH/ROE

NGC3718

©STScI/AURA/CALTECH/ROE

NGC3718

H221[1]	11:32.6	+53 04	GLX
SB(s)a pec	10.5′ × 5.5′	11.52B	

This face-on spiral galaxy has a very peculiar morphology and has also been catalogued as Arp 214. The DSS image shows an extremely bright, star-like nucleus in a bright core. The large oval envelope has two diffuse, irregular spiral arms attached. Faint extensions increase the size of this galaxy to about 10.5′ × 5.5′. A complex, thin dust lane crosses ESE to WNW in front of the core and appears tidally disturbed. We may be seeing the result of a recent merger between two galaxies. Visually, NGC3718 is elongated N–S and appears as an oval, ethereal glow that is broadly brighter to the middle. The core is a little brighter than the main envelope and there is no trace of the faint, irregular spiral structure. A clutch of very small, interacting galaxies is visible as an E–W streak of light 7.1′ to the S and is catalogued as UGC6527. Radial

velocity: 1058 km/s. Distance: 47 mly. Diameter: 144 000 ly.

NGC3725

H836[2]	11:33.7	+61 53	GLX
SBc	1.6′ × 1.1′	13.79B	

Photographically, this face-on, barred spiral galaxy has a small, round core and a faint bar. The arms of this four-branch spiral are bright and grainy and a faint field star is 1.0′ WSW. The slightly fainter, face-on spiral galaxy UGC6528 is 7.4′ WSW. NGC3725 is quite faint visually and is best seen at medium magnification as a round ill defined glow which is broadly brighter to the middle. Radial velocity: 3429 km/s. Distance: 153 mly. Diameter: 71 000 ly.

NGC3726

H730[2]	11:33.3	+47 02	GLX
SAB(r)c	6.2′ × 4.3′	10.68B	

This slightly inclined, barred spiral galaxy has a small, elongated core and

a faint bar. Photographs show three principal spiral arms, which are grainy textured with several knotty condensations, especially to the S, and spiral arm fragments emerge from the principal arms. Visually, the galaxy is a large and moderately bright, broad, elongated oval oriented N–S. The envelope is smooth textured and moderately well defined at the extremities. It is broadly brighter to the middle with a field star involved 2.25′ N. Herschel considered this galaxy resolvable. Radial velocity: 909 km/s. Distance: 41 mly. Diameter: 73 000 ly.

NGC3729

H222[1]	11:33.8	+53 08	GLX
SB(r)a pec	2.8′ × 1.9′	11.72B	

The DSS image shows a slightly inclined, barred spiral galaxy with a bright core and a slightly fainter bar attached to a bright partial ring with several knotty condensations. One broad, faint spiral arm emerges from the N and curves around the galaxy, terminating at a broad elongated condensation in the NE. Visually, the galaxy can be seen in the same medium-magnification field as NGC3718; it is elongated SSE–NNW and a field star 0.8′ to the SSW hinders observation. At high magnification it is a bright, oval patch with a bright core and a very faint, diffuse outer envelope. The bright galaxy NGC3718 is 11.5′ WSW and is almost certainly a physical companion. At the apparent distance the core-to-core separation would be about 169 000 ly. Radial velocity: 1126 km/s. Distance: 50 mly. Diameter: 41 000 ly.

NGC3733

H771[3]	11:35.0	+54 51	GLX
SAB(s)cd:	4.5′ × 1.6′	12.97B	

Photographs show an inclined, barred spiral galaxy oriented NNW–SSE with a very small core and a weak bar attached to low-surface-brightness, grainy-textured spiral arms. Visually, the galaxy is very dim and fairly difficult to

NGC3726

NGC3738

see, owing to the presence of a sixth magnitude field star 4.0′ to the S. It is intermittently seen as a very hazy, roundish patch of light, which is ill defined and broadly brighter to the middle. Herschel came across the galaxy on 14 April 1789 and commented: 'Extremely faint, small, irregularly extended. On account of the brightness of 179 Ursae Majoris of Bode's Catalogue which was in the field of view with it, I had nearly overlooked it.' NGC3737 is 7.5′ to the NE. Radial velocity: 1257 km/s. Distance: 56 mly. Diameter: 73 000 ly.

NGC3737

H772[3]	11:35.6	+54 57	GLX
SB0	0.8′ × 0.4′	13.99B	

In photographs, this lenticular galaxy has a bright core and bright extensions and is elongated NNE–SSW. It has very high surface brightness and seems to be the brightest galaxy in the Abell 1318

galaxy cluster. The DSS image shows the field peppered with faint galaxies of which ten are within a circle of radius 2.0′ around NGC3737. Visually, the galaxy is very small but fairly bright, and is seen at high magnification to be slightly ENE–WSW elongated with a bright, fairly well defined envelope and a bright, star-like core. Radial velocity: 5895 km/s. Distance: 263 mly. Diameter: 61 000 ly.

NGC3738

H783[2]	11:35.8	+54 31	GLX
Irr	2.5′ × 1.9′	11.55B	12.0V

This irregular galaxy is a probable member of the Canes Venatici I Cloud. It has also been catalogued as Arp 234 and the DSS image shows a high-surface-brightness galaxy with an elongated, angular central region. Despite its brightness, the galaxy seems to be a dwarf with several bright condensations bordering the core on all

sides. These condensations resolve at about magnitude 20 and they are probably HII regions. There is an extensive but faint, grainy-textured outer envelope. Visually, the galaxy is large and moderately bright, oval in shape and elongated SE–NW. The envelope is grainy textured, bright and moderately well defined with a small, bright core. Radial velocity: 300 km/s. Distance: 13.4 mly. Diameter: 10 000 ly.

NGC3740

H847[3]	11:36.2	+59 59	GLX
S	0.8′ × 0.4′	14.80B	

Photographs show a highly inclined spiral galaxy oriented ESE–WNW with an elongated core in a fainter outer envelope. The galaxy is quite dim visually and is seen as an oval, hazy patch of light which is fairly even in surface brightness and moderately well defined at the edges. Radial velocity: 3333 km/s. Distance: 149 mly. Diameter: 35 000 ly.

NGC3756

H784[2]	11:36.8	+54 18	GLX
SAB(rs)bc	4.2′ × 2.1′	12.08B	

This galaxy is well seen on the DSS image as an inclined, multi-arm spiral galaxy with a small, brighter core. There are three principal spiral arms, which are bright, fairly massive and laced with nebulous knots and prominent dust lanes delineating the spiral structure. Visually, the galaxy is quite large and is well seen but the surface brightness is fairly low. The disc is a large oval and is broadly brighter to the middle but no core is visible. The edges are hazy and the galaxy is elongated N–S. A field star is 3.6′ NNW. Radial velocity: 1389 km/s. Distance: 62 mly. Diameter: 76 000 ly.

NGC3757

H843[3]	11:37.1	+58 26	GLX
S0?	0.8′ × 0.6′	13.40B	

The DSS image shows a lenticular galaxy with a bright core and faint extensions which may be a bar, embedded in a faint, broad envelope. Although quite small visually, this galaxy is seen as a nebulous 'star' at medium and high magnification; it is round, fairly well condensed and brighter to the middle. A faint field star follows 1.0′ to the E. Radial velocity: 1245 km/s. Distance: 56 mly. Diameter: 13 000 ly.

NGC3762

H837[2]	11:37.5	+61 46	GLX
Sa	2.1′ × 0.5′	13.14B	

This is an edge-on lenticular galaxy. Images reveal a large, bright oval core in a fainter, box-like inner envelope. Thin extensions emerge and a faint field star is embedded immediately NW of the core. This is quite a bright galaxy telescopically; it is a well defined streak of light, oriented NNW–SSE and averted vision brings out the faint extensions, which taper to points. The extensions near the core are fairly bright; the core is condensed, small and quite bright. Radial velocity: 3558 km/s. Distance: 159 mly. Diameter: 97 000 ly.

NGC3769

H731[2]	11:37.7	+47 54	GLX
SB(r)b:	3.0′ × 0.9′	12.53B	

The DSS image shows a highly inclined, barred spiral galaxy with a bright core and a foreshortened bar attached to a possible bright inner ring. The fainter outer envelope has an elongated

brightening NNW of the galaxy core. The companion galaxy NGC3769A is 1.2′ ESE of the core and appears to be a Magellanic-type galaxy. A faint plume of material bridges the two galaxies, suggesting some interaction and the core-to-core separation would be about 12 000 ly at the presumed distance. Together they are also catalogued as Arp 280. Visually, NGC3769 is a small but moderately bright galaxy, elongated SSE–NNW. Its surface brightness is fairly even and its edges are well defined. The magnitude 14.2 companion is not seen. Radial velocity: 785 km/s. Distance: 35 mly. Diameter: 31 000 ly.

NGC3770

H838[2]	11:38.0	+59 37	GLX
SBa	1.1′ × 0.8′	13.64B	

Photographs show an inclined, barred spiral galaxy with a large core in a fainter bar. There are two smooth-

textured and tightly wound spiral arms. The galaxy is moderately bright visually but fairly small; it is visible as an irregularly round, condensed patch of light which is quite well defined and a little brighter to the middle. Radial velocity: 3339 km/s. Distance: 149 mly. Diameter: 48 000 ly.

NGC3780

H227[1]	11:39.4	+56 16	GLX
SA(s)c:	3.2′ × 2.6′	12.13B	

The DSS photograph reveals an almost face-on spiral galaxy with a very small, brighter core and a complex, multi-arm spiral pattern which originates just outside the core. The arms are fragmentary and studded with HII regions as well as discrete dust patches. The galaxy is a large and fairly bright object visually, but quite diffuse. It is moderately well defined at the edges and very broadly brighter towards the middle. The envelope is quite grainy;

NGC3780

high magnification reveals a N–S pair of stellar concentrations in the disc on the E side, though these are only intermittently visible. The oval disc is oriented E–W with a field star 2.0′ ENE of the core. Radial velocity: 2477 km/s. Distance: 111 mly. Diameter: 103 000 ly.

NGC3782

H732[2]	11:39.3	+46 31	GLX
SAB(s)cd:	1.7′ × 1.1′	13.09B	

This dwarfish galaxy is a probable member of the Canes Venatici II Cloud. Photographs show an inclined spiral galaxy with a slightly brighter, elongated inner region in a grainy outer envelope. At high magnification this is a faint and diffuse galaxy, the visibility of which is hindered somewhat by two faint field stars, one immediately W and the other bordering the envelope to the S. The surface brightness is fairly even and the galaxy is elongated N–S with ragged edges. A third field star is to the N, and a fourth, the brightest, is 1.75′ ESE. Radial velocity: 783 km/s. Distance: 35 mly. Diameter: 17 000 ly.

NGC3795

H844[3]	11:40.1	+58 37	GLX
Sc	2.1′ × 0.5′	13.92B	

This galaxy forms a loose grouping with NGC3757, NGC3795A and NGC3795B. Photographs show an edge-on spiral galaxy, elongated NE–SW and slightly brighter in the middle. The galaxy is quite faint visually and is seen only as a dim sliver of light in a star-poor field; it is even in brightness and well defined at the edges with a faint field star 5.0′ to the SW. Radial velocity: 1296 km/s. Distance: 58 mly. Diameter: 35 000 ly.

NGC3796

H839[2]	11:40.6	+60 18	GLX
S?	1.1′ × 0.6′	13.28B	

In photographs, this is an inclined galaxy, probably a spiral, with a large, elongated core in a very faint outer envelope, elongated ESE–WNW.

Visually, only the brighter central region is visible and the galaxy appears as a small and fairly faint object, with an opaque and condensed envelope that surrounds a small, stellar core. It is almost round with well defined edges. Radial velocity: 1330 km/s. Distance: 59 mly. Diameter: 19 000 ly.

NGC3804

H830[2]/H773[3]	11:40.9	+56 12	GLX
SAB(s)d	2.6′ × 1.9′	13.73B	

This galaxy was recorded twice by Herschel: his first observation was on 14 April 1789, when he described the galaxy as: 'Considerably faint, pretty small, a little extended, just following a very small star.' Photographs show an inclined, barred spiral galaxy with a small core. The bar is difficult to see and is overwhelmed by two bright condensations immediately E and W of the core. The grainy, broad spiral pattern is studded with bright patches, particularly on the E side. Visually, the galaxy is situated 0.75′ E of a field star and its outer reaches are just in contact with the star. It is moderately bright and a little more distinct at medium magnification. It is seen as a diffuse patch of light that is very broadly brighter to the middle, slightly oval and oriented E–W with moderately well defined edges. Radial velocity: 1460 km/s. Distance: 65 mly. Diameter: 49 000 ly.

NGC3811

H737[2]	11:41.3	+47 42	GLX
SB(r)cd:	2.2′ × 1.7′	12.93B	

The DSS image reveals a barred spiral galaxy with a round, bright core and a fainter bar. The bar is attached to a knotty ring which extends into a single narrow spiral arm that encircles the galaxy. This arm broadens SE of the galaxy, appearing very faint and fragmentary. Visually, the galaxy is a moderately faint, small object slightly elongated N–S. The surface brightness is fairly even across the well defined envelope with a bright, star-like nucleus at its centre. Radial velocity: 3154 km/s. Distance: 141 mly. Diameter: 90 000 ly.

NGC3813

H94[1]	11:41.3	+36 33	GLX
SA(rs)b:	2.2′ × 1.1′	12.28B	

Photographs show an inclined spiral galaxy with a small core and grainy-textured spiral arms. A large extended dust lane is seen N of the core. Visually, the galaxy is quite bright and moderately large and is well seen at high magnification. It is an oval, opaque, but grainy disc, elongated E–W with well defined edges. A faint field star is located 1.3′ E of the core. Radial velocity: 1474 km/s. Distance: 66 mly. Diameter: 42 000 ly.

NGC3824

H774[3]	11:42.8	+52 47	GLX
SA(s)a? sp	1.3′ × 0.7′	14.44B	

This galaxy forms a physical pair with NGC3829 7.5′ to the ESE. Photographs show a nearly edge-on 'Sombrero'-type galaxy with a large, bright core and bright extensions. It is brightest along its northern flank and a thick dust lane runs the length of the plane of the galaxy. The minimum separation between NGC3824 and NGC3829 would be about 556 000 ly at the presumed distance. Telescopically, both galaxies can be seen in a medium-magnification field, with NGC3824 the larger and brighter of the two. Situated NNE of a triangle of field stars, the disc is oval and extended ESE–WNW, quite well defined and even in surface brightness. Radial velocity: 5712 km/s. Distance: 255 mly. Diameter: 96 000 ly.

NGC3829

H775[3]	11:43.5	+52 43	GLX
SB(s)b:	1.0′ × 0.6′	14.89B	

Photographs show an inclined, barred spiral galaxy with a small core and a very faint bar, which is attached to a mottled spiral pattern. The galaxy is difficult visually and seen as a very hazy patch of light; it is elongated E–W, even in surface brightness and moderately well defined. Radial velocity: 5742 km/s. Distance: 257 mly. Diameter: 75 000 ly.

NGC3838

H831[2]	11:44.2	+57 57	GLX
SA0/a?	1.1′ × 0.4′	13.24B	

This is an almost edge-on lenticular galaxy oriented SE–NW. Photographs show a bright core and bright extensions, with a compact nebulous patch at its NW tip. The galaxy is fairly bright visually, but only its core is visible. It is seen as a very slightly oval and fairly condensed patch of light; the disc is grainy and fairly well defined and is brighter to the middle. Radial velocity: 1405 km/s. Distance: 63 mly. Diameter: 20 000 ly.

NGC3850

H776[3]	11:45.6	+55 53	GLX
SB(s)c:	1.9′ × 0.8′	14.45B	

Images reveal a highly inclined, barred spiral galaxy elongated ESE–WNW with a long narrow bar and faint spiral arms. The galaxy is a very dim object visually; it is a hazy oval of light and is just a little brighter than the sky background. It is very broadly brighter to the middle and the edges are poorly defined. Radial velocity: 1234 km/s. Distance: 55 mly. Diameter: 31 000 ly.

NGC3870

H833[3]	11:45.9	+50 12	GLX
S0?	1.0′ × 0.9′	13.49B	

Photographs show a dwarfish lenticular galaxy with a bright, elongated core in a fainter outer envelope. In a moderate-aperture telescope, this is a small, moderately bright and fairly well-condensed galaxy. The central region is bright and the galaxy is slightly elongated NNE–SSW with a very hazy secondary envelope. Radial velocity: 815 km/s. Distance: 37 mly. Diameter: 11 000 ly.

NGC3877

H201[1]	11:46.1	+47 30	GLX
Sc	5.4′ × 1.3′	11.79B	

Photographs show this member of the Ursa Major Cluster is an almost edge-on spiral galaxy with a very small core, bright spiral arms, and no central bulge. There are several dust patches along the SE flank. Visually, this is a bright, edge-on galaxy in a moderate-aperture telescope, located about 15′ S of Chi Ursae Majoris. The galaxy is a flat, well defined streak of light elongated NE–SW with a slightly brighter central region; it is slightly thicker at the centre but no core is visible. A bright field star is 3.8′ NNW. Radial velocity: 945 km/s. Distance: 42 mly. Diameter: 67 000 ly.

NGC3888

H785[2]	11:47.6	+55 58	GLX
SAB(rs)c	1.8′ × 1.3′	12.68B	

Photographs show that this is an inclined spiral galaxy featuring two main spiral arms of the grand-design type, which have dust lanes involved near their roots before becoming luminous. This galaxy and NGC3898 can be seen together in a medium-magnification field. Visually, the galaxy appears as an elongated oval oriented E–W that is brighter to the centre. A faint field star is immediately to the W. Radial velocity: 2487 km/s. Distance: 111 mly. Diameter: 58 000 ly.

NGC3891

H723[2]	11:48.0	+30 23	GLX
Sbc	2.4′ × 2.1′	13.41B	

The DSS image shows an inclined spiral galaxy with a bright, elongated core and

NGC3877

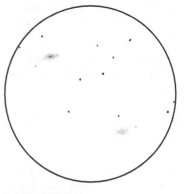

NGC3888 NGC3898
15-inch Newtonian 146x

two principal spiral arms. Subarms emerge from each arm after half a revolution around the core and extend to faint, narrow, outer spiral arms. Visually, the galaxy is a moderately bright, broad oval of light elongated E–W with a condensed core embedded in a bright envelope which fades gradually into the sky background. Radial velocity: 6208 km/s. Distance: 277 mly. Diameter: 193 000 ly.

NGC3893

H738[2]	11:48.6	+48 43	GLX
SAB(rs)c:	4.5′ × 2.8′	10.82B	10.7V

This is a grand-design spiral galaxy viewed almost face-on with a large, bright core and two bright, narrow spiral arms. The DSS image shows that several knotty condensations are located along the spiral arms and the galaxy is elongated N–S. Very faint, asymmetrical outer spiral structure

increases the apparent size to about 7.0′ × 4.0′. In moderate-aperture telescopes this Ursa Major Cluster member can be seen with its companion NGC3896, located 3.7′ SE. It is a large, bright galaxy, seen as a mottled oval of light with a well-condensed, round envelope which drops off gradually in brightness at the edges. A magnitude 12 field star lies just outside of the envelope to the NW and a field star is 3.0′ SW. A small, compact galaxy may be visible in large apertures 5.0′ ENE. Radial velocity: 1022 km/s. Distance: 46 mly. Diameter: 93 000 ly.

NGC3894

H248[1]	11:48.8	+59 25	GLX
E4-5	2.0′ × 1.4′	12.62B	

In photographs, this elliptical galaxy has a bright, elongated core and an extensive outer envelope and is elongated NNE–SSW. NGC3895 is situated 2.1′ to the ENE and is very

probably a physical companion. The minimum core-to-core separation between the two galaxies would be about 90 000 ly at the presumed distance. Both galaxies can be seen together in a high-magnification field; NGC3894 appears larger and brighter and is seen as a roundish, mottled patch of light which is brighter to the middle with fairly well defined edges. Radial velocity: 3314 km/s. Distance; 148 mly. Diameter: 86 000 ly.

NGC3895

H832[2]	11:49.1	+59 26	GLX
SB(rs)a:	1.2′ × 0.7′	13.98B	

Photographs show this inclined, barred spiral galaxy has a bright, elongated core and a fainter bar attached to a faint inner ring. Two spiral arms emerge from the ring and the disc is elongated ESE–WNW. It is a little fainter and smaller than NGC3894, which can be seen in the same field, and only the core of NGC3895 can be seen visually. It appears as a very slightly oval patch of light, well defined at the edges. Radial velocity: 3249 km/s. Distance; 145 mly. Diameter: 51 000 ly.

NGC3896

H739[2]	11:48.9	+48 41	GLX
SB0/a: pec	1.5′ × 1.0′	13.17B	13.6V

This blue compact dwarf galaxy is a companion to NGC3893 in the Ursa Major Cluster. Photographs show an almost face-on spiral galaxy with a large, irregularly shaped, bright core and two very faint spiral arms. The core-to-core separation between the two galaxies would be about 49 200 ly at the presumed distance and deep photographs show a faint plume of material curving towards NGC3896 from NGC3893's outer spiral arm to the S. Visually, the galaxy is very faint and only confirmed at medium to high magnification. It is a hazy patch of light, brighter to the middle with ill defined edges and located 0.5′ S of a magnitude 12 field star which hinders observation. Radial velocity: 959 km/s. Distance: 43 mly. Diameter: 19 000 ly.

NGC3893/3896

©STScI/AURA/CALTECH/ROE

NGC3897

H408[2]	11:49.0	+35 01	GLX
Sbc	2.2′ × 2.1′	13.67B	

This is a face-on spiral galaxy. Photographs show a small, bright core and two principal spiral arms which branch, fade and broaden away from the core. Visually, this galaxy is moderately bright and appears as an almost round diffuse disc, with a tiny brighter core embedded in an envelope which is even in surface brightness. Radial velocity: 6417 km/s. Distance: 287 mly. Diameter: 183 000 ly.

NGC3898

H228[1]	11:49.2	+56 05	GLX
SA(s)ab	4.4′ × 2.6′	11.26B	

This highly inclined spiral galaxy can be seen in the same medium-magnification field as NGC3888, which is located to the SSW. It is prototypical of its class, with a complex, multi-arm, spiral pattern defined by HII regions and dust. Deep, high-resolution photographs show extremely faint, thin spiral arms defined by HII regions far from the core. At the eyepiece, NGC3898 is by far the brighter of the two, a small, bright galaxy elongated ESE–WNW with a sharp stellar core. Faint extensions are readily visible and they are well defined but the outer spiral arms are too faint to see visually. Herschel's 14 April 1789 description reads: 'Very bright. Very bright irregular nucleus and faint branch 1.5′ wide and 0.75′ broad.' Radial velocity: 1256 km/s. Distance: 56 mly. Diameter: 71 000 ly.

NGC3906

H715[3]	11:49.7	+48 26	GLX
SB(s)d	1.9′ × 1.7′	13.52B	

Photographs show a face-on, barred spiral galaxy. A thin, bright bar is embedded in knotty but fragmentary spiral structure. The spiral arms are asymmetrical, broader to the N, W and S and the galaxy is very slightly elongated N–S. Visually, the galaxy is a faint and fairly diffuse object: it is a round, well defined patch of light, which

NGC3906

©STScI/AURA/CALTECH/ROE

occasionally displays the brighter bar, best seen at high magnification with averted vision. Otherwise, the envelope is even in surface brightness. This dwarfish galaxy is a likely member of the Ursa Major Cluster. Radial velocity: 1015 km/s. Distance: 45 mly. Diameter: 25 000 ly.

NGC3913

H786[2]	11:50.6	+55 21	GLX
(R')SA(rs)d:	3.0′ × 3.0′	13.36B	

This face-on spiral galaxy has low surface brightness and is possibly a dwarf. Its small, round core is surrounded by knotty, low-surface-brightness spiral arms. Visually, the galaxy appears large though pale, with poorly defined extremities. Averted vision reveals a slight brightening to the centre and a suspected gradual elongation of the outer envelope in a N–S direction. Two supernovae have

appeared in this galaxy. The first was a type Ia supernova with a maximum brightness of 13.3 and was recorded on 24 May 1963. The second supernova, also of type Ia, occurred in 1979, reaching magnitude 12.3 on 15 March. For visual supernova hunters, this is evidently a rewarding galaxy to follow. Radial velocity: 1032 km/s. Distance: 46 mly. Diameter: 41 000 ly.

NGC3916

H787[2]	11:50.8	+55 09	GLX
SAb: sp	1.5′ × 0.3′	14.64B	

This is an almost edge-on spiral galaxy with a bright, elongated central region and fainter spiral arms. Photographs reveal a probable dust lane along the NW flank. The peculiar galaxy NGC3921 is 4.4′ SE, while the small spindle MCG+9-19-213 is 5.8′ to the SSW. Visually, NGC3916 is a difficult object, appearing as a small sliver of light, elongated NE–SW, fairly even in surface brightness with no core visible. It can be seen in the same high-magnification field as NGC3921. Herschel's 14 April 1789 observation rated this galaxy as: 'Pretty bright, small.' Radial velocity: 5811 km/s. Distance: 260 mly. Diameter: 113 000 ly.

NGC3917

H824[2]	11:50.8	+51 50	GLX
SAcd:	5.0′ × 1.0′	12.53B	

Photographs show a large, highly inclined spiral galaxy with a very small core and multiple spiral arms. A complex pattern of dust defines the spiral structure, with many knots and condensations. Visually, the galaxy is a long, flat streak of light oriented ENE–WSW. It is sharply defined and even in surface brightness but the small core is not visible. In large-aperture telescopes, a razor-thin companion galaxy may be seen 6.25′ WNW. The Herschel object NGC3931 can also be seen in the field 11.0′ to the N. Radial velocity: 1030 km/s. Distance: 46 mly. Diameter: 67 000 ly.

NGC3921

H788[2]	11:51.1	+55 05	GLX	
(R')SA(s)0/a pec	2.3′ × 1.2′	13.16B	12.6V	

An almost face-on spiral galaxy with a peculiar morphology, this galaxy has also been catalogued as Arp 224.

NGC3916/3921

©STScI/AURA/CALTECH/ROE

Photographs show a large, bright core surrounded by a faint outer envelope. Filamentary arms emerge from the core, including one arm that heads out of the galaxy towards the SSE. The other emerges W of the core and makes at least two abrupt angle changes as it circles the core. The structure suggests a recent interaction but apart from a dim spindle 1.25′ to the SW, there are no obvious candidates nearby. A remote swarm of tiny, faint galaxies is centred about 3.0′ NNE, while the likely companion galaxy NGC3916 is 4.4′ NW. The minimum core-to-core separation between the two galaxies would be 342 000 ly at the presumed distance. Telescopically, NGC3921 is larger, brighter and more distinct than NGC3916. Only the core of this peculiar galaxy is visible; it is a

roundish, fairly high-surface-brightness patch of light with a brighter core and well defined edges. Radial velocity: 5973 km/s. Distance: 267 mly. Diameter: 179 000 ly.

NGC3922

H716[3]/H825[2]	11:51.2	+50 11	GLX
S0/a	2.8′ × 0.7′	13.72B	

Images reveal an almost edge-on lenticular galaxy with a bright, elongated core and slightly fainter extensions. A faint field star is immediately N of the core while a brighter field star is 3.5′ SW. Discovered on 9 March 1788, this galaxy was subsequently recorded on 17 March 1790 as H825[2] and later given a duplicate NGC number, NGC3924, though Dreyer noted the duplication in *The Scientific Papers*. Visually, the galaxy is moderately faint with a fairly bright central region and the envelope is slightly elongated NE–SW. The edges are fairly well defined with faint extensions along the major axis. Radial velocity: 966 km/s. Distance: 43 mly. Diameter: 35 000 ly.

NGC3928

H740[2]	11:51.8	+48 41	GLX
SA(s)b?	1.4′ × 1.3′	13.16B	

Photographs show a face-on galaxy, possibly a spiral, with a large, bright, round core in a fainter outer envelope. There is evidence of a large dust patch

NGC3917

©STScI/AURA/CALTECH/ROE

embedded in the bright core on the southern boundary. A thin lane of dust is in the core on the eastern flank, while a field star is 1.25′ SE. At high magnification, this is quite a bright galaxy; it is small but takes magnification well. It has a small, bright core with a slightly fainter, opaque envelope. There is some graininess in the envelope, which is round and well defined and the object is bracketed NNW and SE by two field stars. Radial velocity: 1043 km/s. Distance: 47 mly. Diameter: 19 000 ly.

NGC3930

H616³	11:51.8	+38 01	GLX
SAB(s)c	4.6′ × 2.6′	13.29B	

This is a slightly inclined barred spiral galaxy with a small core and a faint bar. Photographs show broad, sweeping, S-shaped spiral arms emerging from the bar, with several bright, knotty condensations along the spiral arms. The galaxy is quite faint visually; it is a dim and ill defined glow, very broadly brighter to the middle and irregularly round. Herschel discovered the galaxy on 17 March 1787 and described it as: 'Very faint, considerably large, irregular form, 4′ in diameter. 5′ south of a star of the sixth magnitude.' This star was Groombridge 1830, a famous star of high proper motion. It is now located more than 10.0′ SSE of the galaxy. Radial velocity: 937 km/s. Distance: 42 mly. Diameter: 56 000 ly.

NGC3931

H769³	11:51.2	+52 00	GLX
SA0⁻:	1.1′ × 0.9′	14.37B	

Photographs show a lenticular galaxy with a bright core and a broad, fainter outer envelope. This is a faint oval of light in moderate-aperture telescopes, very gradually brighter to the centre and elongated NNW–SSE. A magnitude 9 field star is 5.1′ E. Large-aperture telescopes may detect a very small, compact, two-arm spiral galaxy 3.5′ SW. Radial velocity: 903 km/s. Distance: 40 mly. Diameter: 13 000 ly.

NGC3938

H203¹	11:52.8	+44 07	GLX
SA(s)c	5.4′ × 4.9′	10.92B	10.4V

The DSS shows a large, bright, face-on spiral galaxy with a bright core in a faint inner envelope. Two spiral arms emerge and branch out quickly into a multi-arm pattern. They are grainy-textured with several knots and condensations along their length. A bright field star is 8.5′ NW. Visually, the galaxy is well seen at medium magnification as a large and fairly bright object with a large, slightly brighter and round core. The surrounding envelope is just a little fainter and quite round with edges that fade into the sky background. Radial velocity: 849 km/s. Distance: 38 mly. Diameter: 59 000 ly.

NGC3941

H173¹	11:52.9	+36 59	GLX
SB(s)0°	3.5′ × 2.3′	11.27B	

This is a typical lenticular galaxy: photographs show a very bright, elongated core, a slightly fainter inner disc and an even fainter outer envelope. High-resolution photographs show a broad, inner bar and, while the core is elongated N–S, the outer envelope is decidedly NNE–SSW. Telescopically, this large and bright galaxy is well seen at high magnification as an oval disc oriented N–S with a brighter, oval central region and a tiny, round and bright core. The core is best seen at medium magnification and the outer disc is moderately well defined. Radial velocity: 942 km/s. Distance: 42 mly. Diameter: 43 000 ly.

NGC3945

H251¹	11:53.2	+60 41	GLX
SB(rs)0⁺	5.2′ × 3.5′	11.70B	

This is a bright, barred lenticular galaxy with a large, bright, elongated core. A faint bar is attached to a bright inner ring and is oriented perpendicular to the core. Photographs show that the ring brightens where it

NGC3945

©STScI/AURA/CALTECH/ROE

touches the bar and a second very faint ring surrounds the galaxy. A prominent dust lane is silhouetted against the bright bar on the W side. In a moderate-aperture telescope, the galaxy is quite bright and takes magnification well. A very small, bright, star-like core is surrounded by a bright inner envelope elongated E–W; this is probably an indication of the bar. This is surrounded by an ill defined, faint secondary envelope which is almost round. Three faint field stars are visible, one to the WNW, one to the SW and one to the S. Herschel discovered the galaxy on 19 March 1790 and called it: 'Very bright, perfectly round, bright nucleus and faint chevelure. Very gradually brighter to the middle. 1.5′ diameter.' Radial velocity: 1355 km/s. Distance: 61 mly. Diameter: 92 000 ly.

NGC3949

H202¹	11:53.7	+47 52	GLX
SA(s)bc	2.7′ × 1.6′	11.37B	

The DSS image shows an inclined spiral galaxy with a slightly brighter, elongated central region in a bright spiral arm complex. A few brighter condensations are visible and the small, compact galaxy NGC3950 is 1.6′ to the N, a galaxy probably only visible in large-aperture telescopes. Visually, NGC3949 is a very bright and quite large galaxy; it is a well defined, mottled oval of high surface brightness, oriented

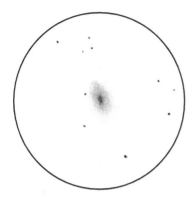

NGC3953 15-inch Newtonian 177x

ESE–WNW. The bright core is a little extended along the major axis and is a little more distinct at medium magnification. Radial velocity: 853 km/s. Distance: 38 mly. Diameter: 30 000 ly.

NGC3953

H45[5]	11:53.8	+52 20	GLX
SB(r)bc	6.9′ × 3.5′	10.48B	

In photographs, this large, bright, inclined spiral galaxy has a brighter core, and a complex, multi-arm spiral system. Many knots, condensations and dust patches are seen along the tightly wound spiral arms. In moderate apertures, this galaxy is a well defined oval of light elongated NNE–SSW. The bright, well-concentrated core is distinctly oval and elongated along the major axis. It is surrounded by a smooth-textured outer envelope which is even in surface brightness. Radial velocity: 1121 km/s. Distance: 50 mly. Diameter: 100 000 ly.

NGC3958

H833[2]	11:54.6	+58 22	GLX	
SB(s)a	1.3′ × 0.5′	13.82B	13.2V	

This is a highly inclined barred spiral galaxy. Photographs show a bright, elongated core and a bright inner ring involving brighter patches. Two fainter spiral arms emerge and wind tightly around the inner disc. The probable companion galaxy NGC3963 is 8.1′

NGC3953

©STScI/AURA/CALTECH/ROE

NNE and, if physically related, the minimum core-to-core separation between the two galaxies would be 363 000 ly. The pair of galaxies can be seen together in a medium-magnification field. Visually, NGC3958 has high surface brightness and appears as a small, well defined sliver of light oriented NE–SW that is even in brightness except for a slightly brighter core. Radial velocity: 3455 km/s. Distance: 154 mly. Diameter: 58 000 ly.

NGC3963

H67[4]	11:55.0	+58 30	GLX
SAB(rs)bc	2.7′ × 2.5′	12.60B	

This is a face-on, barred spiral galaxy of the grand-design type. Photographs show a small, round, bright core in a slightly fainter bar. Two bright spiral arms emerge and they have many knotty condensations along their entire length. Two very faint field stars are involved in the S spiral arm, one SE and the other SSW of the core, and

companion galaxy NGC3958 is 8.1′ SSW. Visually, NGC3963 appears slightly fainter than its smaller companion. It is a moderately bright but fairly diffuse, round, somewhat mottled patch of light and is a little brighter to the middle with ill defined edges. The faint star SSW of the core is just visible. Recorded on 18 March 1790, Herschel placed this galaxy in his Class IV, calling it: 'Pretty bright, pretty large, round. The greatest part of it equally bright, then fading away pretty suddenly; between 2 and 3′ in diameter.' Radial velocity: 3277 km/s. Distance: 146 mly. Diameter: 115 000 ly.

NGC3971

H724[2]	11:55.6	+30 00	GLX	
S0	1.3′ × 1.3′	14.10B	13.1V	

Photographs show a lenticular galaxy with a bright, round core and a very faint outer envelope with a faint edge-on galaxy 3.9′ to the SSE. NGC3971 is moderately faint but is well seen at high magnification as a round, fairly well defined patch of light with a grainy envelope and broadly brighter to the middle. A line of three field stars to the S points at the galaxy. Radial velocity: 6733 km/s. Distance: 301 mly. Diameter: 114 000 ly.

NGC3972

H789[2]	11:55.8	+55 19	GLX
SA(s)bc	3.7′ × 1.0′	12.97B	

Photographs show an almost edge-on spiral galaxy with a small, brighter core in a grainy spiral structure. A prominent dust lane crosses below the core on the SW flank and the spiral structure is asymmetrical. NGC3977 is in the field 5.2′ to the NE but probably does not form a physical pair with NGC3972. In moderate-aperture telescopes NGC3972, NGC3977 and NGC3982 can be seen in the same medium-power field. NGC3972 is a moderately bright, flat streak oriented WNW–ESE. It is well defined and is bright along its major axis. Radial velocity: 931 km/s. Distance: 42 mly. Diameter: 45 000 ly.

NGC3977

H790[2]	11:56.1	+55 24	GLX
(R)SA(rs)ab:	2.0′ × 2.0′	14.41B	

Photographs reveal a face-on spiral galaxy slightly elongated N–S with a brighter core in a fainter, inner envelope. The spiral arms are extremely faint and fragmented on the DSS image. Visually, it is very faint, small and round with well defined edges and is a little brighter to the middle. Only the core is seen visually; the spiral structure is too faint for visual observation. Radial velocity: 5886 km/s. Distance: 263 mly. Diameter: 153 000 ly.

NGC3978

H840[2]	11:56.2	+60 31	GLX
SABbc:	1.7′ × 1.5′	13.34B	

This is a face-on spiral galaxy with a brighter core. Photographs show that two grainy spiral arms emerge, expanding to very faint outer spiral

structure. A field star is 4.4′ E, while a very faint edge-on galaxy is 2.0′ W. Visually, the galaxy is moderately bright but the field star to the E hinders visibility somewhat. The envelope is round and fairly well defined and brightens broadly to the middle, where a tiny, bright core is visible. Radial velocity: 10 074 km/s. Distance: 450 mly. Diameter: 222 000 ly.

NGC3982

H62[4]	11:56.5	+55 08	GLX
SAB(r)b:	2.3′ × 2.0′	11.77B	

Images reveal a face-on spiral galaxy with a brighter core. Two bright, knotty spiral arms emerge, expanding to a fainter outer spiral structure. At medium magnification, the galaxy is very bright, large and round and a little brighter to the centre. The envelope is very opaque and quite well defined though there is some raggedness at the extremities. Herschel discovered the galaxy on 14 April 1789 and assigned it to his Class IV: planetary nebulae, calling it: 'Considerably bright, quite round. A large place in the middle is nearly of an equal brightness. Towards the margin it is less bright.' Radial velocity: 1187 km/s. Distance: 53 mly. Diameter: 36 000 ly.

NGC3985

H707[3]	11:56.7	+48 20	GLX
SB(s)m:	1.0′ × 0.7′	13.13B	

This dwarfish galaxy is a probable member of the Ursa Major Cluster and photographs show a curious, one-armed spiral galaxy with a large, irregular core; the bright spiral arm emerges from the E and curves clockwise, halfway around the core. Herschel commented that he suspected a very faint object S following but no candidate is readily visible in photographs. Visually, this is a small and faint galaxy, an oval patch a little brighter to the middle and well defined. It is slightly elongated E–W. Radial velocity: 1004 km/s. Distance: 45 mly. Diameter: 13 000 ly.

NGC3990

H791[2]	11:57.6	+55 28	GLX
S0˙: sp	1.4′ × 0.8′	13.46B	

Photographs show this lenticular galaxy has a bright, elongated core and a slightly fainter outer envelope elongated NE–SW. NGC3998 is 2.9′ E, but there is a relatively large difference in their radial velocities (about 345 km/s) and they may not be associated. Visually, both galaxies can be seen in a high-magnification field and they are easy objects. NGC3990 is smaller and is dominated by a bright core which overwhelms its faint outer envelope. The galaxy is very slightly elongated and there is no trace of the faint outer envelope. Radial velocity: 776 km/s. Distance: 35 mly. Diameter: 14 000 ly.

NGC3992

H61[4]	11:57.6	+53 23	GLX
SB(rs)bc	7.6′ × 4.7′	10.46B	

This is the dominant galaxy of the NGC3992 group of galaxies, a nearby association which has 41 members. Although now known as M109, this galaxy was not included in Messier's original published catalogue. Herschel first recorded it on 14 April 1789 as a Class IV object, describing it as: 'Considerably bright. Bright nucleus with very faint elongated branches about 30 degrees np/sf, 7′ or 8′ long, 4′ or 5′ broad.' Photographs reveal a large, bright, inclined, barred spiral galaxy with a bright core and a well defined

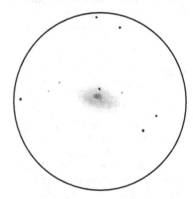

NGC3992 15-inch Newtonian 177x

NGC3992

©STScI/AURA/CALTECH/ROE

NGC4013

©STScI/AURA/CALTECH/ROE

vision helps to bring out the fainter extensions, which taper gradually to points. Radial velocity: 872 km/s. Distance: 39 mly. Diameter: 59 000 ly.

NGC4020

H725[2]	11:58.9	+30 25	GLX
SBd? sp	2.5′ × 1.1′	13.39B	

The DSS image shows a highly inclined spiral galaxy, possibly barred, which is a little brighter along its major axis with two grainy, fainter spiral arms. The galaxy is elongated NNE–SSW and a possible dust lane is seen along the eastern flank. Though moderately faint visually, the galaxy is well seen at medium magnification as an elongated, even-surface-brightness object with well defined edges and a field star 3.1′ to the WSW. Radial velocity: 753 km/s. Distance: 34 mly. Diameter: 24 000 ly.

bar. The bar brightens where it is attached to the two major spiral arms. The spiral arms are very regular and well developed. Bright along a narrow path and dotted with HII regions, they broaden to a fainter envelope. Visually, this is a rewarding object in moderate apertures. It is a bright, extensive galaxy, though the outer envelope is poorly defined. The central region is very bright with an extensive, oval outer envelope surrounding it, elongated ENE–WSW. A magnitude 12 star lies about 45″ NNW of the core; two other faint stars border the outer envelope to the W and the NE. Radial velocity: 1121 km/s. Distance: 50 mly. Diameter: 110 000 ly.

NGC3998

H229[1]	11:57.9	+55 27	GLX
SA(r)0°?	2.8′ × 2.3′	11.41B	

Images reveal a lenticular galaxy with a bright, slightly elongated core and an extensive outer envelope. It is elongated SE–NW and NGC3990 is located 2.9′ W. Visually, it is quite similar to its photographic appearance: it is a compact, well defined object with an intense core surrounded by a bright, grainy outer envelope which is round. Radial velocity: 1120 km/s. Distance: 50 mly. Diameter: 41 000 ly.

NGC4013

H733[2]	11:58.5	+43 57	GLX
SAb	5.2′ × 1.0′	12.38B	

This is an edge-on spiral galaxy with a large, elongated core and fainter extensions. Images reveal a narrow, prominent dust lane extending along the entire major axis, which is oriented ENE–WSW. A field star is directly in front of the dust lane immediately ENE of the centre. At high magnification this edge-on galaxy is difficult to see, owing to the field star located near the core. This star greatly hinders observation. Averted

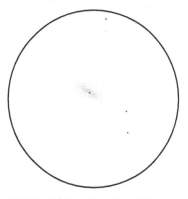

NGC4013 15-inch Newtonian 272x

NGC4025

H617[3]	11:59.2	+37 48	GLX
SB(s)cd	2.6' × 1.6'	14.11B	

Images show a face-on, barred spiral galaxy, with a long, thin, bright bar attached to two faint spiral arms. There are several knots and condensations along the arms as well as hints of dust patches. Visually, the galaxy is very dim, poorly defined and just slightly brighter to the middle. Slightly oval in shape, it is elongated NNE–SSW. A bright field star is 7.0' to the NW. Radial velocity: 3231 km/s. Distance: 144 mly. Diameter: 109 000 ly.

NGC4026

H223[1]	11:59.4	+50 58	GLX
S0	5.2' × 1.3'	11.70B	

In photographs, this edge-on lenticular galaxy has a bright, elongated central region and bright extensions in an overall fainter envelope. The galaxy is large and very bright visually, and is well seen at high magnification. The large oval central region bulges very slightly and has an intense, star-like core embedded. The edge-on extensions are well seen, brightening and lengthening considerably with averted vision. The texture is grainy and somewhat even in brightness, being slightly fainter near the central bulge. The extensions taper slightly and the galaxy is oriented N–S. A bright field star is 7.1' NNE. Radial velocity: 995 km/s. Distance: 44 mly. Diameter: 67 000 ly.

NGC4036

H253[1]	12:01.4	+61 54	GLX
S0⁻	4.3' × 1.7'	11.54B	

This is a highly inclined lenticular galaxy elongated E–W with a bright, elongated core and bright extensions. The DSS image shows four distinct brightness gradients, including a very faint outer envelope, and there is a

NGC4026

suspected dust patch E of the core in the third gradient. Visually, this galaxy is very bright and is well seen at high magnification as an elongated oval object elongated E–W and is bright along its major axis with a good deal of mottling throughout the disc. A faint secondary envelope is visible, which is fairly well defined. Radial velocity: 1546 km/s. Distance: 69 mly. Diameter: 87 000 ly.

NGC4041

H252[1]	12:02.2	+62 08	GLX
SA(rs)bc	2.8' × 2.3'	11.64B	

Photographs show a face-on spiral galaxy with a large, bright, round core and fainter spiral arms. Five distinct branches of the spiral structure are visible. The arms are thin and knotty with several condensations and are quite symmetrical. Visually, only the brighter core is seen, though it is fairly bright. It is round, quite mottled, well defined and brighter to a nonstellar core. The galaxy is probably physically related to NGC4036. Radial velocity: 1336 km/s. Distance: 60 mly. Diameter: 49 000 ly.

NGC4047

H741[2]	12:02.9	+48 38	GLX
(R)SA(rs)b:	1.7' × 1.2'	12.88B	

Photographs show a slightly inclined spiral galaxy of the multiple-spiral-arm type; it has a large, bright core and knotty spiral arms and is elongated E–W. Visually, this is a small and faint galaxy; it is a small oval patch which is a little brighter to the middle and well defined. Radial velocity: 3469 km/s. Distance: 155 mly. Diameter: 77 000 ly.

NGC4051

H56[4]	12:03.2	+44 32	GLX	
SAB(rs)bc	5.2' × 4.6'	10.79B	10.2V

This bright galaxy was one of the first Seyfert galaxies identified: these galaxies feature a small, intensely bright nucleus, now known as an 'active galactic nucleus'. NGC4051 is a principal member of the Canes Venatici

NGC4051

©STScI/AURA/CALTECH/ROE

similar radial velocities. The galaxy has a small, bright core surrounded by a slightly fainter, inner disc and a faint outer envelope. The small companion galaxies are 0.25′ SE and 0.25′ NE of the core. At the presumed distance, the minimum separation between each of the companion galaxies and the larger spiral would be about 34 000 ly. The SE galaxy has high surface brightness and a bright plume, or possibly a fourth galaxy, extends N into the principal galaxy's spiral arm. NGC4054 is very faint visually, however, and is dimly seen at medium magnification as a small, irregular patch of light with fairly even surface brightness and moderately well defined edges. Radial velocity: 10 324 km/s. Distance: 461 mly. Diameter: 67 000 ly.

NGC4062

H174[1]	12:04.1	+31 54	GLX
SA(s)c	4.8′ × 2.0′	11.89B	

This galaxy is the principal member of a group of 19, variously known as the NGC4062 group or the Coma I group of galaxies. Photographs show a highly inclined spiral galaxy with a small core and multiple, bright and knotty spiral arms. A dust lane is visible in the spiral pattern SW of the core and a faint field star is involved 1.0′ W of the core. Visually, the galaxy is a bright and large object, elongated E–W and brightest along its major axis. It is quite grainy along this axis and is embedded in a fainter, diffuse outer envelope. Radial velocity: 770 km/s. Distance: 34 mly. Diameter: 48 000 ly.

NGC4068

H781[2]	12:04.0	+52 35	GLX
IAm	3.3′ × 1.7′	13.43B	

This irregular galaxy has a very small radial velocity and may be a nearby dwarf irregular. The DSS image shows a low-surface-brightness system with a faint, grainy envelope and several brighter, star-like condensations. Telescopically, this galaxy appears as a faint haze surrounding a magnitude 11 field star. It is best seen at medium

II Cloud of galaxies. Photographs show a slightly inclined spiral galaxy with a small, bright core and a fainter, curved possible bar. The bar transmutes easily to massive, bright spiral arms. Several knotty condensations are seen among the spiral arms, the largest of which may be up to 650 ly across. The spiral

structure is asymmetrical, being more pronounced to the NNE, where several dust patches are involved. In a moderate-aperture telescope this is a large and bright galaxy. The envelope is elongated ESE–WNW and quite broad. The bright core is a little fainter than the field star located 2.4′ WSW. The core appears offset to the NE. Herschel discovered this galaxy on 6 February 1788 and placed it in his Class IV, describing the object as: 'Considerably bright, irregularly round. Considerably bright nucleus in the middle with extensive chevelure, 5.0′ in diameter.' Radial velocity: 745 km/s. Distance: 33 mly. Diameter: 50 000 ly.

NGC4054

H794[3]	12:03.2	+57 54	GLX
0.5′ × 0.3′	15.58B		

The DSS image reveals a slightly inclined spiral galaxy that is part of a triple system, all three members having

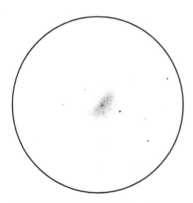

NGC4051 15-inch Newtonian 166x

magnification as an irregular patch of light elongated NNE–SSW which is a little more distinct S of the star. The edges are very ill defined and the surface brightness uneven. A brighter field star is 3.25′ WNW. Recorded on 12 April 1789, Herschel described the object as: 'A pretty small star involved in nebulosity of no great extent; the star does not seem to belong to it.' Radial velocity: 282 km/s. Distance: 12.5 mly. Diameter: 12 000 ly.

NGC4085

H224[1]	12:05.4	+50 21		GLX
SAB(s)c:?	2.8′ × 0.8′	12.30B	12.8B	

This highly inclined spiral galaxy has a brighter, elongated central region and photographs show the spiral arms are bright and tightly wound with several dust patches and a prominent dust lane involved. Bright field stars are 6.5′ ESE and 5.5′ SW. Companion galaxy NGC4088 is 12.0′ N and both galaxies

are members of the Ursa Major Cluster. In a medium-magnification field, NGC4085 and the much larger NGC4088 are both visible. NGC4085 is a bright, almost edge-on galaxy, elongated ENE–WSW, very gradually brighter to the centre and well defined along the edges. Radial velocity: 811 km/s. Distance: 36 mly. Diameter: 30 000 ly.

NGC4088

H206[1]	12:05.6	+50 33	GLX
SAB(rs)bc	6.0′ × 2.3′	11.26B	

This galaxy has also been catalogued as Arp 18 and photographs show that this Ursa Major Cluster member is a highly inclined spiral galaxy with a brighter, elongated core and a dusty bar. The spiral arms are bright, massive and tightly wound, with many knots and condensations. A tremendous amount of dust is visible, much of it in patches, and a significant dust lane starts E of the core and runs NE. The smaller

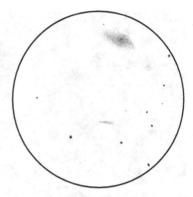

NGC4085 NGC4088
15-inch Newtonian 146x

companion galaxy NGC4085 is 12.0′ S and at the apparent distance, the core-to-core separation between the two galaxies would be about 130 000 ly. Visually, NGC4088 features a bright, dense envelope with a bright, nonstellar core. The extensions exhibit incipient, curved spiral structure. The eastern arm appears a little longer than the western one and the galaxy is elongated NE–SW. Despite the fact that both galaxies are fairly bright objects, they were discovered by Herschel one year apart: NGC4088 on 9 March 1788 and NGC4085 on 12 April 1789. Radial velocity: 823 km/s. Distance: 37 mly. Diameter: 65 000 ly.

NGC4096

H207[1]	12:06.0	+47 29	GLX
SAB(rs)c	5.8′ × 1.7′	11.09B	

This is a highly inclined spiral galaxy with a very small, brighter core. Photographs show a dusty, fragmentary spiral structure with many knots, dust patches and lanes and a highly asymmetrical structure: the outer envelope is much more extensive towards the SSW. Visually, this is a large and fairly bright galaxy; it is a long streak oriented NNE–SSW, which is quite bright along its major axis and has a slightly brighter core, and somewhat diffuse extremities, particularly to the SSW. Radial velocity: 622 km/s. Distance: 28 mly. Diameter: 47 000 ly.

NGC4088

©STScI/AURA/CALTECH/ROE

NGC4096

Its visibility is hindered somewhat by the presence of a field star 0.75′ to the W. The disc is slightly oval and elongated NE–SW. Herschel's 12 April 1789 observation described the galaxy as: 'Very bright, pretty large, with a brighter nucleus just following a considerable star.' Radial velocity: 920 km/s. Distance: 41 mly. Diameter: 37 000 ly.

NGC4141

H795[3]	12:09.7	+58 51	GLX
SBcd:	1.5′ × 0.8′	14.72B	

Images show a slightly inclined, barred spiral galaxy with a narrow bar and two fainter, knotty spiral arms, elongated E–W. The galaxy is quite faint visually and is best seen at medium magnification as a diffuse, roundish patch of light with poorly defined edges that is very slightly brighter to the middle. Radial velocity: 1991 km/s. Distance: 89 mly. Diameter: 39 000 ly.

NGC4142

H814[3]	12:09.5	+53 06	GLX
SB(s)d:	2.2′ × 1.2′	13.91B	

This is a slightly inclined, barred spiral galaxy. Images reveal a narrow bar and fragmentary, ill defined spiral arms. The galaxy is a faint and difficult object visually; it is a smallish, ill defined and very faint patch of light which is a little extended N–S and very broadly brighter to the middle. Radial velocity: 1232 km/s. Distance: 55 mly. Diameter: 35 000 ly.

NGC4097

H400[3]	12:06.1	+36 52	GLX
S0	1.2′ × 0.8′	14.63B	

Photographs show a lenticular galaxy oriented E–W with a bright and elongated core in a fainter outer envelope. Telescopically, this galaxy is moderately faint, appearing as a roundish, somewhat diffuse patch of light with hazy edges. It is very slightly brighter to the middle with a field star situated 1.6′ SSW. Radial velocity: 6310 km/s. Distance: 282 mly. Diameter: 98 000 ly.

NGC4100

H717[3]	12:06.2	+49 35	GLX
(R')SA(rs)bc	5.4′ × 1.8′	11.71B	

This possible member of the Ursa Major Cluster is a highly inclined spiral galaxy with a very small, brighter core. Photographs reveal narrow, grainy, tightly wound spiral arms emerging from the core with many patches of dust and narrow dust lanes delineating the spiral structure. A bright field star is 7.2′ NNW. Visually, this is a moderately bright, greatly elongated object, oriented NNW–SSE. The brightness is almost even along the entire major axis and no brighter core is visible. Radial velocity: 1137 km/s. Distance: 51 mly. Diameter: 80 000 ly.

NGC4102

H225[1]	12:06.4	+52 43	GLX
SAB(s)b?	3.1′ × 1.8′	12.08B	

This is an inclined spiral galaxy. Photographs show a bright, slightly elongated and large core which may be a bar. The overlapping spiral arms suggest a ring-like structure with prominent knots evident N of the core and dust lanes visible along the inside edge of the spiral arms. Telescopically, the galaxy is very bright and is well seen at all magnifications. The core is very bright and nonstellar and is embedded in a grainy though ill defined envelope.

NGC4102

NGC4157

H208[1]	12:11.1	+50 29	GLX
SAB(s)b? sp	9.0′ × 1.0′	12.09B	11.3V

The DSS image shows that this Ursa Major Cluster member is a large, almost edge-on spiral galaxy which is bright along its major axis with grainy extensions. A large, prominent dust lane extends along the major axis on the NNW flank. A field star is at the WSW tip, 3.9′ from the core. A brighter field star is 4.4′ NNW. In a moderate-aperture telescope this is a bright, well defined streak of light elongated ENE–WSW, with even surface brightness along its entire length. The tips taper to blunt points and averted vision reveals hazy extensions beyond these points. Radial velocity: 842 km/s. Distance: 38 mly. Diameter: 99 000 ly.

NGC4161

H803[2]	12:11.4	+57 44	GLX
S?	1.4′ × 0.8′	13.81B	

Images reveal a slightly inclined spiral galaxy with a small, round, brighter core and bright, grainy spiral arms emerging from the core. Telescopically, this is a moderately bright galaxy with an opaque and fairly condensed envelope of even surface brightness. The edges fade fairly abruptly into the sky background and the oval disc is elongated NE–SW. A field star is 2.0′ to the NNW. Radial velocity: 4990 km/s. Distance: 223 mly. Diameter: 91 000 ly.

NGC4144

H747[2]	12:10.0	+46 27	GLX
SAB(s)cd? sp	7.3′ × 1.7′	12.05B	

Photographs show an almost edge-on spiral galaxy with an elongated, knotty core. The outer envelope is quite grainy with many minute condensations. The spiral structure is broader and blunter to the E; W of the core it gradually narrows and tapers. The galaxy is a member of the Canes Venatici I Cloud and is relatively nearby; high-resolution photographs begin to resolve stars at about blue magnitude 20. Although moderately faint visually, this galaxy stands out well as a flat streak of light, oriented ESE–WNW. It is brighter along the major axis and the extremities are somewhat diffuse and fade uncertainly into the sky background. A magnitude 9 field star is 7.5′ SW. Radial velocity: 319 km/s. Distance: 14.3 mly. Diameter: 30 000 ly.

NGC4149

H802[2]/H845[3]	12:10.4	+58 19	GLX
S?	2.0′ × 0.3′	13.94B	

This galaxy was inadvertently recorded twice by Herschel, first on 17 April 1789 and again on 18 March 1790. The description and position given were nearly identical each time. Photographs show an edge-on, high-surface-brightness spindle-shaped galaxy, possibly lenticular, with a bright, elongated core and bright extensions. There is evidence of a possible narrow dust lane crossing the core. Visually, the galaxy is a small, faint sliver of light elongated E–W, displaying even surface brightness, though a star-like core is visible at medium magnification. A bright field star is 7.1′ NE. Radial velocity: 3148 km/s. Distance: 141 mly. Diameter: 82 000 ly.

NGC4172

H792[2]	12:12.2	+56 11	GLX
S?	1.3' × 1.2'	14.22B	

Despite the uncertain classification, the DSS image shows an almost face-on spiral galaxy, with a large, bright, slightly extended core. Faint, broad spiral arms emerge from the core with some evidence of dust patches throughout. A sizable, bright condensation is involved in the spiral arm immediately S of the core. Visually, this fairly faint galaxy is visible as a faint, ethereal disc of light with a bright, stellar core. The disc is even in surface brightness, round and well defined at the edges. Radial velocity: 9360 km/s. Distance: 418 mly. Diameter: 158 000 ly.

NGC4181

H777[3]	12:12.8	+52 54	GLX
E	0.7' × 0.4'	14.86B	

Photographs show an elliptical galaxy with an oval core and a fainter disc oriented N–S. The galaxy is very faint visually and appears as a hazy oval of light which is a little brighter to the middle with a field star situated 1.5' SSW. The galaxy is plotted immediately N of NGC4187 in the first edition of *Uranometria 2000.0*, but this is an error. Radial velocity: 9746 km/s. Distance: 435 mly. Diameter: 89 000 ly.

NGC4194

H867[2]	12:14.2	+54 32	GLX	
IBm pec	2.7' × 1.5'	12.95B	13.5V

This peculiar system has also been catalogued as Arp 160. The DSS photograph shows an irregular galaxy with a large, bright, oval core connected to a slightly fainter spiral arm fragment immediately S. A narrow plume loops out of the core in the E, and a faint, broad, knotty patch of matter fans out towards the N. Faint patches are seen throughout, especially to the S and the galaxy is elongated N–S. This is quite a bright galaxy telescopically; it is a roundish, opaque, fairly well defined disc with a small, stellar, bright core. Radial velocity: 2582 km/s. Distance: 115 mly. Diameter: 91 000 ly.

NGC4195

H796[3]	12:14.2	+59 37	GLX
SB(s)cd	1.6' × 1.4'	15.24B	

The DSS photograph shows a face-on, barred spiral galaxy with a narrow, brighter bar and two narrow spiral arms emerging into a broad, fainter envelope. There are nebulous knots throughout the outer disc. The galaxy is quite faint and diffuse visually and is best seen at medium magnification, though averted vision is needed to see the galaxy consistently. It is a hazy, even-surface-brightness patch, round and somewhat ill defined at the edges. Radial velocity: 4447 km/s. Distance: 199 mly. Diameter: 92 000 ly.

NGC4198

H793[2]	12:14.3	+56 01	GLX
S0/a	1.0' × 0.7'	14.85B	

Images show a slightly inclined spiral galaxy with a large, bright, extended core. Two narrow, smooth-textured spiral arms emerge from the bright central disc. A condensation or compact object is attached to the tip of the spiral arm in the NW. Telescopically, the galaxy is quite faint and appears as a hazy oval disc of light which is fairly well defined and oriented NW–SE with a tiny, star-like, brighter core. The galaxy is situated midway between two field stars on a NNW–SSE orientation. Radial velocity: 9433 km/s. Distance: 421 mly. Diameter: 122 000 ly.

NGC4199

H797[3]	12:14.8	+59 54	GLX
E	0.7' × 0.4'	15.29B	

This elliptical galaxy is the brighter of a physical pair: photographs show the slightly fainter companion 0.5' ENE and a faint field star 0.3' N of the principal galaxy. These two galaxies are the brightest members of the Abell 1507 galaxy cluster and the minimum separation between them at the presumed distance would be about 120 000 ly. NGC4199 is among the most remote entries in the Herschel catalogue. Visually, it is quite faint and is seen as a small, nebulous patch of light, round and fairly condensed, with an even-surface-brightness disc and quite well defined edges. Radial velocity: 18 198 km/s. Distance: 813 mly. Diameter: 166 000 ly.

NGC4271

H804[2]	12:19.6	+56 45	GLX
S0⁻:	1.5' × 1.3'	13.67B	

Visually and photographically, this lenticular galaxy presents similar appearances. The core is bright and embedded in a faint outer envelope, which is very slightly elongated E–W. A threshold star is visible 30″ NNE. Radial velocity: 4842 km/s. Distance: 216 mly. Diameter: 95 000 ly.

NGC4284

H798[3]	12:20.2	+58 06	GLX
Sbc	2.6' × 1.3'	14.53B	

In photographs this highly inclined spiral galaxy has a small, bright core and two bright, narrow spiral arms expanding into a broader spiral structure. It forms a close pair with the galaxy NGC4290 4.6' to the E, but it is unlikely that the two galaxies are physically related. NGC4284 is quite faint visually; it is a small, evenly bright, oval patch of light elongated E–W with a field star 1.0' to the E, and another is 1.25' S. Radial velocity: 4298 km/s. Distance: 192 mly. Diameter: 145 000 ly.

NGC4290

H805[2]	12:20.8	+58 06	GLX
SB(rs)ab	2.3' × 1.7'	12.85B	

Images show an inclined, barred spiral galaxy with a bright, elongated core attached to a bright bar. The bar is attached to a pseudo-ring which is especially bright where it meets the bar.

NGC4284/4290

The ring is surrounded by a broad, faint outer envelope, which has many knotty condensations. Visually, this galaxy is quite bright, is broadly brighter along its major axis, shows some mottling of the well defined envelope and appears elongated NNE–SSW. The fainter galaxy NGC4284 is 4.6′ W. A very faint edge-on galaxy is 3.75′ E. Radial velocity: 3127 km/s. Distance: 140 mly. Diameter: 94 000 ly.

NGC4335

H806[2]	12:23.0	+58 27	GLX
E	1.9′ × 1.5′	13.63B	

This elliptical galaxy has a bright, elongated core and a fainter inner disc surrounded by a much fainter outer envelope. Photographs show that a compact object or faint star borders the bright core to the NW. Telescopically, the galaxy is fairly bright with a well-condensed envelope and an extremely faint outer envelope.

It is slightly brighter to the core. Two Herschel objects (NGC4358 and NGC4362) are 9.0′ SE. Radial velocity: 4718 km/s. Distance: 211 mly. Diameter: 116 000 ly.

NGC4358

H799[3]	12:24.0	+58 24	GLX
S0	1.1′ × 0.9′	16.35B	

Photographs show a lenticular galaxy with a bright core and a fainter outer envelope. A companion galaxy is 0.5′ to the WSW and was identified as NGC4364 in the first edition of *Uranometria 2000.0*, but this is almost certainly incorrect. NGC4362 is 1.8′ to the SE. Visually, NGC4358 is a little easier to see than NGC4362; it is a round, fairly well-condensed patch of light which is a little brighter to the middle. The tiny companion to the WSW is not seen. Radial velocity: 4593 km/s. Distance: 205 mly. Diameter: 66 000 ly.

NGC4362

H800[3]/H801[3]	12:24.2	+58 22	GLX
S?	1.1′ × 0.4′	15.01B	

Photographs show a highly inclined spiral galaxy with a large, bright core and fainter extensions. Possible dust patches are in the outer envelope SW and NE of the core. Visually, only the core of NGC4362 is visible; it is small and round, fairly well condensed and located N of a faint field star. Herschel claimed to have seen two nebulae at this position on 17 April 1789, describing them as: 'Two. Both considerably faint, considerably small, round.' There is a very slim possibility that Herschel mistook the faint field star 0.75′ to the S for a nebula, but this is doubtful. J. L. E. Dreyer's interpretation appeared in *The Scientific Papers* as: 'III. 800-801. Very probably the word 'two' refers to III. 799 and III. 800, as nobody seems to have seen three nebulae in the place.' There is, of course, the tiny galaxy immediately W of NGC4358 but it is highly unlikely that Herschel could have seen this object. Radial velocity: 4625 km/s. Distance: 207 mly. Diameter: 66 000 ly.

NGC4384

H879[3]	12:25.2	+54 30	GLX
Sa	1.3′ × 1.0′	13.52B	

This is a slightly inclined spiral galaxy. Photographs show a bright, irregular core and two bright, stubby spiral arms emerging into a fainter outer envelope. The galaxy is somewhat faint visually, but is well seen at both medium and high magnification as a small, oval and well defined patch of light oriented E–W. The disc is a little grainy and very slightly brighter to the middle. Radial velocity: 2598 km/s. Distance: 116 mly. Diameter: 44 000 ly.

NGC4500

H234[1]	12:31.4	+57 58	GLX
SB(s)a	1.6′ × 1.0′	13.12B	

Photographs show an inclined, barred spiral galaxy with a bright core and oval bar attached to two narrow,

S-shaped spiral arms. A bright field star is 1.0′ E. Visually, the galaxy is small but moderately bright and well defined. It is a narrow oval, elongated SE–NW with a small, bright core. The envelope is only slightly fainter and somewhat grainy. The bright field star hinders visibility of the galaxy. Radial velocity: 3208 km/s. Distance: 143 mly. Diameter: 67 000 ly.

NGC4511

H834[3]	12:32.2	+56 29	GLX
1.0′ × 0.7′	14.64B		

This galaxy is unclassified though it is probably an Sb spiral. Photographs show a small, slightly inclined spiral galaxy with a small, bright core and two slightly fainter spiral arms. Best seen at medium magnification, this is a very faint galaxy, a small, nebulous patch of light, very slightly elongated N–S. It is fairly well defined with even surface brightness across the envelope. A faint field star is 2.4′ S. Radial velocity: 4716 km/s. Distance: 211 mly. Diameter: 61 000 ly.

NGC4547

H802[3]	12:34.8	+58 55	GLX
1.0′ × 0.8′	15.16B		

Although unclassified, this is probably an elliptical or lenticular galaxy. Photographs show that it has a faint outer envelope and a brighter core. A slightly fainter galaxy is 0.5′ ESE and these two galaxies are the brightest of a compact swarm of faint galaxies in the field, including the Herschel object NGC4549 4.25′ ENE. Visually, NGC4547 appears as a faint, elongated glow which can just be resolved into a pair of galaxies with averted vision at medium magnification. The image is extended ESE–WNW with a very faint stellar core intermittently visible in the halo to the W, while the eastern member of the pair appears smaller and fainter. Both are fairly well defined and even in brightness. Radial velocity: 12 918 km/s. Distance: 577 mly. Diameter: 159 000 ly.

NGC4549

H807[3]	12:35.3	+58 57	GLX
0.7′ × 0.3′	16.13B		

Photographs show a highly inclined, probably barred, spiral galaxy, situated ENE of NGC4547. Visually, three galaxies are seen at medium magnification; NGC4547 has a faint companion immediately ESE, while NGC4549 appears as a very faint, flat streak oriented E–W, situated midway between a pair of field stars aligned N–S. Radial velocity: 13 527 km/s. Distance: 609 mly. Diameter: 124 000 ly.

NGC4566

H880[3]	12:35.9	+54 13	GLX	
Sb	1.1×0.9′	13.84B	13.3V	

The DSS image shows an inclined, barred spiral galaxy with a bright elongated core and a stubby bar leading to a fainter ring. At high magnification the galaxy is faint but visible as a roundish, nebulous patch of light with a small, brighter core. The envelope is hazy and ill defined. Radial velocity: 5438 km/s. Distance: 243 mly. Diameter: 78 000 ly.

NGC4605

H254[1]	12:40.0	+61 37	GLX
SB(s)c pec	6.3′ × 2.7′	10.82B	

This highly inclined galaxy, probably a spiral, has a bright, elongated major axis embedded in a broad, fainter outer envelope. Photographs show many knotty condensations in the inner envelope, particularly to the ESE and the outer envelope is very grainy. Many dust patches are involved, especially along the NE flank. At high magnification, the galaxy is large and quite bright. Bright across its major axis and fairly well defined, the envelope is convex towards the SSW and concave along its NNE flank, though no direct evidence of the prominent dust patch is indicated. The envelope tapers sharply towards the ESE and is blunter towards the WNW. A faint field star is

NGC4605

©STScI/AURA/CALTECH/ROE

immediately S and the galaxy is elongated ESE–WNW. Radial velocity: 253 km/s. Distance: 11.3 mly. Diameter: 21 000 ly.

NGC4644

H794[2]	12:42.7	+55 09	GLX
Sb	1.7′ × 0.4′	11.60B	

Images reveal a highly inclined barred spiral galaxy elongated NE–SW with a bright core and two sweeping, open spiral arms. A fainter field galaxy, designated NGC4644B, is 1.4′ E; it shows a similar radial velocity and is a likely companion. At the presumed distance the core-to-core separation between the two galaxies would be about 92 000 ly. Visually, only the brighter galaxy is visible in a moderate-aperture telescope; it is a faint, oval patch of light, moderately well defined and a little brighter to the middle. Radial velocity: 5029 km/s. Distance: 225 mly. Diameter: 111 000 ly.

NGC4646

H910[2]/H794[2]	12:42.9	+54 51	GLX
S0?	1.0′ × 0.4′	14.34B	

Photographs show an inclined lenticular galaxy oriented NNE–SSW with a bright, elongated core. A faint field star is immediately NE of the core. Due to the identification difficulties experienced by John Herschel and other nineteenth-century observers, NGC4646 and NGC4644 were once thought to be duplicate observations of the same object, H794[2] (recorded on 14 April 1789), but NGC4646 was not actually discovered until 24 March 1791, when it received the designation H910[2]. Visually, the galaxy is moderately bright though small, a narrow sliver of light, fairly well defined and with a slightly brighter, small core. Radial velocity: 4739 km/s. Distance: 212 mly. Diameter: 62 000 ly.

NGC4675

H778[3]	12:45.6	+54 44	GLX
SBb:	1.6′ × 0.5′	14.84B	

In photographs, this is a highly inclined, barred spiral galaxy with a bright core

in a bright, inner ring. Two thin, very faint spiral arms emerge from the ring and a faint edge-on galaxy is 6.7′ NNE. At medium magnification the galaxy is a faint and nebulous, slightly elongated patch of light which is a little brighter to the middle and extended E–W. The edges are hazy and ill defined. Radial velocity: 4849 km/s. Distance: 216 mly. Diameter: 101 000 ly.

NGC4686

H795[2]	12:46.7	+54 33	GLX
Sa	2.0′ × 0.6′	13.60B	

This is a highly inclined spiral galaxy with a bright elongated core and tightly wound spiral arms defined by narrow dust lanes, which photographs show well, S of the core. Visually, the galaxy is small but moderately bright, elongated due N–S with well defined edges and a bright, star-like core. A faint field star is 0.8′ E. Radial velocity: 5109 km/s. Distance: 229 mly. Diameter: 133 000 ly.

NGC4695

H796[2]/H985[3]	12:47.5	+54 23	GLX
S	1.1′ × 0.7′	13.94B	

Photographs show an inclined spiral galaxy with a bright elongated core and two main, clumpy spiral arms. A faint star or compact object is involved immediately WSW of the core. At medium magnification, this faint galaxy is a small, somewhat diffuse, oval patch of light elongated E–W. The surface brightness is even across the envelope. This galaxy seems to have been recorded twice: on 14 April 1789 and again on 24 March 1791. Radial velocity: 4966 km/s. Distance: 222 mly. Diameter: 71 000 ly.

NGC4732

H814[2]	12:50.1	+52 52	GLX
E	1.0′ × 0.6′	14.86B	

Photographs show a very small, elliptical galaxy elongated N–S with a brighter, elongated core and faint extensions. Visually, the galaxy is faint and small but is readily visible as a round patch of light with even surface

brightness and diffuse edges. Radial velocity: 10 022 km/s. Distance: 448 mly. Diameter: 130 000 ly.

NGC4801

H816[3]	12:54.6	+53 06	GLX
0.9′ × 0.6′	15.19B		

This is one of the more remote systems in Herschel's catalogue. Though unclassified, it appears to be an elliptical or lenticular galaxy. Photographs show it features an elongated core and fainter extensions elongated ESE–WNW. It is far and away the largest and brightest of a swarm of very faint galaxies in the field which are probably beyond the capabilities of even the largest amateur telescopes. At medium magnification NGC4801 is a very faint galaxy, difficult to hold with direct vision. It is a small, almost round patch of light, broadly brighter to the middle with a diffuse outer envelope. Radial velocity: 16 232 km/s. Distance: 725 mly. Diameter: 190 000 ly.

NGC4814

H243[1]	12:55.4	+58 21	GLX
SA(s)b	3.3′ × 2.3′	12.83B	

The DSS image shows an inclined spiral galaxy with a large, elongated core surrounded by bright, tightly wound spiral arms. The arms fade after half a revolution around the core, broadening slightly and becoming loosely wound. Visually, this is a large and moderately bright galaxy that is well seen at high magnification. The brightest part is an opaque, well defined oval, elongated ESE–WNW with a small, bright core. A fainter secondary envelope is intermittently visible which broadens the visual appearance of the galaxy. Radial velocity: 2623 km/s. Distance: 117 mly. Diameter: 112 000 ly.

NGC4964

H779[3]	13:05.4	+56 18	GLX
Sa	1.0′ × 0.7′	13.93B	

Photographs show an inclined spiral galaxy with an elongated core and bright, clumpy spiral arms. The arms are asymmetrical and more extensive on the NE flank. Visually, this is a small, faint galaxy; it is a well defined oval patch of light which is broadly brighter to the middle and elongated SE–NW. Radial velocity: 2623 km/s. Distance: 117 mly. Diameter: 34 000 ly.

NGC4967

H783³	13:05.7	+53 34	GLX
E	0.7′ × 0.7′	15.02B	

The DSS image shows a small elliptical galaxy with a bright core and a fainter outer envelope. This is a rewarding field for observers with large-aperture telescopes, as the galaxy marks the centre of a faint galaxy cloud including Herschel objects NGC4973 and NGC4974. About two dozen galaxies are seen on the 15.0′ DSS image. Visually, the three Herschel objects can be seen in a medium-magnification field. NGC4967 is the faintest of the three; it is round and well defined with even surface brightness and can be seen with direct vision once

found. Radial velocity: 8949 km/s. Distance: 400 mly. Diameter: 81 000 ly.

NGC4973

H781³	13:05.5	+53 41	GLX
E/S0	0.8′ × 0.7′	14.75B	

Images show a small elliptical galaxy with a bright core and a fainter outer envelope. It is the brightest of the three Herschel objects seen in a faint galaxy cloud, the other two being NGC4967 and NGC4974 and is seen as a round, well defined patch of light. Radial velocity: 9011 km/s. Distance: 402 mly. Diameter: 94 000 ly.

NGC4974

H782³	13:05.9	+53 40	GLX
E/S0	0.7′ × 0.4′	14.31B	

Photographs show an elliptical galaxy oriented ESE–WNW with an oval core and a faint envelope. This galaxy is ESE of NGC4973; a faint field star lies between the two objects. It is a little fainter than the preceding galaxy but otherwise similar in appearance

visually. Radial velocity: 8921 km/s. Distance: 398 mly. Diameter: 81 000 ly.

NGC4977

H780³	13:06.2	+55 39	GLX
SA(r)b	2.0′ × 2.0′	14.58B	

The DSS image shows a face-on spiral galaxy with a small, very bright, round core in a somewhat fainter inner pseudo-ring. A multi-arm spiral pattern made up of narrow, low-surface-brightness spiral arms emerges from the pseudo-ring. At medium magnification the galaxy is faint, though readily visible as a small, moderately well defined patch of light with a brighter central region in a slightly diffuse outer envelope. Radial velocity: 8431 km/s. Distance: 377 mly. Diameter: 219 000 ly.

NGC5007

H848³	13:09.3	+62 10	GLX
S0⁻:	1.0′ × 0.8′	14.20B	

Situated 5.25′ SW of a bright field star, photographs show a face-on lenticular galaxy with a large, bright, round core and extremely faint, broad spiral arms or outer disc structure. A very thin dust lane crosses the core from SE to NW. Likely companion galaxy UGC8234 is 6.75′ to the NNW, while MCG+10-19-44 is 2.0′ ESE. NGC5007 is quite faint visually and difficult to see owing to the bright field star to the NE. Best at medium magnification, it appears as a small, round patch of light which is even in surface brightness and fairly well defined at the edges. Radial velocity: 8465 km/s. Distance: 378 mly. Diameter: 110 000 ly.

NGC5109

H808³/H826²	13:20.9	+57 39	GLX
Sbc	1.7′ × 0.5′	14.18B	

Photographs show an almost edge-on spiral galaxy, probably barred, with a thin, long, brighter bar in a grainy envelope. A field galaxy, plotted as NGC5113 in the first edition of *Uranometria 2000.0*, is 5.1′ NE. Visually, NGC5109 is the brighter of the two galaxies visible in a high-

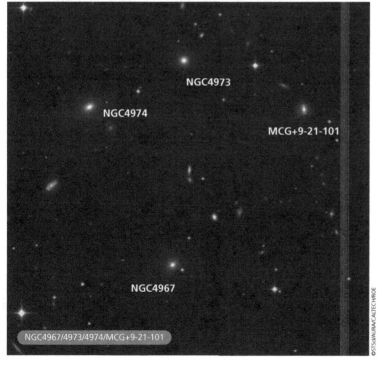

NGC4973
NGC4974
MCG+9-21-101
NGC4967
NGC4967/4973/4974/MCG+9-21-101
©STScI/AURA/CALTECH/ROE

magnification field; it is a faint nebulous streak elongated SSE–NNW and is a little brighter along its major axis with well defined edges. This galaxy was listed twice by Herschel: it was first observed and recorded as H808³ on 24 April 1789 and subsequently as H826² on 17 March 1790. Radial velocity: 2243 km/s. Distance: 100 mly. Diameter: 50 000 ly.

NGC5163

H821³	13:26.9	+52 45	GLX
E	1.0′ × 0.7′	14.77B	

The visual and photographic appearances of this elliptical galaxy are similar. It is a small, faint patch of light, fairly well defined and brighter to a very small, condensed core. Visually, it appears round but photographs show it elongated N–S. The field star 6.0′ to the ENE hinders observation and should be kept out of the field. Radial velocity: 9024 km/s. Distance: 403 mly. Diameter: 117 000 ly.

NGC5164

H784³	13:27.2	+55 29	GLX
SBb	1.0′ × 1.0′	14.45B	

Images reveal an almost face-on barred spiral galaxy with a small, bright, elongated core in a fainter bar. The bar attaches to two short, brighter arcs embedded in broad, grainy spiral arms elongated NNE–SSW. Visually, the galaxy is very faint; it is a round patch of light with even surface brightness and fairly well defined edges. Radial velocity: 7326 km/s. Distance: 327 mly. Diameter: 95 000 ly.

NGC5201

H797²	13:29.2	+53 04	GLX
S?	1.6′ × 1.0′	14.19B	

Despite the uncertain classification, the DSS photograph shows a definite, almost face-on spiral galaxy with a peculiar morphology. The bright, slightly elongated core is embedded in a slightly fainter inner envelope. There are two faint condensations at the border of the inner envelope to the SE and NW which may be evidence of a

bar. A narrow half ring emerges from the inner envelope on the NE side. A single, disturbed spiral arm emerges from the SE and wraps completely around the galaxy counterclockwise, broadening and fading in the S. The galaxy is elongated SE–NW and a bright field star is 5.75′ N. Visually, however, only the bright core of the galaxy is seen, appearing as a moderately faint, irregularly round patch of light, well defined but with a suspected hazy, secondary envelope. The central region is fairly opaque and a little brighter to the middle. Radial velocity: 8901 km/s. Distance: 397 mly. Diameter: 185 000 ly.

NGC5204

H63⁴	13:29.6	+58 25	GLX
SA(s)m	5.0′ × 3.0′	11.73B	

Herschel discovered this remarkable galaxy on 24 April 1789 and placed it in

his Class IV. His description reads: 'Considerably bright, considerably large, irregularly round, easily resolved. Very gradually much brighter to the middle, 4′ diameter. I suppose, with a higher power, I might have seen the stars.' The DSS image shows a large, almost face-on spiral galaxy. The core is large, bright and irregular in form and the spiral structure is fragmentary and difficult to trace, but the entire envelope is peppered with bright, knotty HII regions, some of them quite large. High-resolution photographs begin to show individual stars at about blue magnitude 20.5. Visually, the galaxy is large and fairly bright, but quite diffuse. It is an oval, extended glow, oriented N–S, broadly brighter to the middle with poorly defined edges. Very faint field stars bracket the galaxy to the E and W south of the core. Radial velocity: 318 km/s. Distance: 14 mly. Diameter: 21 000 ly.

NGC5204

©STScI/AURA/CALTECH/ROE

NGC5216

H841[2]	13:32.1	+62 42	GLX
E0 pec	2.0' × 1.0'	13.68B	

The DSS image shows a probable elliptical galaxy with a peculiar morphology. This galaxy is interacting with NGC5218 which is 4.1' N and connected by a long, narrow filament which appears to emerge from the

**NGC5216 NGC5218
15-inch Newtonian 272x**

NGC5216/5218

©STScI/AURA/CALTECH/ROE

northern galaxy. NGC5218 has a round, bright core in a fainter envelope which is more prominent to the S. There is a long, brighter filament emerging from the core in the S and curving SW. Together with NGC5218 this is known as Keenan's System and it has also been catalogued as Arp 104. Telescopically, the galaxy forms quite a bright pair with NGC5218. NGC5216 is a round, fairly concentrated patch of light, which is a little brighter to the middle with well defined edges. Radial velocity: 3067 km/s. Distance: 137 mly. Diameter: 80 000 ly.

NGC5218

H842[2]	13:32.2	+62 46	GLX
SB(s)b? pec	1.8' × 1.5'	13.10B	

This inclined spiral galaxy, probably barred, has a peculiar morphology. Photographs show the galaxy is interacting with NGC5216 which is 4.1' S and connected by a long, narrow filament. The two galaxies undoubtedly form a physical pair and the minimum distance between the two would be 163 000 ly at the presumed distance. NGC5218 has a bright, irregular core in a blunt, slightly fainter bar. Spiral arms emerge, but the structure is chaotic. There is a short, bright arm which emerges in the N and curves back quickly into the bar. There is a second, larger spiral arm emerging from the SW and curving S of the galaxy. A third, disconnected, arm is in the W and curves N where it broadens and fades, extending a further 1.7'. Visually, NGC5218 is larger and brighter than its companion. It is an oval object, elongated E–W, broadly brighter to the middle with somewhat well defined edges. Radial velocity: 3061 km/s. Distance: 137 mly. Diameter: 72 000 ly.

NGC5250

H817[2]	13:36.0	+51 14	GLX
S0	1.0' × 0.7'	13.92B	

The appearance of this lenticular galaxy is quite similar both photographically and visually. At the eyepiece, it is a small, round and condensed patch of light; it is brighter to a bright core and its edges fade quickly into the sky background. Photographs show it elongated SE–NW. A bright field star is 5.75' to the NE. Radial velocity: 4626 km/s. Distance: 207 mly. Diameter: 60 000 ly.

NGC5255

H803[3]	13:37.3	+57 06	GLX
	0.8' × 0.2'	15.11B	

Though this galaxy is unclassified, the DSS photograph shows an edge-on lenticular galaxy with a bright, slightly elongated core in a fainter elongated envelope, oriented NE–SW. Visually, this exceedingly faint galaxy is seen at medium and high magnification as a hazy, oval patch of light. Intermittently, extensions are seen at high magnification and the galaxy is very small, though somewhat well defined. A bright field star is 1.75' east. Radial velocity: 7074 km/s. Distance: 316 mly. Diameter: 74 000 ly.

NGC5256

H673[3]	13:38.3	+48 17	GLX
Compact pec	1.4' × 1.3'	14.1B	

The DSS image shows a pair of galaxies in contact, but it is very difficult to separate the two components. The principal galaxy appears to have a large and very high-surface-brightness core with disturbed spiral arms, or tendrils emerging from the core. The secondary galaxy appears to be a fainter system immediately N with a highly distorted structure and a core offset to the W, bordered by a dark cavity or dust patch. A plume of material emerges and is seen in silhouette in front of the core of the larger galaxy. A bright double star is 5.75' WSW, while a faint field galaxy is 1.2' SSE of the double star. Visually, the galaxy is small but fairly bright and is well seen at high magnification. Round and very condensed, it is brighter to the middle with well defined edges. Radial velocity: 8368 km/s. Distance: 374 mly. Diameter: 152 000 ly.

NGC5278

H798[2]	13:41.6	+55 40	GLX
SA(s)b? pec	0.9' × 0.7'	14.71B	

The DSS image shows an almost face-on spiral galaxy with a peculiar morphology, which forms an interacting pair with NGC5279 0.5' E. Together the galaxies are also catalogued as Arp 239. NGC5278 has a round, bright core and one thick, bright spiral arm emerging, encircling the core and extending to NGC5279. The companion's northern spiral arm curves directly into NGC5278's spiral arm. The southern arm of NGC5279 has a bright condensation at its tip. The surface brightness of both galaxies is quite high. The field galaxy UGC8671 is 2.7' WSW. Visually, NGC5278 is a patchy, irregular glow, oval in shape and a little brighter to the middle. The edges are ill defined and the galaxy is oriented E–W. NGC5279 is visible as a very faint, nebulous point of light immediately E. The magnitude 7 field star 7.75' ENE interferes with observation. Herschel recorded the galaxy on 14 April 1790,

describing it as: 'Pretty bright, extended, 1.5' long, 0.5' broad', indicating that he probably saw both galaxies as an unresolved glow. Radial velocity: 7655 km/s. Distance: 342 mly. Diameter: 89 000 ly.

NGC5294

H785[3]	13:45.2	+55 17	GLX
0.7' × 0.3'	15.17B		

The classification of this dwarfish galaxy is uncertain. Photographs show that it is possibly spiral, with a patchy, irregular core elongated ESE–WNW and a very faint star immediately W. Visually, the galaxy is very dim and is seen only as a hazy and very gradual brightening in the sky background; it is patchy and irregularly round with poorly defined edges. Radial velocity: 2076 km/s. Distance: 93 mly. Diameter: 19 000 ly.

NGC5308

H255[1]	13:47.0	+60 58	GLX
S0[−]	3.7' × 0.7'	12.24B	

Images show an edge-on lenticular galaxy with a large, bright, elongated core in a bright, elongated spindle. There is a very small, compact galaxy located between the galaxy's ENE tip and a faint field star which is almost due E of the core. Visually, this is a very beautiful, bright, spindle-shaped galaxy, oriented ENE–WSW with a bright, bulging core. The extensions are bright and well defined, especially on the SSE flank. The extensions taper to soft points. Radial velocity: 2169 km/s. Distance: 97 mly. Diameter: 104 000 ly.

NGC5322

H256[1]	13:49.3	+60 12	GLX
E3-4	5.3' × 3.9'	11.05B	

The DSS image reveals a large, bright, elliptical galaxy with a large, elongated core in a fainter, extended envelope. Visually, this is a very bright and large galaxy, much brighter to a well-condensed core. The outer envelope is quite bright and mottled but the extremities are quite diffuse and ill

defined. At high magnification two extremely faint stars are suspected, one just beyond the outer envelope to the S, the other to the E. The DSS image supports this observation. The galaxy is oriented E–W. Radial velocity: 1881 km/s. Distance: 84 mly. Diameter: 130 000 ly.

NGC5342

H849[3]	13:51.4	+59 53	GLX
S0	1.1' × 0.4'	14.23B	

Photographs reveal a lenticular galaxy with a bright, elongated core and slightly fainter extensions, elongated SSE–NNW. In a moderate-aperture telescope this is a small, moderately bright galaxy with a well-condensed, bright, though nonstellar, core. The core is surrounded by a small secondary halo which is slightly oval. Radial velocity: 2376 km/s. Distance: 106 mly. Diameter: 34 000 ly.

NGC5368

H786[3]	13:54.5	+54 20	GLX
(R')SABab:	1.2' × 0.8'	14.08B	

This is an almost face-on spiral galaxy, very probably barred, with a bright, elongated central region in a fainter outer disc, which suggests possible spiral structure. Visually, the galaxy is moderately bright; it is a small, roundish object with a very small, bright core embedded in a fainter outer envelope. The envelope is quite well defined; a faint field star is 1.5' to the NNE. Radial velocity: 4772 km/s. Distance: 213 mly. Diameter: 74 000 ly.

NGC5370

H843[2]	13:54.1	+60 41	GLX
SB(r)0[0]	1.3' × 1.3'	14.08B	

The DSS image reveals a beautiful, face-on, theta-shaped, barred lenticular galaxy, with a bright, elongated core and a slightly fainter bar. The bar is attached to a bright inner ring, while the ring is surrounded by a very faint, broad outer envelope. A field star is 1.4' NNE. Visually, however, this galaxy is a faint, diffuse, almost round patch of light with moderately well defined extremities. It is gradually brighter to the middle though no bright core is visible. Radial velocity: 3187 km/s. Distance: 142 mly. Diameter: 54 000 ly.

NGC5372

H809[3]	13:54.7	+58 41	GLX
S?	0.6′ × 0.4′	15.44B	

Photographs show a very small, inclined galaxy, probably a spiral, with a large, bright core in a fainter outer· envelope, elongated SE–NW. A narrow dust lane is suspected SW of the core. At high magnification the galaxy is small but moderately bright and appears round with a very small, brighter core. The edges are quite well defined and the disc is opaque. Radial velocity: 1843 km/s. Distance: 82 mly. Diameter: 14 000 ly.

NGC5376

H238[1]/H844[2]	13:55.3	+59 30	GLX
SAB(r)b?	1.8′ × 1.1′	12.97B	

Images reveal an inclined spiral galaxy with a brighter core and two knotty, bright spiral arms which expand to broader arms. Several dust patches are involved. The galaxy UGC8859 is in the field 7.1′ NNE. Bright Herschel companions NGC5379 and NGC5389 are about 15.0′ NNE. Visually, this is a moderately bright galaxy which is elongated ENE–WSW, with moderately well defined extremities. A faint stellar core is intermittently visible at high magnification. In *The Scientific Papers*

the designations H844[2] and H844[3] are assigned to NGC3795, but the position provided for H844[2] is almost identical to that of NGC5376; the identification of H844[2] with NGC3795 is therefore an error. NGC5376 was recorded twice by Herschel: first on 24 April 1789 and again on 19 March 1790. Radial velocity: 2214 km/s. Distance: 99 mly. Diameter: 52 000 ly.

NGC5379

H239[1]	13:55.6	+59 45	GLX
S0	2.3′ × 0.6′	14.03B	

Photographs show an inclined lenticular galaxy with a bright, elongated core. A fainter inner envelope surrounds the core and expands to a bright ring, which is embedded in a still fainter outer envelope. A dust lane may be involved and seems to border the inner portion of the bright ring for almost a complete revolution. Bright Herschel companion galaxy NGC5376

is 15.0′ SSE, while NGC5389 is 4.1′ E. The diffuse spiral galaxy UGC8859 is 7.5′ S. In moderate-aperture telescopes, NGC5379 is a hazy patch of light at best and is most prominent at high magnification; it is a little brighter to the middle and elongated ENE–WSW. Radial velocity: 1735 km/s. Distance: 78 mly. Diameter: 52 000 ly.

NGC5389

H240[1]	13:56.1	+59 44	GLX
SAB(r)0/a:?	4.7′ × 1.0′	12.88B	

This is a large, bright, highly inclined spiral galaxy with a large, bright, elongated core. Photographs show a bright spiral arm with two condensations in the S. Two narrow dust lanes are W of the core. A magnitude 9 field star is 3.9′ ENE. The bright Herschel companion NGC5376 is 15.0′ SSE and NGC5379 is 4.1′ W, while the faint spiral galaxy UGC8859 is 8.2′ SSW. Visually, both this galaxy and

NGC5376

NGC5379/5389

NGC5379 can be seen in the same high-power field of view. NGC5389 is the brighter of the two and has a prominent star-like core at medium magnification and diminishes somewhat as magnification is increased. The outer envelope is a little diffuse and oval and tapers to points; the galaxy is elongated N–S. Radial velocity: 1970 km/s. Distance: 88 mly. Diameter: 120 000 ly.

NGC5402

H810[3]	13:58.2	+59 49	GLX
S?	1.3′ × 0.3′	14.50B	

The visual and photographic appearances of this edge-on spiral galaxy are quite similar. It is extremely faint at medium magnification and is confirmed with averted vision at high power. It is a thin sliver of light which is moderately well defined and a little brighter at the core; it is oriented almost due N–S. Radial velocity: 3150 km/s. Distance: 141 mly. Diameter: 53 000 ly.

NGC5422

H230[1]	14:00.7	+55 10	GLX
S0	3.9′ × 0.7′	12.81B	

Photographs reveal a classic, edge-on lenticular galaxy elongated SSE–NNW with a bright, round core embedded in an oval, slightly fainter envelope and even fainter, tapered extensions. A field star is 2.3′ E. In moderate-aperture telescopes, it is a relatively easy object, appearing as a small, thin streak with a slightly brighter core. The galaxy is well defined and several bright field stars are visible in a medium-magnification field. Radial velocity: 1940 km/s. Distance: 87 mly. Diameter: 98 000 ly.

NGC5430

H827[2]	14:00.8	+59 20	GLX
SB(s)b	2.2′ × 1.7′	12.82B	

The DSS photograph shows an almost face-on, barred spiral galaxy with a peculiar morphology. The small, bright, round core is in an irregular bar. To the SE is a bright condensation, while two fainter knots are SE of this condensation. One thin, faint spiral arm emerges from the bar to the W, curving S. The other spiral arm emerges from the E curving N; it is much broader than the western spiral arm and a second spiral arm fragment is visible on the E side, closer to the core. A faint lenticular galaxy, MCG+10-20-50, is 6.75′ WSW. Visually, this galaxy is quite bright and is well seen at high magnification. It features a prominent bar oriented SE–NW with a small, round, bright core embedded. The whole is surrounded by a fainter, diffuse and ill defined envelope. A faint, tight double star is located 1.75′ to the ENE. Radial velocity: 3090 km/s. Distance: 138 mly. Diameter: 56 000 ly.

NGC5443

H799[2]	14:02.2	+55 49	GLX
Sdm:	3.0′ × 1.0′	13.35B	

Photographs show a highly inclined spiral galaxy with a bright core embedded in a fairly bright inner disc and fainter outer spiral extensions. A faint field star is involved in the envelope to the WSW. In a moderate-aperture telescope this is a large and quite bright galaxy; it is an elongated streak, oriented NE–SW and is well defined at the edges and well seen at high magnification. A small, bright core is embedded in an extended envelope that is somewhat mottled and uneven in surface brightness. Two faint field stars are seen along the SE flank of the galaxy, one S of the core, the other NE of the core. Radial velocity: 1925 km/s. Distance: 86 mly. Diameter: 75 000 ly.

NGC5447

H787[3]	14:02.5	+54 17	EN
1.0′ × 0.2′			

NGC5447, NGC5461 and NGC5462 are bright knots in the spiral arms of NGC5457 (M101). NGC5447 is located

NGC5430

NGC5447 NGC5461 NGC5462
15-inch Newtonian 146x

7.75′ WSW of the core of M101 and the DSS image shows it as a resolved network of more than half a dozen nebulous patches at the end of the outermost spiral arm to the W. Visually, it can be seen as an elongated, detached portion of the galaxy oriented SSE–NNW. It is moderately bright along its major axis and moderately well defined. Radial velocity: 271 km/s. Diameter: 3500 ly.

NGC5448

H691[2]	14:02.8	+49 10	GLX
(R)SAB(r)a	4.0′ × 1.8′	12.12B	

Images reveal a highly inclined barred spiral galaxy with a large, bright core and a bright, foreshortened bar. Two bright, massive spiral arms emerge from the bar, forming an inner ring, with small knots involved. A dust lane is

NGC5448

evident to the SE of the core and the outer spiral arms are very faint and broad. Visually, the galaxy is well seen at medium magnification as a moderately bright object elongated ESE–WNW. The envelope is fairly well

defined and the core region is brighter, mottled and appears elongated along the major axis. Radial velocity: 2130 km/s. Distance: 95 mly. Diameter: 110 000 ly.

NGC5461

H788[3]	14:03.7	+54 19	HII/EN/OC
0.75′ × 0.2′	10.85B		

NGC5447, NGC5461 and NGC5462 are bright knots in the spiral arms of NGC5457 (M101). NGC5461 is located 4.5′ ESE of the core of M101 and is well resolved in photographs. Visually, this object is best seen when the eye is well dark adapted and the spiral structure of M101 is clearly visible. Direct vision shows it as an oval brightening in the eastern spiral arm. Herschel located these objects on one of the most productive nights of his observing

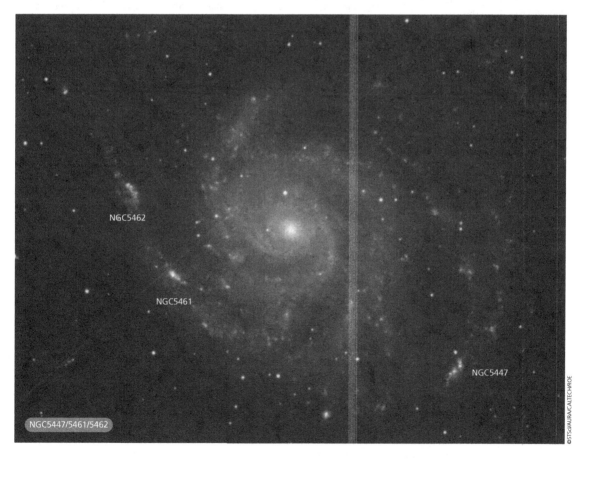

NGC5462

NGC5461

NGC5447

NGC5447/5461/5462

career, 14 April 1789, when he recorded no less than 50 new nebulae and galaxies.

NGC5462

H789[3]	14:03.9	+54 22	HII/EN/OC
1.7' × 0.8'	9.81B		

NGC5462, a bright knot in the spiral arm of M101 is located 6.0' ENE of the core and is clearly resolved in photographs. At the eyepiece this nebulous knot is seen as an elongated, brighter patch near the end of M101's eastern spiral arm. It is larger than NGC5461, which is visible to the SW.

NGC5473

H231[1]	14:04.7	+54 54	GLX
SAB(s)0⁻:	2.3' × 1.7'	12.37B	

Photographs show an almost face-on barred lenticular galaxy with a bright, elongated core and a short bar in a slightly fainter, elongated envelope. The bar is perpendicular to the envelope and is oriented E–W. A faint field star is involved in the envelope immediately ENE of the core and the galaxy is elongated SSE–NNW. Telescopically, this small and bright galaxy is dominated by an intensely bright, large core with the edges fading suddenly into the sky background. The outer envelope is mottled in texture and round, but there is no sign of the bar. Radial velocity: 2147 km/s. Distance: 96 mly. Diameter: 64 000 ly.

NGC5474

H214[1]	14:05.0	+53 40	GLX
SA(s)cd pec	4.8' × 4.3'	11.48B	10.8V

This galaxy is part of the group dominated by M101 and photographically appears as a spiral galaxy of peculiar morphology. The core is large, bright and irregular with a bright spur to the S. The spiral structure is strong with many nebulous knots to the S, but almost nonexistent to the N, giving a highly asymmetrical appearance. In moderate-aperture telescopes, this is a fairly bright galaxy which is brighter to the middle and large compared with other galaxies in the

region. Best seen at low to medium magnification, the core is clearly displaced towards the N. NE of the core is a magnitude 14 star and a magnitude 12 star precedes the galaxy 3.5' to the NW. Herschel noted the galaxy on 1 May 1788, saying: 'Considerably bright, considerably large. North ends abruptly. South very gradually.' Radial velocity: 392 km/s. Distance: 17.5 mly. Diameter: 24 000 ly.

NGC5475

H800[2]	14:05.2	+55 45	GLX
Sa? sp	2.0' × 0.7'	13.42B	

The classification of this almost edge-on galaxy is uncertain. It displays characteristics of both a spiral and a lenticular galaxy: the core is bright and elongated and is embedded in a fairly bright disc that may involve dust lanes, particularly in the southern extension. Visually, the galaxy is elongated almost due N–S; it is small with very high surface brightness and well defined edges. Bright at high magnification, the galaxy features a small, bright core and bright extensions which taper to points. Radial velocity: 1839 km/s. Distance: 82 mly. Diameter: 48 000 ly.

NGC5477

H790[3]	14:05.6	+54 28	GLX
SA(s)m	1.7' × 1.3'	14.38B	14.0V

Photographs show a face-on, dwarf spiral galaxy with a very small, brighter core in a grainy spiral envelope. There are bright condensations in the spiral

arms to the E and S. Telescopically, this very diffuse, irregular patch of light is very slightly brighter to the middle and a little elongated E–W with very ill defined edges. The galaxy is probably a satellite of M101. Radial velocity: 424 km/s. Distance: 19 mly. Diameter: 9300 ly.

NGC5480

H692[2]	14:06.4	+50 43	GLX
SA(s)c:	1.7' × 1.7'	12.81B	

The DSS photograph shows a slightly inclined spiral galaxy with a small, round core and complex, bright spiral structure. The two brightest arms form an S-shaped pattern emerging from the core; there are at least three fainter minor arm fragments, and several knots and condensations as well as dust patches are seen in the arms. The spiral pattern is asymmetrical, being a little more extensive to the N, and a very faint outer envelope is seen on the eastern flank. Companion galaxy NGC5481 (in Boötes) is 3.2' E. Visually, this is an attractive pair of galaxies; NGC5480 though bright is the more diffuse; it is an oval patch of even surface brightness, elongated N–S. Although it fades gradually at the edges, it is fairly well defined. Radial velocity: 1968 km/s. Distance: 88 mly. Diameter: 44 000 ly.

NGC5484

H791[3]	14:06.8	+55 02	GLX
E2	0.4' × 0.3'	15.80B	

Photographs show a very small elliptical galaxy with a brighter, elongated core and a fainter outer envelope elongated N–S. Several galaxies are in the DSS field. Companion galaxy NGC5485 is 3.75' ESE, while NGC5486 is 6.8' NE. Visually, NGC5484 is by far the faintest of the three in the field; it is extremely difficult and is seen only intermittently as a very small, almost star-like patch, round with even surface brightness. Radial velocity: 2152 km/s. Distance: 96 mly. Diameter: 11 000 ly.

NGC5485

H232[1]	14:07.2	+55 00	GLX
SA0 pec	2.3′ × 1.6′	12.40B	

This is the principal member of a small group of eight galaxies. Photographs reveal a lenticular galaxy of peculiar morphology, with a large, bright central region and an extensive, fainter outer envelope. Deep photographs show a dust lane or perhaps a dust ring perpendicular to the major axis of the galaxy. Several faint galaxies are in the field, possibly dwarf companions. Telescopically, the galaxy is bright and round, appearing well defined with a bright core. A faint field star is 1.8′ SE of the core. Neighbouring galaxy NGC5484 is 3.75′ WNW and NGC5486 is 6.5′ NNE. Radial velocity: 2122 km/s. Distance: 95 mly. Diameter: 63 000 ly.

NGC5486

H801[2]	14:07.4	+55 06	GLX
SA(s)m	1.7′ × 1.0′	14.23B	

This slightly inclined spiral galaxy is a low-surface-brightness dwarfish galaxy with a very small, bright core and fainter, knotty spiral arms. It is quite faint visually; it appears as a small uncondensed smudge of light of even surface brightness which is elongated E–W. Radial velocity: 1513 km/s. Distance: 67 mly. Diameter: 33 000 ly.

NGC5526

H804[3]/H835[3]	14:13.9	+57 46	GLX
Scd	2.3′ × 0.3′	14.74B	

This object was catalogued twice by Herschel: first on 17 April 1789 and again on 17 March 1790. The descriptions are nearly identical and Herschel thought this galaxy was resolvable. Photographs show an edge-on spiral galaxy with a narrow, bright major axis and fainter extensions. There is a dust lane along the SW flank. A small, faint galaxy is 0.5′ WSW of the core and may be visible in large apertures. Its radial velocity (12 119 km/s) is far higher than that of the larger galaxy, and it is presumed to be a background object. A faint field star is 0.75′ N. NGC5526 is a faint and difficult galaxy visually, visible as an ethereal blur, flat and elongated SE–NW. The surface brightness is even across the major axis and the edges are well defined. Radial velocity: 2135 km/s. Distance: 95 mly. Diameter: 64 000 ly.

NGC5540

H805[3]	14:14.8	+60 00	GLX
0.7′ × 0.6′	14.74B		

This galaxy is unclassified but photographs indicate that it may be a slightly inclined spiral galaxy with a large, elongated, brighter core and a faint outer envelope which includes a brighter knot to the S. Several faint galaxies are in the field, including one at 2.3′ NNW, a second at 3.0′ WNW and a third at 2.75′ SSW. They may be visible in very-large-aperture telescopes. Visually, the galaxy is quite faint and small but is readily apparent and holds magnification surprisingly well. It is round, moderately well defined and broadly brighter to the middle. Radial velocity: 11 117 km/s. Distance: 497 mly. Diameter: 101 000 ly.

NGC5585

H235[1]	14:19.8	+56 44	GLX
SAB(s)d	5.8′ × 3.7′	11.39B	11.2V

The DSS image shows a large, slightly inclined spiral galaxy with an elongated, brighter core embedded in a broad spiral pattern. The spiral arms are studded with many knotty condensations and dust lanes are visible to the SE and NW of the core. It is one of the brighter companion galaxies to M101 and deep

NGC5585

©STScI/AURA/CALTECH/ROE

photographs show that the brightest stars of this galaxy begin to resolve at about blue magnitude 20. In a moderate-aperture telescope, the galaxy is large, fairly bright and broadly concentrated towards a brighter middle. The edges of the galaxy are diffuse and poorly defined and the envelope is somewhat grainy. It is very slightly elongated NNE–SSW with a quite faint star bordering the envelope to the SSW. Radial velocity: 435 km/s. Distance: 19.5 mly. Diameter: 33 000 ly.

NGC5631

H236[1]	14:26.6	+56 35	GLX
SA(s)0°	3.0′ × 3.0′	12.48B	

This lenticular galaxy has a large, bright, slightly elongated core in an extensive, slightly asymmetrical outer envelope. The DSS image shows a curving dust lane NE of the core and suspected thin dust lanes are evident in the outer envelope throughout, so spiral structure may exist. The galaxy is small but very bright visually and holds high magnification well. The envelope is round and condensed, opaque but with a grainy texture. A bright, concentrated core is visible and the edges are fairly well defined. Radial velocity: 2112 km/s. Distance: 94 mly. Diameter: 82 000 ly.

Ursa Minor

The 24 Herschel objects in this most northerly constellation were observed in two principal periods: 1785 and in the years between 1791 and 1802. All are galaxies and, except for NGC6217, none is a bright or compelling object.

NGC6331 is a very challenging object and a marker for the remote galaxy cluster Abell 2256, which may be visible in large-aperture telescopes.

NGC5034

H909[3]	13:12.2	+70 39	GLX
S?	0.9′ × 0.7′	14.43B	

Despite its uncertain classification, the DSS image shows that this is obviously an almost face-on spiral elongated NNE–SSW with a bright, elongated core and two spiral arms. Telescopically, the galaxy is faint and small but is readily visible at both medium and high magnification. It appears as a round patch of light, even in surface brightness and fairly well defined. Radial velocity: 8836 km/s. Distance: 395 mly. Diameter: 103 000 ly.

NGC5144

H70[4]	13:22.9	+70 31	GLX
SAc? pec	1.2′ × 0.9′	13.47B	

This face-on spiral galaxy displays a peculiar morphology. Photographs show three short, brighter spiral arms in a faint, broad outer envelope. A bright, detached condensation 0.4′ S is a likely companion galaxy seen in silhouette. The galaxy is well seen at high magnification but appears as an ethereal, ghostly patch of light which is a little brighter to the middle and a little extended NNE–SSW, situated midway between two field stars. The edges are not well defined. Recorded on 6 May 1791, Herschel called it: 'Considerably bright, round, almost equally bright throughout, resembling a very ill defined planetary nebula about 0.5′ in diameter.' Radial velocity: 3299 km/s. Distance: 147 mly. Diameter: 51 000 ly.

NGC5323

H899[2]	13:45.5	+76 51	GLX
Sab	1.4′ × 0.4′	14.52B	

Images reveal an almost edge-on spiral galaxy with a bright, elongated core in a faint outer envelope. Visually, the galaxy is fairly faint and is located 4.6′ SSE of a tight double star with equally bright components. The galaxy is very slightly elongated N–S with a small, faint core embedded in a faint envelope with moderately well defined edges. Radial velocity: 2136 km/s. Distance: 96 mly. Diameter: 39 000 ly.

NGC5452

H947[3]	13:54.5	+78 13	GLX
SAB(s)d	1.7′ × 1.1′	14.42B	

Photographs show a slightly inclined spiral galaxy slightly elongated E–W with a slightly brighter, small core and faint spiral arms. Visually, this is a small and quite faint galaxy located 2.1′ almost due S of a faint field star. It is a round, uncondensed spot with fairly well defined edges and even surface brightness across the envelope. Radial velocity: 2226 km/s. Distance: 99 mly. Diameter: 49 000 ly.

NGC5547

H948[3]	14:09.8	+78 36	GLX
S	0.5′ × 0.4′	14.83B	

Images reveal a very small, face-on spiral galaxy with a brighter core and two main spiral arms. IC4404 is 0.7′ N, but this field galaxy is not seen in moderate-aperture telescopes.

NGC5547 is extremely small and faint and high magnification is required for its identification. It is slightly oval with well defined edges, is extended N–S and is slightly brighter to the middle. An extremely faint field star immediately S is visible with averted vision. Radial velocity: 11 863 km/s. Distance: 530 mly. Diameter: 77 000 ly.

NGC5607

H331[2]/H319[3]	14:19.4	+71 35	GLX
Pec	0.9′ × 0.8′	13.91B	

The DSS image shows a slightly inclined spiral galaxy with a brighter, elongated core and faint, asymmetrical spiral structure, which is more extensive to the W. A possible dust lane borders the core to the S. Visually, this galaxy is only moderately faint though small and is well seen at high magnification. The galaxy appears round and is brighter to a star-like core. The edges are fairly well defined and the envelope appears grainy. H319[3] has been separately identified as NGC5620 at right ascension 14:22.5, declination +69 34, but Herschel's positional measurements for the two objects lead to the location of NGC5607. Radial velocity: 7751 km/s. Distance: 346 mly. Diameter: 91 000 ly.

NGC5671

H882[3]	14:27.7	+69 38	GLX
SB(r)b	1.7′ × 1.0′	14.35B	

Photographs show a slightly inclined barred spiral galaxy with a small, round core and a fainter bar. Two narrow spiral arms emerge from the bar, forming a pseudo-ring with only one outer arm extending for a complete revolution around the galaxy. The galaxy is elongated NE–SW and a field star is 3.8′ SW. Telescopically, only the core and bar are visible and the galaxy is a difficult object. It is best seen with averted vision at high magnification as a narrow, ill defined ESE–WNW streak with fairly even surface brightness. Radial velocity: 9171 km/s. Distance: 410 mly. Diameter: 203 000 ly.

NGC5712

H950[3]	14:29.7	+78 51	GLX
E	0.7′ × 0.7′	15.30B	

This is probably an elliptical galaxy. Photographs show a small, round core and an extensive, fainter outer envelope. Visually, the galaxy is very faint but is revealed as a round, well-condensed patch and is evenly bright across a well defined, sharp-edged envelope. A faint field galaxy is 4.1′ WNW and appears brighter than NGC5712; it is immediately WSW of a faint field star. Photographs show that this galaxy is a high-surface-brightness, edge-on lenticular galaxy. Radial velocity: 6958 km/s. Distance: 311 mly. Diameter: 63 000 ly.

NGC5819

H311[3]	14:54.0	+73 08	GLX
SAB(rs)bc	1.0′ × 1.0′	14.37B	

Photographs show a face-on barred spiral galaxy with a bright, elongated core and bright, knotty spiral structure. A faint galaxy is 3.3′ NE. Visually, NGC5819 is faint but is readily visible as a small, round and fairly well defined patch of light with even surface brightness, situated just NE of a line joining two faint field stars 4.1′ apart. Herschel's 16 March 1785 description reads: 'Very faint, small, irregularly round. Between 2 pretty small stars 6′ apart.' Radial velocity: 7413 km/s. Distance: 331 mly. Diameter: 96 000 ly.

NGC5832

H332[2]	14:57.8	+71 41	GLX
SB(rs)b?	3.7′ × 2.2′	13.60B	

The DSS image shows an almost face-on spiral galaxy with a star-like, slightly brighter core. The spiral structure is faint and poorly defined and the galaxy is elongated NE–SW. Visually, the galaxy is large, though quite dim, appearing almost round and broadly brighter towards the middle. The envelope is fairly smooth-textured and the edges fade gradually into the sky background. Discovered on 16 March 1785 Herschel found this object: 'Pretty bright, considerably large, brighter towards the preceding side.' Radial velocity: 611 km/s. Distance: 27 mly. Diameter: 28 000 ly.

NGC5836

H312[3]	14:59.5	+73 53	GLX
SB(rs)b	1.2′ × 1.0′	14.71B	

Images show a slightly inclined, barred spiral galaxy with a small, round core and a fainter bar. The bar is attached to two narrow spiral arms and a faint field star is involved 0.4′ S of the core. The galaxy is best seen at high magnification as a very ethereal, ghostly glow, almost round and diffuse. It has a bright, very compact and star-like core and the presence of the faint field star makes it look like a double nuclei. Herschel recorded the galaxy on 16 March 1785 and interpreted the bright core as a star, his description reading: 'Extremely faint, very small, a little extended. Two very small stars in it.' Radial velocity: 7332 km/s. Distance: 327 mly. Diameter: 114 000 ly.

NGC5909

H943[3]	15:11.5	+75 22	GLX
Sb	1.1′ × 0.5′	14.69B	13.6V

The DSS image shows a highly inclined spiral galaxy with a slightly elongated core and asymmetrical spiral structure that is oriented ENE–WSW and more extensive to the WSW. A dust lane is seen along the SE flank and immediately E of the core. It forms a pair with

NGC5912, which is 0.8′ E. The two galaxies are probably physically related and the core-to-core separation would be about 75 000 ly at the presumed distance. A faint galaxy is 4.5′ SSE. Visually, both NGC5909 and NGC5912 can be observed in the same high-magnification field, although only the core of NGC5909 is seen. It is a small, round patch with well defined edges. Radial velocity: 7221 km/s. Distance: 322 mly. Diameter: 103 000 ly.

NGC5912

H944[3]	15:11.7	+75 22	GLX
E?	0.7′ × 0.7′	14.51B	13.0V

The classification of this galaxy is uncertain and it may be an S0 galaxy. Photographs show a bright, slightly-elongated core in a fainter outer envelope, elongated E–W. A faint star or very possibly a compact companion galaxy is immediately E of the core and is seen in silhouette against the outer envelope. At high magnification, this galaxy is very slightly brighter than NGC5909 and appears as a small, well defined, slightly elongated oval oriented E–W, with even surface brightness. Radial velocity: 7402 km/s. Distance: 331 mly. Diameter: 67 000 ly.

NGC6011

H313[3]	15:46.6	+72 09	GLX
Sb	2.0′ × 0.7′	14.36B	

Photographs reveal a highly inclined spiral galaxy with a small, bright core and slightly fainter spiral arms. At high magnification, the galaxy is faint but readily visible as an elongated patch of light, even in surface brightness, and elongated almost due E–W with diffuse edges. A faint field star is 0.9′ ENE of the core. Radial velocity: 7652 km/s. Distance: 342 mly. Diameter: 199 000 ly.

NGC6048

H873[2]	15:57.6	+70 42	GLX
E	2.0′ × 1.4′	13.57B	

This elliptical galaxy has a bright, elongated core in a fainter, elongated envelope, oriented SE–NW.

Photographs show that a faint neighbouring galaxy is 2.4′ ESE. At high magnification NGC6048 is small, though moderately bright. The inner envelope is bright and brightens to a star-like core. The envelope is round with moderately well defined edges. Radial velocity: 7886 km/s. Distance: 352 mly. Diameter: 205 000 ly.

NGC6068

H973[3]	15:55.4	+79 00	GLX
SBbc?	1.3′ × 1.3′	12.72B	

The DSS image shows a face-on, barred spiral galaxy with a bright, elongated core and two bright spiral arms. The N spiral arm fades suddenly and broadens as it curves W, then S. Dark zones N and S of the core may be dust patches. The edge-on companion galaxy NGC6068A is 1.9′ W and is probably physically related. The core-to-core separation at the observed distance would be about 103 000 ly. In a moderate-aperture telescope, NGC6068 is best viewed at medium magnification. It appears as a hazy, oblong patch of light extended N–S and is a little brighter to the middle with poorly defined edges. NGC6068A is intermittently seen as a faint, star-like point SE of a faint field star. Herschel discovered NGC6068 on 6 December 1801 and thought it resolvable. Radial velocity: 4152 km/s. Distance: 188 mly. Diameter: 52 000 ly.

NGC6071

H883[3]	16:01.1	+70 37	GLX
0.9′ × 0.7′	15.17B		

This object is unclassified, though very probably it is an elliptical galaxy. Its visual and photographic appearances are quite similar. Visually faint, the galaxy is a small, round, nebulous patch that is best seen at medium magnification. It is broadly brighter to the middle and has ill defined edges. Photographs show a faint field galaxy 3.8′ WSW. Radial velocity: 7736 km/s. Distance: 346 mly. Diameter: 90 000 ly.

NGC6079

H884[3]	16:04.5	+69 40	GLX
E	1.4′ × 1.0′	13.85B	

This elliptical galaxy has a brighter, elongated core and a faint outer envelope, elongated SSE–NNW. Photographs show a faint spiral galaxy 7.5′ SE. At high magnification NGC6079 is small, though fairly condensed, and almost round. The envelope is opaque and even in brightness with well defined edges. Radial velocity: 7627 km/s. Distance: 341 mly. Diameter: 139 000 ly.

NGC6094

H314[3]	16:06.6	+72 30	GLX
S0	1.8′ × 1.4′	14.35B	

Images show this lenticular galaxy has a bright core in a fainter inner envelope and an irregular outer envelope. Visually, this faint galaxy stands out best at high magnification. It features a faint star-like core, which is only intermittently visible, and a moderately well defined disc oriented ESE–WNW. Herschel discovered this galaxy on 16 March 1785 and thought it was easily resolvable. Radial velocity: 7781 km/s. Distance: 348 mly. Diameter: 182 000 ly.

NGC6217

H280[1]	16:32.6	+78 12	GLX
(R)SB(rs)bc	3.0′ × 2.5′	11.79B	

Photographs show this face-on, barred spiral galaxy to have spiral arms emerging from the bar, but it also has the peculiar feature of a third spiral arm that emerges from the E of the bright core. This third arm then curves S to join the spiral arm that emerges from the S of the bar. In moderate-aperture telescopes, the galaxy is bright and well defined, though the edges are ragged. The envelope is mottled and the galaxy brightens suddenly to a sharp stellar core. SE of the core along the major axis is a bright condensation or star, which is brighter than the core itself. The galaxy is elongated NNW–SSE. Radial velocity: 1544 km/s. Distance: 69 mly. Diameter: 61 000 ly.

NGC6217

©STScI/AURA/CALTECH/ROE

NGC6251

H974[3]	16:32.5	+82 33	GLX
E	1.8′ × 1.5′	13.90B	

Photographs show an elliptical galaxy with a bright, slightly elongated core in a fainter outer envelope. It is paired with NGC6252 2.4′ N, though the spread in radial velocities suggests that the two may not form a physical pair. Visually, both galaxies are visible but NGC6251 is easier to see; it is a small, moderately bright, well defined glow which brightens to a small, brighter core. The main envelope is somewhat grainy in texture. Radial velocity: 7587 km/s. Distance: 339 mly. Diameter: 177 000 ly.

NGC6252

H975[3]	16:32.7	+82 36	GLX
S?	0.9′ × 0.5′	15.27B	

The DSS image shows an inclined galaxy, probably a spiral, with a small, bright core and a smooth-textured disc which is elongated ENE–WSW. At the eyepiece, this galaxy is very faint; it is a slightly oval patch and is little more than an ill defined glow due N of NGC6251. Radial velocity: 6607 km/s. Distance: 298 mly. Diameter: 77 000 ly.

NGC6324

H945[3]	17:05.4	+75 25	GLX
Sc	1.0′ × 0.7′	13.92B	

This is an inclined spiral galaxy with a bright core and two spiral arms.

NGC6331

Radial velocity: 5054 km/s.
Distance: 266 mly. Diameter:
66 000 ly.

NGC6331

H951[3]	17:03.5	+78 38	GLX
E	0.4′ × 0.3′	15.20B	

The DSS image reveals three elliptical galaxies on a line running ESE–WNW; the two galaxies to the E appear to be in contact and share a common envelope. These galaxies are among the principal members of the Abell 2256 galaxy cluster and are some of the most remote objects in the Herschel catalogue. Dozens of faint galaxies are visible in the immediate field of the DSS image. Visually, this is an extremely faint object that forms a diamond-shaped asterism with three very faint field stars. It appears as a very small, elongated glow, which is the unresolved image of the two galaxies in contact. Herschel discovered this object on 20 December 1797, describing it as: 'Extremely faint, small, better with 320 (magnification).' Large amateur telescopes may be able to pick up three additional members of the cluster, situated 1.75′–3.5′ to the ENE. Radial velocity: 17 743 km/s. Distance: 793 mly. Diameter: 92 000 ly.

Photographs show one is short and stubby and the other is long wrapping around the galaxy and broadening into a faint envelope. A bright field star is 5.6′ N. Visually, the galaxy is moderately faint, but holds magnification well; it is an oval patch, elongated E–W and fairly well defined with a small, brighter core. The core is offset somewhat towards the N.

Virgo

Virgo is home to the largest number of objects in Herschel's catalogue, numbering 344 in all and, except for the globular cluster NGC5634, all are galaxies. Of course, this is the location of the Virgo Cluster of galaxies and a high percentage of the listings in this constellation are associated with the cluster. However, there are many lesser groups scattered throughout the constellation, some of them fairly remote, and a significant number of Herschel objects are associated with these groups as well. Many of the objects are found in contrasting pairs, triplets and larger associations, so the observer will find many attractive fields to explore. And Herschel was not necessarily exhaustive in his sweeps through this

constellation; the observer with a moderate- or large-aperture instrument will frequently find faint objects that Herschel missed. A significant number of objects in this constellation have more than one Herschel number. He swept through these fields repeatedly year after year, often using different stars for his positional offsets and many objects were inadvertently recorded more than once. There are many compelling individual galaxies here, including NGC4594 (the Sombrero galaxy) and NGC5746, two of the author's favourites. However, each individual observer is sure to list different objects as worthy of repeated attention and this constellation will provide many seasons of fruitful exploration.

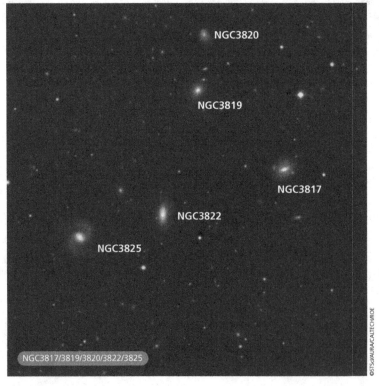

NGC3820

NGC3819

NGC3817

NGC3822

NGC3825

NGC3817/3819/3820/3822/3825

©STScI/AURA/CALTECH/IOE

NGC3818

H284[3]	11:42.0	−6 09	GLX
E5	2.2′ × 1.4′	12.71B	

This elliptical galaxy has a bright core and a fainter outer envelope in photographs, and visually is seen as a bright, well defined, slightly oval object, oriented E–W. The opaque disc displays a slightly brighter core and a bright field star is 7.2′ N. Radial velocity: 1567 km/s. Distance: 70 mly. Diameter: 45 000 ly.

NGC3822

H153[2]/H35[3]	11:42.2	+10 17	GLX
Sb	0.8′ × 0.5′	14.10B	

Photographs show an inclined spiral galaxy with a bright core and a bright inner envelope, surrounded by a fainter outer envelope. It is part of a compact group of galaxies catalogued as Hickson 58, including the Herschel object NGC3825, 3.25′ ESE and three others: NGC3817 (4.75′ WNW), NGC3819 (4.75′ NNW) and NGC3820 (6.75′ WNW). There should be no problem in seing them with large-aperture telescopes. Visually, NGC3822 is a little fainter than NGC3825 but larger; it is elongated N–S and is a little brighter to the middle with no stellar nucleus visible. The galaxy NGC3848 is plotted separately in the first edition of *Uranometria 2000.0* and was thought to be Herschel's H35[3], but it is actually a repeat observation of NGC3822/H153[2]. Radial velocity: 6054 km/s. Distance: 270 mly. Diameter: 63 000 ly.

NGC3825

H154[2]/H36[3]	11:42.4	+10 17	GLX
SBa	1.7′ × 1.0′	13.92B	

This is a face-on, barred spiral galaxy with a bright, round core and a short, faint bar in a faint, round envelope. Two faint, smooth and broad spiral arms are attached and the galaxy is elongated SSE–NNW. In a moderate-aperture telescope only the core of this galaxy is visible, appearing smaller than its neighbour NGC3822, very slightly oval in shape and brighter to the middle with a stellar core. Like NGC3822, this galaxy was recorded twice by Herschel, inadvertently receiving the duplicate designation of NGC3852. Radial velocity: 6419 km/s. Distance: 287 mly. Diameter: 142 000 ly.

NGC3833

H102[3]	11:43.5	+10 10	GLX
Sc	1.4′ × 0.7′	14.54B	

Photographs show an inclined spiral galaxy elongated NNE–SSW with a very small, bright core in a fainter spiral pattern. Telescopically, this galaxy can be seen in the same field as the galaxy erroneously plotted in the first edition of *Uranometria 2000.0* as NGC3848, which is 6.0′ NNE. NGC3833 is a larger but very diffuse, faint nebulous patch of light which is even in surface brightness, fairly round and moderately well defined. Radial velocity: 5952 km/s. Distance: 266 mly. Diameter: 108 000 ly.

NGC3874

H104[3]	11:45.6	+8 34	DS

This object was observed by Herschel only once; it was recorded on 15 April

1784 and described as: 'Very faint, very small, left doubtful. Twilight.' It is now associated with a pair of stars oriented E–W and is very faint visually. The preceding of the two stars is slightly brighter and the pair is best seen at high magnification. The edge-on spiral galaxy NGC3863 is 10.0′ to the SW of this pair and is a little easier to see. In a high-magnification field, the galaxy is moderately faint, but is well seen as a well defined elongated patch of light oriented ENE–WSW which is a little broader at the middle and brighter along a narrow line of the major axis. It is blue magnitude 13.81 and measures 2.7′ × 0.6′ in size. The double star is difficult to see in a dark sky and it would have been even more challenging in twilight. The present author feels this galaxy is a more credible candidate for H104[3] than the pair of stars.

NGC3876

H103[3]	11:45.4	+9 11	GLX
Sab:	1.3′ × 0.7′	13.87B	

Photographs show an inclined spiral galaxy, possibly barred, with a bright, rectangular central region and two main spiral arms attached. The N spiral arm is the brighter; two knotty spiral condensations are visible and the galaxy is elongated E–W. A bright field star is 5.75′ W and a faint edge-on galaxy, UGC6734, is 3.4′ SE. At the eyepiece, this is a very faint, small galaxy and is easily overlooked. The galaxy is almost round with a slight concentration towards the middle and the outer envelope is moderately well defined. Herschel thought the object was resolvable. Radial velocity: 2806 km/s. Distance: 125 mly. Diameter: 47 000 ly.

NGC3914

H90[3]	11:50.6	+6 35	GLX
(R')SB(rs)b	1.3′ × 0.7′	14.43B	

This is an inclined, barred spiral galaxy, with a bright, oval core and two spiral arms. Photographs show that dust patches are involved with the spiral arms, which broaden as they expand outward. A very faint field star is

embedded in the spiral arm N of the core. Telescopically, the galaxy is a moderately bright but diffuse patch of light, best seen at medium magnification. The envelope is ill defined and elongated NE–SW. A faint field star is 1.4′ NNW of the core. Radial velocity: 6020 km/s. Distance: 269 mly. Diameter: 102 000 ly.

IC2963

H113[3]	11:49.4	−5 07	GLX
S0[+]: sp	1.7′ × 0.4′	14.62B	

This object was originally recorded as NGC3915 and was presumed to be nonexistent, but has subsequently been identified as the same object as IC2963. Visually, it is a very faint and slightly elongated patch of light, immediately E of the easternmost star in a faint triangle. The surface brightness is even and the edges are diffuse. Herschel's 24 April 1784 description reads: 'Extremely faint, extremely small, with 240 (magnification) 2 very small stars and nebula.' Radial velocity: 1543 km/s. Distance: 69 mly. Diameter: 34 000 ly.

NGC3952

H612[3]	11:53.7	−4 00	GLX
IBm: sp	2.1′ × 0.7′	13.52B	

Photographs show an irregular galaxy with a bright, crescent-shaped central region and very faint extensions. Telescopically, this is a small but reasonably bright galaxy, a slightly elongated, well defined object with even surface brightness, oriented E–W. A bright field star is 7.5′ NNW. Radial velocity: 1454 km/s. Distance: 65 mly. Diameter: 40 000 ly.

NGC3976

H132[2]	11:56.0	+6 45	GLX
SAB(s)b	3.8′ × 1.2′	12.56B	

In photographs, this is a highly inclined spiral galaxy with a bright oval core in an elongated, fainter envelope. Two bright spiral arms emerge, with many knots and condensations involved, as well as dust lanes and patches.

NGC3976

stars.' Radial velocity: 1354 km/s. Distance: 60 mly. Diameter: 74 000 ly.

NGC4044

H491[3]	12:02.5	−0 11	GLX
E⁺:	1.1′ × 1.0′	14.52B	

Photographs show an elliptical galaxy with a bright core and a faint, extensive outer envelope. A faint star or compact object is immediately NNE of the core. The object is small and faint visually, fairly even in surface brightness and well defined. Radial velocity: 6026 km/s. Distance: 269 mly. Diameter: 86 000 ly.

NGC4045

H276[2]	12:02.7	+1 59	GLX
SAB(r)a	2.9′ × 1.8′	12.73B	

This is a slightly inclined spiral galaxy with a bright, elongated core embedded in a narrow, knotty inner spiral structure. Photographs show that the arms expand to a broad, extensive pattern of smooth-textured and faint outer arms, elongated E–W. Also in the field is NGC4045A, located 1.5′ to the S but evidently not a physical companion as its radial velocity is considerably higher. Telescopically, NGC4045 is a small but fairly bright object, very slightly elongated N–S with a bright, grainy envelope that is a little brighter to the middle with fairly well defined edges. Radial velocity: 1880 km/s. Distance: 84 mly. Diameter: 71 000 ly.

Extremely faint, disturbed plumes are seen to the ENE. At medium magnification this is a fairly bright galaxy, with an extended mottled envelope elongated ENE–WSW and a bright, round core. The outer envelope is hazy and ill defined, but the inner envelope is bright and opaque. Radial velocity: 2408 km/s. Distance: 108 mly. Diameter: 119 000 ly.

NGC4030

H121[1]	12:00.4	−1 06	GLX
SA(s)bc	4.2′ × 3.0′	11.23B	

The DSS image shows a slightly inclined galaxy with a large, bright central region and at least four principal spiral arms emerging; they are fragmentary with several knotty condensations involved. Telescopically, the galaxy is very bright and is situated just E of a N–S line joining two bright field stars. The main envelope is bright, grainy and

broadly brighter to the middle. The edges are diffuse and poorly defined and the galaxy is slightly elongated NNE–SSW. Herschel recorded this galaxy on 1 January 1786 and described it as: 'Very bright, considerably large, a little extended, much brighter to the middle. 3′ long, 2.5′ broad, between 2 pretty bright

NGC4030

NGC4067

H37[3]	12:04.2	+10 51	GLX
SA(s)b	1.4′ × 1.0′	13.20B	

Images reveal a slightly inclined, barred spiral galaxy with a bright, elongated core and a narrow bar. Two principal spiral arms emerge from the bar and are tightly wound. In a moderate-aperture telescope the galaxy is not bright, and only the core and the bar are readily visible as a well defined oval elongated NE–SW with a slight brightening towards the middle. Radial velocity: 2353 km/s. Distance: 105 mly. Diameter: 43 000 ly.

NGC4073

H277[2]	12:04.5	+1 54	GLX
E5/S0⁻	2.5′ × 1.9′	12.51B	13.2V

In photographs this elliptical or lenticular galaxy is elongated ESE–WNW and has a large bright core embedded in an extensive, fainter outer envelope. A compact possible companion is involved in the outer envelope ESE of the core. Several fainter galaxies are nearby to the S, including Herschel object NGC4077 7.0′ SSE and NGC4063, but there is a considerable spread in radial velocity and the galaxies may not form a physical group. Visually, NGC4073 forms a bright pair with NGC4077 in a medium-magnification field. It is a little brighter than NGC4073, and appears as a slightly extended, well defined oval oriented E–W with a bright, star-like core. Radial velocity: 5771 km/s. Distance: 258 mly. Diameter: 187 000 ly.

NGC4077

H258[3]	12:04.7	+1 48	GLX
SB(r)0°	1.2′ × 0.9′	14.29B	

Images reveal a lenticular galaxy elongated N–S with a large core embedded in an extensive, slightly fainter outer envelope. A faint star or compact companion is involved in the outer envelope N of the core. Several fainter galaxies are nearby, including NGC4139 (IC2989) 1.7′ to the NW. At medium magnification, NGC4077 is visible as an elongated sliver of light, oriented NNE–SSW and is fairly well defined. Radial velocity: 7009 km/s. Distance: 313 mly. Diameter: 109 000 ly.

NGC4123

H4[5]	12:08.2	+2 53	GLX
SB(r)c	4.6′ × 3.7′	11.93B	

This is an almost face-on, barred spiral galaxy with a small, bright core, embedded in a faint, dusty bar.

Photographs show four principal spiral arms; they are faint, tightly wound and studded with brighter condensations. Visually, this galaxy can be seen in a medium-magnification field with the barred spiral NGC4116, which is situated 14.0′ to the SSW. NGC4123 is a large but very diffuse patch of light, broadly brighter to the middle with poorly defined edges. NGC4116 presents a similar appearance and the galaxies probably form a physical pair with a minimum core-to-core separation of 225 000 ly at the presumed distance. Radial velocity: 1232 km/s. Distance: 55 mly. Diameter: 74 000 ly.

NGC4124

H33[1]/H14[2]/H60[2]	12:08.2	+10 23	GLX
SA(r)0⁺	4.3′ × 1.3′	12.39B	

Photographs show a bright, highly inclined lenticular galaxy with a large, bright core surrounded by a slightly fainter inner envelope and a much fainter outer envelope. Visually, the galaxy is a moderately bright, elongated sliver of light oriented ESE–WNW. The surface brightness is fairly even, though it does brighten and thicken slightly in the middle and the edges are fairly well defined. Herschel inadvertently recorded this galaxy twice. It was first observed on 15 March 1784 and classed as a faint nebula with the terse comment: 'Faint, small.' One month later on 15 April, Herschel recorded it as a bright nebula, calling it: 'Bright, large, much extended, much brighter to the middle. Resolvable.' There is also a possibility that it was recorded a third time, as H14[2], on 18 January 1784, and called: 'a little extended, not cometic' but this identification is uncertain. Radial velocity: 1608 km/s. Distance: 72 mly. Diameter: 90 000 ly.

NGC4129

H548[2]	12:08.9	−9 02	GLX
SB(s)ab: sp	2.3′ × 0.6′	13.08B	

This almost edge-on spiral galaxy has a very bright, elongated central region with faint outer spiral structure in the DSS image, but high-resolution

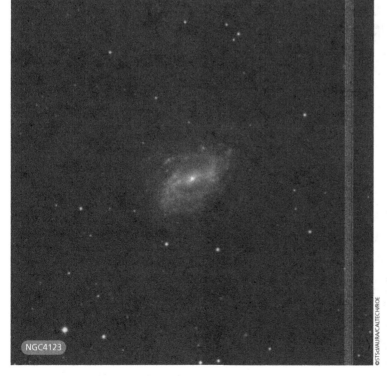

NGC4123

photographs resolve the inner disc into several bright clumps. The galaxy is bright and well seen at high magnification as an elongated, high-surface-brightness disc oriented E–W with extensions that taper to points. It is broadly brighter along its major axis and the edges are very well defined. Radial velocity: 1047 km/s. Distance: 47 mly. Diameter: 32 000 ly.

NGC4168

H105[2]	12:12.3	+13 12	GLX
E2	2.1′ × 1.6′	12.11B	

This galaxy is the principal member of the NGC4168 group of galaxies, a subcluster in the Virgo Cluster numbering about 44 members. It is an elliptical galaxy with a large, bright core and an extensive outer envelope. Fainter companion galaxy NGC4164 is 2.8′ W and NGC4165 is 2.7′ NNW. Visually, NGC4168 is bright, round and brighter to a nonstellar core. The large, fainter outer envelope fades indefinitely

into the sky background. Radial velocity: 2155 km/s. Distance: 96 mly. Diameter: 59 000 ly.

NGC4179

H9[1]	12:12.9 +1	18	GLX
S0⁻: sp	4.0′ × 1.1′	11.84B	

Photographs show a spindle-shaped, edge-on lenticular galaxy with a large, bright, elongated core and slightly fainter, narrow extensions. This is a bright, opaque galaxy visually, elongated SE–NW with slightly fainter extensions that taper to points. The core is bright and bulges and is elongated along the major axis. This is one of Herschel's early Class I discoveries; recorded on 24 January 1784, his description reads: 'Considerably bright, extended north preceding, south following [with a] nucleus and 2 branches, 3′ long.' Radial velocity: 1158 km/s. Distance: 52 mly. Diameter: 60 000 ly.

NGC4179

NGC4180

H133[2]	12:13.1	+7 02	GLX
Sab:	1.6′ × 0.5′	13.32B	

This is an almost edge-on spiral galaxy with a bright, elongated central region and slightly fainter outer arms. Photographs show a probable dust lane along the eastern flank. Visually, the galaxy is moderately bright, well defined at the edges and uniformly bright along its major axis, which is oriented NNE–SSW. Radial velocity: 2016 km/s. Distance: 90 mly. Diameter: 42 000 ly.

NGC4193

H163[2]	12:13.9	+13 10	GLX	
SAB(s)c:?	2.1′ × 1.0′	13.19B	12.3V	

Images show an inclined spiral galaxy with a small, brighter core and grainy, slightly fainter spiral arms. Telescopically, the galaxy is moderately bright; it is an E–W elongated, moderately well defined oval, with a brighter elongated central region and a very slightly brighter core. The disc features a grainy texture and is well seen at medium magnification. Radial velocity: 2421 km/s. Distance: 108 mly. Diameter: 66 000 ly.

NGC4197

H134[2]	12:14.6	+5 48	GLX
Scd(f)	4.2′ × 0.6′	13.47B	

The DSS image shows an edge-on spiral galaxy with a bright, elongated central region and extensive dust patches along its NW flank. Knotty condensations are prominent in the NE extension. Visually, the galaxy is moderately faint; it appears as a flat streak that is fairly well defined along its edges. It is elongated NE–SW and its surface brightness is even along its major axis. Radial velocity: 1981 km/s. Distance: 88 mly. Diameter: 108 000 ly.

NGC4200

H164[2]	12:14.8	+12 11	GLX
S0	1.4′ × 0.8′	13.91B	

Photographs show a lenticular galaxy elongated E–W with a bright, elongated core and a slightly fainter and extensive outer envelope. Visually, the galaxy is quite faint; it is a small, round nebulous patch which requires averted vision to see clearly. It has even surface brightness across the face of the disc and its edges are not well defined. Radial velocity: 2314 km/s. Distance: 103 mly. Diameter: 42 000 ly.

NGC4206

H165[2]	12:15.3	+13 02	GLX
SA(s)bc:	5.2′ × 0.8′	12.85B	12.1V

Images reveal an almost edge-on spiral galaxy with a very small, brighter core embedded in a grainy spiral pattern. A dust lane is visible on the eastern flank and dust patches are seen throughout the spiral pattern. Visually, this galaxy, NGC4216 and NGC4222 all just fit in the field of a medium-magnification eyepiece. NGC4206 appears faint, though well defined; it is a flat streak oriented N–S and greatly elongated. Radial velocity: 643 km/s. Distance: 29 mly. Diameter: 44 000 ly.

NGC4215

H135[2]	12:15.9	+6 24	GLX
SA(r)0+: sp	1.9′ × 0.7′	13.01B	

The DSS image shows an edge-on lenticular galaxy with a bright, oval core embedded in a fainter, rectangular envelope. The tapered extensions are brighter than the rectangular envelope. The galaxy is a probable member of the W Cloud of the Virgo Cluster. In a medium-aperture telescope this is a small but fairly bright galaxy elongated N–S, with a bright, condensed core. Well defined along its edges, the envelope is opaque along the major axis. Radial velocity: 1993 km/s. Distance: 89 mly. Diameter: 49 000 ly.

NGC4216

H35[1]	12:15.9	+13 09	GLX
SAB(s)b:	8.1′ × 1.8′	10.95B	10.0V

Photographs reveal the classic form of an almost edge-on spiral galaxy with a bright core embedded in a tightly wound spiral pattern. There is a

NGC4216

©STScI/AURA/C.ALTECH/ROE

prominent dust lane along the ESE flank and a great deal of dust defines the entire spiral pattern. The galaxy is elongated NNE–SSW and there is a faint field star 0.5′ E of the core. A compact possible companion galaxy is 4.0′ N of the core and may be visible in large-aperture telescopes. Visually, this is a remarkable object even in a small-aperture telescope. The nucleus is sharp and stellar and is surrounded by a small but very bright core. The envelope is bright and well defined and tapers to sharp points. The galaxy is located about 2° W of the centre of subcluster A in the Virgo Cluster and most likely is a member. Radial velocity: 73 km/s. Distance: 55 mly (estimate). Diameter: 130 000 ly (estimate).

NGC4223

H137[2]	12:17.5	+6 42	GLX
SA(s)0+:	3.6′ × 2.0′	13.02B	

The DSS image reveals an inclined spiral galaxy with a bright, elongated core in a

slightly fainter envelope, surrounded by a faint spiral pattern with some evidence of dust, particularly to the SE. A magnitude 8 field star is 5.5′ to the S. The first edition of *Uranometria 2000.0* erroneously identifies this galaxy as NGC4241. In the atlas, the designation NGC4241 should be attached to IC3115. Visually, NGC4223 is a small but moderately bright galaxy; it is a well defined elongated object which is a little brighter to the middle and oriented ESE–WNW. NGC4241 is 8.75′ ESE. Radial velocity: 2156 km/s. Distance: 96 mly. Diameter: 101 000 ly.

NGC4224

H136[2]	12:16.6	+7 28	GLX
SA(s)a: sp	3.3′ × 1.0′	12.89B	

Images reveal a highly inclined spiral galaxy with a bright, elongated core in a slightly fainter envelope. A narrow, sharply defined dust lane extends the length of the galaxy along its NW flank with a very faint field star embedded

NGC4224

NGC4241

H480[3]	12:18.0	+6 39	GLX
SB(s)cd	2.5′ × 2.3′	13.77B	

The DSS image shows a face-on, barred spiral galaxy with a very slightly brighter core in a straight, narrow bar. A very faint field star is embedded in the bar to the SE and a second faint star is in the spiral arm to the ESE. An S-shaped spiral pattern emerges from the bar. In the first edition of *Uranometria 2000.0* the galaxy is identified as IC3115 but the *New General Catalogue* identification is more accurate. Visually, this galaxy is moderately large but quite dim and is a little more distinct at medium magnification. It is a round, diffuse patch of light that is a little brighter to the middle and has hazy edges. Radial velocity: 655 km/s. Distance: 29 mly. Diameter: 21 000 ly.

NGC4246

H91[3]	12:18.0	+7 11	GLX
SA(s)c	2.5′ × 1.3′	13.41B	

Photographs show an inclined spiral galaxy with a very small, slightly brighter core and two major spiral arms attached. The arms are grainy with evidence of dust and they broaden and fade as they expand outward, branching quickly into a multi-arm pattern. The peculiar galaxy NGC4247 is 5.4′ N. At medium magnification, NGC4246 is quite a faint galaxy, an oval hazy patch of light slightly elongated E–W. The edges are very poorly defined and the brightness of the main envelope is uniform. Radial velocity: 3643 km/s. Distance: 163 mly. Diameter: 118 000 ly.

NGC4260

H138[2]	12:19.4	+6 06	GLX
SB(s)a	3.3′ × 1.6′	12.69B	

This probable member of the W Cloud of the Virgo Cluster is an inclined barred spiral galaxy with a bright, elongated core and a fainter bar. Photographs show that the bar appears to be slightly curved and uneven in brightness; this is possibly a consequence of the presence of dust patches. There is a narrow dust

near the end of the lane to the WSW. A magnitude 8 field star is 6.75′ SSW. This is a probable member of the W Cloud of the Virgo Cluster. Visually, NGC4224 can be seen in the same medium-magnification field as NGC4233. It is moderately bright, fairly evenly bright along its major axis and broadly brighter to the middle. It is fairly well defined and is elongated ENE–WSW. Herschel considered this galaxy resolvable. Radial velocity: 2527 km/s. Distance: 113 mly. Diameter: 108 000 ly.

NGC4233

H496[2]	12:17.1	+7 37	GLX
S0°	2.3′ × 1.0′	12.99B	

This galaxy, located to the NE of NGC4224, is also probably a member of the W Cloud of the Virgo Cluster. The DSS image reveals a lenticular galaxy oriented N–S with a bright, round core and fainter, elongated extensions. In a moderate-aperture telescope little more than the bright core is seen; the galaxy appears as a small, fairly faint, fairly well condensed oval patch which is slightly elongated N–S. Radial velocity: 2295 km/s. Distance: 102 mly. Diameter: 68 000 ly.

NGC4235

H17[2]	12:17.2	+7 11	GLX
SA(s)a	3.8′ × 0.9′	12.61B	

Photographs show an almost edge-on spiral galaxy with a bright, slightly elongated core in a fainter, dusty envelope. In a moderate-aperture telescope, this thin, well defined and bright galaxy is moderately condensed along the major axis and somewhat brighter in the middle. It is elongated NE–SW and its ends taper slightly. This is a probable member of the W Cloud of the Virgo Cluster. Radial velocity: 2333 km/s. Distance: 104 mly. Diameter: 115 000 ly.

NGC4260

small well-concentrated, almost round spot and, though faint, it is readily visible with direct vision. Radial velocity: 2452 km/s. Distance: 110 mly. Diameter: 39 000 ly.

NGC4267

H166[2]	12:19.8	+12 48	GLX
SB(s)0⁻?	3.2′ × 3.0′	11.84B	

Photographs reveal a face-on lenticular galaxy with a bright core in an extensive, fainter outer envelope. There is a sharp and sudden drop off in brightness between the core and the envelope and the outer envelope fades very slowly into the sky background. The appearance of the galaxy in small-aperture telescopes is very similar to that in photographs: a bright, almost stellar core is surrounded by a much fainter, round outer envelope. Radial velocity: 952 km/s. Distance: 42 mly. Diameter: 40 000 ly.

NGC4270

H568?[2]	12:19.8	+5 28	GLX
S0	2.0′ × 0.9′	13.10B	

This galaxy is part of a subgrouping of galaxies in the Virgo Cluster, which, if their radial velocities are any indication, are on the far side of the cluster as seen from our perspective. The photographic and visual appearances of NGC4270 are quite similar: it is an inclined lenticular galaxy oriented ESE–WNW with a large, bright, elongated core in an extended, slightly fainter envelope. There is an extremely faint, possible dwarf companion galaxy immediately to the WSW, which is dimly seen in photographs but is well beyond visual range. Along with NGC4270, 17 NGC objects are crowded into an area of about 1 square degree and Herschel, who observed the region on both 17 April 1786 and 23 April 1786, observed at least four, and perhaps five, of the member galaxies. He was somewhat vague in his description, lumping the four galaxies together, thus: 'Four nebulae. They are scattered about. The place [coordinates] is that of the last [NGC4281].' The four

patch in the bar to the NE and a narrow dust lane borders the core to the SE. Two broad, smooth-textured spiral arms emerge from the bar. In a medium-magnification eyepiece, this moderately bright galaxy is elongated NE–SW and is a little brighter along the bar. The edges of the bar are fairly well defined and evidence of the spiral arms is seen in the faint, elongated haze surrounding the bar. Large-aperture amateur telescopes may reveal up to four other galaxies within a 15.0′ circle centred on NGC4260, including NGC4269 8.25′ SE, IC3155 7.75′ SE and IC3136 7.8′ NW. Only NGC4269 is noted in a moderate-aperture telescope. Radial velocity: 1878 km/s. Distance: 84 mly. Diameter: 81 000 ly.

NGC4261

H139[2]	12:19.4	+5 49	GLX
E2-3	4.1′ × 3.6′	11.35B	

The DSS image shows that this elliptical galaxy has a large, bright, slightly

elongated core in an extensive, fainter outer envelope. It is a probable member of the W Cloud of the Virgo Cluster and, as with NGC4260, several fainter galaxies are nearby, including the Herschel object NGC4264 3.3′ ENE. An interesting target for large-aperture amateur telescopes is the compact, M32-like dwarf elliptical VCC334, which is 1.75′ S. Visually, NGC4261 is a bright, almost round galaxy with fairly well defined edges and a brighter, concentrated core. The envelope appears grainy. Radial velocity: 2158 km/s. Distance: 97 mly. Diameter: 115 000 ly.

NGC4264

H140[2]	12:19.6	+5 51	GLX
SB(rs)0⁺	0.9′ × 0.7′	13.76B	

Photographs show an inclined, barred lenticular galaxy with a bright, elongated core and a slightly fainter bar embedded in a faint outer envelope. Visually, the galaxy is a

galaxies are identified by Dreyer as NGC4270, NGC4273, NGC4277 and NGC4281, though there are a number of other galaxies present which were certainly within the capabilities of both Herschel and his instrument, but were not catalogued. One of them, NGC4268, which is SSW from NGC4273, could very likely be one of the four and is brighter than NGC4277, the faintest of the galaxies identified by Dreyer. NGC4268 is a highly inclined lenticular galaxy oriented NE–SW and the present author suggests that H571[2] = NGC4268. Discovery of this galaxy is credited to Eduard Schönfeld, who used a 6-inch refractor to record the galaxy in 1862. Radial velocity: 2276 km/s. Distance: 102 mly. Diameter: 59 000 ly.

NGC4273

H569?[2]	12:19.9	+5 21	GLX
SB(s)c	2.3′ × 1.5′	12.32B	

In photographs this is an almost face-on, barred spiral galaxy with an asymmetrical form. A bright bar and a core are attached to two bright, dusty spiral arms. The arms expand to a fainter but more extensive, multi-arm pattern, which is offset towards the S. Visually, the galaxy is quite bright; it is a well defined oval of light with a brighter, though nonstellar core. Elongated N–S, the companion galaxy NGC4277 is 1.9′ E. and both are members of the W Cloud subgrouping of the Virgo Cluster. Radial velocity: 2296 km/s. Distance: 103 mly. Diameter: 69 000 ly.

NGC4259/4266/4268/4270/4273/4277/4281/4282

NGC4277

H570[2]	12:20.1	+5 21	GLX
SAB(rs)0/a:	1.1′ × 0.8′	14.55B	

Photographs show a face-on, barred spiral galaxy with a small, bright, round core in a weak N–S bar. The faint, smooth-textured outer envelope is elongated N–S and companion galaxy NGC4273 is 1.9′ W. This is the smallest and faintest of the four galaxies Herschel recorded in the area and visually it is a small, faint, almost round, concentrated patch of light with well defined edges. It is immediately N of a faint field star. Radial velocity: 2114 km/s. Distance: 95 mly. Diameter: 30 000 ly.

NGC4281

H573[2]/H571?[2]	12:20.4	+5 23	GLX
S0+: sp	3.0′ × 1.6′	12.30B	

NGC4281 and NGC4273 are the principal members of the W Cloud subgrouping in the Virgo Cluster, though the radial velocity of NGC4281 is somewhat higher than many of its neighbour galaxies. It is a highly inclined lenticular galaxy with a bright, elongated core and an extensive outer envelope and high-resolution photographs show a thin, curving dust lane bordering the core to the N. Visually, it is the largest and brightest of the galaxies visible in a medium-magnification field (six galaxies are readily visible); it is an oval, well defined glow oriented due E–W with a brighter, elongated core. J. L. E. Dreyer believed that Herschel catalogued this galaxy twice: on 17 April 1786, he noted it with three other galaxies he discovered in the field (see NGC4270); six nights later, he described it as: 'A nebula, but cloudy.' Radial velocity: 2630 km/s. Distance: 117 mly. Diameter: 103 000 ly.

NGC4294

H61[2]	12:21.3	+11 31	GLX
SB(s)cd	3.2′ × 1.2′	12.35B	

This is a highly inclined barred spiral with a very small core embedded in a grainy bar. High-resolution

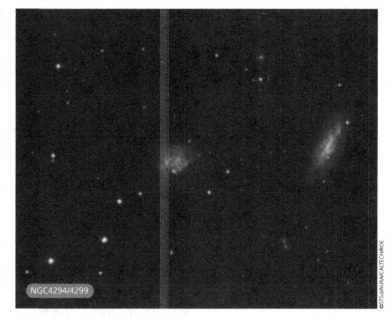

NGC4294/4299

©STScI/AURA/CALTECH/ROE

photographs show that there are numerous HII regions in both the bar and the spiral arms. A faint field star is 1.1′ NNW of the core at the edge of the outer envelope and the companion galaxy NGC4299 is 5.5′ E. Visually, both galaxies fit comfortably in a medium-magnification field. NGC4294 is the larger and brighter of the two, a very elongated galaxy, oriented SSE–NNW, bright along is major axis and a little brighter to the middle. The central portion is a little broader than the extremities, which taper slightly. Radial velocity: 261 km/s. Distance: 55 mly. Diameter: 51 000 ly.

NGC4296

H92[3]	12:21.5	+6 40	GLX
S0	1.1′ × 1.1′	14.50B	

Photographs show that this lenticular galaxy has a bright core in a slightly fainter inner envelope. A very faint and extensive secondary envelope, elongated NNE–SSW, surrounds the core. Companion galaxy NGC4297 is 1.1′ NNW. Visually, NGC4296 is readily visible as a small, elongated galaxy that is fairly condensed along its major axis

but with hazy edges. Radial velocity: 4022 km/s. Distance: 180 mly. Diameter: 89 000 ly.

NGC4297

H93[3]	12:21.5	+6 40	GLX
S0?	0.5′ × 0.2′	13.86B	

This lenticular galaxy is elongated N–S and has a bright core in a fainter outer envelope. The DSS image reveals possible tidal interaction between this galaxy and NGC4296. Visually, this is an extremely difficult object to view. Herschel's discovery description (for both NGC4296 and NGC4297), recorded on 13 April 1784, states: 'Two. One very faint, very small. The other just by, extremely faint, extremely small, left doubtful.' A later observation by Herschel in 1785 seems to have revealed only NGC4296. As on both occasions Herschel was using his 18.7-inch reflector in its less efficient, Newtonian configuration, it is uncertain if he actually saw this extremely faint galaxy. Radial velocity: 3954 km/s. Distance: 177 mly. Diameter: 26 000 ly.

NGC4299

H62[2]	12:21.7	+11 30	GLX
SAB(s)dm:	1.7′ × 1.6′	12.70B	

This is a face-on spiral galaxy with a very small core and poorly defined spiral structure. The DSS image shows several knotty condensations, particularly to the SE in an otherwise faint and grainy envelope. Visually, the galaxy appears irregularly round; the main envelope is uneven in brightness, a little brighter to the S with some condensations visible. Radial velocity: 171 km/s. Distance: 55 mly. Diameter: 27 000 ly.

NGC4300

H572[2]	12:21.7	+5 23	GLX
Sa	1.7′ × 0.6′	13.74B	

Photographs show a highly inclined galaxy, possibly a ring, with a bright elongated core and smooth outer spiral arms or ring. There are large dust patches or a cavity on each side of the core. In a moderate aperture, the galaxy is faint but readily visible as an elongated patch of light which is even in surface brightness along its major axis and well defined at the edges. It is elongated NE–SW. Radial velocity: 2229 km/s. Distance: 99 mly. Diameter: 49 000 ly.

NGC4303

H139[1]	12:21.9	+4 28	GLX	
SAB(rs)bc	6.5′ × 5.8′	10.14B	9.7V	

This is Messier's object number 61 in his catalogue and was discovered on 11 May 1779. Interestingly, he observed

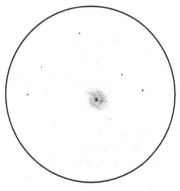

NGC4303 15-inch Newtonian 146x

NGC4303

©STScI/AURA/CALTECH/ROE

the galaxy previously on 5 May and 6 May, each time mistaking it for a comet that he was observing that year. Why Herschel included this galaxy in his catalogue is a bit of a mystery, as he generally avoided recording Messier's objects in his own listing. Observed on 17 April 1786, Herschel described the galaxy as: 'Extremely bright, very bright nucleus. Resolvable, 6′ or 7′ diameter.' Though Sandage and Bedke (1994) classify this galaxy as a grand-design spiral, the brightness and complexity of the spiral pattern seem to indicate that it is more likely a multi-arm spiral. It is a very bright and large, face-on, barred spiral galaxy with a bright core in a fainter bar. Massive, bright spiral arms emerge from the bar and are notable for making sharp and abrupt angle changes. Several dust patches and bright HII regions are involved. The main spiral arms broaden to faint, outer spiral structure. In moderate- and large-aperture telescopes this galaxy is a very rewarding sight. It appears very bright

with an intensely bright, round and nonstellar nucleus. While spiral structure is not visible with certainty, the outer envelope is very patchy with dark zones visible SW and E of the core. Three blunt, straight, bright bars surround the core. These are segments of the spiral arms. The outer envelope, which seems round on initial inspection, is, in fact, slightly elongated N–S. A threshold star is visible WSW of the core in the envelope and another is SW of the outer envelope. Radial velocity: 1483 km/s. Distance: 66 mly. Diameter: 125 000 ly.

NGC4313

H63[2]	12:22.6	+11 48	GLX
SA(rs)ab: sp	4.2′ × 1.0′	12.60B	

Images reveal a highly inclined spiral galaxy with a very small, bright core in a fainter, outer envelope. Several dust patches are involved, including a prominent one bordering the core to the SW. In a moderate-aperture telescope, this is a bright, very elongated galaxy; it is quite bright along its major axis with a small, condensed and bright core. The galaxy is well defined and elongated SE–NW, and the outer disc tapers slightly. Radial velocity: 1384 km/s. Distance: 62 mly. Diameter: 76 000 ly.

NGC4325

H38[3]	12:23.1	+10 37	GLX
E4	1.1′ × 0.7′	14.33B	

This elliptical galaxy appears as a bright, oval core in a fainter envelope on photographs. Visually, the galaxy is moderately faint and is best seen at medium magnification as a roundish, broadly concentrated patch of light with an ill defined, hazy outer envelope. NGC4320, a peculiar interacting pair of galaxies, can be seen 4.5′ to the SSW as an irregularly round, ill defined glow. Radial velocity: 7646 km/s. Distance: 341 mly. Diameter: 109 000 ly.

NGC4326

H141[2]	12:23.2	+6 04	GLX
SAB(r)ab:	1.3′ × 1.1′	14.31B	

This is the first of a triplet of Herschel objects visible in a high-magnification field; the others are NGC4333 and NGC4339. In photographs, this is an almost face-on, barred spiral galaxy with a small, bright core and very short, tapered extensions. Extremely faint, narrow spiral structure surrounds the bar. Companion galaxy NGC4333 is 3.25′ SE and has a similar radial velocity to NGC4326 so they may be physically related. NGC4339, 5.75′ ENE, has a much lower radial velocity than NGC4326 and is probably a closer system. Visually, NGC4326 is an extremely faint, almost star-like patch of light, situated between two field stars oriented N–S. Only the core and the bar are visible in a moderate-aperture telescope. Radial velocity: 7043 km/s. Distance: 315 mly. Diameter: 119 000 ly.

NGC4333

H142[2]	12:23.4	+6 02	GLX
SB(s)ab	0.9′ × 0.7′	14.36B	

Photographs show a face-on, barred spiral galaxy with a round, bright core

in a slightly fainter bar. Very faint spiral arms emerge from the bar. Companion galaxy NGC4326 is 3.25′ NW; the radial velocities of the two galaxies are similar and they may form a physical pair. At the presumed distance the core-to-core separation would be about 294 000 ly. The brighter galaxy NGC4339 is 3.8′ NE. Visually, NGC4333 is a small, concentrated patch of light which requires averted vision to see clearly. It is almost round and located along a line joining two field stars oriented ENE–WSW. Radial velocity: 6962 km/s. Distance: 311 mly. Diameter: 81 000 ly.

NGC4339

H143[2]	12:23.6	+6 05	GLX
E0	2.4′ × 2.3′	12.31B	

NGC4339 forms an apparent triplet with NGC4326 and NGC4333 and is by far the easiest of the three to see. This galaxy is classified as elliptical, though the

extensive fainter outer envelope visible in photographs along with a possible dust lane along the eastern flank suggest that this is a lenticular galaxy. Visually, the main envelope is high in surface brightness with a small, bright core. The galaxy is grainy and diffuse at the edges with some suggestion of a fainter secondary envelope. Herschel located the trio on 13 April 1784, describing them as: 'Three nebulae. The last [NGC4339] is the largest.' Radial velocity: 1212 km/s. Distance: 54 mly. Diameter: 38 000 ly.

NGC4341

H95[3]	12:24.0	+7 07	GLX
SAB(s)0°	1.6′ × 0.5′	13.43B	

The DSS image shows a highly inclined lenticular galaxy oriented E–W with a bright, elongated core in a fainter inner envelope. It is plotted as IC3260 in the first edition of *Uranometria 2000.0*. Only the core of this E–W edge-on galaxy is

NGC4326/4333/4339

NGC4341/4342/4343

©STScI/AURA/CALTECH/ROE

and has even surface brightness and slightly tapering extensions. Possible companion galaxy NGC4342 is 6.0′ N. Radial velocity: 939 km/s. Distance: 42 mly. Diameter: 30 000 ly.

NGC4348

H625[2]	12:23.9	−3 27	GLX
SAbc: sp	3.1′ × 0.6′	13.30B	

Photographs show a highly inclined spiral galaxy with a large, bright, elongated central region in a fainter envelope. A field star is 3.4′ S. Visually, the galaxy is a flat streak oriented NE–SW with well defined edges; it is bright along its major axis with extensions that taper to soft points. A very faint field star is immediately W. Radial velocity: 1899 km/s. Distance: 85 mly. Diameter: 76 000 ly.

NGC4352

H64[2]	12:24.1	+11 13	GLX
SA0 sp:	1.9′ × 0.8′	13.55B	

The photographic and visual appearances of this lenticular galaxy are quite similar. Visually, the galaxy is small but moderately bright; it is an elongated oval oriented E–W with a brighter, condensed core. The extensions are hazy and ill defined and taper slightly to blunt points. Radial velocity: 2010 km/s. Distance: 90 mly. Diameter: 49 000 ly.

NGC4356

H481[3]	12:24.3	+8 33	GLX
Sc	3.2′ × 0.4′	14.13B	

Images show an edge-on spiral galaxy with a narrow dust lane running the length of the major axis of the galaxy, which is elongated NE–SW. A faint field star is immediately E. In a moderate-aperture telescope this is an extremely faint and low-surface-brightness galaxy appearing only as a faint haze around the field star. Larger apertures may reveal the edge-on character. Herschel only described this galaxy as: 'Very faint.' Radial velocity:

visible in a moderate-aperture telescope and it appears as a round, very small, well defined patch of light. It is the most northerly of four galaxies seen in a medium-magnification field; the others are IC3267, NGC4342 and NGC4343. Radial velocity: 848 km/s. Distance: 38 mly. Diameter: 18 000 ly.

NGC4342

H96[3]	12:23.6	+7 03	GLX
S0⁻	1.3′ × 0.6′	13.46B	

This object may be a member of the Virgo Cluster subcluster B. In the DSS image, it is an edge-on and spindle-shaped, bright ellipsoid elongated almost due N–S with a faint outer envelope. It is plotted in the first edition of *Uranometria 2000.0* as IC3256. Companion galaxy NGC4343 is 6.0′ S, NGC4341 (IC3260 in *Uranometria 2000.0*) is 4.6′ NE and IC3267 is 6.5′ ESE. A tiny, compact galaxy is 0.5′ SE of

the core. Visually, NGC4342 appears as a very small, hazy patch of light, very slightly elongated N–S. This field is a difficult one for moderate-aperture telescopes. Herschel discovered NGC4341, NGC4342 and NGC4343 on 3 April 1784 and commented: 'Three. All extremely faint, very small, round. In the 2nd observation two of them were overlooked.' Radial velocity: 677 km/s. Distance: 30 mly. Diameter: 11 000 ly.

NGC4343

H94[3]	12:23.7	+6 57	GLX
SA(rs)b:	2.5′ × 0.7′	13.76B	

Photographs show a highly inclined spiral galaxy with a bright, elongated core in a fainter inner envelope. Faint spiral extensions are involved with several faint dust patches. This is the largest and brightest object in a difficult field. Visually, it is moderately bright and well defined; it is extended SE–NW

1069 km/s. Distance: 48 mly. Diameter: 44 000 ly.

NGC4365

H30[1]	12:24.5	+7 19	GLX
E3	6.9′ × 5.0′	10.50B	

This elliptical galaxy is a probable member of the Virgo Cluster subcluster B, which is centred on M49, and is the second brightest member of the subcluster. It has a large, bright, elongated core and an extensive outer envelope. Companion galaxy NGC4366 is 5.0′ ENE, while NGC4370 is 10.0′ NE. In a moderate-aperture telescope all three galaxies can be seen together in a medium-magnification field. NGC4365 is a large, bright galaxy, slightly oval in shape with a bright, grainy envelope and a large, brighter core. The boundaries of the bright, inner envelope are ragged and a faint secondary haze is visible. The galaxy is elongated NE–SW. Radial velocity: 1170 km/s. Distance: 52 mly. Diameter: 105 000 ly.

NGC4366

H97[3]	12:24.8	+7 21	GLX
dE6	0.8′ × 0.3′	14.81B	

Recorded on 13 April 1784 and evidently seen at the same time as NGC4370, Herschel described this object as: 'The smallest of 2, extremely faint. The other is H144[2] [NGC4370].' Unfortunately, Herschel did not provide a location for this object in reference to the brighter galaxy and on a subsequent sweep on 28 December 1785, Herschel only noted NGC4370. The object was not seen by John Herschel, Heinrich d'Arrest or the observers at Birr Castle. However, there is a very faint galaxy, designated NGC4366 which is situated 5.75′ SSW from NGC4370 and 5.0′ ENE of NGC4365. Photographs show a likely dwarf elliptical galaxy elongated NE–SW with a brighter core in a faint outer envelope. Visually, this galaxy is seen only intermittently, but repeatedly, at medium magnification. It is a very small and faint, though somewhat

concentrated, patch of light, round and fairly well defined. Guardedly, this may be accepted as the object that Herschel saw. Radial velocity: 1203 km/s. Distance: 54 mly. Diameter: 12 500 ly.

NGC4370

H144[2]	12:24.9	+7 27	GLX
Sa	1.6′ × 1.0′	13.53B	

Photographs show a probable spiral galaxy elongated E–W, with a peculiar dust lane along the major axis. The dust lane is extremely narrow at the core but broadens as it extends towards the galaxy's outer envelope. There is an extensive halo around the bright core and NGC4366 is 5.75′ SSW. Visually, NGC4370 is a small, round and faint nebulous patch of light of even surface brightness, adjacent to an equilateral triangle of field stars situated 3.0′ to the ENE. Radial velocity: 710 km/s. Distance: 32 mly. Diameter: 15 000 ly.

NGC4371

H22[1]	12:24.9	+11 42	GLX
SB(r)0+	4.0′ × 2.2′	11.82B	10.8V

This inclined, barred lenticular galaxy has a bright core and a broad bar oriented N–S. Images reveal the bar is attached to a very faint, ill defined ring showing the brightness flare at the attachment point typical of this type of galaxy. Both the core and ring are elongated E–W. Visually, the galaxy is bright and almost round with a bright, large core in a fairly bright envelope. The edges are poorly defined and a hazy

NGC4365/4366/4370

©STScI/AURA/CALTECH/ROE

NGC4371

©STScI/AURA/CALTECH/ROE

secondary envelope, slightly elongated E–W, is visible. Radial velocity: 885 km/s. Distance: 39 mly. Diameter: 46 000 ly.

NGC4376

H530[2]	12:25.3	+5 44	GLX
Im	1.4′ × 0.8′	13.86B	

Photographs show a possible spiral galaxy with a brighter, knotty central region, possibly involving a bar, in a grainy outer envelope, oriented SSE–NNW. The galaxy is not plotted in the first edition of *Uranometria 2000.0* and care should be taken to note the correct declination which is given above. The galaxy is moderately faint visually, but is well seen at medium magnification as a roundish patch of light, fairly well defined and broadly brighter to the middle. Radial velocity: 1058 km/s. Distance: 47 mly. Diameter: 19 000 ly.

NGC4378

H123[1]	12:25.3	+4 55	GLX
(R)SA(s)a	3.3′ × 2.7′	12.60B	

The DSS image shows an almost face-on spiral galaxy with a bright, slightly elongated core and very faint spiral structure, which is possibly of the multi-arm type with an outer pseudo-ring, elongated very slightly N–S. Bright field stars are 4.0′ N and 3.5′ ESE. In a moderate-aperture telescope this is a bright, almost-round galaxy, with a small, bright core. The envelope is bright, somewhat grainy and moderately well defined, surrounded by a faint haze. Radial velocity: 2479 km/s. Distance: 111 mly. Diameter: 107 000 ly.

NGC4387

H167[2]	12:25.7	+12 49	GLX	
E5	1.8′ × 1.1′	12.97B	12.1V	

The photographic and visual appearances of this elliptical galaxy are quite similar. The galaxy features a bright, elongated core and a fainter outer envelope and is elongated SE–NW. The field is a compelling one as it is

situated between and S of M84 and M86, and in a low-magnification field several other galaxies can be viewed without much trouble. Neighbouring galaxy NGC4388 is 9.0′ S. The radial velocities of the galaxies in this field vary widely and they may not actually form a coherent subgroup in the Virgo Cluster. NGC4387's radial velocity is only 417 km/s, well below the Virgo Cluster mean (as are several other nearby galaxies), indicating a distance of about 19 mly and a diameter of only 9300 ly.

NGC4388

H168[2]	12:25.8	+12 40	GLX	
SA(s)b: sp	5.6′ × 1.3′	11.9B	11.0V	

Photographs reveal a highly inclined spiral galaxy with a bright, elongated inner core and fainter extensions. The extension to the W is broader and larger than the one to the E and is possibly disturbed. Prominent, large dust patches are immediately E and W of the

core but do not seem to form a continuous ring in front of the core. Visually, the galaxy is moderately bright; it is a flat, mottled streak elongated E–W. Situated between and S of M84 and M86, the radial velocity is high compared with other galaxies nearby in the field. Radial velocity: 2469 km/s. Distance: 110 mly. Diameter: 180 000 ly.

NGC4390

H39[3]	12:25.8	+10 27	GLX
Sbc(s) II	1.6′ × 1.2′	13.32B	

Photographs show this face-on spiral galaxy has a very small, bright core and fragmentary spiral arms dotted with bright knots in a fainter outer disc. Visually, the galaxy is faint, hazy, very slightly oval and slightly elongated E–W. It is very gradually brighter to the middle with poorly defined edges. Radial velocity: 1041 km/s. Distance: 47 mly. Diameter: 22 000 ly.

NGC4388

NGC4403

H755[3]	12:26.3	−7 41	GLX
SA(r)ab pec sp	1.8′ × 0.5′	13.75B	

Despite a slight difference in their radial velocities, this galaxy forms a likely physical pair with NGC4404 0.8′ to the E. Photographs show a highly inclined galaxy, possibly a spiral, with a very bright, elongated central region and fainter extensions, elongated NNE–SSW. Telescopically, both galaxies are moderately bright and best seen at medium magnification. Only the core of NGC4403 is visible, appearing as a round patch of light which is broadly brighter to the middle with ill defined edges. Radial velocity: 5083 km/s. Distance: 227 mly. Diameter: 119 000 ly.

NGC4404

H756[3]	12:26.4	−7 41	GLX
SA(r)0⁻ pec:	1.2′ × 0.9′	14.25B	

Photographs show a slightly inclined lenticular galaxy elongated SE–NW with a bright, slightly elongated core and a fainter outer envelope. The galaxy forms a close pair with NGC4403 and the minimum separation would be about 53 000 ly if they are at the presumed distance. Visually, this galaxy appears a little smaller than NGC4403; it is a round disc which is broadly brighter to the middle with hazy edges. Radial velocity: 5468 km/s. Distance: 244 mly. Diameter: 85 000 ly.

NGC4412

H34[2]	12:26.6	+3 58	GLX
SB(r)b? pec	1.4′ × 1.3′	13.16B	

Images show a face-on, barred spiral galaxy with a small, round core in a faint bar. Two narrow, grainy spiral arms of fairly high surface brightness emerge from the bar and the one emerging from the W appears disturbed. Visually, the galaxy is a moderately faint object; it is a round, poorly defined haze and is broadly brighter to the middle. Radial velocity: 2212 km/s. Distance: 99 mly. Diameter: 40 000 ly.

NGC4413

H169[2]	12:26.5	+12 37	GLX	
(R′)SB(rs)ab:	2.3′ × 1.5′	12.87B	12.3V

Photographs show a slightly inclined, barred galaxy with a mottled inner spiral pattern and smooth-textured, faint outer spiral structure. In a medium-magnification field, both this galaxy and NGC4425 can be seen; NGC4413 is a moderately bright, large, oval patch of light which is somewhat ill defined at the edges and broadly brighter to the middle. It is elongated NE–SW with a magnitude 10 field star 1.3′ N. Both galaxies are situated to the SE of the bright elliptical galaxies M84 and M86. The radial velocity is only 47 km/s and no reliable information is available regarding either distance or size. Distance: 55 mly (estimate). Diameter: 37 000 ly (estimate).

NGC4415

H482[3]	12:26.6	+8 25	GLX
S0/a	1.3′ × 1.2′	13.30B	

Photographs show a face-on lenticular galaxy with a brighter core in a faint outer envelope. This object is faint and very small in a moderate-aperture telescope; it is a hazy oval patch of light which is slightly elongated N–S and broadly brighter to the middle with hazy, ill defined edges. Radial velocity: 842 km/s. Distance: 38 mly. Diameter: 14 000 ly.

NGC4417

H155[2]	12:26.8	+9 35	GLX
SB0: sp	3.4′ × 1.3′	12.08B	

This edge-on lenticular galaxy displays a symmetrical box- or peanut-shaped central region, depending on the resolution quality of the photograph, and a slightly fainter disc component. Visually, this galaxy is very bright with a condensed, well defined and somewhat grainy envelope. The core is bright and condensed and the fainter extensions taper slightly. The galaxy is elongated NE–SW and is probably a member of subcluster B of the Virgo Cluster. Radial velocity: 756 km/s. Distance: 34 mly. Diameter: 33 000 ly.

NGC4418

H492[3]	12:26.9	−0 51	GLX
Sa	1.5′ × 0.7′	14.03B	

Photographs show an inclined spiral galaxy with a large, bright core, elongated ENE–WSW. An irregular dwarfish, possible companion galaxy, MCG+0-32-13, is 3.0′ ESE. Visually, only the core of NGC4418 is visible. The galaxy is faint and is seen as a slightly oval patch of light which is even in surface brightness and moderately well defined at the edges. Radial velocity: 2082 km/s. Distance: 93 mly. Diameter: 41 000 ly.

NGC4420

H23[2]/H17[3]	12:27.0	+2 30	GLX
SB(r)bc:	2.0′ × 1.0′	12.86B	

This is a highly tilted galaxy; photographs indicate possible spiral structure in a grainy disc that does not show a brighter core. Visually, the galaxy is a moderately bright, elongated streak oriented N–S with fairly even surface brightness along its major axis and well defined extremities. This object was inadvertently recorded twice by Herschel. Radial velocity: 1598 km/s. Distance: 71 mly. Diameter: 41 000 ly.

NGC4422

H114[3]	12:27.3	−5 49	GLX
SA0⁻ pec?	1.3′ × 1.3′	14.59B	

This lenticular galaxy presents similar appearances photographically and visually. The galaxy is fairly faint and is best seen at medium magnification as a small, round patch of light, even in surface brightness and fairly well defined. Radial velocity: 7491 km/s. Distance: 335 mly. Diameter: 126 000 ly.

NGC4423

H145[2]	12:27.1	+5 52	GLX
Sdm:	2.3′ × 0.4′	14.10B	

Photographs show a spindle-shaped galaxy, possibly a spiral or an irregular, with a brighter central region offset towards the S in irregular, asymmetrical extensions. Visually, the galaxy is quite faint but is seen at medium magnification as a narrow sliver of light, even in surface brightness, well defined and oriented NNE–SSW, with a distorted, diamond-shaped asterism to the S. Radial velocity: 1041 km/s. Distance: 47 mly. Diameter: 31 000 ly.

NGC4425

H170²	12:27.2	+12 44	GLX	
SB0⁺: sp	2.5′ × 0.7′	12.98B	11.8V	

The DSS image shows an almost edge-on lenticular galaxy, but higher-resolution images reveal a galaxy with a boxy or X-shaped core with asymmetrically aligned extensions, somewhat reminiscent of NGC5403 but without the central dust lane. Visually, NGC4425 is bright, elongated NNE–SSW and immediately E of a faint field star. It is fairly well defined along its major axis with a flat profile and a small, bright core. Radial velocity: 1812 km/s. Distance: 81 mly. Diameter: 59 000 ly.

NGC4429

H65²	12:27.4	+11 07	GLX
SA(r)0⁺	5.9′ × 2.8′	11.05B	

In photographs, this lenticular galaxy has a bright, elongated central region with a ring-like, ill defined outer spiral structure. High-resolution images show a narrow, well defined dust lane immediately S of the core, which is almost burnt out in the DSS image. Visually, the galaxy is large and quite bright, featuring an extended, very grainy, oval envelope with a condensed, brighter core, which is best seen at medium magnification. Although the primary envelope is fairly well defined, a hazy outer envelope is suspected. A bright field star is 2.0′ NNE of the core. Radial velocity: 857 km/s. Distance: 38 mly. Diameter: 66 000 ly.

NGC4430

H26²/H146²	12:27.4	+6 15	GLX
SB(rs)b:	2.3′ × 1.9′	12.76B	

This is a face-on, barred spiral galaxy with a very small and faint core in a short bar. The DSS image shows three spiral arms but only one extends for more than half a revolution; it broadens, fades and almost entirely

encircles the galaxy, providing a very asymmetric profile. A small spiral galaxy, NGC4432, is 2.4′ SE. At the eyepiece, this is quite a faint galaxy; it is a dim, very diffuse, oval patch of light with poorly defined edges which is broadly concentrated towards the middle and elongated roughly E–W. This galaxy was recorded twice by Herschel: first on 28 January 1784 (as H26² = NGC4453) and again on 13 April 1784 (as H146² = NGC4430). The 28 January observation describes NGC4430 perfectly: 'Pretty bright, considerably large, brighter towards the following side', but at the position given there is only a blank patch of the sky. Dreyer correctly postulated that NGC4453 was actually NGC4430. The designation NGC4453 is currently given to a very small, remote and faint spiral galaxy that Herschel never recorded. Radial velocity: 1376 km/s. Distance: 62 mly. Diameter: 41 000 ly.

NGC4431/4436/4440

NGC4431

H171[2]	12:27.5	+12 18	GLX
SA(r)0	1.7′ × 1.1′	13.80B	12.9V

This is the first of a trio of galaxies visible in a high-magnification field and situated about 1° W of M87. They are probably members of subcluster A of the Virgo Cluster of galaxies and though there is a significant spread in the measured radial velocities of the three galaxies (400 km/s) it is fairly likely that they form a physical system. NGC4431, a lenticular galaxy, presents similar photographic and visual appearances, being an elongated oval oriented N–S which is fairly well defined and broadly brighter to the middle. NGC4436 is 3.6′ to the ENE, NGC4440 is 6.25′ to the E and all three galaxies are about 5.0′ N of a magnitude 9 field star. Radial velocity: 879 km/s. Distance: 39 mly. Diameter: 19 000 ly.

NGC4434

H497[2]	12:27.5	+8 09	GLX
E0/S0(0)	1.4′ × 1.4′	13.01B	

Preceding M49 to the WNW, this elliptical galaxy is a probable member of subcluster B of the Virgo Cluster of galaxies. Its visual and photographic appearances are similar: it is a moderately bright, small and well-concentrated galaxy which brightens to a bright stellar core. It is perfectly round

with well defined edges; only one faint field star, which is ESE of the galaxy, is visible in a high-magnification field. Radial velocity: 1002 km/s. Distance: 45 mly. Diameter: 18 000 ly.

NGC4435

H28[1]	12:27.7	+13 05	GLX
SB(s)0°	3.3′ × 2.2′	11.51B	10.8V

Herschel recorded this galaxy and NGC4438 in the same entry: 'One of two, at 4′ or 5′ distance. Bright. Considerably large.' The two galaxies were christened 'The Eyes' by Leland Copeland, and are also known as Arp 120, an interacting pair in Markarian's Chain. The centre-to-centre separation between NGC4435 and NGC4438 is 4.4′. If they are indeed companions, the distance separating the two galaxies would be a minimum of about 42 000 ly. The visual and photographic appearances of this lenticular galaxy are quite similar. Though small visually, the galaxy is quite bright; it is an elongated oval oriented almost due N–S with an intense core and very well defined edges. Deep photographs show narrow, blunt extensions in the oval envelope and a possible bridge of material extending eastward from NGC4435 to the northern extension of NGC4438. Radial velocity: 749 km/s. Distance: 34 mly. Diameter: 32 000 ly.

NGC4436

H172[2]	12:27.7	+12 19	GLX
dE6/dS0,N	1.9′ × 0.7′	13.98B	13.0V

This is part of a trio of galaxies that includes NGC4431 and NGC4440. The classification is somewhat uncertain but this lenticular galaxy is very likely a dwarf as are its companion galaxies. Visually, the galaxy is a quite faint, broad patch of light immediately ESE from a faint field star. Its appearance is similar to that in photographs: it is slightly oval and elongated ESE–WNW. NGC4431 is 3.6′ to the WSW, while NGC4440 is 3.2′ ESE. Presuming that all three galaxies are at approximately the distance of NGC4436 (whose radial velocity most closely matches that of M87, the dominant galaxy in the region), then the minimum centre-to-centre separation between NGC4436 and NGC4440 would be about 44 000 ly, while the separation of NGC4436 and NGC4431 would be about 50 000 ly. Radial velocity: 1069 km/s. Distance: 48 mly. Diameter: 26 000 ly.

NGC4438

H28[1]	12:27.8	+13 01	GLX
SA(s)0/a pec:	9.0′ × 3.7′	10.93B	10.2V

This large bright galaxy is seemingly paired with NGC4435. Images show that NGC4438 is a peculiar galaxy with a disrupted outer envelope marked by heavy, distorted dust clouds bordering the bright inner core to the W. The outer envelope is asymmetrical and distended,

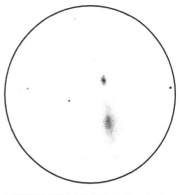

NGC4435 NGC4438 15-inch Newtonian 146x

NGC4435/4438

twisting away from the core towards the SSW. Visually, the bright core is large, elongated and intense, and is embedded in the outer envelope which is ill defined but easy to see, elongated NNE–SSW. The cause of the distortions seen in this galaxy is unknown. Its neighbour NGC4435 shows no disruption in its form, though an apparent bridge of material seems to join the two galaxies. There is also a wide difference in their radial velocities: that of NGC4438 is only 18 km/s. As a result, a reliable size and distance cannot be determined for this galaxy. Distance: 34 mly (estimate). Diameter: 88 000 ly (estimate).

NGC4440

H173[2]	12:27.9	+12 18	GLX
SB(rs)a	1.9′ × 1.7′	12.84B	11.7V

This is the easternmost member of a small group which includes NGC4431 and NGC4436, and photographically it is the most interesting of the group. It is an almost face-on, barred spiral galaxy with a large core and a short bar attached to a partial pseudo-ring which generates two spiral arms. As is frequently seen in systems of this type, there is a brightness flare where the end of the bar makes contact with the ring or spiral arm. In a moderate-aperture telescope, NGC4440 is easily the brightest of the three galaxies; it is a

round, fairly well defined patch of light with a bright, condensed core. Radial velocity: 669 km/s. Distance: 30 mly. Diameter: 16 000 ly.

NGC4442

H156[2]	12:28.1	+9 48	GLX
SB(s)0°	4.6′ × 1.8′	11.39B	

Photographs show this is a large and bright lenticular galaxy with a bright central region and a faint outer envelope. Visually, the galaxy is quite bright and is well seen at high magnification. The bright, well defined envelope is grainy and brighter to a small core. Beyond the bright inner envelope is a very faint haze. The galaxy is elongated E–W. A faint field star borders the outer envelope to the E. Herschel considered this galaxy resolvable. This is a probable member of subcluster B of the Virgo Cluster centred on M49. Radial velocity: 469 km/s. Distance: 21 mly. Diameter: 28 000 ly.

NGC4452

H23[1]	12:28.7	+11 45	GLX
S0(9)	2.8′ × 0.6′	12.87B	12.0V

Morphologically, this galaxy is very similar to the more well known NGC4762 and is an edge-on lenticular galaxy. In photographs, NGC4452 has a very bright major axis and only a very

slight central bulge and the whole is embedded in an extremely faint secondary envelope. The faint elliptical galaxy IC3381 is 7.0′ to the WNW, 2.3′ S of a bright field star. Visually, this is a bright galaxy; it is a sliver elongated NNE–SSW which is well defined along its extremities and evenly bright along its major axis. There appears to be a very slight 'bump' or broadening in the middle on the NW flank. Radial velocity: 109 km/s. Distance: 55 mly (estimate). Diameter: 45 000 ly.

NGC4454

H180[2]	12:28.8	−1 56	GLX
(R)SB(r)0/a	2.3′ × 2.3′	12.85B	11.9V

This is a very slightly tilted, theta-shaped, barred galaxy. Photographs show a large bright core and a faint bar pattern. A pseudo-ring surrounds the core and is attached both directly to the bar and by faint streamers emerging from the core, particularly on the E side. The whole is surrounded by a very faint, broad outer arc, which is most easily seen on the eastern flank. Visually, the galaxy is faint, diffuse and broadly brighter along its major axis with poorly defined edges. It is elongated NNE–SSW and roughly oval in shape. Radial velocity: 2308 km/s. Distance: 103 mly. Diameter: 69 000 ly.

NGC4457

H35[2]	12:29.0	+3 34	GLX
(R)SAB(s)0/a	2.8′ × 2.3′	11.71B	

The DSS image reveals a spiral galaxy of complex and unusual structure. The core of the galaxy is large and bright with short, narrow, high-surface-brightness spiral arms emerging into a slightly fainter inner spiral pattern. The arm emerging from the S is particularly bright. Surrounding this inner pattern is a very faint, broad pseudo-ring formed by two spiral arms emerging from the inner disc. Visually, this is a bright galaxy which takes magnification well. It has a small, bright core which is somewhat elongated E–W. The inner envelope is bright and grainy and the edges are not well defined: they fade to a hazy outer envelope. Overall, the galaxy appears round. Radial velocity: 800 km/s. Distance: 36 mly. Diameter: 29 000 ly.

NGC4458

H121[2]	12:29.0	+13 15	GLX
E0	1.6′ × 1.6′	12.88B	12.1V

Photographs show an elliptical galaxy with a bright core and a round outer envelope. Paired with NGC4461 in Markarian's Chain, this elliptical galaxy is small but fairly bright visually; it is round and well condensed with well defined edges. Radial velocity: 584 km/s. Distance: 26 mly. Diameter: 12 000 ly.

NGC4461

H122[2]/H174[2]	12:29.0	+13 11	GLX
SB(s)0+	3.5′ × 1.4′	11.88B	11.2V

On deep photographs this galaxy has a large, oval core embedded in a smooth-textured, broad and very faint spiral structure. The arms are devoid of dust. Only the central region of this bright, elongated galaxy is visible at medium magnification. Paired with NGC4458, it is larger and brighter than its presumed companion; it is a well defined oval galaxy, elongated almost due N–S. Its brightness increases slowly to the core. Herschel catalogued this galaxy twice: on 8 April 1784 he coupled it with NGC4458, calling them: 'Two. Both pretty faint. Small. Brighter to the middle'; nine nights later, on 17 April, he noted only NGC4461 calling it: 'Faint'. Presumably, observing conditions on this occasion were considerably inferior to those during the previous session. There is an almost 1300 km/s difference in radial velocity between the two galaxies and they may not form an actual physical pair. Radial velocity: 1880 km/s. Distance: 84 mly. Diameter: 86 000 ly.

NGC4464

H483[3]	12:29.4	+8 10	GLX
E3	1.1′ × 0.8′	13.56B	

Photographs show an oval-shaped elliptical galaxy with an ill defined outer

envelope. The galaxy is a close companion of M49 and at high magnification appears as a fairly condensed object; it appears small but moderately bright and much brighter to the middle to a star-like core. It is almost round with abruptly defined edges. A faint field star is WNW of the galaxy. Herschel described the galaxy as very faint. Radial velocity: 1175 km/s. Distance: 52 mly. Diameter: 17 000 ly.

NGC4469

H157[2]	12:29.5	+8 45	GLX
SB(s)0/a? sp	3.5′ × 1.2′	12.37B	

The DSS image reveals an almost edge-on spiral galaxy with a broad but faint, 'peanut'-shaped central region with tapered fainter extensions. There is a considerable amount of dust along the central plane: to the E of the core in the form of elongated dust lanes and to the W of the core in discrete, circular patches. Visually, this quite bright galaxy is large and extended E–W. The

edges are fairly well defined but the inner envelope is uneven in brightness. A very slight haze is visible around the main envelope. These structural features led Herschel to believe the galaxy was resolvable. This is a probable member of subcluster B of the Virgo Cluster, associated with M49. Radial velocity: 521 km/s. Distance: 23 mly. Diameter: 24 000 ly.

NGC4470

H18[2]/H19[2]/H498[2]	12:29.6	+7 49	GLX
Sa?	1.6′ × 1.0′	13.03B	

This galaxy, which S precedes M49, was recorded twice by Herschel: first on 23 January 1784 and again on 28 December 1785. The 23 January observation used different offsets but both lead to the same object. There was some confusion about the identity of this galaxy as Herschel erroneously identified M49 as M61 in his notes but a field sketch he made clarified this issue. NGC4470's radial velocity implies that

NGC4469

©STScI/AURA/CALTECH/ROE

it is probably a background object and not associated with M49. Photographs show a possible spiral galaxy with a bright core surrounded by a thin ring of dust and a bright inner spiral pattern, with a very faint outer envelope. In moderate-aperture telescopes, the galaxy is faint, but readily visible as an oval of light oriented N–S, uniform in brightness except for a gradually brighter core and well defined at the edges. Radial velocity: 2271 km/s. Distance: 101 mly. Diameter: 47 000 ly.

NGC4472

H7[1]	12:29.8	+8 00	GLX
E2/S0	10.2′ × 8.3′	9.27B	

The brightest of the Virgo Cluster galaxies and the dominant galaxy of subcluster B, this elliptical galaxy (also known as M49) has a bright core and extensive, fainter outer envelope. Photographs show a number of small, fainter and probably dwarfish companion galaxies surrounding this major galaxy. There are many globular clusters involved, but not nearly as many as surround the dominant galaxy of subcluster A, M87. Visually, the galaxy is large and very bright; its appearance is very similar to that in photographs. The field star immediately E is well seen, bordering the visual outer envelope. This was one of Herschel's early Class I discoveries, recorded on 23 January 1784 and M49's inclusion in his catalogue is curious, due to his propensity to avoid recording objects previously discovered by Messier. In the note section in *The Scientific Papers* (in a letter to the astronomer Bode), Herschel indicates that he was uncertain of the identity of the reference star (49 Leonis) used to indicate the position of the bright nebula; he also confesses: 'My apparatus in 1784 was not so complete as it is now . . .' He undoubtedly thought he was recording a new nebula, realizing later that it had been discovered previously by Messier. It has also been catalogued as Arp 134. Radial velocity: 929 km/s. Distance: 41 mly. Diameter: 124 000 ly.

NGC4476

H123[2]	12:30.0	+12 21	GLX
SA(r)0−:	1.7′ × 1.2′	13.16B	12.2V

This galaxy forms an apparent pair with NGC4478 4.5′ to the ESE and appears in photographs as an elongated lenticular galaxy with a bright oval core and an extensive envelope oriented NNE–SSW. Its radial velocity is significantly higher than that of its neighbour, however, and it may be a background object. Both galaxies can be seen in the same high-magnification field, with NGC4476 appearing as a small, slightly oval patch of light which is broadly brighter to the middle with slightly hazy edges. Radial velocity: 1916 km/s. Distance: 86 mly. Diameter: 42 000 ly.

NGC4478

H124[2]	12:30.3	+12 20	GLX
E2	1.9′ × 1.6′	12.16B	11.5V

This galaxy is 8.75′ WSW of the giant elliptical galaxy M87 and forms an apparent pair with NGC4476. The DSS image shows an elliptical galaxy with a large, bright central region and a slightly fainter outer envelope. A bright, star-like condensation is immediately N of the core. Visually, NGC4478 is a fairly high-surface-brightness object and is broadly brighter to the middle with a grainy texture to the oval disc, which is oriented NNW–SSE. The edges are fairly well defined. Radial velocity: 1296 km/s. Distance: 58 mly. Diameter: 32 000 ly.

NGC4480

H531[2]	12:30.4	+4 15	GLX
SAB(s)c:	2.1′ × 1.1′	13.11B	

Images show a slightly tilted spiral galaxy with a small core and grainy spiral arms of the multi-arm type. Visually, the galaxy is a quite faint, diffuse, elongated patch of light with hazy edges. The envelope is almost uniform in brightness and elongated N–S. Radial velocity: 2360 km/s. Distance: 105 mly. Diameter: 65 000 ly.

NGC4472

©STScI/AURA/CALTECH/ROE

NGC4482

H40[3]	12:30.2	+10 47	GLX
dE,N:	1.6′ × 0.9′	13.72B	

Photographs show that this elliptical galaxy has a modest-sized, bright core and an extensive outer envelope. This is a fairly faint galaxy visually; it is a hazy patch of light with fairly even surface brightness across its disc. The edges are fairly well defined and the galaxy is elongated SE–NW. Radial velocity: 1812 km/s. Distance: 81 mly. Diameter: 38 000 ly.

NGC4487

H776[2]	12:31.1	−8 03	GLX
SAB(rs)cd	4.2′ × 2.8′	11.83B	

The DSS photograph reveals a slightly inclined spiral galaxy with a bright core and a possible bar. The spiral structure is defined principally by dust lanes; the arms are broad and grainy with several nebulous knots involved. Telescopically, the galaxy is large and fairly bright but a little diffuse and is seen as a very slightly oval disc which is fairly well defined and brighter to a sizable core. A field star is just beyond the disc to the N, while another follows to the ESE. Radial velocity: 919 km/s. Distance: 41 mly. Diameter: 50 000 ly.

NGC4488

H484[3]	12:30.9	+8 22	GLX
SB(s)0/a pec:	4.7′ × 1.2′	13.08B	

This galaxy has a unique morphology and has been classed as a barred lenticular. Photographs show a galaxy with a bright elongated core embedded in a much fainter bar which broadens as it expands away from the core, giving it a 'wasp-waisted' appearance. Two very faint spiral plumes curve away from the bar's extremities. Visually, this is a small, faint and diffuse object; it appears as a hazy bar of light elongated SE–NW with fairly even surface brightness and ill defined edges. Radial velocity: 906 km/s. Distance: 40 mly. Diameter: 55 000 ly.

NGC4488

©STScI/AURA/CALTECH/ROE

NGC4491

H41[3]	12:31.0	+11 29	GLX	
SB(s)a:	1.7′ × 0.9′	13.47B	12.6V	

This galaxy and NGC4497 can be seen together in a medium-magnification field. Photographs show an inclined, barred spiral galaxy with two short spiral arms emerging from the bar and quickly broadening to a faint envelope. The galaxy is very probably a dwarf. Visually, the galaxy is a faint, thin sliver of light elongated NW–SE so that only the bar is readily visible. It is fairly even in surface brightness along the major axis with somewhat diffuse tips. Radial velocity: 441 km/s. Distance: 20 mly. Diameter: 9700 ly.

NGC4492

H499[2]	12:31.0	+8 05	GLX
SA(s)a?	1.9′ × 1.7′	13.21B	

Photographs show that this face-on spiral galaxy has a round core and a faint spiral pattern, well defined by prominent dust lanes. Visually, it is a diffuse and difficult object, located immediately SSW of a prominent field star which hinders visibility. A second, slightly brighter field star is located ESE. The galaxy is only weakly concentrated to the middle and is slightly oval in form, oriented NNW–SSE. Best seen at medium magnification, this is probably a background object despite its proximity to M49. Radial velocity: 1707 km/s. Distance: 76 mly. Diameter: 42 000 ly.

NGC4487

©STScI/AURA/CALTECH/ROE

NGC4496

H36²/H18³	12:31.6	+3 56	GLX
SB(rs)m	4.0′ × 3.2′	11.94B	

There are two entries in Herschel's catalogue for this object, both recorded on 23 February 1784, and the *New General Catalogue* listed a second galaxy, NGC4505, for the H18³ entry although there is no object at the designated position. Part of the confusion lies in the fact that Herschel used two different reference stars for location: Sigma Virginis for NGC4496 and 16 Virginis for NGC4505. The description for NGC4496 reads: 'Faint, very large, irregularly round, brighter to the middle. 6′ long, 4′ broad.' While for NGC4505, Herschel remarked: 'Very faint, considerably large, resolvable.' The DSS image shows two galaxies, one of which is very obviously a satellite and is superimposed on NGC4496; this is sometimes referred to as NGC4496B. The principal galaxy is a low-surface-brightness, barred spiral with loosely wound multiple spiral arms, populated by many brighter knots and condensations. The companion, located 0.8′ to the S, may be a Magellanic type; it seems to have a short bar with low-surface-brightness possibly spiral structure. Visually, NGC4496 is a faint and diffuse galaxy, broadly brighter toward the S with very hazy and ill defined edges. The galaxy appears slightly elongated N–S and while not resolved as two separate objects, the brightening towards the S suggests that the companion is visible in moderate-aperture telescopes. Radial velocity: 1650 km/s. Distance: 74 mly. Diameter: 86 000 ly.

NGC4497

H42³	12:31.5	+11 37	GLX
SAB(s)0⁺:	2.0′ × 0.9′	13.27B	12.5V

Images reveal that this is very possibly an inclined, barred lenticular galaxy with a bright elongated core and faint traces of a bar in a slightly fainter envelope. Visually, the galaxy is a faint, elongated oval which is broadly brighter to the middle. The galaxy is oriented ENE–WSW and its edges are hazy. Radial velocity: 990 km/s. Distance: 44 mly. Diameter: 26 000 ly.

NGC4503

H66²	12:32.1	+11 11	GLX
SB0⁻:	3.5′ × 1.7′	12.08B	11.1V

This lenticular galaxy presents similar appearances both visually and photographically. In the DSS image there seems to be evidence of a possible dust lane bordering the bright core to the NE. A small, face-on lenticular galaxy is 6.75′ to the NNE. Visually, the galaxy is fairly bright, well condensed and well defined along its major axis. It has a brighter core; the inner envelope is quite dense and fainter extensions are visible. The galaxy is elongated almost due N–S. Radial velocity: 1286 km/s. Distance: 57 mly. Diameter: 58 000 ly.

NGC4504

H771²	12:32.3	−7 34	GLX
SA(s)cd	4.4′ × 2.7′	12.17B	

Photographs show an inclined spiral galaxy with an elongated central region and asymmetrical spiral structure: the eastern spiral arm is defined by a string of likely HII regions. A broad and prominent dust lane borders the inner spiral arm on the western flank and very faint outer spiral structure can be traced outward to an apparent diameter of about 6.5′. Telescopically, the galaxy is a large but quite dim and diffuse object, appearing irregularly round and even in surface brightness across the disc. The edges are a little hazy and the disc is a little grainy in texture. Radial velocity: 885 km/s. Distance: 39 mly. Diameter: 63 000 ly.

NGC4517

H5⁴	12:32.8	+0 07	GLX
SA(s)cd: sp	10.7′ × 1.5′	11.09B	

The DSS image shows a large, edge-on galaxy which lacks a core or central

NGC4496

©STScI/AURA/CALTECH/ROE

NGC4517

©STScI/AURA/CALTECH/ROE

bulge. The relatively low-surface-brightness disc is studded with small knots and laced with complex dust patches and lanes. Visually, the galaxy is diffuse and quite faint, and is hindered by the presence of a bright field star which borders the galaxy near the middle on the N flank. The galaxy is flat and much extended almost due E–W; it is a little brighter on the W side with moderately well defined edges. Discovered on 22 February 1784, Herschel described the galaxy as: 'A pretty bright star with a milky ray south parallel, 15′ or 20′ long.' Radial velocity: 1038 km/s. Distance: 46 mly. Diameter: 144 000 ly.

NGC4519

H158[2]	12:33.5	+8 39	GLX
SB(rs)d	2.9′ × 2.3′	12.36B	

This barred spiral galaxy is located in subcluster B of the Virgo Cluster, which is dominated by the giant elliptical galaxy M49. Photographs reveal a galaxy with spiral arms emerging from an incomplete ring surrounding the bar; several HII regions are found in the spiral arms. Visually, however, the galaxy is moderately faint, large, diffuse and gradually brighter to the middle to a nonstellar core. The edges are very poorly defined and the galaxy is very slightly oval in form, oriented NW–SE. Herschel thought this galaxy resolvable. Radial velocity: 1161 km/s. Distance: 52 mly. Diameter: 44 000 ly.

NGC4520

H757[2]	12:33.8	−7 23	GLX
SA0⁻ pec sp	1.3′ × 0.6′	14.80B	

Photographs show a lenticular galaxy with a dominant, oval core and fainter, tapered extensions. Visually, this object is quite faint and is seen as a diffuse, oval patch of light which is fairly even in surface brightness and elongated E–W. Radial velocity: 7516 km/s. Distance: 336 mly. Diameter: 127 000 ly.

NGC4526

H31[1]/H38[1]/ H119?[1]	12:34.0	+7 42	GLX
SAB(s)0°:	7.2′ × 2.4′	10.61B	

The visual and photographic appearances of this lenticular galaxy are quite similar. Situated mid-way between two bright field stars, NGC4526 is a bright, well defined object with a mottled envelope and is well seen at high magnification. It brightens to a well-developed, sizable and nonstellar core and is oriented ESE–WNW. Herschel recorded the galaxy at least twice: on 13 April 1784 he called it: 'Very bright, extended, much brighter to the middle, resolvable, between 2 bright stars'; on 18 April 1784 his description reads: 'Bright, very large, much extended, much brighter to the middle.' A somewhat mysterious entry on 28 December 1785, designated NGC4560 by Dreyer, may have been a third recording of the galaxy, though Herschel's description reads: 'Very bright, pretty small' and the position is 2.1 minutes in time E of NGC4526. There is nothing at the position, though it is possible that Herschel observed a telescopic comet. This is a possible member of the Virgo Cluster; however, the radial velocity is much lower than the cluster mean. Radial velocity: 381 km/s. Distance: 17 mly. Diameter: 35 000 ly.

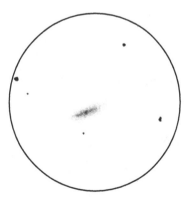

NGC4526 15-inch Newtonian 313x

NGC4527

H37[2]	12:34.1	+2 39	GLX
SAB(s)bc	6.2′ × 2.1′	11.48B	

This is a bright, highly tilted spiral galaxy and photographs show two thin, principal spiral arms in a disc studded with knots and condensations. A large dust patch is seen on the NNW flank. In a moderate-aperture telescope, this is a large and fairly bright galaxy. The inner region is bright and very elongated along the major axis and is surrounded by a diffuse glow which is somewhat well defined. The galaxy is elongated ENE–WSW. Radial velocity: 1654 km/s. Distance: 74 mly. Diameter: 133 000 ly.

NGC4528

H67[2]	12:34.1	+11 19	GLX
S0°:	1.7′ × 1.0′	12.93B	12.1V

The photographic and visual appearances of this lenticular galaxy are quite similar. Visually, this is a small, moderately bright, oval galaxy; it is well condensed with even surface brightness across the disc. It is very slightly brighter to the middle with well defined extremities and is elongated N–S. Radial velocity: 1287 km/s. Distance: 57 mly. Diameter: 28 000 ly.

NGC4531

H175[2]	12:34.3	+13 05	GLX
SB0+:	3.1′ × 2.0′	12.48B	

The DSS image shows an inclined spiral galaxy, possibly barred, with a bright inner pseudo-ring and suspected fragmentary spiral structure. Visually, the galaxy is moderately bright; it is a large, oval object with a fairly bright, well defined envelope that gradually brightens to the middle. At the centre is a slightly brighter core. Radial velocity: 146 km/s. Distance: 42 mly (estimate). Diameter: 38 000 ly (estimate).

NGC4532

H147[2]	12:34.3	+6 28	GLX
IBm	2.8′ × 1.0′	12.28B	

This is a possible member of subcluster B of the Virgo Cluster and morphologically the galaxy seems to be an irregular system. The DSS image shows an elongated galaxy which is fairly bright along its major axis though no core is visible. An elongated dust patch, oriented SSE–NNW, is in the northern portion and a significant, detached probable HII region is immediately SSE of the centre of the galaxy. This HII region measures about 8″ in diameter and would actually be almost 3400 ly in diameter if the presumed distance to this galaxy is correct. Visually, this galaxy is fairly bright and is elongated SSE–NNW; it is even in surface brightness across much of its length, though quite diffuse at its SSE extension. A very faint field star is 0.5′ E of the centre and the galaxy is blunt at its NNW extremity. A magnitude 8 field star is 5.4′ S. Herschel considered the galaxy resolvable. Radial velocity: 1942 km/s. Distance: 87 mly. Diameter: 71 000 ly.

NGC4535

H500[2]	12:34.3	+8 12	GLX
SAB(s)c	6.9′ × 5.3′	10.52B	

This magnificent, S-shaped spiral galaxy was christened the 'Lost Galaxy' by Leland Copeland, owing to its faintness in small telescopes, especially when compared with the nearby NGC4526. The DSS photograph shows a face-on spiral galaxy with a small, bright core in a short bar which spawns the two principal spiral arms. After less than half a revolution, secondary arms form, which are both broader and fainter than the principal pattern. The arms are dotted with HII regions and unwind gracefully into a complex outer pattern. Telescopically, the galaxy is very large and quite bright, but it is also diffuse and most distinct at medium magnification. It is situated within a triangle of field stars and the disc is an oval, ill defined glow oriented N–S and extending between the N–S pair of stars in the

NGC4535

triangle. The large, round central region is a little brighter than the disc and appears off-centre to the N in the disc. Radial velocity: 1896 km/s. Distance: 85 mly. Diameter: 170 000 ly.

NGC4536

H2[5]	12:34.5	+2 11	GLX
SAB(rs)bc	7.6′ × 3.2′	11.11B	10.6V

The DSS image shows a highly inclined, large spiral galaxy of the grand-design type with a possible very short bar emerging from the bright core. The spiral arms are loosely wound, fairly massive, studded with knots and laced with dust. The galaxy may be an outlying member of subcluster B in the Virgo Cluster. Visually, this is a large and moderately bright galaxy, with a small, bright core embedded in a fairly bright but uneven envelope, which is elongated ESE–WNW. The envelope is surrounded by a very faint, ill defined haze. Herschel discovered the galaxy on

24 January 1784, placing it in his Class V: very large nebulae. He may have seen something of the spiral structure as his description reads: 'Considerably bright, much extended north preceding, south following. Much brighter to the middle, easily resolvable, 9′ or 10′ long with a branch towards the north preceding.' This matches the location of one of the spiral arms. Radial velocity: 1725 km/s. Distance: 77 mly. Diameter: 171 000 ly.

NGC4541

H493[3]	12:35.2	−0 13	GLX
(R′)SAB(r)bc:	1.6′ × 0.7′	13.92B	

In photographs, this highly inclined spiral galaxy shows two principal spiral arms emerging from an inner, probable pseudo-ring structure. The galaxy is moderately faint visually; it is an oval patch of light which is very slightly brighter to the middle, elongated E–W

with somewhat diffuse edges. Radial velocity: 6795 km/s. Distance: 304 mly. Diameter: 141 000 ly.

NGC4546

H160[1]	12:35.5	−3 48	GLX
SB(s)0⁻:	3.3′ × 1.4′	11.35B	10.3V

The photographic and visual appearances of this lenticular galaxy are quite similar. Visually, this is a bright and quite well defined galaxy which takes magnification well. Oval in shape and oriented ENE–WSW, the envelope is slightly grainy and a bright, compact core is embedded. Radial velocity: 949 km/s. Distance: 42 mly. Diameter: 41 000 ly.

NGC4550

H36[1]	12:35.5	+12 13	GLX
SB0°:	3.3′ × 0.9′	12.48B	11.7V

This galaxy forms a very attractive high-magnification pair with NGC4551, located 3.2′ to the NE. Photographs show a very bright, edge-on lenticular galaxy with a bright, elongated central region embedded in a fainter secondary envelope. The visual appearance is quite similar: the galaxy is very bright and elongated N–S, with a well-condensed, well defined, opaque envelope, the ends of which taper to points. The core is small and condensed. The two galaxies appear to be a physical pair but the radial velocities are very dissimilar, with that of NGC4550 being much the lower. If the two galaxies are at the same distance, the core-to-core

NGC4536

NGC4550/4551

separation between them would be about 46 000 ly. Radial velocity: 330 km/s. Distance: 50 mly (estimate). Diameter: 48 000 ly.

NGC4551

H37[1]	12:35.6	+12 16	GLX
E:	1.8′ × 1.4′	12.89B	12.0V

This galaxy's classification is somewhat uncertain; if it is indeed an elliptical it would be about E4. Visually and photographically, the galaxy is a bright oval, elongated ENE–WSW with a condensed envelope; it is well defined, opaque and a little brighter to the middle. Radial velocity: 1121 km/s. Distance: 50 mly. Diameter: 26 000 ly.

NGC4564

H68[2]	12:36.4	+11 26	GLX
E6	3.5′ × 1.5′	11.97B	11.1V

Visible in the same medium-magnification field as NGC4567/68, this galaxy is a prototypical E6 system lacking a disc component. It is part of subcluster A in the Virgo Cluster. Visually, the core is small and bright and extends along the major axis to the slightly fainter outer envelope and overall the galaxy is well defined and oriented NE–SW. Radial velocity: 1089 km/s. Distance: 49 mly. Diameter: 49 000 ly.

NGC4567

H8[4]	12:36.5	+11 15	GLX
SA(rs)bc	3.0′ × 2.0′	12.10B	11.3V

NGC4567 and its close companion NGC4568 were named the 'Siamese Twins' by Leland Copeland, who popularized astronomy in the 1950s. At first glance, photographs show the two galaxies to be in apparent contact, but the lack of evidence of tidal distortion suggests that these galaxies may be well separated from each other. It would seem that NGC4567 is in front of the larger galaxy, as its spiral pattern is complete where the two galaxies appear to join. The galaxy features a small core and multiple spiral arm fragments which are laced with dust and brighter

knots. Herschel discovered these two galaxies on 15 March 1784, describing them as: 'A double nebula. The chevelure run into each other. Close, not very faint.' Visually, the core of NGC4567 is brighter than that of its companion and the outer envelope appears broader as well. The envelope is bright, mottled and elongated E–W. Both galaxies are presumed to be members of subcluster A in the Virgo Cluster, though their radial velocities are quite high. Radial velocity: 2220 km/s. Distance: 99 mly. Diameter: 86 000 ly.

NGC4568

H9[4]	12:36.6	+11 14	GLX
SA(rs)bc	5.0′ × 2.6′	11.69B	11.2V

The morphology of this galaxy is almost identical to that of its close companion, NGC4567, though it is seen more obliquely and is much larger, and features a broader, fainter outer spiral structure. The diffuse outer arm on the

western flank seems to disappear behind the outer disc of NGC4567, suggesting that NGC4568 is a little more distant. It features a multiple spiral arm structure dominated by bright knots and several dust patches, including a long, broad dust lane N of the core. At the eyepiece, NGC4568 has a bright central region, although the core is less prominent than that of its companion. Much extended NNE–SSW, the envelope is mottled and moderately well defined and the two galaxies appear fused on the NE side. Radial velocity: 2201 km/s. Distance: 98 mly. Diameter: 145 000 ly.

NGC4570

H32[1]	12:36.9	+7 15	GLX
S0(7)/E7	3.8′ × 1.1′	11.74B	

The visual and photographic appearances of this bright lenticular galaxy are strikingly similar. Well seen at high magnification, the galaxy features a bright, elongated core with a very slight bulge and bright, well

NGC4567/4568

defined extensions which taper to points. The galaxy is oriented NNW–SSE and seems to be a pure lenticular system rather than an elliptical one. Herschel recorded the galaxy on 13 April 1784 and remarked: 'Considerably bright, small, bright nucleus and two very faint branches.' Radial velocity: 1664 km/s. Distance: 74 mly. Diameter: 82 000 ly.

NGC4578

H15[2]	12:37.5	+9 33	GLX
SA(r)0°:	3.0′ × 2.2′	12.36B	

Visually and photographically this lenticular galaxy presents very similar appearances. In a moderate-aperture telescope this is a very bright galaxy, with a large, bright core and a hazy outer envelope; it is slightly oval in shape and elongated NNE–SSW with hazy, ill defined edges. Radial velocity: 2214 km/s. Distance: 99 mly. Diameter: 86 000 ly.

NGC4580

H124[1]	12:37.8	+5 22	GLX
SAB(rs)a pec	2.0′ × 1.4′	12.72B	

This spiral galaxy is fairly bright on the DSS image but high-resolution photographs reveal its somewhat peculiar nature. The DSS image shows a bright central core and a pseudo-ring made up of two overlapping spiral arms. High-resolution photographs show a series of HII regions along the spiral arms, indicating high levels of star formation. The outer arms are faint and very smooth textured. Visually, this is quite a bright galaxy which is broadly brighter to the middle and quite grainy in texture over much of its envelope. The main envelope is quite opaque and surrounded by a much fainter haze which is slightly elongated N–S. Radial velocity: 962 km/s. Distance: 43 mly. Diameter: 25 000 ly.

NGC4586

H125[1]	12:38.5	+4 19	GLX
SA(s)a: sp	3.9′ × 1.2′	12.58B	

Images reveal an inclined spiral galaxy with a bright core and grainy spiral structure which features a prominent, thin dust lane crossing N of the galaxy's core. The galaxy is quite large and moderately bright visually but somewhat diffuse as well. It is broadly brighter along its major axis, which is a little grainy in texture. The envelope is oval, broader in the middle, fairly well defined and oriented ESE–WNW. This galaxy appears to be associated with a Virgo Cluster subgroup centred on NGC4636. Radial velocity: 719 km/s. Distance: 32 mly. Diameter: 37 000 ly.

NGC4588

H98[3]	12:38.8	+6 46	GLX
Sc(s)	1.2′ × 0.4′	14.99B	

Photographs show a highly inclined spiral galaxy oriented ENE–WSW with an elongated central region and a probable multi-arm spiral pattern. Telescopically, the galaxy is quite dim and is visible intermittently as a hazy, roundish patch of light, which is even in brightness and best seen at medium magnification. Radial velocity: 5543 km/s. Distance: 247 mly. Diameter: 86 000 ly.

NGC4591

H504[3]/H13[3]	12:39.3	+6 01	GLX
Sb	1.6′ × 0.8′	13.96B	

This is an inclined spiral galaxy, though the DSS image does not show the spiral pattern well. There are three principal brightness gradients: the core and central region, the inner envelope and the fainter outer envelope. Visually, the galaxy is a small and fairly faint oval extended NE–SW. It is broadly brighter to the middle and the edges are somewhat ill defined. This galaxy may also be Herschel's H13[3], which was identified as NGC4577 by Dreyer. Though there was no offset in declination published by Herschel (clouds intervened and he lost the object) his measured offset from the star 11 Virginis puts the object in the vicinity of NGC4591 and there are no other obvious candidates nearby. Radial velocity: 2353 km/s. Distance: 105 mly. Diameter: 49 000 ly.

NGC4592

H31[2]	12:39.3	−0 32	GLX
SA(s)dm:	5.8′ × 1.6′	12.56B	

Photographs show this highly inclined spiral galaxy to have many HII regions resolved in an envelope of uneven surface brightness. It is a difficult object visually for small-aperture telescopes (though Herschel thought it quite bright and perhaps even resolvable); it is extended E–W and is smooth textured with no concentration to the middle. Radial velocity: 980 km/s. Distance: 44 mly. Diameter: 74 000 ly.

NGC4593

H183[2]	12:39.7	−5 21	GLX
(R)SB(rs)b	3.9′ × 3.3′	12.16B	

This is an inclined, barred spiral galaxy. Photographs reveal that the spiral arms emerging from the bar are tightly wound and form a pseudo-ring before expanding to a more relaxed outer pattern. An anonymous possible dwarf companion is situated 4.0′ ESE of the core. Telescopically, the galaxy can be seen in a medium-magnification field with NGC4602, which is probably a physical companion as both galaxies have similar radial velocities. NGC4593 is moderately bright and only the bar and brighter core are visible. The bar is elongated NE–SW and is fairly well defined. Radial velocity: 2595 km/s. Distance: 116 mly. Diameter: 131 000 ly.

NGC4594

H43[1]	12:40.0	−11 37	GLX	
SA(s)a	9.4′ × 6.7′	9.15B	8.0V

This is the magnificent 'Sombrero' galaxy, one of the most striking galaxies for amateur observation in the heavens. It was discovered by Pierre Méchain on 11 May 1781, though the observation was not published at the time. Herschel recorded the galaxy on 9 May 1784 and described it as: 'Extended, very bright to the middle. 5 or 6′ long. The bright place in the middle is pretty large, but breaks off abruptly.' On the DSS image the galaxy is shown to have one of the largest central bulges of any galaxy in the heavens, with a bright, well defined disc and a prominent dust lane extending the length of the major axis on the S side of the disc. An extensive faint halo surrounds the galaxy and dozens of globular clusters are visible. Visually, this galaxy is a remarkable object in medium- and large-aperture instruments. The bright bulge and disc, both elongated E–W, are well seen and a sharply defined dust lane extends across the disc. It is narrow near the core but slowly broadens as it extends away to the E and W. The bulge is less extensive to the S of the dust lane than it is to the N and a bright, nonstellar core is visible immediately N of the dust lane. Radial velocity: 904 km/s. Distance: 40 mly. Diameter: 110 000 ly.

NGC4596

H24[1]	12:39.9	+10 11	GLX
SB(r)0+	4.1′ × 3.6′	11.44B	

In photographs this is a bright, theta-shaped, barred spiral galaxy featuring a large, bright core and a prominent bar which flares and brightens at its contact point with the faint inner ring. Smooth spiral arms emerge from the ring and broaden to a very faint and large outer envelope. At high magnification, the galaxy is bright in moderate-aperture telescopes, though only the core and the bright bar are visible. The galaxy is oval in shape and elongated ENE–WSW, and the bar is moderately well defined but the edges are not sharp. The core is bright and well defined but seems to lack a stellar nucleus. This Virgo Cluster galaxy has a fairly high radial velocity given its size and brightness. Radial velocity: 1815 km/s. Distance: 81 mly. Diameter: 97 000 ly.

NGC4597

H636[2]	12:40.2	−5 48	GLX
SB(rs)m	3.7′ × 1.4′	12.87B	

This barred spiral galaxy has a somewhat unusual structure: in

photographs, the bar is weak with little evidence of a core and the arms attached to the bar appear fairly straight, giving the object the appearance of a letter 'H' seen obliquely. There are several HII candidates, including a particularly bright one at the southern tip of the western arm. Telescopically, the galaxy is a very diffuse, dim and difficult object and is best seen at medium magnification as a somewhat large but very faint patch of light which is fairly even in surface brightness and elongated NNE–SSW. The edges are very poorly defined and the overall outline is rectangular. Radial velocity: 935 km/s. Distance: 42 mly. Diameter: 45 000 ly.

NGC4598

H105[3]	12:40.3	+8 23	GLX
SB0	1.5′ × 1.2′	13.50B	

The DSS image shows a barred lenticular galaxy with a small bright core and a bar which is straight as it leaves the core but which begins to curve at its extremities. The whole is embedded in a faint, broad outer envelope. Visually, the galaxy is quite faint: it appears as a hazy patch of light which is very slightly elongated N–S, with a broadly brighter central region. The edges are ill defined. Radial velocity: 1900 km/s. Distance: 85 mly. Diameter: 37 000 ly.

NGC4599

H509[3]	12:40.5	+1 11	GLX
SA0/a	2.0′ × 0.7′	13.80B	

Photographs show that this edge-on lenticular galaxy has a bright, bulging core embedded in a bright inner disc with fainter outer extensions. Although quite small visually, the galaxy is moderately bright and only the brighter inner region is well seen. Brighter to the core, the galaxy is a fairly well defined object which is slightly elongated SE–NW with a grainy envelope and extensions that taper to points. Radial velocity: 1753 km/s. Distance: 78 mly. Diameter: 45 000 ly.

NGC4600

H577[2]	12:40.4	+3 08	GLX
S0:	1.3′ × 0.9′	13.51B	

The visual and photographic appearances of this lenticular galaxy are quite similar. At high magnification, it is a moderately faint galaxy which precedes a widely separated pair of bright field stars. The galaxy is an oval glow; it is elongated ENE–WSW, fairly even in surface brightness and moderately well defined at the edges. Radial velocity: 775 km/s. Distance: 35 mly. Diameter: 13 000 ly.

NGC4602

H184[2]	12:40.6	−5 08	GLX
SAB(rs)bc	3.4′ × 1.2′	12.52B	

Photographs show a highly inclined barred spiral galaxy with a bright, multi-arm structure involving several knotty condensations. Visually, the galaxy appears a little brighter and larger than NGC4593, which can be seen in the same medium-magnification field. The galaxy is seen as an elongated, flattened oval oriented almost due E–W, with even surface brightness along the major axis, and well defined edges. Radial velocity: 2437 km/s. Distance: 109 mly. Diameter: 108 000 ly.

NGC4606

H43[3]	12:41.0	+11 55	GLX
SB(s)a	3.2′ × 1.6′	12.72B	

On the DSS image, this galaxy appears to be an inclined, barred spiral galaxy with a bright core and hazy, curving spiral arms broadening from the faint bar. Visually, it is a small but moderately bright galaxy which is quite flat and extended NE–SW. The surface brightness is fairly even along its major axis and a brighter core is embedded. Two faint field stars are involved: the first is 0.4′ to the SSW in the outer envelope and the second is 1.1′ SSW beyond the envelope. The companion galaxy NGC4607 is 3.8′ ESE and is visible as a very faint, hazy sliver of light elongated N–S. The core-to-core separation between the two galaxies at the suspected distance would be about 80 000 ly. Radial velocity: 1615 km/s. Distance: 72 mly. Diameter: 67 000 ly.

NGC4608

H69[2]	12:41.2	+10 09	GLX
SB(r)0°	3.2' × 2.8'	12.05B	

This galaxy is located between subclusters A and B of the Virgo Cluster. Like NGC4596, this is a bright, theta-shaped, barred spiral galaxy, but it is seen more nearly face-on and the ring-like structure stands out more as a result. The bar does not flare where it touches the inner ring, however, and the ring is surrounded by a very faint outer envelope which does not display spiral structure except for a dark, curving lane on the NW flank. Visually, only the central region and bar are visible and the galaxy is well seen at high magnification, which also keeps nearby Rho Virginis out of the field. It appears as an elongated bar; the bulging central region is not seen though the

galaxy appears brighter to a stellar nucleus. Herschel considered this galaxy resolvable. Radial velocity: 1809 km/s. Distance: 81 mly. Diameter: 75 000 ly.

NGC4612

H20[2]/H148[2]	12:41.5	+7 19	GLX
(R)SAB0°	2.5' × 1.9'	12.07B	

The DSS image shows a bright lenticular galaxy with a bright core and a fainter outer envelope. High-resolution images show at least three brightness gradients as well as the central bar. In a moderate-aperture telescope, the galaxy is small but very bright, and responds well to high magnification. The core, which is bright and slightly elongated E–W, is embedded in a dense and grainy envelope, surrounded by a very faint outer haze. A magnitude 10 field star is 1.0' E of the core. This galaxy was recorded twice by Herschel, who thought he was recording two different objects: first on 23 January 1784 (using

31 Boötis as his position reference) and again on 13 April of that same year (this time using 31 Virginis). The characteristic star pattern to the E and NE and its relation to the galaxy were described on at least two separate occasions, leaving little doubt that the same object was seen both times. Radial velocity: 1811 km/s. Distance: 81 mly. Diameter: 59 000 ly.

NGC4623

H149[2]	12:42.2	+7 41	GLX
SB0[+]: sp	2.2' × 0.5'	13.24B	

This lenticular galaxy presents very similar appearances photographically and visually. In a moderate-aperture telescope the galaxy is small, moderately faint and elongated N–S. It is brighter along its major axis and broadly brighter in the centre, which shows a very slight bulge. The edges are well defined. In *The Carnegie Atlas of Galaxies* (Sandage and Bedke, 1994), consideration is given to classifying this galaxy as a very rare E7 elliptical. Radial velocity: 1830 km/s. Distance: 82 mly. Diameter: 53 000 ly.

NGC4626

H772[2]	12:42.4	−7 02	GLX
SB(s)bc: sp	1.3' × 0.4'	13.91B	

This highly inclined spiral galaxy forms a probable physical pair with NGC4628 4.3' to the N. The minimum separation between the two galaxies would be about 156 000 ly at the apparent distance. Photographs show an irregular spindle with a possible dust patch on the western flank. Visually, the galaxy appears a little more diffuse than NGC4628; it is an oval patch of light which is a little brighter to the middle and oriented NE–SW. Radial velocity: 2790 km/s. Distance: 125 mly. Diameter: 47 000 ly.

NGC4628

H773[2]	12:42.4	−6 57	GLX
SA(s)b: sp	1.7' × 0.5'	14.01B	

Photographs show this spindle-shaped spiral galaxy is oriented at almost

NGC4608

©STScI/AURA/CALTECH/ROE

exactly the same position angle as its companion to the S, NGC4626, though it is larger and has a prominent central bulge with ill defined spiral arms. Telescopically, both galaxies can be seen in a high-magnification field, but they are quite faint. NGC4628 appears a little larger and more distinct; it is a well defined sliver of light with tapered extensions which is evenly bright across the disc and oriented NE–SW. Radial velocity: 2722 km/s. Distance: 122 mly. Diameter: 60 000 ly.

NGC4630

H532[2]	12:42.5	+3 58	GLX
IB(s)m?	1.8′ × 1.3′	13.18B	

This is probably a dwarf irregular galaxy. Photographs show a slightly inclined galaxy with a brighter, elongated and irregularly shaped core, surrounded by a faint but lumpy outer envelope which shows a vague spiral pattern. The galaxy appears moderately bright visually but is diffuse, broadly brighter to the middle and very slightly elongated N–S with ill defined edges. Radial velocity: 663 km/s. Distance: 30 mly. Diameter: 15 000 ly.

NGC4632

H14[1]	12:42.5	−0 05	GLX
SAc	3.1′ × 1.2′	12.28B	

This is an inclined spiral galaxy with a slightly brighter central region. Photographs show that the spiral pattern is knotty and asymmetrical, being brighter and more extensive towards the ENE. The galaxy is fairly bright visually with a somewhat grainy envelope which is a little brighter to the middle. It appears fairly well defined but the NNW flank is irregular in outline. The galaxy may be physically related to NGC4666 and NGC4668, which form a pair about 45.0′ to the SE. Radial velocity: 1637 km/s. Distance: 73 mly. Diameter: 66 000 ly.

NGC4636

H38[2]	12:42.8	+2 41	GLX
E/S0	6.5′ × 4.9′	10.43B	

The DSS image shows a galaxy oriented SE–NW with a bright, round, elliptical core and an extensive, elongated outer disc, which is typical of a lenticular galaxy. The galaxy's retinue of globular clusters are seen as small, hazy patches arrayed throughout the outer disc and beyond. Visually, this is quite a large and bright galaxy. It is round and the edges are diffuse. A very diffuse and faint outer envelope surrounds the condensed inner disc which brightens to a star-like core. A bright field star is 7.25′ SSW. This appears to be the principal member of a subgroup in the Virgo Cluster which includes NGC4600, NGC4630 and NGC4665. Radial velocity: 861 km/s. Distance: 39 mly. Diameter: 73 000 ly.

NGC4638

H70[2]/H176[2]	12:42.8	+11 26	GLX	
S0[−]	2.2′ × 1.4′	12.09B	11.2V	

This object was recorded twice by Herschel: on 15 March and 17 April 1784. Photographically, it is a well defined, edge-on lenticular galaxy with a tapered disc surrounded by a much fainter outer envelope. Telescopically, although the galaxy appears small, it has a high-surface-brightness disc which is very well defined and oriented ESE–WNW. A stellar, bright core is well seen at high magnification and the faint, dwarfish galaxy NGC4637, 1.75′ to the E is intermittently seen as a small, hazy patch of light. Radial velocity: 1115 km/s. Distance: 50 mly. Diameter: 32 000 ly.

NGC4639

H125[2]	12:42.9	+13 15	GLX
SAB(rs)bc	2.8′ × 1.9′	12.19B	

This is a very slightly inclined barred spiral galaxy with the bar joined to a pseudo-ring made up of spiral arms. High-resolution photographs show several bright HII regions at these two junctures, as well as two very faint, open spiral fragments beyond the inner arms. An extremely faint irregular companion is 2.8′ to the W. Visually, NGC4639 is quite a bright galaxy which

holds magnification well. The main envelope is fairly dense with a bright core and the bar appears as a thin bright band running N–S. Beyond the inner envelope is a faint secondary haze which extends almost to a faint field star 0.9′ SE of the core. Herschel considered this galaxy resolvable. Radial velocity: 974 km/s. Distance: 43 mly. Diameter: 36 000 ly.

NGC4642

H494[3]	12:43.3	−0 39	GLX
SAB(rs)bc pec sp	1.8′ × 0.5′	13.76B	

Photographs show a highly inclined spiral galaxy with an asymmetrical spiral pattern and a dust patch NE of the core. A faint field star is immediately ENE of the core. Visually, the galaxy can be seen in the same medium-magnification field as NGC4653, which is 9.8′ to the ENE and probably a physical companion. NGC4642 is an elongated streak, a little diffuse at the edges and oriented NNE–SSW. The faint star near the core can be mistaken for the actual core, particularly at low magnification. Radial velocity: 2557 km/s. Distance: 114 mly. Diameter: 60 000 ly.

NGC4643

H10[1]	12:43.3	+1 59	GLX
SB(rs)0/a	3.2′ × 2.8′	11.68B	

The DSS image shows a large, barred galaxy with a theta-shaped inner pattern, formed by a bright core and a fairly bright bar surrounded by a faint, inner ring. Typically, the bar flares in brightness where it touches the ring and high-resolution images show evidence of recent star formation here. Beyond the inner ring is a smooth-textured outer disc which shows faint spiral structure and is elongated NE–SW and at least 4.0′ in extent. Although bright, only the core of this barred galaxy is visible at high magnification. The envelope is round and grainy and brightens to a condensed core. A very faint, quite diffuse secondary halo surrounds the inner core but the bar is not seen. Radial velocity: 1256 km/s. Distance: 56 mly. Diameter: 65 000 ly.

NGC4647

H44[3]	12:43.5	+11 35	GLX
SAB(rs)c	2.9′ × 2.3′	12.09B	12.5V

This galaxy is evidently a companion of the bright lenticular galaxy M60, which appears to the NW in photographs and is possibly in contact with this galaxy. Although there is no distortion in the disc of M60, NGC4647 is a somewhat peculiar looking, slightly inclined spiral galaxy that is also catalogued as Arp 116. The bright core has a very small bar involved with a bright inner ring of uneven surface brightness. The outer spiral structure is not well defined and is very clumpy, fragmentary and asymmetrical in form, being more extensive towards the NW. Visually, the galaxy is well seen with M60 at both medium and high magnification. The core of the galaxy is prominent, while its outer envelope is hazy and ill defined, and it appears to be in contact with the larger galaxy. At high magnification, the envelope is mottled and brighter to a nonstellar core. Radial velocity: 1373 km/s. Distance: 61 mly. Diameter: 52 000 ly.

NGC4653

H662[3]	12:43.9	−0 34	GLX
SAB(rs)cd	3.0′ × 2.7′	12.81B	

Despite their similar brightnesses and proximity on the sky, NGC4653 and NGC4642 were discovered separately, with NGC4653 being recorded on 11 April 1787, while NGC4642 was first seen on 1 January 1786. Photographs show that NGC4653 is a face-on spiral galaxy with two principal inner spiral arms and a faint, fragmentary outer spiral pattern. The minimum core-to-core separation between the two galaxies would be 324 000 ly at the presumed distance. Visually, NGC4653 is very slightly brighter than its companion but fairly diffuse; it is a roundish patch of light which is very

gradually brighter to the middle. Radial velocity: 2544 km/s. Distance: 114 mly. Diameter: 99 000 ly.

NGC4654

H126[2]	12:44.0	+13 08	GLX
SAB(rs)cd	5.2′ × 3.0′	11.03B	

The DSS image shows an inclined spiral galaxy with a very small bar and complex spiral fragments emerging, dotted with HII regions, while the arms are laced with dust patches and lanes. In a moderate-aperture telescope, the galaxy is a large and quite bright, extended oval oriented ESE–WNW. There is a brighter, thin band running along the major axis, identifiable with the bar and some of the brightest spiral arm fragments. The main envelope is quite opaque, but the edges are poorly defined and hazy. Herschel thought this object was resolvable. A magnitude 10 field star is 3.25′ WNW.

Radial velocity: 1003 km/s. Distance: 45 mly. Diameter: 68 000 ly.

NGC4658

H558[2]	12:44.6	−10 05	GLX
SB(s)bc	1.8′ × 1.0′	12.94B	

Situated 2.5′ W of a magnitude 8 field star, photographs show this highly inclined, barred spiral galaxy, which has a large, bright bar and two short, slightly fainter spiral arms that fade suddenly into a fainter outer envelope. Telescopically, this is a moderately bright, elongated galaxy oriented N–S which precedes a fairly bright field star. The envelope is bright and well defined and brightens broadly to the middle with a very faint field star in the envelope NNW of the core. NGC4663 is also visible in the field, 7.2′ to the SSE, and is seen as a small, round, condensed and faint galaxy with a star-like core. Radial velocity: 2279 km/s. Distance: 102 mly. Diameter: 53 000 ly.

NGC4654

©STScI/AURA/CALTECH/ROE

NGC4660

H71[2]	12:44.5	+11 11	GLX
E5	2.2′ × 1.6′	12.07B	

In photographs, this elliptical galaxy appears as a bright disc surrounded by a faint halo. At high magnification, the galaxy is very small but has a high surface brightness and is seen as a well defined oval disc oriented E–W with tapered ends and a bright, intense stellar core. The surrounding field appears starless. Radial velocity: 1033 km/s. Distance: 46 mly. Diameter: 30 000 ly.

NGC4665

H142[1]/H39[2]	12:45.1	+3 03	GLX	
SB(s)0/a	5.0′ × 3.0′	11.52V	10.5V

In photographs, this barred spiral galaxy has a bright, elongated central region and a slightly fainter bar. The smooth-textured spiral arms emerge from the ends of the bar but are quite faint and difficult to trace; each arm extends for a little more than half a revolution around the bar. Visually, the galaxy is bright and large with a bright central region which is elongated N–S; this is a manifestation of the bar. The core is a little brighter but nonstellar and there is evidence of a very faint and diffuse secondary envelope. A bright field star is situated 1.75′ to the SW. The galaxy is probably part of the Virgo Cluster subgroup which includes NGC4600, NGC4630 and NGC4636. Radial velocity: 710 km/s. Distance: 32 mly. Diameter: 46 000 ly.

NGC4666

H15[1]	12:45.1	−0 28	GLX
SABc:	4.6′ × 1.3′	11.45B	

Photographs show a large, highly inclined spiral galaxy with complex dust patches and lanes which border the central region particularly on the SE flank. Even in small-aperture telescopes, this is a bright galaxy elongated NE–SW with a prominent, slightly bulging central region embedded in a slightly fainter but well defined envelope. One of Herschel's early Class I discoveries, it was recorded on 22 February 1784. The

fainter companion galaxy NGC4668 is only 7.4′ to the SE. Radial velocity: 1444 km/s. Distance: 65 mly. Diameter: 86 000 ly.

NGC4668

H663[3]	12:45.5	−0 32	GLX
SB(s)d	1.4′ × 0.7′	13.54B	

Although it is considerably smaller than its neighbour NGC4666, photographs show that this dwarfish spiral galaxy has a fairly high surface brightness and its somewhat chaotic structure is partially resolved into HII regions. In small-aperture telescopes it is very faint and requires averted vision to be well seen as a circular glow which is slightly brighter to the centre. A faint star triplet is 3.0′ W. Herschel discovered this galaxy three years after NGC4666, on 11 April 1787, and though he found it very faint and small, he was able to detect its irregular form. Radial velocity: 1550 km/s. Distance: 69 mly. Diameter: 28 000 ly.

NGC4671

H774[2]	12:45.8	−7 04	GLX
E[+] pec:	1.6′ × 1.3′	13.61B	

The DSS photograph shows an elliptical galaxy with a bright core, a fainter outer disc and a broad, blunt plume which extends away from the galaxy to the NNW. Telescopically, the galaxy is fairly bright but it appears quite small; it is a round, opaque patch of light which is fairly well defined and a little brighter to the middle. Radial velocity: 2876 km/s. Distance: 128 mly. Diameter: 60 000 ly.

NGC4682

H523[3]	12:47.3	−10 04	GLX
SAB(s)cd	2.5′ × 1.3′	13.11B	

The DSS image shows an inclined spiral galaxy with the principal, narrow spiral arms emerging from a small inner disc. Faint arms emerge from the principal pattern after about a quarter revolution. Visually, the galaxy is a very faint and diffuse, largish oval patch of light elongated E–W with a faint field star following 1.5′ to the ENE. Radial velocity: 2223 km/s. Distance: 99 mly. Diameter: 72 000 ly.

NGC4684

H181[2]	12:47.3	−2 43	GLX
SB(r)0[+]	2.8′ × 1.0′	12.08B	

In the DSS image, this almost edge-on lenticular galaxy appears to have a dust lane crossing the bright, elongated central region. There are three brightness zones, which is typical of the class, and the faint outer envelope tapers slightly. Telescopically, the galaxy is quite bright, large and oval in outline, elongated NNE–SSW. The core is bright and a faint field star is involved in the envelope 0.75′ to the NNE. The edges are moderately well defined and the galaxy is broadly brighter to the middle. Radial velocity: 1479 km/s. Distance: 66 mly. Diameter: 54 000 ly.

NGC4688

H543[3]	12:47.8	+4 20	GLX
SB(s)cd	4.0′ × 4.0′	13.51B	

This is a low-surface-brightness, barred spiral galaxy. Images show an object with a weak bar and two relaxed spiral arms with knotty condensations throughout, including two bright ones in the western spiral arm. The faint peculiar galaxy MCG+1-33-14 is 6.75′ to the NNE. Telescopically, the galaxy is large but faint and fairly diffuse, irregularly oval in shape and very slightly extended NNE–SSW. A very faint and small core is intermittently visible in the poorly defined envelope. Radial velocity: 917 km/s. Distance: 41 mly. Diameter: 48 000 ly.

NGC4690

H664[3]	12:47.9	−1 38	GLX
(R′)SA0[−]?	1.2′ × 0.9′	13.81B	

This lenticular galaxy is similar in appearance visually and photographically. At medium magnification it is moderately faint and small, almost round and a little brighter to the middle. The edges are fairly well defined and photographs show that the galaxy is elongated SSE–NNW. Radial velocity: 2717 km/s. Distance: 121 mly. Diameter: 42 000 ly.

NGC4691

H182[2]	12:48.2	−3 20	GLX
(R)SB(s)0/a pec	2.8′ × 2.3′	11.68B	11.1V

This galaxy appears to be a barred spiral, based on its appearance on the DSS image, but it is somewhat peculiar as the form of the bar is irregular with several indentations along the southern flank which may be dust patches. The spiral pattern is faint and smooth and forms a pseudo-ring around the bright bar. The galaxy is moderately bright visually and is well seen at high magnification. The central bar is quite bright and is elongated E–W with a small, brighter core, best seen at medium magnification. A diffuse secondary envelope is visible immediately N and S of the bar. Radial velocity: 1018 km/s. Distance: 65 mly. Diameter: 37 000 ly.

NGC4694

H72[2]	12:48.2	+10 59	GLX
SB0 pec	3.2′ × 1.5′	12.29B	

In photographs, several small dark patches are visible in the envelope of this lenticular galaxy, principally N and S of the brighter core; otherwise there is little evidence of spiral structure. The presence of an absorption spectrum indicates recent, intense star formation. At the eyepiece, the galaxy is fairly bright and elongated NW–SE. It has a bright core but the core is off centre, displaced to the NE in the galaxy's envelope. This makes the NW extremity appear blunter than the SE one, which is tapered to a sharp point. Radial velocity: 1127 km/s. Distance: 50 mly. Diameter: 47 000 ly.

NGC4697

H39[1]	12:48.6	−5 48	GLX
E6	7.2′ × 4.7′	10.25B	

Photographs show a large elliptical galaxy with a bright, elongated core and an extensive, faint outer envelope. A small, possibly spiral galaxy with a bright central region is 6.0′ to the WNW. In a small-aperture telescope, the oval core is well seen and is sharply defined with a smooth-textured disc and

a brighter middle. Moderate apertures show an almost stellar core in a grainy disc with a very faint, secondary haze surrounding the disc, which is oriented ENE–WSW. Radial velocity: 1142 km/s. Distance: 51 mly. Diameter: 107 000 ly.

NGC4698

H8[1]/H6[3]	12:48.4	+8 29	GLX
SA(s)ab	7.6′ × 3.6′	11.56B	

The DSS image shows a multi-arm spiral galaxy with a large, bright core and significant dust lanes, which define the spiral pattern. There is a thin inner lane and a much larger and broader lane in the outer envelope, barely seen in silhouette. There is an extremely faint outer spiral pattern which greatly increases the apparent size of the galaxy. High-resolution images reveal just how odd this galaxy is: the core is much like an elongated elliptical galaxy whose major axis is perpendicular to the galaxy's disc. Visually, the galaxy is very bright and well seen at high magnification. It is situated between two bright field stars. The core is sizable and quite intense and is surrounded by a fairly bright envelope, extended almost due N–S. This envelope is quite grainy in texture and moderately well defined. The galaxy was recorded twice by Herschel and curiously was classified once as a bright nebula (23 January 1784) and once as a very faint one (18 January 1784). Radial velocity: 946 km/s. Distance: 42 mly. Diameter: 94 000 ly.

NGC4699

H129[1]	12:49.0	−8 40	GLX
SAB(rs)b	3.9′ × 2.9′	10.38B	

In photographs this slightly inclined spiral has an extensive, bright, oval central region, surrounded by a flocculent, fragmentary inner spiral pattern and broad, ill defined outer spiral structure. Telescopically, the galaxy is moderately well condensed, with a brighter, almost star-like core located not quite in the centre of the grainy-textured disc. It is slightly brighter to the S and E and the disc is surrounded by a faint, ill defined haze

oriented NNE–SSW. Radial velocity: 1287 km/s. Distance: 57 mly. Diameter: 65 000 ly.

NGC4700

H524[3]	12:49.1	−11 25	GLX
SB(s)c? sp	3.6′ × 0.6′	12.04B	

On the DSS image the morphology of this galaxy is difficult to ascertain as much of the galaxy appears overexposed. It is an edge-on galaxy and may be either an Sc spiral or irregular type: high-resolution photographs show many brighter knots along the major axis and some dust patches but little that indicates spiral structure. Visually, the object appears as a beautiful and moderately bright edge-on galaxy, featuring a high-surface-brightness envelope which tapers to points. It is well defined throughout. There is no central bulge and only intermittently does the galaxy appear brighter to the middle. Oriented NE–SW, a dust lane was suspected along the western flank but this is not supported by photographs. Radial velocity: 1295 km/s. Distance: 58 mly. Diameter: 61 000 ly.

NGC4701

H578[2]	12:49.2	+3 23	GLX
SA(s)cd	3.6′ × 3.0′	12.82B	

Viewed almost face-on, this spiral galaxy is dominated by a bright, irregularly round core surrounded by a faint, multi-arm spiral pattern dotted with knotty condensations. Visually, however, only the bright core is visible and thus the galaxy appears as a small, circular, well defined patch of light. Radial velocity: 650 km/s. Distance: 29 mly. Diameter: 30 000 ly.

NGC4703

H514[3]	12:49.3	−9 07	GLX
Sb	3.5′ × 0.5′	14.40B	

This edge-on spiral galaxy is situated 8.0′ NNE of a magnitude 8 field star. The DSS image shows that its morphology is similar to that of brighter examples of the Sb class, such as NGC4565 and

NGC5746. It appears as an edge-on galaxy with a large but flattened central bulge with faint spiral extensions bisected by a prominent dust lane. Visually, the galaxy is a moderately faint and diffuse, elongated oval object of even surface brightness oriented SSE–NNW with fairly well defined edges. A fairly large-aperture telescope is required to show the full extent of the galaxy. Radial velocity: 4351 km/s. Distance: 194 mly. Diameter: 198 000 ly.

NGC4705

H610[3]	12:49.4	−5 12	GLX
SAB(s)bc: sp	4.0′ × 1.1′	12.52B	

In photographs, this is a highly inclined spiral galaxy with a large, elongated central region and spiral arms which curve below the plane of the galaxy to the ESE and above the plane to the WNW. A prominent dust lane is visible below the core along the southern flank. Visually, the galaxy is a faint, small, elongated patch of light which is ill defined, quite diffuse and oriented ESE–WNW. Radial velocity: 4201 km/s. Distance: 188 mly. Diameter: 218 000 ly.

NGC4708

H722[3]	12:49.7	−11 06	GLX
SA(r)ab pec?	1.2′ × 0.9′	13.91B	

In photographs, this inclined galaxy has a bright core and two bright spiral arms. A small galaxy is 0.4′ WNW of the core. The minimum separation between the two galaxies would be about 21 000 ly at the presumed distance if they are physically related. Telescopically, NGC4708 is a very faint galaxy; it is a diffuse, oval patch of light with poorly defined edges which is very gradually brighter to the middle. It is elongated ENE–WSW. Radial velocity: 4053 km/s. Distance: 181 mly. Diameter: 63 000 ly.

NGC4713

H140[1]	12:50.0	+5 19	GLX
SAB(rs)d	2.4′ × 1.9′	12.22B	

The DSS shows a slightly inclined spiral galaxy with a bright, bar-like central

NGC4731

region in a knotty, chaotic and fragmentary spiral disc. It is a fairly large and moderately bright galaxy telescopically, though a little diffuse. It appears as a fat oval of light oriented E–W with a fairly well defined, mottled envelope that is broadly brighter to the middle. Radial velocity: 587 km/s. Distance: 26 mly. Diameter: 18 000 ly.

NGC4720

H611[3]	12:50.7	−4 07	GLX
S0−	1.0′ × 0.6′	14.30B	

Photographs show an inclined lenticular galaxy with a bright, elongated central region and a very faint outer envelope. The galaxy is moderately faint visually; it is a small, oval object of even surface brightness, though condensed to the core. The edges are quite well defined and the galaxy is elongated ESE–WNW. Radial velocity: 1411 km/s. Distance: 63 mly. Diameter: 18 000 ly.

NGC4731

H41[1]	12:51.0	−6 24	GLX
SB(s)cd	6.6′ × 3.2′	11.89B	

This barred spiral galaxy has an interesting unique morphology: photographs show a bright narrow bar with no core and very relaxed, narrow spiral arms, which curve slightly, broaden and then fade at the ends. The eastern arm in particular appears disturbed: a broad, faint plume extends to the N. Visually, the galaxy is large but quite diffuse with only the bar visible as a moderately bright, elongated patch of light oriented ESE–WNW. It is situated on the W side of a bright triangle of field stars. Radial velocity: 1390 km/s. Distance: 62 mly. Diameter: 119 000 ly.

NGC4733

H73[2]	12:51.1	+10 55	GLX
E+:	1.9′ × 1.6′	12.70B	

Although it is classified as elliptical, examination of the DSS image of this galaxy reveals a face-on barred spiral galaxy with a bright core and a bar embedded in a smooth-textured, fainter spiral structure, defined by narrow dust lanes, particularly to the S of the core. Visually, it is a moderately faint galaxy, oval in form with a hint of a bright, star-like core. The outer envelope is elongated E–W and a magnitude 14 field star precedes the galaxy immediately W. Radial velocity: 881 km/s. Distance: 39 mly. Diameter: 22 000 ly.

NGC4739

H515[3]	12:51.6	–8 25	GLX
E+ pec:	1.5′ × 1.3′	14.02B	

Photographs show an elliptical galaxy with a bright core and a grainy outer envelope. Visually, the galaxy is moderately faint, but is well seen as a round disc that is very broadly brighter to the middle and fairly well defined at the edges. Radial velocity: 3668 km/s. Distance: 164 mly. Diameter: 72 000 ly.

NGC4742

H133[1]	12:51.8	–10 27	GLX
E4:	2.2′ × 1.4′	12.11B	

Situated 9.2′ SE from the bright double star Struve 1682, this elliptical galaxy is quite similar in appearance visually and photographically. The galaxy is very bright telescopically but very small, featuring a sizable, bright core embedded in a diffuse and ill defined envelope. The galaxy is elongated ENE–WSW with a field star 1.4′ to the SE. Radial velocity: 1160 km/s. Distance: 52 mly. Diameter: 33 000 ly.

NGC4753

H16[1]	12:52.4	–1 12	GLX
I0	6.0′ × 2.8′	10.85B	9.9V

This is apparently a member of the NGC4643 subgroup in the Virgo Cluster (numbering about 15 galaxies arrayed around Gamma Virginis) and is difficult to classify owing to the complex pattern of thin and irregular dust lanes which cross in front of the bright, lenticular disc of the galaxy. Photographs show an extensive outer disc which increases the length of the major axis to about 8.5′. The galaxy is well seen even in small-aperture telescopes; it has a bright core and a well defined central region. Elongated E–W, an extensive, fainter outer envelope fades gradually into the sky background. Radial velocity: 1156 km/s. Distance: 52 mly. Diameter: 128 000 ly.

NGC4754

H25[1]/H74[2]	12:52.3	+11 19	GLX
SB(r)0⁻:	4.6′ × 2.5′	11.5B	

Photographically, this lenticular galaxy has a large, bright core and a faint disc, oriented NNE–SSW. Visually, only the bright core is likely to be detected in moderate-aperture telescopes. The galaxy appears round and well condensed with a bright core. The outer envelope is mottled at high magnification and a few faint sparkles are visible. At medium magnification, it forms an attractive, contrasting pair with NGC4762. Both galaxies were discovered on 15 March 1784; curiously, Herschel assigned two numbers to this galaxy, as well as two descriptions. Under H25[1] he called the galaxy: 'Bright, small, in a line with 2 stars'. As H74[2], he described the galaxy as part of a pair with NGC4762 and called it: 'Pretty bright, nearly round'. Radial velocity: 1302 km/s. Distance: 58 mly. Diameter: 77 000 ly.

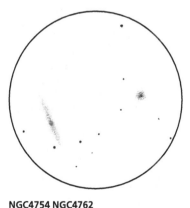

NGC4754 NGC4762
15-inch Newtonian 146x

NGC4759

H559[2]	12:53.1	–9 11	GLX
S0+/S0°	1.7′ × 1.4′	14.03B	

The DSS image shows three tightly packed galaxies, two of which are extremely close, face-on lenticulars which share a common envelope. The brighter of the pair is NGC4761 and is situated 0.4′ ESE of the fainter of the pair, NGC4759. The third galaxy, NGC4764, is a smaller, edge-on lenticular 1.1′ ENE of the brightest galaxy. NGC4759 has a slightly pear-shaped bright core but otherwise there does not seem to be any indication of interaction between the two galaxies and they exhibit different radial velocities: NGC4759's radial velocity is 3455 km/s, while that of NGC4761 is 4153 km/s. Visually, NGC4764 is not detected, but NGC4759 and NGC4761 appear as a large, somewhat mottled oval glow elongated ESE–WNW. The oval appears bi-nuclear, which is evidence of two galaxies in contact. A field star 1.5′ S hinders the observation, making the edges of the oval diffuse. Herschel may have seen NGC4759 and NGC4761 as one unresolved object, though his 25 March 1786 description reads only: 'Faint, small.' Radial velocity: 4153 km/s. Distance; 182 mly. Diameter: 76 000 ly.

NGC4762

H75[2]	12:52.9	+11 14	GLX
SB(r)0° sp	8.7′ × 1.7′	11.10B	

Photographically, this lenticular galaxy is a stunning object, amongst the flattest galaxies known. The bright, bulging core is oval and attached to bright extensions with rounded tips. Beyond this is a very faint outer envelope, not seen in moderate-aperture telescopes, which appears disturbed and broadens before fading into the sky background. The disruption may be due to the proximity of NGC4754, though there is no disturbance visible in that galaxy. Visually, NGC4762 is bright, well defined and very much elongated NNE–SSW. It is a beautiful needle of light: the core is round and bright and the outer envelope is smooth with ends which taper to sharp points. Three prominent field stars to the S bracket the galaxy. Radial velocity: 939 km/s. Distance: 42 mly. Diameter: 106 000 ly.

NGC4765

H544[3]	12:53.2	+4 28	GLX
S0/a?	1.1′ × 0.8′	14.82B	

Photographs show a very small galaxy, very likely a dwarf, which is broadly brighter to the middle with a grainy texture and an irregular, oval form. The galaxy is small but quite bright visually with a condensed, well defined envelope, a little brighter to the middle and almost round. Radial velocity: 650 km/s. Distance: 29 mly. Diameter: 9000 ly.

NGC4770

H525[3]	12:53.5	−9 31	GLX
S0/a:	1.6′ × 1.0′	13.76B	

The photographic and visual appearances of this lenticular galaxy are quite similar. Situated almost due W of the bright star Psi Virginis, which hinders observation somewhat, the galaxy is small and has a brighter core embedded in a rather dim envelope, oriented E–W. Radial velocity: 3541 km/s. Distance: 158 mly. Diameter: 74 000 ly.

NGC4771

H535[2]	12:53.4	+1 16	GLX
SAd? sp	4.6′ × 0.8′	12.93B	

This highly tilted galaxy appears to be a multi-arm spiral in photographs. The pattern is quite fragmented and there are many dust patches and a prominent dust lane along the NE flank. There is no central bulge and the inner region varies considerably in surface brightness. At moderate magnification, this is a moderately faint galaxy, but it is well defined and tapers very slightly to blunt points. The galaxy is elongated SE–NW and the texture is somewhat mottled along the major axis. Radial velocity: 1059 km/s. Distance: 47 mly. Diameter: 63 000 ly.

NGC4772

©STScI/AURA/CALTECH/ROE

NGC4772

H24[2]	12:53.5	+2 10	GLX	
SA(s)a	4.0′ × 2.3′	11.99B	11.0V	

Photographs show that this tilted spiral galaxy has a large, round, bright core surrounded by a faint, elongated inner envelope which features a prominent dust lane that can be followed almost completely around the galaxy. A very faint, broad outer envelope is also visible. Visually though, the galaxy is somewhat faint, just gradually brighter to the middle and elongated NNW–SSE. The dust lane probably restricts the apparent size of the galaxy at the eyepiece. Radial velocity: 967 km/s. Distance: 43 mly. Diameter: 50 000 ly.

NGC4773

H516[3]	12:53.6	−8 38	GLX
E+ pec:	1.3′ × 1.0′	13.67B	

The DSS photograph shows an elliptical galaxy with a bright elongated central region in a very faint and limited outer

envelope. A small, compact galaxy is 0.4′ immediately S and may be a physical companion. At the apparent distance, the minimum separation between the two galaxies would be about 18 000 ly. Visually, the galaxy forms a fairly prominent triangular triplet with the Herschel object NGC4777 and the face-on spiral galaxy NGC4780, which appears to have been overlooked by Herschel. NGC4773 is marginally the brightest, a condensed, roundish disc, well defined and even in surface brightness. There is no sign of the smaller companion to the S. Radial velocity: 3379 km/s. Distance: 151 mly. Diameter: 57 000 ly.

NGC4775

H186[2]	12:53.8	−6 37	GLX
SA(s)d	2.2′ × 2.0′	11.58B	

Photographs show a face-on spiral galaxy with a bright, irregular core and ill defined, fragmentary spiral arms studded with bright nebulous knots. The galaxy is bright and fairly large visually, an irregularly round, mottled disc of fairly even surface brightness, quite well defined at the edges with a faint spur to the NE. Radial velocity: 1469 km/s. Distance: 66 mly. Diameter: 42 000 ly.

NGC4777

H517³	12:54.0	−8 47	GLX
(R)SAB(s)a:	1.9′ × 0.8′	13.98B	

Photographs show that this inclined, possible spiral galaxy has a bright, elongated core in a smooth-textured disc which may involve faint dust lanes on the eastern flank and bordering the core to the W. A bright field star is 4.7′ to the WNW. Only the bright central region of NGC4777 is seen visually, appearing as a well defined, condensed, round glow with a tiny star-like core. Radial velocity: 3420 km/s. Distance: 153 mly. Diameter: 84 000 ly.

NGC4779

H106³	12:53.8	+9 44	GLX
SB(rs)bc	2.1′ × 1.8′	13.20B	

The DSS image reveals a face-on, barred spiral galaxy with a pseudo-ring structure made up of spiral arm fragments. There are several condensations along the spiral arms and the structure is somewhat chaotic towards the periphery. The bar is oriented N–S. The galaxy is a faint object visually, not much more than a round, nebulous patch that is a little brighter to the middle, with ill defined edges. Herschel thought the galaxy was resolvable. Radial velocity: 2781 km/s. Distance: 124 mly. Diameter: 76 000 ly.

NGC4781

H134¹	12:54.4	−10 32	GLX
SB(rs)d	3.5′ × 1.5′	11.39B	

This galaxy is part of a small group of galaxies of similar radial velocities, including NGC4742 to the W, NGC4784 and NGC4790. In the DSS image the galaxy appears as an inclined barred spiral with two fairly short principal arms in a fragmentary spiral envelope. Many of the fragments are defined by brighter HII regions in high-resolution photographs, which also begin to resolve the brightest stars at blue magnitude 21.5. In a medium-magnification field the galaxies NGC4781, NGC4784 and NGC4790 form a triplet, of which the largest and brightest is NGC4781. It appears as a large, elongated oval with a mottled envelope and moderately well defined edges. It is oriented ESE–WNW and a faint field star appears embedded in the outer envelope 0.75′ W of the core. NGC4784 is 5.5′ to the SE and the minimum core-to-core separation between the two galaxies would be 97 000 ly at the presumed distance. Radial velocity: 1151 km/s. Distance: 52 mly. Diameter: 52 000 ly.

NGC4784

H526³	12:54.6	−10 37	GLX
S0	1.6′ × 0.4′	14.40B	

In photographs, this is a very attractive, edge-on lenticular galaxy with a prominent, elongated bulge and fainter extensions oriented ESE–WNW. Visually, the galaxy appears as a small, faint and moderately concentrated spot; the fainter extensions may be visible in large-aperture telescopes. A field star is 2.0′ WSW of the core. Herschel's 25 March 1785 description reads: 'Extremely faint, extremely small, some little doubt.' Radial velocity: 1151 km/s. Distance: 52 mly. Diameter: 24 000 ly.

NGC4786

H187²	12:54.5	−6 52	GLX
E⁺ pec	1.8′ × 1.4′	12.72B	

This elliptical galaxy appears similar visually and photographically. Telescopically, it is a small but high-surface-brightness object, well seen at high magnification as a slightly oval disc elongated N–S with a bright central 'bar' along the major axis. A slightly fainter, secondary halo is a little hazy at the edges. Radial velocity: 4548 km/s. Distance: 203 mly. Diameter: 106 000 ly.

NGC4790

H560²	12:54.9	−10 15	GLX
SB(rs)c:?	1.6′ × 0.8′	12.77B	

Although this is classified as a barred spiral galaxy, high-resolution photographs show no bar and only a small, round core in this slightly inclined galaxy. It is probably a spiral, but the pattern is very difficult to trace: there are many nebulous knots scattered throughout the oval envelope. It is situated 18.0′ NNE of the bright galaxy NGC4781 of which it is probably a companion: the resolution of stars and HII regions is similar for both galaxies. Telescopically, NGC4790 is an oval, fairly even-surface-brightness object, moderately well defined and oriented E–W. Faint field stars are 1.0′ N and 1.4′ E of the core. Radial velocity: 1236 km/s. Distance: 55 mly. Diameter: 26 000 ly.

NGC4795

H21²	12:55.0	+8 04	GLX
(R′)SB(r)a pec	2.2′ × 1.9′	13.21B	

This is the principal member of a somewhat peculiar double system, which seems to involve a small companion galaxy undergoing an encounter with the larger galaxy. NGC4795 has a bright core and a suspected bar elongated NNE–SSW. Two main spiral arms emerge but they are smooth textured and show no evidence of dust or star formation, despite the evident close encounter. The fainter companion, probably an elliptical or lenticular, is immediately E of the core and is involved with the tip of NGC4795's eastern spiral arm. The core-to-core separation between the two galaxies is 0.5′, which at the presumed distance, is only 17 700 ly. *Uranometria 2000.0* (first edition) identifies this object as NGC4796 but calls it a star. The field galaxy NGC4791 is 4.75′ to the W. Visually, NGC4796 is not seen; NGC4795 is a moderately bright, almost round galaxy with a bright core. The edges are somewhat ill defined and the envelope is fairly bright. NGC4791 is visible in the W at medium magnification. Discovered on 23 January 1784, Herschel's description for NGC4795 reads: 'Pretty bright, pretty large, brighter towards the preceding side.' Radial velocity: 2727 km/s. Distance: 122 mly. Diameter: 78 000 ly.

NGC4799

H548[3]	12:55.3	+2 54	GLX
S?	1.3′ × 0.4′	14.13B	

The classification of this faint galaxy is somewhat uncertain; photographically it appears to be an almost edge-on spiral with a bright central region and very faint outer spiral structure. Visually, it is a faint, roundish patch with even surface brightness, 45″ NNW of a faint field star. Herschel used a magnification of 240 to ascertain that the galaxy was elongated. It is oriented E–W. Radial velocity: 2723 km/s. Distance: 122 mly. Diameter: 46 000 ly.

NGC4808

H141[1]	12:55.8	+4 18	GLX
SA(s)cd:	2.8′ × 1.1′	12.30B	

Photographs show this galaxy to be a multi-armed spiral seen obliquely. There is no core that is distinctly visible and the arms are laced with a chaotic pattern of dust patches. In moderate-aperture telescopes, the galaxy is moderately faint; it is a fat oval of light elongated ESE–WNW. The edges are well defined and the envelope appears smooth textured and very gradually brighter to the middle. Radial velocity: 695 km/s. Distance: 31 mly. Diameter: 25 000 ly.

NGC4813

H777[2]	12:56.6	−6 48	GLX
S0°	1.6′ × 0.8′	14.10B	

Photographs show a lenticular galaxy with a brighter, oval central region in a fainter outer halo. A compact anonymous galaxy is 5.2′ to the N. Telescopically, NGC4813 is small and quite faint but is well seen at high magnification as a slightly oval, well defined object that is a little brighter to

the middle. Radial velocity: 1297 km/s. Distance: 58 mly. Diameter: 27 000 ly.

NGC4818

H549[2]	12:56.8	−8 31	GLX
SAB(rs)ab pec:	4.3′ × 1.5′	12.00B	

In photographs, this large, tilted spiral galaxy features a bright and large, elongated central region embedded in a grainy spiral pattern. Visually, the galaxy is somewhat faint and oriented N–S, but the core is visible as a roundish glow with faint, ill defined extensions. Radial velocity: 963 km/s. Distance: 43 mly. Diameter: 54 000 ly.

NGC4825

H563[2]	12:57.2	−13 40	GLX
E3/SA0⁻	2.5′ × 1.5′	12.88B	

This lenticular galaxy also has characteristics of an elliptical galaxy, principally the bright oval core. Photographs show the galaxy in a field of smaller galaxies and it may be the dominant member of a small group. A compact, probable dwarf companion is 0.75′ NNW of the core in the outer halo of the principal galaxy, which is elongated SE–NW. Visually, NGC4825 appears small and fairly faint; it is a round object with diffuse edges and is a little brighter to the middle. Also visible in the field is the fainter edge-on galaxy NGC4820, situated 4.0′ to the SW. Radial velocity: 4336 km/s. Distance: 194 mly. Diameter: 141 000 ly.

NGC4843

H613[3]	12:58.0	−3 37	GLX
SA(r)0/a?	2.1′ × 0.5′	14.19B	

The DSS image shows a bright and very highly inclined spiral galaxy with a bright, extended central region but no obvious core. The galaxy is fairly faint visually, quite small and best seen at medium magnification. It is a fairly flat, diffuse and elongated object oriented E–W. The galaxy is very broadly brighter to the middle and a faint field star is visible 0.75′ ESE of the core. Radial velocity: 4829 km/s. Distance: 215 mly. Diameter: 132 000 ly.

NGC4845

H536[2]/ H37[5]	12:58.0	+1 35	GLX
SA(s)ab sp	5.0′ × 1.2′	12.13B	

In photographs, this highly tilted spiral galaxy features a small core, no central bulge and a conspicuous dust lane. A faint two-armed spiral pattern emerges from the bright central disc. Visually, NGC4845 is a fairly bright object for moderate-aperture telescopes; it is a large, well defined streak of light, much extended ENE–WSW and framed by three stars with magnitudes between 11 and 13. The central region is broad and the envelope is bright along the extended major axis with the ends tapering to blunt points. There is a remote possibility that this object is a duplicate observation of H3[5] = NGC4910, made by Herschel on 24 January 1784 but his description of the object: 'Extremely faint, very large, easily resolvable, round, 7 or 8′ in diameter' is a poor description of NGC4845. Herschel further mentioned: 'The place of this nebula is not determined with accuracy.' There is nothing in the vicinity that remotely matches the description and NGC4910 is now considered nonexistent. Radial velocity: 1160 km/s. Distance: 52 mly. Diameter: 75 000 ly.

NGC4856

H68[1]	12:59.3	−15 02	GLX
SB(s)0/a	4.2′ × 1.6′	11.44B	

This is a bright and highly inclined lenticular galaxy. Photographs show a bright core and an extensive outer envelope and high-resolution images show two short, stubby spiral arms emerging from the bright core before blending into the outer envelope. The faint, edge-on galaxy MCG−2-33-80 is 6.0′ to the NE and a bright field star is 7.0′ ESE. In a medium-magnification field, NGC4856 is the brightest of three galaxies, the others being NGC4877 and MCG−2-33-82. It is quite bright with a large, bright, oval core. It is a little brighter at the edges and oriented NE–SW with a faint star bordering the envelope to the ESE of the core. Radial

velocity: 1235 km/s. Distance: 55 mly. Diameter: 67 000 ly.

NGC4866

H162[1]	12:59.5	+14 10	GLX
SA(r)0[+]: sp	6.3′ × 1.4′	12.02B	

The DSS image shows an almost edge-on spiral galaxy with a bright core and a bright inner ring or spiral pattern; it is smooth textured except for significant dust lanes E and W of the core. The image shows an extensive, very faint, outer spiral pattern with dust involved. Visually, this is a very bright, fairly large galaxy that is well seen at high magnification; it appears as a flat streak, elongated E–W with tapering extensions. It is well defined along the edges with a field star 0.6′ WNW of the core, just bordering the inner envelope. The core is bright, condensed and elongated E–W. Discovered on 17 January 1787, Herschel called it: 'Very bright, extended south preceding north following. Small star in it 0.5′ preceding nucleus.' Radial velocity: 1957 km/s. Distance: 87 mly. Diameter: 160 000 ly.

NGC4877

H299[2]	13:00.4	−15 17	GLX
SA(s)ab:	2.3′ × 0.9′	13.15B	

This is a highly inclined spiral galaxy. Photographs show a large, oval central region surrounded by a fainter spiral envelope. A bright field star is 2.8′ to the NW, another is 5.5′ to the SE and the fainter edge-on galaxy MCG–2-33-82 is 7.0′ to the SW. Visually, NGC4877 is elongated almost due N–S with a brightish major axis; it is a little brighter to the middle and well defined at the edges. MCG–2-33-82 can be seen in the field as a small, elongated patch of light. Radial velocity: 4803 km/s. Distance: 215 mly. Diameter: 144 000 ly.

NGC4878

H758[3]	13:00.3	−6 06	GLX
SB(r)0[+]	1.5′ × 1.4′	13.96B	

This galaxy is a face-on, barred lenticular. Photographs show an oval central region and a short bar attached to a ring-like outer disc.

NGC4879/H759[3], 1.4′ to the ESE, is a single faint star and therefore this object can be considered nonexistent. Herschel in his 23 March 1789 description described both NGC4878 and NGC4879 as: 'Two nebulae both very faint, very small' but provided only one position. There is only one other galaxy in the immediate vicinity (NGC4888/H778[2] is 4.5′ to the ENE). Visually, NGC4878 is fairly well seen at high magnification as a round, moderately bright object that is broadly brighter to the middle with hazy edges. NGC4979 is a single, magnitude 14 star to the ESE which is intermittently visible. Radial velocity: 3735 km/s. Distance: 167 mly. Diameter: 73 000 ly.

NGC4879

H759[3]	13:00.5	−6 06	SS
nonexistent			

NGC4879 is a faint star ESE of NGC4878, not a nebula, so this Herschel object can be considered nonexistent.

NGC4880

H83[3]	13:00.2	+12 29	GLX
SA(r)0[+]:	3.0′ × 2.3′	13.11B	

In photographs this almost face-on spiral galaxy has a large oval central region and very faint spiral structure. Visually, it is a very faint, hazy object that is slightly extended N–S and a little brighter to the middle. The edges are very poorly defined. Radial velocity: 1341 km/s. Distance: 60 mly. Diameter: 52 000 ly.

NGC4888

H778[2]	13:00.6	−6 04	GLX
Sa: sp	1.4′ × 0.5′	13.5B	

Photographs show that this spindle-shaped lenticular galaxy follows a bright pair of field stars by 1.4′. Recorded on the same night as NGC4878 and NGC4879, there can be no mistaking its identity, as Herschel calls it: 'Faint, small, south following a double star.' Nevertheless, some sources consider that NGC4888 is nonexistent,

NGC4878/4879/4888

or that it should be considered equal to NGC4879. It forms a likely physical pair with NGC4878, the minimum core-to-core separation at the apparent distance would be about 218 000 ly. Visually, NGC4888 can be seen following a pair of bright stars, which hinders visibility somewhat, but the galaxy is an oval, even-surface-brightness object, well defined and oriented ESE–WNW; its position angle is almost identical to that of the double star. Radial velocity: 3773 km/s. Distance: 169 mly. Diameter: 69 000 ly.

NGC4890

H614[3]	13:00.6	−4 34	GLX
SB(r)b	1.0′ × 0.8′	13.84B	

Photographs reveal a high-surface-brightness galaxy of peculiar morphology. It is a bright, irregular oval with uneven surface brightness, fading towards the NNE. A faint and irregular outer envelope exists and there is no clear spiral pattern. Visually, the galaxy can be seen in the same medium-magnification field as NGC4915, appearing faint and fairly small but well defined, very slightly elongated E–W and a little brighter to the middle. Radial velocity: 2939 km/s. Distance: 131 mly. Diameter: 38 000 ly.

NGC4899

H300[2]	13:00.9	−13 57	GLX
SAB(rs)c	2.2′ × 1.3′	12.66B	

This is a slightly inclined spiral galaxy. Photographs show a complex, knotty spiral pattern which is fragmentary and difficult to trace. Visually, the galaxy is moderately bright but somewhat diffuse; it is an extended oval oriented NNE–SSW. The galaxy is broadly brighter to the middle and the edges are fairly well defined. Radial velocity: 2544 km/s. Distance: 114 mly. Diameter: 73 000 ly.

NGC4900

H143[1]	13:00.6	+2 30	GLX
SB(rs)c	2.2′ × 2.1′	11.91B	

Images show a face-on barred galaxy featuring a short, blunt bar in a

grainy disc which lacks a clear spiral pattern. Visually, the galaxy is fairly bright but quite diffuse with somewhat ill defined edges. A bright field star borders the galaxy 0.6′ to the SE and the envelope is very broadly brighter to the middle. At high magnification a very small and slightly brighter core is intermittently visible. Radial velocity: 892 km/s. Distance: 40 mly. Diameter: 25 000 ly.

NGC4902

H69[1]	13:01.0	−14 31	GLX
SB(r)b	2.5′ × 2.2′	11.81B	

Photographs show that this bright, barred spiral galaxy has a pseudo-ring made up of nearly overlapping, bright internal arms, which are fairly tightly wound and expand to a fainter, broader outer spiral pattern. High-resolution photographs show many bright HII knots in the inner spiral structure. Telescopically, the galaxy is quite bright but a little diffuse and almost round with a large round core. The brightness of the envelope falls sharply at the border of the core and the envelope has a mottled texture. The edges are diffuse and there are bright field stars 2.0′ to the NW and 2.25′ to the WSW. Radial velocity: 2557 km/s. Distance: 114 mly. Diameter: 83 000 ly.

NGC4904

H517[2]	13:01.0	−0 02	GLX
SB(s)cd	2.2′ × 1.4′	12.69B	

This is an almost face-on, barred spiral galaxy. Photographs show a bright bar with no discernible core and two spiral arms expanding quickly to a broad pattern. Visually, the galaxy is moderately bright, irregularly round and fairly well defined with a suspected bar involved, elongated ESE–WNW. The envelope is quite grainy with the suspicion of dark patches involved. Radial velocity: 1114 km/s. Distance: 50 mly. Diameter: 32 000 ly.

NGC4915

H47[4]	13:01.5	−4 33	GLX
E0	1.7′ × 1.4′	12.89B	

Photographs show that this galaxy has a bright and round elliptical core and a fainter outer envelope and is elongated NE–SW. The fainter galaxy NGC4918 is situated 6.0′ to the ENE. Visually, NGC4915 is quite a bit brighter than NGC4890, which is visible in the same eyepiece field and is a likely physical companion. NGC4915 is a well-condensed and opaque object with well defined edges and a brighter core. Discovered on 11 March 1787, Herschel described it as: 'Pretty bright, stellar, resembles a star with a bur all around.' Radial velocity: 2945 km/s. Distance: 131 mly. Diameter: 65 000 ly.

NGC4925

H779[2]	13:02.1	−7 42	GLX
SA:(s)0° pec	1.1′ × 0.8′	14.41B	

The classification of this galaxy is somewhat uncertain and it may be an elliptical as it has a bright, dominant central disc surrounded by a faint halo. A possible companion galaxy IC4071 is 6.0′ N. Visually, NGC4925 is best seen at medium magnification as a moderately faint, very slightly oval patch of light, with an evenly bright disc and moderately well defined edges. Radial velocity: 3377 km/s. Distance: 151 mly. Diameter: 48 000 ly.

NGC4928

H190[2]/H760[3]	13:03.0	−8 05	GLX
SA(s)bc pec	1.3′ × 1.0′	12.92B	

This face-on galaxy was recorded twice by Herschel: on 3 March 1786 and again on 23 March 1789. The DSS image shows a probable spiral galaxy with a bright, featureless central region and two short, stubby spiral arms dotted with knotty patches. Visually, the galaxy is fairly bright and is seen as a slightly elongated patch of light that is brighter along a N–S axis and set in a diffuse, fairly round envelope, ill defined at the edges. Radial velocity:

1622 km/s. Distance: 72 mly. Diameter: 27 000 ly.

NGC4933

H191[2]	13:03.9	−11 29	GLX
S0/a pec	1.8′ × 1.1′	13.2B	

This is a peculiar triple system which shows definite gravitational interaction between two of the members. The system has also been catalogued as Arp 176. The two principal galaxies are well seen in the DSS image; the larger, central lenticular galaxy is in contact with the smaller elliptical galaxy which is 0.9′ to the SW. A tidal plume extends away to the WSW from the elliptical, and the outer disc of the lenticular is distorted towards the elliptical. High-resolution photographs show a prominent dust lane along the lenticular's eastern flank which is highly distorted by the encounter. The third galaxy, a dwarfish spiral or irregular, does not seem to be affected

and is located 1.2′ to the ENE. All the galaxies have similar radial velocities. The fainter barred spiral galaxy IC4134 is 4.2′ to the ESE. Telescopically, NGC4933 is moderately bright and appears as an extended ray of light oriented NNE–SSW. It is fairly well defined and even in surface brightness except for a slightly brighter patch in the NNE, which is the core of the principal galaxy. The overall dimensions of the interacting pair are about 2.6′ × 1.6′. Radial velocity: 3132 km/s. Distance: 140 mly. Diameter: 73 000 ly.

NGC4939

H561[2]	13:04.2	−10 20	GLX
SA(s)bc	5.5′ × 2.8′	11.89B	

This slightly inclined spiral galaxy is well seen in the DSS image. The bright, elongated core is surrounded by tightly wound spiral arms, which are narrow and dotted with nebulous patches. The arms broaden somewhat and fade near

NGC4939

©STScI/AURA/CALTECH/ROE

the extremities and the faint, outer spiral arms extend the apparent diameter of this galaxy to about 6.2′. The galaxy is somewhat faint visually and only the brighter central region is visible as an oval, even-surface-brightness patch of light with ill defined edges. Radial velocity: 3007 km/s. Distance: 134 mly. Diameter: 243 000 ly.

NGC4941

H40[1]	13:04.2	−5 33	GLX
(R)SAB(r)ab:	4.3′ × 3.3′	11.90B	

Images show an oblique, multi-arm spiral galaxy with a bright core and bright, fragmented spiral arms featuring many nebulous knots. Surrounding the bright inner spiral pattern is a very faint outer shell which does not show any clear spiral form. At high magnification in moderate-aperture telescopes, this galaxy is fairly bright and greatly elongated N–S. It features a much brighter, round and detached core with a faint stellar nucleus visible. Though bright, the outer envelope is diffuse and poorly defined. A field star is 2.7′ S. Radial velocity: 1019 km/s. Distance: 46 mly. Diameter: 57 000 ly.

NGC4942

H761[3]	13:04.3	−7 39	GLX
SAB(s)d:	1.7′ × 1.2′	13.66B	

Photographs show a face-on spiral galaxy with a probable short bar oriented N–S in a fragmentary spiral pattern dotted with nebulous knots. The galaxy is an irregularly round object

NGC4933

©STScI/AURA/CALTECH/ROE

visually and best seen at medium magnification as a somewhat hazy patch of light that is very broadly brighter to the middle with poorly defined edges. Radial velocity: 1649 km/s. Distance: 74 mly. Diameter: 37 000 ly.

NGC4951

H188[2]	13:05.1	−6 30	GLX
SAB(rs)cd:	3.8′ × 1.3′	12.58B	

In photographs this is an oblique spiral galaxy with a large, bright core and faint outer spiral structure. At medium magnification, the galaxy is a prominent oval object oriented E–W that is a little brighter to the middle but lacks a stellar nucleus. The main envelope is fairly well condensed but the edges are diffuse. Herschel considered this galaxy resolvable. It may be a companion of NGC4941. Radial velocity: 1085 km/s. Distance: 49 mly. Diameter: 54 000 ly.

NGC4958

H130[1]	13:05.8	−8 01	GLX
SB(r)0? sp	4.3′ × 1.2′	11.49B	

The DSS photograph shows a bright, edge-on lenticular galaxy with a spindle-shaped central region and a diffuse and extensive outer halo. Visually, the galaxy is very bright and is well seen at high magnification as an irregularly oval, grainy-textured patch of light oriented NNE–SSW. There is a bright core and the edges are fairly well defined, but the outer halo is not seen. Radial velocity: 1360 km/s. Distance: 61 mly. Diameter: 76 000 ly.

NGC4981

H189[2]	13:08.8	−6 47	GLX
SAB(r)bc	2.8′ × 2.0′	12.07B	

In photographs, this slightly oblique spiral galaxy has a large, bright central region; two narrow spiral arms emerge, broadening to fainter, more diffuse outer spiral structure. Visually, the galaxy is very bright with a bright field star 1.0′ SSE. The core is sizable and quite bright and is embedded in a fairly bright, oval envelope with diffuse edges. The galaxy is oriented SSE–NNW. Radial velocity:

1590 km/s. Distance: 71 mly. Diameter: 58 000 ly.

NGC4984

H301[2]	13:09.0	−15 31	GLX
(R)SAB(rs)0+	3.1′ × 2.3′	12.20B	

The DSS image of this galaxy suggests a face-on barred galaxy with an inner disc and a faint outer ring, but high-resolution photographs show that the galaxy is actually a face-on spiral with multiple inner spiral arms defined by dust and a large dust patch ENE of the core. The outer spiral pattern consists of two, tightly wound, smooth-textured spiral arms and a much fainter outer arm that may be interpreted as a ring. Visually, this is a large and moderately bright galaxy with an oval-shaped, diffuse envelope, fairly well defined edges and a small, brighter core. A short bar is suspected running roughly N–S; it is a manifestation of the oval central region of the galaxy. A wide pair of field stars is 2.5′ ENE. Radial velocity: 1165 km/s. Distance: 52 mly. Diameter: 47 000 ly.

NGC4989

H185[2]	13:09.4	−5 25	GLX
S0°:	1.5′ × 0.9′	14.40B	

Photographs show this lenticular galaxy has a peculiar, dwarfish companion located 1.4′ S. A very faint plume of material extends from NGC4989 towards the smaller galaxy, suggesting interaction. The companion may be visible in very large amateur telescopes. Visually, NGC4989 is an oval patch of light oriented almost due N–S with a faint stellar core. A field star is 2.75′ SSW and the naked-eye star Psi Virginis is located to the SE. Also visible in the field in moderate apertures is the elliptical galaxy NGC4990, located 7.3′ due N. Radial velocity: 3016 km/s. Distance: 135 mly. Diameter: 59 000 ly.

NGC4995

H42[1]	13:09.7	−7 50	GLX
SAB(rs)b:	2.5′ × 1.6′	11.92B	

In photographs this slightly oblique spiral galaxy has a high surface

brightness; the spiral pattern is principally of the grand-design type with two main spiral arms emerging from the core. In moderate-aperture instruments, this is a bright galaxy located 3.4′ SSE of a magnitude 8 field star. The envelope is oval and oriented E–W with fairly well defined edges and is a little brighter to the middle. Radial velocity: 1650 km/s. Distance: 74 mly. Diameter: 54 000 ly.

NGC4999

H537[2]	13:09.6	+1 40	GLX
SB(r)b	2.3′ × 1.9′	13.12B	

This is an almost face-on barred spiral galaxy of almost perfect symmetry. Photographs show that the bright core is embedded in a strong bar with the two principal inner arms forming a pseudo-ring. The ring branches out to a narrow but well defined multi-branch spiral pattern with dust lanes involved. Telescopically, the galaxy is moderately large but faint and diffuse, broadly brighter to the middle and almost round. Radial velocity: 5582 km/s. Distance: 249 mly. Diameter: 167 000 ly.

NGC5015

H637[2]	13:12.4	−4 19	GLX
(R)SB(r)a:	2.1′ × 1.8′	13.45B	

Photographically, this almost face-on, barred spiral galaxy has a large, elongated and bright core with short, narrow bars. The inner spiral pattern is

NGC4999

tightly wound for half a revolution before quickly fading to a broad outer pattern. Telescopically, the galaxy is moderately bright but only the bright central region is visible as a slightly oval, fairly well defined patch of light with even surface brightness, oriented ENE–WSW. A prominent field star is 4.25′ to the NW. Radial velocity: 3068 km/s. Distance: 137 mly. Diameter: 84 000 ly.

NGC5017

H669[3]	13:12.9	−16 46	GLX
E[+]?	1.6′ × 1.3′	13.54B	

This elliptical galaxy is probably an outlying member of the NGC5044 group of galaxies. Its visual and photographic appearances are quite similar. At medium magnification the galaxy is a small but moderately bright object with a condensed and moderately well defined disc that is very slightly oval and oriented NE–SW with a small, bright core. Radial velocity: 2429 km/s. Distance: 109 mly. Diameter: 50 000 ly.

NGC5018

H746[2]	13:13.0	−19 31	GLX
E3:/S0	3.4′ × 2.4′	11.69B	

The classification of this galaxy is somewhat uncertain but it would appear to be a lenticular galaxy with an extensive outer envelope. High-resolution photographs show at least four distinct dust patches in the outer halo which borders the N of the bright central region. Visually, the galaxy is quite bright, featuring a bright elongated central region, oriented E–W, which resembles a bar. Surrounding it is a mottled, faint outer envelope, which is ill defined. The edge-on lenticular galaxy NGC5022 is 7.2′ ESE and is visible as a faint sliver of light, oriented NNE–SSW and well defined with a faint pair of stars 1.75′ W. The two galaxies probably form a physical pair with a minimum projected separation of 253 000 ly at the presumed distance. Radial velocity: 2695 km/s. Distance: 120 mly. Diameter: 119 000 ly.

NGC5019

H545[3]	13:12.8	+4 44	GLX
SB?	0.8′ × 0.7′	14.54B	

This appears to be a slightly inclined spiral galaxy with a possible bar and asymmetrical spiral structure. Photographs show it to have a brighter and more definite pattern on the W side. Visually, the galaxy is quite faint and quite diffuse though its edges are moderately well defined. The surface brightness is fairly high but the E edge is a little fainter. Radial velocity: 6375 km/s. Distance: 285 mly. Diameter: 66 000 ly.

NGC5020

H129[2]	13:12.6	+12 36	GLX
SAB(rs)bc	3.2′ × 2.7′	13.15B	

The DSS image shows a face-on barred spiral galaxy with a bright core and a very faint, wide bar which broadens to a pseudo-ring. Five spiral arms emerge from the pseudo-ring, three in the E and two in the W. Visually, although bright, this galaxy is a small, hazy patch of light, slightly elongated N–S with very poorly defined edges. A small bright core is embedded in the grainy-textured envelope. Recorded on 12 April 1784, Herschel called this galaxy: 'Faint, pretty large, a little brighter to the middle. Round, resolvable.' Radial velocity: 3333 km/s. Distance: 149 mly. Diameter: 138 000 ly.

NGC5037

H510[2]	13:15.0	−16 35	GLX
SA(s)a:	2.2′ × 0.6′	13.11B	

This galaxy is probably a member of the NGC5044 galaxy group, though its radial velocity is somewhat lower than that of the principal galaxy of that group. Photographs show a highly tilted spiral galaxy with a large central region and tightly wound spiral arms which are involved with significant amounts of dust, particularly to the NE of the core. At medium magnification, the galaxy can be seen in the field with three other Herschel objects: NGC5044, NGC5047 and NGC5049. It is a moderately bright, extended, flat streak of light, well defined and oriented NE–SW with a field star at its NE tip. Radial velocity: 1774 km/s. Distance: 79 mly. Diameter: 51 000 ly.

NGC5044

H511[2]	13:15.4	−16 23	GLX
E0	3.0′ × 3.0′	11.59B	

This is the principal galaxy of a fairly rich and moderately bright group of galaxies which contains a mix of large, bright galaxies and many smaller dwarf ones. About a dozen galaxies in an area of 1° × 1° have NGC designations, though not all the galaxies in the field are group members; for example, NGC5047 has a significantly higher radial velocity. The DSS image show that NGC5044 is a large, perfectly round elliptical galaxy with a bright core and an extensive outer envelope with seven smaller, though significant, galaxies in the field, including MCG–3-34-33 4.0′ to the S. Visually, it is far and away the brightest galaxy in a medium-magnification field that includes NGC5037, NGC5047, and NGC5049; it is a large moderately bright and round galaxy with a large, bright core, a grainy envelope and hazy edges. Radial velocity: 2592 km/s. Distance: 116 mly. Diameter: 101 000 ly.

NGC5047

H670[3]	13:15.8	−16 31	GLX
S0	2.7′ × 0.5′	13.60B	

This is a bright, spindle-shaped lenticular galaxy and its photographic

NGC5047

and visual appearances are quite similar. It is well seen visually as a small, bright streak of light, elongated ENE–WSW, well defined and high in surface brightness along its major axis with a slight bulge and extensions which taper to points. Though situated 10.0′ SSE of the principal galaxy of the NGC5044 group, it is evidently not a member as its radial velocity is much higher than the group average. Radial velocity: 6218 km/s. Distance: 278 mly. Diameter: 218 000 ly.

NGC5049

H512[2]	13:16.0	−16 24	GLX
S0	1.9′ × 0.6′	13.75B	

Photographs show this member of the NGC5044 group of galaxies to be a likely edge-on, spindle-shaped lenticular galaxy with a very large, elongated central region and a much fainter, slightly tapered disc. Visually, only the core of the galaxy is visible; it is small, round, well defined and fairly bright. Herschel made at least two sweeps of this area of the sky, picking up the brighter galaxies in the region on 31 December 1785, while fainter galaxies like NGC5017 and NGC5047 were noted on 7 May 1787. Radial velocity: 2632 km/s. Distance: 118 mly. Diameter: 65 000 ly.

NGC5054

H513[2]	13:17.0	−16 38	GLX
SA(s)bc	5.1′ × 3.0′	11.64B	

This large spiral galaxy may be an outlying member of the NGC5044 group of galaxies, though its radial velocity is lower than the average for the group. The galaxy is well seen in the DSS image, which shows a slightly inclined spiral galaxy with four principal spiral arms which each fade quickly after about half a revolution to a fainter and broader outer spiral pattern. A small, edge-on galaxy 2.7′ N may be a background object; if it is a physical companion the minimum separation would be 57 000 ly at the presumed distance. Visually, NGC5054 is a large and fairly bright galaxy but it is diffuse

and poorly defined with a bright core embedded asymmetrically towards the S in a somewhat grainy envelope. The envelope is irregularly round with a faint star bordering the core to the ENE. Radial velocity: 1629 km/s. Distance: 73 mly. Diameter: 108 000 ly.

NGC5068

H312[2]	13:18.9	−21 02	GLX
SB(s)d	7.1′ × 6.6′	10.64B	9.9V

This barred spiral galaxy appears to be relatively nearby and the DSS image shows it well as a face-on specimen with a fragmentary, multi-arm pattern, which is well resolved into brighter HII regions and possible stellar associations. A large and particularly complex nebulosity is situated in the outer envelope 2.4′ NNW of the bar. High-resolution photographs indicate that the brightest stars may be resolved at blue magnitude 18.5. Telescopically, the galaxy is an interesting object. It is a

large, bright but quite diffuse galaxy; the central region is large and slightly oval, and is a little brighter than the main envelope. The bar is well seen, running SSE–NNW, while the outer envelope is diffuse and a patch of dust is suspected on the eastern flank. A faint field star is in the outer envelope to the NNE and another is to the W. Radial velocity: 546 km/s. Distance: 24 mly. Diameter: 50 000 ly.

NGC5073

H282[3]	13:19.4	−14 52	GLX
SB(s)c? sp	4.1′ × 0.6′	13.32B	

The DSS image shows an edge-on galaxy, probably a spiral, with a bright central region and fainter extensions. A probable dust lane is suspected along the WSW flank of the galaxy. Two small, compact galaxies of high surface brightness are in the field but their high radial velocities indicate that they are background objects. One is 5.4′ to the

NGC5068

NNE, preceding a faint field star, and the other is 6.75′ to the WSW. Telescopically, NGC5073 is a fairly bright galaxy; it appears as a well defined sliver of light, which is fairly even in surface brightness and greatly elongated SSE–NNW. The ends taper to points. Herschel's 8 February 1785 description reads: 'Very faint, much extended south following, north preceding, very narrow.' Radial velocity: 2638 km/s. Distance: 118 mly. Diameter: 140 000 ly.

NGC5076

H117[3]	13:19.4	−12 45	GLX
SB(rs)0⁺	1.3′ × 0.9′	13.91B	

The DSS image shows a barred spiral galaxy with a bright, oval core and a slightly fainter bar with two faint, smooth-textured spiral arms attached. Its radial velocity is similar to the dominant galaxy in the field, NGC5077, and the minimum core-to-core

separation between the two galaxies would be about 176 000 ly at the presumed distance. Visually, it is a small, faint and hazy patch, almost round and moderately well defined. Radial velocity: 2891 km/s. Distance: 129 mly. Diameter: 49 000 ly.

NGC5077

H193[2]	13:19.5	−12 39	GLX
E3-4	2.3′ × 1.9′	12.33B	

This galaxy is the brightest of a small clutch of about a dozen galaxies in a 30′ circle which probably form a physical group. The photographic and visual appearances of NGC5077 are quite similar. In a medium-magnification field, the galaxy can be seen with NGC5076, NGC5079, NGC5072 and NGC5088 and appears as a large, bright oval, broadly brighter to a bright core, fairly well defined and oriented almost due N–S. The galaxy may actually be an S0, owing to its extensive outer

envelope and a very faint field star is seen in photographs bordering the bright central region to the S. The magnitude 7 star to the W interferes somewhat with observation. A compact possible satellite galaxy is only 45″ to the ESE. Neighbouring Herschel galaxies NGC5076 and NGC5079 are 5.0′ S and 2.75′ SSE, respectively. Radial velocity: 2717 km/s. Distance: 121 mly. Diameter: 81 000 ly.

NGC5079

H118[3]	13:19.6	−12 42	GLX
SB(rs)bc pec:	1.6′ × 1.0′	13.04B	

This galaxy is the middle member of a trio of galaxies, the others being NGC5076 and NGC5077, though its radial velocity is considerably smaller than that of the other two. All three galaxies were discovered by Herschel on 11 May 1784. Photographically, it is a slightly inclined, possibly barred spiral galaxy with a clumpy ill defined spiral structure dominated by brighter HII regions. If it is a physical companion of NGC5077, the minimum core-to-core separation between the two galaxies would be 97 000 ly at the presumed distance. Visually, the galaxy is large, faint, a little brighter to the middle, elongated NNE–SSW and a little hazy at the edges. Radial velocity: 2142 km/s. Distance: 96 mly. Diameter: 44 000 ly.

NGC5084

H313[2]	13:20.3	−21 50	GLX
S0	11.0′ × 2.5′	11.60B	

In photographs this edge-on lenticular galaxy is quite remarkable and displays some peculiar traits. The inner portion displays a large, elongated central bulge with bright, thin extensions and a fainter halo, but there is a considerable, very faint outer extension on each side. High-resolution photographs show that extensive bands of dust exist in these outer arms and are particularly prominent on the E side. A likely companion galaxy, ESO576-31, is 8.0′ to the WSW. Not surprisingly, only the bright inner region of NGC5084 is seen visually, but it is a bright, large and well

NGC5076/5077/5079

©STScI/AURA/CALTECH/ROE

NGC5084

©STScI/AURA/CALTECH/DOE

Radial velocity: 6386 km/s. Distance: 285 mly. Diameter: 124 000 ly.

NGC5100

H22[2]	13:21.0	+8 59	GLX
0.9′ × 0.5′	14.89B		

Though it is unclassified, photographs show that this remote galaxy is possibly a slightly inclined spiral with a bright, elongated central region surrounded by an asymmetrical disc, oriented NNE–SSW. It may form a physical pair with a fainter, face-on spiral 0.5′ to the NW. Recorded on 23 January 1784, it was described by Herschel as: 'Faint, very small' but he lost the object before he was able to make an accurate determination of its position. J. L. E. Dreyer suggested that this object is the same as one discovered by Albert Marth at this position in the nineteenth century, and that is the identification accepted here. Visually, the galaxy is quite faint but is consistently seen as a small, round patch of light which is quite well defined and situated about 1.5′ WSW of a faint field star. Radial velocity: 9541 km/s. Distance: 426 mly. Diameter: 112 000 ly.

NGC5111

H119[3]	13:22.8	−12 58	GLX
E2/S0−	1.9′ × 1.6′	13.45B	

Photographs show that this elliptical galaxy is slightly elongated E–W with a bright central region and an extensive outer envelope. A field star is 2.5′ WNW and several small and very faint galaxies are in the immediate field, including one 1.25′ WNW of the core. Telescopically, NGC5111 is moderately bright and almost round; it is a small and moderately well-condensed object with a brighter core and well defined edges. Radial velocity: 5417 km/s. Distance: 242 mly. Diameter: 134 000 ly.

NGC5118

H925[3]	13:23.5	+6 23	GLX
Scd:	0.9′ × 0.8′	14.52B	

This very slightly inclined spiral galaxy features a small, bright core and two

defined oval, oriented E–W with a faintish, moderately well defined envelope. The bright central region is elongated along the major axis with a star-like core embedded. The inner disc bulges slightly and a faint field star is just beyond the outer disc to the ENE. Radial velocity: 1599 km/s. Distance: 71 mly. Diameter: 228 000 ly.

NGC5087

H724[3]	13:20.4	−20 37	GLX
E+/S0	3.0′ × 2.1′	12.23B	

This galaxy may be either elliptical or lenticular in form; high-resolution images of the core show the presence of a dust lane. Otherwise, the visual and photographic appearances are fairly similar. Telescopically, the disc has high surface brightness and well defined edges. A bright core is embedded in the envelope, which is slightly mottled and oriented N–S. Radial velocity:

1700 km/s. Distance: 76 mly. Diameter: 66 000 ly.

NGC5094

H539[3]	13:20.7	−14 05	GLX
E2/SA0−	1.5′ × 1.1′	14.27B	

The DSS image shows that this lenticular galaxy has two probable physical companions, situated about 1.25′ to the WSW; one of these galaxies is identified as MCG–2-34-36. These companions may be visible in large-aperture telescopes and if they are physically related to the larger galaxy, the minimum separation would be about 104 000 ly at the presumed distance. Visually, NGC5094 is moderately faint, small and round with a condensed disc, even surface brightness and well defined edges. Two faint field stars follow 1.0′ to the ESE and neither of the faint companions is detected.

high-surface-brightness spiral arms embedded in a fainter envelope. Photographs show the smaller, fainter galaxy CGCG044-079 2.4' to the ENE. Visually, NGC5118 is a very small, faint and almost round galaxy with even surface brightness across the envelope and well defined edges. Radial velocity: 6934 km/s. Distance: 310 mly. Diameter: 81 000 ly.

NGC5129

H653[2]	13:24.1	+13 59	GLX
E	1.7' × 1.4'	13.03B	

This galaxy has been classified as simply elliptical, but it is considerably elongated and has an extensive outer envelope and could therefore be classed as E3 or E4. The DSS image shows that the galaxy is situated in a compelling field which would be a challenge for large-aperture telescopes. The theta-shaped, barred galaxy NGC5132 is 8.0' NNE and three tiny galaxies flank the magnitude 10 field star 3.1' to the SE. In a medium-magnification field, NGC5129 is a small and fairly faint galaxy, elongated N–S with a bright, condensed core and well defined edges. A magnitude 10 field star is 1.6' E. Discovered on 19 March 1786, Herschel described the galaxy as: 'Pretty bright, very small, much brighter to the middle, just preceding a pretty considerable star.' Radial velocity: 6869 km/s. Distance: 307 mly. Diameter: 152 000 ly.

NGC5134

H314[2]	13:25.3	−21 08	GLX
(R)SAB(r)a	4.0' × 4.0'	12.45B	

The DSS image shows a bright, inclined spiral galaxy with a slightly oval core and bright, knotty spiral arms. Brighter HII regions are evident near the edge of the disc towards the SSE and NNW and a smooth-textured, faint outer ring encircles the oval inner region. Telescopically, the galaxy is a large and quite bright, extended oval with a large, elongated core. Oriented SSE–NNW, the envelope is somewhat mottled but fairly well defined, although it seems slightly

dimmer along the western flank. In a medium-magnification field, the galaxy IC4237 is easily visible 10.0' to the W. Radial velocity: 1639 km/s. Distance: 73 mly. Diameter: 85 000 ly.

NGC5136

H84[3]	13:24.8	+13 44	GLX
0.7' × 0.6'	14.51B		

This unclassified but very probably elliptical galaxy is similar visually and photographically, appearing very small and faint in a medium-magnification eyepiece. It is a round, nebulous patch of light with even surface brightness and well defined edges. A star-like, brighter core is intermittently visible. Herschel used a magnification of 240 to confirm this galaxy's nonstellar nature. Radial velocity: 6635 km/s. Distance: 296 mly. Diameter: 60 000 ly.

NGC5146

H115[3]	13:26.5	−12 19	GLX
S0° pec	2.1' × 1.4'	13.95B	

Photographs show that this lenticular galaxy has a bright, oval core and a faint outer envelope. There are several faint and small galaxies in the field, the most prominent being an anonymous galaxy situated 7.3' NNW. NGC5146 is a fairly faint galaxy visually; it is a small, slightly oval patch which is well defined at the edges and slightly elongated NE–SW. Radial velocity: 6586 km/s. Distance: 294 mly. Diameter: 180 000 ly.

NGC5147

H25[2]	13:26.3	+2 06	GLX
SB(s)dm	2.1' × 1.6'	12.28B	

Photographs show an almost face-on spiral galaxy which is a little brighter to the middle with a possible faint bar and four knotty, principal spiral arms. The galaxy is moderately bright at high magnification and fairly well seen; it has a diffuse envelope that is moderately well defined at the edges. A faint field star, which is offset slightly to the NW in the envelope, can be mistaken for a

bright core. Radial velocity: 1035 km/s. Distance: 46 mly. Diameter: 28 000 ly.

NGC5170

H22[5]	13:29.8	−17 58	GLX	
SA(s)c: sp	9.9' × 1.2'	12.07B	10.8V	

This is a classic, edge-on spiral galaxy whose morphological cousins include galaxies such as NGC4565 and NGC5746. Photographs show a spiral galaxy that is tilted very slightly, so that its small central bulge can be seen over the prominent dust lane. The bulge is surrounded by a bright inner region but the surface brightness drops off quickly to an outer disc dominated by dust patches and lanes. Visually, the galaxy is a bright, flat and well defined object, elongated ESE–WNW with a brighter core. The brightness fades gradually from the centre towards the slightly tapered tips. Radial velocity: 1393 km/s. Distance: 62 mly. Diameter: 180 000 ly.

NGC5174

H45[3]/H46[3]	13:29.3	+11 00	GLX
Scd:	3.2' × 1.7'	13.67B	

Recorded on 15 March 1784, Herschel called this object, 'Two, mistaken for one; but 240 [magnification] shewed them both. Considerably large. Very faint.' In fact, this is a single galaxy. The second object (NGC5175 = H46[3]) is a single foreground star 0.8' S of the galaxy core. The DSS photograph shows a slightly inclined, multi-arm spiral galaxy with a small core in a probable bar. The spiral arms are bright, narrow and relaxed with several knotty condensations involved. Visually, the galaxy is obviously a single object, appearing as a large and very diffuse oval patch of light, very broadly brighter to the middle with a field star to the NNE. It is best seen at high magnification; the bright star 71 Virginis, located S of the galaxy, greatly hinders observation and should be kept out of the field. Radial velocity: 6815 km/s. Distance: 305 mly. Diameter: 284 000 ly.

NGC5170

NGC5183

H679[2]	13:30.1	−1 44	GLX
Sb pec	2.1′ × 0.9′	13.99B	

This galaxy forms a probable physical pair with NGC5184, situated 3.5′ to the NNE on the DSS image. The minimum core-to-core separation between them would be about 192 000 ly at the presumed distance.

The galaxy is a highly inclined spiral with a brighter core and two knotty spiral arms. Of the pair, NGC5183 appears smaller and slightly fainter visually, with an envelope that is elongated ESE–WNW and brighter to the middle with fairly well defined edges. Radial velocity: 4228 km/s. Distance: 189 mly. Diameter: 115 000 ly.

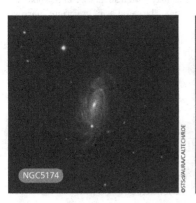

NGC5174

NGC5183/5184

NGC5184

H680[2]	13:30.2	−1 41	GLX
SAB(r)c	2.0′ × 1.1′	13.64B	

Photographs show an almost face-on spiral galaxy, which may be barred; the core is small and bright and there appear to be three sets of spiral arms, each composed of parallel sets of arms with dust lanes involved. Visually, it is larger and brighter than NGC5183; it is a little extended ESE–WNW with somewhat ill defined edges and a little brighter to the middle. Radial velocity: 3929 km/s. Distance: 175 mly. Diameter: 102 000 ly.

NGC5185

H642[3]	13:30.0	+13 25	GLX
Sb	1.9′ × 0.7′	14.48B	

Despite its classification, photographs show that this highly inclined spiral galaxy is very likely barred. The bright core is embedded in a straight, bright bar and there is a tightly wound spiral pattern. This galaxy is elongated NE–SW and dust lanes and patches can be seen. Visually, NGC5185 is quite faint and somewhat diffuse and is best seen at medium magnification as an oval patch of light, very slightly brighter to the middle and situated immediately ENE of a linear asterism of four evenly bright stars. The edges are somewhat diffuse and a medium-magnification field also shows NGC5181 as a round, well defined, concentrated patch of light 8.0′ to the SSW. Radial velocity: 7356 km/s. Distance: 329 mly. Diameter: 182 000 ly.

NGC5203

H507[3]	13:32.2	−8 48	GLX
S0−	1.5′ × 0.8′	14.14B	

Photographs show this lenticular galaxy has a brighter, lens-shaped central region and a fainter outer envelope. Two small, condensed objects, one 0.4′ NNW the other 0.4′ SSW of the core, are probably dwarf satellites of the principal galaxy. Telescopically, the galaxy is a very faint, oval patch of light, fairly even in surface brightness, well

defined at the edges and elongated E–W.
Radial velocity: 6674 km/s. Distance:
298 mly. Diameter: 130 000 ly.

NGC5207

H643[3]	13:32.2	+13 54	GLX
SAB(r)b	1.7′ × 1.0′	14.35B	

This is a slightly inclined spiral galaxy.
Photographs show a bright inner ring
formed by overlapping spiral arms which
expand to a fainter, broader outer spiral
pattern. Visually, the galaxy is faint and
most easily seen at medium
magnification as a diffuse, oval patch of
light extending SE from a faint field star,
which hinders observation. The galaxy is
oriented SE–NW and is broadly brighter
along its major axis with hazy edges.
Radial velocity: 7638 km/s. Distance:
341 mly. Diameter: 169 000 ly.

NGC5208

H9[3]	13:32.5	+7 19	GLX
S0	1.7′ × 0.6′	14.50B	

This edge-on lenticular galaxy is part of
a triple system of Herschel objects that
very probably form a physical group.
The DSS image also shows several
smaller, fainter objects which may be
part of the group, including the non-
Herschel object NGC5212. Visually, this
is an attractive triplet in a medium-
magnification field. NGC5208 is a well
defined, elongated streak, oriented
SSE–NNW; it is quite even in surface
brightness and moderately bright with
slightly tapered ends. Radial velocity:
6728 km/s. Distance: 301 mly.
Diameter: 149 000 ly.

NGC5209

H10[3]	13:32.7	+7 20	GLX
E	0.8′ × 0.7′	14.59B	

This elliptical galaxy is 3.75′ ENE of
NGC5208. Photographs show a bright,
round core in an elongated halo,
oriented ESE–WNW. Visually, the
galaxy appears round and well
condensed and is a little brighter to the
middle with hazy edges. Radial velocity:
7016 km/s. Distance: 313 mly.
Diameter: 73 000 ly.

NGC5210

H99[3]	13:32.9	+7 10	GLX
Sa	1.5′ × 1.3′	14.03B	

Photographs show a face-on spiral
galaxy with a bright core and one well-
developed spiral arc S of the core,
defined by a probable dust lane along its
inner edge. Beyond is a faint outer halo.
Telescopically, NGC5210 appears larger
than either NGC5208 or NGC5209; it is
a round, somewhat diffuse object,
evenly bright with hazy edges and a
faint field star following 2.0′ ENE of the
core. Radial velocity: 6833 km/s.
Distance: 305 mly. Diameter:
133 000 ly.

NGC5221

H86[3]	13:34.9	+13 50	GLX
Sb:	3.1′ × 0.8′	14.36B	

This galaxy is part of a loose grouping of
galaxies that includes NGC5222 and

NGC5230 which share similar radial
velocities and are probably physically
related. All three objects were recorded
by Herschel on 12 April 1784. This
galaxy and the galaxy pair NGC5222
are also catalogued as Arp 288. The DSS
image reveals that NGC5221 is a highly
inclined spiral galaxy which has
recently undergone a collision with
another galaxy, though it seems that
this other galaxy may have been
absorbed by the large spiral. NGC5221
features a bright core surrounded by a
broken, ring-like structure probably
formed by overlapping spiral arms. The
core itself is irregular and appears
disturbed and streamers of material are
being ejected from the spiral arm to
the S. A much longer, broad plume
extends well away from the galaxy to
the WNW. NGC5221 is separated
from the galaxy pair NGC5222 by 5.5′
and the minimum separation between
the two would be about 500 000 ly

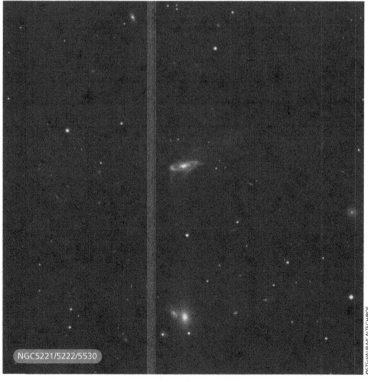

NGC5221/5222/5530

years at the apparent distance. The plumes extend the apparent dimensions of the galaxy to about 3.6′ × 0.8′. Visually, in a medium-magnification field NGC5221 forms an attractive triplet with NGC5222 and NGC5230. It is the northernmost of the group, a well defined, elongated streak of light oriented E–W and very broadly brighter to the middle. Radial velocity: 6968 km/s. Distance: 311 mly. Diameter: 326 000 ly.

NGC5222

H85[3]	13:34.9	+13 44	GLX
E:	1.3′ × 1.0′	14.02B	

Images show this elliptical galaxy forms a possible interacting pair with a spiral galaxy 0.4′ to the ENE. There is also a smaller, fainter galaxy 0.9′ to the ESE. The elliptical has a large oval core and an extensive outer disc, oriented N–S. While the elliptical appears intact, the spiral has an asymmetrical spiral structure with two broad arms curving away from the core to the N. These two galaxies and NGC5221 are also catalogued as Arp 288. Visually, NGC5222 is unresolved at medium magnification, but appears as an E–W elongated patch of light which is moderately well defined with a tiny, stellar, brighter core. Radial velocity: 6846 km/s. Distance: 306 mly. Diameter: 116 000 ly.

NGC5224

H926[3]	13:35.2	+6 28	GLX
0.6′ × 0.6′	14.98B		

Although this galaxy is unclassified, photographs reveal a face-on, probably lenticular galaxy with a round, bright core in an extensive, fainter outer halo. Visually, the galaxy is situated between two bright field stars aligned ENE–WSW. It is a concentrated spot which is quite faint but well seen at both medium and high magnification as a round, evenly bright object with well defined edges. Radial velocity: 6787 km/s. Distance: 303 mly. Diameter: 53 000 ly.

NGC5227

H928[3]	13:35.3	+1 25	GLX
(R′)SB(r)b	2.1′ × 1.6′	13.84B	

Images show that this face-on, barred spiral galaxy has a small core and a bar surrounded by tightly wound spiral arms forming an inner pseudo-ring before expanding to a broad, faint, four-branch spiral pattern. Telescopically, the galaxy is quite faint and blurry but is seen consistently as an ill defined, roundish patch of light which is very slightly brighter to the middle. Radial velocity: 5191 km/s. Distance: 232 mly. Diameter: 142 000 ly.

NGC5230

H87[3]	13:35.5	+13 40	GLX
SA(s)c:	2.2′ × 1.9′	13.14B	

This is a face-on, multi-arm spiral galaxy. Photographs show a small, bright core in a complex, asymmetrical spiral pattern made up of narrow, knotty arms which are fairly tightly wound. Visually, it is the largest and brightest of a group of three galaxies, the others being NGC5221 and NGC5222; it appears as a round, somewhat diffuse patch of light which is very broadly brighter to the middle and well defined. Radial velocity: 6841 km/s. Distance: 306 mly. Diameter: 195 000 ly.

NGC5235

H100[3]	13:36.1	+6 34	GLX
SB	1.2′ × 0.5′	14.81B	

Photographs show an inclined spiral galaxy with a bright core surrounded by a bright pseudo-ring. The northern spiral arm is narrow, well defined, and arcs towards the ESE, while the southern arm is broader and fainter as it sweeps W, evidently above the rotation plane of the galaxy. Visually, however, this galaxy is quite faint and only the core is visible as a round, well defined, small and evenly bright glow. Both this galaxy and NGC5224 are probably outlying members of the galaxy group surrounding NGC5208. Radial velocity: 6547 km/s. Distance: 292 mly. Diameter: 102 000 ly.

NGC5247

H297[2]	13:38.1	−17 53	GLX
SA(s)bc	5.6′ × 4.9′	10.77B	

Like NGC5170, this large, bright galaxy is isolated in the field and their similar radial velocities suggest that the two galaxies may form a physical pair, despite their rather wide separation on the sky. The DSS image shows a classic, face-on spiral of the grand-design type, featuring two bright principal arms emerging from a brighter core as well as two more diffuse arms which separate from the brighter arms after about half a revolution. Visually, the galaxy is very large and diffuse, and is seen as a round, hazy patch of light with a brighter, slightly elongated central region. The envelope has a somewhat grainy texture and the edges are moderately well defined but there is no evidence of spiral structure. Radial velocity: 1253 km/s. Distance: 56 mly. Diameter: 91 000 ly.

NGC5252

H505[3]	13:38.3	+4 31	GLX
S0	1.3′ × 0.8′	14.25B	

This is a lenticular galaxy. Photographs show a lens-shaped object oriented NNE–SSW with a brighter core and faint extensions. Visually, only the core of this faint galaxy is detected, appearing as a small, slightly extended patch of light. It is fairly well defined and a little brighter to the middle. Radial velocity: 6850 km/s. Distance: 306 mly. Diameter: 116 000 ly.

NGC5257

H895[2]	13:39.9	+0 50	GLX	
SAB(s)b pec	1.6′ × 0.9′	13.66B	13.0V	

This galaxy forms an interacting pair with NGC5258 and together they have also been catalogued as Arp 240. The galaxies have similar radial velocities and the core-to-core separation would be about 110 000 ly at the presumed distance. The DSS image shows a pair of spiral galaxies joined by an E–W tidal plume. NGC5257 is a slightly inclined spiral galaxy with a possible small bar

NGC5247

and faint inner spiral structure. The outer spiral arms feature two bright arcs which fade after less than half a revolution to fainter, broader outer arms. Faint plumes are visible S of the galaxy. Visually, both galaxies can be seen in a high-magnification field. NGC5257 has a high-surface-brightness envelope, is fairly well defined and is slightly brighter to the

NGC5257/5258

middle. It is slightly elongated E–W and a little smaller than its companion. Radial velocity: 6750 km/s. Distance: 302 mly. Diameter: 140 000 ly.

NGC5258

H896[2]	13:40.0	+0 50	GLX	
SA(s)b: pec	1.6′ × 1.3′	13.97B	12.8V

The DSS image shows this inclined, S-shaped spiral has a complex inner structure with several bright condensations but no definite core. There is an extensive dust patch just N of a bright condensation near the core and a second to the S, which appears to sever the southern spiral arm. Like NGC5257, there are bright arcs in the spiral structure, though they are not as prominent. Telescopically, the galaxy is moderately bright with a high-surface-brightness envelope which is elongated NNE–SSW. It is well defined at the edges and a little brighter to the middle. Radial velocity: 6790 km/s. Distance: 300 mly. Diameter: 140 000 ly.

NGC5300

H533[2]	13:48.3	+3 57	GLX
SAB(r)c	3.5′ × 2.2′	12.94B	

In photographs, this almost face-on galaxy has low surface brightness and a very small, brighter core is embedded in the somewhat fragmentary, multi-arm spiral structure. The galaxy is very diffuse and fairly faint visually; it is slightly elongated SSE–NNW and broadly brighter to the middle. It appears as a broad oval patch of light with ill defined edges and a prominent field star is 8.25′ to the SW. Radial velocity: 1138 km/s. Distance: 51 mly. Diameter: 52 000 ly.

NGC5306

H306[2]	13:49.1	−7 12	GLX
E1	1.3′ × 1.0′	13.10B	

In photographs, this elliptical galaxy can be seen to form a triple system with the galaxy MCG–1-35-12 0.9′ NNE and a compact anonymous galaxy 0.5′ to the SSW. A third object, the edge-on galaxy MCG–1-35-13 is 3.25′ WNW. A large-aperture telescope is required to detect all of them. There is evidence of a tidal plume extending from MCG–1-35-12 to the principal galaxy, which has a bright, round central region and an extensive outer envelope oriented NNE–SSW. NGC5306 is a fairly bright galaxy visually; it is well seen at high magnification as an irregularly round object with a small bright core in a hazy envelope. A faint field star is 0.75′ NW of the core, while a bright star is 5.5′ S. Herschel recorded the galaxy on 5 March 1785, calling it: 'Faint, very small, irregular form, resolvable.' This suggests that he may have detected one or other of the associated galaxies. Radial velocity: 7147 km/s. Distance: 319 mly. Diameter: 121 000 ly.

NGC5324

H307[2]	13:52.1	−6 03	GLX
SA(rs)c	2.3′ × 2.1′	12.62B	

Photographs show a face-on spiral galaxy with a small core and a multi-arm pattern with a faint field star 0.5′ E of the core. The galaxy is large and somewhat diffuse visually; it is fairly bright, almost round and slightly brighter to the middle, while the edges are hazy and ill defined. Radial velocity: 2981 km/s. Distance: 133 mly. Diameter: 89 000 ly.

NGC5327

H685[2]	13:52.1	−2 12	GLX
SAB(rs)b:	1.9′ × 1.6′	13.76B	

This is a face-on barred spiral galaxy with two principal inner spiral arms expanding to a fainter, broader outer pattern. Smaller galaxies are in the field: CGCG017-079 is 4.75′ N and CGCG017-075 is 6.6′ to the NNW. Visually, NGC5327 is a moderately faint galaxy, almost round and a little brighter to the middle. The edges are fairly well defined with a faint but attractive pair of stars 7.25′ to the E. Radial velocity: 4315 km/s. Distance: 193 mly. Diameter: 107 000 ly.

NGC5329

H549[3]	13:52.2	+2 20	GLX	
E:	1.3′ × 1.3′	13.40B	12.4V	

Photographically, this elliptical galaxy has a bright core and extensive outer envelope. A smaller, possible satellite galaxy is situated 0.75′ SE, and other galaxies lie 3.5′ ENE, 2.3′ NNE and 4.75′ almost due N. Visually, NGC5329 can be seen in the same medium-magnification field as the galaxy pair NGC5331. It is quite faint but fairly opaque, appearing as a round, well defined patch of light with a condensed core. Radial velocity: 7073 km/s. Distance: 316 mly. Diameter: 120 000 ly.

NGC5331

H929[3]	13:52.4	+2 06	GLX
Sb pec:	1.3′ × 0.4′	14.3B	

The DSS photograph reveals a pair of interacting spirals which may be in contact on the W side. The pair are oriented N–S with the southern one appearing more nearly edge-on with a prominent, distorted dust lane and a faint, extended plume oriented

ESE–WNW. The northern component is probably a barred spiral galaxy with a bright, elongated core and three prominent spiral arms. There is an extremely faint, broad plume, which is difficult to detect, immediately N of the galaxy. The components cannot be resolved in a moderate-aperture telescope; NGC5331 appears as a small, diffuse, ill defined patch of light of fairly even surface brightness. Radial velocity: 9870 km/s. Distance: 441 mly. Diameter (including plumes): N and S components about 167 000 ly each.

NGC5334

H665[3]	13:52.9	−1 07	GLX
SB(rs)c:	3.9′ × 2.9′	12.95B	

In photographs this face-on barred spiral galaxy has very low surface brightness; the core is extremely small and is set in a small bar with a tightly wound pseudo-ring that expands to a broad, multi-arm spiral pattern with several condensations involved. Visually, the galaxy is dim and moderately large and appears as a fat oval of light with diffuse edges and fairly even surface brightness. The disc is elongated N–S with a field star 3.25′ S of the core. Radial velocity: 1340 km/s. Distance: 60 mly. Diameter: 68 000 ly.

NGC5343

H308[2]	13:54.0	−7 34	GLX
SA(r)0−?	1.7′ × 1.2′	14.26B	

Images show a lenticular galaxy oriented NE–SW with a bright, elongated central region in a fainter envelope. Telescopically, this is a fairly bright galaxy which is well seen at high magnification as a round, high-surface-brightness object with a grainy envelope, well defined edges and a small, brighter core. Radial velocity: 2572 km/s. Distance: 115 mly. Diameter: 57 000 ly.

NGC5345

H686[2]	13:54.3	−1 25	GLX
SA(r)a pec:	1.8′ × 1.8′	13.92B	

This face-on spiral galaxy has a brighter core surrounded by a bright, pseudo-

ring structure, which is fainter and poorly defined towards the SW. This pseudo-ring is surrounded by a very dim outer spiral structure. Photographs show a large condensation 0.25′ S of the core, which is probably a satellite galaxy seen in silhouette. Telescopically, this galaxy is fairly well seen despite the presence of the bright star 90 Virginis 8.0′ to the ESE. The galaxy appears as a round, moderately high-surface-brightness object that is fairly well defined at the edges and slightly brighter to the middle. Radial velocity: 7222 km/s. Distance: 322 mly. Diameter: 169 000 ly.

NGC5356

H506[3]	13:55.0	+5 20	GLX
SABbc: sp	3.7′ × 0.7′	13.61B	

This edge-on spiral galaxy presents similar appearances visually and photographically. It is best seen at medium magnification as a greatly extended sliver of light with even surface brightness except for a very slightly brighter core and fairly well defined edges. The galaxy is oriented NNE–SSW and its ends are tapered. Radial velocity: 1349 km/s. Distance: 60 mly. Diameter: 65 000 ly.

NGC5363

H6[1]	13:56.1	+5 15	GLX
I0?	4.9′ × 3.2′	11.10B	

In photographs, this lenticular galaxy has a large bright core and an extensive faint outer envelope. Telescopically, it is a very bright galaxy with a very faint, intermittently visible secondary shell. The core is bright and gradually brighter to the middle. The galaxy is elongated SE–NW and a magnitude 8 field star is 3.75′ NE. Radial velocity: 1115 km/s. Distance: 50 mly. Diameter: 71 000 ly.

NGC5364

H534[2]	13:56.2	+5 01	GLX
SA(rs)bc pec	6.8′ × 4.4′	11.19B	

Photographically, this superb spiral galaxy is notable for the moderately bright ring which encircles the core.

NGC5364

velocity: 4361 km/s. Distance: 195 mly.
Diameter: 96 000 ly.

NGC5382

H546[3]	13:58.2	+6 16	GLX
S0	1.5′ × 1.1′	13.51B	

This galaxy forms a probable physical
pair with NGC5386 which is located
5.0′ to the NNE. Photographs show a
lenticular galaxy with a bright oval
core in a fainter outer disc. The pair is
moderately bright and is well seen in a
high-magnification field. NGC5382
appears a little smaller than its
neighbour; it is a very slightly oval
disc that is a little brighter to the
middle, fairly well defined and
oriented NNE–SSW. Radial velocity:
4291 km/s. Distance: 192 mly.
Diameter: 84 000 ly.

NGC5386

H547[3]	13:58.2	+6 21	GLX
S0/a	1.2′ × 0.5′	13.60B	

Photographs reveal a probable inclined
spiral galaxy with a bright core and a
bright half arc along the NW flank
with a fainter, outer spiral disc.
Visually, NGC5386 appears a little
more elongated than NGC5382 to the
SSW, is oriented NE–SW and has a
bright, well defined disc and a tiny,
bright core. A faint field star is at the
SW edge of the disc. Radial velocity:
4270 km/s. Distance: 191 mly.
Diameter: 67 000 ly.

High-resolution images show two short
spiral arcs emerging from the core to
form the ring. The spiral arms are thin
and somewhat faint, but each can be
followed for more than a full rotation
around the galaxy core. The dwarfish
galaxy NGC5360 is 7.7′ to the WSW
and may be associated with the brighter
NGC5364. Visually, NGC5364 is a
diffuse, though moderately bright
galaxy and is quite large. The galaxy has
poorly defined extremities and is only a
little brighter to the middle. It is very
slightly oval and is oriented NNE–SSW
with a faint field star to the NNW. Radial
velocity: 1216 km/s. Distance: 54 mly.
Diameter: 108 000 ly.

NGC5369

H285[3]	13:56.8	−5 29	GLX
E0?	1.0′ × 0.9′	14.45B	

The photographic and visual
appearances of this elliptical galaxy are

quite similar. Though moderately faint,
the galaxy is well seen at high
magnification as a small, round patch of
light with a brighter core and well
defined edges. Radial velocity:
2915 km/s. Distance: 130 mly.
Diameter: 38 000 ly.

NGC5374

H889[2]	13:57.5	+6 06	GLX
SB(r)bc?	1.7′ × 1.5′	13.63B	

Photographs show a face-on spiral
galaxy that has a boxy inner ring
structure made up of overlapping spiral
arms surrounding a small core. Three
fainter spiral arms emerge and are fairly
tightly wound. Visually, this is a
moderately bright galaxy, situated
immediately E of a field star and located
within a bright isosceles triangle of field
stars. The galaxy is almost round and
fairly even in surface brightness, with a
mottled, fairly well defined disc. Radial

NGC5392

H666[3]	13:59.4	−3 12	GLX
S0[+]?	1.1′ × 0.8′	14.99B	

The DSS image shows a bright
lenticular galaxy with a very faint
outer envelope. A small, bright, oval
condensation borders the central
region on the NE side and may be a
satellite galaxy seen in silhouette.
Visually, the galaxy is a very faint and
difficult object; it is very small, round
and even in surface brightness. Radial
velocity: 7291 km/s. Distance:
326 mly. Diameter: 104 000 ly.

NGC5400

H667[3]	14:00.6	−2 51	GLX
(R')S0⁻?	1.4′ × 1.2′	14.31B	

Photographs show that this lenticular galaxy is the brightest of a tight group of six galaxies in the field, three to the S and two to the N; all of the group are located within 3.25′ of NGC5400. Visually, this lenticular galaxy is a small, very faint, and extremely difficult object, appearing as a dim, round patch of light with even brightness and moderately well defined edges. Radial velocity: 7391 km/s. Distance: 330 mly. Diameter: 134 000 ly.

NGC5426

H309[2]	14:03.4	−6 04	GLX
SA(s)c pec	3.0′ × 1.6′	12.71B	

The DSS image shows that this elongated spiral galaxy and NGC5427 2.3′ N form an interacting pair; the galaxies are also catalogued as Arp 271. NGC5426 features a large, elongated core and thin spiral arms dotted with HII regions. The arms on the W side are somewhat warped and two filaments are drawn towards NGC5427. The core-to-core separation between the two galaxies at the projected distance would be about 75 000 ly. Visually, NGC5426 is quite small and is seen as a hazy oval of light with a condensed middle. Herschel had an exceptional view of the two galaxies, recording them on 5 March 1785. His description reads: 'Two nearly in the meridian. Distance 4′, small star in between. Chevelure touching. The north [nebula] pretty bright, considerably large, much brighter middle. The south [nebula] faint, small.' The description suggests that Herschel was able to detect the interaction between the two galaxies. Radial velocity: 2519 km/s. Distance: 112 mly. Diameter: 98 000 ly.

NGC5427

H310[2]	14:03.4	−6 02	GLX
SA(s)c pec	2.8′ × 2.4′	11.06B	

This bright, face-on spiral galaxy features a round core and two principal

NGC5426/5427

©STScI/AURA/CALTECH/ROE

spiral arms of high surface brightness. Curiously, photographs fail to show any sign of interaction with NGC5426 to the S: the spiral pattern is completely normal and undisturbed. Visually, NGC5427 is brighter and larger than its companion; it is quite diffuse and only very gradually brighter to the middle. The envelope is round and there is no evidence of a core. Radial velocity: 2565 km/s. Distance: 114 mly. Diameter: 93 000 ly.

NGC5468

H286[3]	14:06.6	−5 27	GLX
SAB(rs)cd	2.6′ × 2.4′	12.94B	

The DSS image shows a face-on spiral galaxy with two principal arms emerging from the central region and quickly fragmenting into outer spiral arm fragments; at least five fragments are well seen in the image. The almost edge-on spiral NGC5472 is 4.8′ to the E and a bright field star is 4.2′ SSE. Visually, the galaxy is a moderately

large but dim and diffuse object; it is almost round and is broadly brighter to the middle. The edges are poorly defined and the bright field star hinders visibility. Radial velocity: 2793 km/s. Distance: 125 mly. Diameter: 94 000 ly.

NGC5476

H287[3]	14:08.2	−6 05	GLX
SAB(rs)dm?	1.4′ × 1.3′	13.36B	

Photographs show a very slightly inclined spiral galaxy with grainy, fragmentary spiral structure. Visually, the galaxy is a small, moderately faint and irregularly round object; it is fairly even in surface brightness and quite well defined. Radial velocity: 2609 km/s. Distance: 116 mly. Diameter: 47 000 ly.

NGC5478

H762[3]	14:08.2	−1 42	GLX
SAB(s)bc	1.1′ × 0.7′	14.46B	

This is a slightly inclined spiral galaxy. Photographs show a large, bright central region, which may contain a bar with broad, fainter spiral arms emerging. Visually, the galaxy is small and quite dim; it is seen as a round gradual brightening which is fairly diffuse and a little brighter to the middle with fairly well defined edges. Radial velocity: 7502 km/s. Distance: 335 mly. Diameter: 107 000 ly.

NGC5491

H890[2]	14:10.9	+6 22	GLX
S?	1.4′ × 0.8′	13.93B	

While the classification is uncertain, the DSS image clearly shows a slightly inclined spiral galaxy with a dusty, mottled spiral structure. The galaxy forms a physical pair with the compact galaxy NGC5491B 0.4′ to the N. Visually, only the larger spiral is seen; it is quite faint and appears as a roundish, diffuse patch of light with hazy edges and is only very broadly brighter to the middle. Radial velocity: 5879 km/s. Distance: 262 mly. Diameter: 107 000 ly.

NGC5493

H46[4]	14:11.5	−5 03	GLX
S0 pec sp	2.1′ × 1.7′	12.29B	

The DSS image shows a lenticular galaxy with a bright, boxy core and blunt extensions oriented ESE–WNW, embedded in an extensive, faint and irregular halo. In a moderate-aperture telescope, the galaxy appears as a small, high-surface-brightness object which is brighter to a condensed core with faint extensions. Recorded on 22 February 1787, Herschel's description reads: 'Pretty bright, almost considerably bright, very small, stellar, like a star with burs.' The last comment led him to include the object in his Class IV. Radial velocity: 2666 km/s. Distance: 116 mly. Diameter: 73 000 ly.

NGC5506

H687[2]	14:13.2	−3 13	GLX
Sa pec sp	3.2′ × 0.8′	12.81B	12.0V

Images reveal a bright, almost edge-on galaxy, possibly a spiral, with a prominent dust lane along the major axis, cutting in front of the core. Another dust lane is suspected along the southern flank and faint tendrils emerge all around the outer disc. NGC5507 is 3.6′ to the NNE, and is very probably a physical companion: the minimum core-to-core separation at the presumed distance would be about 85 000 ly. Visually, both galaxies can be seen in a high-magnification field. NGC5506 is large, bright and elongated E–W. It exhibits fairly even surface brightness across its major axis, but is somewhat grainy and has irregular, slightly ill defined edges. Radial velocity: 1815 km/s. Distance: 81 mly. Diameter: 76 000 ly.

NGC5507

H49[4]	14:13.3	−3 09	GLX
SAB(r)0°	1.7′ × 0.9′	13.43B	12.5V

This galaxy is paired with NGC5506. Photographs show an inclined lenticular galaxy with a large bright core and an extensive outer envelope. Telescopically, the galaxy is a little smaller but brighter than its neighbour; it is a round, condensed and well defined object that is a little brighter to the middle. Recorded as a Class IV object, Herschel's 15 April 1787 entry described it as: 'Pretty bright, stellar, like a star with a small bur all around.' Radial velocity: 1814 km/s. Distance: 81 mly. Diameter: 40 000 ly.

NGC5549

H552[3]	14:18.5	+7 22	GLX
S0	1.6′ × 0.7′	14.05B	

In photographs this spindle-shaped lenticular galaxy has a bright, elongated core with faint extensions oriented ESE–WNW. The DSS image shows a triangular group of fainter galaxies to the SW and a bright field star 7.2′ to the E. Visually, only the core of this galaxy is seen; it is moderately bright, round and broadly brighter to the

NGC5506/5507

middle. The edges are moderately well defined and the galaxy is just S of a line joining two field stars oriented E–W. Radial velocity: 7702 km/s. Distance: 344 mly. Diameter: 160 000 ly.

NGC5560

H579[2]	14:20.1	+4 00	GLX
SB(s)b pec	3.7′ × 0.7′	13.18B	12.4V

Photographs show this highly inclined spiral galaxy to have a bright, elongated inner region involving a probable bar and fainter, extensive spiral arms. High-contrast images show these arms to be slightly warped and there is almost certainly tidal interaction with the brighter galaxy NGC5566 located 5.2′ to the SE. In moderate apertures, NGC5560 is fainter than NGC5566 but is quite prominent; it is a smooth, elongated and fairly well defined galaxy, oriented ESE–WNW, with a faint field star NNW of the core. Radial velocity:

1717 km/s. Distance: 77 mly. Diameter: 83 000 ly.

NGC5566

H144[1]	14:20.3	+3 56	GLX
SB(r)ab	6.6′ × 2.2′	11.41B	10.6V

This galaxy is the brightest member of an interacting triplet catalogued as Arp 286, the other galaxies being NGC5560 and NGC5569. In photographs, NGC5566 is very interesting structurally, displaying a large, bright core in a theta-shaped bar/ring structure. Two broad, extensive and faint spiral arms emerge from the ring; they are fairly smooth textured but each arm has a dust lane, the one in the northern spiral arm being more prominent. High-contrast photographs show a warp in this dust lane and a faint plume of material extending to the smallest member of the trio, NGC5569, which is not a Herschel object. Visually,

NGC5566 is an intensely bright galaxy, the main body of which is quite round and very opaque with a bright stellar nucleus. Averted vision brings out hints of a faint outer envelope oriented NNE–SSW, but the companion galaxy NGC5569 is not visible in a moderate-aperture telescope. At the presumed distance, the core-to-core separation between the centres of NGC5566 and NGC5569 would be about 80 000 ly. The separation between NGC5560 and NGC5566 would be about 100 000 ly. Radial velocity: 1495 km/s. Distance: 67 mly. Diameter: 128 000 ly.

NGC5574

H145[1]	14:20.9	+3 14	GLX
SB0⁻? sp	1.6′ × 1.0′	13.30B	12.4V

This galaxy is part of a triplet that includes NGC5576 and the larger but

NGC5574/5576

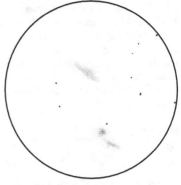

NGC5574 NGC5576 NGC5577
15-inch Newtonian 146x

NGC5560/5566

fainter NGC 5577 (not a Herschel object). Images show this lenticular galaxy has a bright, oval core embedded in a stubby extended envelope. A faint, elongated shell surrounds the brighter inner part of the galaxy. In a medium-magnification field, NGC 5574 is a little fainter than its close companion NGC 5576 and appears much elongated ENE–WSW with a slightly brighter core. The galaxy is quite well defined. The fainter NGC 5577 to the N is hazy but readily visible and it is surprising that Herschel missed it. Discovered on 30 April 1786, NGC 5574 and NGC 5576 were jointly described as: 'Two; the preceding pretty bright, pretty large, extended. Distance 3′ or 4′ south preceding, north following. The following considerably bright, round, pretty large. Place of 2nd.' Radial velocity: 1645 km/s. Distance: 73 mly. Diameter: 34 000 ly.

NGC5576

H146[1]	14:21.1	+3 16	GLX
E3	4.6′ × 3.3′	11.79B	11.0V

The visual and photographic appearances of this elliptical galaxy are quite similar. It is very slightly oval in shape, is oriented E–W, and is moderately condensed with fairly well defined edges. It is sharply concentrated to a bright core. A magnitude 12 field star is located 1.25′ NW of the core. The DSS image shows a faint plume emerging from the E and curving clockwise to the N. Only 2.75′ separates this galaxy from NGC 5574 to the SW. At the presumed distance, the core-to-core separation would be 53 000 ly. Radial velocity: 1473 km/s. Distance: 66 mly. Diameter: 88 000 ly.

NGC5599

H927[3]	14:23.7	+6 34	GLX
Sb	1.4′ × 0.5′	14.62B	

Images show a highly inclined spiral galaxy with an elongated, brighter central region in a faint spiral envelope. Six small and very faint galaxies are

within 3.0′ to the W. Telescopically, the galaxy is fairly dim and is seen as an oval, even-surface-brightness patch of light elongated SSE–NNW with fairly well defined edges. Radial velocity: 7219 km/s. Distance: 322 mly. Diameter: 131 000 ly.

NGC5604

H668[3]	14:24.7	−3 13	GLX
Sa	1.9′ × 1.0′	13.57B	

Photographs of this slightly inclined spiral galaxy show a large and bright central region embedded in a very faint disc that has traces of grainy spiral structure. Visually, only the bright central region is detected and it is moderately faint, almost oval in shape, elongated N–S and very gradually brighter along the major axis. The edges are quite diffuse and poorly defined. Radial velocity: 2717 km/s. Distance: 121 mly. Diameter: 67 000 ly.

NGC5618

H763[3]	14:27.2	−2 16	GLX
SB(rs)c	1.7′ × 1.3′	14.14B	

Images show a face-on barred spiral galaxy with two principal and fairly massive spiral arms emerging from the bar: the arm emerging from the S is the larger and, extends for more than a full revolution, while that from the N fades after half a revolution. Visually, though, this is a faint and quite diffuse patch of light; it is a little brighter to the middle and round with ill defined edges. Radial velocity: 7106 km/s. Distance: 317 mly. Diameter: 157 000 ly.

NGC5634

H70[1]	14:29.6	−5 59	GC
CC4	5.5′ × 5.5′	10.7B	10.1V

The DSS image shows a compressed and bright globular cluster, round, with a dense core which transmutes to a much less dense and very well resolved outer halo. The compressed core is about

NGC5634

1.75′ in diameter. There is a magnitude 7.5 field star 1.4′ ESE of the centre of the globular cluster and a magnitude 11 field star is 1.6′ NW of the core. Visually, although quite small, this globular cluster is very bright and condensed, growing brighter to a sizable core. There is a fairly sharp drop-off in brightness at the edge and hazy evidence of outliers on the threshold of resolution at medium magnification. Spectral type: F5. Radial velocity in approach: 63 km/s. Distance: 81 500 ly.

NGC5636

H580[2]	14:29.7	+3 16	GLX
SAB(r)0⁺	1.3′ × 0.9′	13.84B	

This galaxy forms a triplet with NGC5638 and the non-Herschel object UGC9310, which should be visible in large-aperture telescopes. Photographically, NGC5636 is a theta-shaped, barred ring galaxy seen at a slightly oblique angle. In medium-aperture telescopes, it is a hazy, indeterminate patch of light which is a little brighter to the middle and best seen with averted vision. NGC5638 is 2.0′ almost due S and at the presumed distance the core-to-core separation between the two galaxies would be 45 000 ly. Radial velocity: 1737 km/s. Distance: 78 mly. Diameter: 29 000 ly.

NGC5638

H581[2]	14:29.7	+3 14	GLX	
E1	2.3′ × 2.1′	12.16B	11.2V	

This galaxy is the brightest member of the triplet that also includes NGC5636 and UGC9310 and its visual and photographic appearances are similar. It features a bright envelope surrounded by a fainter secondary glow and the core is fairly large and bright. UGC9310 is 5.2′ almost due E. Radial velocity: 1668 km/s. Distance: 75 mly. Diameter: 50 000 ly.

NGC5645

H150[2]	14:30.7	+7 17	GLX
SB(s)d	2.7′ × 1.3′	12.90B	

This is an inclined, barred spiral galaxy. Photographs show a bright bar without

a core in an ill defined spiral pattern with brighter knots distributed in the disc, mostly along the northern flank. A small, oval galaxy, possibly a satellite, borders the galaxy ESE of the bar. Telescopically, the galaxy appears fairly bright at medium magnification and dims somewhat at high magnification. It appears as an irregularly round, smooth-textured patch of light which is broadly brighter to the middle with diffuse edges. Radial velocity: 1375 km/s. Distance: 61 mly. Diameter: 48 000 ly.

NGC5652

H891[2]	14:30.9	+5 58	GLX
SABbc	2.2′ × 1.7′	14.51B	

This is a slightly inclined spiral galaxy. Photographs show a small, bright core and a possible bar. The spiral pattern is complex and multi-arm, made up of bright, narrow, tightly wound arms dotted with nebulous knots. Visually, the galaxy is large and fairly bright; it is a fat oval of light elongated E–W which is somewhat grainy in texture and broadly brighter to the middle. The edges are quite well defined. Radial velocity: 7497 km/s. Distance: 335 mly. Diameter: 214 000 ly.

NGC5661

H892[2]	14:31.9	+6 14	GLX
SBb:	1.7′ × 0.5′	14.36B	

Photographs show a highly inclined, barred spiral galaxy with two narrow, bright, spiral arms. Fainter, possible companion galaxies are 4.5′ SE and 6.3′ SSW. Visually, NGC5661 is moderately faint and is best seen at medium magnification as an oval patch of light which is a little brighter to the middle, well defined and oriented NNE–SSW. Radial velocity: 2357 km/s. Distance: 105 mly. Diameter: 52 000 ly.

NGC5668

H574[2]	14:33.4	+4 27	GLX
SA(s)d	3.7′ × 3.7′	12.26B	

This is a face-on galaxy with a small core and thin, fragmented spiral arms.

Photographs show several condensations, particularly in the inner spiral structure, which are significant HII regions. At medium magnification, this is a moderately bright, though diffuse object; it appears slightly oval in shape with a patchy-looking outer envelope and is a little brighter to the SW. A faint field star is in contact to the ENE. Photographs show this star to be the brighter of a very close pair of field stars situated between two HII regions in the spiral arms. Radial velocity: 1580 km/s. Distance: 70 mly. Diameter: 76 000 ly.

NGC5674

H893[2]	14:33.8	+5 26	GLX
SABc	1.1′ × 1.0′	14.20B	

This face-on, barred spiral galaxy has a bright, oval bar in photographs, but appears to have only one spiral arm which emerges from the bar in the S and wraps tightly around the bar for a little more than one full revolution. The arm has many brighter, knotty condensations along its length. Visually, the galaxy is quite bright and very well defined, appearing as a boxy patch of light with high surface brightness with no visible core. Radial velocity: 7476 km/s. Distance: 334 mly. Diameter: 107 000 ly.

NGC5679

H894[2]	14:35.0	+5 20	GLX
Sb	1.1′ × 0.6′	14.10B	

This galaxy is part of a triple system which has also been catalogued as Arp 274. The DSS image reveals an inclined spiral galaxy in apparent contact with a slightly smaller spiral 0.7′ to the WSW and a small, compact object 0.7′ to the ESE. A faint double star is immediately N of the WSW spiral. At the apparent distance, the core-to-core separation between NGC5679 and each of the two galaxies would be about 79 000 ly. The extremely dim, face-on spiral galaxy UGC9385 is 6.0′ to the SE. Visually, NGC5679 is small and moderately faint but well seen at both medium and high magnification as a somewhat condensed, irregularly oval patch of light which is a little brighter to the middle with fairly well defined edges. Neither companion galaxy was detected visually. Radial velocity: 8666 km/s. Distance: 387 mly. Diameter: 124 000 ly.

NGC5690

H582[2]	14:37.7	+2 17	GLX
Sc? sp	3.4' × 1.0'	12.48B	

In photographs, significant dust lanes are visible in silhouette against the brighter inner disc of this almost edge-on spiral galaxy. There is no central bulge or brighter core and a very faint star can be seen in the outer envelope to the SE. Telescopically, the galaxy is situated 3.25' ENE of a bright field star. The galaxy appears large and fairly bright and is well seen despite the presence of the star. It is a flat, elongated sliver of light, oriented SE–NW with a very faint field star involved at its SE tip. The surface brightness is highest along the major axis with a slight, elongated brightening at the core. Radial velocity: 1748 km/s. Distance: 78 mly. Diameter: 77 000 ly.

NGC5691

H681[2]	14:37.9	−0 24	GLX
SAB(s)a: pec	1.9' × 1.5'	12.90B	

The DSS image suggests a slightly inclined, barred spiral galaxy with one principal arm winding around the galaxy for more than one revolution. There may be a second arm but if so it is very weak and short. Higher-resolution photographs show many knotty condensations in the spiral arm near the bar and significant dust patches, while the outer spiral structure is very broad and faint. In a moderate-aperture telescope, the galaxy is bright and well seen at high magnification as a broad, oval patch of light elongated ESE–WNW with fairly well defined edges. The envelope is very slightly brighter along the major axis. Radial velocity: 1857 km/s. Distance: 83 mly. Diameter: 46 000 ly.

NGC5701

H575[2]	14:39.2	+5 22	GLX	
(R)SB(rs)0/a	4.3' × 4.1'	12.08B	10.9V	

This face-on, barred spiral galaxy is well seen on the DSS image. The bright core has a fainter, N–S oriented bar whose ends flare in brightness where they meet the inner disc. The faint spiral arms emerge from the disc rather than from the bar and appear to form a pseudo-ring. The arms broaden quickly as they spiral outward with several faint, knotty condensations involved. A compact, fainter galaxy can be seen 1.0' SW of the core within the spiral pattern, but this may be a background object. Visually, only the bright core of this barred spiral is visible and it is well seen at high magnification. It appears as a round, grainy patch of light which is well defined and brighter to the middle. A hazy secondary envelope surrounds the core. Radial velocity: 1510 km/s. Distance: 67 mly. Diameter: 84 000 ly.

NGC5713

H182[1]	14:40.2	−0 17	GLX	
SAB(rs)bc pec	2.8' × 2.5'	11.62B	11.2V	

This is one of the principal members of a loose E–W association of half a dozen galaxies, the others being NGC5691, NGC5705, NGC5719, NGC5733 and NGC5750. Photographs show a barred spiral galaxy of high surface brightness with short, bright spiral segments which transmute into a single, broad, faint outer spiral arm. Visually, the galaxy appears as a bright, even-surface-brightness, oval object elongated E–W with well defined extremities. Companion galaxy NGC5719 fits comfortably in the same medium-magnification field to the ESE. Radial velocity: 1888 km/s. Distance: 84 mly. Diameter: 68 000 ly.

NGC5718

H550[3]	14:40.8	+3 28	GLX
S0⁻:	1.4' × 1.0'	14.18B	

This galaxy forms an interacting pair with IC1042, situated 0.9' WNW and the pair has been catalogued as Arp 171. Both galaxies appear to be lenticular and there is a bridge of material connecting the two; the minimum core-to-core separation at the presumed distance would be 96 000 ly. The DSS image shows that these galaxies are probably the principal members of a small cluster of galaxies; at least eight of them appear to be bright enough to detect in large amateur instruments and there are many more fainter galaxies in a 15.0' field. IC1039 is 3.75' WSW, while IC1041 is 5.5' SSW. In a moderate-aperture telescope, only NGC5718 is visible and it is a very dim and difficult object, best seen at medium magnification as a very diffuse and ill defined brightening in the field. A field star only 1.25' to the NE hinders the view somewhat; at high magnification the galaxy is seen only intermittently as a round patch which is very broadly brighter to the middle. Radial velocity: 8207 km/s. Distance: 366 mly. Diameter: 149 000 ly.

NGC5719

H682[2]	14:40.9	−0 19	GLX
SAB(s)ab pec	3.2' × 1.2'	13.31B	

This is a highly tilted spiral galaxy with a bright central bulge in a slightly fainter outer envelope. Photographs reveal a thick and extensive dust lane which is strongly inclined to the plane of the galaxy and associated with a spiral segment which shares the inclination. In a moderate-aperture telescope, only the bright core of the galaxy is noted; the dust lane hides a significant percentage of the outer spiral structure. The galaxy is a high-surface-brightness oval elongated E–W with a brighter core and well defined edges. Radial velocity: 1722 km/s. Distance: 77 mly. Diameter: 71 000 ly.

NGC5740

H538[2]	14:44.4	+1 41	GLX
SAB(rs)b	3.0' × 1.5'	12.60B	

The DSS image reveals a slightly inclined spiral galaxy with a short bar attached to a broken inner ring structure which expands into fragmentary spiral arms. In a moderate-aperture telescope, the galaxy appears as an oval glow which is oriented SSE–NNW and broadly concentrated towards the middle with hazy, poorly defined edges. Radial velocity: 1569 km/s. Distance: 70 mly. Diameter: 61 000 ly.

NGC5746

H126[1]	14:44.9	+1 57	GLX
SAB(rs)b? sp	8.0′ × 1.3′	11.34B	

This galaxy forms a physical pair with NGC5740 to the SSW and is part of a bright and extensive group of galaxies numbering at least 30 members, located to the W of the bright field star 109 Virginis. Photographs show this galaxy to be a classic, edge-on spiral with a moderate-sized and slightly bulging core embedded in a bright spiral pattern. A massive and extensive dust lane is visible along the entire length of the galaxy on the eastern flank. Visually, the galaxy is very bright; the central bulge is well seen as a slight broadening

along the major axis and a minute, brighter core is visible which is especially well seen at high magnification. The extensions are bright and taper gradually while fading into the sky background and the western flank is more sharply defined. Radial velocity: 1723 km/s. Distance: 77 mly. Diameter: 178 000 ly.

NGC5750

H183[1]	14:46.2	−0 13	GLX
SB(r)0/a	3.3′ × 1.5′	12.55B	

Photographically, this inclined, barred spiral galaxy has a bright, thin ring attached to the bar. High-resolution photographs show the ring to be two separate, overlapping spiral arms. Massive dust patches and short, thin dust lanes are silhouetted against the inner envelope in the NE and an asymmetrical outer spiral pattern is seen extending to the ENE. Visually, the galaxy is a hazy, oval patch of light, elongated NE–SW with a brighter,

condensed central region in a slightly fainter envelope. The edges of the envelope fade gradually into the sky background. Radial velocity: 1680 km/s. Distance: 75 mly. Diameter: 72 000 ly.

NGC5770

H576[2]	14:53.4	+3 57	GLX
SB0	1.7′ × 1.3′	13.18B	12.3V

The DSS image shows a theta-shaped, barred galaxy with a round, bright core and a very faint bar attached to a faint ring. A peculiar condensation immediately W of the core in the bar is either a foreground star or a compact object seen in silhouette against the galaxy. Visually, the galaxy is small, bright and grainy and is well seen at high magnification as a slightly elongated, opaque, well-condensed and well defined object oriented E–W. There are two condensations in the envelope, with the eastern one being the core of the galaxy. Recorded on 29 April 1786, Herschel called this object: 'Faint, small, a little extended like 2 stellar. joined closely.' Radial velocity: 1501 km/s. Distance: 67 mly. Diameter: 33 000 ly.

NGC5775

H554[3]	14:54.0	+3 33	GLX
Sb(f)	4.2′ × 1.0′	12.26B	11.3V

Seen as a nearly edge-on spiral galaxy in the DSS image, this galaxy forms a physical pair with NGC5774, a very dim face-on spiral galaxy 4.3′ to the NW; the minimum separation between them would be 94 000 ly at the presumed distance. Many condensations are visible in the flattened disc of NGC5775 and dust lanes are prominent along its NE flank. Visually, NGC5775 is quite bright and distinct; it is an edge-on galaxy which is well defined along the edges and bright along its major axis. The disc tapers to blunt points. Three faint field stars bracket the galaxy, while NGC5774 is also visible and is best seen at medium magnification as a diffuse patch of light. Radial velocity: 1690 km/s. Distance: 76 mly. Diameter: 92 000 ly.

NGC5746

©STScI/AURA/CALTECH/ROE

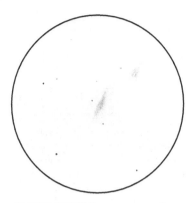

NGC5774 NGC5775
15-inch Newtonian 272x

NGC5806

H539[2]	15:00.0	+1 54	GLX	
SAB(s)b	3.1′ × 1.6′	12.35B	11.7V	

This inclined spiral galaxy features a moderate-sized core embedded in a grainy, fairly high-surface-brightness spiral pattern. The DSS image shows very faint, outer spiral structure beyond the inner spiral pattern. In medium-sized instruments the galaxy is moderately faint, oval and extended N–S. It is brighter along its major axis, though no core is visible. Only the well defined, bright inner envelope is visible. The galaxy is considered an outlying member of the NGC5846 galaxy group. Radial velocity: 1368 km/s. Distance: 61 mly. Diameter: 55 000 ly.

NGC5813

H127[1]	15:01.2	+1 42	GLX	
E1-2	4.2′ × 3.0′	11.52B	10.5V	

The photographic and visual appearances of this bright elliptical galaxy are quite similar. Visually, the galaxy is quite bright and round and features a brighter core. It is situated within a magnitude 11–13, diamond-shaped asterism and is due S of a magnitude 9 field star. The outer envelope is fairly well defined and NGC5814, which may be visible in larger-aperture telescopes, is 4.5′ to the SE. Photographs show an extensive outer envelope, which is oriented SE–NW. NGC5813 is part of the NGC5846 group of galaxies, which are arrayed in an E–W line S of 110 Virginis. Radial velocity: 1981 km/s. Distance: 88 mly. Diameter: 108 000 ly.

NGC5831

H540[2]	15:04.1	+1 13	GLX	
E3	2.2′ × 2.0′	12.43B		

This galaxy is a member of the NGC5846 galaxy group. Photographs show an elliptical object with a bright, round central region and an extensive, fainter outer envelope elongated SSE–NNW. Visually the galaxy is a bright, round, well-condensed object with well defined edges and a grainy envelope. A brighter, nonstellar core is visible. Radial velocity: 1665 km/s. Distance: 74 mly. Diameter: 48 000 ly.

NGC5838

H542[2]	15:05.4	+2 06	GLX	
SA0⁻	4.2′ × 1.5′	11.79B		

Photographs show a quite bright lenticular galaxy with a large central core embedded in a faint, but extensive outer disc. There is evidence of dust patches in the disc all around the core. Visually, this is a bright compact galaxy located 5.3′ NNE of a magnitude 9 field star. The galaxy is much brighter to the middle with an oval, well defined envelope elongated NE–SW. It is a probable member of the NGC5846 galaxy group. Radial velocity: 1372 km/s. Distance: 61 mly. Diameter: 75 000 ly.

NGC5839

H541[2]	15:05.5	+1 38	GLX	
SAB(rs)0°:	1.4′ × 1.3′	13.69B	12.2V	

Images reveal a lenticular galaxy seen face-on with a bright, oval core and a very large outer envelope. Visually, the

NGC5775

©STScI/AURA/CALTECHROE

galaxy is quite small but has high surface brightness and responds well to high magnification. It is a condensed patch of light slightly elongated E–W with well defined edges. It is probably a member of the NGC5846 group of galaxies, although its radial velocity is somewhat lower than that of the principal galaxy of the group. Radial velocity: 1237 km/s. Distance: 55 mly. Diameter: 22 000 ly.

NGC5845

H511[3]	15:06.0	+1 38	GLX
E:	0.8′ × 0.5′	13.44B	11.2V

The visual and photographic appearances of this galaxy are quite similar. Visually, it is a very small, oval-shaped object which is slightly elongated SE–NW and faint but quite condensed with well defined edges. Photographs indicate that this galaxy should probably be classified as an E4 dwarf elliptical. Radial velocity: 1462 km/s. Distance: 65 mly. Diameter: 15 000 ly.

NGC5846

H128[1]	15:06.4	+1 36	GLX
E0-1:	4.1′ × 3.8′	11.09B	

The DSS image shows a large, bright elliptical or lenticular galaxy; the core is massive and large and there is an extensive fainter envelope. A smaller companion galaxy, NGC5846A, is 0.6′ S of the core and is seen in silhouette against the fainter outer envelope. It is quite likely that NGC5846 and NGC5846A form a physical pair, though there is no evidence of interaction between the two galaxies. NGC5845 is 7.2′ WNW. NGC5846 is the principal galaxy of a small group with at least ten members, including NGC5806, NGC5813, NGC5838, NGC5839 and NGC5845. There is a considerable spread in the radial velocities of the member galaxies. Visually, NGC5846 is a large, bright galaxy which is quite round with a bright central region. The envelope appears pear shaped, tapering to a slight

concentration to the S, which is the companion galaxy NGC5846A. The companion is resolvable as a separate object, and is very small, concentrated and well defined. Curiously, Herschel did not note this companion galaxy. Radial velocity: 1726 km/s. Distance: 77 mly. Diameter: 92 000 ly.

NGC5850

H543[2]	15:07.1	+1 33	GLX
SB(r)b	5.0′ × 5.0′	11.89B	

The DSS image shows a classic, theta-shaped, barred spiral galaxy, the inner ring of which is formed by two overlapping spiral arms emerging from the ends of the straight bar. The inner ring expands to two very faint, broader spiral arms, loosely wound around the galaxy. Visually, the galaxy is fairly prominent but hazy and poorly condensed. The core is well seen, is gradually brighter to the middle and sits

in a very faint, ill defined hazy envelope. Radial velocity: 2568 km/s. Distance: 115 mly. Diameter: 167 000 ly.

NGC5854

H544[2]	15:07.8	+2 34	GLX
SB(s)0+	2.7′ × 0.8′	12.65B	

This lenticular galaxy is probably an outlying member of the NGC5846 galaxy group. Photographs show a classic, nearly edge-on galaxy with the three principal brightness zones, typical of the class. The bright core is surrounded by a slightly fainter inner disc and a faint outer envelope which seems speckled with discrete dust patches. The galaxy is small but quite bright visually; it is an elongated lens-shaped object oriented ENE–WSW with a brighter core and sharply defined edges. Radial velocity: 1753 km/s. Distance: 78 mly. Diameter: 61 000 ly.

NGC5850

©STSCI/AURA/CALTECH/ROE

NGC5864

H585[2]	15:09.6	+3 03	GLX
SB(s)0°? sp	2.8′ × 0.9′	12.68B	

The DSS image shows a highly inclined lenticular galaxy of somewhat unusual structure. The bright, elongated core appears to be bisected by a thin, low-contrast dust lane and brighter, stubby arms emerge from the core before blending into the fainter outer disc. The galaxy is probably an outlying member of the NGC5846 galaxy group. Visually, the galaxy is small but bright; it is an elongated, oval object oriented ENE–WSW with well defined edges and fairly even surface brightness. A very faint field star borders the galaxy 0.5′ ESE of the core. Radial velocity: 1904 km/s. Distance: 85 mly. Diameter: 69 000 ly.

NGC5865

H684[2]	15:09.9	+0 31	GLX
SAB0⁻	0.7′ × 0.7′	14.75B	

This galaxy forms a pair with the larger and brighter galaxy NGC5869, situated 3.5′ to the S. They are not physical companions, however, as NGC5865 has a much higher radial velocity. Photographs show a face-on lenticular galaxy with a bright, oval core and an extensive outer envelope. In the first edition of *Uranometria 2000.0*, chart 243 in volume 1 identifies the galaxy as NGC5865 while chart 243 in volume 2 calls the galaxy NGC5868, but both are the same object. Curiously, neither chart plots the much brighter NGC5869. Visually, NGC5865 can be seen in the same high-magnification field as NGC5869 and appears as a fainter, more diffuse object. It is round and broadly brighter to the middle with somewhat ill defined edges and is more distinct at medium magnification. Discovered on 11 April 1787, Herschel came upon this object almost 14 months after he had found NGC5869, describing it as: 'Two. The 2nd pretty bright, small, irregularly extended. For the 1st see II. 545 [NGC5869].' Radial velocity: 11 725 km/s. Distance: 524 mly. Diameter: 107 000 ly.

NGC5869

H545[2]	15:09.9	+0 28	GLX
S0°:	2.3′ × 1.7′	13.15B	

This is an almost face-on lenticular galaxy. Photographs show a large, elongated core in an extensive fainter outer envelope oriented ESE–WNW. Telescopically, the galaxy is much larger and brighter than NGC5865, which is seen in the same high-magnification field. NGC5869 appears as a round, well defined and well condensed patch of light with even surface brightness. Recorded on 24 February 1786, Herschel called it: 'Pretty bright, small, irregularly extended, a little brighter to the middle.' Radial velocity: 2096 km/s. Distance: 94 mly. Diameter: 63 000 ly.

Vulpecula

Except for NGC6793 (recorded in 1789) all the objects discovered here were seen in 1784. Vulpecula is located in the summer Milky Way; star clusters are dominant here and only one external galaxy is included. The most interesting object visually is the rich star cluster NGC6940.

NGC6793

H81[8]	19:23.2	+22 01	OC
III2p	7.0′ × 7.0′		

The DSS shows a poor, scattered cluster of bright stars in a fairly rich field. Visually, this coarse cluster is easily identified at low magnification, as it stands out well from the rich Milky Way background. The cluster is made up primarily of magnitude 10 and 11 stars, scattered over a 10′ field. Two coarse triangles stand out at centre. The northern triangle is the brighter of the two and its northernmost member is a tight double star. Fifteen stars are recognized as members of this cluster. Discovered on 18 July 1789, Herschel called the group: 'A scattered cluster of considerably large stars. Irregular form, pretty rich, above 15′ in extent.'

NGC6800

H21[8]	19:27.2	+25 08	OC
IV1p	12.0′ × 12.0′		

The photographic and visual appearances of this large scattered cluster are similar. Visually, it is a large, bright cluster in a rich Milky Way field. The brightest members are about magnitude 10 and form a distinct, ring-shaped structure surrounding a prominent starless zone about 5′ across. Only 20 stars are recognized as members. Spectral type: A.

NGC6802

H14[6]	19:30.6	+20 16	OC
I1m	5.0′ × 1.5′	10.1B	8.8V

The DSS image shows a rich cluster of faint stars with a narrow brightness range that are fairly well separated from the rich Milky Way field. In small telescopes, this cluster is a nebulous haze, bar shaped, and oriented N–S. It is framed by six stars in the field which are unconnected to the cluster: a bright pair to the NE and another bright pair to the NW and single stars SE and SW of the cluster. Medium to high magnification helps to resolve a handful of magnitude 13 cluster members; resolution is better

in medium-aperture telescopes at high magnification where the group is quite compressed and the brightness range of the members is small. Herschel recorded the cluster on 11 July 1783, calling it: 'A cluster of extremely small and very compressed stars. A parallelogram of 4′ long, 2′ broad in the direction of the meridian.' Cluster membership is about 200 stars. Spectral type: B6. Age: about 1.7 billion years. Distance: 3300 ly.

NGC6823

H18[7]	19:43.1	+23 18	OC
I3m n	7.0′ × 7.0′	7.7B	7.1V

Small apertures show a distinct cluster of bright and faint stars of magnitude 8–13 centred on a compressed knot of four bright stars with faint threshold stars involved. The field is rich but the cluster is well separated from the sky background and is elongated NNE–SSW. Photographs show the cluster embedded in faint nebulosity

designated NGC6820, which is difficult to see visually. About 80 stars are considered to be members of this very young cluster, which is part of the Vulpecula OB1 association. Spectral type: O7. Age: 2 million years. Distance: 8550 ly.

NGC6830

H9[7]	19:51.0	+23 04	OC
II2p	6.0′ × 6.0′	8.4B	7.9V

Photographs show a somewhat compressed cluster of brighter stars in a rich Milky Way field. Visually, this is a small, compressed cluster of magnitude 10 or fainter stars in a fairly rich field. The cluster stands out well and the brighter members form a cruciform structure; indeed it looks like a little 'Cygnus' with the brightest star in the 'Deneb' position. Cluster membership is about 80 stars. Spectral type: B0. Age: about 100 million years. Radial velocity in recession: 22 km/s. Distance: about 5500 ly.

NGC6847

H202[2]	19:57.0	+29 20	MWP?

This was recorded as a faint nebula on 17 July 1784, when Herschel's description was: 'A resolvable nebulous patch; there are great numbers of them in this neighbourhood like forming nebulae, but this is the strongest of them; they are evidently congeries of small stars.' Visually, this probable Milky Way patch is located on the Cygnus–Vulpecula border; medium magnification reveals a slightly condensed patch of resolved stars N of a magnitude 9 field star. The stars are magnitude 12 or fainter, and about 17 are resolved in an elongated patch oriented NE–SW, about 10′ × 6′ in size.

NGC6885

H20[8]/H22[8]	20:12.1	+26 29	OC
III2m	20.0′ × 20.0′	8.7B	8.1V

This is a large, bright and scattered cluster of stars surrounding the bright field star 20 Vulpeculae; most of the members lie N, S or SW of the bright

NGC6802

©STScI/AURA/CALTECH/ROE

NGC6823

©STScI/AURA/CALTECH/ROE

This object was recorded by Herschel on 18 July 1784, as: 'A cluster of many large scattered stars.' At low magnification, the field shows an elongated, fairly rich grouping of stars of moderate brightness range about 20.0′ × 10.0′ in size, oriented E–W. At least 35 stars are involved but this may not actually be a true cluster.

NGC6940

H8[7]	20:34.6	+28 18	OC
III2r	25.0′ × 25.0′	7.0B	6.3V

This cluster appears similar visually and photographically. In a moderate-aperture telescope this rich cluster is best seen at low magnification and is well resolved against the rich Milky Way background. Over 100 cluster members are visible in a 30′ × 30′ area and the cluster appears slightly elongated E–W with the majority of members around magnitude 11–12. The total membership of the cluster is 170 stars. Spectral type: B8. Age: 1.1 billion years. Radial velocity in recession: 5 km/s. Distance: 2700 ly.

NGC7052

H145[3]	21:18.6	+26 27	GLX
E	2.1′ × 1.1′	13.90B	

The DSS image shows a galaxy with a bright, elongated core and a substantial, fainter outer envelope. Visually, the galaxy is rather faint and is best seen at high magnification. It is slightly oval, elongated ENE–WSW and gradually brighter to the middle, while the edges are poorly defined. A magnitude 9 field star is 1.5′ E. Radial velocity: 4888 km/s. Distance: 218 mly. Diameter: 134 000 ly.

star. A smaller, fainter concentration of stars NW of 20 Vulpeculae is not NGC6882 but a group first recognized by R. J. Trumpler and later catalogued as Collinder 416. Herschel described the cluster as: 'A cluster of coarse scattered stars, not rich.' He presumably saw the stars making up Collinder 416 and may have included them in his description. At least 50 stars are visible in a 20′ × 20′ area; about 35 stars are accepted as members of NGC6885, while up to 40 stars are associated with Collinder 416. According to Archinal and Hynes in the exhaustive catalogue *Star Clusters* (2003), H22[8] (NGC6882) is probably a duplicate observation of H20[8] (NGC6885), the clusters being in almost identical positions and observed on successive nights in September 1784. Spectral type: B5. Distance: 1950 ly.

NGC6938

H17[8]	20:34.8	+22 15	OC?
20.0′ × 10.0′			

Photographs show an irregular grouping of bright stars in a rich field.

General References

Allen, D. A., Barker, E. S. and Jones, K. G., editors (1979). *Webb Society Deep-Sky Observer's Handbook*, Volume 2: *Planetary and Gaseous Nebulae*. Hillside, New Jersey, USA: Enslow Publishers.

Archinal, B. A. (1993). *The "Non-existent" Star Clusters of the RNGC*. Portsmouth, UK: The Webb Society.

Archinal, B. A. and Hynes, S. J. (2003). *Star Clusters*. Richmond, Virginia, USA: Willmann-Bell, Inc.

Ashbrook, J. (1984), *The Astronomical Scrapbook*. Cambridge, UK: Cambridge University Press; and Cambridge, Massachusetts, USA: Sky Publishing Corporation.

Barker, E. S. and Jones, K. G., editors (1980). *Webb Society Deep-Sky Observer's Handbook*, Volume 3: *Open and Globular Clusters*. Hillside, New Jersey, USA: Enslow Publishers.

Barker, E. S. and Jones, K. G., editors (1981). *Webb Society Deep-Sky Observer's Handbook*, Volume 4: *Galaxies*. Hillside, New Jersey, USA: Enslow Publishers.

Cragin, M., Rappaport, B. and Lucyk, J. (1993). *The Deep Sky Field Guide to Uranometria 2000.0*, first edition. Richmond, Virginia, USA: Willmann-Bell, Inc.

de Vaucouleurs, G., de Vaucouleurs, A., Corwin, H. G. Jr *et al.*, (1991). *Third Reference Catalogue of Bright Galaxies*, three volumes. New York, USA: Springer-Verlag Inc.

Dreyer, J. L. E. (1888) A New General Catalogue of Nebulae and Clusters of Stars, *Mem. Roy. Soc. London*, L9.

Falco, E. E., Kurtz, M. J., Geller, J. J. *et al.* (1999). The updated Zwicky catalog (UZC). *Publ. Astron. Soc. Pac.*, **111**, 438–452.

Gil De Paz, A., Boissier, S., Madore, B. F. *et al.* (2007). The GALEX ultraviolet atlas of nearby galaxies. *Astrophys. J., Suppl. Ser.*, **173**, 185–255.

Hernandez-Toledo, H. M. and Ortega-Esbri, S. (2008). Broad-band BVRI photometry of isolated spiral galaxies. *Astron. Astrophys.*, **487**, 485–502.

Herschel, Sir J. F. W. (1847). *Results of Astronomical Observations Made During the Years of 1834, 5, 6, 7 & 8, at the Cape of Good Hope*. London, UK: Smith, Elder and Co.

Herschel, Sir J. F. W. (1864). Catalogue of nebulae and clusters of stars, *Phil. Trans. Roy. Soc. London*, **154**, 1.

Herschel, Mrs. J. (1879). *Memoir and Correspondence of Caroline Herschel*, second edition. London, UK: John Murray.

Herschel, Sir W. and Dreyer, J. L. E. (1912). *The Scientific Papers of Sir William Herschel*, two volumes. London, UK: The Royal Society and the Royal Astronomical Society.

Hirshfeld, A. and Sinnott, R. W., editors (1985). *Sky Catalogue 2000.0.*, Volume 2: *Double Stars, Variable Stars and Nonstellar Objects*. Cambridge, Massachusetts, USA: Sky Publishing Corporation; and Cambridge, UK: Cambridge University Press.

Hoskin, M. (1963). *William Herschel and the Construction of the Heavens*. London, UK: Oldbourne Book Co. Ltd.

Hoskin, M. (2003). *The Herschel Partnership as Viewed by Caroline*. Cambridge, UK: Science History Publications Ltd.

Hynes, S. J. (1991). *Planetary Nebulae, A Practical Guide and Handbook for Amateur Astronomers*. Richmond, Virginia, USA: Willmann-Bell Inc.

Hynes, S. J. and Jones, K. G., editors (1987). *Webb Society Deep-Sky Observer's Handbook*, Volume 7: *The Southern Sky*. Hillside, New Jersey, USA: Enslow Publishers.

Jones, K. G. (1975). *The Search for the Nebulae*. Buckinghamshire, UK: Alpha Academic, Science History Publications Ltd.

Jones, K. G. (1991). *Messier's Nebulae and Star Clusters*, second edition. Cambridge, UK: Cambridge University Press.

King, H. C. (1955). *The History of the Telescope*. Buckinghamshire, UK: Charles Griffin and Company.

Lauberts, A. and Valentijn, E. A. (1989). *The Surface Photometry Catalogue of the ESO-Uppsala Galaxies*. Garching, Germany: European Southern Observatory.

Lubbock, C. A. (1933). *The Herschel Chronicle, The Life-story of William Herschel and his Sister Caroline Herschel*. London, UK: Cambridge University Press.

Luginbuhl, C. B. and Skiff, B. A. (1989). *Observing Handbook and Catalogue of Deep-Sky Objects*. Cambridge, UK: Cambridge University Press.

Mallas J. H. and Kreimer, E. (1978). *The Messier Album*. Cambridge, Massachusetts, USA: Sky Publishing Corporation.

Maurer, A. (1998). *A Compendium of All Known William Herschel Telescopes*. Cumming, Georgia, USA: The Antique Telescope Society.

Maurer, A. (2005–2006). The Replica of William Herschel's 25-foot Spanish Telescope, *The Speculum: The Journal of the William Herschel Society*, **4**, No. 2, 36–37.

Paturel, G., Petit, C., Prugniel, P. *et al*. (2003). HYPERLEDA. I. Identification and designation of galaxies. *Astron. Astrophys*. **412**, 45–55.

We acknowledge the usage of the HyperLeda database (http://leda.univ-lyon1.fr).

Petrosian, A., McLean, B., Allen, R., Kunth, D. and Leitherer, C. (2008). The north galactic pole +30° zone galaxies. I. A comparative study of galaxies with different nuclear activity. *Astrophys. J., Suppl. Ser.*, **175**, 86–96.

Sandage, A. and Bedke, J. (1994). *The Carnegie Atlas of Galaxies*, two volumes. Washington, DC, USA: Carnegie Institution of Washington with The Flintridge Foundation.

Serio, G. F., Indorato, L. and Nastasi, P. (1985). G. B. Hodierna's observations of nebulae and his cosmology. *J. History Astron*. XVI, 1–36.

Sidgwick, J. B. (1953). *William Herschel Explorer of the Heavens*. London, UK: Faber and Faber Ltd.

Skrutskie, M. F., Culri, R. M., Stiening, R. *et al*. (2006). The Two Micron All Sky Survey (2MASS) *Astron. J.*, **131**, 1163–1183.

Tirion, W., Rappaport, B. and Lovi, G. (1987) *Uranometria 2000.0*, two volumes, first edition. Richmond, Virginia, USA: Willmann-Bell, Inc.

Tully, R. B. (1988). *Nearby Galaxies Catalog*. Cambridge, UK: Cambridge University Press.

Van Altena, W. F., Lee J. T. and Hoffleit, E. D. (1995). *The General Catalogue of Trigonometric Stellar Parallaxes*, fourth edition. New Haven, Connecticut, USA: Yale University Observatory.

Van den Bergh, S. (2000). *The Galaxies of the Local Group*. Cambridge, UK: Cambridge University Press.

Whiston, G. S. and Jones, K. G., editors (1982). *Webb Society Deep-Sky Observer's Handbook*, Volume 5: *Clusters of Galaxies*. Hillside, New Jersey, USA: Enslow Publishers.

Further reading

Interested readers may learn additional information about the Herschels and their observing careers by consulting the following books:

Branchett, B., Branchett, D. and Jones, P. (1980). *Observe: The Herschel Objects*. Pittsburgh, Pennsylvania, USA: The Astronomical League.

Clerke, A. (1895). *The Herschels and Modern Astronomy*. London, UK: Cassell and Company.

Cole, C. and Pratt C. (1997). *Observe: The Herschel II Objects*. West Burlington, Iowa, USA: The Astronomical League.

Holden, E. S. (1881). *Sir William Herschel*. New York, USA: Charles Scribner's Sons.

Lemonick, M. (2009). *The Georgian Star*. New York, USA: W. W. Norton and Company, Inc.

Mullaney, J. (2007). *The Herschel Objects and How to Observe Them*. New York, USA:Springer-Verlag Inc.

Mullaney, J. and Tirion, W. (2010). *The Cambridge Atlas of Herschel Objects*. Cambridge, UK: Cambridge University Press.

O'Meara, S. J. (2007). *Steve O'Meara's Herschel 400 Observing Guide*. Cambridge, UK:Cambridge University Press.

Steinicke, W. (2010). *Observing and Cataloguing Nebulae and Star Clusters: From Herschel to Dreyer's New General Catalogue*. Cambridge, UK: Cambridge University Press.

Index

Printed in the United States
by Baker & Taylor Publisher Services